THE
FACTS ON FILE
ENCYCLOPEDIA
OF SCIENCE,
TECHNOLOGY,
AND SOCIETY

Editorial Board and Contributors

THE
FACTS ON FILE
ENCYCLOPEDIA
OF SCIENCE,
TECHNOLOGY,
AND SOCIETY

VOLUME 3 O–Z

Rudi Volti

Facts On File, Inc.

The Facts On File Encyclopedia of Science, Technology, and Society

Facts On File, Inc.
11 Penn Plaza
New York, NY 10001

Library of Congress Cataloging-in-Publication Data

Volti, Rudi.
    The facts on file encyclopedia of science, technology, and society / Rudi Volti.
        p.   cm.
    Includes bibliographical references and index.
    ISBN 0-8160-3459-1 (v.1)
    ISBN 0-8160-3460-5 (v.2)
    ISBN 0-8160-3461-3 (v.3)
    ISBN 0-8160-3123-1 (set) (alk. paper)
    1. Science—Encyclopedias.   2. Technology—Encyclopedias.
3. Science—Social aspects—Encyclopedias.   4. Technology—Social aspects—
Encyclopedias.   I. Title.
Q121.V65   1998
503—dc21   98–39014

THE
FACTS ON FILE
ENCYCLOPEDIA
OF SCIENCE,
TECHNOLOGY,
AND SOCIETY

Occupational Safety and Health Administration

In 1970, The U.S. Congress passed the Occupational Safety and Health Act (amended in 1990), which consolidated a myriad of state and federal rules and laws addressing worker health and safety into one major legislative act. That law created the Occupational Safety and Health Administration (OSHA), which is housed in the Department of Labor and headed by an assistant secretary for Safety and Health. Prior to the passage of this legislation, the rules regarding the safety and health of American workers varied considerably from state to state and across the diversity of agricultural and industrial workplace settings. OSHA's mission was to bring some regularity to safety and health standards, and thereby to save lives, prevent injuries, and protect the health of all 100 million American workers. OSHA's task is to work with the $6\frac{1}{2}$ million employers affected by the legislation and with counterpart agencies (there are currently 25) at the state level. OSHA receives scientific and technical assistance from the National Institute for Occupational Safety and Health (NIOSH), one of the institutes of the NIH.

In its first 25 years, OSHA claims some credit for the following workplace safety and health improvements: Workplace deaths have been reduced by 50 percent; OSHA's cotton dust standard has virtually reduced brown lung disease in the textile industry; deaths from construction trench cave-ins have declined by 35 percent; and OSHA's lead standard has reduced blood poisoning in battery plant and smelter workers by 67 percent. In industries where OSHA inspections have been focused, there has been an average reduction of in-

jury and illness of 22 percent; where OSHA has been less vigilant, injury and illness has remained at similar levels or has increased. In spite of these improvements, problems remain. About 6,000 American workers die annually from workplace injuries; an estimated 50,000 die from illnesses caused by workplace chemical exposures; and 6 million suffer nonfatal workplace injuries each year.

OSHA undertakes two tasks. It sets regulations deemed necessary for employee safety and health—e.g., machine parts must be guarded, stairways must have handrails, employees must not be exposed to harmful levels of toxic chemicals—and then inspects workplaces to determine compliance. Not surprisingly, the agency has been most successful in setting standards where the causes and effects of injury or illness are easily established and for which there are clear and practicable remedies. Where causes are more ambiguous, where effects show up only decades later, and where there is disagreement among scientists about safety thresholds (as is this case with many chemical exposures), OSHA has encountered stiff resistance from employers and costly legal challenges. Enforcement at millions of workplace settings has been compromised by a relatively small staff of inspectors (2,100) attached to 200 offices across the country.

The work of the agency has also been hampered by the public view that OSHA's rules too often are unnecessarily complex and unreasonable when applied uniformly to the variety of workplace settings. In some quarters, OSHA is viewed as an agency that has lost sight of its primary goal of protecting workers and is instead hopelessly snarled in red tape and excessive paperwork. In an era of weaker labor unions and with government and business elites focused on "downsizing" as a means to enhance American global competitiveness, OSHA has had to rethink its strategies for better compliance. In an effort to encourage employers to develop their own strong and effective health and safety programs, OSHA offers them two options for meeting their obligations under the law: partnership or the usual means of enforcement. Partnership rewards employers who establish their own comprehensive health and safety programs with the promise of onsite inspections, penalty reductions, and highest priority for assistance when needed. Firms that do not opt for partnership and set up their own programs will face strong and traditional enforcement procedures. An experiment with this approach has been initiated in Maine, where 200 firms with the highest incidence of injuries were offered the choice of partnership or consensual enforcement mechanisms. All but two chose partnership, and the success to date has encouraged OSHA to develop ways to nationalize the "Maine 200" concept. OSHA has also initiated a process to streamline and rationalize current regulations so that they are up-to-date and sensible for today's workplace environment. Moreover, efforts are underway to reduce paperwork requirements in order to focus on safety and health results, not red tape. Time will tell whether these changes improve the public's confidence in OSHA's ability to protect worker safety and health.—T.I.

See also LEAD POISONING

Octane Number

When the air-fuel mixture used by an internal combustion engine is not burned properly, the result can be a condition known as *pinging* or *knocking*. More than an unpleasant sound, it can be an indication that excessive forces are being applied to piston crowns and other parts. If this condition persists, the result can be serious engine damage. In an internal combustion engine, the ignition of the air-fuel mixture normally produces a flame front that generates an even pressure on the piston crown. If portions of the air-fuel mixture ignite too rapidly, they can produce high-frequency pressure variations that are audible as knock. A number of things can make this happen. One of them is improper ignition

timing that causes the spark to occur too early in the engine's operating cycle. Another cause is the presence of hot spots in the combustion chamber that ignite the air-fuel mixture prematurely. Knocking can also be produced by gasoline that is unsuited to the engine.

In 1919, Charles Kettering (1876–1958) and Thomas Midgley, Jr. (1889–1944), began an inquiry into the causes of knock. They took up the issue in order to counter the accusation that knock was being caused by the battery ignition system produced by their employer, the Delco division of General Motors. In 1922, after much experimentation, they discovered that tetraethyl lead (four CH_3CH_2 groups attached to a lead atom) significantly reduced the tendency of an engine to knock. This discovery allowed the design of more powerful and efficient engines through the use of higher compression ratios by counteracting the greater propensity of high-compression engines to knock (compression ratio is the ratio of the volume of the cylinder and combustion chamber when the piston is at its lowest point to the volume at its highest point).

At this time, there was no standardized measurement of the knock resistance of different blends of gasoline. Such a measure was developed in 1926 by Graham Edgar, an employee of the Ethyl Corporation, a firm jointly founded by General Motors and the Standard Oil Company of Indiana. Edgar created a scale based on the propensity of different components of gasoline to induce knocking. One of them, *n*-heptane, caused a great deal of knocking, whereas another, isooctane, produced none. Accordingly, the first was given a rating of zero, and the latter was given a rating of 100. Mixtures of the two were used for intermediate numbers; e.g., gasoline with an octane number of 90 was the equivalent of a mixture of 10 percent *n*-heptane and 90 percent isooctane.

Through the addition of tetraethyl lead and other additives, it is possible to produce gasoline with an octane number above 100, as is the case with some aviation fuels. On the other hand, because of the removal of lead due to its harmful effect on catalytic converters, the octane rating of automotive gasoline is generally lower today than it was in the 1960s and early 1970s. Improvements in combustion chamber design and the use of fuel injection and computerized ignition controls have reduced the knocking tendency of car engines, allowing the use of lower octane gasoline in high-compression engines.

See also CATALYTIC CONVERTER; ENGINE, FOUR-STROKE

Oedipus Complex

The Oedipus complex is a hypothetical set of problems faced by any adolescent boy in his psychological development. Sigmund Freud (1856–1939) recognized elements of this complex as early as 1897, when undergoing his own self-analysis, and described it in greater detail, beginning with the second edition of his book *The Interpretation of Dreams* and also in his later work *Totem and Taboo*. As a true complex, it is a group of psychological problems that all derive from the boy's situation and that challenge his continuing development.

Freud assumed that a boy's first relationship is with his mother; thus, a boy begins life, like a girl, modeling his behavior to this maternal relationship. However, unlike a girl, the boy must eventually transfer his role-modeling development to a relationship with his father. At some point, the boy must give up intimacy with his mother and embrace a masculine relationship with his father. In most indigenous societies this process of development was achieved through rites of initiation.

Since Freud believed that sexual concepts and feelings begin at a very early age, he believed that the Oedipus complex possessed powerful sexual barriers and lessons as well. An adolescent boy has begun to establish sexual feelings within his maternal relationship and, consequently, will experience sexual jealousy of his father's relationship as husband. This jealousy may provoke anger when his father is seen asking him to break away from his mother to join him in the masculine world. Indeed, to the boy who is comfortable within his mother's intimacy, the masculine world of the father may look uncomfortable and threatening. In the extreme, the boy may harbor fantasies in which he tries to eliminate his father so that he can stay with his mother forever.

This complex of sexual feelings and problems is, indeed, the source of the name. Oedipus was the tragic figure of Greek drama who, having been exposed as a baby, was inadvertently returned to Thebes as a young man. Along the way and propelled by the hubris of the

immature male, Oedipus kills the king of Thebes (his own father) and, finally, marries the queen (his own mother). This succession of accidents eventually brings down a natural disaster upon the people of Thebes. The accidents are uncovered and Oedipus, in full awareness of what he has done, sinks into remorse and self-destruction.

The Oedipus complex is a psychologically potent situation in the development of a boy. In normal development, with understanding parents and correct management, the boy makes it through these situations, with all of their powerful feelings, and emerges psychologically healthy, in the masculine world and with sexual interests redirected outside of the family. However, as Freud observed, a misguided passage through this situation may bring about a host of abnormal psychological difficulties in later life.—T.B.

Office of Technology Assessment

Most of the governments in the industrialized world have set up agencies to predict, promote, monitor, and regulate technological change. Unlike many other countries, however, the United States does not have anything resembling a ministry of science and technology. Congress and its various committees and subcommittees often are involved with scientific and technological issues, and the executive branch usually includes a presidential science adviser. However, until 1995, the most extensive work on the possible consequences of technological change is done by an agency attached to the Congress, the Office of Technology Assessment (OTA).

The Office of Technology Assessment was established in 1972 as a result of the efforts of Congressman Emilio Daddario, a Democratic congressman from Connecticut and at the time chairman of the House Subcommittee on Science, Research, and Development. OTA was not to be a policymaking agency but one that improved legislation by providing information to Congress regarding various aspects of technology and technological change. As the legislation establishing the Office put it, OTA was to be ". . . an aid in the identification and consideration of existing and probable impacts of technological application. . . ." Although it was seen as a neutral agency, the formation of OTA also had a political aspect. President

Richard Nixon was known to be uninterested in receiving advice on matters involving science and technology, and in fact he abolished his own Office of Science and Technology in 1973. OTA was therefore seen both as a source of information and as a way for Congress to gain some leverage over the President when policies involving technology came to the fore.

Although the formation of OTA owed something to political maneuvering, it always attempted to avoid partisanship. OTA was overseen by a board comprised of six members of the House of Representatives and six members of the Senate, with Republicans and Democrats having equal representation. The chairmanship of the board alternated between a member of the House and a member of the Senate. The board appointed OTA's director, who served as its chief executive officer.

OTA took up issues that had been suggested by congressmen and their staff, but it also initiated projects on its own. Compared to most federal agencies, OTA was a tiny organization; in 1995, it had a staff of about 150 and an annual budget of $22 million. It was able to conduct some research in house, but much of the research was done by outside contractors, usually academics in universities or research organizations.

The Office of Technology Assessment investigated a broad range of subjects, everything from an examination of the biological factors underlying drug addiction to an evaluation of the Social Security Administration's plans to upgrade its computer facilities. In the years immediately following its creation, OTA was criticized for not tackling controversial topics, and for its infrequent involvement with the two largest components of the federal science and technology budget, defense and space. In recent years, however, it tackled some highly controversial issues, most notably the Strategic Defense Initiative or "Star Wars" program. Here, however, it ran into trouble on a number of fronts. The publication of the OTA report on SDI required a classification review of 9 months, as various agencies concerned themselves with the possible divulgence of secret information. The three chapters that dealt with survivability of the system were deleted as a result of Defense Department objections.

OTA's review of the Strategic Defense Initiative demonstrated that technology assessment cannot always be conducted in a politically neutral atmosphere. More than this, OTA's critical evaluation of some tech-

nologies threatened the agency's very existence. Being on the "wrong side" in some controversial issues has undercut OTA's political support in Congress, especially after the Republicans attained a majority in both houses of Congress in 1994. The Republicans targeted OTA for extinction, arguing that its functions could be eliminated or transferred to other government agencies, such as the National Academy of Sciences. Despite strong bipartisan support outside Congress, the Office of Technology Assessment was closed down in 1995.

See also NATIONAL ACADEMY OF SCIENCE; STRATEGIC DEFENSE INITIATIVE

Oil and Gas Exploration

In the early days of the oil industry, the presence of oil in an underground reservoir was indicated when it seeped to the surface or showed up in brine wells. Wildcatters also drilled for oil guided by little more than intuition and hope. These efforts met with success only occasionally; on average, 29 out of 30 wells ended up as dry holes.

These odds were improved considerably as geological knowledge began to be applied to oil and gas exploration. Since oil and gas deposits tend to appear only in certain kinds of underground structures, an understanding of local geological conditions could be put to good use. Geologists can gain a great deal of information by observing the local terrain and by collecting rock samples. Knowledge gained in this way is augmented by the examination of aerial photographs that provide detailed views of an area's geomorphology, vegetation, fracture patterns, and soil characteristics; all of these can provide important clues about the likely presence of oil or gas. In recent years, photographs taken from airplanes have been complemented and in some cases replaced by highly detailed images that are relayed from orbiting satellites.

Exploration for oil and gas also makes use of a number of on-the-ground technologies. One of these is based on the detection of minute variations in the Earth's gravitational attraction that are caused by different geological formations beneath the Earth's surface. For example, the greater density of igneous rocks causes them to exert more gravitational pull than sedimentary rocks. By using an instrument known as a *gravimeter* (a device that dates back to the end of the 19th century), geologists can construct a subsurface profile that may indicate the presence of relatively dense rocks close to the surface, a possible indication that a dome has been formed by an upward-thrusting sedimentary layer. Domes of this sort are often found to contain oil or gas. The use of a gravimeter may also lead to the discovery of a salt dome, another structure that often indicates the presence of oil or gas.

Magnetism, another one of the Earth's natural features, is the basis of another method of oil exploration. As with variations in gravitational attraction, differences in local magnetic fields indicate the presence of different kinds of subsurface formations. An analysis of magnetic variations can therefore provide important information about what lies underground. Practical magnetometers date back to 1870, but they were not commonly used for mapping geological structures until the mid-1920s. Unlike a gravimeter, which produces a series of discrete readings, a magnetometer provides a continuous record of variations in a region's magnetic field. One common practice is to trail a magnetometer behind an airplane, allowing a large area to be surveyed in a short space of time.

More precise information about underground features can be obtained through the use of seismography. This is a procedure that entails initiating tiny artificial earthquakes and then recording the speed of the resultant sound waves. Since different kinds of rocks reflect, transmit, and refract sound waves at different speeds, the measurement of these speeds provides important clues about the geology of an area. Seismic tests sometimes begin with the detonation of an explosive, but mechanical percussion is more commonly used today. However it is conducted, seismic testing is an expensive process that is largely confined to areas that already have given some evidence of the presence of oil or gas reservoirs.

Drilling continues to be a key part of the exploration process, with much more involved than simply sinking a borehole and waiting to see if oil shows up. One important aspect of exploratory drilling is the examination of material brought up from a well in order to determine the geological profile of the surrounding area. This may entail an analysis of the drill cuttings by geologists, geophysicists, geochemists, paleontologists, and radiologists, all of whom use their expertise to provide information about the area being drilled. If the formation allows it, a special bit can be used for the extraction of solid core of considerable length. In this way a significant portion of the geological strata can be examined. One often-used procedure is the production of a borehole log. Developed in France by Conrad (1878–1936) and Marcel (1884–1953) Schlumberger in the early 20th century, it measures the electrical resistances of different segments of the bore as a means of mapping the types of rocks residing at different layers of the subsurface.

In the past, the data obtained through these various methods were recorded as graphs on long sheets of paper. Today, the data are fed into computers that analyze a large number of variables in order to determine the likely occurrence of oil or gas. Yet for all their sophistication, modern exploration technologies do not guarantee success; even today only 1 exploratory well out of 10 will yield oil or gas in commercial quantities. To a significant degree, oil and gas exploration is still a wildcat enterprise.

See also ARTIFICIAL INTELLIGENCE; REMOTE SENSING

Oil and Gas Pipelines

By the 4th-century C.E., the Chinese were using hollow bamboo tubes to transport natural gas to the places where it was used. Much later in the United States, the first successful attempts to drill for petroleum were followed by the laying of pipelines. In 1865, a short line constructed of 5.1-cm (2-in.) diameter cast-iron pipe began to carry oil from a field in Pennsylvania to a link with the Oil Creek Railroad. The first pipeline directly connecting an oil field with a refinery was built in the Pittsburgh area in 1874, and 5 years later the Allegheny Mountains were traversed by a 177-km (110-mi) pipeline.

The expanded demand for petroleum products stimulated by the growth of automobile ownership led to accelerated pipeline construction in the 1920s. At that time, however, pipeline construction was impeded by the high cost of connecting pipe segments and burying the pipe in the ground. The development of improved welding techniques in the 1930s reduced the time and expense of joining pipe segments, while the use of mechanized earth-moving equipment greatly facilitated the excavation and filling of pipeline trenches. During World War II, fears about the vulnerability of tanker ships motivated the U.S. government to construct the "Big Inch" (it was actually 24 inches [61 cm] in diameter) and "Little Inch" (20 in. [50.8 cm] in diameter) pipelines to connect fields in Texas with the East Coast. The accelerated construction of pipelines in the postwar years did much to reduce transportation costs for petroleum and petroleum products. Pipelines were even more important for the natural-gas industry, for the pipeline transmission of gas largely put an end to the wasteful practice of flaring off gas at the wellhead because there was no way to get it to potential users.

An oil pipeline is not confined to the transmission of only one product at a time. Provided that two or more fluids are not dissimilar (different grades of gasoline, for example), it is possible to send them in sequence, with the interface between two different batches determined by a recording gravitometer, or delineated by dye or radioactive isotopes. When different products, such as gasoline and diesel fuel, are sent through a pipeline, special plugs or inflatable rubber spheres are used as a mechanical separation between the two batches.

Pressure in petroleum pipelines is maintained by pumping stations, while compressor stations do the same thing for gas pipelines. These stations are placed every 130 to 240 km (80–150 mi), depending on the type of terrain being traversed. Pumping pressures and flow rates are monitored and automatically controlled from a central station. Since the products flowing through pipelines often leave residue behind, from time to time it is necessary to clean the interior walls with tools known as *pigs*. These enter the pipeline through special traps and allow the cleaning to be done even while products are flowing.

Pipelines have long been regulated by the government. In the United States, the control of pipelines by Standard Oil helped that company to nearly monopolize the oil industry in the early years of the 20th century. In order to open up the use of pipelines, the Hepburn Act of 1906 declared that pipelines moving products across state lines are "common carriers" and must carry the products of any shipper. Prices charged for using a pipeline were set by the Interstate Commerce Commission until 1977. Shorter, privately owned pipelines that carry products between two plants or terminals are not subject to the common-carrier rule. Government authority has also been important to pipeline constructors, for they can invoke the right of eminent domain, the government-sanctioned power to secure a right-of-way in exchange for the payment of a negotiated fee.

Lying underground for the most part, pipelines tend to be out of the public eye. Occasionally, however, a pipeline becomes a major political issue. The Alaska oil pipeline, completed in 1977 after the expenditure of $9 billion, was bitterly opposed by some environmentalists. The pipeline traversed 1,290 km (800 mi), from Prudhoe Bay to Valdez, half of it over permafrost and tundra. Concerns were voiced that the pipeline would cause irreversible damage to the permafrost and disrupt a fragile ecosystem. The political struggle that ensued delayed the pipeline's completion for nearly 5 years and was resolved only after some extraordinary measures were taken. Most importantly, unlike most land pipelines that are buried beneath the surface, the portion of the trans-Alaska pipeline that traverses the permafrost is supported by heat-resistant trestles so that the pipeline's heat does not damage the ground below it.

See also NATURAL GAS; WELDING, ELECTRICAL; WELDING, GAS

Oil Refining

Petroleum, or "crude oil," is a mixture of a variety of hydrocarbons. Petroleum was used in ancient times as a medicine, and in China it was used as a fuel and a lubricant for cart axles as early as the 3d-century B.C.E. None of these applications was very important, and petroleum became a valuable substance only in the latter half of the 19th century, when kerosene began to displace whale oil as a fuel for illuminating lamps. At first, petroleum was refined through distillation, a physical process that separated its various components. The petroleum was heated in a reaction vessel, causing each substance or "fraction" to boil off in sequence, substances with the lowest boiling points first, such as gasoline, followed by kerosene and other substances. The vaporized fractions were then returned to a liquid form by running them through a condenser. After all of the volatile substances were removed, a residue of asphalt was all that was left in the reaction vessel.

At the time that petroleum was being refined by distillation, the gasoline it produced was a nuisance. It was a dangerous, highly incendiary substance that, because it had few uses, was often dumped into a conveniently situated body of water. But the situation began to change early in the 20th century. Automobiles with internal combustion engines fueled by gasoline were being produced in increasing numbers, while at the same time kerosene lamps were being displaced by electric lights. As a result, oil refiners were faced with the need to change the mix of products being extracted from crude oil.

Gasoline is not a single substance; it is a mixture of various hydrocarbons that are distinguished by the number of carbon and hydrogen atoms in each molecule and the way that these atoms are arranged. Gasoline thus consists of an assortment of paraffins, pentanes, hexanes, heptanes, octanes, isooctanes, and alkenes, along with various additives the prevent vapor lock, improve starting, and so on. At first, gasoline could only be produced through distillation. It is now possible to convert heavier fractions into gasoline. This is done by running these fractions through a second distillation process, this time at a partial vacuum. This allows these fractions to be vaporized at a lower temperature, thus reducing their tendency to decompose into elemental carbon and hydrogen. Among the products of this stage is a substance known as *gas oil*; this is put through a breakdown process that turns it into gasoline and other substances, a process known as *cracking*.

During the early decades of the American oil industry, control over the refining process was one key to monopolizing the industry and reaping enormous wealth as a result. In 1873, John D. Rockefeller's Standard Oil Company controlled 10 percent of refining capacity in the United States; 7 years later it controlled 90 percent. It used its near-monopoly position to extract massive profits, making Rockefeller one of the world's richest men. Standard Oil held its dominant position until 1910, when the U.S. Supreme Court ruled that the firm was in violation of the Sherman Anti-Trust Act. As a result, Standard Oil was broken up into a number of smaller companies.

See also CATALYTIC CRACKING; FUELS, FOSSIL; THERMAL CRACKING

A portion of a large oil refinery in Southern California (courtesy ARCO).

Oil Shale

The great majority of the world's energy is obtained from fossil fuels. One fossil fuel that is little used today but may be of great importance in the future is oil shale. Oil shale is a sedimentary rock that contains a mixture of various high-molecular-weight fatty acids and their salts known as *kerogen*. An early stage in the formation of petroleum, the kerogen in oil shale has remained in its present state because the sediments did not remain buried long enough or deep enough to be converted to petroleum. Kerogen is mostly insoluble in petroleum solvents, but it can be heated and then subjected to catalytic hydrogenization to produce an oil that can be refined like ordinary petroleum. The amount of oil that can be extracted from the shale is variable, ranging from 25 liters per metric ton (6 gal per ton) to more than 220 liters per metric ton (52 gal per ton). In the latter case, more than half of the rocks' volume is organic matter.

The amount of energy that might be extracted from oil shale is enormous. The organic material estimated to be locked up in sedimentary rocks has been estimated to include 1.1×10^{16} metric tons (1.2×10^{16}) tons of carbon. This is 1,000 times the amount of carbon contained in known coal reserves. In terms of petroleum equivalent, the world's oil shales contain 5.2×10^{11} m^3 (3.3×10^{12} barrels) of oil, more than all of the world's proven and likely reserves of petroleum.

It is, however, not certain that significant amounts of energy will be obtained from shale oil in the years to come. Although the total resources of shale oil are vast, they are scattered over large geographic regions. In many of these places, the amount of oil is small relative to the rock that contains it. Moreover, all existing methods of extracting oil from the shale entail costs that are greater, sometimes considerably greater, than those associated with drilling for petroleum.

The extraction of oil from the shale deposits is done in two ways. The first begins with mining the shale, crushing it, and then extracting the oil by heating the crushed shale at 460° to 480°C (860°–914°F) in a large retort. One of the problems with this method is that it produces vast deposits of rock that may disfigure the natural environment. This is not a problem when the oil is extracted *in situ*. The extraction can be done in two ways. In the first, a hole is drilled into the shale, fracturing it. The oil is then extracted through the application of heat; the shale at the top is ignited, and as it burns it heats the rocks below. Oil flows to the bottom, from where it is pumped to the surface. In the second method, some of the shale is mined out, and the remaining shale is broken and put into the mined-out area. Again, heat is used to extract the oil. Both methods reduce the amount of waste brought to the surface, but there remains the danger of polluting underground water.

The financial and environmental costs of producing energy from shale oil will be difficult to overcome. But should the price of petroleum reach high levels for a prolonged period, shale oil may become an attractive option for a world accustomed to consuming vast amounts of energy.

See also COAL; DRILLING, ROTARY; FUELS, FOSSIL; OIL REFINING

Oil Spills

Oil spills are one of the hazards of living in a petroleum-dependent civilization. The damage that can be done by an oil spill was dramatically illustrated in 1989, when the tanker *Exxon Valdez* ran aground in Prince William Sound, Alaska. The accident caused the spillage of 38.2 million liters (10.1 million gal) of oil and produced a major environmental disaster. Oil eventually polluted 1,450 km (900 mi) of coastline and killed 330,000 to 390,000 waterfowl and 3,500 to 5,500 sea otters. Many years will pass before the affected areas return to their previous state.

Yet for all the damage it did, the *Exxon Valdez* spill was by no means the largest. An incomplete list of notable oil spills caused by tanker accidents includes the following:

> Hull failure of the *World Glory* off the coast of South Africa (1968): 51 million liters (13.5 million gal)
> Grounding of the *Sea Empress* at St. Ann's Head, Wales (1996): 72 million liters (19 million gal)
> Grounding of the *Braer* off the Shetland Islands (1993): 98 million liters (26 million gal)

Grounding of the *Urquiola* at La Coruna, Spain (1976): 111.3 million liters (29.4 million gal)

Grounding of the *Torrey Canyon* off Land's End, England (1967): 132.1 million liters (34.9 million gal)

Grounding of the *Amoco Cadiz* near Portsall, France (1978): 248.3 million liters (65.6 gal)

Collision of the *Atlantic Empress* and the *Aegean Captain* off the coast of Trinidad and Tobago (1979): 333.9 million liters (88.2 million gal)

While tankers figure prominently in the history of oil spills, the largest oil spill was the result of an accident at an offshore platform in 1979. Before it was capped, 556.5 liters (147 million gal) of oil flowed into the Gulf of Mexico. Although they make the headlines, tanker accidents on average account for only 12.5 percent of the oil spilled throughout the world, while oil rigs account for 1.5 percent. Of all the oil spilled in the ocean, 37 percent results from municipal and industrial drainage and runoff, while natural seepage accounts for another 8 percent. Intentional dumping of waste oil by ships at sea is another major cause of ocean pollution.

Although spills from tankers and offshore platforms account for a relatively small portion of the oil spilled into the ocean, the concentration of spills in relatively small areas greatly amplifies their impact. The extent of the damage caused by a spill depends on a number of variables: the size of the spill, the type of oil (light oil is more likely to evaporate and disperse before it hits the shoreline, but it is more toxic than heavier portions of oil), the weather, and the time of year. Winds can help to disperse spilled oil, but they can also whip it into an emulsion that is hard to pump and lasts for weeks.

A number of techniques have been developed to deal with oil spills. One makes use of containment devices known as *booms*; these confine the spill and allow the oil to be vacuumed up or soaked up by absorbent materials. Alternatively, chemical dispersants can be used to break up the oil into tiny droplets. The value of dispersants continues to be debated. They can prevent an oil spill from migrating to a coastline, but they do not remove the oil; instead, they produce a highly toxic concentration of oil that can do considerable damage to marine life. No matter what technique is used, action must be taken within a few hours of the spill.

These efforts result in the recovery of no more than 15 percent of the oil lost in a large spill. A sizable portion of the remainder usually ends up along a coastline, where it can pose a severe threat to plant and animal life. Some cleanup measures, such as spraying hot water at high pressures, may cause more harm than good. One promising technique attacks spilled oil through the use oil-consuming bacteria. An oiled area is sprayed with fertilizer, which stimulates the growth of naturally occurring bacteria, which in turn convert the oil to fatty acids, carbon dioxide, and water. However, this technique works only if the oil has already settled evenly over the shoreline, a process that may take a fair amount of time.

In response to the *Exxon Valdez* disaster and other previous oil spills, the U.S. government enacted the Oil Pollution Act of 1990. Its provisions include the phasing in of double hulls for large oil tankers, so that by 2015 the entire fleet will be so equipped. The act also requires more comprehensive contingency planning for massive spills, more federal oversight of tanker operations, and the coordination of federal efforts with those mounted by the industry-sponsored Marine Spill Response Corporation. These are useful measures, but given the volume of tanker traffic and the inherent imperfections of human creations, future oil spills are all but inevitable.

See also FOSSIL FUELS; "NORMAL ACCIDENTS"; TANKERS

Operations Research

Managers in the public and private sector constantly face decisions in circumstances in which they have some information but also a great deal of uncertainty. The field of operations research (OR) emerged as a way of providing managers with tools that allow them to analyze such decisions. The field began in the early part of the 20th century with the scientific approach to management developed by Frederick W. Taylor (1856–1915). The field underwent much development during World War II when the military found that it had many problems in need of quantitative analysis. The operations research techniques developed during the war proved to be extremely useful when industry and government began to incorporate them into organizational decision-making process. Today, one finds these techniques used quite widely. In many textbooks, the field is referred to as operations research, management science, or quantitative analysis. All three terms describe the attempt to use systematic analysis in organizational decision making.

To illustrate one kind of OR, consider the following example. Imagine that you manage a small food concession stand in a city park. Each day you must make a decision about the amount of food to prepare for sale. The main factor that affects your decision is weather. If the weather is good, attendance at the park will be high, and you will sell lots of food. If the weather is bad—say it rains—attendance will be low, and you will sell a small amount of food. Unfortunately, you cannot be sure about the weather. The information you have is an estimate from the weather bureau as to the likelihood of rain. Using that information, you must somehow estimate what the best decision would be: Prepare a large amount of food costing $1,000 or a small amount of food costing $400.

We will look at this process in general terms and then return to our example to illustrate one simple technique used in quantitative analysis. OR researchers will typically outline a process that includes the following steps: Define the problem, develop a model that illustrates the process, gather relevant data, analyze the data to identify a possible solution, test the solution, analyze the results to evaluate the solution, and implement the solution. This process makes OR seem to be a neat, linear process, but it must be kept in mind that the textbook description of the process is rarely found in the real world. What this description does alert us to are the various elements of the process, elements that are utilized in one way or another in the OR process.

Let us now return to our small city park. As park concession manager, you have two choices: Prepare a large amount of food or a small amount. But you also have two possible states of nature to anticipate, rain or no rain. Using historical data from the concession stand, you are able to construct the following table in which the table entries were your net revenues in each condition.

Food amount	No rain	Rain
$1,000	$200	$ – 150
$ 400	$ 50	$ – 25
Probability of no rain/rain	.70	.30

This table tells you that historically when you spent $1,000 for food, you realized a net revenue of $200 when it did not rain, but you lost $150 (from food that had to be thrown out) when it did rain. Alternatively, when you spent $400 for food, your net revenue was $50 when it did not rain but –$25 when it did rain. Because you had prepared less food in anticipation of rain, you had to throw out less food. Using the information in this table, we can calculate what OR researchers call the "expected value" (EV) of each alternative. The general formula for EV is:

EV = (payoff alternative 1) × (probability alternative 1) + (payoff alternative 2) × (probability alternative 2) + . . . + (payoff alternative N) × (probability alternative N)

Thus, you would have a term in the formula for each alternative you identified for your problem.

EV($1,000) = (payoff no rain) × (probability no rain) + (payoff rain) × (probability rain)

EV($400) = (payoff no rain) × (probability no rain) + (payoff rain) × (probability rain)

As the table above indicates, the weather bureau has announced that there is a 30 percent chance for rain today. So, in your case, the two alternatives, no rain and rain, have the following expected values:

EV($1,000) = (200)(.70) + (– 150)(.30) = $95.00
EV($400) = (50)(.70) + (– 25)(.30) = $27.50

The results of this calculation tell us that, when the chance of rain is 30 percent, our expected net revenue will be larger if we spend $1,000 for food than if we spend $400 for food. Your decision on this particular day, then, will be to spend $1,000 on food.

In this example, you started with a simple problem: Should I prepare a large or a small amount of food for my food stand? Taking into account the probability of rain today and the history of net revenue on rainy and clear days, you were able to determine that today, with a 30 percent chance of rain, you could expect more net revenue if you prepared a large amount of food. You should note that, as the chance of rain increases above 30 percent, the expected values will ultimately favor preparing a small amount of food. See if you can figure out what chance of rain would lead you to choose the small alternative.

In the simplest terms, OR consists of a set of tools for aiding managers in making better decisions. In addition to calculations based on expected value, OR researchers utilize statistics, forecasting, inventory control techniques, linear programming, queuing techniques, and simulation to analyze problems.—J.D.S.

See also SCIENTIFIC MANAGEMENT; STATISTICAL INFERENCE

Further Reading: Barry Bender and Ralph M. Stair, Jr., *Quantitative Analysis for Management*, 1991.

Opium

Opium is extracted from *Papaver somniferum*, a poppy that is widely grown in parts of Asia and elsewhere. The opium is obtained by cutting into the plant's seed capsule soon after the petals have fallen. The juice that oozes from the cuts is allowed to harden and is then collected. In this form, opium contains at least 20 chemical compounds known collectively as *alkaloids*, including codeine and morphine. Use of opium leads to addiction, and breaking the habit requires a difficult and painful process of withdrawal.

When smoked or ingested, opium acts as a narcotic (the term *opiate* is also used) that reduces pain, diminishes anxiety, and creates a sense of euphoria along with slight drowsiness. It produces these effects by acting on the central nervous system. In the early 1970s, it was learned that some of the components of opium bind to receptors in cells located in the spinal column and the medial thalamus of the brain. It has been hypothesized that opiates produce their characteristic effects by slowing the rate at which these cells can transmit pain messages. It was later found that the body produces substances that also lock on to these receptors, thereby producing a natural means of alleviating pain.

Opium's effects have been known for centuries. Opium was used by the Egyptians during the 2d-millennium B.C.E., was known to the ancient Greeks, and was introduced into India and China by Arab traders during the 7th-century C.E. During the 19th century, opium use in China was promoted by British and other foreign traders, resulting in widespread addiction. One Chinese official hoped to stop the importation of opium by importuning Queen Victoria: "We have heard that in your country the people are not permitted to inhale the drug. If it is admittedly so deleterious, how can seeking profit by exposing others to its malefic power be reconciled with the decrees of Heaven?" These words did nothing to stem the traffic in opium. Efforts by the Chinese government to put an end to the importation of "foreign mud" in two "opium wars" (1839–1842 and 1856–1858) ended in China's defeat and a humiliating loss of some elements of national sovereignty.

Opium use was also rampant in the West at this time. Many physicians prescribed it for treatment of diarrhea, fever, pain, sleeplessness, and the symptoms of tuberculosis. It was also widely used as anesthetic for surgical operations. Opium was consumed in widely available medications like laudanum, a saffron-flavored combination of opium and sherry. Laudanum was especially popular with parents who used it to treat cranky children. Casual use of opium decreased as its addictive hazards became known and stricter government regulations came into play. Efforts by the U.S. government to control the opium trade were at least partially motivated by the anti-Chinese sentiment that was rampant in the late 19th century. Many lurid tales were told about Chinese opium dens and the lost souls that frequented them. In 1887, the importation of opium by Chinese was forbidden, but no such strictures applied to whites. In 1909, the importation of smoking-grade opium was forbidden, and in 1914 penalties for the illegal use of opium were increased. By this time, more potent opium-derived drugs, first morphine and then heroin, had largely replaced opium as drugs of choice.

See also ANESTHESIA; HEROIN; MORPHINE

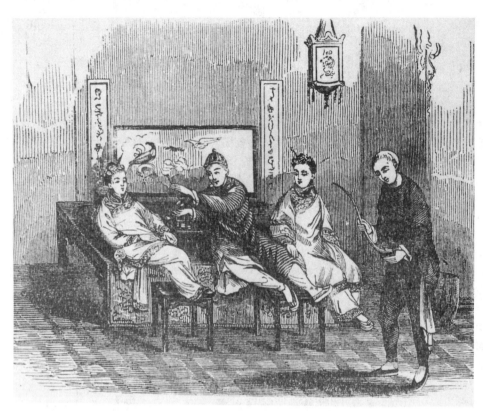

A 19th-century depiction of opium use in China (courtesy National Museum of Medicine).

Oral Rehydration Therapy

Dehydration caused by diarrhea has killed vast numbers of people, claiming 150 million in the last 4 decades alone. In many parts of the world, this condition is brought on by cholera, which claims 5 million lives each year, but it can also be caused by a number of other waterborne pathogens. For many years, the standard treatment for dehydration was the intravenous injection of a saline solution. Intravenous fluid injection is effective, but it requires equipment and skills not readily available in many places where the need is the greatest.

In the early 1960s, Robert A. Phillips, a retired U.S. Navy physician, experimented with a rehydration therapy that only required the patient to drink a solution of glucose, sodium bicarbonate, and potassium bicarbonate. Not only was this "cholera cocktail" easier to administer than an intravenous treatment, it quickly returned to the body the vital electrolytes that had been lost due to diarrhea. Phillips went on to head a cholera research laboratory in Dhaka, then a part of Pakistan and now the capital of Bangladesh. The laboratory became the base of operations for two other American

doctors, David Nalin and Richard Cash, who became strong advocates of oral rehydration therapy. Early tests of oral rehydration therapy produced mixed results until the researchers learned how to precisely coordi-

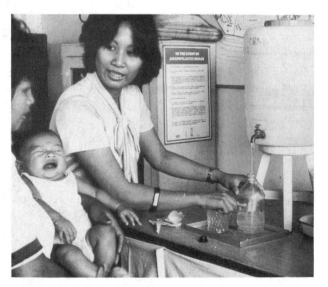

Preparation of a solution for oral rehydration (courtesy National Library of Medicine).

nate fluid intake with fluid loss. Subsequent tests conducted in the Bangladeshi town of Matlab produced highly encouraging results. Oral rehydration therapy then demonstrated its effectiveness when it was used to treat cholera victims during Bangladesh's war of independence against Pakistan.

Since that time, the United Nations Children's Fund (UNICEF) has estimated that the lives of 1 million children have been saved by oral rehydration therapy, and that diarrheal diseases are no longer the leading cause of death among children. However, with oral rehydration therapy being used worldwide to treat only 25 to 30 percent of those suffering from diarrhea, a quarter of the deaths of children under the age of 5 are still being caused by cholera and other diarrheal diseases.

Organ Transplantation

Ancient surgeons dabbled in skin grafting, limb transplantation, and enough attempts at xenografting (transplants between species) to give the mythical chimeras the tiniest basis for belief. Later ages recorded successes, particularly with skin autografts (transplants from one location to another on the same patient), but successful whole-organ transplantation awaited developments in blood typing, surgical procedure, and immunology during the 20th century.

As with organ transplants, blood transfusions had a long and mostly unsuccessful history until Karl Landsteiner (1868–1943) demonstrated the existence of the first grouping system in 1900. Subsequently, 23 other blood-grouping systems were identified, allowing more precise phenotype matching for both blood and organ transfer. Shortly after discovering the ABO groupings, the Austrian-born Landsteiner moved to the United States and continued his career at the newly founded (1901) Rockefeller Institute for Medical Research, which also soon attracted the pioneering French vascular surgeon Alexis Carrel (1873–1944). Carrel first became interested in the problem of suturing blood vessels when French President Sadi-Carnot was stabbed to death in 1894. While it was possible then to stitch ends of vessels together, any affixed exposure of the outside surface of one end to the inside of another caused clots to form. Over a number of years, Carrel experimented

on animals using extremely fine needles and paraffin-coated thread, before discovering that he could split the ends of both vessels, fold them back like shirtsleeve cuffs, and suture them to each other so that only their inside surfaces touched. He announced his procedure in 1902, arrived at the Rockefeller Institute 2 years later, and between 1904 and 1908 turned his experimental procedure, improved with aseptic techniques, into accepted practice.

Carrel realized that the new surgical techniques opened possibilities for allografting (between different individuals of the same species) whole organs, but his experiments with kidney transfer in dogs showed the same kind of engraftment failure several European researchers had observed (then mainly attributed to surgical mistakes). A German surgeon performed the first keratoplasty (cornea transplant) in 1906, but the absence of blood vessels in the cornea removed the need for ABO matching and reduced immune-response rejection. Until the 1950s, then, organ transplantation in human beings remained limited to skin and venous autografting and to cornea allografting.

Knowledge that engraftment failure actually resulted from an immune response began with Australian immunologist Macfarlane Burnet's (1899–1985) theory that, during embryonic life, a body develops tolerance to potentially antigenic compounds, whether its own or from a foreign source. After birth, that tolerance constituted the nonimmunogenic "self," which was distinguished from "nonself" substances that stimulated an immune response. Soon after Burnet put this notion forward in 1949, British zoologist Peter Medawar (1915–1987) tested it with skin grafts on mice, ultimately showing that an inoculation of cells from a future donor into fetal mice generated tolerance when those mice received the donor's graft later in life. Further unraveling the genetics of this phenomenon, George Davis Snell (1903–1996) of the Jackson Laboratory in Maine proposed that certain histocompatibility genes existed, producing, as Jean Dausset (1916–) in Paris showed, lymphocyte antigens that controlled the acceptance or rejection of organ transplants. In human beings, the major histocompatibility complex was found on the short arm of chromosome 6, encoding highly polymorphic expressions of seven basic antigens. By matching these antigens between donors and recipients, surgeons tremendously improved the chances of a graft or transplant being accepted.

Matching antigens, however, was not a problem Joseph E. Murray (1919–) faced when he performed the first successful kidney transplant in December 1954 at Boston's Peter Bent Brigham Hospital; his patients, Richard and Ronald Herrick, were identical twins. Murray heterotopically implanted one of Ronald's kidneys in his brother's pelvis (earlier attempts had sited the armpit and groin); Richard survived for 8 years. Murray realized that to extend this procedure to less-closely related patients, he needed a way of suppressing the host's immune response, as had first been done in 1951 with cortisone to prolong skin grafts. Antimetabolite drugs, such as 6-mercaptopurine and azathiorine, which disrupted DNA synthesis, inhibited reproduction of lymphocytes or any other rapidly expanding cells. Murray used these as preoperative immune suppressants. By 1980, cyclosporine, a drug discovered by Swiss immunologist Jean Borel that selectively targets the growth cycle of Helper T cells (consequently causing fewer side effects), became the preferred chemotherapy in organ transplantation—increasing, for example, the success rate of kidney transplants from 50 to more than 80 percent.

By the spring of 1963, all the elements existed for successful allotransplants of visceral organs in human beings, proceeding with the liver and the lung. At the Denver Veterans Administration Hospital, Thomas Starzl removed the whole liver of a white man who had died of a brain tumor and implanted it in a black man suffering from an incurable hepatoma. Infusing large amounts of fibrinogen (to promote clotting) prevented the patient from bleeding to death on the operating table, as had happened in an earlier attempt. But the infusion also promoted clots that passed into the lungs, causing abscesses that resulted in the patient's death 22 days later. Technically, the transplant succeeded, though it was another 20 years, during which the critical blood-handling problems were resolved, before liver transplantation became a serviceable, rather than an experimental, operation.

A little more than a month after Starzl's liver transplant, James Hardy at the University of Mississippi Hospital performed the first human lung allotransplant. His team took the left lung from a cadaver and sewed it into the chest of a patient whose left lung was removed due to an advanced bronchogenic carcinoma. Because of an undetected metastases and chronic renal disease, the patient survived for only 18 days, but the surgery showed that a lung could be transplanted and caused to function immediately by rejoining only four passages: the pulmonary artery, both pulmonary veins, and the bronchus. Infection at the suture sites, caused by recipients breathing nonsterile air postoperatively, posed serious problems (e.g., limiting use of immunosuppressants). Gradually, physicians resolved the infection difficulties and other complications, allowing even heart and lung transplantations, as first performed by Denton Cooley (1920–) at the Texas Heart Institute in September 1969.

No organ transplantation captured worldwide attention more than Christiaan Barnard's (1923–) heart homograft in December 1967 at the Groote Schuur Hospital in Cape Town, South Africa. Three years earlier, James Hardy, following his pioneering lung transplant, had transferred a chimpanzee heart to a human being. It supported the patient for about $1\frac{1}{2}$ hours after a pulse defibrillator normalized the beat. This demonstrated the technical possibilities, but an important legal and ethical problem remained. For a human to human heart transplant to succeed, the beating heart of a braindead donor had to be stopped, while the recipient's still beating heart was removed. Besides all the other risks associated with transplant surgery, then, cardiac allotransplants required that both patients "die," according to the heartbeat definition of being alive. This issue was debated but not resolved by the time Barnard transferred the heart of a male accident victim to 54-year-old Louis Washkansky, who lived for 18 days before a pulmonary infection ended his life. Surgical centers all over the world performed over 100 heart transplants in the following year, most with disappointing results, though Barnard's second patient, transplanted a month after Washkansky, survived for 19 months.

In 1968, Robert Good, at the University of Minnesota, performed the first allograft of bone marrow, from the sister to a boy suffering from Severe Combined Immune Deficiency. In that case, graft vs. host disease posed a greater risk than host vs. graft rejection, but the experiment succeeded, and that patient remains alive and healthy in 1996. Gradually, improved means of resection, histocompatibility testing, controlling immune response, and infection suppression opened a wide range of organ transplantations—of the pancreas, cartilage, bone, small bowel, parathyroid, and even adrenal medullary tissue (from fetal donors).

The National Organ Transplant Act in 1984 established a federally funded network of organ procurement organizations (of which Boston's Interhospital Organ Bank, founded in 1968, was first). That legislation also prohibited the buying and selling of human organs for transplantation. Many legal and ethical questions remain, however, in a field of medicine where thousands of prospective organ recipients outnumber available donations by about five to one.—G.T.S.

See also BLOOD GROUPS; BRAIN DEATH; DEFIBRILLATOR; SURGERY, ANTISEPTIC

Organic Fertilizer—see Fertilizer, Organic

Orphan Drugs

Many human afflictions have been prevented, cured, or alleviated by substances such as penicillin and polio vaccines. Many other illnesses are susceptible to some form of chemical treatment, but until recently pharmaceutical companies were reluctant to develop, test, and market drugs that would be used by only a few people. The unwillingness to get involved with these drugs was based on unassailable economic logic; a small market generated revenues that did not justify the cost of bringing such drugs to market. Because they have a small base of commercial support, these drugs are commonly known as *orphan drugs*.

In 1983, the United States Congress passed the Orphan Drug Act in order to encourage the development of drugs used for the treatment of "rare" disorders. This legislation defined a rare disorder as one affecting fewer than 200,000 people. This is a small number relative to a population in excess of 260 million, but it includes people who are suffering from illnesses that are by no means exotic: Multiple sclerosis, hemophilia, and cystic fibrosis all qualify as rare disorders. However, nearly half the illnesses covered by the act are quite uncommon, afflicting fewer than 25,000 people.

The Orphan Drug Act contains a number of provisions intended to stimulate the development of drugs aimed at a small clientele. It includes annual grants of $12 million for research into the treatment of rare disorders, and it provides assistance to small pharmaceutical companies as they seek to win approval from the Food and Drug Administration to market a new drug. The act also contains a substantial tax credit; a firm intending to market an orphan drug is eligible for a tax credit that offsets 50 percent of the costs of conducting clinical trials for that drug.

Orphan drugs can be patented if they are new substances. However, many orphan drugs are already-known substances for which a new use has been discovered. To give pharmaceutical firms an incentive to discover and test substances that might have new medical applications, the Orphan Drug Act grants a temporary monopoly to firms putting orphan drugs on the market. The act gives a firm an exclusive right to market an orphan drug for a period of 7 years, presumably a sufficient time to recoup the costs of developing and testing the drug.

The Orphan Drug Act has undoubtedly contributed to an increased availability of drugs used for the treatment of rare disorders; between the passage of the act and early 1995, 108 orphan drugs were approved by the FDA. On the other hand, drug manufacturers have been accused of reaping inordinately high profits on some of the orphan drugs they have brought to market. Moreover, on occasion some drug manufacturers have gained windfall profits on substances like the human growth hormone, which initially qualified as an orphan drug but ended up with a very large market.

See also FOOD AND DRUG ADMINISTRATION, U.S.; HUMAN GROWTH HORMONE; PATENTS; PENICILLIN; VACCINES, POLIO

Orthodontics

Different cultures have different standards of beauty, and people in these cultures often have to undergo significant discomfort in order to achieve these standards. In Western culture, straight, even teeth are considered to be aesthetically pleasing. Since not everyone is born with dentition of this sort, many children and adults have undergone the tooth-straightening procedures collectively known as *orthodontics*.

In the 18th century, the pioneering French dentist Étienne Bourdet (1722–1789) used arch-shaped pieces of ivory that were tied to the teeth in order to pull them into better alignment. But orthodontics was little practiced at this time; many people had lost the majority of their teeth by early adulthood, so the pressing need was for false teeth, not improved natural teeth. Beginning in the 19th century, better oral hygiene and improved dentistry allowed more people to retain their teeth. Since people kept more of their teeth into adulthood, they now had a motivation to make them more attractive. At the same time, rising personal incomes allowed more people to bear the expenses of orthodontic procedures.

An important step forward in orthodontics was the publication in 1880 of Norman W. Kingsley's (1829–1913) *Treatise on Oral Deformities as a Branch of Mechanical Surgery*. Along with systematizing orthodontic practices, Kingsley's text presented a number of practical procedures, some of which are still in use today. Orthodontics became a recognized specialty within dentistry through the efforts of Edward Hartley Angle (1855–1930). In addition to setting up his own school of orthodontia, Angle used relationship of the first molars as the basis for a typology of malocclusion that is still in use today. Angle also was instrumental in the formation of the American Society of Orthodontists and served as the organization's first president.

In addition to learning from past experience, today's orthodontia has benefited from the application of new materials. Stainless steel alloys are used for the brackets and the bands, although better-appearing plastic brackets also are used. Brackets are bonded onto the tooth surface by resin adhesives, while the bands encircling the teeth in the rear of the mouth are held in place by a cement based on phosphoric acid and zinc oxide. The archwire that applies the force to move teeth to their desired locations may be fashioned from stainless steel, but a number of alloys also are used for this purpose. The archwire is most commonly secured to the brackets by elastic bands. Some orthodontic procedures do not use fixed appliances; instead, the teeth are moved or held in place by stainless steel wires attached to a removable acrylic fixture.

See also ADHESIVES; FALSE TEETH; STEEL IN THE 20TH CENTURY

Oscilloscope

An oscilloscope is an instrument used to display variations in the voltage of a signal. In the most common kind of oscilloscope, known as a *cathode-ray oscilloscope*, a point of light moves from left to right across a screen, tracing a line as it moves. The line's form indicates voltage variations in the circuit generating the signal. If, for example, an oscilloscope is connected to a source of alternating current, the line on the screen take the form of a sine wave, the characteristic "signature" of alternating current. On the face of an oscilloscope's screen is a grid known as a *graticule*. This grid is marked out with 8 vertical divisions and 10 horizontal ones, with a distance of 1 cm (0.4 in.) usually separating two division lines. The divisions on the vertical axis indicate the voltage of the signal fed into the oscilloscope, while those on the horizontal axis are a function of time.

An oscilloscope's screen, like that of a television set, is one end of a cathode-ray tube. The history of the cathode-ray tube begins with William Crookes (1832–1919), who was able to produce an invisible discharge of electricity within an evacuated tube by connecting grids at opposite ends of the tube to a power source. The current traveled from a negative electrode (the cathode) to a positive one (the anode); for this reason, the discharges were known as *cathode rays* (it is now known that they are electrons). One of the interesting properties of these "rays" was that they produced a glow in fluorescent materials. In 1897, the German physicist Karl Ferdinand Braun (1850–1918) found that a spot of fluorescence produced by a cathode-ray discharge could be moved by a shifting electromagnetic field. In effect, Braun had created the first oscilloscope.

The simplest oscilloscopes provide a graphic indication of the waveform of a circuit while it is in operation. A more versatile instrument known as a *storage oscilloscope* does exactly what its name implies: It retains a waveform after the cessation of the signal that produced it. This allows a thorough examination of the waveform and a better understanding of the circuit that produced it.

The ability to store waveforms is an inherent quality of digitizing oscilloscopes. Unlike traditional analog oscilloscopes that feed the signal directly into the instrument, digitizing oscilloscopes first convert the signal to a digital form. In addition to facilitating waveform storage, digitizing oscilloscopes easily accommodate computerized analyses of input signals.

See also ALTERNATING CURRENT; ELECTROMAGNET; ELECTRON

Further Reading: Ian Hickman, *Oscilloscopes: How to Use Them, How They Work*, 4th ed., 1995.

Oven, Microwave

Microwaves are short (0.1 mm), high-frequency (3×10^9 Hz) waves that occupy the portion of the electromagnetic spectrum between radio waves and visible light. They are widely used today for the transmission of communications signals. In the late 1930s, microwaves began to be used for the detection of far-off objects through a process known as "radio detection and ranging," or *radar* for short. Early radar sets were powered by a device known as a *cavity magnetron*. In 1946, an engineer named Percy L. Spencer stepped in front of an operating magnetron and was surprised to find that the candy bar in his shirt pocket was beginning to melt. Intrigued, Spencer then tried some kernels of popcorn; held in front of the magnetron they soon popped. The microwaves were causing the water molecules of the food to vibrate rapidly, and the resultant heat caused the food to cook.

Spencer's employer, the Raytheon Company, was deeply involved in the production of radar sets, an enterprise that faced a shrinking market due to the end of the World War II. Cooking with microwaves seemed to offer a new commercial niche for radar technology. In 1953, the company was granted a U.S. patent for a "high frequency dielectric heating apparatus." In that year, Raytheon marketed its first microwave oven, which had been given the name Radar Range (later shortened to Radarange). It was not intended for household kitchens; weighing 340 kg (750 lb), standing 1.7 m (5.5 ft) high, and costing $3,000, it was purchased for use in restaurants, railroad dining cars, and ocean liners.

In 1955, Tappan, a manufacturer of conventional ovens and ranges, marketed a microwave oven intended for households, but it was still quite bulky. In Japan, Keishi Ogura made the miniaturization of the microwave oven possible by developing an electron tube to replace the magnetron. Amana, a domestic appliance

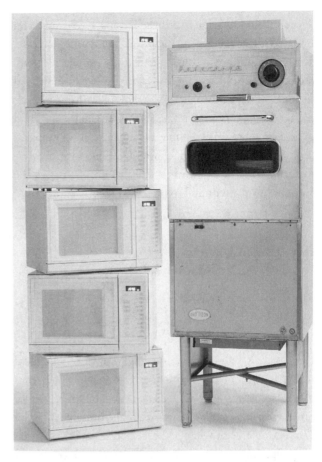

A microwave oven from the late 1940s alongside a stock of modern models (courtesy Raytheon).

maker that had been acquired by Raytheon in 1965, used this tube for a compact microwave oven it began to sell in 1967 at a price of $495. At first sales were slow and consumers wary. Many foods cooked in microwave ovens lacked the color and texture of foods prepared through conventional methods. Potential purchasers were also concerned about radiation. Some of this was misplaced; microwaves are nonionizing, which means that they cannot cause food to become radioactive by detaching charged particles from atoms and molecules. Still, there were justifiable concerns about microwave leakage. In 1971, the U.S. government required that all microwave ovens had to meet or exceed federal standards for leakage. They also had to be fitted with interlocks that prevented their doors from being opened while food was being cooked.

A surprising discovery that was followed by years of

development made the microwave oven a kitchen fixture; by 1990 three-quarters of American households had one. No less important to the microwave's popularity were the economic and social changes that affected industrial societies in the postwar era. Women's labor force participation increased, commuting distances lengthened, and in general there was less time available for food preparation. Growing numbers of consumers came to appreciate the ability of the microwave oven to cut down on the amount of time required to prepare a meal.

See also MICROWAVE COMMUNICATIONS; RADAR

Ozone Layer

Ozone (O_3) is an allotrope of oxygen, a molecule formed from three (instead of the normal two) atoms of oxygen. It is one of the constituents of photochemical smog, to which it contributes the pungent smell that makes smog so unpleasant. The ozone in smog, sometimes known as *ground level ozone*, is thought to be the product of a photochemical reaction involving oxygen and some impurities. Ozone is also created in the upper atmosphere by the action of solar ultraviolet radiation, which splits O_2 molecules into separate oxygen atoms that then combine with an oxygen molecule to form O_3. Most atmospheric ozone is concentrated in the stratosphere in a layer that begins at about 15 km (9.3 mi) and contains a peak volume of ozone at 30 km (18.6 mi); At this level, ozone is more than 1,000 times more abundant than it is near the surface of the Earth. The ozone layer is not uniform; it is thinnest at the equator and generally increases in thickness in both directions away from the equator.

Unlike the ground-level ozone found in smog, ozone in the upper atmosphere plays a highly beneficial role, for it screens out much of the ultraviolet radiation that would otherwise reach the Earth's surface. Ultraviolet radiation (short-wavelength light that is often rendered simply as UV) is important for certain life processes such as the synthesis of vitamin D in the body, but in excessive amounts it can cause skin cancer, melanoma, and glaucoma, and disrupt crucial natural processes such as plant photosynthesis and the growth of phytoplankton.

In recent years there has been much concern that the ozone layer is being eroded by the action of chlorine atoms. This happens when chlorine atoms combine with one of an ozone molecule's oxygen atoms, resulting in the formation of one chlorine monoxide (ClO) molecule and one O_2 molecule. It is believed that this process may continue for approximately 10,000 reactions, because chlorine atoms are released from chlorine monoxide when these molecules encounter free oxygen atoms. This results in the formation of an O_2 molecule and a chlorine atom, which is now free to strip individual oxygen atoms from ozone molecules.

In 1974, researchers at the University of California at Irvine published the results of a study that indicted chlorofluorocarbons (CFCs) as a major source of chlorine atoms in the ozone layer. CFCs can release chlorine atoms through photochemical processes initiated by UV light. At the time that the study was conducted, CFCs were being used extensively as refrigerants, most notably those sold under the brand name Freon, as propellants for aerosols, and for the manufacture of Styrofoam. Subsequent studies of the atmosphere conducted by the National Air and Space Agency and the National Oceanic and Atmospheric Administration have lent support to the indictment of CFCs as a major source of ozone layer depletion.

The most troubling indication of the erosion of the ozone layer has come from observations of the atmosphere above Antarctica, which have found a "hole" in the ozone layer during certain months. Ozone depletion is particularly serious in this region because of the presence of strong circular winds, known as the *polar vortex*, that funnel ozone-depleting gases over Antarctica. Reinforcing the ozone-depleting process are ice crystals that act as platforms for the reaction through which chlorine breaks down ozone molecules.

While observations of the Antarctic ozone hole have been troubling, not all scientists agree that ozone depletion is anything more than a temporary phenomenon, a manifestation of normal fluctuations in ozone levels over the Antarctic. Nor is there a consensus that CFCs are the prime cause of ozone depletion. Volcanic eruptions also produce large amounts of chlorine, perhaps 20 times as much as the 750,000 tons (682,000 tonnes) generated by the production of CFCs (on the other hand, recent research seems to indicate that most volcanic chlorine never reaches the level of the stratosphere). It has also been noted that since ozone concentrations vary by latitude, the effects of a depleted ozone layer are no greater than those caused by moving a few miles north or south of the equator. Finally, the claim is made that ozone is continually being created. Oxygen molecules are continually split by photodisassociation, and the resulting oxygen atoms are more likely to join O_2 molecules to form O_3 than they are to unite with chlorine atoms.

According to a 1996 report by the director of the National Oceanic and Atmospheric Administration's climate-monitoring laboratory, during the late 20th century about 4 percent of the world's ozone layer was lost every decade. Not all atmospheric scientists agree with this estimate; assessing the extent and nature of the problem is difficult because of the great complexity of the phenomena involved. Nearly 200 chemical reactions in upper stratosphere have been identified, many of which interact with one another in a highly complex fashion.

Despite these reservations, strong measures to protect the ozone layer have been taken by the world's governments. In 1978, the United States banned the use of CFCs in spray cans. In 1987, 43 nations signed the Montreal Protocol, which stipulated holding CFC production at 1986 levels by 1990, followed by a 20 percent reduction by 1994, and a further 30 percent reduction by 1999. Since poorer nations would have greater difficulty meeting these targets, their timetable was extended.

The role of CFCs in the depletion of the Earth's ozone layer remains a matter of debate, and their phasing out is an example of how far-reaching policies may have to be implemented in the absence of complete scientific certainty. Efforts to prevent CFC-induced ozone depletion are similar to proposed restrictions on fossil-fuel consumption in order to avert hypothesized global warming. Although differences remain regarding the severity of the ozone depletion problem, to many policymakers and members of the general public, the possible consequences of inaction are so serious that taking firm measures in the face of inadequate information is justified.

See also CHLOROFLUOROCARBONS; CLIMATE CHANGE; NATIONAL AERONAUTICS AND SPACE ADMINISTRATION (NASA); PHOTOSYNTHESIS; SMOG, PHOTOCHEMICAL; VITAMINS

P

**Pacemaker—see
Cardiac Pacemaker**

Paint

Paint is a coating that adheres to the surface of a material. This distinguishes it from dyes or stains, which are absorbed into the material being treated. The simplest paint consists of two basic elements: a pigment that provides the color, and a medium (also known as a *vehicle* or *binder*) in which the pigment is suspended. In most cases, paint also includes a solvent that thins it and helps it to cover the surface to which it is applied, as well as a drying agent. A paint "dries" when the medium and the solvent evaporate, or when it undergoes a chemical reaction that causes it to harden. In many cases, both processes occur. To illustrate, a simple paint might consist of a pigment, linseed oil (the medium), and turpentine (the solvent). After the paint has been applied, the volatile turpentine evaporates, leaving a layer of linseed oil and suspended pigment. As the oil absorbs oxygen it polymerizes, hardening the paint.

Painting is one of the oldest of human activities; cave paintings of animals and other subjects were produced about 25,000 years ago. The paint used by early artists was based on yellow, black, and red pigments derived from magnesium or iron oxide; the medium was likely hot fat. The ancient world made use of a variety pigments: various ochres for red, brown, and yellow; copper carbonate (malachite) for green; and lapis lazuli for blue. Alternatively, blue could be obtained by heating together silica, calcium carbonate, copper carbonate, and sodium carbonate. White was obtained from chalk and gypsum, or by using white lead. For paint media, people used water mixed with an adhesive made from such diverse substances as milk casein, starch, gums, and egg whites. "Encaustic" paints in a medium of hot wax were used from the 5th-century B.C.E., onwards. Oil paints, which used linseed oil as a medium, had been around since the 6th-century C.E., but did not come into prominence until the 15th century, when first Flemish and then Italian artists used them to create works of unprecedented richness and depth.

On a more prosaic level, paints have long been used on structures, vehicles, utensils, and furniture for decorative purposes. Equally important, the application of one or more coats of paint helps to seal the surface of the painted object. This retards the deterioration brought on by corrosion, mildew, and the effects of weather. Until fairly recently, the pigments used for common paint were based on common, natural materials. Barns usually were red and freight cars had a coat of "boxcar red" because these colors could be cheaply derived from common oxides. White was a popular color for houses because white lead was readily available. Only in recent decades has it become apparent that lead can pose a serious health hazard.

The paints used today fall into two general categories: solvent thinned and water thinned. Solvent-thinned paints have a pigment suspended in a volatile compound like turpentine or acetone. Some solvent-thinned paints dry through the evaporation of the solvent, while others dry by oxidation. In the former case, acrylic resins, vinyl resins, or cellulose derivatives serve as the medium (paints that use cellulose media are known as *lacquers*). Solvent-thinned paints that dry by oxidation generally use an oil medium like the aforementioned linseed oil. In this case, "drying" is actually the polymerization of the oil when it is exposed to air.

Water-thinned paints have been around for centuries. The temperas used by artists are one example of a water-thinned paint. In recent decades a new type of water-thinned paint based on latex, an emulsified polymer, has been developed. These paints are convenient to use since no special solvents are required. In addition, water-thinned paints are now required in many localities because they do not put volatile, pollution-generating solvents into the air.

Paints can be applied in a variety of ways. They are commonly put on a surface by a brush or roller, or they can be sprayed. In the latter case, the paint is atomized into tiny droplets and blown at the surface to be covered. Effective spray painting requires careful consideration of the viscosity of the paint, the choice of a properly sized nozzle, the selection of the right spray pressure, and a correct orientation of the spray gun. The paint should hit the surface at a right angle, and the distance between the gun and the surface should not vary as the spray gun is moved. It is also important to keep the gun at the right distance from the surface. If it is too far away, the paint droplets will partially dry before hitting the surface, resulting in a dull, rough finish. On the other hand, if the gun is held too close, too much paint will be delivered, causing runs and sags.

Spray painting has been made more convenient by the use of spray cans containing aerosol propellants. Aerosols have been a mixed blessing, however, for until recently the propellant used was a fluorocarbon that eroded the Earth's ozone layer. Today's spray paints use environmentally safe propellants, but they can still be a problem when they are used by graffiti artists. In many communities, aerosol paints are kept in locked cases and their sale is restricted.

While some kinds of paint are applied as liquids and others as sprays, still others take the form of a powder. Often used when maximum durability is required, powder coatings are applied by taking advantage of electrostatic attraction. The powder is given an electrical charge opposite to the object to be painted, causing the powder to cling tightly to the surface. The coated object is then heated, causing the powder to liquefy; it then hardens like a conventional paint.

See also AEROSOL SPRAYS; CHLOROFLUOROCARBONS; LEAD POISONING

Pap Test

The Pap test, Pap smear, or cervical cytologic smear test is a screening procedure used to distinguish abnormal from normal cells of the cervix, vagina, and uterus. It is used primarily to prevent cervical cancer through early detection of dysplasia, i.e., abnormal cell growth. Such changes usually take years to develop into cervical cancer, and regular Pap tests can reveal precancerous lesions and cancer at early stages when they are most easily cured. The test can also detect other female reproductive tumors, albeit much less effectively, as well as the presence of certain infections, including yeast, fungus, herpes, and the human Papilloma virus (HPV).

The Pap test is named after George N. Papanicolaou (1883–1962), a Greek-American physician trained at the University of Athens and the University of Munich who moved to New York after serving as a medical officer in the Balkan War of 1912–13. While working at the Pathology Laboratory at the New York Hospital, Papanicolaou recognized that cervical cancer is accompanied by the shedding of malignant cells into vaginal fluid. By scraping a few cells from the cervix, smearing them on a slide, and examining them under a microscope, cell changes associated with both cancer and infection could be readily identified. Others before Papanicolaou also had used vaginal smears for cancer diagnosis. Aurel Babes in particular has been recognized for his concurrent discovery of the value of smears for the diagnosis of uterine cancer. However, Papanicolaou was the first to describe the evolutionary steps in the development of cancer at the level of the individual cell, and he also became known for his method of preservation of these cells by wet fixation and precise staining. Finally, he is distinguished for his persistence in winning acceptance for cytologic examination of the cervix.

When the Pap test was first introduced, cervical cancer had caused more deaths in women than any other type of malignancy. The test became a regular part of gynecologic examinations in the United States during the 1950s, and over the following decades it contributed to a 70 to 75 percent reduction of the incidence of invasive cervical cancer. Today, cervical cancer ranks well below cancers of the breast, lungs, colon, and rectum in U.S. women.

To obtain a specimen, a clinician holds the vaginal

walls open with a speculum, inserts a cotton swab, spatular, or fine brush into the vagina, and scrapes cells from the outside of the cervix and just inside the cervical canal. The clinician then places the gathered cells on a slide, "fixes," and then stains them to enhance features that can be used in diagnosis. The slide is sent to a cytology laboratory for microscopic analysis. The sample is examined to find if and how much of the surface tissues (or epithelium) of the cervix is affected, and to determine the kind and degree of cell changes. Women who receive an "abnormal" result from a Pap test should have it repeated because the condition may improve naturally with time and because erroneous test results are a possibility. A "false positive" error occurs when the test reports an abnormality when, in fact, there is none. This mistake may cause emotional distress and unnecessary and expensive follow-up testing.

A "false negative" error occurs when the test reports no disease when, in fact, cervical cancer is present, an outcome that may needlessly cost women their fertility, bodily integrity, and lives. Due to the inherent limitations of the test, every negative Pap test result has a 5 percent chance of being a false negative, but 20 to 40 percent of tests may be falsely read as negative. Errors result from several causes: failure to take samples correctly, failure to apply chemical fixatives correctly, and mistakes in a cytologist's or cytotechnician's examination of a slide due to habituation, fatigue, or incomplete screening of a specimen. False results may stem from errors made by incompetent cytotechnologists who frequently are found in so-called "Pap test mills." Women and practitioners can take steps to reduce the risk of errors by checking to see if laboratories have been cited for violation of standards (under the Clinical Laboratories Improvement Act enforced by the Department of Health and Human Services) and by a program of regular testing. New techniques to automate and improve slide preparation and assessment have recently become available, but whether these more costly procedures will be deemed cost effective in actually improving medical outcomes remains to be seen.

It is generally recommended that screening begin when a woman reaches 18 to 20 years of age, or is sexually active, whichever occurs first. Some practitioners recommend annual Pap tests for all their patients, but others suggest that monogamous women at low risk can decrease the frequency after having had three normal annual Pap smears. Factors that put women at higher risk of abnormalities include: beginning sexual intercourse at a young age, having multiple sex partners, intercourse with a high-risk male, such as one with cancer of the penis, a history of HPV, exposure to synthetic hormones (such as DES), smoking, and an occupation involving carcinogenic substances (as used in metal or chemical industries). Women of lower socioeconomic status have a much higher incidence of CIN and cervical cancer than middle-class women and, unfortunately, are less likely to be screened on a regular basis for a variety of reasons, including limited access to regular preventive care.—A.H.S.

See also CANCER; DES

Paper

People in literate cultures have written on a variety of media: papyrus (the interleaved and pressed fibers taken from the plant of the same name), bark, stone, clay, wood, wax, vellum (the prepared skin of calves, lambs, and young goats), and parchment (the prepared skin of goats and sheep). All of these were less than ideal, being either high in cost, difficult to write on, or hard to store and carry.

In China during the later Shang dynasty (about 3,500 years ago), flat animal bones were used for the inscription of divinations. At a later date the Chinese used bronze and stone for inscriptions, followed by writing materials made from bamboo strips and silk. These were precursors of one of China's most important contributions to the world, paper. The invention of paper is conventionally attributed to a Chinese official named Cai Lun (d. 121), who is said to have produced high-quality paper and presented it to the emperor in 105 C.E. In fact, archeological excavations have uncovered samples of hemp-based paper that date back to the 1st-century B.C.E.

Paper had become the standard writing medium in China by the 4th-century C.E. The availability of paper made possible the expansion of the Chinese printing industry and the production of books in massive numbers, reinforcing China's status as a civilization based on the written word. Along with the pages of books, the Chinese also produced wallpaper, wax paper, tissue paper, and even clothes, shoes, and armor made from paper.

One of the most significant applications of paper in China was paper money, which first appeared during the Song Dynasty (960–1279). Paper currency at first consisted of bank drafts, but in 1023 the government began to issue its own notes. The government's ability to create money was soon abused, leading to unfortunate consequences. As economists have long noted, the amount of money in circulation affects the overall price level, and in China the excessive issuance of paper money led to skyrocketing inflation. And, as might be expected, counterfeiting also became a problem, even though it was punishable by death.

Paper diffused from China into Vietnam and Korea at the beginning of the 3d century, and then made its way into Japan. By the end of the 7th century, paper had reached the Indian subcontinent, and then entered the Arab world in the middle of the 8th century. According to the traditional account, Arab conquerors learned about paper production after conquering Samarkand in 753 and capturing the Chinese paperworkers living in that city. As with many important ideas and inventions, paper was introduced into Europe by the Arabs. Paper mills were established in Spain and France in the 13th century, and by the 16th century paper had largely replaced older media such as parchment.

Although paper was considerably cheaper than older materials, the massive expansion in publishing made possible by printing with movable type outstripped the ability to make paper in sufficient quantities. The production of paper lagged because manufacturing processes had changed little from those first used in China. Rags (which had replaced hemp and other vegetable fibers) were pounded and soaked to form a pulp. The pulp was spread over a mesh frame and then transferred to a sheet of felt or other absorbent material, pressed, and then allowed to dry. The dry sheet would then be dipped in a starch sizing to give it a glaze and body.

Around 1670, the production of paper was speeded up by the invention in the Netherlands of a wind-powered device that used spiked rollers to tear up rags. An even more important invention was a device for making paper in long rolls of continuous sheets that was patented by France's Nicholas Louis Robert (1761–1828) in 1798. This machine formed the paper on an endless belt of woven wire instead of individual frames. It was subsequently improved in England by Brian Donkin (1768–1855) and the brothers Henry (1768–1854) and Sealy (d. 1847) Fourdrinier, whose name was eventually used for the machine, even though the brothers were unable to protect their patent and went bankrupt in 1810.

Improved papermaking machinery sped up the process of production, but the shortage of rags continued to be a serious bottleneck. Taking his inspiration from a wasps' nest, the French scientist René-Antoine Ferchaulot de Réaumur (1683–1757) presented in 1719 a paper for the French Royal Academy that suggested the use of wood as a source of pulp. Réaumur never tried to make paper from wood; that task was undertaken in Germany by Jacob Christian Schäffer, who between 1765 and 1771 published a series of books describing his experiments with wood and many other substances. It was not until 1850 that the first paper mill using wood (along with straw) was set up by a Hollander living in England, Mattias Koops. The paper was of poor quality, however, and Koops's business failed. Improved grinding machines made the process more practical, and wood-pulp paper finally gained a measure of commercial success by the late 1860s. From this point it began to take over the market. Production of wood-pulp paper outstripped that of rag-based paper during the 1880s. Today, 90 percent of the paper comes from wood.

In modern papermaking, pulp is made through mechanical processes that tear up the wood, or by means of chemical process that use steam along with sodium hydroxide (NaOH) and sodium sulfide (Na_2SO_3) to digest the wood chips. A considerable amount of paper is also made from recycled paper. Recycled paper saves many trees, but it has its limits. The processing of the old paper results in shortened fibers, so after a few rounds of recycling, the fibers are too short to be made back into paper.

The production of paper from wood chips allowed a substantial increase in paper production. At the same time, however, it resulted in paper of lower quality. Unless it is specially treated, paper made from wood pulp is slightly acidic; consequently, it slowly yellows and crumbles until it becomes unusable. Improved production processes have made possible the manufacture of acid-free paper, but many books and periodicals published in the past will have much shorter lifespans than much older ones that were printed on rag-based paper.

Although computer enthusiasts used to talk about the advent of a "paperless society," paper consumption continues to grow. Per-capita paper consumption in most industrial countries was about 200 kg (440 lb) per year in the mid-1990s (the United States consumed about 50 percent more), and consumption is expected to be 70 percent higher in 2010 than it was in 1990. Since each 100 kg of paper requires about 0.2 cu m of wood, the world's forests will be heavily stressed if present trends continue.

See also PRINTING WITH MOVABLE TYPE; RECYCLING; WINDMILLS

Paper Clip

One of the keys to efficiency is keeping together the things that belong together. Although not much thought is given to it, the simple paper clip makes a significant contribution to an organized life by keeping sheets of paper in one neat bundle. There are of course other ways to accomplish this: Strings, ribbons, and pins have been used for this purpose for centuries. These methods have their drawbacks. The first two require holes or slits to be made, and they do not allow new papers to be easily added to the stack, or old ones removed. Pins are easily inserted (up to a certain stack thickness), but they can prick the fingers, snag other papers, and leave holes that can become increasingly ragged.

During the last quarter of the 19th century, the growing volume of clerical work motivated inventors to seek a better way of attaching papers. At the same time, improvements in the drawing and bending of wire helped to turn a need into a practical reality. The wire from which a paper clip is made has to have a certain degree of springiness, for the resulting tension is what holds the papers together. At the same time, the wire cannot be too springy, for otherwise it would not hold the shape into which it is bent. The production of the right kind of wire and the development of machinery to automatically bend it made the manufacture of paper clips a practical possibility.

The actual design of the paper clip was the result of considerable experimentation. Credit is often given to a Norwegian, Johan Vaaler, for the design of the first successful paper clip in 1899. But contemporaries of Vaaler produced their own designs, some of which were closer to the design that is most common today. The documentary record is imperfect, but there is some reason to credit William Middlebrook of Waterbury, Conn., with the "Gem" design in common use today. A patent application for a clip-forming machine was granted to him in 1899, and the patent application includes a rendition of a paper clip identical to today's Gem. On the other hand, Middlebrook's patent application concerns the machinery used to make the clip and makes no claim regarding the novelty of the shape of the clip depicted in the application. The Gem clip may in fact have originated with Gem, Limited, a British manufacturing firm.

Although the vast majority of the 20 billion paper clips sold annually are of the Gem design, it has its drawbacks. The Gem clip can be difficult to push onto a stack of papers, especially when there are more than about 10 of them, and its outside leg can snag paper when the clip is pulled off. Today there are paper clips designed to slide on and off more easily, clips expressly designed for holding stacks of index cards, clips with small ridges for gripping the paper more tightly, and clips for holding many sheets of paper together, as well as plastic clips for use where nonmagnetic properties are necessary. No single design perfectly meets all requirements, but the billions of paper clips in service today play an essential role in keeping things organized.

Further Reading: Henry Petroski, *The Origin of Useful Things*, 1992, pp. 51–77.

Parachute

A parachute commonly is an umbrella-like device for slowing down the descent of a body falling through the atmosphere; other shapes and uses will be noted later on. Leonardo da Vinci often is credited with inventing the parachute on the basis of some notebook sketches (1485). However, his notebooks were unknown until little more than a century ago, so Leonardo did not influence the actual development and use of the parachute.

Although somewhat earlier descents of animals were

recorded, the first recorded, successful parachute jump by a person is credited to the Frenchman A.-J. Garnerin (1797). In subsequent years, he frequently gave parachute exhibitions, most notably a jump from a free balloon at 2,500 m (8,200 ft) over England in 1802. Those first parachutes were bulky, canvas devices, although chutes made of Japanese silk became the standard in relatively short order, hence the expression "hitting the silk" to signify making a (usually emergency) jump. The general unavailability of Japanese silk during World War II led to the use of a new synthetic fiber, nylon, for parachutes. Nylon remains the standard material for parachutes of various configurations. Modern, conventional parachutes reduce the rate of descent almost 90 percent, compared to the terminal velocity of a free-falling person, about 58 m (190 ft) per second.

The first parachute jump from an airplane was made by U.S. Army Captain A. Berry at St. Louis, Mo., in 1912. In spite of this demonstration of the feasibility of making such jumps, World War I combat pilots made no use of the compact, pack-type parachutes invented by the American showman L. P. Stevens in 1908. Parachute use was discouraged by commanding officers, who believed that parachute-equipped pilots would be less likely to engage in air combat if they had the option of bailing out. However, during that conflict, the crews of captive observation balloons and dirigibles did use parachutes to escape from their craft when those were attacked and set afire. During World War II, air crews of all belligerents except some fighting for Japan were issued parachutes to escape from disabled aircraft.

Other military uses for parachutes were devised and used during World War II. The most important such use was the dropping of large numbers of fully equipped paratroopers behind enemy lines to outflank and surround enemy forces. The German Army was the first to use this tactic, although the Soviet Army had trained soldiers for such service during the 1930s. Other, less-dramatic uses for parachutes during that war included supplying isolated or surrounded friendly forces, dropping agents behind enemy lines, and even slowing down the rate of descent of flares and bombs.

The ring parachute, named for the geometry of its construction, invented in Germany during World War II, proved to be especially suitable for dropping heavy loads accurately as well as serving as an airbrake to shorten the landing run of airplanes. Since the 1950s, parachutes with highly controllable, if modest, gliding capabilities called *parafoils* or *parawings* have become popular, particularly for use in the increasingly popular sport of skydiving. These hang-gliderlike devices allow teams of parachutists to put on displays suggesting an aerial ballet, while individuals may compete in contests of skill, the simplest of which is spot landing. It is the parafoil that has elevated skydiving to its present status as an international sporting event. However, not all present sky divers engage in these rather sophisticated activities. Throughout the world, there are still thousands of people who do no more than use an airplane as an elevator that allows them to make an exciting descent to earth protected by a conventional parachute.—M.L.

See also NYLON

Further Reading: Lynn White, Jr., "The Invention of the Parachute," *Technology and Culture*, vol. 9, no. 3 (July 1968): 462–67.

Paradigm

In 1962, the physicist-turned-historian of science, Thomas S. Kuhn (1922–1996), published *The Structure of Scientific Revolutions*, a book that presented an account of the nature of scientific activity that diverged radically from those offered by mainstream philosophers of science. At the same time, Kuhn retained the conventional supposition that science differed from most other intellectual activities in its "progressive" character. Kuhn's account, which has played a tremendous role in stimulating the entire social studies of science movement, centered on the role played by paradigms, or "universally recognized scientific achievements that for a time provide model problems and solutions to a community of practitioners" (p. viii; see Further Reading). Like most seminal concepts, a certain vagueness surrounds the term *paradigm*. Margaret Masterman has convincingly argued that Kuhn used the term in 23 different ways in his book. All, however, are related analogically or metaphorically to the central concept.

According to Kuhn, a paradigm is generated when some particular work implicitly determines the legitimate problems and research methods of a research field for a succeeding generation of practitioners. Historical examples of paradigm-setting works include Ptolemy's *Almagest*, Newton's *Principia* and *Optics*, Lavoisier's *Principles of Chemistry*, Lyell's *Geology*, and Bohr's "On the Quantum Theory of Line-Spectra." Dominant paradigms emerge only when the work in question has two critical characteristics: first, its achievement must seem so substantial as to attract an enduring group of followers, and second, it must be sufficiently open-ended to suggest a large number of problems waiting to be solved.

A paradigm is not merely a theory, although it may contain one or more. It is not merely a set of instruments, though it may incorporate some. Nor is it a set of metaphysical or epistemic assumptions, though it may explicitly or implicitly incorporate some of those too. A paradigm is, instead, a concrete artifact or model investigation that produces a particular way of seeing phenomena and which invites elaboration and extension.

Once a community of practitioners forms around a paradigmatic work, the process of "normal science" commences. Since the paradigm both defines what constitutes interesting problems as well as the strategies and instrumentation that should be used in solving them, normal science is basically a puzzle-solving activity, and normal scientists can be viewed as sophisticated puzzle solvers who have been given both the rules they must follow and the problems they must solve. The progressive character of science follows from the fact that normal science rapidly expands the number of problems solved as scientists elucidate the implications of the paradigm they share. Occasionally, however, the puzzle solving leads to a result that violates paradigm-induced expectations, as when the results of the Michaelson-Morley experiment, expected to determine a relatively large velocity for the Earth's motion through an electromagnetic ether, suggested that that value was zero. Kuhn called these novel or unexpected results *anomalies*.

At first, anomalies will be ignored, and elaboration of the paradigm will continue unabated. But if the number of anomalies increases substantially, or if they occur in connection with problems that are considered particularly important and central, scientists begin to question aspects of the dominant paradigm. Eventually, a new paradigm will be offered which: (1) resolves many or all of the anomalies of the old paradigm, (2) is capable of solving most or all of the other problems solved by the old paradigm, and (3) suggests an extended domain of problems to be solved. Then begins a "revolutionary" period in the science, during which adherents of the old and the new paradigms compete for converts. In general, proponents of the old paradigm will be unable to change their way of seeing the world, so the new paradigm must gain its support from newer recruits to the field. In time, supporters of the old paradigm will drop away, and followers of the new paradigm will usher in a new period of normal science in the field. Since the new paradigm will almost certainly solve nearly all of the problems solved by the old, and since it will also have an extended domain of applicability, the change of paradigms is likely to be seen as progressive. Nothing, however, about the process of paradigm shifts guarantees that the process will produce knowledge that comes increasingly close to some perfect knowledge of any "real" natural world. The process of new paradigms replacing old ones is likely to be unending.

Much of the criticism of Kuhn's paradigm theory has focused on the claim that different paradigms must be psychologically and sociologically exclusive so that a mature science cannot exist when two or more paradigms coexist. In contrast, Imre Lakatos (1922–1974), for example, argued that sciences are characterized by the existence of robust "research programmes" that may run in parallel without serious conflict. Other criticisms focus on the notion that ordinary science, which had long been characterized as focusing on attempts to test or falsify theories, is instead a relatively routine and intellectually conservative enterprise. This aspect of the paradigm concept has evoked the hostility of many working scientists, who resent the implication that their work is narrow and routine.

From the perspective of the social studies of science, Kuhn's paradigm concept has been of considerable importance because it focuses attention on the social process of consensus formation in the sciences as the central feature of scientific activity. In so doing, it draws attention to the scientific community's mechanisms for enforcing conformity, processes such as peer review for purposes of funding and access to publication.—R.O.

See also HYPOTHESIS; LIGHT, SPEED OF; RELATIVITY, SPECIAL THEORY OF; THEORY

Further Reading: Thomas Kuhn, *The Structure of Scientific Revolutions*, 2d ed., 1970.

Parallel Processing

When computers first were produced, the usual mode of operation was called the *fetch-execute* cycle: An instruction was fetched from the memory and was executed, then another instruction was fetched, etc. Accordingly, early computers were essentially processing instructions serially, i.e., one after the other. This mechanism, the basis on which all modern computers operate, was sufficient until the demand for faster and faster speed required some modification. Faster computers were constructed by simply using faster technology to accomplish the same task. Fast transistors replaced slower vacuum tubes, and these, in turn, were replaced by faster versions of integrated circuits. There is, however, a limit to how fast any particular technology can operate. When this is reached, engineers attempt to design devices that will do jobs in parallel in order for the total elapsed time to be reduced and thus speed up the job at hand.

The first stage in producing a parallel computer—exemplified in machines designed in the late 1950s like the IBM Stretch and UNIVAC LARC computers—was to duplicate the internal data pathways and some of the control electronics of the computer so that the processor would fetch an instruction from memory and, while this was being executed, it would already be fetching the next instruction in sequence so that the execution electronics would not have to wait when it had finished with the first command. This is often referred to as a *pipelined* computer because many different instructions can be in the execution "pipeline" at one time, each in a different stage of execution. Today, this is the major design idiom of the computer engineer. Only the very simplest computers—those used to control microwave ovens for example—are still based on the older sequential fetch-execute cycle.

Another way in which parallelism can be obtained is to have computers whose instructions will operate on more than one piece of data at a time. While most machines are limited to operations that will, for example, add two numbers together and store the result in memory, many problems require this operation to be performed on hundreds or even thousands of similar numbers at each stage of the solution. Special "vector"-oriented machines have instructions that can perform the

same operation, say an addition, on thousands of numbers—again using a form of pipelined processing.

When computers stopped being produced from individual components, such as transistors, and started being made of integrated-circuit technology, it became possible to consider building and using machines in which there were essentially two or more full processors inside each computer. A computer with multiple processors can operate on a single problem in a variety of ways depending on the problem at hand. If, for example, a computer program was attempting to predict the weather tomorrow over North America, then one processor could be calculating the forecast for Los Angles while a different processor was attempting the same task for Toronto, both using the same weather data from today. On the other hand, some problems do not easily lend themselves to such parallelism. To calculate the weather for several consecutive days, it is first necessary to calculate the forecast for tomorrow, then calculate it for the day after, etc. This is a process that by its very nature is serial rather than parallel.

In general, the programming of these multiple-processor parallel machines is a very difficult task, and strange problems occasionally arise. The most famous of these is known as a *deadlock*, where it can happen that processor A cannot proceed with its calculation until it obtains the result of a calculation being done by processor B. Meanwhile, B is waiting for a result from A, so nothing at all gets done. These types of situations, and how to avoid them, have proved to be fertile grounds for interesting research in computer science.

While the vast majority of computers are simple single-processor machines, many different multiple-path parallel-processing machines have been designed and used for solving particular problems. Foremost among these are problems faced by the military, by geological exploration (such as in the oil industry), and for problems such as weather forecasting. Many of the so-called supercomputers are constructed from processors that are similar, (and often identical) to those found in the more-advanced personal computers. The only difference is that the supercomputers can have hundreds, and sometimes even tens of thousands, of these processors capable of being controlled to operate in parallel on one problem.

The advent of these special-purpose parallel machines has also opened up new methods of thinking about how problems might be solved. For example, rather than designing a sequential, step-by-step, algorithm for a problem, one can specify the initial data items and then simply indicate how these are to be modified (added, subtracted, etc.) to produce other values and then specify the operations on these in turn, until the final answer is reached. Special "dataflow" computers then take this information and perform the indicated operations when the operands become available as the result of other operations. This naturally leads to parallel working, and many different attempts have been made to both simulate and actually design dataflow machines and languages for describing the problems to be solved.

Other paradigms include having machines where many different interconnected processors share one large memory or have their own "local" memory for programs and data. Alternatively, the processors may have both their own memory and access to a shared large memory. Results from each local processor can be either left in a "global" memory for other processors to use, or these results can be passed from processor to processor, with each one modifying them in some way until the final result is produced. Several experiments have even been done using hundreds of computers on the Internet, each one tackling one aspect of the problem and then communicating their results to the others over the network. This scheme has been of particular use in complex number theory computations, such as those involving finding very large prime numbers.—M.R.W.

See also COMPUTER, MAINFRAME; INTEGRATED CIRCUIT; INTERNET; OIL EXPLORATION; THERMOIONIC (VACUUM) TUBE; TRANSISTOR; WEATHER FORECASTING

Parking Meter

On average, cars spend only 4 percent of their existence on the road; the rest of the time they are parked. But a car that is not on the road may still be a source of revenue—provided that it is parked in space controlled by a parking meter.

In addition to producing revenue, parking meters are used to stimulate the circulation of traffic in business districts by preventing cars from occupying the same spaces all day. Parking meters offer an intermedi-

ary position between complete laissez faire for cars and drivers on the one hand and the banning of cars from certain areas on the other. The dangers of the latter situation became evident in 1920 when the city of Los Angeles banned parking in its downtown business district in order to relieve congestion. The result was a disastrous drop in the patronage of downtown businesses as motorists shopped elsewhere. Clearly, there is a need to control parking without choking off commerce altogether.

One answer was provided by Carlton C. Magee. A journalist and president of the Businessmen's Traffic Committee of Oklahoma City, Magee is generally credited with being the inventor of the parking meter. Magee applied for his first parking meter patent in 1932, and the first installation of his meter occurred in Oklahoma City in 1935. Since then, parking meters have helped to control parking, while at the same time, meters bring in a fair amount of revenue to municipalities. The 30,000 meters in operation in Los Angeles, to cite one example, raise $20 million annually, and perhaps 10 times this amount in fines or parking violations.

For decades, parking meters retained the essential configuration of Magee's design, a mechanism activated by the deposit of a coin. In recent years, the application of microelectronics has opened up new possibilities. Meters are now being tested that display the amount of time a car has been parked, and do not increase time beyond a preset limit. This prevents "feeding the meter," and allows other cars to use a space after the time on the meter expires. Also in development are meters equipped with infrared sensors that erase the remaining time when a car vacates the space. If these become universal, one of life's small triumphs will be obliterated—finding a parking space with time still left on the meter.

Particle Accelerators

Particle accelerators are used by physicists to study the nature of physical reality on a subatomic scale. As its name implies, a particle accelerator increases the kinetic energy of charged subatomic particles. After reaching their maximum velocity, beams of particles are split up into several separate beams. These beams then collide with a target, producing other kinds of subatomic particles, many of them not normally found on Earth. Electric and magnetic fields select particular particles for experimentation. Under most circumstances, several experiments may be conducted simultaneously.

The presence of particular subatomic particles is indicated by the distinct track that each makes on a detector. The first detector to be used for research into subatomic particles was the cloud chamber. This instrument revealed the presence of particles by taking advantage of their ionizing properties. The same principle is also used for two other widely used detectors, the bubble chamber and the spark chamber. Detectors are major instruments in their own right, often being as large as a house and costing several million dollars to construct.

The value of accelerators has been considerably enhanced by the use of computers. Collisions occur in an infinitesimally short amount of time, and many things happen at once. Detectors now include "trigger" systems that take only a few millionths of a second to determine which of the signals from each collision are likely to be of interest to researchers. The computer records these events, and also generates online displays of the collisions, giving researchers an immediate picture of the data being produced. A computer graphics system can be used to visually rotate the tracks produced by particles, zoom into places of particular interest, and eliminate extraneous material.

Particle accelerators are the epitome of "big science." The Tevatron of the Fermi National Accelerator Laboratory near Chicago (Fermilab) occupies 2,750 hectares (6,800 acres), employs 2,000 people, and hosts 1,000 particle physicists at any given time. The Tevatron, which consists of two accelerators arranged one on top of the other, uses 60 megawatts of energy, more than enough to meet the needs of a small city of 150,000 people. Protons make thousands of circuits inside the accelerators, each time receiving an accelerating "kick" from electrical fields within. Each accelerator uses 2,000 electromagnets to keep the protons on a circular course until they reach an energy of 1,000 GeV (i.e., 1×10^{12} electron volts. An electron volt is the amount of energy gained by an electron when it accelerates through an electrical potential of 1 volt. The protons travel 3 million km (1.86 million mi) in 20 seconds, and attain a velocity 99.9995 percent of the speed of light.

The first device for accelerating subatomic particles was the cyclotron, built under the direction of Ernest O. Lawrence (1901–1958) in 1931. The first use of an improved version of the cyclotron known as the *synchrocylotron*, was in 1947. The machine was so called because the electrical field producing the acceleration of particles was kept in step with the particles' increasing periods of revolution that resulted from their relativistic increases in mass. By this time, a group of heavy subatomic particles had been discovered in cosmic rays, but even the synchrocyclotron with its 4.6-m (15.1-ft) magnet could not produce enough energy to create them in the laboratory. What was needed was a more powerful accelerator, and in 1947 the U.S. Atomic Energy Commission provided initial funding for building a new kind of accelerator, the Synchrotron. The key component of this device was a large ring containing a series of C-shaped electromagnets. These magnets held the accelerating particles to a proper course as they accelerated inside the ring. In 1952, the first Synchrotron, which had been given the name Cosmotron, went into operation at Brookhaven National Laboratory on New York's Long Island. The second, known as a Bevatron, began to be used in late 1954. Its first major discovery was an antiproton, a negatively charged proton that had already been found in cosmic rays.

At about this time, work was proceeding on the particle accelerator sponsored by the pan-European Conseil Européen pour la Recherche Nucléaire (CERN). Located in Switzerland near the French border, in late 1959 this Synchrotron began to accelerate protons to 24 Gev (giga electron volts), far more than the 10 Gev achieved by the previous record holder, the Dubna Laboratory in Moscow.

The years that followed were the scene of efforts to build increasingly powerful accelerators, including the Fermilab accelerator described above. In early 1967, another type of accelerator, one that accelerated electrons, went into operation. Unlike circular proton accelerators, this one was arranged in a straight line that extended for a distance of 3 km (1.9 mi), hence its name, Stanford Linear Accelerator (SLAC). It had to take this configuration because high-energy electrons radiate energy away from themselves as they move along a circular path (protons also radiate energy, but their greater mass allows them to reach much higher energies before they lose a significant amount of energy). SLAC now is capable of 30 GeV, on a par with the energy imparted to protons by some circular accelerators.

The largest of today's accelerators are *colliders*. These accelerate beams of protons and antiprotons or electrons and positrons in opposite directions until they collide at a predetermined point. The collisions generate enough energy to produce rare particles like pions (particles liberated when a proton's nuclear force field is disrupted). The first collider went into operation at Stanford University in 1965. It consisted of two side-by-side rings, each one similar to a regular accelerator. The rings were joined at a common point where the collisions took place.

Stanford was also the locale of the Stanford Positron Electron Asymmetric Rings (SPEAR). Electrons supplied by the aforementioned Stanford Linear Accelerator merge inside the ring. Positrons also are supplied by the linear accelerator and are accelerated by reversing the electric fields within the accelerator. In 1971, it became possible to produce collisions between protons when CERN put into service its ISR (Intersecting Storage Rings) accelerator. This produced a collision energy of 63 GeV, equal to a stream of protons striking a stationary target at 1,800 GeV.

In addition to being the fundamental instruments of nuclear research, accelerators have a number of practical applications. Radiation therapy for cancer generally uses X-ray radiation, but neutron radiation offers the prospect of more effective treatment. In conventional radiation therapy, X rays liberate electrons that attack cancerous tissue, but these electrons are not always energetic enough to destroy the genetic material in the cancerous cells. In contrast, accelerated neutrons dislodge protons and alpha particles when they collide with atomic nuclei. These heavy, charged particles are capable of destroying the DNA in the cancerous tissue, thereby preventing its further growth.

Accelerators are also used for medical diagnosis through positron emission tomography (PET). A PET scan provides detailed information about metabolic activity in an organ, including the brain. This makes it particularly valuable for studying neurological disorders like epilepsy. To produce a PET scan, an accelerator produces radioactive nuclei by accelerating a beam of protons, alpha particles, or deuterons (the nucleus of a deuterium atom, i.e., a proton and a neutron) and having them col-

lide with the nuclei of a suitable element. One of the products of these collisions are positrons. Because they are a form of antimatter, the positrons annihilate electrons in the vicinity, creating two gamma rays in the process, one for each electron and positron. Detection of gamma ray pairs allows the emitting nucleus to be located and provides the basis of a computer-generated picture of the organ. For example, a radioactive isotope of fluorine attached to a sugar molecule can be used to trace the metabolism of sugar in different areas of the brain, thereby revealing the brain's response to different stimuli. Positrons produced by accelerators can also be used to find incipient metal fatigue by revealing distortions in the atomic lattice of a metal.

In the 1980s, researchers began to plan the next step in accelerators, the superconducting supercollider. In 1984, the U.S. Congress allocated funds for the superconducting magnets that would accelerate subatomic particles, and construction of the collider itself was authorized in 1990. But cost overruns plagued the program, and a Congress increasingly concerned about budgetary deficits killed the program in 1994.

See also ALPHA PARTICLES; ANTIMATTER; CANCER; CLOUD CHAMBER; COMPUTER, MAINFRAME; COSMIC RAYS; CYCLOTRON; DNA; GAMMA RAYS; LIGHT, SPEED OF; METAL FATIGUE; PROTON; RELATIVITY, SPECIAL THEORY OF; SUPERCONDUCTING SUPERCOLLIDER; X RAYS

Further Reading: Frank Close, Michael Marten, and Christine Sutton, *The Particle Explosion*, 1987. Sharon Traweek, *Beamtimes and Lifetimes: The World of High Energy Physicists*, 1988.

Passive Solar Design—see Solar Design, Passive

Pasteurization

Preserving food by applying heat is a technology that goes back to the early 19th century. As with many technologies, the process worked, even though the reason for its effectiveness was unknown. The answer was provided by Louis Pasteur (1822–1895) when he discovered that microscopic organisms caused the spoilage of food and beverages, and that these organisms could be killed by heating them to 55°C (131°F). Pasteur's discovery was not completely novel; in 1768, Lazzaro Spallanzani (1729–1799) had shown that prolonged boiling prevented the emergence of microorganisms that seemed to arise through spontaneous generation. However, Pasteur was the first to use research into microorganisms to solve a practical problem.

The preservation of food through the application of heat, which came to be known as *pasteurization*, was further developed by a native of Ireland, John Tyndall (1820–1893). Tyndall's process of sterilization entailed subjecting the food to alternate periods of heating and chilling, the former taking 45 minutes to an hour, the latter taking 24 hours. Tyndall's process was an improvement over a single application of heat, for it killed not only existing bacteria but also the spores that are part of the reproductive cycle of some bacteria.

Many food products are subjected to pasteurization today, among them wine, beer, pickles, juices, and dairy products. Pasteurization can be done over a range of temperatures and processing times. For example, milk can be pasteurized by applying heat at 63°C (145°F) for 30 minutes or at 100°C (212°F) for .01 second. For milk, the standard for successful pasteurization involves the destruction of *Coxiella burnetti*, a rickettsia distinguished by its resistance to heat.

Pasteurization is not a perfect process. Not all pathogens are destroyed, and even when they are, their spores may survive and mature into new pathogens. Moreover, the process reduces vitamins B_1 and C and breaks down some enzymes and proteins. Adherents of "natural foods" claim that pasteurized foods have diminished nutritional value, and that in many cases pasteurization is used because food production and handling are not as sanitary as they could be.

See also FOOD PRESERVATIVES; SPONTANEOUS GENERATION; VITAMINS

Patents

A patent gives an inventor exclusive control over the commercial use of his or her invention. In the United States, it is granted for a period of 20 years from the date that the application was filed with the U.S. Patent and Trademark Office. The possession of a patent gives the holder a legal monopoly, making it one of the few instances of the government's giving its blessing to restraint of trade. There are two major reasons for this. First, it is believed that without the prospect of this legal monopoly, potential inventors would be less motivated to pursue their inventions, leaving society with a retarded rate of technological advance. Second, a patent renders an invention more visible, for the basic design and specifications of a patented item are open to inspection by anyone willing to pay a small fee. Many avail themselves of this right; 13,000 copies of descriptions of patented items are sold every day. In fact, the word *patent* is derived from the Latin *pateo*, which means "to open." As the U.S. Supreme Court ruled in 1933, an inventor ". . . may keep his invention secret and reap its fruits indefinitely. In consideration of its disclosure and the consequent benefit to the community, the patent is granted." Inventors thus may face a dilemma: They may seek a patent that leads to someone else's gaining the insight necessary to "invent around" their device, or they may eschew a patent and attempt to shroud their invention in secrecy, hoping that no one else figures it out.

The U.S. government's ability and responsibility to grant patents is embedded in the Constitution. Article 1, Section 8, gives Congress the power to ". . . promote the Progress of Science and the Useful Arts by securing for limited Times to Authors and Inventors the Exclusive Rights to their respective Writings and Discoveries." At the same time, however, the procedures through which patents are granted has undergone a fair amount of change. During the early years of the republic, the U.S. Secretary of State, the Secretary of War, and the Attorney General, or any two of them, had to approve each patent. This was awkward and restrictive, so in 1793 a revised patent law removed all governmental oversight and simply granted a patent to anyone who registered for one. It was then up to the courts or even Congress to deal with the inevitable problems that arose. The procedure was put on a firmer legal and organizational footing when the U.S. Patent Office was established in 1836.

The establishment of the Patent Office did not create a completely unambiguous system; as cynics have pointed out, a patent can be little more than a license to sue. Court decisions have been of the utmost importance in determining the validity of many individual patents, often to the dismay of the patent holder. From

1921 to 1973, federal circuit courts nullified nearly two-thirds of the patents brought before them.

The awarding of a patent may be highly contentious because the standards used are broad to the point of vagueness. Over time, the fundamental question, "what makes something patentable?" has been answered in a number of different ways. One thing is clear: An abstract idea cannot be patented. Albert Einstein (himself a former patent office clerk) would have met with frustration had he attempted to patent the General Theory of Relativity. However, in 1980 the patent system became involved in an area that previously was thought inviolate: the patenting of life forms. In the case of *Diamond v. Chakrabarty*, by a 5 to 4 decision the United States Supreme Court ruled that human-made organisms were entitled to full patent protection. Although the bacteria genetically engineered by Dr. Chakrabarty were living creatures, the Court ruled that they were human creations, and therefore patentable.

Currently, all that the Patent Office and the courts require is that the item be "new, useful, and non-obvious." The first two criteria are as old as the first patent law, formulated in 1790; the last dates to 1953. The extent to which a patent has to meet these three criteria has changed over time, as has the significance placed on each of them. The second criterion, usefulness, has been the least problematic, for it has been interpreted broadly. Practicality and commercial success are not taken into consideration, except for the case of alleged perpetual-motion machines. For decades the Patent Office has rejected these out of hand. At times, however, the issue of usefulness to society as a whole has been considered. In the past, inventions that were deemed injurious to society, such as birth-control devices in the 19th century, could not be patented.

The first criterion, newness, has been a source of vexation, since there are no clear standards regarding the extent of novelty required. Anything from a radical innovation to a minor modification of something already in existence could logically be judged as novel. In 1880, a Supreme Court justice sought to clarify matters by ruling that a truly novel invention required "a flash of genius." But this too was a hard-to-determine standard, although it endured until patent legislation was revised in 1952. It was the discarding of the "flash of genius" standard that led to the formulation of the third

criterion currently invoked, that an invention had to be "non-obvious." The problem here is that it is not always obvious whether or not something is obvious. It is necessary to consider such matters as the existing "state of the art," how much the invention diverged from it, and what are the usual levels of intelligence and skill of practitioners of the art. In an attempt to sort things out, in 1982 the Court of Appeals for the Federal Circuit was created to decide on the validity of challenged patents. This court has backed off a bit from using the three established criteria, and instead has tended to view commercial success as strong evidence that an invention is necessarily "new, useful, and non-obvious."

The inescapable vagaries of patent law create a situation perfect for litigation, a circumstance reinforced by the fact that millions of dollars can be riding on a decision. Just getting involved in a patent infringement suit can be very costly; the average cost of one is on the order of $250,000. As was noted above, it may be in an inventor's best interest to not seek a patent and instead attempt to preserve exclusive control over an invention through secrecy. Often, the issue is moot; many patents are quickly rendered irrelevant as technologies advance. Under these circumstances, it may be best to be satisfied with a temporary commercial advantage, for by the time a rival comes up with a successful copy it won't have much value.

Should an inventor seek a patent, he or she may have to make an important decision: whether to use the patent offensively or defensively—as a "sword" or as a "shield," to use a legal metaphor. When the description of a patented invention is made in broad and encompassing terms, the patent can be the basis of infringement charges brought against many other inventors. This was done in the late 19th century by George Selden (1846–1922), who was granted a patent on no less an invention than the automobile. For many years, he and subsequent holders of the patent were able to exact licensing fees from most American car manufacturers, until Henry Ford succeeded in having a court of law restrict the patent to automobiles with a certain kind of (by then obsolete) two-stroke engine. On the other hand, when an inventor has produced something of a more limited nature (these constitute the vast majority of cases), the patent will usually be drafted in much narrower terms. In this way, the patent holders have a "shield" for protection against the wielders of "swords," for the patent specifically applies to the particular improvements they have created.

When it is of enough significance, a patent can serve as the foundation for a monopoly position within a particular industry. Conversely, an established firm may try to suppress innovations that threaten its position or even its very existence. Firms may buy up the patents covering threatening innovations in order to bury them. This happens rarely, and one should not put too much credence in stories of potentially epochal inventions that were suppressed by established firms. In any event, there are no industries where the key technology is covered by a single "master patent" that serves to wall out potential competitors.

To prevent competitors from using innovations to gain a competitive advantage, firms within a particular industry may enter into cross-licensing agreements with these competitors. Under these arrangements, individual firms agree to make their patents available to other firms, and in return gain free use of the inventions patented by others. Although these arrangements may lower the motivation of firms to invent, they serve to increase the diffusion of the inventions that are made. Also, on occasion a cross-licensing agreement may result in a better product and process, because the recipient firm doesn't just use them as they come but makes them better.

When patent laws began to be formulated centuries ago, their prime beneficiary was the individual inventor. Individuals still account for many significant inventions today, and many of them have been protected by patents. But a great deal of inventing is now done in large units, such as corporate research and development facilities. The inventions that result may in turn be highly dependent on work done in other places, such as university laboratories. Under these circumstances, it may be very difficult to determine who is entitled to receive a patent and subsequently profit from it. Also, there are many areas of inventive activity that have been heavily dependent on research funded by the federal government. This raises the question of whether individuals and firms in the private sector should be allowed to receive patents for devices and processes based on publicly funded research.

The considerable expenditure of financial, governmental, and legal resources that goes into the issuing and securing of patents is justified by the assumption that the patent system plays a key role in the promotion of technological advance. But this is only an assumption. Other industrial nations, notably the Netherlands and Switzerland, went for decades with no patent protection for inventors, yet there is no evidence that their rate of technological advance was significantly slower than comparable countries with a patent system. Whether or not the patent system aids, hinders, or is irrelevant to technological advance is not easily resolved. Like many social institutions, it owes its existence as much to inertia as to anything else. When it is impossible to determine the mixture of advantages and disadvantages with any degree of confidence, the status quo usually prevails.

See also PERPETUAL-MOTION MACHINES; RESEARCH AND DEVELOPMENT; SELDEN PATENT

Peer Review

In order to ensure that the highest-quality scientific work gets funded and published, virtually all funding sources and publishers submit proposals and manuscripts to members of the scientific community (the *peers* of the person seeking to get support or to gain access to an audience) for evaluation and to get recommendations. This process, called *peer review*, was initiated in the mid-17th century by the Royal Society of London. When letters were submitted for inclusion in the Royal Society's *Philosophical Transactions*, the society's secretary sent them to members with some expertise in the area for their evaluation. Nearly simultaneously, the French Crown insisted that a primary function of the Academie des Sciences should be to review manuscripts for publication by the state and to propose patent protection of technical procedures and products.

In the most carefully controlled version of the modern peer review process, evaluation is "double blind"— that is, the reviewer in principle does not know who submitted the material to be reviewed, and the submitter does not know who is doing the reviewing. In this way it is hoped that all forms of personal favoritism and animosity can be avoided and proposals or papers can be evaluated solely on their technical merit or on their conformity to whatever other criteria the funding agency or publication might establish and make public.

Recent attacks on peer review procedures are based on the claims that: (1) In some fields the active workers know one another so well that attempts at blind reviewing are ineffective, so personal relationships inevitably play a substantial role; (2) referees may gain advantage by learning from proposals, which they then use for their own purposes after negatively evaluating their competitors; and (3) peer review is not only not useful, it may also be of negative impact in making decisions about relative allocations to different fields, since peer reviewers are likely to try to preserve funding and publishing opportunities in their own fields, even when those fields are producing little high-quality work or little work of value to the funding agencies or society at large.—R.O.

Pen, Ballpoint

A ballpoint pen consists of a small ball rolling inside a socket that is connected to a tube filled with ink. When used for writing, the ball picks up the ink and rolls it onto the page. The basic idea goes back to the 1880s, when an American named John J. Loud invented a rolling-ball marker to be used on bales of leather or fabric. It was not successful, and the idea lay dormant until the late 1930s, when it was revived in Hungary by Ladislao and Georg Biro. The two brothers subsequently fled the Nazis by moving to Argentina, where they met a British financier who helped them set up a ballpoint pen factory in England. The Royal Air Force was the initial customer for the pen, which had the virtue of not leaking at high altitudes, as did conventional fountain pens. Meanwhile, an American named Milton Reynolds began to produce ballpoint pens in the United States, also for use by the armed forces. Although he took his inspiration from Biro pens that he had seen in Argentina, he was able to skirt the Biros's patent by using a different kind of ink feed.

In the years immediately following World War II, the ballpoint pen found millions of customers, even though early ballpoints tended to leak and leave smudges. Improvements in inks and manufacturing processes largely eliminated these drawbacks. Manufacturing improvements also dramatically lowered the cost of ball-point pens, making them the archetypal disposable product; most people seem to lose ballpoint pens long before their ink supply runs out.

The ballpoint pen represents a continuation of a long-established trend in the evolution of writing implements. Moreover, as Lewis Mumford (1890–1990) pointed out in the 1930s, the evolution of the pen is indicative of some general features of technological change and its influence on human skills. In the early 19th century, the standard instrument for writing was the goose quill pen. Organically based and sharpened by the user to meet his or her requirements, the quill pen mirrored the handicraft technologies of its time. It was cheap and simple, and a fair amount of skill and practice were required for effective use. Later in the century, it was supplanted by the steel-nib pen, a typical artifact of the industrial age, standardized and machine made. Mumford's analysis antedated the ballpoint pen, but it is well suited to it. Built to close manufacturing tolerances, the ballpoint pen is the product of highly advanced manufacturing technologies. Yet the precision embodied in the pen often stands in sharp contrast to the sloppy handwriting that emanates from it. Advances in pen technology have not been paralleled by improvements in human skills, and may in fact have contributed to their deterioration.

See also MASS PRODUCTION

Penicillin

The story of the dramatic discovery of penicillin has often been told; however, as with many stories of its kind, it is less than complete. According to the popular account, in 1928 Alexander Fleming (1881–1955), a British bacteriologist working at St. Mary's Hospital in London, noticed that among a set of petri dishes containing colonies of staphylococcus bacteria there was one dish in which the staphylococci were obviously dying. Fleming eventually traced the cause of death to spores of the fungus *Penicillium notatum* that had apparently entered through an open window.

Fleming's discovery was an important one, but it was neither the beginning nor the end of the search for what we today call *antibiotics*. In the decades prior to 1928, many efforts had been made to find substances that killed bacteria but did not affect the organism in which they resided. It had even been noticed that the penicillium mold appeared to check the growth of certain microorganisms, but nothing else was done at the time. Research languished even after Fleming's discovery. Efforts were mounted to isolate the antibacterial material in the penicillium mold, but they ended in failure. Meanwhile, the recently discovered sulfa drugs were demonstrating remarkable curative powers, thereby deflecting interest away from penicillin.

In 1938, 14 years after Fleming's original discovery, three other researchers in Great Britain—Howard Florey (1898–1968), Ernst Chain (1906–1979), and Norman Heatley—began to work on penicillin. One of their first tasks was to extract sufficient quantities of penicillin from the penicillium mold so that it could be used experimentally. Once this was done, tests on mice infected with streptococcus bacteria dramatically demonstrated the antibacterial quality of penicillin. Clinical trials on human beings with severe infections also gave strong indications of penicillin's efficacy.

There still remained the major problem of obtaining adequate quantities of the drug. In 1942, it was discovered that a slightly different mold, *Penicillium chrysogenum*, produced much greater quantities of penicillin. This was a step forward, but what was needed was a technology that allowed penicillin production at an industrial level. In the United States, it was found that the addition of corn-steep liquor (a by-product of corn-starch extraction) to the culture medium increased the amount of penicillin by a factor of 10. Producing penicillin in large quantities was also stimulated by the development of a submerged fermentation process, which in turn required new tank designs, special cooling systems, and turbine mixers. Penicillin needs air to grow, and since the mold was submerged, the medium in which it grew had to be aerated, but this required the development of antifoaming agents. Finally, the extraction of the penicillin itself was done by a freeze-drying process that originally had been developed for preserving blood plasma. As production increased, the cost of penicillin dropped dramatically, from $20 a dose in 1942 to only 55 cents 3 years later.

Even at these low prices, a substantial amount of money was made from penicillin. One American bacteriologist who had worked on the use of corn-steep as a growing medium was able to patent the process, even though the much of the knowledge that led up to it had been provided by British researchers. The latter had deliberately refrained from seeking a patent, having been advised by their government that the public funding of penicillin research precluded individuals from profiting from it.

Penicillin cured a large number of bacterial diseases and was especially valuable in dealing with staphylococcal infections, against which sulfa drugs had little effect. But penicillin has not been a completely unmixed blessing; small numbers of people have severe allergic reactions to penicillin, resulting in death in some cases. A more subtle problem is the rise of drug-resistant bacteria, as the widespread use of penicillin has been accompanied by an increasing number of patients infected by varieties of staphylococci resistant to it. Extensive reliance on penicillin as a "wonder drug" also led to a contraction of other measures necessary for the avoidance of disease. For example, the effectiveness of peni-

Fermentation tanks for the large-scale production of penicillin (courtesy Bristol-Meyers Squibb Company Archives).

cillin against syphilis led to a slacking of efforts aimed at prevention; consequently, cases of the disease began to increase after reaching a low point in 1957. Finally, for many years there was a tendency to prescribe penicillin for everything that seemed remotely like an infection, including viral ailments like hepatitis, even though penicillin has no effect on viruses.

See also ANTIBIOTICS, RESISTANCE TO; SULFA DRUGS

Further Reading: Harry F. Dowling, *Fighting Infection: Conquests of the Twentieth Century*, 1977.

Periodic Table of the Elements

The periodic table of the elements serves as both a dictionary and a quick reference for the chemical and physical properties of the elements. The arrangement of the elements, which led to the present version, was first proposed by the Russian chemist Dmitri Ivanovich Mendeleev (1834–1907) in 1869, who was influenced by the concept of "periodicity." This idea arose from the fact that a number of scientists had made several groupings ("families") of elements according to chemical properties that repeated as one ordered the elements according to increasing atomic weight. For example, the Englishman John Alexander Newlands (1838–1898), with his "law of octaves," had also attempted to organize the elements in a coherent arrangement. In 1870, the German chemist Lothar Meyer (1830–1895) independently devised the system most similar to Mendeleev's table.

Mendeleev based his organization of the 63 known elements on increasing atomic weight and similar chemical properties. His great insight included leaving empty spaces in the table for then-undiscovered elements, thus allowing the table to be a predictive scientific model. For example, he predicted the existence of the elements scandium and gallium, as well as their atomic weights and properties.

In its modern form, the elements are ordered according to increasing atomic number (number of protons in the atom). This pattern is based on Henry Moseley's (1887–1915) studies of the X-ray spectra of elements in 1913. For example, even though nickel has a lower atomic weight than cobalt, cobalt precedes nickel in the table because it has a lower atomic number as deduced from its X-ray emission. (Of course, the number of electrons in an atom is equal to the number of protons.) The elements are normally arranged in 7 rows (periods) that yield 16 columns (groups) of elements with similar chemical properties. This arrangement reveals that the number of outer or "valence" electrons is the theoretical basis of chemical properties. The electronic structure of atoms is derived from the application of quantum mechanics, which governs the types of "orbitals" and their order of filling by electrons. Elements increased in size (radius) going down the groups because they are adding electrons to outer shells, but the radii diminish going to the right across a period because the added protons in the nucleus attract electrons inward.

The table begins with the lightest element, hydrogen, and currently ends with element 109. The elements are divided in several genera types: metals, nonmetals, and metalloids (or semimetals). Metals tend to lose electrons and form positive ions, whereas nonmetals tend to gain or share electrons. The metals are further classified into main group metals, transition metals (in the middle of the table), lanthanides (rare earths), and actinides (heavy radioactive elements). Most elements are metals; the largest block is the transition metals that include many common metals such as iron, copper, and zinc, and precious metals such as silver and gold. The transition metals have importance in metallurgy and as industrial and biochemical catalysts. Although there are only about 16 nonmetals, they are very abundant and include carbon, nitrogen, oxygen, chlorine, and all the rare gases. Silicon and germanium are well-known metalloids (about seven elements), and are especially important in semiconductor technologies. Main group metals include sodium, calcium tin, and lead.

Naturally radioactive elements are common among elements with atomic number 84 (polonium) or greater. The elements beyond uranium (92) in the actinide series include the heaviest elements, and many have very unstable forms (isotopes) that decay rapidly, especially those over atomic number 100. Neptunium (93) was first prepared in 1940, and Glenn Seaborg (1922–) and his colleagues synthesized other heavier actinides in the following years. Lighter elements can also exist as unstable isotopes, such as tritium (a form of hydrogen used as a biological tracer and in nuclear weapons) and

The Periodic Table

1	2	3	4	5	6	7	8	9	10	11	12	13	14	15	16	17	18
1 H																	2 He
3 Li	4 Be											5 B	6 C	7 N	8 O	9 F	10 Ne
11 Na	12 Mg											13 Al	14 Si	15 P	16 S	17 Cl	18 Ar
19 K	20 Ca	21 Sc	22 Ti	23 V	24 Cr	25 Mn	26 Fe	27 Co	28 Ni	29 Cu	30 Zn	31 Ga	32 Ge	33 As	34 Se	35 Br	36 Kr
37 Rb	38 Sr	39 Y	40 Zr	41 Nb	42 Mo	43 Tc	44 Ru	45 Rh	46 Pd	47 Ag	48 Cd	49 In	50 Sn	51 Sb	52 Te	53 I	54 Xe
55 Cs	56 Ba	57* La	72 Hf	73 Ta	74 W	75 Re	76 Os	77 Ir	78 Pt	79 Au	80 Hg	81 Tl	82 Pb	83 Bi	84 Po	85 At	86 Rn
87 Fr	88 Ra	89† Ac															

Transition elements

s–block
d–block
p–block

*Lanthanides

57 La	58 Ce	59 Pr	60 Nd	61 Pm	62 Sm	63 Eu	64 Gd	65 Tb	66 Dy	67 Ho	68 Er	69 Tm	70 Yb	71 Lu

†Actinides

89 Ac	90 Th	91 Pa	92 U	93 Np	94 Pu	95 Am	96 Cm	97 Bk	98 Cf	99 Es	100 Fm	101 Md	102 No	103 Lr

f–block

technetium (which is synthesized for use in medical imaging procedures).—A.Z.

See also ATOMIC NUMBER; CHEMICAL BONDING; ISOTOPE

Further Reading: J. W. von Spronsen, *The Periodic System of the Elements*, 1969.

Perpetual-Motion Machines

One of the dreams of ancient and medieval mechanics was the construction of a perpetual-motion machine, i.e., a machine, which once motion was imparted to it, would never stop moving. Better yet, some hoped to produce a machine from which work might be extracted without diminishing its motion. One such device was proposed by Villard de Honnecourt (born c. 1200), a highly accomplished medieval engineer-architect who described a wheel that was to be perpetually rotated by the falling of mallets loaded with mercury. Medieval schoolmen, including Jean Buridan (c. 1300–c. 1385), discussed thought experiments like this one: What would be the motion of a heavy object dropped down a hole drilled all the way through the Earth in the absence of the retarding friction caused by motion through the atmosphere? Those who thought about it considered that perpetual motion might occur; but they agreed that in every physically realizable case, frictional forces would gradually retard the motion.

In the late 16th century, Simon Stevin (1548–1620), a Dutch engineer, mathematician, and natural philosopher, devised a geometrical proof that one frequently proposed design for a perpetual-motion machine was an impossibility. By the mid-18th century, the idea of perpetual motion was in such disrepute that the members of the Parisian Academie des Sciences announced that applications for patents on perpetual-motion machines would no longer be considered, an edict that caused substantial bitterness among lower-class inventors.

The Academie's action and a general condemnation of perpetual-motion devices by the international scientific community failed to discourage would-be inventors. In the second half of the 19th century, for example, the British patent office received an average of 10 applications per year for perpetual-motion machines. Even

today, creators of such devices enthusiastically continue the quest for perpetual motion, submitting proposals to national governments, patent offices, private investors, technical magazines, and scientific societies around the world.

Although the proposals came in myriad varieties, reflecting the personal styles of their inventors, they typically hinged on a few basic notions, such as the overbalancing wheel described centuries ago by Villard de Honnecourt. In the 19th century, projects involving the new forces of electromagnetism were especially popular, and in the 1970s and the 1980s, the energy crisis stimulated a special interest in permanent magnetism, electromagnetism, and gravity, among other forces, as sources of limitless power.

The dream of perpetual motion has attracted not only sincere individuals who genuinely believe that their contrivances will supply mankind with free energy but also deceivers and frauds. Of the instances of deception, one of the most notorious was John Keely, a charismatic Philadelphia mechanic, who, in 1872, persuaded financial backers to invest millions of dollars in his "Hydro-Pneumatic-Pulsating-Vacuo Engine" that allegedly drew on the power of hidden vibratory forces. Only after his death did his disappointed investors find the real mechanical source of power hidden beneath the floorboards of his house.

Honest attempts to create perpetual motion will always founder on two basic principles of physics. A device that, once started, performs work with no additional input of energy ("perpetual motion of the first kind") runs up against the First Law of Thermodynamics. This law states that no device can release more energy than it receives. In other words, energy efficiency cannot exceed 100 percent. This at least allows for the possibility of perpetual motion if the effects of friction could somehow be negated ("perpetual motion of the third kind"), but such a device could not serve as a source of limitless energy. Another type of perpetual motion ("perpetual motion of the second kind") would produce motion by taking heat from some source and converting it into useful energy. If such a thing were possible, a ship could be powered by a device that extracted heat from the ocean. Unfortunately, this possibility is nullified by the Second Law of Thermodynamics, which requires that for work to be performed heat must flow from a high-temperature source to a low-temperature reservoir. Finally, perpetual motion of a sort actually is achievable when electricity continuously flows through a resistance-free superconductor. But if the electrical current is used as a source of energy, some of that energy will be converted to heat, and the electrical current will eventually dissipate.—A.M. and R.O.

See also CONSERVATION OF ENERGY; ELECTROMAGNET; ENERGY EFFICIENCY; ENTROPY; GRAVITY; SUPERCONDUCTIVITY

Further Reading: Arthur W. Ord-Hume, *Perpetual Motion: The History of an Obsession*, 1980.

Phlogiston

One of the major concerns of 17th-century scientists was the nature of combustion. Experiments with vacuum pumps seemed to indicate that there was something in the air that supported combustion, but it was a very elusive substance. Most puzzling was the fact that when a combustible material was burned in a sealed chamber there was no subsequent loss of pressure inside the chamber. On the other hand, some investigators speculated that air had nothing to do with combustion; rather, it was thought that combustible materials burned because they contained a substance that promoted burning. Johann Becher (1635–1682) called the substance "oily earth." In 1703, another German scientist, Georg Stahl (c. 1660–1734), elaborated on Becher's ideas, and substituted the word *phlogiston* for "oily earth."

According to this theory, materials that are combustible contain phlogiston, while noncombustible substances are lacking in it. Furthermore, it was possible to return phlogiston to the substance that had lost it. A "calx," a substance that we would now call an *oxide*, could be returned to its original state by heating it with a material rich in phlogiston, such as charcoal, resulting in the transfer of the phlogiston from the charcoal to the calx. Air was still needed for combustion, but its purpose was to soak up phlogiston. When air had been saturated with phlogiston, burning ceased. This explained why a candle could burn for only a limited time in a closed container.

All of this made good sense, except for one annoying observation. Experimenters beginning with Robert Boyle

(1627–1691) had weighed substances that had undergone combustion and found that they weighed more than they had prior to being burned. Since phlogiston supposedly was released during the combustion process, this would imply that it had negative weight. Some ingenious thought went into the explanation of this phenomenon. One adherent of the phlogiston theory surmised that the loss of phlogiston led to a "weakening of the repulsion between the particles and the aether," thereby diminishing their mutual gravitational attraction. Another argued that a metal that lost its phlogiston acquired gas at the same time, increasing its weight accordingly. But many simply admitted that there was much that was not understood about phlogiston and the mechanisms through which it worked.

Little could be gained by speculating on the nature of phlogiston; what was needed were more experiments on combustion. Some of the most important of these were performed by the great French chemist Antoine-Laurent Lavoisier (1743–1794). In 1772, he heated lead oxide with charcoal (a process now known as *reduction*) and observed that a significant amount of air was released as the lead oxide turned into pure lead. In his notebook, Lavoisier noted that this and subsequent experiments were "destined to bring about a revolution in physics and chemistry."

He was right. Lavoisier's experiments were the beginning of the end of the phlogiston theory, but they did not solve the puzzle at once, for Lavoisier and his contemporaries did not know that ordinary air is made of a number of distinct gases. Consequently, Lavoisier had no way of knowing that it was oxygen that combined with a substance undergoing combustion, while the gas he observed being given off in the process of reduction was carbon dioxide.

An identification of the gases comprising the atmosphere began in the 1770s. The best publicized discovery came in 1774, the result of Joseph Priestley's (1733–1804) experiments with mercury oxide. When he directed the concentrated light from a burning glass onto mercury oxide, the resultant heat converted it into pure mercury, at the same time liberating a gas. Priestley called this gas "dephlogisticated air," because its ability to support combustion indicated that it contained no phlogiston. Thus, it supported rapid combustion as it absorbed the large amounts of phlogiston given off by a burning substance. In 1779, Lavoisier renamed the gas *oxygen*, meaning "acid former," on the mistaken assumption that all acids contained it. Priestley's contribution was not unique. In Sweden, Carl Wilhelm Scheele (1742–1786) had discovered oxygen a few years before Priestley, but Scheele had not published his findings. Ironically, Priestley, whose discovery of oxygen was the foundation of a new understanding of combustion, remained a firm believer in phlogiston to his dying day.

Further Reading: J. H. White, *A History of the Phlogiston Theory*, 1932.

Phonograph

The principle underlying the phonograph was discovered almost simultaneously by Charles Cros (1842–1888) in France and by Thomas Edison (1847–1931) in the United States. Both realized that variations on a surface could be used to move a diaphragm that generated sound. In April 1877, Cros published an article describing how sounds could be reproduced by cutting grooves in a soot-coated glass cylinder. The grooves were then to be made permanent by photoengraving them on a steel cylinder. Cros's device may have worked, but there is no firm evidence that he ever tried it out. The first instrument that indisputably recorded sound and then played it back was devised by Edison and his associates in July 1877. Edison's original intention was to invent a machine that would record telephone calls, and his device borrowed from some elements of telephone technology. The phonograph, as Edison had called it, consisted of a diaphragm that was made to vibrate by sounds projected into a telephone speaker. The vibrations of the diaphragm produced up-and-down movements of a stylus as it cut into a moving strip of wax-coated paper. The sound was played back by reversing the process. The irregularities in the paper moved the stylus, and these in turn were translated into the movements of a diaphragm that (imperfectly) reproduced the sound. By the end of the year a cylinder covered with tinfoil had replaced the paper strip. Edison's own recording of "Mary Had a Little Lamb" on that cylinder is generally credited with being the world's first phonograph record.

Edison and Cros were not the only ones who had been working on mechanical sound reproduction. In 1880, after considerable experimentation, Chichester Bell (a cousin of Alexander Graham Bell) and Charles Sumner Tainter (1854–1940) obtained several patents that covered a phonograph of their own invention, which, among other things, used a wax-covered cylinder in the place of Edison's tinfoil-covered cylinder. Other inventors dispensed with the cylinder altogether. These included the ancestor of the tape recorder, as well as what eventually emerged as the standard record player, a machine using flat disks. The disk-playing machine was the creation of Emil Berliner (1851–1929), a German who had immigrated to the United States in order to escape induction into the Prussian Army. In addition to using disks instead of cylinders, Berliner's phonograph differed in that the physical variations that produced the sound were cut in a side-to-side direction rather than in the up-and-down pattern used by Edison.

Thomas Edison and the first phonograph (courtesy Smithsonian Institution).

In this early phonograph, sound entering the mouthpiece causes the vibration of a stylus that inscribes indentations on the rotating cylinder (from M. Kranzberg and C. W. Pursell, Jr., *Technology in Western Civilization*, vol. I, 1967, p. 647).

Berliner called his machine a *gramophone*, a term that is still used in some parts of the world.

An important advantage of disk records was that it was easier to stamp them out than it was to mold the cylindrical variety. Edison's company began to produce disks in 1913, although it continued to make cylinder records and their players for a number of years afterwards. Cylinder-playing machines were largely supplanted by disk players in the 1920s, not so much because they were technically deficient but because Edison, guided largely by his own tastes, did not supply the kind of music that consumers wanted to hear. At that time, most phonographs could only play records that had been made by the maker of the phonograph. Under these circumstances, a firm's recorded repertoire was at least as important as the technical quality of its machines.

Although Edison had invented an electrically powered turntable in 1887, the great majority of early phonographs were powered by spring mechanisms that were wound up by turning a crank. Electrical drive

eventually removed this slight inconvenience, but this was not the most important contribution that electricity made to phonograph technology. Until the mid-1920s, phonographs were acoustic devices that reproduced sound solely through the movement of a diaphragm. Some amplification was achieved by putting this sound through the funnel-shaped horns that were a distinguishing feature of early phonographs. Not only was their volume low, they also failed to reproduce sounds of certain frequencies, a deficiency that was also present when the original recording was made by a similar acoustic process.

The electrical augmentation of sound was not a totally new technology; it had long been used in the telephone industry, and for devices like microphones and radios. The basic principles of electronic recording and playback were developed at Western Electric, the manufacturing division of American Telephone and Telegraph. AT&T already had wide experience with the amplification of telephone messages, and much of the knowledge that had been gained could be directly applied to the phonograph. The recording technology developed at AT&T used a condenser microphone to convert sounds into electrical current. The current was strengthened by a vacuum-tube–based amplifier and then was used to power the electromechanical cutter that produced the record grooves. In this way, variations in sound were converted into variations in current, which in turn were converted into profile variations within the record's grooves. The process was essentially reversed when the record was played: As the phonograph's needle moved in the groove, its motion caused variations in the magnetic field of a pickup located in the player's tone arm. These produced a changing electrical current that, after being amplified, drove the speaker.

Electrical recording and playback allowed higher volumes, and equally important, they provided for uniform responses throughout the sound spectrum and caused much less distortion. Sound quality was further improved by the use of loudspeakers that reproduced sound by means of a diaphragm attached to a coil that was actuated by a changing magnetic field. By the late 1920s, acoustic record players were technologically obsolete, although many of them continued to be used for many years thereafter.

By this time the record industry had gone into a temporary commercial eclipse. The industry had flourished in the early years of the 20th century, but in the 1920s the sales of records and phonographs tumbled due to the widespread availability of free music and other programming on radio. The economic hard times brought on by the Great Depression of the 1930s made a bad situation worse. Fortunately for the record industry, the jukebox had emerged as a major consumer of phonograph records; in 1936 more than half the records produced were destined for jukeboxes.

Post–World War II prosperity helped the industry to revive, as did a series of technological advances: high-fidelity long-playing records and stereophonic sound. But by the 1980s, the conventional phonograph had reached the end of the road, as audiocassettes and compact disks emerged as the prime media for music and other recordings.

In their heyday, phonographs and the records played on them did more than make prevailing musical styles available to a wide audience; they also stimulated the diffusion of new kinds of music. In particular, they played a key role in the growth and widespread popularity of two key products of 20th-century America, jazz and rock-and-roll. In ways unimagined by its original inventors, the phonograph helped to bring fundamental changes to the cultural landscape.

See also AMPLIFIER; COMPACT DISK; JUKEBOX; MICROPHONE; MUSIC, HIGH FIDELITY; STEREOPHONIC SOUND; TELEPHONE

Further Reading: Andre Millard, *America on Record: A History of Recorded Sound*, 1995.

Photographs, Instant

From its inception in the early 19th century, a general trend in photography was to take the effort and drudgery out of taking pictures. Roll film, easy-to-use cameras, exposure meters, and electric flashes all contributed to making photography a less-daunting task. But there was still an important shortcoming: After the shutter was snapped, a lengthy series of steps were necessary before the photograph could be viewed. First the negative had to be developed and fixed, and then the actual photograph had to be developed and printed. The

process could take hours, and most consumers had to wait days to get their pictures back.

In 1947, Edwin H. Land (1909–1991) transformed photography by marketing a camera and film that produced finished photographs. Prior to working on the new photographic system, Land had done extensive work in optics. After a year at Harvard, he left that institution to conduct research on the polarization of light. Two years later he invented a polarization filter for camera lenses that reduced glare and reflections. His Polaroid corporation went on to manufacture a number of products based on the principle of polarization: sunglasses, lamps, and filters.

The original Polaroid model-95 camera used two separate film rolls, one negative and the other positive, that were loaded at opposite ends of the camera and then put together. After exposure, the ends of the two were drawn through rollers that forced a developing reagent between them. After a minute's wait, a door on the back of the camera was opened, allowing the positive print to be pulled away from the negative, which was then discarded. The photograph was then made permanent by applying a clear, fast-drying liquid that had been supplied with the film. The process produced instant pictures, but it required a fair amount of skill for loading the film and getting the right exposure. In 1963, these problems were surmounted by the model-100 Polaroid camera, the first instant camera capable of producing color photographs. Loading was simplified through the use of a film pack that contained several kinds of paper layered together, each contributing some of the colors of an individual photograph. The new process still required that the print be separated from the negative, which then had to be discarded with a caustic solution still adhering to it. In 1972, Polaroid solved this problem by introducing the SX-70 camera, which used a 10-exposure film pack containing a thin 6-volt battery. Each picture unit contained several emulsion layers and some developing reagent that was squeezed between a clear cover sheet and the emulsion layers. The picture emerged from the camera dry and required a few minutes exposure to ordinary light to develop completely.

Edwin Land was a rarity in that he was both an accomplished inventor and a successful entrepreneur. Polaroid cameras were a great commercial success in the years immediately following Land's introduction of instant photography, and by 1960 the value of Polaroid camera sales had risen even with those of Kodak. But from 1978 to 1988, sales fell by half. While the quality of photographs taken with conventional cameras steadily improved, Polaroid photographs continued to lack sharpness. The Polaroid camera brought instant gratification, but it was not enough to offset the technical superiority of conventional photography.

Photography, 35-mm

In the late 19th and early 20th centuries, serious photography was done with large photographic plates, usually 8 × 10 in. (20.3 × 25.4 cm) for studio work, and 4 × 5 in. (10.2 × 12.7 cm) when some portability was required. For the general public, the Kodak camera using roll film that produced a negative $2\frac{5}{8}$ in. (6.7 cm) in diameter was the source of many snapshots, most of indifferent quality. What photography needed was a film and camera combination that coupled the picture quality of large cameras with the portability and ease of use of the Kodak.

Many miniature cameras were produced at this time, but they were hindered by the need to use separate plates for each shot. By the end of the 19th century, it became possible to create cameras that used roll film with no loss of picture quality. What allowed this was the film used by Thomas Edison (1847–1931) for early motion pictures. It produced a negative 35 mm (1.38 in.) in width, and had sprocket holes on each side to transport the film from spool to shutter chamber to take-up spool. The 35-mm width was the result of dividing a strip of standard 70-mm Kodak negative stock into two pieces. A single frame's height of 26.25 mm (1.03 in.) may have been motivated by a desire to have a 4 to 3 ratio of width to height—close to the aesthetically pleasing "golden section" (or "divine proportion") of 1.618 to 1. The first camera to effectively employ this film was the Leica, a device so successful that it set the standard for decades to come. Invented in 1914 by Oskar Bernack (1879–1936), the Leica began to be commercially produced by the German Leitz company in 1925.

After World War II, serious photographers turned to 35-mm cameras in increasing numbers because they combined portability with good picture quality. A further boost for 35-mm photography came from the development of the single-lens reflex (SLR) camera that eliminated the problem of parallax. While professionals and serious amateurs overwhelmingly used 35-mm cameras, casual photographers were also drawn to them, as increasingly sophisticated cameras automatically set the f-stop (which governs lens aperture), shutter speed, and focus. Later generations of these cameras offered automatic loading, automatic reading of film speed, and other features that made adequate photographers out of people who had little interest in acquiring technical competence. In 1975, only 1 in 10 pictures taken worldwide was exposed on 35-mm film; by 1987 the ratio had risen to 2 out of 3.

See also FILM, ROLL; LIGHT, POLARIZED; MOTION PICTURES, EARLY; PHOTOGRAPHY, EARLY

One of the first examples of a Leica camera (courtesy Leica Camera, Inc.).

Photography, Color

From the very inception of photography, people were disappointed that only black-and-white (actually, shades of gray) images were produced. Early photographs were sometimes colored by hand, a laborious process that usually included only the tinting of certain parts of the photo, such as adding a touch of red to the cheeks of a portrait's subject. An early color photographic process based on the interference of light waves in a thin film was invented in 1891 by Gabriel Lippman (1845–1921), a French professor of physics who won a Nobel Prize in 1908, but practical difficulties prevented its commercial application.

The foundation for an effective means of taking color photographs was laid in 1861 by another professor of physics, James Clerk Maxwell (1831–1879), who demonstrated that light of any color can be created by mixing three primary colors: red, green, and blue. Colors are formed through an additive process: Yellow is made by mixing blue and green, magenta is made by mixing blue and red, and so on (this holds only for colored light; a mixture of paint pigments may produce very different effects). In 1892, Maxwell's discovery was employed commercially for the production of stereoscopic slides, which when seen through a special viewer presented realistically colorful three-dimensional scenes.

Although this was fine for parlor entertainment, people wanted color pictures to carry in their wallets and mount on their walls. This was achieved in 1893 by a Dublin resident, John Joly (1857–1933). The technique entailed the use of a screen that contained tiny, evenly distributed areas of red, green, and blue. A photographic plate was exposed by light passing through the screen. The developed plate then was made into a transparency that was bound to the screen. In this finished photograph the transparency's areas of black, gray, and white passed different amounts of light that then blended the colors on the screen to reproduce the colors of the original scene or object.

Additive processes eventually gave way to subtractive ones. These are based on the principle that different colored objects reflect and absorb different parts of the light spectrum. For example, a blue object reflects only the blue portion of the spectrum, and subtracts (that is, absorbs) the other colors. The first applications of this method required separate exposures of a scene through red, blue, and green filters. The resulting negatives were in turn used to make transparencies tinted in complementary colors: cyan for red, yellow for blue, and magenta for green. When the three transparencies were stacked together the result was a rendition of the scene in its original colors.

In the history of science and technology, there are numerous examples of simultaneous discovery and invention; color photography using the subtractive method was one of them. It was invented in 1869 by two Frenchmen, Louis Ducos de Hauron (1837–1920) and Charles Cros (1842–1888), who presented descriptions and samples of their methods to the French Photographic Society at virtually the same time. Cros did not go much beyond his original invention, but Ducos du Hauron continued to experiment. After Herman Vogel (1834–1898) showed that silver bromide and other photographic emulsions could be sensitized to the entire spectrum, Ducos du Hauron began to make good color prints. His success subsequently inspired a number of similar processes.

Subtractive processes produced good, even excellent, color photographs, but the need to make three separate exposures could be a nuisance. A much more convenient method of taking color photographs emerged in 1935 with the marketing of Kodachrome movie film and the introduction of 35-mm color film for still cameras 2 years later. Kodachrome was invented by Leopold Mannes (1899–1964) and Leopold Godowsky, two amateurs who received some technical assistance from the Eastman Kodak company. Kodachrome film was made from four layers of emulsion, each sensitive to a specific part of the color spectrum. Exposure produced a latent image of three primary colors, which was then developed to create a negative. Reversal processing put complementary colors of yellow, cyan, and magenta in the appropriate areas in order to turn the film to a positive.

Kodachrome photographs had the same drawback as daguerreotypes: Only one copy was produced. In 1941, Kodacolor, another subtractive process, made possible the use of negatives from which any number of prints could be made. Prints were made through a dye-coupling process; just as a monochrome negative turns light into dark and vice versa, a Kodacolor negative also turned photographed colors into their complements: red to green, yellow to blue, and so on. These colors were then restored in the positive print.

Color photography continues to improve; new films allow fine-grained images at slower shutter speeds, while some Polaroid films produce instant photographs with muted, subtle colors. Color prints and slides do not give a precise rendition of the actual colors of the object or scene that was photographed, but color is not perceived with absolute objectivity; many individuals see color differently from other people. Finally, the almost universal use of color photography may have dulled an appreciation of shades, forms, and textures—properties often best explored through the medium of black-and-white photography. Color photography has provided new capabilities, but sometimes at the expense of other aesthetic virtues.

See also PHOTOGRAPHS, INSTANT

Further Reading: Beaumont Newhall, *The History of Photography*, 1982.

Photography, Early

For hundreds of years it was known that light entering a darkened room through a pinhole projected a reduced, inverted image on the opposite wall. At the end of the 16th century, the image was made brighter by fitting the aperture with a lens. A special room or box constructed for this purpose was called a *camera obscura* (literally "dark room") and was widely used by artists for tracing images. All that was necessary to obtain a permanent image was to discover a substance that changed when light fell upon it. The first step in this direction was taken in 1727, when a German doctor, Johanne Heinrich Schulze (1687–1744), dissolved chalk in nitric acid that contained some silver nitrate impurities; to his surprise the resulting substance turned purple when exposed to light. Although Schulze produced designs and inscriptions by masking vessels containing the substance, the phenomenon was little noticed until the early 19th century when Thomas Wedgwood (1771–1805) used screens treated with silver nitrate or silver chloride in a camera obscura to produce nonpermanent photographic images.

The first permanent photographs were produced in France by Joseph Nicéphore Niepce (1765–1833) in the second decade of the 19th century. Niepce made permanent photos by coating a plate with bitumen of Judea, a thick hydrocarbon that hardened when exposed to light. Niepce used this quality to reproduce engravings by placing them on a plate coated with the bitumen. Light passing through the blank areas of the engraving hardened the corresponding parts of the bitumen-coated plate, rendering it impervious to its usual solvent, oil of lavender. Only the dark areas were washed away by the solvent, leaving behind bare metal portions of the plate. These were then etched with acid so that the plate could be used for printing reproductions of the original engraving. Niepce also tried to photograph outdoor scenes through the use of this method; his crude image of a barn that stood outside his window may be considered the first photograph.

In 1829, Niepce formed a partnership with Louis Jacques Mandé Daguerre (1789–1851). Although Niepce died 4 years later, Daguerre continued to work on photography. He eventually created a photographic plate by putting a silver-coated copper sheet on top of a box containing particles of iodine; these reacted with the silver to form a coating of light-sensitive silver iodide. After exposure, the plate was placed over a box containing heated mercury, which formed an amalgam with the silver. A bath of sodium chloride prevented further exposure, leaving a permanent photograph. Daguerre had difficulty in reaping commercial benefits from his technique for making photographs, which were called *daguerreotypes*. But in 1839 with the help of some eminent scientists, he succeeded in getting the French government to buy the invention in return for a small lifetime stipend for Niepce's son and himself.

While Daguerre continued to work on his method of photography, in England William Henry Fox Talbot (1800–1877) was also attempting to chemically capture images on paper. He eventually invented a photographic process that would render the Daguerreotype obsolete within 2 decades. Fox Talbot's first attempts at photography produced direct prints that were made by treating a piece of paper with a weak solution of sodium chloride (common table salt), drying it, and then bathing it with a silver nitrite solution. The ensuing chemical reaction caused the paper to be impregnated with light-sensitive silver chloride. After light from a subject struck the paper, it formed a photographic image that was fixed, although not permanently, by a bath of potassium iodide or sodium chloride. A much better fixing process was invented by John Herschel (1792–1871), an astronomer and inventor of the word *photography*, who used a solution of sodium thiosulphate (known to generations of photographers as "hypo") to wash away the silver salts that had not been exposed to light.

Fox Talbot's greatest contribution to photography came in 1840, when he invented a photographic process that is essentially the same as the one used today: the exposure of a negative that is subsequently used to print one or more positive prints. This gave it a great advantage over the daguerreotype, which was limited to a single photograph. Fox Talbot's process began with paper that had been treated with potassium iodide and silver nitrate to produce a coating of silver iodide, and then was bathed in a mixture of gallic acid and silver nitrate. After exposure, the paper was developed in the same solution to produce a negative image. This was then waxed to make it translucent, and to allow a print to be made on paper coated with silver chloride.

Daguerre's process produced photographs of stunning clarity and detail, but Fox Talbot's negative-positive process was much more economical. And by the middle of the 19th century, sharper negatives, and consequently sharper prints, were being produced by using glass instead of waxed paper. A process invented by an English sculptor, Frederick Archer (1813–1857), became the standard photographic method for several decades. A glass plate was coated with potassium iodide and collodion (a solution of nitrocellulose, alcohol, and ether), then dipped in a bath of silver nitrate right before a picture was to be taken. This produced a light-sensitive plate that had to be exposed while still moist, and developed immediately afterwards, resulting in a sepia-tinted photograph characteristic of this era.

The process was somewhat simplified by the commercial production of photosensitive collodion solutions that could be applied directly to the plate, but the plate still had to be developed immediately after exposure. The late 1870s saw the appearance of dry plates made by combining the photosensitive chemical with gelatin. This gave photographers more time between exposure and development. Even so, photographers still had to prepare and develop their plates, making photography an activity best pursued by professionals and dedicated amateurs, a situation that changed only with the introduction of roll film and factory developing.

All of these inventions and discoveries made pho-

tography a technical possibility. At the same time, its rapid spread during the 19th century was stimulated by the social changes that were then taking place. Photography emerged in the era of the rising bourgeoisie, a time when merchants, manufacturers, and professionals grew in number and influence. Members of this social class had the desire and the financial means to flaunt their affluence by taking on some of the attributes of the old landed gentry. For the latter, a key artifact of their status had been the painted portrait. But even for relatively wealthy bourgeoisie, this was an expensive acquisition, and in any event there were not enough skilled painters to meet the potential demand. Middle-class aspirations were therefore met in the photographic studio, where the subject could pose with all of the trappings of upper-class status, such as rich draperies and elegant furniture. In later years, cheaper methods of photography emerged to meet the needs of poorer yet increasingly affluent people, such as immigrants to America who wanted portraits to send home. Meanwhile, millions of amateur and professional photographers were recording what they saw in their camera's viewfinders. Only a few decades after its invention, photography had created a vastly expanded permanent record of people, places, and events.

See also ROLL FILM

Further Reading: Beaumont Newhall, *The History of Photography: From 1839 to the Present*, 1982.

Photography, Stroboscopic

A stroboscope is a device that makes a moving object appear to stop. The simplest stroboscope, the origins of which go back to 1832, is a disk with a number of slots arranged radially from the edge to the axis. As the disk is rotated, an observer looking at a moving object through the slots will see the object intermittently. When the rotation of the disk is coordinated with the movement of the object, the object may appear to be stationary, or it may appear to move in a manner different from its actual motion. This effect can be seen in motion pictures where the wheels of a car seem to be rotating backwards as the car travels forward. This occurs because succeeding frames of the film were not exposed at the instant that the wheels had rotated through some multiple of 360 degrees. If each frame had been taken a bit earlier than the time required for a complete rotation, the wheels will appear to be going backwards.

Successful photographs of moving objects are almost as old as photography itself. William Henry Fox Talbot, the coinventor of photography was able to photograph a clipping from the *New York Times* on a revolving disk by briefly illuminating it with an electric spark generated by a Leyden jar.

Other photographers had considerable success with electric sparks, but it is fair to say that the modern era of stroboscopic photography begins with the work of Harold E. Edgerton (1903–1990). While a graduate student in electrical engineering at the Massachusetts Institute of Technology, Edgerton was studying the stability of power systems, particularly the ability of generators and motors to remain in step after being disturbed by events like lighting strikes on transmission lines. While using mercury arc rectifiers as the power source for a generator, Edgerton noticed that the generator's rotor seemed to be oscillating over a small arc. As Edgerton was well aware, the rotor was not oscillating; rather, the rectifiers were supplying brief flashes of light that illuminated the rotor as it turned. This inspired Edgerton to build a stroboscope, which he described in the journal *Electrical Engineering* in 1931.

Aided by Kenneth Germeshausen and Herbert Grier, Edgerton went on to develop the art and technology of stroboscopic photography. With the capability to take still photographs in less than 1/10,000 of a second, and movies at 300 frames per second, Edgerton, Germeshausen, and Grier investigated such phenomena as the flight of a hummingbird and the swing of a golfer. Many of the photographs have become classics, not just as scientific records but also as works of art.

A major challenge for Edgerton and his colleagues was to set off brief flashes of light at the proper moment or moments. This might be done by having the object being photographed break a beam of light from a photoelectric cell. Alternatively, the sound of the object could be used to trigger the circuitry. For moving pictures, they used a sequence of lights flashing at very brief intervals. The flashing of the lights obviated the need for a conventional camera shutter, as a frame was taken every time that a light flashed. This occurred so quickly that

no blurring was observed when the developed film was projected.

Most strobe photography uses a bulb filled with inert xenon gas as the source of illumination. The light is emitted when the bulb receives a sudden discharge of electricity from a capacitor. Strobe photography often produces striking effects by revealing what hitherto had been unseen. In addition to their aesthetic appeal, strobe photographs can be valuable tools for the study of such phenomena as the propagation of shock waves and the movement of fluids.

See also CAPACITOR; PHOTOGRAPHY, EARLY

Further Reading: Harold E. Edgerton and James R. Killian, Jr., *Moments of Vision: The Stroboscopic Revolution in Photography*, 1979.

Photosynthesis

Photosynthesis is the process through which plants use sunlight to convert carbon dioxide and water into carbohydrates. Photosynthesis is the basis of most life on Earth, for it provides food for plants, and these plants directly or indirectly nourish most other organisms. Moreover, most of the oxygen in our atmosphere had its origin in the oxygen that was released through photosynthesis. Photosynthesis is crucial to the stability of the atmosphere since it uses up carbon dioxide (CO_2) and releases oxygen, helping to offset the buildup of CO_2 caused by the burning of fossil fuels and other human activities.

Photosynthesis can be summarized by the equation $CO_2 + 2H_2O \rightarrow [CH_2O] + H_2O + O_2$. Many reactions are nested inside this equation, however, and the actual process involves many complex transformations. Photosynthesis has never been artificially replicated. If it could be, it would be one of the greatest breakthroughs of all time, for as Otto Warburg (1883–1970) demonstrated, photosynthesis operates with close-to-perfect thermodynamic efficiency.

The quantitative study of plant growth began with Johannes van Helmont (1579–1644), a Flemish physician. Seeking to prove his theory that water was the material from which everything else was made, Helmont grew a willow tree in a tub for 5 years, during which time it received no additions other than water. At the end of the 5-year period, Helmont showed that the weight of the soil in the tub hardly changed, but the tree had gained 74 kg (164 lb). Helmont also studied the carbon dioxide given off by burning wood, which he called *gas sylvestre* (gas from wood), not realizing that this gas, and not water, was the source of the tree's growth.

The importance of carbon dioxide for the life of plants began to be perceived when Stephen Hales (1677–1761) discovered that the decomposition of plants resulted in the release of a gas that did not support combustion, which he called *fixed air*, what we now know as CO_2. In 1772, Joseph Priestley (1733–1804) found that green plants can replenish air by removing phlogiston from it. Ironically, Priestly, the codiscoverer of oxygen, did not realize that phlogiston was a wholly imaginary substance, and that it was the oxygen released from the plants that was replenishing the air. The experiments of Hales and Priestly were an inspiration to Jan Ingen-Housz (1730–1799), a Dutch physician who had interested himself in plant physiology. In 1779, Ingen-Housz published *Experiments on Vegetables*, which made the important observation that living plants absorb carbon dioxide and release oxygen, and that the visible light of the sun (but not its heat) was necessary for the process to occur.

Ingen-Housz also had noted that only the green parts of plants are involved in the release of oxygen, and subsequently Jean Senébier (1742–1809) showed that a whole plant was not needed for the production of oxygen. Its leaves, even after they were chopped up, would change "fixed air" into "dephlogisticated air" (i.e., liberate the oxygen from carbon dioxide). An important clue as to why this happened came in 1817, when chlorophyll was chemically isolated by Pierre-Joseph Pelletier (1788–1842) and Joseph Bienaimé Caventou (1795–1877), French chemists whose research into plant alkaloids later paved the way for the synthesis of antimalarial drugs. At the time, Pelletier and Caventou did not realize the significance of chlorophyll's role in photosynthesis. In 1865, the German botanist Julius von Sachs (1832–1897) published a treatise that demonstrated how the chlorophyll in the cells of green plants was concentrated in organelles that came to be called *chloroplasts*. Von Sachs also determined that chlorophyll catalyzes (i.e., increases the speed of a reac-

tion) the chemical reactions that occur during photosynthesis, and he hypothesized that the carbon dioxide taken in by plants is incorporated into the carbohydrates that are produced through photosynthesis.

Von Sachs's hypothesis did in fact present an accurate, if schematic, picture of photosynthesis. In general terms, photosynthesis begins when water enters a plant through its roots, and carbon dioxide enters through tiny holes (known as *stomata*) in its leaves. Within the chloroplasts, the water molecules are broken down into oxygen and hydrogen. These are then involved in a series of reactions that result in the production of glucose and fructose, the simple sugars that serve as the plant's nutrients.

The exact mechanisms through which these reactions occur began to be understood as research into photosynthesis continued into the 20th century, aided by the development of new processes of chemical analysis. One of the most important of these was chromatography, a process of separating chemicals that was created during the first decade of the 20th century by Mikhail Tsvett (1872–1920), a Russo-Italian botanist. Tsvett's techniques were refined by Richard Willstätter (1872–1942), who used chromatography to investigate the nature of chlorophyll. Willstätter found that there were two major types of chlorophyll in land plants: a blue-green "a type" and a yellow-green "b type." Willstätter also discovered that the molecular structure of chlorophyll was similar to that of hemoglobin, the oxygen-transporting constant of red blood cells. The molecules of both hemoglobin and chlorophyll have ringlike structures that surround a central atom, iron in the case of hemoglobin and magnesium in the case of chlorophyll.

The use of chromatography provided important insights into photosynthesis. Of equal importance was the use of radioactive isotopes to trace the actions of the elements involved in photosynthesis. The first of these studies was conducted by Martin Kamen (1913–), who used oxygen-18 to determine that the oxygen generated by photosynthesis was derived from the dissolution of water molecules, and not from carbon dioxide. Kamen's subsequent isolation of carbon-14 made possible the study of the biochemical reactions as they occurred in plant cells. Of particular importance was the research of Melvin Calvin (1911–1997). Using green algae cells, Calvin studied the process of photosynthesis by following carbon-14 molecules as they went through a plant's sequence of chemical reactions. Calvin stopped the reactions at different stages by treating the algae with alcohol, and then used paper chromatography to analyze the chemicals that had been produced. He eventually discovered that photosynthesis entails the creation within a few seconds of at least 10 intermediate chemical products. The process begins as the chloroplasts absorb energy from light (the "light-dependent" or "light" reaction). This is followed by a series of reactions that require no light and terminate with the production of carbohydrate molecules (the "light-independent" or "dark" reaction). The metabolic pathway through which these light-independent reactions occur is known as the Calvin cycle.

See also CARBON DIOXIDE; CELL; CHROMATOGRAPHY; CLIMATE CHANGE; ENERGY EFFICIENCY; FOOD CHAIN; ISOTOPE; PHLOGISTON; RADIOACTIVE TRACERS; RADIO-CARBON DATING

Photovoltaics

In photovoltaic cells, unlike other methods of producing electricity with solar energy, the sun's light rather than its heat generates the electricity. This photovoltaic effect was discovered in 1839 by Edmund Becquerel, a French scientist. During the 1880s, American inventor Charles Fritts made the first solar cells. The cells were small, thin selenium wafers covered with a transparent gold film; they were extremely inefficient, only 1 percent of the sunlight striking the cells actually being converted to electricity. Fritts's discovery was forgotten, rediscovered during the 1930s, and then used to develop light-sensitive devices such as photometers. Selenium, however, was not capable of producing real power.

During the 1940s, silicon was discovered to work well as a rectifier, changing alternating current into direct current. Russell Ohl, while working on rectifiers at Bell Telephone Laboratories, also discovered that silicon provided a good photovoltaic response. He and other Bell scientists continued working with silicon rectifiers into the 1950s, when Gordon Pearson, director of the rectifier project, discovered accidentally that a significant electrical current was generated when a silicon rectifier was exposed to sunlight. Pearson took his discovery to Darryl Chapin, leader of another Bell project that was seeking to

develop a dependable alternative power source for rural telephone systems. Chapin, who had been trying unsuccessfully to improve the selenium solar cell, soon was working with Pearson and scientist Calvin Fuller on a silicon solar cell. Within a few months, they succeeded in producing a silicon cell that converted 6 percent of available sunlight into electricity, and they soon increased the conversion ratio to 15 percent.

Although Bell Lab's so-called "sunshine battery" received an extraordinary response—pundits claimed that solar cells soon would power everything from lawn mowers to automobiles—the cost of producing cells was prohibitive; even for rural telephone systems, conventional electric power was cheaper. For the National Aeronautics and Space Administration's satellite program, however, solar cells proved very cost effective. In 1958, NASA installed them on Vanguard I, America's second satellite, and adoption of photovoltaics for the space program led to steady improvement in solar cell design for the next several years. Knowledge of this work and other worldwide solar research activity was shared at world energy symposiums, the first held in Phoenix, Ariz., in 1955, and through the Association of Applied Solar Energy. But at the outset of the 1970s, few people knew of these activities.

The 1973 energy crisis renewed interest in terrestrial use of photovoltaics. A U.S. government-funded research program at the Jet Propulsion Laboratory in Pasadena, Calif., accelerated development of silicon photovoltaics, and the utility-funded Electric Power Research Institute began photovoltaic research at a time that new solar-cell markets were appearing. Annual world sales of photovoltaic modules increased from 0.5 megawatts to 55 megawatts (mW) between 1976 and 1991. Photovoltaics provided electric power for remote railroad-crossing lights, communication systems, harbor buoys, livestock water pumps, and monitoring equipment, as well as consumer products such as calculators and watches. The appropriate technology movement embraced photovoltaics along with other solar technologies, eulogizing rooftop solar cell modules as a way for individuals to break the grip of central-station electric power companies. These units, costing from $2,000 to $5,000, found a modest market with vacation home-owners in the United States and Europe, and in villages of less-developed countries.

Meanwhile, during the 1980s, utilities began investigating large-scale application of photovoltaics. Pacific Gas and Electric Company and ARCO Solar opened the largest demonstration photovoltaic generating facility in 1984. Their 71.7-ha (177-acre), 6.4 mW photovoltaic farm in San Luis Obispo County, Calif., contained several hundred 10.4-by-11-m (34-by-36-ft) rectangular solar cell panels mounted on 6.1-m- (20-ft-) high pedestals, "mechanical toadstools" that automatically tracked the sun. With oil prices climbing and photovoltaic cell developments steadily reducing the cost of power generation to $1.50 per kWh in 1980, it seemed a prudent project. By 1988, however, oil prices had declined even faster than the kilowatt-hour cost of photovoltaics, which had fallen to 35 cents. As a result, plans to expand the plant were dropped.

Nevertheless, PG&E and other utilities discovered that photovoltaics were cost effective where they reduced peak loads in situations that otherwise would require expensive system upgrading. In 1993, the firm opened a 500-kW solar system near Fresno to handle peak load. The Southern California Edison Company installed a similar system in a South Pasadena neighborhood, and a consortium of 86 utilities, known as the Utility Photovoltaic Group, announced plans to install 50 mW of new solar capacity before the end of the century. The efficiency of photovoltaics has increased steadily since the 1950s, while the cost of cells, silicon sheets, thin-film technologies, and concentrator systems has declined. Photovoltaic technology surely will fill an important place in the world's electric power future.—J.W.

See also ALTERNATING CURRENT; APPROPRIATE TECHNOLOGY; CALCULATOR; DIRECT CURRENT; NATIONAL AERONAUTICS AND SPACE ADMINISTRATION (NASA); RECTIFIER; SOLAR DESIGN, ACTIVE; SPINOFFS

Further Reading: Thomas B. Johansson et al., eds., *Renewable Energy: Sources for Fuels and Electricity*, 1993.

Phrenology

The field of phrenology began with the work of Franz Joseph Gall (1758–1828). Gall was a successful anatomist especially interested in the physiology of the brain. He showed that the brain and spinal cord consisted of two kinds of substance: gray and white matter. From a series of studies comparing healthy and damaged human brains, as well as the brains of animals, children, adults, and the elderly, he concluded that the workings of the mind were related to the operation of the brain, and that the mental faculties depended on the integrity of the brain, especially the cortex. However, Gall is best remembered for his claims that personality and intelligence can be measured from external characteristics of the skull. Gall's theory of phrenology was based on the notion that personality and intelligence were reducible to 42 powers or functions, and that each of these was localized in specific area of the brain. These areas were identified with phrenological charts of the skull. Using these charts, one could measure intellectual ability and personality through palpation of the skull, with well-developed powers creating small bumps on the skull and less-developed powers leading to indentations in the skull.

Phrenology attracted both devoted followers and powerful enemies. His enemies included the Catholic Church and the Emperor of Austria, as well as scientists and physiologists. His most important supporter was a theologian and physician, Johann Caspar Spurzheim (1776–1832). In 1832, after Gall's death, Spurzheim was invited to lecture in the United States. His initial lectures in Boston created a sensation and were followed by a series of very successful and popular lectures at hospitals and universities. However, as phrenology gained in popularity in the United States, it lost any claims to scientific method and became little more than a money-making enterprise. Three American entrepreneurs, Orson and Lorenzo Fowler, and Samuel Wells, capitalized on phrenology's popularity and marketed neatly labeled phrenological busts with manuals that detailed how to undertake a complete phrenological self-analysis. They, and other phrenologists, opened phrenological offices where the gentry could have their heads read. Some employers included phrenological examinations as a condition of employment, and phrenologists counseled young people on the selection of appropriate marriage partners.

Although popular among the general public, phrenology from its start was regarded by most scientists as, at best, as a pseudoscience, such as astrology or alchemy. In 1843, a leading investigator of brain functions, Pierre Flourens (1794–1867), published a devastating critique of the theory, *An Examination of Phrenology*. He included in his critique findings from his own studies demonstrating that the contours of the skull did not correspond to those of the brain. Thus the basic assumption of phrenology was wrong.

Despite its serious flaws as a scientific theory, phrenology nonetheless made indirect but important contributions to psychology. It bolstered the belief that mental functions could be localized in the brain. In contending that they could measure personality and intellectual differences between people, phrenologists also reinforced the notion of individual differences and the ability to measure them. This notion later became the basis for personality theorists, psychological testing, and the measurement of intelligence.—M.M.

See also INTELLIGENCE, MEASURES OF

Piggyback Transportation

From the second half of the 19th century to the early 20th century, most of the freight transported any significant distance overland traveled by rail. In the 1920s, the railroad's preeminent position began to erode as the motor truck developed into an effective means of moving freight. Much of the advantage of using a truck lay in its ability to provide flexible, door-to-door service. But when goods had to be hauled long distances, railroads remained the cheapest mode of transport. Piggyback trains combine the best of both forms of transportation by the long-distance haulage of truck trailers loaded onto railroad cars. These trailers are moved from a shipper to a piggyback terminal by conventional truck tractors, which also haul them from their arrival terminal to their final destination. Piggyback service is sometimes called as TOFC, for "trailer on flat car."

The basic idea underlying piggyback transportation is not new. Toward the end of the 19th century, the Long Island Rail Road initiated a kind of piggyback service when it hauled produce-carrying farm wagons into New York City. In the early 20th century, a few electric interurban railroads transported truck trailers on flatcars. The first mainline railroad to provide this service was the Chicago Great Western; in 1935, it began to transport trailers between Chicago and Dubuque. After World War II, railroads began to turn to piggyback as a partial answer to truck competition and declining revenues.

A 19th-century phrenologist at work (courtesy National Library of Medicine).

Piggyback technology was fairly straightforward at first, involving little more than the use of existing flatcars and the construction of end-loading ramps. Considerably more problematic was the political environment. Railroading was a highly regulated industry, and the railroads had to wait for rulings by the Interstate Commerce Commission (ICC) regarding the setting of freight rates and the legality of cooperation between individual railroads and trucking companies. The resolution of these issues in 1953 was followed by the rapid expansion of piggyback service.

After being given the green light by the ICC, the railroads and truck lines had to work out how they would share responsibility for piggyback service. A number of different programs emerged. Some had shippers dealing only with truckers, who then made their own arrangements with the railroads. Another program had the shipper using the services of a railroad that ran its own trucks. Others were based on shippers providing their own trailers, while still others involved the services of third-party freight forwarders that provided everything but the rail portion of the trip and the loading and unloading at each end.

While regulatory and tariff arrangements were largely resolved in the 1950s, advocates of piggyback service also had to contend with internal opposition. Some railroad executives feared that piggybacks would simply cannibalize their own boxcar traffic. Also, the railroads' mechanical and operating departments often saw piggyback service as something that would further complicate their lives. Successful initiation of piggyback service therefore required a fair amount of entrepreneurial effort both inside and outside railroad management.

From the 1950s onwards, the efficiency of piggyback service increased as a result of significant improvements to the equipment used. Early piggybacking was done with ordinary freight cars, which necessitated a cumbersome and time-consuming process of tying down trailers with chains. The process of securing trailers was considerably simplified through the development of special-purpose piggyback flatcars. In particular, the fitting of flatcars with retractable hitches made it much easier to load and unload trailers. By the 1980s, piggyback cars had evolved into 27-m (89-ft) special-purpose cars capable of carrying one 12-m (40-ft) trailer and one 14-m (45-ft) trailer.

No less important for the advance of piggyback service was the development of improved facilities for the loading and unloading of trailers. For many years, this was accomplished through the use of loading ramps situated at the end of a railroad spur. Flatcars were temporarily connected with bridge plates, allowing trailers to be pushed from the loading ramp to the furthermost empty car. This arrangement had the advantage of low capital cost, but it could be highly inconvenient. For example, if a trailer arriving at its destination happened to be in the middle of the train, it was necessary either to switch the cars or to unload all of the trailers on one side of the car to be unloaded. In the early 1960s, the railroads improved piggyback operation through the use of cranes for loading and unloading trailers. These devices straddled a track or set of tracks and ran on hydraulic or electric power. Cranes of this sort are still in use, although some piggyback facilities now use sidelift loaders that resemble giant forklift trucks.

The deregulation of the railroads that began in 1980 was followed by a substantial increase in piggybacking. By this time, however, the trailer-on-flatcar was rivaled by an alternative means of moving freight by rail, container-based intermodal transport. Even so, piggybacking continues to be extensively used for hauling freight in trailers that might otherwise be clogging up the highways.

See also RAILROAD; TRANSPORTATION, INTERMODAL; TRUCKS

Further Reading: David J. DeBoer, *Piggyback and Containers: A History of Rail Intermodal on America's Steel Highway*, 1992.

Piltdown Man

In the years that followed Charles Darwin's *The Origin of Species*, a number of scientists were engaged in an effort to find the ancestors of our species, *Homo sapiens*. In the late 19th and early 20th centuries, the fossil record was scanty, and theories about human evolution were difficult to prove or disprove. One of the most contentious issues of the day had to do with the sequence of human development. Some scientists were of the opinion that upright posture and bipedal locomotion preceded the development of high levels of intelligence, while others argued that intelligence came first.

Support for the latter theory came in the years 1911 and 1912 when excavations at the English village of Piltdown conducted by Charles Dawson yielded a remarkable collection of fossils and artifacts. Unearthed were fragments of a skull that included parts of a cranium that appeared to have a brain capacity similar to that of a modern human being. Also found was a massive lower jaw that was far less human in form, although the surviving teeth had wear characteristics similar to those of early human beings. In addition to the human remains, the site contained stone implements and a pointed tool that had been made from an elephant bone. These remains appeared to date back to the Pleistocene era or perhaps to the preceding Pliocene era that extended from 5.2 to 1.65 million years before the present.

The discoveries lent support to the "brain first" theory of human evolution held by many physical anthropologists. The existence of Piltdown Man, or *Eoanthropus dawsoni* to give its scientific name, also resonated with the nationalistic feelings of the time, for there were those who took pride in the apparent fact that the earliest known human being was an Englishman. Even so, many scholars of human origins remained unconvinced by the incongruous association of a human cranium and an apelike jaw.

In the 1950s, the availability of new dating methods made it possible to determine the age of the Piltdown Man with considerable precision. After testing the remains for the accumulation of fluorine and the loss of nitrogen, Kenneth Oakley (1911–1981) announced in 1952 that all of the remains were of recent origin. The thicker-than-normal skull was probably that of an individual with Paget's disease, while the jaw was determined to have come from an orangutan. On careful inspection, it was discovered that the teeth had been filed to shape, while all the remains were found to have been artificially stained.

Piltdown Man was thus revealed as one of the greatest scientific hoaxes of all time. The identity of the perpetrator or perpetrators is not known, but whoever was involved had a good understanding of human anatomy. In addition to embarrassing a number of physical anthropologists who had believed in the authenticity of Piltdown Man, the fraud impeded efforts to trace human origins. When the first *Australopithecus* was discovered in 1924, many physical anthropologists were convinced that its lineage belonged to the apes and not to human beings. For those who had been taken in by the fraud, the combination of bipedal locomotion with a brain that was considerably smaller than Piltdown Man's marked *Australopithecus* as an unlikely ancestor of modern human beings.

See also AUSTRALOPITHICINES; EVOLUTION, HUMAN; NATURAL SELECTION

Pipe Organ

Reduced to its most basic components, an organ is set of pipes, an apparatus to provide air under pressure to these pipes, and a number of devices to control the supply of that air. As with wind instruments in general, vibrating columns of air in the pipes produce the instrument's sounds. The pitch and tone quality of these sounds is largely determined by the kind of material used for the pipes, along with their length and shape.

Until the 19th century, air was supplied by bellows. Sliders actuated from a keyboard controlled the admission of air into the pipes.

The ancestry of the organ can be traced back to a simple wind instrument known as the pan-pipes; the organ was born when air was provided by a source other than the player's lungs. In the 2d-century B.C.E., the Alexandrine engineer Ctesibius (fl. c. 270 B.C.E.) is said to have designed a technically sophisticated organ that used a hydraulic mechanism to keep the air supply at a constant pressure. By the Middle Ages, the pipe organ was the most technologically sophisticated device in existence, and for centuries only the mechanical clock occupied the same technological plane. One of the largest medieval organs, the one installed at Winchester Cathedral at the end of the 10th century, had 400 pipes and 26 bellows to provide air for them. The sound from organs of this sort could be heard a mile away, although the musical quality suffered from the fact that the pipes were not always in tune with one another.

The medieval church organ was a massive device that sometimes required as many as 70 people to work the foot-operated bellows supplying the air. Playing these organs called for a fair amount of physical strength, since the keys worked valves that had to overcome the resistance of air pressure. The pipes were grouped in "ranks" that generally corresponded to a single note, although a rank often included pipes several octaves above and below the primary note, as well as a few pipes at one or more octave plus a fifth or third to add harmonic color. As many as 10 pipes were used for each note, and all sounded when a key was depressed. What the instrument lacked in precision and subtlety of musical expression, it made up in power and volume.

By the 14th century, organ builders added keyboards (including a pedal board worked by the feet) that allowed playing separate pipes instead of the entire rank. In the 15th century, organs began to be equipped with stops; these allowed air to be directed into a selection of pipes that were similar in timbre but with different pitches (from this comes the common expression "pulling out all the stops"). In the 17th century, the organ was equipped with a number of refinements that did not change its basic character: tremolo stops, pipes that simulated a string tone, and the swell, a box with movable shutters that provided crescendos and diminuendos.

At about this time, organs began to be equipped with air reservoirs that eliminated the pulsations produced by air that came directly from a bellows. In the 19th century, the volume of air supplied to the pipes increased substantially as first steam and hydraulic, and then electric motors began to be used for this purpose. This allowed the construction of even larger organs but greatly increased the pressure that had to be overcome when depressing a key. It was therefore necessary to invent pneumatic, electric, or electro-pneumatic devices that opened a valve or set of valves when a key was depressed to sound a note or chord.

These new means of actuating the keys made it possible to physically separate the keys from the pipes. As a result, the physical appearance of organs often had little or nothing to do with their size or their musical resources. Since their way of connecting keys to pipes is not mechanical, electric-action organs are considered by some to be imitations of the real thing. The most thoroughgoing example of such imitation is, of course, the electronic organ, which uses computer chips with sounds of real organs recorded on them. Today's electronic organs range in size from small portables to full-featured instruments that have taken the place of many conventional pipe organs.

See also CLOCKS AND WATCHES; MICROPROCESSOR; MOTOR, ELECTRIC; MUSIC, ELECTRONIC

Planetarium

A planetarium is a device that simulates the appearance and motions of celestial bodies. The first known planetariums were celestial globes constructed under the direction of the Greek (or Egyptian) astronomer Ptolemy (c. 75–?), although none of these have survived. Similar globes were made by Arab and Persian astronomers, as well as by the Danish astronomer Tycho Brahe (1546–1601). These early devices had the disadvantage of putting observers outside the celestial sphere, a vantage point that required them to imagine themselves looking in at the celestial bodies. This defect was remedied by constructing celestial globes that were large enough to allow spectators to sit inside them. The first of these was designed by Adam Oelschlager in the mid-17th century, and was followed by several others, including one built by Wallace Atwood in the early 20th century.

Celestial globes did a good job of reproducing the movements of the stars, but they were unable to do the same with "wandering" celestial objects: the sun, moon, and planets. The Roman author Cicero (106–43 B.C.E.) described a device that replicated the motion of the planets. It had been removed following the Roman conquest of Syracuse in 212 B.C.E., and Archimedes (c. 287–c. 212 B.C.E.) may have been its designer, but historical details are scanty. The invention of the mechanical clock in the late Middle Ages was paralleled by the development of mechanisms that showed planetary motions; in fact, the two functions were sometimes combined in one instrument. In the 17th century, Christian Huygens (1629–1695) designed a clockwork-driven planetarium that depicted the six known planets circling the sun in elliptical orbits and at proper nonuniform speeds.

Mechanical models of the solar system are sometimes called *orreries*, a term that goes back to a device that John Rowley built around 1712 for Charles Boyle, the Fourth Earl of Cork and Orrery. A number of orreries were built in the years that followed. One of the most impressive is the one built by a Dutch wool-comber named Eise Eisinga (1744–1828). Still in operation today, the planets move on the ceiling of a room in Eisenga's home, animated by weight-driven gears and regulated by a pendulum. The planets move in "real time"; Saturn, for example, takes 29.5 years to make a complete orbit of the ceiling.

Today's planetariums use light projected on a dome to simulate celestial bodies. The projection planetarium was conceived by Walter Bauersfeld (1879–1959) at the Zeiss Optical Company in 1919, with construction commencing the following year. Bauersfeld was chief engineer at Zeiss, and was well-situated to carry out the project. In August 1923, the projector was tested at the factory, and in October of that year it was temporarily installed at the Deutsches Museum in Munich. It was then returned to the Zeiss works, where it was used for many demonstrations. It went back to the Deutsches Museum in 1925, where it operated until its retirement in 1960.

This first Zeiss projector was capable of showing stars only at northern latitudes between 49 degrees and 68 degrees. In 1926, a Zeiss planetarium capable of showing the entire celestial sphere was installed in Bar-men, Germany. A number of improved models followed. The largest of these is the Zeiss Mark VI planetarium, which was introduced in 1968. Designed for domes 18 to 25 m (60–80 ft) in diameter, it is 3.8 m (12.5 ft) long and weighs 2,500 kg (5,500 lb). The large globes at either end each have 16 starfield projectors. The star holes, which are produced by a photochemical etching process, have apertures ranging in size from half the diameter of a pin to one-eighth the diameter of a human hair. The sun, moon, and planet projectors are positioned in the framework between the globes. Illumination for the projection of celestial objects is provided by mercury vapor lamps, and the light is passed through special planar lenses. The projector even simulates the twinkling of the stars by beaming light through a rotating device that resembles a small birdcage. The movement of the projector can be controlled to allow a simulated 24-hour period to take place in as little as 30 seconds or as much as 36 minutes.

Smaller, less-expensive planetarium projectors were created by Armand Spitz in the late 1940s. Widely used in schools and museums, the Spitz resembles the Zeiss planetarium in that it is mechanically operated. In contrast, the Digistar planetarium uses a high-resolution cathode-ray tube in conjunction with a wide-angle lens to replicate the firmament. Like some models of the Zeiss and Spitz planetariums, it is controlled by a computer.

Although the primary purpose of planetariums is the simulation of celestial phenomena, planetariums also are used for a number of varied purposes. They have been employed on occasion for historical research, for example to check the position of celestial objects mentioned in a text. They have also been used for navigation training, including space navigation for astronauts. Planetariums have even allowed ornithologists to study how migrating birds are able to fly vast distances without getting lost.

There are about 2,000 fixed planetariums worldwide, visited by about 20 million people every year. In addition, there are many portable planetariums that are used by schools and other educational institutions. These devices project their images on inflatable dome.

See also CLOCKS AND WATCHES

Further Reading: Charles F. Hagar, *Planetarium: Window to the Universe,* 1980.

Planing Machine

A planing machine employs revolving metal cutters to shave or chip the rough surface of a board or a timber in order to reduce its dimensions or to produce a smoother surface. The piece to be planed can be fed by hand or by a self-feeding mechanism and can pass over or under the cutters. By 1800, Samuel Bentham received various patents for planing machines in Britain. These patents established an important principle of using rotating cutters to allow for the continuous cutting of wood. In theory, this process was much faster than the ancient method of using various hand planes to smooth boards.

The earliest successful application of this principle has been associated with a patent granted to William Woodworth of Poughkeepsie, N.Y., in 1828. The Woodworth planer used feed rolls and a rotary cutting cylinder. Boards placed on edge were clamped to a moving carriage and were passed through metal rollers until they met with a rotating cutting cylinder that was mounted vertically. By 1831, this type of machine could plane 2.5 to 2.7 m (8–9 ft) a minute, producing 400 to 500 planks a day. It held a virtual monopoly over the large and profitable market for tongue-and-groove floor boards used in the construction of buildings. But the early machine vibrated considerably because of its wooden frame and because it required better bearing designs. Most improvements in the machine followed the expiration of Woodworth's patent in the 1850s, when the machines became larger and operated at higher speeds, with powered, spring-pressured rollers, chip breakers, and more cutting knives in the cylinder. By 1853, these machines operated at 4,000 revolutions a minute, planing about 15 m (50 ft) of flooring a minute. Cutters required sharpening about once an hour. Improved models were able to dress all four sides of a board in a single pass.

The success of the Woodworth planer increased tremendously the quantity of dressed lumber, providing incentive to develop machines capable of boring, mortising, tenoning, and shaping, especially when better steel allowed for more durable cutters. Occasionally, the principles of the Woodworth planer were adapted for special purposes. By 1880, a planer could smooth 500 doors in a day. However, the Daniels planer was more commonly used in heavier work required by the railroads and carriage manufacturers. The Daniels machine employed a vertical revolving shaft with horizontal arms, the cutters being placed at the ends of those arms. A traveling bed delivered the work to the cutters, which operated above the work at a very high speed. Most of these machines were very large and heavily engineered, with cast-iron frames.

Throughout the 19th century, most furniture and cabinet shops used a variety of smaller, general-purpose machines known as surfacers, jointers, matchers, and moulders. The simplest of these machines were fed by hand and could be extremely dangerous, calling for great skill on the part of their operators. The general design of planing machines has changed very little in the 20th century. However, electric motors eliminated the need for complex shafting with leather belts in larger woodworking shops. Increasingly, safety devices have become common in even the simplest machines for woodworking enthusiasts.—J.C.B.

See also MORTISING AND TENONING, MACHINES FOR; MOTOR, ELECTRIC; SHAPING MACHINES

Plant Hybridization

The 19th-century French naturalist Jean Henri Fabre (1823–1915) pointed out that history "knows the names of the King's bastards, but cannot tell us the origin of wheat." But almost certainly, bread wheat (*Triticum vulgare*) was the creation of the ancient inhabitants of present-day Iraq, who, perhaps accidentally, hybridized emmer (*Triticum dicoccum*) and a species of *Aegilops*, goat grass. Similarly, modern corn appears to be descended from archaic Americans crossing teosinte and primitive corn, or by their taking advantage of natural hybridization and propagating the offspring. The term *hybrid* derives from the Latin for "half breed."

Certainly, a great deal of plant hybridization occurred between the dawn of agriculture and the 18th century, but in 1735, Carolus Linnaeus (1707–1778) published his *System Naturae*, establishing binomial classification and prompting enormous interest in botany. Ironically, Linnaeus supposed his classification system would reveal God's pattern of creation, based on fixed "species." But very soon, naturalists observed variations that gave rise to the idea that crossbreeding generated new species. For example, William Bartram in Philadelphia mentioned in 1739 that he had crossed different species of the same genus and produced unusual flower colors never seen before. Linnaeus himself eventually abandoned the belief that all species descended unchanged since the Garden of Eden. And, by then, agricultural "improvers" had begun simple crossing experiments they called "scientific breeding."

In part due to Linnaeus, and partly because of Louis Antoine de Bougainville's introduction of high-sucrose sugarcane from Tahiti to Martinique, naturalists accompanied the European voyages of exploration, starting with Captain James Cook's circumnavigation in 1768–1771. Ostensibly, their role was to collect specimens of economic importance, but in the case of Charles Darwin on *H.M.S. Beagle* (1831–1836), the real discovery lay in observing the diversity of species and wondering about their origin.

Darwin's great work, *On the Origin of Species* (1859), included a chapter on "Hybridism," in which he explored the selection advantages crossbreeding (e.g., increased fertility) had over inbreeding (e.g., decreased vigor). Later, he amplified this point, based on his own plant experiments, in *Effects of Cross and Self Fertilisation in the Vegetable Kingdom* (1876), and in *Different Forms of Flowers on the Plants of the Same Species* (1877). By the time these books appeared, Darwin had a legion of followers, including Asa Gray (1810–1888) at Harvard, who sought to test, if not demonstrate, the particulars of evolution by natural selection. Gray's former student, William J. Beal at Michigan Agricultural College, contacted Darwin, and in 1877, Beal carried out a controlled field trial with maize that showed that hybrid vigor from intervarietal breeding could increase yields by nearly 25 percent. (Beal simply planted two varieties of the same race in alternating rows and then detasseled one, ensuring that any corn the emasculated stalks produced had been pollinated by the other variety). At the time, most corn growers supposed soil nutrients and environmental conditions explained yield differences. But Beal's results were too impressive to be ignored, and soon researchers at other agricultural experiment stations pursued "hybrid vigor" in different crops and situations.

By 1900, plant scientists knew a great deal about

inbreeding to obtain high expression of certain desirable traits, and about crossbreeding to restore vigor, but they knew nothing about the means by which traits passed from one generation to the next. Most assumed that progeny had an intermediate or blended inheritance from the parental stock; they generally assumed that the environment could exert influence on the transmission of characteristics. But then three European botanists—Hugo De Vries (1848–1935), Carl Correns (1861–1933), and Erich von Tschermak (1871–1962)—independently found, almost simultaneously, that traits seemed to follow patterns in their inheritance, and indeed, expression was mathematically predictable. This observation suggested that distinct "factors" existed, countering the blended inheritance notion. Further, on investigation, these researchers discovered that an Austrian monk, Gregor Mendel (1822–1884), had reached the same conclusions and published them as *Research on Plant Hybrids* in 1866. In 1909, Danish botanist Wilhelm Johannsen (1857–1927) named the Mendelian factors *genes*, a shortened form of De Vries's term, *pangenes*.

Mendel's rediscovered work allowed plant researchers to quantify inheritance for certain traits, further rationalizing breeding programs. Among those testing the new principles was George Harrison Shull (1874–1954), who in 1904 began an 8-year series of experiments with maize at the Station for Experimental Evolution in Cold Spring Harbor, N.Y. Shull first studied segregation of starchy and sugary kernels from the same ear, leading to his observation that ordinary field corn actually consisted of numerous hybridizations. He then investigated the number of kernel rows, where some species had as few as 8, and others as many as 22. Inbreeding these lines weakened both, but crossing species showed enormous increase in vigor, producing some ears of 24 rows of kernels. It had not been Shull's goal to show how interspecific breeding could increase productivity for agriculture, but that resulted after Edward Murray East (1879–1938) and Donald F. Jones (1870–1963), working at the Connecticut Agricultural Experiment Station, showed further improvements, and after Henry A. Wallace (1888–1965) pioneered the commercial hybrid corn business in 1926. Between 1933 and 1945, hybrid corn acreage in the United States rose from some 400,000 acres to nearly 78 million.

In 1974, "recombinant DNA" entered scientific language, meaning the splicing together of genes from different species by means that bypassed sexual reproduction. At first, the new technology appeared to be limited to bacterial genes, but in April 1983, Josef Schell in the Netherlands and Marc Van Montagu in Germany reported that the crown gall pathogen *Agrobacterium tumefaciens* could serve as a vector for gene transfer in plants. In the same year, the Agrigenetics Company used the bacterial vector to deliver the gene of a storage protein, phaseolin, from beans to sunflowers, in effect creating a transgenic "sunbean." Subsequently, researchers have used several means to produce transgenic tobacco, cotton, corn, and wheat that are resistant to virus, insect, and herbicide damage, as well as plants expressing human genes of pharmaceutical value.

In addition to the application of recombinant DNA technology to intergeneric plant hybridization, molecular biologists since 1989 have pursued an organized genome mapping project that includes identification of all the genes of the tiny mustard plant, *Arabadopsis thaliana*. Concurrently, other researchers are investigating genes responsible for specific traits in virtually all commercial crop plants. The *Arabadopsis* map should provide a guide to the main highways of the plant kingdom, while the specialized studies may show the streets and lanes. The combination of identifying genes of economic importance and being able to transfer them from one species to another suggests that the greatest age of plant hybridization, for better or worse, has only recently begun.—G.T.S.

See also CORN, HYBRID; DNA; EVOLUTION; GENE; GENETICS, MENDELIAN; NATURAL SELECTION; NEOLITHIC AGRICULTURAL REVOLUTION; SEEDS; SUGAR; TAXONOMY

Plasma

A plasma is produced by heating a gas to a temperature that causes it to become completely ionized; that is, it consists of ionized molecules and free electrons. A plasma behaves differently from other states of matter (solid, liquid, and gas) and is therefore considered to be a fourth state of matter. In the everyday world, gases are frequently ionized, but they do not become plasmas. One example is a the gas in a fluorescent tube. The individual molecules are ionized for only a short time, for they cool and recombine into ordinary molecules when they bump into the walls of the tube. Maintaining a plasma requires the application of a strong magnetic field to prevent the ions from coming into contact with the walls of the containment vessel.

The study of plasmas goes back to the early days of electrochemistry, when in the 1830s Michael Faraday (1791–1867) passed electrical currents through various gases at low pressures and produced some plasmalike materials. William Crookes (1832–1919) continued this line of experimentation in the 1870s, and was the first to suggest that a plasma might be considered a fourth state of nature. In 1923, the name *plasma* was coined by the American chemist Irving Langmuir (1881–1957), who came upon it in the course of studying neon lighting. For Langmuir, the ionized molecules were reminiscent of corpuscles in blood, hence the name *plasma*.

Since the 1920s, plasmas have been of particular interest to astronomers, for most stellar material exists in the form of a plasma. Some of the most important research into plasmas was conducted by the Swedish astrophysicist Hannes Alfvén (1908–1995), who theorized how plasmas behaved in the presence of magnetic fields, becoming one of the founders of the science of magnetohydrodynamics in the process. Alfvén was able to construct an entire cosmology in which the creation of galaxies was the result of the working of electromagnetic currents within plasmas. In this way, Alfvén proposed an alternative to the more popular Big Bang theory of the universe's creation.

Plasmas are also of critical importance to fusion research. In conducting research into the process that is the source of the energy provided by the sun and other stars, experimenters found that the best means of achieving fusion reactions entailed heating a gas (up to temperatures of tens or even millions of degrees centigrade) and using electromagnets to confine the resulting plasma. Much of today's fusion research is conducted with the use of a device known as a *tokamak*, a doughnut-shaped tube that contains the plasma in a powerful magnetic field. Plasmas have also been used on a small scale for propulsion in space. The plasma is created and then accelerated by a high-current electrical discharge, producing exhaust velocities that are considerably higher than those generated by chemical rockets. Plasma propulsion has been used for the positioning of orbiting satellites, but up to now it has not been used as a rocket propellant.

See also BIG BANG THEORY; FUSION ENERGY; LIGHTS, FLUORESCENT; NEON; ROCKETS, LIQUID PROPELLANT; ROCKETS, SOLID-PROPELLANT

Plate Tectonics

Plate tectonics holds that the Earth's continents rest on discrete plates that float on magma (liquid rock). Specifically, the solid part of the Earth (the lithosphere) is divided into seven major plates and several smaller ones. Plate tectonics classifies a number of different interactions between plates. A *subduction zone* is a place where two plates collide, e.g., the west coast of South America. A *transform fault* is a place where two plates slide past each other, e.g., the San Andreas fault in California. A *constructive plate margin* occurs when two plates move apart, e.g. the mid-Atlantic ridge. When three plates interact together it is known as a *triple junction*.

The theory of plate tectonics developed in the 1960s. However, the idea that the Earth has moveable plates is not new. Sir Francis Bacon commented on the similarities of the Atlantic coastline of Africa and the Pacific coastline of South America in his book *Novum Organum* in 1620. Bacon is often given the credit of being the first to observe the coastline similarities, but he may have been only the first person to commit the idea to writing. In France, François Placet published *La Corruption du grand et petite Monde* in 1666, in which he discussed the subject of sin and its consequences, among which he included Noah's flood. "Avant le Deluge l'Amerique n'estoit point separée des autres parties de la terre, et il n'y avoit point d'Isles" ("Before the deluge America was not separated from the other parts of the Earth, and there were no islands") was the title of the chapter dealing with the physical consequences of the flood. Placet thought that the landmasses may have been formed by a conjunction of islands or by the destruction of Atlantis, the sinking of which caused the uncovering of new continents. This was a "catastrophic" theory of the Earth's development, one that was subsequently challenged by the "uniformitarian" principle.

The idea that the Atlantic ocean was formed by the

destruction of the Atlantis landmass was very popular in the 17th and 18th centuries. In 1749, George-Louis Leclerc, Comte du Buffon (1707–1773), in his *Théorie de la Terre,* postulated that the Atlantic Ocean might have been formed by the subsidence of an intervening landmass—Atlantis—and by subsequent erosion of the American continent by sea currents. Buffon stands out from his peers as he denied the Atlantic fit of Africa and South America by pointing out that the continents had protrusions that faced each other. Theodor Christoph Lilienthal, professor of theology at Königsberg in Germany, published a series of books under the title of *Die gute Sache der Göttlichen Offenbarung* in 1756. He was perhaps the first to suggest that the continents of South America and Africa had once been side by side. In his view, a strict translation of the Old Testament implied that the division of the Earth after its creation occurred between the continents and oceans.

The geometric and geologic similarities of the opposing continents were noted in papers published by the German explorer Alexander von Humboldt (1769–1859); he too speculated that the Atlantic was formed by a catastrophic event, but not the Noachian flood. According to Humboldt, "a flow of eddying waters . . . directed first towards the north east, then towards the north west, and back again to the north east. . . . What we call the Atlantic Ocean is nothing more than a valley scooped out by the sea." In 1858, an American named Antonio Snider-Pellergrine published *La Création et ses Mystères dévoilés.* He put forward the same idea as Humboldt but also included multiple catastrophism in his schema, i.e., there were many catastrophes of which Noah's flood had been the last. Pellergrine also produced a map of the world as he thought it appeared before the separation of the continents.

During the 19th century, geological theory shifted in the direction of uniformitarianism. In 1822, Oswald Fisher postulated that the separation of the continents might be associated with the origin of the moon out of the Pacific Ocean, an idea that persisted until well into the 20th century. In 1910, F. B. Taylor, an American physicist, supplied the first uniformitarian concept of continental drift. But the real pioneer of the theory of continental drift was Alfred Wegener (1880–1980), who devoted his life to its development. Wegener's ideas began to receive empirical support in the 1950s with exploration of the seafloor and the charting of its topography, which included features such as midocean ridges. In 1962, Harry Hess of Princeton University put forward the idea that the seafloor moved sideways, away from the oceanic ridges. This mechanism was also postulated by R. S. Dietz. Hess could not explain what made the crust move away from the midocean ridge, but he did suggest that new crust formed along it. It was suggested that oceanic crust was formed from the Earth's mantle at the crest of a midocean ridge, a sort of submarine volcano. Dietz postulated that the lateral motion of the crust was driven by convection currents in the upper mantle.

F. J. Vine and D. H. Matthews published a paper in the British journal *Nature* in 1963 that did not attract much attention at the time but proved to be one of the most important contributions to the theory of plate tectonics. The paper confirmed the theory of seafloor spreading and noted that the magnetic lineations of the seafloor might be explained in terms of seafloor spreading. This meant that the oceanic crust recorded the occasional reversal of the Earth's magnetic field.

In 1965, J. Tuzo Wilson recognized a new class of fault-transform faults that connect linear belts of tectonic activity. The view of the Earth as seven major plates and several ones in relative motion was put forward on a rigorous geometrical basis by D. P. McKenzie, R. L. Parker, and W. J. Morgan in 1967–1968. This theory was confirmed by the earthquake seismology work of B. Isaacks, J. Oliver, and L. R. Sykes.—J.G.

See also CATASTROPHISM AND UNIFORMITARIANISM; CONTINENTAL DRIFT

Further Reading: Philip Kearey and Frederick J. Vine, *Global Tectonics,* 1990.

Platform Construction

Since the beginning of American civilization, carpenters streamlined methods of building in order to simplify their craft and allow for more rapid construction with standardized materials. No revolution occurred in simple construction; the process was slow and irregular. But taken together, innovations in building techniques made it possible to build houses and other structures at a pace that is measured in days rather than months.

Construction methods can be sorted into distinct periods. The oldest period employed timber frame construction, which involved joinery of large timbers to form a frame that at times was complex and even esoteric. During the 19th century, the balloon frame replaced timber construction as the dominant method of building. Balloon frames eliminated complex joinery by using mass-produced nails to join dimensional lumber cut at saw and planning mills. Platform frames derive from the balloon frame and have become the dominant method for building common homes in the 20th century.

There are three basic steps to building a platform frame. The first step involves construction of a foundation and box sill for the ground floor. In the 19th century, foundations were built of stone, brick, or wooden posts. However, platform frames usually begin with construction of a foundation made from concrete blocks or poured concrete. Next, one or more steel or wooden girders are positioned to span the foundation. Dimensional-lumber sills are then anchored to the top of the foundation wall. The sills and girders support a box frame built of header, end, and floor joists nailed together with bridging added between floor joists for additional strength. Initially, carpenters placed boards diagonally over this frame and nailed them to the floor joists. Today, 2-by-8 (1.2-m by 2.4-m) plywood sheets cover the joists to form a platform for the first floor.

The second step involves construction of interior and exterior walls. In a balloon frame, studs for exterior walls ran from the sill to the rafter. In platform construction, they do not; instead, the walls rest on the plywood platform, allowing for a simpler erection of walls. To build these walls, sole plates, precut 2-by-4 studs (nominally 5 cm by 10 cm, but actually smaller) and top plates are measured, marked, laid out, nailed together, and raised into position, section by section for each of the exterior walls. Specially framed openings called *headers* are built for windows and doors, a practice not used in balloon frames. Once raised, the exterior walls are plumbed and nailed to the subfloor, braced temporarily until all exterior walls are in position and joined. Repeating the process, the interior partition walls are then laid out, raised, and nailed to the platform. A second top plate connects interior and exterior walls to each other. If a house contains a second floor, a similar box frame is built, with plywood subfloor and

framed exterior walls. In a balloon frame, the second-floor joists rested on a ledger board, a 1-by-6 board (nominally 2.5 cm by 15 cm) notched into and nailed to studs that ran the height of the building.

Third, the roof is framed and the exterior is sheathed. At first, the roofs of platform frames were much the same as those for balloon frames. Rafters were cut to run from the top plates to a ridge board. They were nailed at the ridge board and to the ceiling joists. However, since the studs did not run continuously from bottom to top, the gable ends of the house required studs cut to run at varying lengths from the top of the ceiling to the ascending rafter. More recently, factory-built roof trusses have eliminated much of the work of the carpenter at the site.

The platform frame affected common architecture, for it was easier to build a split-level home with a platform than with a balloon frame. But platform and older balloon frames were distinguished as well by the materials and tools employed in their construction. Platform frames were built to accommodate the great advances in plumbing, heating, and electrical services that began to emerge in the second half of the 19th century. They used modern sheathing, insulation, and shingles that often had been developed by chemical companies. Since 1920, nearly all windows, doors, and trim has been made and assembled in factories, and merely installed at the construction site. Portable electric circular saws, routers, drills, screwdrivers, and nail guns eliminated much of the physical work in carpentry and increased the speed in which a home can be built. Carpenters no longer require large tool chests. Once common, the numerous chisels, hand planes, and marking gauges are now obsolete as far as most carpenters are concerned.—J.C.B.

See also CENTRAL HEATING; CONSTRUCTION, BALLOON FRAME; ILLUMINATION, ELECTRICAL; NAILS, PLANING MACHINE; PLYWOOD; SAWS, WOOD; TOILET, FLUSH

Plow

Few tools are more important to civilization than the plow. Almost all forms of agriculture require some kind of soil preparation prior to the planting of seeds. By cutting into the soil, a plow moves vegetation aside, aerates the soil, improves moisture retention, and brings nutri-

TWELVE OXEN PLOUGH OF THE EIGHTEENTH CENTURY.

Plow used in 18th-century England. As many as 12 oxen supplied the power to pull it (from S. D. Chapman and J. D. Chambers, *The Beginnings of Industrial Britain,* 1970).

ents closer to the surface. At the dawn of agriculture, this was accomplished with a simple digging stick or a hoe. Draft animals eventually were harnessed to these early devices, adding greatly to their efficacy. This use of an external source of power turned the digging stick into a simple plow.

A plow consists of a number of specific components. The task of cutting horizontally into the soil is performed by the *share.* At first made of wood like the rest of the plow, the share was soon tipped with flint or metal to increase its durability (one oft-cited passage in the Bible refers to "beating swords into plowshares"). Many plows also are equipped with a *coulter,* a blade that makes a vertical cut into the soil prior to the cut made by the share, thereby facilitating the latter's penetration. The *moldboard,* the third basic element of most plows, turns over the turf and soil cut by the share, and after inverting the slice deposits it into the furrow. Be-

fore the invention of the moldboard, this was accomplished by holding the plow at an angle to the soil being plowed. The moldboard was used in China by the 9th-century C.E. and appeared in Europe no later than the 16th century.

The shape of the moldboard determines the extent to which the soil is pulverized and turned over. A curved surface works better than a straight one, although the best shape was not immediately obvious. The creation of an optimal design occupied a number of scientifically inclined inventors, including Thomas Jefferson, who claimed to have discovered a mathematical theory of moldboard design. A more empirical technique was used at about the same time in England by James Small. His method consisted of making a moldboard of soft wood, and then using it to turn over many furrow slices until the natural scouring action produced a properly curved surface.

Many plows consisted of nothing more than a

frame, share, coulter, and moldboard, but during the Middle Ages a significant number began to be equipped with wheels. A wheeled plow was easier to transport from field to field, and made it easier for the plowman to regulate the depth of the cut into the soil. But plows of this sort were unwieldy in comparison with simpler plows. It is possible that difficulties in turning the plow at the end of the field led to the reconfiguration of farm acreage into long strips in order to minimize the number of U-turns.

Plows in Europe were drawn by as many as 12 oxen, although teams of 2 to 4 were far more common. After the medieval invention of the horsecollar, it became possible to effectively substitute horses for oxen. This was a significant change, because a horse can exert about the same pull as an ox, but it moves twice as quickly. It has also been theorized that the cost of the wheeled plow and associated draft animals necessitated cost-sharing arrangements among peasants, resulting in cooperative agricultural communities and an open-field system of cultivation.

For centuries, plows had been made by local blacksmiths, carpenters, and wheelwrights. Factory production began in 1789 when Robert Ransome, who had patented a method for tempering cast-iron plowshares in 1755, began to manufacture plows in Ipswich, England. By 1808, his factory was using standardized parts, one of the key elements of mass production. Standardization also facilitated the replacement of worn or broken parts. In the United States, farmers in the recently settled Midwest were frustrated by the tendency of plowed soil to stick to the moldboard. To counteract this, in 1837 an Illinois blacksmith named John Deere (1804–1886) built a plow with a moldboard faced with a sheet of steel taken from a saw blade. So successful was the basic design that the Deere steel plow came into widespread use; within 20 years of Deere's original invention his factory was producing 10,000 plows annually.

The plow made possible the westward expansion of America, but its use has had unfortunate consequences there and in other places. Failure to consider local conditions has at times led to severe erosion. This has been a particularly serious problem in parts of Africa, where heavy rainfall and a hilly terrain create conditions that are not well suited to the use of the plow.

See also HORSECOLLAR

Plywood

Plywood, a composite material made by gluing together thin layers of wood (called *veneers*) into large panels, is one of the 20th century's most ubiquitous materials. Modular in concept and standardized in dimension, plywood was a material expression of the growing interest in standardization and uniformity that arose in the late 19th and early 20th centuries. Like many other standardized materials that were developed during the same time period, plywood depended on a combination of technical and bureaucratic support, on public and private cooperation, to nurture a uniform standardized product. Accordingly, plywood should be viewed as an institutional as well as a technological development.

The practice of gluing veneers to a solid foundation or core was ancient in origin. Prior to the 20th century, the development of veneered work was driven by a combination of economic and aesthetic motivations. Sawing wood in certain ways revealed attractive decorative patterns, and applying those patterns in thin layers to cheaper substrates of solid wood conserved the more expensive and exotic surface-quality materials. The decorative (or hardwood) veneer industry grew during the 19th century, spurred by the development of power woodworking machinery that could saw, slice, and press together wood with ever greater speed, accuracy, and power.

Though veneering grew rapidly, the image of veneered goods suffered. Inconsistent manufacturing methods, erratic natural-based glues, and the inherent instability of wood itself combined to make the word *veneer* a synonym for poor quality, for the presence of a slick surface masking underlying inferiority.

While modern plywood developed out of this decorative veneer industry, the ubiquitous mass-produced 4-by-8-ft (1.2-by-2.4-m) softwood panels filling contemporary lumberyard racks differ in structure and history from the furniture-oriented hardwood plywood industry, although the two branches of the industry remain closely related in method and machinery, if not in product or market.

In the late 19th century, the veneer industry's trade association journals bemoaned quality problems plaguing their members, at the same time complaining of the unfairness of the poor reputation of "veneered goods."

In 1915, the veneer industry decided to adopt the word *plywood*, derived from "multi-ply wood," in a conscious attempt to create a new image. The word had been in use in Europe where mills in northern Russia had begun producing rough commodity-type panels, but *plywood* was considered an obscure technical term until formally adopted by the American industry.

A new government agency, the Forest Products Laboratory (FPL), actively lobbied for the name change at veneer trade association meetings. The FPL, a division of the United States Department of Agriculture, was established in 1911, a late outgrowth of the 19th-century forestry conservation movement. The FPL focused its early efforts on wood preservation techniques, conducting simple tests and field studies, but the timber industry viewed the FPL with suspicion, refusing to share statistics and methods. The FPL hoped to use its research and coordinating role to advance the development of a useful wood product, at the same time enhancing its own prestige and influence.

The FPL struggled with limited funding and meager facilities until the onset of World War I. During the war, wood was a strategic military material, and the fledgling profession of "wood products engineer" suddenly became an important component of the war effort. Military agencies with unlimited budgets turned to the FPL for assistance, among them the Army Signal Corps, where the beginnings of an air force was forming. Its airplanes, made of waxed linen and spruce beams, had serious performance limitations; consequently, the Signal Corps sought planes based on more reliable engineering materials, and plywood seemed to offer an answer.

The FPL conducted hundreds of thousands of tests on every aspect of plywood manufacture, from species of wood, formulations of glues, thickness and number of veneers, and more. Unfortunately for the status and aspirations of the wood products engineers, airplane designers soon embraced metals, superior to wood in their homogeneity and predictability, as the materials of the future. But wartime research had successfully transformed plywood into an adequate, if not ideal, engineering material. This was a significant accomplishment, because plywood turned out to be a complex and difficult material to tame because of the nature of wood.

Wood is a relentlessly variable, heterogeneous, unstable, and somewhat unpredictable material that changes dimension constantly as it gains or loses moisture from the surrounding atmosphere. Plywood represented an attempt to overcome and neutralize the inherent instability of wood and to create a quasi-homogenous wood-based material. During moisture changes, wood swells or shrinks considerably across the grain but changes dimension very little parallel with the grain. Plywood is assembled with the grain of each veneer laid perpendicular to that of its neighbor, thereby using the longitudinal stability of each layer to restrain the neighboring pieces from swelling or shrinking. Internal stresses result, however, reducing the overall load-bearing abilities of plywood in ways difficult to predict or measure with precision. Engineering design based on plywood therefore involved unknowns and required large safety factors. As a result, plywood's uses grew in areas where a certain amount of overbuilding was acceptable, especially in house construction.

In the 1920s, as the airplane design and fabrication drifted toward metal, the FPL began a concerted study of the house, applying knowledge gained on airplanes. The FPL found an aggressive ally in the Douglas Fir plywood industry in the Pacific Northwest. The Douglas Fir loggers had virgin forests of enormous, clear-grained logs ideal for peeling into veneers on large lathes. With the FPL providing technical support and the Douglas Fir Plywood Association adding product development and marketing, plywood entered a period of dynamic growth that was further aided by the development of truly waterproof phenol-formaldehyde glues in the 1930s. Plywood of fir and eventually other similar softwoods became a mass-produced, everyday commodity. By the 1960s, standardized 4-by-8-ft plywood sheets sheathed houses, filled lumberyards, and had a familiarity that seemed timeless.—C.H.

See also ADHESIVES; STANDARDIZATION; WOOD

Further Reading: Charles Haines, *The Industrialization of Wood: The Transformation of a Material,* 1990.

Poison Gas—see Gas, Poison

Polarized Light—see Light, Polarized

Polio Vaccines—see Vaccines, Polio

Pollution Charges and Credits

One of the unfortunate consequences of technological advance has been environmental pollution. Pollution-control technologies can counteract the problems created by other technologies, but there have to be incentives for their installation and use. In most cases, the installation and use of pollution-control equipment entails expenditures that do not directly benefit the firm that has cleaned up its operations. Left to their own devices, individual firms will continue to pollute because, in the language of economics, pollution is a "negative externality." Firms produce goods that they sell for a profit to customers, who in turn benefit from having these goods. But external to these market transactions are costs borne by society as a whole: the pollution that is generated in the course of making these goods. A public-spirited firm might voluntarily reduce its sources of pollution, but this will have little effect if other firms in the industry do not follow suit.

Under these circumstances, governments have to "internalize the externalities" by providing incentives for firms to reduce their pollutants. The traditional means of achieving this has been regulation; maximum emissions standards are established, and failure to comply results in fines or even imprisonment. This method is commonly referred to as "command and control," and it continues to be widely used by governments everywhere. In recent years, an alternative strategy has been formulated and applied, one that uses positive economic incentives in the place of punitive regulations.

Under an economic incentives approach, firms have to pay for the pollution they generate. However, the payment is not a fine; rather, it is a charge that internalizes what would otherwise be an externality. Ideally, the charge should equal the costs of the damage done by the pollution. In reality, ascertaining a precise cost is impossible, but the exercise itself can be valuable. In setting charges, government officials have to take into consideration the costs of specific pollutants instead of trying to regulate all of them, irrespective of the actual damage each one does.

A more radical approach to pollution control centers on the issuance of pollution credits to individual firms. Under a program of this sort, a firm is allowed to emit pollutants up to a certain maximum. It receives a credit if its emissions of pollutants fall below this level. For example, a power plant may receive one credit for every 1,000 kg of sulfur dioxide below its maximum allotment. It then can sell these credits to another power plant, one that produces pollutants over its maximum. The price of the credit will be set by the interaction of supply and demand. When there are many polluters exceeding standards, these polluters will bid up the price of the credits, making it more expensive for them to buy credits instead of cleaning up their operations. This will provide a strong incentive to reduce pollution. At the same time, a high price for the credits will motivate firms that are already clean to accumulate more credits by becoming even cleaner. Conversely, a low price for pollution credits means that the credits have been doing their job; there is now less pollution, and consequently less need to obtain credits.

In the United States, the first use of pollution credits occurred in the early 1970s. At that time, the Nixon administration put into operation a program that allowed a single plant to increase its emissions of one pollutant if it reduced emissions of another. In 1975, the U.S. Environmental Protection Agency authorized regional air-quality regulators to issue pollution credits, and to allow the buying and selling of these credits. The 1990 Clean Air Act went one step further, mandating a pollution-credit program as a means of reducing sulfur dioxide emissions by power plants. By the mid-1990s, pollution credit programs had developed to such an extent that a market for pollution-credit futures had been established. This market operates on the same principles as futures' markets for wheat or pork bellies, whereby speculators make purchases and sales in accordance with their expectations of future price movements.

Some environmentalists have criticized pollution credits, claiming that they amount to "licenses to pollute." But in fact the same thing can be said of traditional regulations; unless the regulations require zero pollution, they too give firms the right to pollute up to a certain maximum. It has also been argued that an effective pollution-credit market requires precise monitoring of the emissions of individual firms. But command-and-control programs also require monitoring in order to determine if a firm is in compliance with the regulations. A more serious criticism of pollution-credit programs is that they might allow lightly polluting firms in a low-pollution region to sell their credits to heavy polluters that are located in regions with already high levels of pollution, thereby making a bad situation worse. Under these circumstances it may be necessary to restrict the use of pollution credits to specific areas. The protection of particular areas will also require that firms adhere to local pollution standards even though they may have bought a requisite number of pollution credits.

See also ACID RAIN; ENVIRONMENTAL PROTECTION AGENCY; SMOG, PHOTOCHEMICAL

Polyethylene

Polyethylene (*polythene* to users of British English) is the most commonly produced type of plastic. Polyethylene is a polymer that is produced from ethylene ($CH_2=CH_2$), a gaseous hydrocarbon. There are a number of different types of polyethylenes; these differ in their degree of crystallinity and hence their density. High-density polyethylenes are used when products require strength and stiffness, as well as resistance to penetration by liquids and gases. These virtues are partially offset by their lower resistance to impact and environmental stress cracking, and the higher temperatures and pressures required for molding.

The discovery and subsequent development of polyethylene exemplifies the significance of fortunate accidents for scientific and technological advance. In 1933, two British chemists working for Imperial Chemical Industries (ICI), M. W. Perrin and J. C. Swallow, were looking into how very high pressures affected a number of chemical reactions. One of these reactions entailed reacting ethylene with benzaldehyde at 170°C at an ethylene pressure of 1,400 atmospheres. The reaction left a waxy solid on the walls of the reaction vessel. Subsequent efforts to produce this substance from ethylene alone failed until on one occasion an accidental leak necessitated the introduction of more ethylene into the reactor. This fresh charge carried with it just the right amount of oxygen to push the reaction forward and produce polyethylene.

By 1939, polyethylene was being used to insulate underwater communications cables. Its most important early use was as an insulating material for one of the most important military innovations of the World War II era, radar. Polyethylene was particularly important for the development of radar carried on aircraft. According to Robert Watson Watt (1892–1973), one of the most important figures in the early development of radar, "the availability of polyethylene transformed the design, production, installation, and maintenance problems of airborne radar from the almost insoluble to the comfortably manageable."

High-density polyethylene was developed in the 1950s. It too was the result of an accident, in this case the contamination of a reaction vessel by a nickel residue. This fortuitous discovery led to further experiments with other metals and their compounds used as catalysts, culminating in the development of a manufacturing process that came to be known as the Muelheim atmospheric polyethylene process. This process, which uses titanium tetrachloride and aluminum triethyl as catalysts, allowed the production of high-density polyethylene at relatively low pressures and temperatures.

In 1959, polyethylene became the first plastic to have an annual production of 1 billion pounds (453,600,000 kg). Less than 30 years later, nearly 7 billion pounds (3,175,520,000 kg) were being produced annually in the United States alone. Polyethylene's combination of strength, toughness, durability, and low production cost allows it to be widely used for a broad range of products: bottles and other containers, packaging films, pipes, garbage bags, coatings, monofilament fibers, electrical insulating materials, and all sorts of molded articles.

See also POLYMERS; RADAR; SUBMARINE CABLE

Further Reading: Royston M. Roberts, *Serendipity: Accidental Discoveries in Science*, 1989.

Polymers

Polymers are molecules constructed of aggregations of smaller molecules (monomers). Often structured as long chains, polymers have very high molecular weights; for this reason they are sometimes called *macromolecules* or just "big molecules." Many natural substances are polymers: wood, cotton, muscle fibers, and the molecule that transmits inherited characteristics, DNA. Artificial polymers go back to the 1860s, and they are now essential components of modern industrial society.

Two artificial polymers in common use in the early 20th century were celluloid and bakelite. These found widespread uses, but as often happens with technologies, practical application was considerably in advance of scientific understanding. During the first 2 decades of the 20th century, chemists did not believe that big molecules existed. The prevailing view was that what appeared to be giant molecules were in fact aggregates of small molecules that were joined together by weak intermolecular forces. This was the "micellar theory" first proposed in 1858 by the Swiss botanist Karl von Nägeli (1817–1891) to explain the structure of cellulose. According to this theory, large chemical compounds lacked molecular organization and would simply fall apart once they reached a certain size and level of complexity.

In the early 1900s, the micellar theory began to be challenged by anomalous experimental results. The molecular weight of starch was determined to be 38,000, and that of rubber to be over 100,00 (in fact, it is more like 500,000), far too large for the established theory. The publication in 1920 of Hermann Staudinger's (1881– 1965) article "Uber Polymerisation" ("Concerning Polymerization") may be taken as the opening salvo in the assault of the micellar theory. The recently developed technique of X-ray crystallography seemed to affirm the micellar theory, but subsequent work by Herman Mark, an Austrian chemist, convinced him of the reality of polymers. Support for the polymer theory was also provided by the development of the ultracentrifuge, which the Swedish chemist The Svedberg (1884–1971) had used in 1924 to determine the molecular weights of a number of substances, including hemoglobin (its molecular weight of 66,800 was four times earlier estimates). In the next decade, techniques to determine the atomic weight of substances were refined, while on a theoretical level the new electronic theory of valence posed a challenge to the micellar theory. The electronic theory of valence dealt a severe blow to the idea of partial valence, which had been used to explain the aggregation of small molecules into larger entities.

The debate over the existence of polymers at first aroused little interest in the United States. But in the late 1920s, two American chemists, Carl Marvel (1894–1988) and Wallace Carothers (1896–1937), began their investigations of the structure and properties of polymers. One early result was a synthetic rubber, given the name neoprene, that was discovered by Arnold Collins in Carothers's research team at the DuPont Chemical Company. It went into commercial production in 1933. At this time, the world's first synthetic rubber industry had been established in the Soviet Union, while the German I. G. Farben chemical company was also a leading producer of synthetic rubber. In all three cases, the rubber was used for applications where oil and solvent resistance was needed for products such as fuel lines. World War II gave a powerful impetus to the development of synthetic rubber as many combatants had their supplies of natural rubber sharply curtailed. Significant quantities of rubber were produced in Germany, the United States, and the Soviet Union, and equally important, a great amount of basic research in polymer science was funded by this enterprise. Although synthetic rubber was for many years inferior in some respects to natural rubber, the same cannot be said of another product of polymer chemistry, nylon. Another product of Wallace Carothers and his research team at DuPont, nylon, which was commercially marketed in 1939, was superior to the silk it replaced.

The years following World War II saw continued explorations into the chemistry of polymers as well as improved methods of production. One important development was the use of catalysts for the manufacture of polymers, pioneered in Germany by Karl Ziegler (1898–1973) and in Italy by Giulio Natta (1903–1979). Research and development has produced a vast number of polymer-based plastics, fibers, and rubbers. Beginning with a few basic feedstocks—oil, chlorine or fluorine, oxygen, nitrogen, and hydrogen—industrial chemistry now turns out a vast number of polymer-based products, everything from fabric for business suits to plastic for trash cans.

See also BAKELITE; CELLULOID; CHEMICAL BONDING; CRYSTALLOGRAPHY; DNA; NYLON

Further Reading: Peter J. Y. Morris, *Polymer Pioneers*, 1990.

Polypropylene

Polypropylene is one of the most commonly used plastics; in excess of 10 billion kg (22 billion lb) are produced in the world every year for such items as food wrappers, medical syringes, toys, auto parts, and even carpet fibers and upholstery fabrics. Polypropylene is a polymer-based propylene ($H_2C=CH-CH_3$). It is similar in some respects to polyethylene, but unlike polyethylene, its atoms are arranged in straight, unbranched chains. As a result, polypropylene is stronger, harder, and has a higher melting point than polyethylene.

Polypropylene was discovered at about the same time by a number of researchers working independently. In Italy, a group of chemists headed by Giulio Natta (1903–1979) produced polypropylene in 1954 by using catalysts that had been developed for polyethylene production by Karl Ziegler (1898–1973) at the Max Planck Institute for Coal Research in Germany. Ziegler himself produced the polypropylene at about the same time, as did another group of researchers at the Hoechst Chemical Company. In the United States, a team of scientists working at the research center of the Hercules chemical company produced polypropylene in early 1955, but credit for being the first to produce it goes to John T. Hogan and Robert Banks, who accidentally made polypropylene in 1951 and 1952 while working on new gasoline additives for their employer, Phillips Petroleum. In 1953, the company filed for a patent on polypropylene. Finally, between 1953 and 1955, polypropylene was produced in the laboratories of research chemists in the employ of Standard Oil of Indiana, DuPont, and Standard Oil of New Jersey.

With so many people and organizations capable of claiming credit for the discovery of polypropylene, the substance became the subject of extensive patent litigation. After several court cases and subsequent appeals, the patent covering polypropylene was awarded to Phillips in 1983. Although this gave Phillips a legal claim over the substance, it is important to note that polypropylene could be produced in large quantities because prior improvements in oil refining had guaranteed adequate supplies of feedstocks. Also, polypropylene emerged at a propitious moment, for ever since the creation of nylon in the 1930s, the buying public had been primed to accept and even demand a seemingly endless procession of new synthetic substances.

See also OIL REFINING; POLYETHYLENE; POLYMERS

Further Reading: David B. Sicilia, "A Most Invented Invention," *American Heritage of Invention and Technology*, vol. 6, no. 1 (Spring/Summer 1990).

Polystyrene

Polystyrene belongs to a category of plastics known as *thermoplastics*. It melts when heated and sets when cooled; consequently, it can be recycled many times. The manufacture of polystyrene begins with the alkylation of benzene with ethylene (both derived from petroleum) and hydrogen chloride. The resultant ethyl benzene is then dehydrogenated at 630°C in the presence of a catalyst to form polystyrene. Polystyrene is hard and transparent, although dyes are often added for the production of colored items. Although it has an unfortunate association with toys and cheap consumer goods, polystyrene is a versatile material with a multitude of uses.

Discovered in the 1830s, polystyrene was not commercially produced until the 1930s. Making this possible was the catalytic process that was developed in the late 1920s by Herman Mark, an Austrian chemist in the employ of the German I. G. Farben Chemical Company, which put polystyrene on the market in 1930. A similar process was devised by Iwan Ostromislensky of the Naugatuck chemical division of U.S. Rubber in 1933. In 1937, the Dow Chemical Company developed a less-expensive process that allowed the production of colorless polystyrene; industrial production at Dow began a year later.

Polystyrene products can be made through extrusion and a variety of molding processes. Because it can be formed in molds having sharp relief, polystyrene can be used for the manufacture of items with complex surfaces. The ability of polystyrene to be molded into products with right-angle corners, such as TV cabinets, helped to spur a shift away from bulbous "streamlined" designs that had been popular in the 1930s and 1940s.

Polystyrene can be blended with a rubbery polymer such as butadiene-styrene rubber to produce a plastic with much more impact resistance than polystyrene by itself. Polystyrene is also blended with butadiene and acrylonitrile to produce ABS, a tough, hard plastic used for many purposes. The most controversial use of polystyrene entails its conversion into styrofoam. A solid, lightweight foam that is manufactured by pressurizing liquid polystyrene, styrofoam has been widely used as an insulating material and for food packaging. Much of it is intended to be used once and then thrown away; as a result, it has caused a substantial disposal problem because it does not decay like organic materials. Making matters worse, some former methods of styrofoam production required the use of chlorofluorocarbons (CFCs), gaseous substances that have been implicated in the erosion of the Earth's ozone layer. The mandated phasing-out of CFC production has given rise to the use of alternative production methods. The disposal problem, however, still remains.

See also CHLOROFLUOROCARBONS; POLYMERS; POLYPROPYLENE; RECYCLING; RUBBER; WASTE DISPOSAL

Polyvinyl Chloride

Polyvinyl chloride (or PVC for short) is a polymer that belongs to a family of polyvinyl resins that also includes polyvinyl acetate and polyvinyl alcohol. It is a thermoplastic, which means that it softens when heat is applied. This allows it to be easily molded, extruded, or otherwise formed. PVC is in many industrial, military, and consumer goods, making it one of the three most widely used plastics, along with polystyrene and polyethylene.

PVC was discovered in 1926 by Waldo L. Semon, an employee of the B. F. Goodrich Co., while he was trying to develop a new adhesive. The rubber shortages incurred during World War II encouraged a number of applications of PVC, most notably its use as an insulator for electrical wires. After the war, it was used for many products: pipes, raincoats, shower curtains, and films, to name only a few. It was also the material basis of an important postwar consumer product: the long-playing record.

Although PVC has made new products possible and brought down the price of many existing ones, its production entails some potential environmental hazards. A major problem is the use of chlorine and chlorine compounds in its manufacture. PVC is 57 percent chlorine, and it accounts for 34 percent of chlorine usage in the United States. Two of the chemicals used to make PVC, ethylene dichloride and vinyl chloride, are hazardous. Moreover, their production results in some toxic by-products, most notably a family of chlorinated aromatic hydrocarbons known as *dioxins*. Exposure to dioxins has been linked to a number of health problems, including skin lesions, arthritis, and nerve and liver damage.

More dioxins can be produced when PVC is incinerated (as sometimes occurs with the disposal of medical wastes) or accidentally burned. Many PVC additives and stabilizers, such as lead and cadmium, are also hazardous. Efforts to recycle PVC are often thwarted because most commercially used PVC is not labeled (PVC that is labeled is designated by the number 3). Keeping PVC separate from other plastics to be recycled is often necessary because PVC's high chlorine content may spoil the recycled product.

See also POLYETHYLENE; POLYMERS; POLYSTYRENE; RECYCLING

Polywater

From time to time, scientists discover something that promises to completely upset existing ways of understanding the natural world. On some of these occasions, however, the discovery turns out to be utterly bogus. Such was the case with "polywater." In 1961, a scientist working at a provincial technical institute in the Soviet Union announced that he had discovered an entirely new form of water. The research was quickly pushed forward by a well-respected physical chemist at the Soviet Academy of Sciences and trumpeted to the outside world.

Polywater was thought to be a new substance, polymerized water. Due to a unique and previously unknown bonding pattern of hydrogen and oxygen atoms, polywater was unlike ordinary water, with, for example, different boiling and freezing points. The newly discovered substance was hailed by one eminent British scientist as "the most important physical-chemical discovery of this century." The U.S. Department of Defense was so impressed that it put millions of dollars into polywater research. One scientist even expressed the fear that the displacement of ordinary water by polywater could lead to the complete extinction of life on Earth.

Not all scientists were so easily impressed. Their skepticism was well justified, for by the early 1970s research showed that the distinctive characteristics of polywater were caused by the presence of a variety of contaminants. In the end, polywater failed to pass the empirical tests necessary for the support of a new scientific discovery. But before this happened, a great deal of time, money, and effort had been expended on polywater research because many scientists wanted to be on the cutting edge of an exciting new development. In the competitive climate of modern scientific research, it was all too easy for some researchers to see what they wanted to see and to temporarily succumb to a kind of mass delusion.

See also POLYMERS

Further Reading: Felix Frank, *Polywater*, 1981.

Porcelain

The greatest technical achievement of Chinese potters was the development of the hard, white material known as *porcelain*. As early as the Shang Dynasty (c. 1550–1025 B.C.E.), Chinese potters were producing the world's first stoneware, a type of high-fired brown or buff-colored ceramics covered with glazes high in feldspar. Experience with the high-temperature firing used for stoneware provided Chinese artisans with the technical wherewithal for experimenting with more refined clays and glazes during the prosperous Tan Dynasty (960–1279 C.E.).

Porcelain is made by the fusion of a white clay known as *china clay* or *kaolin* to feldspar, a type of crystalline rock. The fusion is done at high temperatures, about 1,300°C (2,370°F). Objects made from porcelain clay bodies—which are comprised roughly of five parts kaolin, three parts feldspar, and two parts flint or silica—are often covered with glazes with a high percentage of feldspar. During the firing, elements in the kaolin and feldspar fuse, or vitrify, creating a strong bond between body and glaze and resulting in brilliant and often-translucent artifacts. The material's shiny

whiteness led westerners to call it *porcelaine*, after the especially glossy cowry shell. Porcelain is frequently brittle, but it also is impervious to water and resistant to electricity and heat.

Chinese potters of the Song Dynasty (960–1279) and especially the Ming Dynasty (1368–1644) used porcelain to make an array of household and ceremonial artifacts for domestic and foreign consumption. The wondrous material provided these Asian potters with new avenues for creativity—and profits. Asian kaolins were especially plastic, making porcelain a sculpture medium especially suited to relief decoration, while the material's unique color—its pure whiteness—provided potters with a literal *tabula rasa* for decoration. Chinese potters, followed by Korean and Japanese craftsmen, used oxides such as iron and copper to develop a new art of colored glazes, from subtle *celadons* in pale blues, greens, grays, and browns to the deep oxblood red known as *sang-de-boeuf*. They also articulated several new modes in painted decoration, first mastering the application of cobalt blue underglaze motifs, and later developing the art of overglaze polychrome enameling. By the end of the Ming period, China unequivocally boasted the world's most advanced ceramics industry, thanks largely to the development of porcelain.

China led the world in porcelain production, but by the late 15th and early 16th centuries, Asian potters were growing increasingly dependent on the patronage of European consumers whose pocketbooks had been fattened by the prosperous mercantile economies. Rich westerners yearned for novel dining and drinking accessories, particularly after the introduction of the exotic drinks of coffee and tea, and they were dazzled by the stupendous blue-and-white and enameled Asian porcelains brought to the West by Portuguese and Dutch merchants. By the 17th century, the pastime of collecting porcelain—"china"—was a passion among European royalty, who established private factories in an attempt to manufacture the new luxury product. But the alchemy of porcelain manufacture was a closely guarded eastern secret, and European potters struggled unsuccessfully to produce a material with the strength, translucence, and whiteness of china. Italian potters, working under Medici patronage and French artisans St. Cloud, made proto-porcelains in the 16th and 17th centuries, but because their soft-paste porcelain contained no kaolin, the wares were easily scratched. As western potters lagged behind eastern artisans in high-temperature ceramics, imported porcelain became a symbol of power, status, and wealth among European kings and princes. Porcelain production frustrated European ceramists until the early 18th century, when a young German alchemist inadvertently discovered that kaolin was the mysterious ingredient of Chinese potters.

The European rediscoverer of porcelain was Johann Böttger (1682–1719), a prodigious young alchemist whose greatest ambition was the transmutation of base metals into gold. Augustus the Strong, King of Poland and Elector of Saxony, arrested and imprisoned Böttger at Albrechsburg Castle near Meissen (now Dresden), where the alchemist undertook gold synthesis on behalf of his jailer. Although he did not create precious metals, by trial and error Böttger eventually learned the fundamentals of making refined high-temperature ceramics, producing fine, brown stoneware in 1710 and porcelain in 1708. Riding on Böttger's success, Augustus erected a porcelain manufactory in 1710, and his Meissen works enjoyed a virtual monopoly on German china production until the mid-18th century. By the century's end, continental potters laboring under royal or noble auspices, English artisans working in factories operated by middle-class entrepreneurs, and a few American craftsmen in Philadelphia had uncovered and built on Chinese technical secrets and were making an array of porcelain products, including bone china.

In the early 20th century, porcelain's unique characteristics were harnessed for technical and industrial applications. Porcelain had long been used to make accessories for chemists and druggists, including mortars and pestles, but the advent of the electrical age and the automobile age created new used for this old heat-resistant material, most notably as insulators and spark plugs. Today, porcelain's role as the vitreous supermaterial has been usurped by high-temperature glass materials. Porcelain is still the dominant medium for sanitary plumbing fixtures and for household consumer products, such as tableware. That fine china is coveted by consumers rich and poor is a fitting 20th-century legacy to the china mania among the wealthy during the early modern era.—R.L.B.

See also ENGINEERING CERAMICS; GLASS, HIGH-TEMPERATURE; SANITARY PLUMBING FIXTURES; STONEWARE

Postindustrial Society

In viewing the broad sweep of history, some scholars have argued that the most fundamental transformations of human society occurred during the Neolithic Agricultural Revolution and the Industrial Revolution. To these, a third revolution may currently be unfolding, a revolution that is taking humanity beyond the industrial age. The concept of a postindustrial society was first fully articulated by sociologist Daniel Bell in his book *The Coming of Post-Industrial Society*. Since then, the term has been used in a variety of contexts, but there have been few comprehensive studies of the nature of postindustrial societies.

The term *postindustrial society* should not be taken literally, for industrial production does not vanish. Manufacturing continues to be a key part of the economy; its decline is only relative. Indeed, it is the success of industrial manufacture that has made postindustrial society possible. Although scarcity has not been abolished, the ability of modern industry to disgorge a plethora of goods has allowed the expansion of economic sectors that produce nothing tangible. As a result, the "tertiary" or service sector (healthcare, entertainment, education, sales, etc.) occupies the largest part of the economy. Manufacturing was responsible for 40 percent of the U.S. gross national product (GNP) in the late 1940s. In the 1990s, it had slipped to well below 25 percent, and the service sector accounted for around 70 percent of the GNP, with agriculture taking the remainder.

While postindustrial societies typically produce industrial goods in large quantities, they way that they do it has changed. The advance of technologies loosely grouped under the term *automation* has sharply lowered the need for human workers in the manufacturing sector. In 1948, nearly 21 million workers were directly involved in the manufacture of products in the United States. In 1990, only about 19 million workers were employed in manufacturing, even though output had risen substantially. The sharp relative decline of manufacturing work has meant that as far as employment is concerned, all of the action has been in the service sector. In 1990, more than 84 million Americans were employed in services, and this sector has been responsible for virtually all of the net employment gains in recent years.

Another significant characteristic of postindustrial society is that the most important productive resource is not land, resources, or capital, but knowledge. Knowledge-based occupations have taken on great importance as technologies have become more sophisticated and systems have become larger and more complicated. In recent decades, it has been necessary to develop and apply disciplines like operations research, mathematical modeling, and cybernetics to cope with the complexities of modern economies and societies. In short, analysis and planning are key activities in postindustrial society. Accordingly, the evolution of postindustrial society seems to represent the triumph of technocracy, governance by a technological elite.

Many of the key positions in a postindustrial society are occupied by professionals: physicians, lawyers, professors, engineers, and administrators. The importance of professionals underscores the centrality of knowledge, for advanced training and the possession of specialized knowledge are hallmarks of the professions. Knowledge also is a major source of power. In a postindustrial society, the possession of property and the occupancy of an elected office do not necessarily confer power. According to the theory of postindustrial society, the experts in government and the private sector are the ones who exercise guidance and control. Accordingly, the key cleavage in postindustrial society is not between capitalists and workers but between knowledge workers and everybody else, or, more specifically, between professionals with expert knowledge and the nonexpert "laity."

Finally, because postindustrial societies have sharply reduced material scarcity, work may no longer be central to society or the individuals in it. Less time may be expended in work, leaving more time for leisure activities. As a result of the diminished importance of work, a "leisure ethic" may be taking the place of the "work ethic" that has been a central cultural attribute of industrial societies.

The theory of postindustrial society has been subject to a number of criticisms, especially when the theory is presented in the rather crude fashion that appears above. In regard to the point made in the previous paragraph, it is evident that the importance of work has not been declining. Some advanced societies have high levels of unemployment, but joblessness has largely been the result of government policies and labor market rigidities rather than because of the presence of highly productive technologies. And as far as people with jobs are concerned, in aggregate there has been no voluntary reduction in working hours; if anything, the opposite has been the case.

Critics of the theory of postindustrial society also have noted that even knowledge-based societies are embedded in a capitalist matrix. Expert knowledge is important, but the ownership of the means of production is still a key source of wealth, income, and power. In similar fashion, modern governments are not technocracies. Experts and professionals may influence the formation of policy and its implementation, but their authority has been highly variable, and it often is minimal in regard to the most important issues faced by governments.

Finally, sociologists and other scholars who have studied professionals have called into question the notion that specialized knowledge is the basis of professional authority. As these critics have noted, professionals use a number of strategies to gain a monopoly over the practice of a particular occupation. Knowledge is a key element of professional status, but so are licensing procedures, governance of the profession and its members by members of the profession, and the legal right to draw a sharp line between professionals and the "laity."

In sum, the theory of postindustrial society provides valuable insights into the nature of economically and technologically advanced societies. However, the transition to a postindustrial society is far from complete, and many of the basic elements of traditional industrial societies will endure well into the 21st century.

See also AUTOMATION; CYBERNETICS; INDUSTRIAL REVOLUTION; MATHEMATICAL MODELING; NEOLITHIC AGRICULTURAL REVOLUTION; OPERATIONS RESEARCH; SYSTEMS ANALYSIS; TECHNOCRACY; UNEMPLOYMENT, TECHNOLOGICAL

Further Reading: Daniel Bell, *The Coming of Post-Industrial Society: A Venture in Social Forecasting*, 1973.

Pottery

Pottery is one of the world's oldest crafts, with the earliest earthenware vessels dating to the 5th-millennium B.C.E. Initially, clay was used to fashion figurative sculptures of human beings and animals for use in magic rituals and religious ceremonies. However, the need for containers for storing, cooking, drinking, and eating soon led ancient artisans to construct simple gourd-shaped objects, and these remained the basic forms of household pottery from approximately 3500 B.C.E. until the end of the Late Bronze Age, about 1000 B.C.E.

Pottery is clay formed into a shape and hardened in a fire. The first pottery objects were molded by hand, using one of several techniques. The easiest way to form a vessel is to squeeze, or pinch, a piece into a hollow shape from a ball of clay held in one's hand. Another simple method entails layering successive strips or coils of clay on top of each other and pressing them together to build a form. Eventually, potters learned to mount the clay ball on a turntable that was slowly rotated by an assistant. This method generated centrifugal force on the vessel, allowing the potter to apply pressure that formed the piece with minimal effort. The potter's wheel was refined over the centuries, but it remains the fundamental clayworking tool in the East and West. Potter's wheels and modified potter's wheels, known as *jiggers*, are still used in craft workshops and batch-production ceramics factories throughout the world.

Once shaped by the potter, the vessel is set aside to dry. When it reaches a bone-dry state, it must be baked, or fired, in an oven or kiln, to a temperature of approximately 600° to 700°C (1,100°–1,300°F). This initial firing is called a *bisque* (or biscuit), and its primary function is to expel water that is chemically combined in the clay. While bisque-fired vessels are adequate for the storage of dry foods, they often are porous and cannot hold liquids. To overcome the problem of porosity, potters developed the technical process of *glazing*, which involves covering a clay item with a vitreous envelope of glass. Glazing is a multiple-step process, which involves empirical formulation, application, and a second firing at a higher temperature.

Glazes consist of fluxes (e.g., lead, borax, feldspar, and potassium) often mixed with siliceous materials (e.g., sand, flint, and quartz) that are poured or dusted on to the surface of a pot. A clay and glaze must possess the same coefficient of expansion, otherwise the glaze will crack or peel off. The fine glazing that characterizes some ceramics results from this technical imperfection. Several distinct glazing traditions had evolved by Roman times. Two of these—lead glazing and alkaline glazing—undergirded western pottery technology until the 20th century. Mesopotamian potters developed glazes containing a high percentage of lead oxide. Lead is a powerful flux, so that its addition to glazes allowed potters to fire glazed wares at low temperatures—that is, between 900° and 1,000°C (1,650°–1,830°F)—to create vessels that were watertight and easy to clean. Near-Eastern potters created high-alkaline glazes, which were opacified, or whitened, with tin oxide. Glazes in this family, fired at temperatures up to 1,000°C (1,830°F), were compatible with myriad coloring oxides, and potters working with tin glazes developed skills in decorative painting. These high-alkaline glazes were the precursors to modern majolica in Italy, faience in France, and delftware in northern Europe.

Early potters decorated their products in several simple ways: by adding relief decorations made from strips of clay, by incising figurative or abstract motifs with sharp instruments, or by tinting glazes with colors. They also manipulated the atmospheres in their kilns to achieve decorative effects with clays and glazes. Copper oxide, for example, can produce a bright green color in an oxidizing kiln atmosphere, and a deep red under a reducing flame. Near-Eastern, Spanish, and Italian potters of the late Middle Ages and Renaissance skillfully manipulated reduction to produce lustrous metallic glazes.

Western potters depended on this array of relatively simple technologies until the 18th century, when English ceramists began developing new methods of clay preparation, vessel formation, glaze calculation, and ceramic decoration. Staffordshire potters, in fact, were among the pioneers of English industrialization. The best known of these entrepreneurs was Josiah Wedgwood (1730–1795), who not only lobbied Parliament for a expanded canal system that would link the isolated Midlands ceramics manufacturing district to major ports but who also developed modern sales methods that placed his moderately priced earthenware within the reach of the English, continental European, and American middle classes. Wedgwood's greatest technical achievement was the development of cream-colored earthenware, or creamware, a refined, off-white tableware designed to compete with more-expensive English, Asian, and European porcelain.

The new technologies and methods of work organization pioneered by 18th-century English ceramists reshaped the western pottery industry. Continental European and North American potteries increasingly concentrated on the manufacture of refined white or off-white wares, using modified British methods. Between the late 18th and early 20th centuries, innovations were few and far between, falling primarily into the realm of decoration. In this period, western potters developed new methods for embellishing pottery with prints, photographs, and decals. It was not until the 20th century that managers in American potteries broke with English traditions. By the 1930s, some experimented with mass-production methods as they labored to meet the demands of the burgeoning mass market. Notable technical developments in the 20th century include the development of tunnel kilns, or continuous ovens; the introduction of quality control in raw materials handling; and the selective mechanization of various production processes, from ware forming to glazing.

Today, pottery has been largely surpassed by high-temperature porcelain as the preferred material for the commercial production of household ceramics. Low-fired earthenware is still manufactured by craft and batch producers throughout the world, as demonstrated by the brilliantly colored Italian majolica and decorative household accessories made in Mexico, Spain, Portugal, and Italy. Perhaps as a fitting testimony to the craft potters of the preindustrial world, today's studio ceramists have turned away from high-temperature stoneware and porcelain to embrace pottery as their preferred medium, taking advantage of the wide range of colors easily achieved at low temperatures.—R.L.B.

See also KILNS; PORCELAIN; STONEWARE

Power Lines and Electromagnetic Fields

Appliances and the power lines that supply them with electricity have been a boon to modern life, but they also have been identified as a cause of human health problems. In particular, the electromagnetic fields (EMFs) created by appliances and power lines have been linked to cancer and other illnesses. Concern has focused on the health effects of the electromagnetic fields created particularly by high-voltage power lines and whether or not EMFs cause cancer, miscarriages, and birth defects.

Power lines and electrically driven devices emit fields generally in the 60-Hz (cycles per second) range; these emissions consist of an electric and a magnetic field. The electric component generally is attenuated by metal in homes and other structures (and even in animals and people). Consequently, most attention has been directed to the magnetic component of these fields. In general, these fields have low energy and hence interact with the human body differently from other types of radiation such as X rays or those emitted by microwaves. These power frequency fields do not break chemical bonds, but concern has arisen over the possibility that they might create currents in tissue or have effects on magnetic particles found in some living tissue. Human exposure to such fields is generally lower than the field levels produced by the human heart and other parts of the body. Short-term human exposure may, however, reach higher levels, but those would most likely come from electrical equipment and power tools rather than power lines.

In an attempt to shed some light on this issue, the National Research Council commissioned a study that examined over 500 research efforts in this area. The NRC report looked at three different areas: *in vitro* studies or studies on individual cells and tissues, *in vivo* studies or studies on whole animals, and epidemiological studies or field studies. In its survey, the NRC concluded that both the *in vitro* and *in vivo* studies showed no effects in situations in which exposure was at the levels found around human residences. If the exposure levels were increased by 1,000 to 100,000 times normal levels, some studies found changes in cells. As far as the epidemiological studies were concerned, the correlations that were found were very weak.

A great deal of the concern about EMFs derives from casual observation and epidemiological studies in which it is noted that certain kinds of health problems, such as leukemia in children, seem to cluster in areas where power lines are placed. What generally happens is that a count of children with cancer in a power line area will be compared to a similar count from some control group. In cases in which children near power lines seem to exhibit higher rates of the disease, an inference is made that the power lines may be causing the disease. These higher rates raise a public-policy issue: How can children be protected from the effects of the EMFs generated by power lines? Proposed solutions involve rerouting the power lines or burying them. In order to assess this policy recommendation, it is necessary to look at the nature of the evidence that points to power lines as the cause of the problem.

Questions can be raised about the epidemiological studies. Some epidemiological studies have rated current in wiring from "high" to "low" as surrogate measure of the EMFs, and have not directly measured of current levels. In addition, the epidemiological studies have not generally confronted the problem of confounding factors associated with power lines that might be the real culprits. For example, some researchers have suggested that the kinds of defoliant used to keep the ground clear about the base of power line towers might be the problem. Others have suggested that power lines often go through older neighborhoods and that homes built long ago and not up to current construction standards might have problems that could lead to the development of cancer in children. Others have suggested that traffic patterns in different kinds of neighborhoods will vary, which might contribute to the health problems. Yet another problem has to do with the selection of the control group. It is important that the control group represent an unbiased sample from the population in the area so that the results will not be biased in one direction or the other.

In attempting to assess the relationship between power lines and childhood leukemia, researchers have used a qualitative measure of the current carried by a wire ("wire code") that generally ranges from high to low. The studies show that the highest wire code category (i.e., wire with very high current configurations) is associated with leukemia at 1.5 times the expected rate for children. Thus, some studies seem to show a relationship, but it is quite weak. Many epidemiologists would conclude that such a low relationship is likely to be unreliable. In addition, it should be stressed that wire code is a qualitative measure (very low to very high) and not a direct measure of current carried by the wire.

The NRC report (p. 1) concludes that "no conclu-

Powerlines near Searchlight, Nevada (courtesy National Archives).

sive and consistent evidence shows that exposures to residential electric and magnetic fields produce cancer, adverse neurobehavioral effects, or reproductive and developmental effects." In general, then, knowledge at present does not allow for a clear determination that emissions from power lines do, in fact, cause cancer in children or adults. Nonetheless, the public concern about this issue is high and, coupled with the somewhat weak epidemiological results, warrants in the minds of some a commitment to continue research and, in particular, to mount the kinds of careful epidemiological studies that would allow for firmer conclusions to be drawn.—J.D.S.

See also ELECTROMAGNETIC INDUCTION; MICROWAVE COMMUNICATIONS; NATIONAL ACADEMY OF SCIENCE; PUBLIC HEALTH; X RAYS

Further Reading: National Research Council, *Possible Health Effects of Exposure to Residential Electric and Magnetic Fields*, 1997.

Precision-Guided Munitions

The desire for precision in the aiming of projectiles antedates the coming of airpower by millennia. The various approaches to this have been to perfect the aiming before launch, to dispatch multiple projectiles in one discharge in the hope that some or at least one will have an effect, or to fire a single projectile with either a large or a fragmenting warhead to increase the area of lethality around the impact. But only rarely in the days before aviation were attempts made to affect the trajectory of the projectile after it had been launched.

During World War I, in the United States an early effort was made to develop an effective guided missile, which came to be known as the "Kettering Bug" because its creator was the well-known General Motors chief engineer, Charles Kettering (1876–1958). A pilotless airplane with an explosive payload, it was launched along a preplanned trajectory with a gyroscopic autopilot system that would keep it on course and erect until the timer ran out and detached the wings for the final plunge on the target. To have been successful, it would have required precision measuring and control devices beyond the technology of the day—not to mention excellent intelligence on the exact location of the target.

During the 2 decades of the interwar period, some

progress was made in related technologies. One example was the development of realistic target drones for antiaircraft units. These were a series of remotely piloted airplanes with a radio link to control them as targets, enabling much more realistic antiaircraft target practice than with manned aircraft, where flying safety was a controlling consideration. Obviously, the same technology could be used to improve the trajectories of glider or free-fall bombs, or even powered missiles, after they had been launched from the ground, ships, or aircraft.

At the outset of World War II, the air forces of the Allies found soon after the onset of their campaigns against Germany that prewar predictions of accuracy had been far too optimistic. Equally optimistic were the American notions that they could fly above the effective range of enemy antiaircraft fire and that they could bomb inland targets with acceptable losses. One result of these discoveries were substantial wartime efforts to develop accurate, large weapons that could be dispatched from greater ranges. Only a few of these matured to the stage where they could be combat tested. Navy, Army, Air Forces, and the National Defense Research Committee (NDRC) had various programs that anticipated all of today's technologies except laser guidance. One of the most impressive was the NDRC's proximity fuse, which worked very well and which responded to a great need, especially in antiaircraft operations. The U.S. Navy developed a radar- guided glider bomb known as the Bat that was actually deployed to the Pacific and made about 20 kills—one of them a Japanese destroyer. This weapon was launched from a range of about 20 miles and had the very desirable feature of "launch-and-leave": The aircraft crew could lock it on to its target, drop it, and then leave the area without further concern with the weapon.

As for the Army Air Forces (AAF), there were numerous guidance programs afoot, including those using television, infrared, radar, gyroscopic systems, and even pigeons for their direction. Some employed glider bombs and other free-fall weapons. Some even used war-weary bombers loaded with explosives and guided by autopilots; these achieved limited success in combat, for the accuracy was insufficient and the effort too costly for the results achieved.

The most notable AAF programs were the free-fall weapons known as the AZONs, which came in 1,000-

lb (454-kg) and 2,000-lb (907-kg) sizes. They were employed in combat in northern Italy and in Burma with encouraging results, especially in the latter. The AZON was visually guided by a bombardier who watched a flare in its tail as it fell. Gyroscopes stabilized it in the vertical plane, and the bombardier sent left-right orders via a radio data link. He thus had control in azimuth, but not range, hence the acronym AZON for "azimuth only." Another AAF program was called RAZON for "range and azimuth," but although it was fairly well along in development, it never did get into combat.

As the war ended, the Allies made major efforts to acquire the results of German research and development, which subsequently gave a major boost to American guided-missile technology. However, rapid postwar demobilization caused all the AAF and Navy guided-munitions programs to languish during the late 1940s for want of funding.

When the United States was involved in the Korean War, both the Air Force and Navy suffered greater-than-expected losses to ground fire. AZON and a larger (5,443 kg [12,000 lb]) derivative were employed on a limited basis, and the statistics were encouraging but not overwhelming. The experience did revive interest in research and development, and more work was done in the mid-1950s. In the wake of Korea, for example, the Navy developed the Bullpup missile that employed the same sort of guidance as the RAZON but which had a rocket motor that afforded the pilot a little more safety as it extended his launch range.

At the time of the Cuban Missile Crisis (1962), President Kennedy was disappointed in the armed forces because they could not guarantee a "surgical strike" on the Russian sites. However, the principle of laser light had recently been unveiled, and there were programs afoot to employ it in weapon guidance. This was rapidly done, so that before the Vietnam War was over, the United States had the capability to reliably hit point targets from medium altitudes with a minimal number of attacks and some standoff range. The use of laser-guided bombs in LINEBACKER I of the spring of 1972 was a decisive factor in halting the Communist invasion. Another notable precision-guided missile (PGM) came along in those days: the Maverick missile, which now comes with a laser, television, or infrared seeker. Television bombs were also used in Vietnam, and soon after

they were equipped with infrared seekers so that they could also be used in darkness.

By the time of the war against Iraq in 1991, laser-guided bombs (LGBs) were still the mainstay, and there was little problem with that because the initial attacks had been so successful in bringing down Saddam Hussein's air defense system. Still, the American services already had on line Maverick missiles along with television- and infrared-guided bombs, some of them with rocket assist, that had both surgical precision and a "launch-and-leave" option. However, they were much more expensive than were the LGBs, so except for the Maverick, they were not used except where necessary.

Laser bombs led the way in achieving precision, and infrared further closed off an enemy sanctuary of darkness. Bad weather was still somewhat of a handicap in the war against Iraq, but there are new programs afoot that even in the short term promise to penetrate that sanctuary, and in the longer term do it even against moving targets. The shorter-term weapons aim to employ the Global Positioning System combined with inertial guidance to produce an all-weather weapon with nearly as great precision as a laser bomb. More distant are varieties of very short-wavelength micro radars, sometimes combined with other kinds of guidance, that will be able to identify and hit moving targets in all sorts of weather with surgical precision beyond that of laser bombs. Many authorities now argue that PGMs are a main pillar of a technical revolution that will radically change future warfare.—D.R.M.

See also GLOBAL POSITIONING SYSTEM; GYROSCOPE; LASER; MISSILE, INTERCONTINENTAL BALLISTIC

Further Reading: Bill Gunston and Mike Spick, *Modern Air Combat*, 1988.

Pressure Cooker

A pressure cooker takes advantage of the fact that the temperature at which water boils depends on the pressure of the surrounding air. When a lid is clamped tightly to a pot holding boiling water, the accumulation of steam raises the pressure, and with it the temperature. Of course, the pressure cannot be allowed to increase beyond a certain safe level, and pressure cookers are therefore equipped with safety valves to release excessively high pressures. The higher temperatures made possible by the pressure cooker allow food to be cooked more rapidly or thoroughly. In addition to being more flavorful, food cooked in a pressure cooker generally retains more of its nutritional value than food cooked in a conventional manner.

Credit for the invention of the pressure cooker goes to Denis Papin (1647–1712), a French scientist who had worked as an assistant to one of the pioneers of the study of air pressure, Robert Boyle (1627–1691). Papin invented his "steam digester" in 1679, and a year later he was admitted to Britain's Royal society on the strength of his invention. On at least one occasion he used the pressure cooker to prepare a meal for the members of the society, and he pressure-cooked an especially elaborate meal for King Charles II.

In addition to its practical utility, Papin's pressure cooker impressed upon its inventor the power of air pressure. This led Papin to envisage a device that would use air pressure to do useful work. In his engine, steam would lift a piston that moved inside a cylinder. After the piston had reached the top of its travel, the steam would be condensed, thereby producing a partial vacuum that pulled the piston downwards. This was only a hypothetical invention, however, and it remained for others to translate Papin's ideas into actual machinery.

See also AIR PRESSURE; STEAM ENGINE

Prestressed Concrete—see Concrete, Prestressed

Price-Anderson Act

Although disasters and near-disasters have occurred at nuclear power plants in the past, atomic power has not been an especially dangerous means of generating electricity. Even so, a serious accident at a nuclear power plant would likely do a great deal of damage to life and property. The first systematic effort to predict the extent of this damage was provided by report that came to be known as WASH-740. Undertaken in 1956 by the Brookhaven National Laboratory for the U.S. Atomic Energy Commission, the report indicated that in the worst possible case a nuclear accident would result in 3,400 deaths, 43,000 injuries, and $7 billion in property damage.

Damage on this scale would have been financially ruinous to the fledgling nuclear power industry. Equally important, the mere threat of being held liable for damages would have deterred most investors from putting their money in the nuclear industry. In order to encourage the development of the nuclear industry, the U.S. Congress passed the Price-Anderson Act in 1957. When first passed, the act limited liability for a nuclear accident to $560 million. Moreover, in the event of an accident, only $60 million would be paid by private insurance companies. By 1977, the figure would increase to $140 million. The remaining amounts would be the responsibility of the federal government; in other words, American taxpayers would be financially responsible for the consequences of a nuclear accident.

The act was amended a number of times in the years that followed. A 1966 amendment provided for accelerated payment of claims, while adding some elements of no-fault liability. In 1975, the act was amended to allow an upward revision of the $560 million limit. It also provided for the phasing out of government indemnification by assessing each reactor with a payment that would have to be made in the event of an accident anywhere in the industry. The size of the payment would vary inversely with the number of reactors in operation; in 1984, when there were 74 operating reactors, each would have been liable for $5 million. In 1988, the act was renewed for 15 years. At this point each reactor was liable for $63 million in retroactive assessments, of which no more than $10 million would have to be paid in a single year. These retroactive assessments were complemented by $160 million in insurance, for a total of $7 billion in coverage.

Although $7 billion is a great amount of money, critics of the nuclear-power industry have pointed out that the costs of an actual accident could far exceed this figure. It has also been noted that putting any limit on total liability indicates a fundamental lack of confidence that the industry will always be able to function safely. All in all, the development of the nuclear-power industry in the United States was given a significant boost by the original act. Although government funds have comprised a decreasing share of liability insurance for the nuclear industry, commercial nuclear power would have developed at a much slower rate had it not been for government support of nuclear power in the form of the Price-Anderson Act.

See also CHERNOBYL; NUCLEAR REACTOR; THREE-MILE ISLAND

Printed Circuit

A printed circuit allows the wireless connection of individual electronic components like transistors, resistors, and capacitors. This is done by connecting the components to a conductor path that has been laid down on a board made from an insulating material, most commonly a combination of fabric and phenolic plastic. The conductor paths may be laid out on one or both sides of a board. For complex circuits, boards are made with a number of internal layers in addition to the two surfaces. Printed-circuit technology is also used to form passive electronic components: resistors, capacitors, and inductors.

The actual design of a printed circuit is done with careful consideration of the location of individual components and the routing of the conductor path. Increasingly, this job is done through computer-aided design, or CAD. The design is then converted into a master pattern that is used for the manufacture of printed circuit boards. The circuit boards are formed by starting with an insulating board, laminating it with copper foil, and using an etching process to remove the areas that will not form part of the circuit. The etching is done after the master pattern is used for a photomasking process that determines which areas of copper are removed. After the board has been completed, individual components are soldered or otherwise connected to the circuit board. The elimination of wires reduces production costs, increases reliability, and allows considerable reductions in the size and weight of electronic products. For this reason, printed circuits are used on everything from transistor radios to mainframe computers. Printed circuits have the disadvantage of being more difficult to test and repair than wired circuits, but this is a slight drawback compared to their many benefits.

The origins of the printed circuit go back to 1903, when William Hansen received a British patent for a flat cable made by depositing a conducting material on an insulated surface. In the 1940s, a Milwaukee firm known as Centralab was making circuits that used a ceramic base supporting resistors and capacitors joined by printed silver connections. The firm also developed a method of directly printing carbon-based resistors. First used during World War II, by the late 1940s printed circuits were being employed for consumer products such as portable radios. Today, virtually all electronic devices are based on a printed circuit of some sort.

While printed circuits had an immediate commercial and military significance, they were also important in a conceptual sense. The printed circuit gave rise to the idea that a circuit module should be considered as an integral unit rather than a collection of individual components wired together. This idea was taken to its logical conclusion in the late 1950s with the invention of the integrated circuit. Significantly, Jack Kilby (1923–), the coinventor of the integrated circuit, had worked as an engineer for Centralab in the late 1940s. The processes developed for the production of printed circuits also came to be the basis of many of the operations used to manufacture integrated circuits, such as photo masking, selective etching, and photo reduction.

See also COMPUTER-AIDED DESIGN; INTEGRATED CIRCUIT; RADIO, TRANSISTOR

Printing Press, Rotary

The basic technique of printing remained unchanged for centuries after the invention of the printing press: A flat plate containing the type was inked and then manually pressed down on a sheet of paper. This was an adequate technology for printing books and other materials that did not require great production volume or speed. It was also sufficient for the printing of early newspapers, for at the end of the 18th century even large cities supported the printing of only a few hundred copies of a newspaper that might come out once a week. But in the course of the following century, newspapers began to reach a larger audience, putting a premium on large-scale production. Much of the increase in the newspaper-reading public was due to the expansion of literacy, but government policies also could be significant. In the United States the growth of

A contemporary illustration of Hoe's mid-19th-century rotary printing press.

readership was encouraged by the Post Office Act of 1792, which set low postage rates for periodicals. Along with volume, speed took on a growing importance, for news wasn't news if it took days or weeks to reach the reader, and conventional presses were a time-consuming bottleneck.

The 19th century was a time when many products and processes were being transformed in the course of the Industrial Revolution. Key features of industrialization were mass production and the use of new sources of power. Both aspects were evident as the printing of newspapers was industrialized. A new source of power was used for printing when Friedrich Koenig and Friedrich Bauer invented the steam-powered press in 1812. Two years later it was used to print the *Times* of London. Instead of pressing paper against a flat bed of type by hand, the press employed a steam-powered cylinder that rolled over the paper after another cylinder had done the inking. Every hour, 1,100 copies of a newspaper came off the presses, four times the rate of conventional press production. In the United States, the rotary principle was taken to the next level by Richard Hoe (1812–1886) in 1844. His press used a rotating cylinder to hold the type itself, with each half of the cylinder printing a complete sheet on every rotation. By 1856, Hoe's press was capable of 20,000 impressions each hour. The rotary principle was then applied to the use of continuous rolls of paper, which allowed completely automatic printing in the 1860s. Other improvements included presses that simultaneously printed on both sides of a sheet of paper and then cut off individual pages, as well as better inking processes. In addition, the use of wood pulp paper lowered production costs. With rotary presses capable of massive output, the slowest phase of the printing process was the setting of type, a problem that was not resolved until the end of the century.

Several presses could be kept going through the use of the stereotype process. After type was set in a flat bed, a heavy paper mat was then placed on top and pressure applied, creating a mold. This mold was then used to cast a cylindrical stereotype plate that was used for the actual printing. The stereotype process antedates the use of rotary printing by many decades, but it had met with the opposition of printers. As has often been the case in the history of technology, a new technology was seen as a threat to the livelihood of a particular group. Perhaps more importantly, there was little need to produce large numbers of printing plates until the spread of mass-circulation newspapers and magazines made it necessary.

Once all of the equipment was in place, the marginal cost of producing a newspaper was minimal. Under these circumstances, a publisher had every incentive to increase circulation, especially since advertising revenues were a function of the size of the readership. Accordingly, the size of the newspaper-reading public skyrocketed during the last few decades of the 19th century. With a circulation of 10,000, the *Times* of London was the largest newspaper in the early 19th century, yet by the end of the century New York had two newspapers with a readership of more than a million each. Unfortunately, this vast number was obtained by appealing to the lowest common denominator. Stories about crime, scandal, and bizarre events of every description were prominently featured. Along with pandering to some of the basest tastes, newspapers could also influence events, as could be seen in the role of William Randolph Hearst's *New York Journal* and Joseph Pulitzer's *New York World* in whipping up a public hysteria that contributed to the outbreak of the Spanish-American War.

It would be simplistic indict the rotary printing press and its associated innovations as the cause of wars and debased public tastes. The inventions that made the mass-circulation newspaper possible came about at a time when many trends had combined to create a mass readership. The spread of public education increased literacy, urbanization produced concentrated populations, and commercial expansion provided large advertising revenues. In a more democratic political and social climate, the needs and interests of the "common man" took on a growing importance. As the first of the mass media, newspapers contributed to the emergence of mass society, while at the same time they were products of it. And all the while, technological innovations were an integral part of the cultural changes that newspapers both reflected and created.

See also INDUSTRIAL REVOLUTION; LINOTYPE MACHINE; MASS PRODUCTION; PAPER; PRINTING WITH MOVABLE TYPE; UNEMPLOYMENT, TECHNOLOGICAL

Printing with Moveable Type

Even today, most of the world's languages do not exist in written form. Those that do convey great advantages to their users, for even in an electronic age nothing equals the written word as a cheap, portable, and rapid means of storing and transmitting information. Beyond this, literate societies are different from nonliterate societies in many ways, for the use of writing affects everything from forms of government to individual psychology. Yet until a few centuries ago, even where written languages existed, literacy was largely confined to a small elite. In more recent times the spread of public education has boosted literacy, but of even greater importance has been the set of technological innovations that made written materials available to almost everyone.

For many centuries, the only way to produce books was by painstakingly hand-copying them. Output was low, making printed works rare and expensive. Even centers of learning like monasteries and universities possessed only a few hundred or at the most a few thousand volumes. Books and other printed works became available in large numbers only when printing began to supersede hand copying. The first successful effort to do this took place in China. In the 4th-century C.E. the Chinese were using paper rubbings to copy stone inscriptions. By the 7th century, they were using carved wooden blocks to print pages of books. This allowed the publication of vast numbers of books; 400,000 copies of one 10th-century Buddhist collection still exist. The first use of movable type dates to the middle of the 11th century, when Bi Sheng (c. 990–1051) printed books from type made out of fired clay and imbedded in a wax matrix. When a sufficient number of pages was printed, the wax was melted and the pieces of type were available for reuse. Two hundred years later, Wang Zhen invented a typesetting system based on large rotating frames that made it much easier to select individual pieces of type. The Chinese also used metal type, although the most successful applications were developed in Korea in the 14th century.

While printing was being done on a widespread basis in the East, Europe was still engaged in hand copying. By the early 15th century, wooden blocks were being used for the printing of playing cards and pictures of saints, but they were not suited to the printing of large numbers of book pages. (China had successfully used wooden-block printing for centuries, but alphabetic European scripts required more characters per page than the ideographic Chinese written language.) At this time, there were some trading connections between Europe and China, and it is possible that at least the concept of printing with movable type may have spread from East to West. Whatever the source of inspiration, in the Netherlands Laurens Janszoon Coster (1370–1440) and probably others experimented with movable type, albeit with limited success. Credit for being the first to make effective use of moveable type is usually given Johann Gutenberg (c. 1400–1468) of Mainz, Germany. In producing his Bible, the first European book made from moveable type, Gutenberg used copper molds for casting individual pieces of type,

(upper left) The man in the background is sewing together folded sheets that have been printed. The man in the foreground is trimming the uneven edges of sewn sheets.

(upper right) The papermaker is dipping a frame or mold of interwoven wires into a vat of smooth pulp.

(lower left) The typefounder is pouring molten metal into a mold to form the type.

(lower right) A 16th-century print shop (from J. Thorpe, *The Gutenburg Bible*, The Huntington Library, 1975).

which were then set into a frame prior to the actual printing. Keeping the faces of individual pieces of type in the same plane was a major challenge, for any deviation would result in some characters not making an impression on the paper. Gutenberg solved this problem by casting each piece with a ridge on one side and a corresponding groove in the other. When the type was set, the ridge of one piece would link with the groove of the adjacent piece, thereby keeping everything in alignment.

The use of movable type was the key idea in Gutenberg's method of printing, but as is always the case with major technological changes, many other parallel innovations were also necessary. A press of some sort was required; here, an adaptation of the ancient wine press

served the purpose. Printing needed a heavy ink, which was compounded from lampblack and linseed oil. Type that was used over and over again had to be durable; an alloy of tin, lead, and antimony proved to have the right properties. Finally, if the inherent economy of printing with moveable type was to be realized, it was necessary to replace costly parchment with paper, a Chinese invention that Europeans began to produce in the early 14th century after it made its way to Europe by way of the Arab world.

Gutenberg's Bibles were the result of an early form of mass production, although their production was by no means rapid: The typesetting required 2 years and the actual printing another 2 years. While hand-copied books were works of art available only to a few, printed works reached a large and growing audience. The approximately 185 copies of the Bible that Gutenberg completed in 1455 were a harbinger of what was to come. According to one estimate, more than 10 million books, encompassing 40,000 titles were printed in 50 years that followed Gutenberg's Bible. In the following century, as many as 200 million books were printed in Europe.

Such a vast outpouring of printed works had many consequences. Europe was the scene of an information explosion, as everything from medical textbooks to treatises on astrology were widely distributed. With so much information circulating, and much of it contradictory, people came to be less willing to accept the judgment of a few authorities, and a more critical spirit emerged. It is no coincidence that the great European voyages of discovery commenced in the years immediately following the invention of moveable type, for printing disseminated a great amount of geographical information in the form of maps and travelers' tales. This served the double purpose of helping to motivate new explorations, while new printed works served to correct long-established errors.

Some explorations were intellectual rather than geographical. The great advances in scientific knowledge that began to emerge in the 16th century were nurtured by printing. Scientific advance requires accurate data and coherent theories, but before the age of printing much of what passed for knowledge of the physical world was incomplete, quirky, and inaccurate. Of course, printing disseminated a great deal of bogus information, but because it reached a much larger audience, information was subject to informed criticism, which promoted greater accuracy. Printing also liberated scientists from the drudgery of copying and rote memorization. Freed from these tasks, investigators could devote their time and energy to theorizing, collecting new data, and devising new experiments.

While printing helped to foster a scientific revolution, it had a hand in a religious revolution as well. Significantly, the Protestant Reformation began less than 2 generations after the printing of the Gutenberg Bible. In the centuries before printing, unorthodox interpretations of Christian beliefs had cropped up from time to time, but they remained confined to a particular locality. Beginning in the 16th century, the circulation of printed books and tracts unified new challenges to the traditional Catholic order and turned a number of separate revolts into the Protestant Reformation. The growth of literacy and Protestantism reinforced each other. Protestants believed that a person's faith should be based on the individual reading and interpretation of the Bible rather than on the pronouncements of priests, bishops, and popes. The rise of Protestantism stimulated the printing of Bibles as well as the torrent of religious tracts that carried the words of religious reformers throughout Christendom. For good reason, Martin Luther (1483–1546) praised printing as "God's highest and extremest act of grace, whereby the business of the Gospel is driven forward." (It should be noted, however, that he lost his enthusiasm for unguided Bible reading when it became apparent that it was leading to beliefs that contradicted his own.)

While the desire to read the Bible was a powerful incentive for literacy, reading skills could be applied to many other uses. A literate person could comprehend contracts, keep accurate records and accounts, understand the law, and learn about new ways of doing things. The often-noted connection between the rise of Protestantism and accelerated economic growth was at least partly due to the greater degree of literacy in Protestant lands.

In the cultural realm, a Europe unified by adherence to a single Christian church and the use of Latin and Greek by its scholars began to give way to a collection of national states, each with its own established church and a growing printed literature in the vernacular. Printing also helped to standardize these languages, while dialects unsupported by a printed literature slowly faded away, supplanted by common French, English, or German.

Although it is difficult to prove, reading also may have created a number of changes in the human psyche. Reading is usually a solitary activity, and a culture based on print media may be more individualistic than one based on more collective means of communications, such as church sermons or the oral recitation of sagas. Reading also produces a tendency to see things from a single point of view, the fixed position of the reader. This is reflected in the way scenes were depicted by the visual arts of the print era. Medieval painting had a characteristic visual flatness, with figures portrayed not according to the rules of perspective but in correspondence with their place in the spiritual hierarchy. When a reading population began to be conditioned to see the world from a single vantage point, they began to expect to see subjects depicted in perspective.

Although printing with movable type was accompanied by vast economic, political, and cultural consequences, it was not the sole source of these changes. The spread of the printed word was the result not just of a few key technological innovations but of a receptive social environment as well. Printing was invented at a time when emergent capitalism was generating a growing demand for such things as legal contracts and treatises on accounting. In East Asia, where printing was first invented but capitalism did not take hold, the effects of printing were far less evident. Printing helped to create a new world, but simultaneous changes produced an immense demand for the books and other written materials disgorged by the presses.

Finally, it is important to bear in mind that large numbers of people remained illiterate for many decades after the invention of printing. As late as the middle of the 18th century, an estimated 40 percent of English men and 60 percent of English women were unable to read, even though England had a higher literacy rate than most countries. For large portions of the populace, the consequences of printing were remote at best. Literacy increased in 18th- and 19th-century Europe and North America, due in part to social policies like the provision of public education. But perhaps of equal importance were technological changes that increased the volume of printed materials, lowered costs, and helped to create an expanded reading public. Printing with movable type was only the first phase of an "information revolution" that continues to this day.

See also LINOTYPE MACHINE; MASS PRODUCTION; PRINTING PRESS, ROTARY

Further Reading: Elizabeth Eisenstein, *The Printing Revolution in Early Modern Europe*, 1984.

Priority Disputes

New knowledge or devices almost always build on the current state of knowledge and technology, and they often depend for their inspiration on ideas that are widespread in society. Thus, it is frequently the case that two or more scientists or engineers will produce nearly identical knowledge or devices almost simultaneously In rare cases, such as that of the nearly simultaneous formulation of the concept of biological evolution through the process of natural selection by Charles Darwin (1809–1882) and Alfred Russel Wallace (1823–1913), codiscoverers are willing to share the credit for a new discovery. But since both science and technology are domains in which practitioners are rewarded for doing original work, simultaneous discoveries often lead to conflict among the claimants to originality, or to what are called *priority disputes*. Among the more significant disputes of this kind were the disputes between Isaac Newton (1642–1727) and Gottfried Wilhelm Leibniz (1646–1716) over the invention of differential calculus; among the supporters of Antoine Lavoisier (1743–1794), Joseph Priestly (1733–1804) and Carl Scheele (1742–1786) over the discovery of oxygen; among the supporters of Julius Robert Mayer (1814–1878), James Prescott Joule (1818–1889), and Hermann von Helmholtz (1821–1894) over the discovery of the conservation of energy; and between Alexander Graham Bell (1847–1922) and Elisha Gray (1835–1901) over the invention of the telephone.

In the case of useful inventions, the patent system was developed as a means for resolving priority disputes, and the person whose claim is properly registered with a nation's patent office is given the right to profit from the use of his or her invention in that nation for a prescribed period of time. Thus, because Bell's initial patent application was registered a matter of hours before that of Gray's, the latter received no financial reward for his work.

No comparable formal system exists for the adjudication of priority disputes over scientific discoveries, in part because it is more difficult to establish criteria for judging the establishment of a new scientific principle than to establish the creation of a particular material object. In the energy conservation case, for example, supporters of Mayer argue that his enunciation of the general principle in 1842 predates all others by at least 3 years. Supporters of Joule, on the other hand, argue that Mayer simply stated his principle without any experimental evidence to support it, arguing that Joule's 1845 experiment demonstrating the equivalence of mechanical work and heat provided the first adequately grounded proof of the conservation of energy. Then there are the supporters of Helmholtz, who argue that his 1847 paper deserves to be viewed as the first competent expression of the law of conservation of energy because Joule's experiments, while demonstrating conservation in two particular cases, did not prove the principle generally. Conversely, though Mayer's statement may have been completely general, it failed to provide an adequate proof. To Helmholtz goes the credit, they argue, because he was able to provide the mathematical formulation that allows one to follow the transformations among kinetic energy (energy expressed in the motion of all bodies in a system) and potential energies (energy stored in a system by virtue of forces that depend on the spatial configuration of the bodies in the system) in any transformation.

Scientists have used a number of different strategies to establish their priority with respect to some discovery without at the same time making it so public that others might preempt their use of that discovery to make still others. Newton, for example, first "announced" his discovery of differential calculus in the form of an anagram that simply substituted for the sentence announcing the discovery a list of the number of times each letter of the alphabet appeared in the sentence. Lavoisier deposited a sealed letter announcing his discovery of oxygen with the Paris Academy of Sciences, so that if anyone else claimed priority, Lavoisier could have the letter opened to defend his claim. Present scientific practices are based on the notion that scientific knowledge does not really exist until it is made public in the open scientific literature; as a result, practices like those of Newton and Leibniz would not be accepted as legitimate.

At the same time, however, current practices often leave questions of priority ambiguous, for at least two kinds of reasons. First is the question of how openly a discovery needs to be communicated. Because a substantial amount of scientific work today is either classified for reasons of national defense or is given limited circulation to protect private or national commercial advantage, many discoveries are not openly communicated, although they are available to significant communities. Second, the problem of establishing unambiguous criteria for deciding when a principle is adequately established, or when some formulation is substantially more general than or otherwise superior to another, is still not completely solved and it probably never can be.—R.O.

See also CALCULUS, DIFFERENTIAL AND INTEGRAL; CONSERVATION OF ENERGY; EVOLUTION; NATURAL SELECTION; PATENTS; SCIENTIFIC PAPER; TELEPHONE

Further Reading: Thomas Kuhn, "Energy Conservation as an Example of Simultaneous Discovery," *in* Marshall Clagett, ed., *Critical Problems in the History of Science*, 1969.

Production Concurrency

Historically, military forces have acquired their weapons in discrete stages. A weapon is designed, prototypes are built and tested, and modifications are made on the basis of these tests. Only after the weapon has demonstrated its effectiveness, reliability, and affordability is it put into production. During World War II, these stages began to be compressed; weapons went into production before prototype testing was completed, or in some cases before a prototype was even built. In other words, production was "concurrent" with earlier phases of weapon development. The process was particularly evident with aircraft, as airplanes like the B-24 Liberator and P-47 Thunderbolt started rolling down the production lines while their prototypes were still undergoing tests. Production concurrency was justifiable, given the need for rapid rearmament and the availability of funds provided by large military budgets. But the end of World War II did not bring the end of concurrency; if anything it became more common, bringing many problems in its wake.

A number of factors have motivated concurrent production in the postwar era. One of them is the understandable desire to keep up with a rapid pace of technological advance; no military force wants to be equipped with second-rate weapons. Reinforcing the ambition to stay ahead of the technological curve is the habit military planners have of thinking in terms of worst-case scenarios where potential adversaries have a decided technological edge. On the other hand, military planners tend to be technological optimists who believe that technical obstacles facing a new weapon will be overcome with the expenditure of sufficient ingenuity and effort. Accordingly, any problems that crop up will be vanquished at an early stage of a weapon's development.

Technological optimism coupled with a lack of confidence in present weapons thus leads military planners to compress a new weapon's development through concurrent production. But in many cases, concurrency has not been a short cut. The airplanes, tanks, and other weapons that have gone into concurrent production usually have a number of defects that emerge only in the course of extensive testing. By this time, however, the weapon is in production, making it necessary to engage in time-consuming redesigns of the weapon and its production facilities. In the long run, it may take more time to produce the weapon.

In addition to slowing down the pace of production and increasing costs, concurrency may render weapons less effective under actual service conditions. Problems with reliability, durability, and serviceability are often not evident until a weapon has been in service for a fair amount of time. Again, an extensive program of modification and retrofitting may be necessary. Concurrent production also adds to in-service problems by putting the emphasis on expected performance gains and neglecting reliability. Expected performance gains are the justification for a new weapons program; in contrast, reliability only becomes an issue long after the contracts have been signed.

The desire to make use of new, performance-enhancing technologies as soon as possible is not the only reason that concurrency is practiced; the needs of the military and its contractors are also involved. Defense budgets have always gone through periods of feast and famine, so it is in the interest of the makers and users of weapons to get long-term governmental commitments to new weapons when budgets are generous. A process that includes rigorous prototype testing actually may work against the long-term funding of a weapon, because the weapon's inevitable flaws will be revealed, giving its critics a reason for killing it. Concurrency avoids this hazard, for in the early stages of a weapon's development its supporters can stress the expected advantages, while its critics have nothing solid to point to. By the time it becomes evident that these advantages have not appeared or are offset by unforeseen problems, the program has gained a powerful constituency of political and military officials, businessmen, and workers who will resist efforts to terminate it.

Concurrency has been a leading cause of cost over-runs, stretched-out production schedules, and inadequate performance of military programs such as the Sergeant York antiaircraft gun, the SSN-21 submarine, the M-551 tank, the C5A cargo plane, and the B-70 bomber, to name only a few. Recent advances in computers have led to the claim that the past failures of concurrency can now be overcome through the use of superior design and manufacturing technologies. At present, however, no set of technologies can remove all of the unknowns that surround the creation of sophisticated new weapons. Putting weapons into production before they have been extensively tested will always be an inherently risky course of action.

See also COMPUTER-AIDED DESIGN

Further Reading: Michael E. Brown, *Flying Blind: The Politics of the U.S. Strategic Bomber Program*, 1992.

Propeller, Aircraft

An airplane's propeller uses the torque provided by an engine to move the airplane forward. When rotated, the blades of the propeller generate a low-pressure area on their forward surfaces, much as air flowing over the top of the airplane's wing produces lift. The low-pressure area pulls the airplane forward at a velocity proportional to the propeller's rotational speed.

A propeller's performance is strongly influenced by the angle of the blades to the hub, a measurement known as *pitch*. The effect of pitch on an airplane's forward travel can be visualized as the movement of a screw through a solid material, and in fact, in Great Britain, a propeller is sometimes referred to as an *airscrew*.

Since their invention in the 1930s, some airplanes have used variable-pitch propellers. Instead of being fixed at one angle, the pitch of these propellers can be changed by mechanical, hydraulic, or electrical means. This makes it possible to achieve an optimal blade angle: a low angle for full-power applications, as in takeoff, and a high blade angle for efficient cruising. It also makes it possible to "feather" the propeller in the event of an engine failure, that is, putting it at a 90-degree angle so that it creates minimal drag. This allows a twin-engine aircraft to fly on only one engine. The pitch of a propeller can also be changed automatically through a governing device. Propellers with this feature are known as *constant-speed propellers*, for by automatically changing their pitch they keep the airplane's engine operating at its most efficient speed.

Although propellers have been used since the beginning of powered flight, they have an important limitation. As the rotational speed of a propeller increases, the tips begin to approach the speed of sound. When this happens, a "compressibility burble" is generated, resulting in sonic shock waves that severely degrade an airplane's performance. Despite the development of increasingly powerful engines, the speed of propeller-driven aircraft reached a maximum in the mid-1940s. From that time onward, jet engines were used for high-performance aircraft.

See also BERNOULLI EFFECT; TURBOJET

Propeller, Marine

Beginning in the late 18th century, the use of steam power created great opportunities to build faster, larger, and more far-ranging ships. What was not immediately obvious was how to harness that power. John Fitch's (1743–1798) steamboat, the first to be put into commercial service in the United States, used 12 vertical oars that were directly connected to the engine. The mechanism took up space that might better have been used for cargo, and frequent breakdowns limited the ship's utility. More promising was Robert Fulton's (1765–1815) use of paddle wheels, a method of propulsion that came to be the standard in the ensuing decades. In 1858, the largest ship built up that time, the British *Great Eastern*, used side-mounted paddlewheels to move it through the water, but like other oceangoing paddlewheelers, she handled badly when the seas got rough.

The propeller, an idea that dates back at least to Leonardo da Vinci (1452–1519), offered a better means of propulsion. The first ship to successfully combine high-pressure steam power with a propeller was built in 1802 by John Stevens (1749–1838), a renowned New Jersey inventor. Also pioneering the use of propellers were John Ericsson (1803–1889) in the United States and Francis P. Smith in England. The superiority of the propeller over the paddlewheel was dramatically demon-

strated in 1845, when the British navy conducted a test involving two equally powerful ships, the paddlewheel steamer *Alecto* and the propeller-equipped *Rattler*. After cables had connected the ships stern-to-stern, *Alecto's* paddlewheel thrashed ineffectively as *Rattler* pulled her stern-first at a steady 2.5 knots. The superiority of propellers was further demonstrated during the Crimean War (1853–1856), when the British Navy found that exposed paddlewheels were highly vulnerable to enemy fire. Within a short time, paddlewheel craft were built only for plying coastal and inland waters, if at all.

Whereas a paddlewheel just pushes against the water, a propeller makes use of two basic physical principles. One of them is Newton's third law of motion. As it rotates, the propeller accelerates the water passing through it; this produces a reaction in the opposite direction that moves the ship forward. This effect can be produced by using a large, low-speed propeller to accelerate a great amount of water, or a smaller, rapidly rotating propeller that accelerates a small mass of water. Either way, reaction contributes to propulsion, but more important is the thrust produced when the rotating blades generate areas of low pressure on one side of a blade and high pressure on the other, much as an airplane's wing provides aerodynamic lift.

Complexities are introduced because the stream of water in which the propeller turns does not travel at a constant speed. Consequently, the forces developed by the blades are cyclic, leading to problems with vibration and a phenomenon known as *cavitation*. This occurs when the pressure on one side of the blade is so low that the water vaporizes in places, causing inefficient operation. Equally problematic, the collapse of the cavities leads to vibration, noise, and erosion of the blades' surfaces. Marine propellers look different from aircraft propellers because they are designed to prevent cavitation.

Along with the general shape of the blades, the most important parameter of a propeller's design is its pitch. Pitch is the distance that the propeller would travel as it made one rotation while moving through a solid material. In practical terms, pitch defines the angles of the blades to the propeller's axis. In most cases, a blade's pitch is not uniform; it varies to suit average water speeds along different portions of the blade, thereby minimizing cavitation. Propellers also may be fitted with mechanisms that allow the pitch to be changed while the propeller is in motion. As with an aircraft propeller, this allows the engine to work at its optimal speed under a variety of operating conditions. Controllable pitch propellers also allow a ship to be easily reversed by changing the propeller's blade angle.

Marine propellers can have from two to seven blades. They vary in size from a few centimeters to giants that are more than 9 m (30 ft) in diameter and weigh in excess of 70 tons. The best examples are made from nickel-aluminum, manganese bronze, or stainless steel. Large propellers are cast in a sand mold and then finished to a smooth, precisely shaped surface.

See also BERNOULLI EFFECT; MECHANICS, NEWTONIAN; PROPELLER, AIRCRAFT; STEAMBOATS

Prosthesis

A prosthesis is an artificial device that substitutes for a missing or defective part of the body. In the widest sense of the term, prostheses include hearing aids, heart pacemakers, and artificial kidneys and hearts. Artificial joints, such as the ones used for hip replacements, are also considered to be prosthetic devices. More narrowly, the word *prosthesis* refers to replacements for external bodily parts such as breasts, hands, limbs.

Artificial limbs were among the earliest prostheses. For centuries they amounted to little more than wooden legs and the like, although some medieval armorers were skilled at fabricating metal replacements for warriors who had lost limbs in combat. In the 19th century, improvements in surgical techniques allowed for the survival of many people whose limbs had been amputated, and helped to create an expanding market for prostheses. This trend was accelerated by the numerous and highly sanguinary wars of the 20th century.

Today, nearly 4 million Americans, 70 percent of them over the age of 55, use artificial limbs. Most of them are not war veterans; leg amputations are most commonly performed due to poor circulation as a result of diabetes or vascular disease. In some parts of the world, however, the extensive use of land mines has left large numbers of people, many of them quite young, in need of artificial legs. Losses of upper limbs are much less common, comprising fewer than 10 percent of all amputations.

While an artificial leg often serves as an adequate substitute for an amputated one, replacements for arms and hands have been less useful. An artificial hand allows only simple grasping motions, and many amputees prefer a prosthetic hook because it is lighter and more versatile. However, the ongoing development of computerized controls offers hope of eventual restoration of hand functions by prosthetic devices. Some prostheses have already made use of myoelectric devices that use small, skin-mounted electrodes to pick up electrical signals produced by contracting muscles in the remaining part of the arm. These signals are then electronically processed and used to control small electric motors that power an artificial hand. Further advances in miniaturized circuits and computerized controls should result in increasingly useful artificial hands.

Another source of improvement in prosthetic devices stems from the use of new materials. Plastics and composite materials allow for the construction of prosthetic devices that are lighter, stronger, and more durable than their predecessors. Better materials also improve the compatibility of devices and living tissues. This is particularly important when a prosthetic device is affixed to a bone, as is the case with joint replacements.

Improved prosthetic devices are necessarily expensive propositions. However, even if humanitarian concerns are set aside, it is necessary to take into consideration the increased economic productivity of individuals equipped with advanced prostheses. Their contributions may go a long way toward offsetting the costs incurred in developing, manufacturing, and fitting these devices.

See also CARDIAC PACEMAKER, COMPOSITE MATERIALS; HEARING AID; HEART, ARTIFICIAL; HIP REPLACEMENT; KIDNEY DIALYSIS; LAND MINES; SURGERY, ANTISEPTIC

Proton

J. J. Thomson's (1846–1940) discovery of the electron in 1897 demonstrated that the atom was not an indivisible mass. The years that followed brought much speculation about the structure of the atom. Physicists agreed that the negative charges of the electrons had to be matched by equivalent positive charges in order to maintain the electrical neutrality of the atom as a whole, but how the atom was structured remained a matter of conjecture. Thomson himself formulated what came to be called the "plum pudding" model of the atom, which depicted it as a positively charged sphere embedded with individual electrons.

In 1911, the British physicist Ernest Rutherford (1871–1937) discovered that atoms were not unitary but contained an inner nucleus about which the electrons orbited. He came to this discovery through a classic experiment that entailed firing alpha particles at very thin sheets of gold leaf. To his amazement, some of the alpha particles swerved markedly, and a few came almost straight back. Rutherford concluded that some of the positively charged alpha particles were being deflected from their initial trajectory by the electrical repulsion produced by the atom's positively charged core.

A few years after this discovery, Henry Moseley (1887–1915), another British physicist, rearranged the periodic table of the elements according to the amount of charge in each element's atomic nucleus. This produced a more-consistent periodic table and at the same time provided a reason for the similar chemical characteristics of groups of elements.

The source of the positive charge in the *core*, or *nucleus* as it was soon labeled, still remained a matter of speculation. In 1919, Rutherford pursued a new set of experiments using alpha particles to bombard the nuclei of light elements. This resulted in the detachment of hydrogen nuclei from a few of the atoms of the bombarded elements, a fact made evident by their deflection in a magnetic field. Rather than refer to these particles as hydrogen nuclei, Rutherford gave them a name of their own: *proton*. This name was derived from the Greek word *protos*, meaning "first," but Rutherford also wanted to give some recognition to William Prout (1785–1850), a British physician who in the early 19th century had put forth the theory that all elements were aggregations of hydrogen atoms.

The atom was now conceptualized as having a nucleus of protons surrounded by orbiting electrons. This was not the end of the matter, however, for all elements other than hydrogen had atomic weights greater than their nuclear charge. In other words, something in addition to protons was occupying the nucleus. Many physicists were of the opinion that the atom's nucleus contained a particle that weighed as much as a proton but

had no charge. In 1932, this hypothesis was confirmed when James Chadwick (1891–1974) provided experimental evidence for the existence of the neutron.

For many years the proton was taken to be an elementary particle, i.e., one that was not made up of other particles. But in the early 1960s, Robert Hofstadter, an American physicist, used high-energy electron beams to bombard atomic nuclei. The resultant scattering patterns led him to the conclusion that protons and neutrons consisted of central cores surrounded by two shells of mesons. A further decomposition of the proton (and neutron) was proposed by Murray Gell-Man (1929–) in 1964. Gell-Man postulated the existence of even more fundamental particles he called *quarks*. According to this theory, which has had considerable experimental validation, a proton consists of two "up" quarks and one "down" quark. Until recently there was general agreement that a proton cannot decay into separated quarks, but this has been challenged by a class of theories that come under the rubric of "grand unification theories." According to these theories, a proton will decay in the 10^{30} years. This is 10^{20} times the age of the universe, but it may be possible to observe proton decay through appropriate experiments. If proton decay is observed, it will resolve some of the conundrums that remain unexplained in the Big Bang theory of the creation of the universe.

See also ALPHA PARTICLES; ATOMIC NUMBER; BIG BANG THEORY; ELECTRON; NEUTRON; PERIODIC TABLE OF THE ELEMENTS; QUARK

Ptolemaic System

A number of mathematical astronomical systems were created in Hellenic Greece, all based on the assumption that the fixed and "wandering" stars (i.e., planets) moved on some linked network of spheres that were centered on the Earth. The systems of Callipus and Eudoxus, created in the 4th-century B.C.E., were capable of accounting qualitatively for most of the obvious features of the motion of heavenly bodies, including the regularity of the motion of the fixed stars, the different periods of motion of the moon and "wandering stars," the fluctuations of planetary and lunar latitudes (the distances above and below the plane of motion of the sun), and even the occasional retrograde motions of the planets (sometimes the planets seemed to slow down and even reverse their usual direction of motion through the heavens). However, as the Babylonian records of long-term astronomical observations reached Greek astronomers after Alexander the Great's (356–323 B.C.E.) invasion, it became increasingly difficult to fit all observational data to any concentric spherical system. This problem became increasingly urgent during the Hellenistic period as belief in astrology spread and as the casting of horoscopes became increasingly important. Clearly, astrologers had to be able to determine both the position of the heavenly bodies at a person's birth, as well as what the position would be at the time of some future action.

The first major attempt to modify concentric spherical models for calculating the positions of heavenly bodies seems to have been made by Appolonius of Perga (c. 230 B.C.E.). Appolonius proposed that a planet can be represented by a point on a small circle (called an *epicycle*) that rotates with constant velocity about its center, which in turn moves with constant angular velocity along a circle (called the *deferent*) whose center is the center of the Earth. Using such a model—by adjusting the relative radii of the epicycle and deferent, the relative angular velocities of the planet about the center of the epicycle and the center of the epicycle around the Earth, and the initial position of the planet on the epicycle—Appolonius was able to more accurately represent the longitudinal motions of the planets. Somewhat later, Hipparchus of Nicaea (c. 162–126 B.C.E.) became interested in explaining why the different seasons have different durations. The period from the winter solstice to the spring equinox is 94.50 days, that from spring equinox to summer solstice is 92.73 days, and that from summer solstice back to winter solstice again is 178.03 days. He was able to do so by supposing that the circle on which the sun moves is not centered on the Earth but rather slightly displaced (we say it is eccentric to the Earth).

Far and away the most important Hellenistic astronomical system was that of Claudius Ptolemy (85–165 C.E.), who combined the epicyclic system of Appolonius, the eccentric technique of Hipparchus, and one additional mathematical device, the *equant*, in which the point of equiangular motion along a circular path is

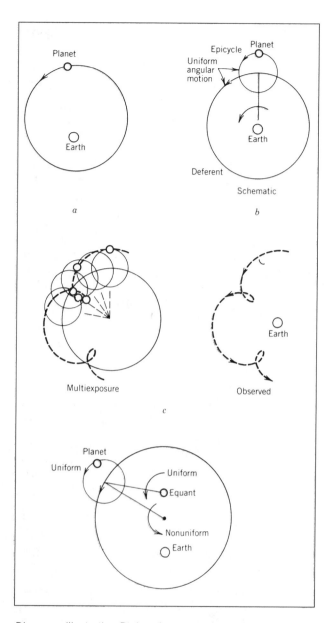

Diagrams illustrating Ptolemy's geocentric model of the solar system. The two illustrations in *c* show how an epicycle is used to explain retrograde motion (from N. Spielberg and B. D. Anderson, *Seven Ideas That Shook the Universe*, 1987; reprinted with permission).

displaced from the spatial center of the path. As a result, he was able to produce a system that was fully capable of predicting the observed positions of all planets, the sun, and the moon accurately, given the precision of naked-eye observations. The Ptolemaic system was most completely presented in Ptolemy's *Mathematical*

Concordance of Astronomy, which came to be known as the *Almagest* because its first Latin version was translated from an 8th-century Arabic version whose title had been corrupted to the equivalent of "the greatest." The longitudinal motion (motion along the path of the sun, or ecliptic) of most planets was successfully replicated by establishing an epicycle moving on a deferent whose center was in turn eccentric to the Earth. In addition, the equiangular motion of the center of the epicycle along the deferent was about an equant that was chosen to be on the opposite side of the Earth from the center of the deferent and at an equal distance from the Earth. In a few cases, including that of the motion of the moon, the model had to be extended by using a double epicycle—i.e., by having the center of one epicycle move along a second epicycle whose center in turn moved along a deferent. Latitude motions were reproduced using a similar system of motions orthogonal to (at right angles to) the plane of the ecliptic.

Ptolemy departed from the systems of Appolonius and Hipparchus in one important way. Most of his Greek predecessors measured the motions of the planets against the background of the fixed stars, so one period of motion was the time that it took a planet to pass from immediately next to a certain star to the time that it was found immediately next to that star again. (We now call this a *sidereal period*.) Ptolemy chose instead to measure the motion of other bodies relative to the position of the vernal equinox, which, as he knew from Hipparchus, moves (precesses) at an angular velocity of about 1 degree per century through the fixed stars. Why he chose to change the framework of measurements is not entirely clear, although it may have been that the only direct way of measuring the length of the solar year was from solstice to solstice. (Since the fixed stars are always blotted out in the vicinity of the sun, no direct sidereal measurement can be made.)

For Greek mathematical astronomers, whose sole goal was to predict stellar and planetary positions, the Ptolemaic system worked well, but it came under attack in Islam and in Europe during the Renaissance for three reasons. First, since the accuracy of any astronomical calculating system must depend on the precision and accuracy of the initial choices for periods, even a tiny error in an assumed period will be amplified over time,

because the amount of the initial error will be added to the calculated position each time the body completes a single cycle. Over time, then, predictions of positions based on the Ptolemaic system began to diverge increasingly from observed positions. Second, Islamic observations seemed to show that there was a complex relationship between the periods of the solar and sidereal years (the pace of precession seemed to fluctuate over time), and the Ptolemaic system offered no widely accepted way to explain this relationship. Finally, both within Islam and among some Renaissance astronomers and natural philosophers was a strong desire to unify astronomical calculations with a physically realizable and aesthetically satisfying system.

In the early 16th century, crises based on predicted positions had produced a desire to modify the Ptolemaic system by both the Catholic Church, which wanted a more satisfactory calendar, and by astrologers, who were disturbed by major inaccuracies. At the same time, numerous scholars, motivated in part by Neoplatonic interests in mathematical harmonies, sought a more elegant system than Ptolemy's. In particular, some were disturbed that the Ptolemaic system offered no way to relate the sizes of the "inferior" planets (Mercury, Venus, and the sun) to those of the "superior" planets (Mars, Jupiter, and Saturn). Moreover, it was disturbing that there was no regular progression of orbital periods as one moved out from the center. These concerns were removed by Nicholas Copernicus (1473–1543), who in his *De Revolutionibus* of 1543 introduced a new sun-centered system of astronomy that measured periods against the background of the fixed stars.—R.O.

See also SOLAR SYSTEM, HELIOCENTRIC

Public Health

Public health is an organized effort by society to protect and promote the health of its members. *Health* as defined in 1948 by the World Health Organization (WHO) includes physical, mental, and social well-being, not just the absence of disease or infirmity. This organized effort takes place on the local, national, and international levels, and involves the prevention or control of infectious and noninfectious diseases as well as injury, the provision of healthcare and adequate sanitation, the ensurance of wholesome food and clean water, the rehabilitation of sick and disabled persons, and the formulation of laws regarding health.

Various disciplines and professions make key contributions to public health. These disciplines include the clinical, biomedical, and environmental sciences, the social and management sciences, law, and ethics. The disciplines most specific to the practice of public health are epidemiology and biostatistics. Among the professions active in public-health work are physicians, nurses, veterinarians, sanitary engineers, microbiologists, industrial chemists, nutritionists, statisticians, laboratory technicians, behavioral scientists, dentists, economists, lawyers, and educators.

Disease-causing organisms, the physical environment, our interactions with them, and our knowledge about them, all change with time, and so do the ideas and practices related to public health. In the ancient world, cleanliness and personal hygiene were important to people all over the world, and notions of purity were often tied to religious practices. When people began living together in larger groups and cities emerged, sanitation and public hygiene became a much greater challenge. Ancient-world engineers devised ingenious ways of providing water to the cities, built underground drains for sewage, and constructed public baths, some with public toilets. In addition to cleanliness and personal hygiene, religious practices and medical theories of the day provided information, advice, and strictures regarding diet, exercise, proper conduct and self-discipline, and the treatment of minor ailments and injuries.

Another challenge for people living together in larger groups was the threat of epidemics. In fact, infectious disease played an important role in shaping world history well into the 19th century and is still an important factor in many parts of the world, especially in underdeveloped and poor countries. For example, the bubonic plague, which was pandemic in 1347 to 1351, and was called the Black Death in Europe, killed millions throughout the world and about a third of Europe's population. No one at the time knew its true cause, only that it was probably decreed by God and the result of the conjunction of the planets and the corruption of the atmosphere. Methods of dealing with other infectious diseases such as leprosy, smallpox, and measles were also applied to the plague: isolation and quarantine of the infected, the burning of sulfur, and scented woods to help clear the corrupted air. The sick were isolated from the community in special isolation areas, such as leprosaria, some of which later became public hospitals.

Quarantine and disinfection were the main preventive means of dealing with infectious diseases until 1798, when the English physician Edward Jenner (1747–1823) introduced the practice of vaccination as a way to prevent the spread of smallpox and eventually to totally eradicate it. Vaccination proved a powerful tool for immunizing nonimmune populations against many other communicable diseases. The advent of the germ theory of disease in the latter half of the 19th century spawned the discovery of many new vaccines and antitoxins. Mass immunization programs were initiated and most infectious diseases such as diphtheria, measles, tetanus, whooping cough, and poliomyelitis came under control. At the same time, however, malaria, tuberculosis, influenza, and infant diarrhea remain as major health problems in many countries. The development of bacteriology and immunology also led to the establishment of public-health laboratories, which in addition to diagnosing contagious diseases also tested the purity of water, food, and milk.

Until the 19th century, most public-health efforts were on the local level, controlled by local councils or boards of health. With the growth of nation-states in Europe starting in the 16th century and the advent of the Industrial Revolution in the 18th, both of which needed large and healthy populations to maintain power and wealth and to supply labor, health-promoting activities became the interest of nations as well. Focus on public health shifted to health problems associated with industrialization and urbanization: poor working conditions in factories and mines, crowding in tenements and factory towns, lack of a clean water supply, and effective sewage disposal, poor food, and inadequate medical services.

These conditions gave rise to sanitary reform movements in Europe and the United States. In England, where the Industrial Revolution began, the sanitary reform movement led to the establishment of national public-health institutions. The Public Health Act of 1848 created a general board of health to help local authorities, especially in matters of sanitation. Following Britain's example, many cities and states in the United States established public-health departments at the turn of the 20th century. The growing health needs of an industrial and urban United States resulted in increased support for public health from the private and public sectors, as reflected in the rapid growth of voluntary health agencies, maternal and child welfare, as well as programs in school health and occupational health, and public-health nursing.

Since disease knows no boundaries and with increased trade and travel throughout the world, especially since the mid-19th century, international efforts have been needed to help stop the spread of disease as well as provide health education, training, and assistance to those in need. International health conferences in the latter half of the 19th century developed conventions for preventing the spread of epidemic diseases and implementing effective quarantine measures. These efforts led to the establishment of the Pan American Sanitary Bureau in 1902, the International Office of Public Health in Paris in 1909, and the World Health Organization in 1948, which eventually absorbed the other two organizations. As a specialized agency of the United Nations, the WHO maintains close ties with other UN agencies, particularly the United Nations International Children's Emergency Fund (UNICEF), the Food and Agricultural Organization (FAO), and international labor organizations.

The provision of healthcare throughout the world today occurs through public assistance, health insurance, and national health service. In the United States, healthcare is provided primarily through the health insurance system. The principal federal government agency in charge of health in the United States is the Public Health Service (PHS), a significant part of the Department of Health and Human Services. The PHS grew out of the Marine Hospital Service, which was established in 1798 to care for sick and disabled seamen. In order to fulfill its very broad mission of promoting health in the United States and the world, the PHS has designed programs and created agencies that help control and prevent diseases; conduct and fund biomedical research that will eventually lead to better treatment and prevention of diseases; protect us against unsafe food, drugs, and medical devices; improve mental

health and deal with drug and alcohol abuse; expand health resources; and, provide healthcare to people in medically underserved areas and to those with special needs. Some of the major agencies of the PHS which carry out these functions are the Centers for Disease Control and Prevention (CDC), the National Institutes of Health (NIH), the Food and Drug Administration (FDA), the Substance Abuse and Mental Health Services Administration (SAMHSA), and the Indian Health Service (IHS).

Although the emphasis of public-health work has shifted somewhat, especially in developed countries, from concern with infectious disease control to other areas such as cancer, heart disease, stroke, health in the workplace, and the impact on health of environmental problems such as toxic waste disposal, we need to remain vigilant in all areas of public health. The tragedy of the AIDS pandemic and the threat of other newly emerging viruses and drug-resistant microorganisms are a constant reminder of that need.—R.K.

See also ACQUIRED IMMUNE DEFICIENCY SYNDROME (AIDS); ANTIBIOTICS, RESISTANCE TO; AQUEDUCT; BUBONIC PLAGUE; CANCER; EPIDEMICS IN HISTORY; FOOD AND DRUG ADMINISTRATION, U.S.; GERM THEORY OF DISEASE; INDUSTRIAL REVOLUTION; INFLUENZA; NATIONAL INSTITUTES OF HEALTH; ORAL REHYDRATION THERAPY; SMALLPOX ERADICATION; VACCINES, POLIO; WASTE DISPOSAL; WATER SUPPLY AND TREATMENT

Pulleys

A pulley is one of the so-called "six basic machines" (the others are the inclined plane, the lever, the wedge, the wheel, and the screw). In its simplest form, a pulley does nothing more than change the direction of a force, as when it allows a weight to be hoisted upwards through the application of a downwards force. However, a single pulley can also provide mechanical advantage. When a worker lifts a load with a rope, the worker is supporting the entire weight of the load, and for each meter of rope pulled up, the load rises 1 meter. But if a pulley is attached to the load, and one end of the rope is tied to a fixed point, threaded through the pulley, and then run back to the worker, we have a mechanical advantage of 2 to 1. The weight of the load is now divided between the end of the rope tied to the fixed point and the end that is held by the worker. When the worker pulls up 1 meter of rope, the load only rises ½ meter, but now the worker only has to pull with half the force.

Even more mechanical advantage is supplied by using more than one pulley. First described by Archimedes (287–212 B.C.E.), these devices (commonly called a *block and tackle*) allow virtually any degree of mechanical advantage to be gained. The basic rule governing pulleys is that they trade speed for lifting capacity, so using the right arrangement of pulleys allows the lifting of heavy loads,

albeit very slowly. Of course, each set of pulleys introduces friction losses that can become significant when using wooden pulleys and bearings, but these losses are small for systems with 3 or 4 to 1 advantage. Through the centuries, block and tackles have been extensively used on construction sites and sailing vessels, and they have been integral elements of many cranes.—E.R.

See also BEARINGS; CRANES

Pumps

An immense variety of technological devices is subsumed under the term *pump*. Pumps may be designed to raise fluids ("lift pumps") or induce them to flow from one place to another more quickly than they might do naturally ("force pumps"). Pumps may be powered by hand or by water, steam, wind, or electricity. Pumps are also used to compress air for such purposes as filling pneumatic tires, supercharging engines, powering pneumatic drills, or spraying liquids like paint or insecticides. Air compressors are basically a variant of the force pump, impelling a fluid—air in this case—through a closed system.

A lift pump partially evacuates the air above a column of water by means of a piston, the most familiar examples being the lever-operated household "suction" pump and the pump driven by a windmill, usually for irrigation. Since they depend on the pressure of the atmosphere, their practical limit is about 8.5 m (28 ft) of height at sea level. A force pump amplifies the pressure available from the atmosphere by means of a powered plunger. Some pumps combine both functions, using lift to raise fluid to a certain height, and then pushing it further through the application of pressure.

Pumps may be designated as "single-acting" or "double-acting." There are rotary pumps in addition to reciprocating pumps. There are positive-displacement pumps and nonpositive-displacement pumps of various types, the most common being the centrifugal pump that employs a closed impeller that throws off a liquid at an angle to the plane of its axis. Other noteworthy varieties of pump include ejectors and injectors, the latter being used to force water into steam boilers against pressure.

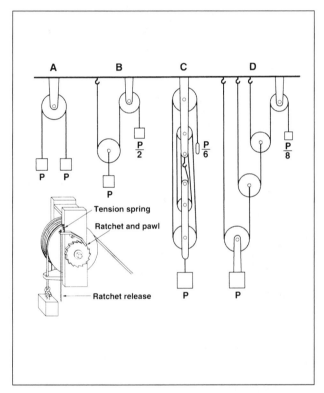

Four pulley systems. In A, no mechanical advantage is gained, and the same weight P is required to keep the other weight, also P, in equilibrium. In B, the weight P is suspended from two equal-length ropes; consequently, only half the weight (P/2) is required to maintain an equilibrium. In C, the same principle allows a weight of P/6 to maintain an equilibrium. In similar fashion, the arrangement of pulleys in D allows a weight P to be lifted by the application of a weight of only P/8, but the latter has to move 8 times the distance. The ratchet-and-pawl mechanism allows the lifting to be stopped when necessary (from V. Smil, *Energy in World History*, 1994; used with permission).

Force pumps date back to ancient Greece and Rome, where they were used for irrigation, fire fighting, and removing bilgewater from ships. Lift pumps are a more recent invention, going back to the 15th century. By 1588, the Italian engineer Agostino Ramelli (1531–1600) was able to illustrate 100 different pumps in his *Le Diversi et Artificiose Machine*, although he did not differentiate between lift and force pumps. Agricola's (1494–1555) *De Re Metallica*, published 30 years before Ramelli's treatise, describes and pictures a pump driven by men treading inside a large wooden wheel and using bored-out logs to expel water from a mine shaft 20 m (66 ft) deep. During the 17th and 18th centuries, French engineers became particularly adept at designing pumping machinery. Beginning in the 1680s, the Marly machine, one of the technological wonders of its day, pumped river water in three separate lifts to the top of a hill above Versailles, from which point aqueducts conveyed it to the fountains of the royal gardens. These pumps were driven mechanically through rods linked to waterwheels. In the 1760s, while a pier was under construction on the Pont d'Orleans, the site was enclosed by a cofferdam excavated to below river level and kept dry by means of a pump driven by a waterwheel and a battery of hand pumps.

Although the Marly machine demonstrated French superiority in pump technology, the British subsequently took the lead when they began developing steam power in conjunction with mining operations. Indeed, the first practical application of steam power was pumping out mines, nearly all of which have problems with water incursion. In 1698, the military engineer Thomas Savery (1650–1715) patented a "water raising engine" that harnessed atmospheric pressure. Four years later he described it in *The Miner's Friend* as suitable for supplying towns with drinking water as well as pumping out mines. However, no engine of the sort that Savery described is known to have been put to work in a mine, and it remained to Thomas Newcomen (1663–1729) and particularly James Watt (1736–1819) to render the idea practical. By 1778, some 70 Newcomen pumping engines were working in the mines of Cornwall, but within a decade nearly all of these had been supplanted by James Watt's improved design, which featured a separate condenser. After forming a partnership with businessman Matthew Boulton (1728–1809),

Watt made his first installation of an atmospheric mine pump in 1776, and, following that, hundreds more such pumps were in operation, making possible the extraction of coal and other minerals at ever-greater depths. The relationship between pumps and the steam engine was symbiotic. While the steam engine made possible more effective pumps, the tools and techniques used in their manufacture also could be applied to the production of better steam engines.

In the United States, the earliest significant pumping operation dates to 1801 in Philadelphia, the first city to have an adequate water supply. The city's Fairmount Works, on the Schuykill River, were built between 1819 and 1822, and throughout the rest of the 19th century, most other cities and towns installed pumping systems to supplement gravity feed. Typically, the latter was accomplished by pumping water into a tank mounted on stilts, providing a "head" that had the same effect as putting the water under pressure.

Pumps are ubiquitous in the modern world. The majority of water-cooled internal-combustion engines have a pump. Most municipalities have pumps to manage water supply and sewage. Pumps are used for irrigation and for bringing petroleum to the surface, and they are used in power plants and factories, aboard ships, and in countless other technological settings. Pumps move everything from blood to molasses to asphalt. Some of them, particularly those intended for extracting groundwater for irrigation, still rely on suction; that is, they rely on the physical reality that the Earth's atmosphere exerts pressure and will seek to fill any space where the air pressure is lower than the normal 101.3 kPa (14.7 lb/in.^{2}). The majority of water supply systems, however, rely on mechanical power to impel liquid from one place and expel it elsewhere, often at the rate of thousands of liters per minute. The term *pump* covers a lot of territory, from such voracious monsters as these to the prosaic devices that, with the application of some vigorous muscle power, will fill a kitchen sink or a watering trough.—R.C.P.

See also AEROSOL SPRAYS; AGRICULTURE, IRRIGATED; AIR PRESSURE; AQUEDUCT; DRILLS, PNEUMATIC; INJECTOR, STEAM; SEPARATE CONDENSER; STEAM ENGINE; SUPERCHARGER; TIRE, PNEUMATIC; WATER SUPPLY AND TREATMENT; WATERWHEEL; WINDMILLS

Punch Cards

Punch cards were first used in 18th-century France for the automatic weaving of designs. In the 1880s, Herman Hollerith (1860–1929), a Washington engineer, devised a way to use punch cards for recording, tabulating, and processing data. From the 1940s to about 1975, punch cards were also used extensively for entering data and programs into computers. The most common punch card was a 18.7 × 7.9-cm (7 $\frac{3}{8}$ × 3 $\frac{1}{4}$-in.) piece of heavy paper that was punched using special equipment to record 80 columns of data, each column representing a single number, letter, or symbol. The cards stored the data and could be quickly read, tabulated, counted, and sorted by electromechanical machines.

The processing of the 1890 United States Census marked the first large-scale use of punch cards. It had taken 8 years to process and publish the data of the 1880 census, and as bureaucrats had come to believe that good government depended on good data, there

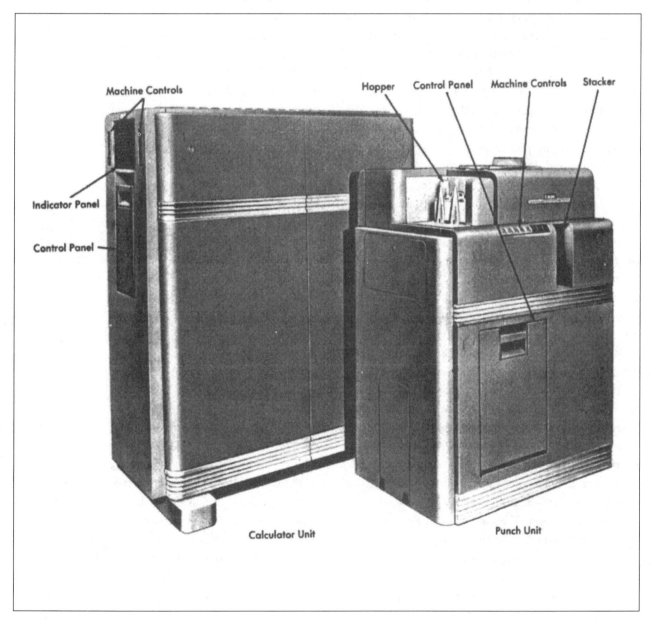

A punch-card reader and its accompanying calculator from the 1950s (from M. R. Williams, *A History of Computing Technology,* 1997; used with permission).

were even more questions scheduled for the 1890 census. The Census Bureau used 96 Hollerith machines to tabulate information on 80,000 people a day.

Large businesses quickly adopted punch card systems to mechanize and deskill some of their bookkeeping and filing tasks. The new machines not only made the job easier but also faster, allowing more and more data to be processed. Businesses as diverse as railroads and insurance companies found the new machines to be very useful for storing information and facilitating decision making. The United States government found increasing uses for punch cards and the machines that read them. The army used hundreds of machines during World War I to keep inventories as well as medical and psychological records. The War Industries Board, which controlled much of the economy during the war, did its accounting on the machines. Fifteen years later, New Deal administrators built on the successful World War I experience, putting punch-card machinery to wide use in mobilizing and directing the economy. The Social Security Administration used hundreds of pieces of punch-card equipment; even the checks that began to be sent in 1937 were in the form of punched cards.

Early Hollerith machines could just sort and count. By the 1930s, advanced punch-card machines could do mathematical calculations—multiplication and even integration of differential equations—and complex tabulations in response to programs. Punch-card technology became more widespread in the 1940s, as libraries began to use punch cards to keep track of books, police departments to track criminals, and businesses for payroll and factory management. But it was only in the 1950s, as businesses began to use computers, that everyone began to see punch cards. Companies sent punch cards out with bills to make tracking payments easier, and then began to use the cards themselves as the bills. By the 1960s, punch cards were familiar, everyday objects. Punch cards became a symbol for the computer, and they also became a symbol of alienation. The cards were, it seemed, a two-dimensional portrait of people that had been abstracted into numbers that machines could use. The cards came to represent a society where it seemed that machines had become more important than people, where people had to change their ways to suit the machines. People weren't dealing with each other face to face, but rather through the medium of the punch card.

In the 1960s, increased anxiety about technology, the information society, "big brotherism," and automation attached themselves to punch cards. As early as the 1930s, punch cards had been marked with the warning: "Do not fold, spindle, or mutilate." In the 1960s, the mutilation of punch cards came to symbolize resistance to the brave new information world. The student protests that began at the University of California at Berkeley in the mid-1960s seized upon punch cards as metaphor, as a symbol of the "system," first the system of registering for classes and then alienating bureaucratic systems in general. "Do not fold, spindle, or mutilate" became a rallying cry for the counterculture. The ecological movement of the early 1970s, itself a child of the 1960s counterculture, picked it up too: a popular poster for Earth Day 1970 showed a picture of the Earth taken from space with the legend "Do not fold, spindle, or mutilate."

Punch cards disappeared in the 1980s. New technologies of optical character recognition (OCR) that allowed computers to read type directly replaced punch cards in bills and most other public uses. The rise of time sharing and personal computers meant that programmers could enter their data and programs directly from a keyboard and that punch cards were no longer necessary. Finally, cheaper magnetic recording devices replaced punch cards for the storage of data.—S.L.

See also CD-ROM; DESKILLING; DISK STORAGE

Further Reading: Steven Lubar, "Do Not Fold, Spindle, or Mutilate: The Cultural History of the Punch Card," *Journal of American Culture* (Winter 1992).

Pyramids

The pyramids of Egypt are among the most stupendous of human creations. Intended as tombs for the pharaohs, their families, and members of the royal court, the pyramids were built with very simple tools and techniques. The pyramids are a powerful example of what can be accomplished with unsophisticated technologies.

More than 70 pyramids can be seen today, and it is likely that more existed in the past but disappeared when their stones were taken for other building projects. The largest of the pyramids is the Great Pyramid of the pharaoh Khufu (or Cheops, to use his Greek name), who reigned for 23 years around 2600 B.C.E. Located at the pyramid complex at Gizeh, it is a magnificent and awe-inspiring edifice. A few dimensions provide some idea of the size and grandeur of the Great Pyramid. Each side is 230 meters (m) (756 ft) in length, and its base occupies 52,600 square meters (13 acres), the equivalent of seven city blocks. The pyramid stands 40 stories high, which made it the world's tallest structure until the 19th century. It was built from 2 million limestone blocks weighing an average of 2.3 metric tons (2.5 tons) each. Altogether the pyramid is estimated to weigh 5.9 million metric tons (6.5 million tons).

The construction of the Pyramid of Khufu was accomplished in 24 years; this implies putting into position one limestone block every 3 minutes. The obvious difficulty of the tasks have provoked intense speculation concerning the construction techniques that were used. Although not completely understood, there is a fair degree of agreement over the basic procedures employed in the building the pyramids. Construction began with the laying out of the site. Each of the pyramids' four sides faces one of the cardinal directions: north, south, east, and west. This was done with a high degree of accuracy using observations of the stars or the sun. The site was leveled and the first course of blocks put in place. Many of the blocks had been quarried from nearby locations, but some were shipped up the Nile from as far away as Aswan, a distance of about 800 km (500 mi). The quarrying of the blocks was a major operation in itself; some of the limestone beds were cut to a depth of 27 m (90 ft). The extraction and shaping of the blocks was done using copper (and possibly bronze) chisels and wooden crowbars, as iron tools were unknown at the time.

Each course of blocks was laid on top of the course immediately below it. In this way the layout of the pyramids followed the design of the step pyramids that had come before them. The sides of each course had to be absolutely parallel to the sides of the course below it; otherwise, the sides of the completed pyramid would have been skewed. The sides of the pyramid were covered with a facing of high-quality white limestone (most of which was stripped off in the succeeding centuries). Moving and elevating the blocks posed formidable challenges for the builders. Sometimes the blocks could be moved on rollers, but the ground around the construction site was usually too soft for them to be used. Instead, the blocks were probably mounted on wooden sledges, and moved along wooden beams that had been mounted transversely in a clay roadbed. In order to decrease friction, water and possibly other fluids were poured under the runners of the sledges.

The most daunting task was lifting the blocks. The ancient Egyptians had no knowledge of pulleys; the only device to provide mechanical advantage was the lever. It is almost certain that the blocks were moved up ramps that led up to the pyramid and then spiraled around the sides. The building of these ramps took as much effort as the construction of the pyramids themselves, for to provide a grade of 1 in 10, it was necessary to build a ramp about 1.6 km (1 mi) long. It is possible that the last few courses of blocks were put in place by levering them up from a lower course. When the blocks were in place, the pyramid was faced with specially cut blocks. In profile, these had the shape of a right triangle. The proper slope for the outside of the blocks was ensured by using simple dimensional ratios: 14 units on the side away from the slope and 11 units at the base.

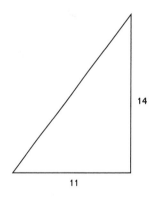

Given the simple tools and techniques of the time, the "secret" of building the pyramids lay in recruiting and organizing large numbers of workers. At any given time, at least 10,000 workers labored on the largest pyramids during their construction. They were not slaves, as is often assumed; rather, they were ordinary folk who were required to work on the pyramids in much the same way that people today are required to pay taxes. Feeding, housing, and managing so many people must have presented many difficulties. Coordinating the efforts of so large a labor force was as significant an accomplishment as the actual building of the pyramids.

The pyramids could be built with simple technologies because there was a powerful motivation to build them. The source of that motivation, however, remains elusive. The pyramids played a central role in ensuring the pharaoh's smooth transition to the world beyond death. Some Egyptologists have postulated that the pyramid's shape was intended to replicate the appearance of sunbeams shining through a break in the clouds, visually connecting heaven and earth. Even if this were so, much remains unexplained about the intentions of the people who built the pyramids. The great mystery of the pyramids is not how they were built but why they were built.

Further Reading: Ahmed Fakhry, *The Pyramids*, 1969.

Quantum Theory and Quantum Mechanics

At the end of the 19th century, physics entered a revolutionary period that would challenge some of the most basic conceptions of physical reality. This revolution began with the study of a phenomenon that seemed to be of rather minor importance at the time, the radiation given off by a black body (i.e., any object that completely absorbs all frequencies of light). In the 1860s, Gustav Kirchoff (1824–1887) had devised methods for empirically studying black-body radiation, but the resulting data resisted mathematical generalization. In Germany, Wilhelm Wien (1864–1928) had worked out an equation that described the distribution of energy at the violet end of the spectrum, while in England John William Strutt (Lord Rayleigh [1842–1919]) and James Jeans (1877–1946) had done the same thing for red end of the spectrum. These were significant but incomplete accomplishments, for the distribution of energy in the whole spectrum could not be captured by a single equation. Moreover, contemporary science could not explain an anomaly known as the *ultraviolet catastrophe*. According to prevailing theory, the maximum intensity of black-body radiation should occur at the higher frequencies. In fact, the intensity was known to increase in the low-frequency range and then reach a maximum. The intensity even decreased with further increases in frequency.

Around 1900, these observations began to make sense when Kirchoff's former student, Max Planck (1858–1947), made a radical assumption. Energy did not exist as a continuum that could take any value; rather, it consisted of discrete "packets," which Planck called *quanta* (the plural of *quantum*, from the Latin "how much?"). Accordingly, the energy emitted by a black body was not radiated as a continuous flow but as an aggregation of discrete quanta.

In Planck's formulation, the size of each quantum was determined by the wavelength of the emitted radiation: The shorter the wavelength of the radiation, the more energetic were the quanta. Since the frequency of a wave is inversely related to its wavelength, the energy of the quantum was directly related to the radiation's frequency. This relationship is expressed in one of the most important equations in the history of physics: $E = h\nu$, where E stands for the energy of a quantum, ν (the Greek letter nu) for the frequency, and h is Planck's constant, 6.626176×10^{-34} joule-second. Given the very small size of h, it follows that the energy of a quantum is also very small, so small that light radiation appears to be continuous, not a stream of individual packets.

As often happens with a radical theory, Planck's formulation met with considerable skepticism at first. Its value became evident in 1905 when Albert Einstein (1879–1955) used quantum theory to explain the photoelectric effect. (Einstein received the 1921 Nobel Prize in physics for this accomplishment, not for the theory of relativity, as might be assumed.) The photoelectric effect—the ability of light to cause the emission of electrons from the surfaces of certain metals—had been discovered by the German physicist Heinrich Hertz (1857–1894), and then was studied intensively by Hertz's former assistant, Philipp Lenard (1862–1947) in the early 1900s. Lenard's experiments revealed the puzzling phenomenon that increasing the intensity of the light did not cause the emitted electrons to have any more kinetic energy. However, as the wavelength of the light decreased (and the frequency increased), the energy of the electrons did increase. This effect was so pronounced that for some metals, light at low frequencies did not cause the electrons to leave the surface at all.

In 1905, Einstein solved this puzzle by applying Planck's quantum theory. Einstein hypothesized that an electron leaves a metal surface only when it is struck by a quantum that is large enough to provide the requisite amount of energy. Since Planck's equation stipulates that the energy of each quantum of light depends on its frequency, a yellow light (low frequency) may not supply quanta capable of dislodging electrons, while the quanta of blue light are up to the task. Moreover, the more energetic the quantum, the faster and more energetic the electron that it dislodges.

Einstein's hypothesis could be tested by measuring the maximum kinetic energy of electrons emitted by different wavelengths of light. In 1914, the American physicist Robert A. Millikan (1868–1953) published the first of several papers that presented experimental validations of Einstein's explanation of the photoelectric effect and the quantum theory on which it was based. Ironically, Millikan had embarked on his experiments in the hope of disproving both, but the experiments were unambiguous, and Millikan was forced to support a set of ideas that continued to cause him discomfort.

A quantum of visible light was given the name *photon* by the American physicist Arthur H. Compton (1892–1962) in 1928. Beyond this, Compton provided experimental evidence that quanta have the characteristics of particles. In his investigations of what came to be called "the Compton effect," he found that the quanta of high-energy electromagnetic radiation, such as X rays, lose energy when they are scattered by free electrons. In physical terms, the quanta of the X-ray radiation have transferred their energy to the electrons, a situation analogous to an elastic collision between two solid bodies—a billiard ball slamming into another ball, for instance. Thus, by the mid-1920s, physicists had to accommodate themselves to the idea that light had the properties of both a wave (as evidenced by such phenomena as refraction and diffraction) and a particle. It was not an easy accommodation to make.

On another front, in 1912 the Danish physicist Niels Bohr (1885–1962) used quantum theory to solve a different problem, the behavior of electrons as they orbit around an atomic nucleus. Contemporary conceptions of atomic structure contained a major flaw; because it is constantly changing direction, an orbiting electron is always accelerating. According to electromagnetic theory, this acceleration should result in the loss of energy and the electron's fall into the atomic nucleus. This, of course, did not happen. The complete explanation of why this did not occur had to await the development of quantum mechanics (see below), but Bohr took an important step forward when he hypothesized that orbiting electrons did not radiate energy so long as they remained in certain orbits. In Bohr's formulation, electrons can move only in particular orbits because each orbit is associated with a specific energy value. According to Bohr, an atom's energy level changes when one of its electrons moves to an orbit of a different energy level. In particular, a quantum of light (a photon) is emitted when an electron "jumps" from its stable orbit to an orbit with a lower energy level. Conversely, an electron is able to move to a higher energy orbit when it absorbs a quantum of energy of just the right size.

Bohr's theory was an elegant one, but it provided no explanation for why orbits were fixed in their particular positions, and Bohr's calculations contained quite a few arbitrary numbers. The behavior of atoms began to make more sense when electrons were conceptualized as having properties similar to those of photons. Since light waves had the characteristic features of particles, it was reasonable to expect that subatomic particles like electrons had wavelike features. This expectation was confirmed in 1927 by the American physicist Clinton Davisson (1881–1958) when he discovered that electron beams were diffracted by a nickel crystal, a phenomenon that could occur only if the electrons had wavelike properties. Similar observations were made in England by George Paget Thomson (1892–1975), who was able to produce diffraction patterns by passing fast electrons through thin gold foil.

The conceptual foundation for the wave-particle duality of electrons was set down in 1924 by the French physicist Louis DeBroglie (1892–1987), who presented a series of equations that described the wave properties of free electrons. The wavelength of a particle moving through space was described by the equation $\lambda = h/mv$, where λ is the wavelength, h is Planck's constant, and mv is the momentum of the particle. For particles of macroscopic size (a golf ball, for example), *mv* is so large that λ is undetectable. But for a small particle like an electron, a velocity of 100 cm/sec gives a wavelength of 0.07 cm.

DeBroglie's concepts were an inspiration to the Austrian physicist Erwin Schroedinger (1887–1961), who in 1926 modified the "planetary" model of electron orbit by theorizing that electrons take the form of standing waves (i.e., waves that consist of an exact number of wavelengths) around the nucleus. Because it took the form of a standing wave, an electron was not an accelerating electric charge, and for this reason it did not radiate light. Schroedinger's conceptualization of the electron as a wave therefore allowed a reconciliation of the 20th-century model of the atom with Maxwell's equations. It also provided a mathematical explanation for Bohr's theory that electrons have to remain in specific orbits, as any orbit that required a fractional number of wavelengths was not in accordance with Schroedinger's equations.

At about the same time that Schroedinger was developing wave mechanics, Werner Heisenberg (1901–1976), Max Born (1882–1970), and Pascal Jordan were working out another mathematical approach, which came to be known as *matrix mechanics*. Nearly 2 decades passed before John von Neumann (1903–1957) was able to demonstrate that wave mechanics and matrix mechanics were mathematically equivalent.

Schroedinger's explorations into the wave nature of electrons was the beginning of quantum mechanics. Just as Newton's laws of motion provided a mathematical description for the motion of particles larger than atoms and molecules (e.g., a hockey puck sliding down the ice or a planet orbiting the sun), Schroedinger's wave equations mathematically described the motions of matter waves like the ones associated with individual electrons. Unfortunately, the situation is not perfectly analogous to the Newtonian situation, for while it is easy to visualize the motion of Newtonian bodies, Schroedinger's equations force us to think in terms of clouds of electrons, rather than the orbits of individual electrons. Even harder to grasp is the idea that the Schroedinger wave is a "probability wave." The wave doesn't literally carry the electron; rather, what the wave is "carrying" is the probability of finding the electron at a particular point in space. Accordingly, an electron orbit associated with a particular energy level is not a well-defined circle

or ellipse, but what has been characterized as a "cloud of probability density" extending through space. The electron has the highest probability of being found where the "cloud" is at its most dense, and least likely to be found where the "cloud" is least dense.

As the preceding paragraph indicates, quantum theory solved some key conundrums about the behavior of light and other forms of electromagnetic radiation, but in so doing it introduced major conceptual problems. Quantum mechanics did not do away with the idea that electromagnetic radiation took the form of waves. Electromagnetic radiation was seen as a duality that encompassed both waves and particles (quanta). Bohr called this "the principle of complementarity," which at least gave a name to a situation that is very hard to accommodate to commonsense views of the world.

Quantum mechanics poses a further challenge to accepted ways of understanding the world by replacing a deterministic approach with a probabilistic one. The necessity to invoke probability destroys the concept, fundamental to much of the science of past centuries, that the universe follows a course that, in theory at least, could be predicted if one knew the velocities and positions of each of its individual components. Quantum mechanics' challenge to determinism has troubled many scientists, including Einstein, who was of the opinion that "God does not play dice with the universe." Some scientists hold to the belief that the physical world contains "hidden variables" that, when discovered, will bring determinacy back to physics. At present, however, they are swimming against a tide that was set in motion by Planck's discovery of the quantum at the end of the 19th century.

See also ATOMIC THEORY; HEISENBERG UNCERTAINTY PRINCIPLE; INFRARED RADIATION; MAXWELL'S EQUATIONS; MECHANICS, NEWTONIAN; ULTRAVIOLET RADIATION; X RAY

Further Reading: Banesh Hoffman, *The Strange Story of the Quantum*, 1959. Nathan Spielberg and Bryon D. Anderson, *Seven Ideas that Shook the Universe*, 1987.

Quark

In the 1930s, all atoms seemed to be formed from only three subatomic particles. Each atom had a nucleus containing one or more protons and one or more neutrons (unless it was an atom of ordinary hydrogen, which lacked the neutrons). Orbiting the nucleus in concentric shells were one or more electrons. However, even before the discovery of the neutron in 1932, physicists were given some indication that the situation might be more complicated, for in 1928 the English physicist Paul A. M. Dirac (1902–1984) had predicted the existence of a positively charged electron, or *positron*. In 1932, the American physicist Carl Anderson (1905–1991) confirmed this prediction while studying cosmic rays. The situation was further complicated by Wolfgang Pauli's (1900–1958) hypothesis in 1931 that a massless, chargeless particle was necessary to explain beta decay (the conversion of a neutron into a proton and electron). His hypothesis was borne out in 1956 when this particle, which had been given the name *neutrino*, was positively identified.

The ability to pry into the interior structure of the atom has been vastly augmented by the development of particle accelerators. These devices have produced a sizable number of subatomic particles; about 200 of them have been identified by now. Although some of these particles are clearly more elementary than many of the others, it has become apparent that the simple proton-neutron-electron model represents only the first step in our comprehending the nature of matter. But the situation is not as messy as it seems, for all of these subatomic particles appear to be made from a relatively small number of even more fundamental particles.

In order to bring some order to the apparent chaos caused by the proliferation of subatomic particles, physicists use a classificatory scheme that divides them into basic groupings according to their mass. The lightest (electrons, muons, neutrinos, and tau particles) are known as *leptons*. These particles are truly elementary, and cannot be decomposed into smaller constituents. The other basic group are the *hadrons*. These in turn divide into two subgroups. The ones with the greatest mass are *baryons* (protons, neutrons, pions, and so forth); a key characteristic of baryons is that they decay into protons. The other subgroup of hadrons are *mesons* (kaons, pions, and psi particles).

While leptons are indivisible, hadrons are composed of even smaller particles. The manner in which this occurs began to be explored in 1961, when Ameri-

can physicist Murray Gell-Mann (1929–) and Yuval Ne'eman, an Israeli physicist, independently devised a new way to classify all elementary particles. This came to be known as "the eightfold way," a play on Buddhism's Eightfold Path to Enlightenment. Like the periodic table, the eightfold way predicted the existence of hitherto unknown entities. The discovery of a predicted subatomic particle in 1964, the omega-minus, provided a strong confirmation of the eightfold way. In 1963, Robert Serber, an American physicist, suggested that this schema could be strengthened by the assumption that three fundamental particles serve as the basis for all subatomic particles. A year later, Gell-Man developed this idea and gave the name *quark* to the three particles. The term *quark* is usually attributed to a passage in James Joyce's *Finnegan's Wake*—"Three quarks for Muster Mark!"—but Gell-Man may have come up with it on his own. A similar three-particle scheme was proposed in the same year by another American physicist, George Zweig, who proffered the name *aces*, a term that failed to catch on.

These theoretical formulations were soon complemented by experimental studies of subatomic particles. Particularly important were experiments conducted using the Stanford Linear Accelerator (SLAC). Beginning in 1967, the behavior of electron beams aimed at atomic nuclei seemed to confirm the quark model. Meanwhile, experiments at the Centre Européen de Researche Nucléaire (CERN) using neutrinos instead of electrons pointed in the same direction. It had become evident that protons were made from quarks.

However, by 1970, physicists realized that the three-quark model was inadequate, and the existence of a fourth quark, which was given the name *charm*, was proposed. This was subsequently confirmed through the use of the accelerators at SLAC and at the Brookhaven National Laboratory. In 1977, a particle discovered at the Fermi National Accelerator Laboratory implied the existence of a fifth quark, which was given the name *bottom* or *beauty*. In 1984, researchers at CERN appeared to have found evidence for a sixth quark, which has been given the name *top* or *truth*. Its existence is still a matter of some controversy.

Although much remains to be learned about quarks, physicists have identified their key features. Quarks never exist in isolation, and all efforts to produce "naked" quarks have been unsuccessful. Quarks carry a charge, but unlike protons and electrons, which have equal but opposite charges, quarks carry charges that are a fraction of a proton's or electron's charge, either $+2/3$ or $-1/3$. These charges are associated with different "flavors" of quarks, giving us the following typology of quarks: "up" $(+2/3)$, "down" $(-1/3)$, charm $(+2/3)$, strange $(-1/3)$, top or truth $(+2/3)$, and bottom or beauty $(-1/3)$. Each of these quark types also has an associated antiquark, which carries an opposite charge (e.g., the "up" antiquark has a charge of $-2/3$).

Each flavor also comes in three possible "colors." As with flavor, this is not meant literally; color is a quality somewhat akin to electric charge, and it is responsible for keeping the quarks stuck together. The theory that explains how color charges hold different quarks together is known as *quantum chromodynamics*. It follows the theory of quantum electrodynamics that emerged in the 1940s as a means of explaining how the electromagnetic force pertains to subatomic particles. Color charge, however, is not completely analogous to electric charge because the latter appears only as positive and negative, whereas there are three types of color charge: red, blue, and green. Moreover, each color can be positive or negative. As with electric charges, oppositely charged quarks attract each other, while the same thing happens with the different colors; for example, blue attracts red and green, but repels blue. Antiquarks have complementary anticolors: cyan, magenta, and yellow. Quarks combine either in threesomes or in quark-antiquark pairs. In the former case, three quarks combine to form a baryon, of which protons and neutrons are the most common examples. Mesons are quark-antiquark pairs. Pairs of antiquarks form pions and kaons, unstable particles that last for only a tiny fraction of a second. The rules of combination for the formation of hadrons require color combinations that result in white. For example, a baryon is made from red, green, and blue quarks. Mesons, being quark-antiquark pairs, contain a primary color and its complementary anticolor.

Quarks are combined in such a way that their resulting particle carries a charge of +1 or 0. A proton, for example, consists of two up quarks and a down quark (hence, $2/3 + 2/3 - 1/3 = +1$). A neutron is made from two down quarks and an up ($2/3 - 1/3 - 1/3 = 0$). Because quarks do not exist in isolation, they cannot be detached from the particle they help to form. Hence, when a proton is accelerated into another proton, the resulting collision produces new quark pairs and triplets.

Along with quarks and antiquarks, matter requires the presence of the "strong force" that holds protons and neutrons together in an atomic nucleus. This force resides in a particle, as was first hypothesized by Japanese physicist Hideki Yukawa (1907–1981). As befits its name, the strong force is presumed to be the most powerful force in the universe, being 100 times stronger than the electromagnetic force. According to contemporary theory, the strong force is mediated by particles known as *gluons*. These are analogous to photons in that both are "gauge bosons," i.e., particles that transmit a fundamental force: electromagnetism in the case of photons, the strong force in the case of gluons. Like photons, gluons have no mass, but unlike photons, which are free to travel anywhere through space, they remain confined to the atomic nucleus, a region with a radius of only 10^{-15} m or less. Gluons are color charged; each one carries a color and an anticolor, making nine color-anticolor pairs, although one combination is excluded since it is equivalent to white. The short range of the force mediated by gluons is the result of these color charges, which cause the gluons to interact strongly with quarks and with each other.

Physicists today refer to the "standard model" of matter, which assumes the existence of six types of quarks and leptons, along with "virtual" or "mediating" particles that carry the four basic forces (gravitation, electromagnetism, and the "strong" and "weak" forces). The quarks and leptons can be arranged in pairs according to their energy level. At energy level 1 are found particles that exist under normal terrestrial conditions: the up and down quarks along with electrons and electron neutrinos. Energy level 2 includes particles occurring in cosmic rays; these also can be produced in particle accelerators. At this energy level are the strange and charm quarks, and two leptons, the muon and muon neutrino. Energy level 3 is attainable on earth only through the use of the most powerful particle accelerators; in nature, these particles are associated with the energy levels that existed when the Big Bang created the universe. The particles associated with energy level

3 are the top (or truth) and bottom (or beauty) quarks, and the tau and tau neutrino leptons.

Much more needs to be learned about the fundamental particles that make up the universe, as well as the forces that hold them together. Had it been built, the Superconducting Supercollider would have provided a great amount of experimental data and new insights into the nature of matter. But even in its absence, research done with existing high-energy colliders will make many important contributions to our knowledge of the constituents of the universe.

See also BIG BANG THEORY; COSMIC RAYS; CYCLOTRON; ELECTRON; NEUTRON; PARTICLE ACCELERATORS; PERIODIC TABLE OF THE ELEMENTS; PROTON; QUANTUM THEORY AND QUANTUM MECHANICS; SUPERCONDUCTING SUPERCOLLIDER

Further Reading: Frank Close, Michael Marten, and Christine Sutton, *The Particle Explosion*, 1987.

Quasars

A quasar is, by definition, a starlike object with a large red shift. Quasars were discovered in the 1960s when catalogues of the heavens were compiled using large radio telescopes. A number of strong radio sources were noted that looked like stars on photographs taken with optical telescopes but possessed spectra that were unique. For several years, astronomers puzzled over the correct interpretation of the unusual spectra of these objects. The breakthrough came on Feb. 5, 1963, when Maarten Schmidt (1929–) realized that he could interpret the spectrum of 3C 273, one of the brightest and best-studied quasars, if he assumed that the object had an enormous red shift.

Normal galaxies appear to be red shifted due to the expansion of the universe. The further away from us a galaxy is, the larger its red shift. If quasars also have red shifts for the same reason, they have to be extremely far away. Throughout the late 1960s and into the 1970s, a debate raged in astrophysics about the correct interpretation of quasar red shifts. The debate was fueled by the fact that many quasars did not give off steady amounts of light but varied erratically, often doubling the amount of light, or energy, they emitted in a matter of months. This rapid variability implies small size. Quasars, or at

least their energy-generating parts, must be not much larger than our solar system. This conclusion shocked scientists, because in order for us to see quasars at the enormous distances implied by their red shifts, they must be putting out more energy than even the largest galaxies. Thus, we are left with the dilemma of an object that outshines an entire galaxy perhaps 100,000 light-years across, but is only a light-month or less in diameter. Faced with this prospect, a number of scientists, most noticeably Halton Arp (1927–), postulated alternative mechanisms to account for the quasars' red shifts, one that would allow the quasars to be much closer to us and thus be less-powerful energy sources.

A number of observations made in the 1980s seem to confirm the huge distances between quasars and our galaxy. Many quasars have been found to lie in clusters of galaxies, and the galaxies are indeed at the distances indicated by the quasars' red shifts. Also, many cases have been found where the light from a distant quasar passes through dust and gas clouds in galaxies that lie between the quasar and ourselves. The large distances to these intervening galaxies confirms that the quasars are indeed very distant.

Quasars do not all appear starlike when seen by the newest generation of telescopes and cameras. Some have faint hazy patches around a central bright core. Some have one or two opposed jets, thin streamers that flow outward from the core at very high speeds. When viewed by a radio telescope, some quasars show extended lobes, many thousands or hundreds of thousands of light-years from the central core. Most quasars have broad emission lines in their spectrum, which are attributed to hot, chaotic gas clouds. Some have narrower "forbidden" lines in their spectrum, probably due to an extended cooler gas cloud surrounding the hotter core. Further complicating the picture is the existence of spectral absorption lines caused by light from the quasar passing through cool gas clouds that lie between the quasar and us. Quasars often put out light at ultraviolet and infrared wavelengths as well as visual and radio wavelengths. A few quasars are X-ray and gamma-ray emitters.

Most scientists today think that quasars are an extreme example of the phenomena seen in "active" galaxies. The fact that all of the quasars have large red shifts implies that they existed long ago but are now extinct. Active galaxies can have smaller red shifts; many currently exist. A comprehensive model of quasars, based on suggestions first advanced by Donald Lynden-Bell (1935–) at Cambridge University, postulates a supermassive black hole as the central energy source for quasars and most active galaxies. This central black hole may have as much as several million times the mass of our sun. Material from surrounding stars and gas clouds is attracted by the black hole's gravity and orbits the supermassive black hole in a thin disk, called an *accretion disk*. As material from the accretion disk eventually spirals in towards the black hole, it is compressed and heated, and emits high-energy radiation. We are able to see some of the radiation that comes from the quasar, but much of it is absorbed by surrounding dust and gas clouds, and then is remitted at longer wavelengths. Some of the material is not drawn into the black hole but instead is thrown out along the rotation axes of the black hole at n4 eearly the speed of light. This is the material that we see as the jets coming from the quasar's center, and it is this material, continuing outward into space for many thousands of years, that eventually collides with intergalactic material to produce the extended lobes seen with radio telescopes. Eventually, the supply of material that replenishes the accretion disk is exhausted, and the quasar ceases to exist as a quasar, perhaps becoming a less-energetic active galaxy.—S.N.

See also BLACK HOLES; DOPPLER EFFECT; EXPANDING UNIVERSE; GAMMA RAYS; INFRARED RADIATION; TELESCOPE, RADIO; TELESCOPE, REFLECTING; ULTRAVIOLET RADIATION; X RAYS

Further Reading: Harry L. Shipman *Black Holes, Quasars, and the Universe*, 2d ed., 1980.

Radar

"Radio ranging and detection," or radar, uses electromagnetic energy (radio waves) to detect distant objects and to find their position and distance. Radar can take two basic forms: pulse and continuous wave (CW). Pulse radar measures the distance to an object by noting the time interval between the transmission of a pulse and the reception of the return signal. CW radar uses modulated signals in conjunction with circuits that identify the time at which a returned signal was originally transmitted.

Radar had its origins in the 1930s, although early experiments on radio wave anticollision devices dated to the early 20th century. In 1900, Nicola Tesla described how radio waves might be used to determine the location of a moving object, and in 1904 Christian Hulsemeyer received a patent for a radio detector that drew on Tesla's idea. The sinking of the *Titanic* in 1912 tragically demonstrated that there was a pressing need for systems that could detect, identify, and locate objects in the water, and the search for U-boats in World War I gave rise to the development of ASDIC (sonar in the United States) systems. In 1922, Guglielmo Marconi (1874–1937) suggested that radio waves could be used to detect ships, and over a decade later the French passenger liner *Normandie* carried a radio detection system. In the meantime, U.S. researchers had observed effects produced by aircraft on short-wave radio transmissions. Radar thus resulted from researchers applying sonar principles to the detection of airplanes.

The advent of large, modern air forces and the fear of bombing by long-range aircraft underscored the importance of effective air defenses. As early as 1934, the British Air Ministry had commissioned studies to determine how to detect enemy aircraft at a distance. Early work began with continuous-wave transmissions, but electromagnetic pulses soon became the basis of early radar. In 1935, Robert Alexander Watson-Watt (1892–1973), a Scottish engineer working in the British National Physical Laboratory, demonstrated a pulse apparatus that detected aircraft at up to 64 km (40 mi). American experiments in 1934–35 used electromagnetic pulses reflected off aircraft and converted to signals that energized a cathode-ray tube display. Impulses could be measured to determine an object's range and, when compared to an earlier and later set of impulses, an object's bearing. However, the curvature of the Earth limited the range of land-based radars.

Many nations began to employ pulse radar, and, as hostilities loomed, radars were specialized for anti-aircraft, gun laying, and other tasks. The first successful use of radar in World War II came in December 1939 when the Germans detected a formation of bombers 115 km (71 mi) offshore. Both the U.S. Navy and the Royal Navy quickly moved to install gun-laying radar on their capital ships. By the outbreak of the war, most German warships were already equipped with radar, and German ships with gun-laying radar shelled Norway with great accuracy in 1940. But German development of radar stalled, due in part to Adolph Hitler's belief that electronics was the province of Jewish scientists and engineers. German radar research resumed in earnest in 1943, but by then Britain and the United States were well out in front.

When the war broke out in 1939, Britain had a chain of radar stations, along with the ability to tell if the blips on their radar screens represented enemy or friendly aircraft. The radar stations were connected to a communications network that provided essential information for interceptor aircraft. Radar played a vital role in the ensuing Battle of Britain; many military historians consider it the decisive factor in Britain's triumph over the Luftwaffe. While radar was helping the Royal Air Force to repel the German air assault, short-range radar apparatus was being fitted to more than 150 Allied ships, allowing them to detect submarines even when their periscope was the only thing above the surface of the water. In the air, the fitting of aircraft with radar was made possible by the invention of the cavity magnetron, a device capable of generating pulses of radio energy equal to those produced by a large transmitter. Airborne radar allowed the development of effective night fighters, and on Sep. 19, 1940, a Bristol Beaufighter scored the first kill by a radar-equipped airplane. Also under extensive development during the war were radar proximity fuses for shells. These eliminated the need for a direct hit on a target, for the shell detonated once it got within killing distance of a target.

Post–World War II advances in radar generally originated with the military, and then were adopted for civilian uses. New types of radars included pulse-Doppler radar, which allows the determination of a target's speed by comparing the frequencies of emitted and reflected radiation. One of the common sights at air bases is the radar dish, a piece of equipment associated with another type of radar, synthetic aperture radar. In early radar, the shape of the antenna sometimes resulted in signals that distorted distance or size. To provide better resolution, researchers developed synthetic aperture (dish-shaped) radar that processed the pulse to provide finer azimuth resolution, or, in other words, a better estimate of an object's size. Subsequently, geographers and other researchers used synthetic aperture radar to form an image over large objects, such as terrain masses.

Another radar, phased array radar (as opposed to single-element radars), consists of multiple stationary antenna elements (transmitter-receivers) and, rather than having a dish shape, the antennae are flat plates that can be installed almost anywhere. These antennae are well suited to military use because they are much more resistant to the overpressures produced by enemy bombs and missiles. Each element of a phased array radar system acts both as a receiver and transmitter, whose combined signal processing allows it to function as a large dish. The entire array thus becomes a single unit whose numerous "sightings" can be correlated into one data set.

During the Cold War, the threat of low-flying, high-speed bombers raised another problem for traditional radar designs, the "ground clutter" that made it difficult to discriminate targets flying below interceptor aircraft. In response, researchers modified radars with sophisticated filters that now provide modern warplanes with "look-down/shoot-down" radar that discriminates between real targets and clutter. Another early weakness of radar, the "line of sight" limitations imposed by the Earth's horizon, have been overcome by "Over the Horizon" (OTH) radar. OTH radar sends high-frequency signals to the ionosphere, which then refracts these signals back towards the surface of the Earth. When an airplane or missile flies through an area where signals are being reflected, it is detected by signals that are reflected back to the ionosphere, and then refracted to the receiving antennae on the ground. Yet another new radar, moving-target indication radars (MTIs), such as those found in JSTARS (Joint Strike Targeting Attack Radar System), can track moving targets, as opposed to fixing a single position. The development of the laser has created new possibilities, as a laser-based radar system can combine small size with good angular resolution.

While radar continues to advance, electronic countermeasures and technologies that render objects resistant to radar detection are also progressing. A significant amount of military research (and expenditure) now centers on a cat-and-mouse game between the designers of radar detection technologies and the designers of technologies intended to foil detection. Meanwhile, radar continues to play an important role in civilian applications like air traffic control and meteorology. Also, in the 1940s, radar technology rather unexpectedly spun off a common household appliance, the microwave oven.—L.S.

See also AIR TRAFFIC CONTROL; BOMBING, STRATEGIC; CATHODE-RAY TUBE; DOPPLER EFFECT; LASER; REMOTE SENSING; SONAR; STEALTH TECHNOLOGY; SUBMARINES; WEATHER FORECASTING

Further Reading: Russell Burns, ed., *Radar Development to 1945*, 1988.

Radial Tires—see Tires, Radial

Radio

In 1865, the British physicist and mathematician James Clerk Maxwell (1831–1879) published a paper entitled "A Dynamical Theory of the Electromagnetic Field" in the journal, *Philosophical Transactions*. In this paper, Maxwell addressed the problem of how an electric charge could be neither created nor destroyed. His solution, which was largely based on mathematical reasoning, was to propose the existence of what he called a "displacement current," something similar to an electric current, except that it did not require a conductor.

Maxwell's theory and accompanying mathematical models remained in the realm of abstraction until 1888, when a young German physicist named Heinrich Hertz (1857–1894) announced that he had successfully generated this current and then detected it with a receiving apparatus. This was the first empirical demonstration of

Heinrich Hertz (courtesy Institute of Electrical Engineers).

radio waves. Hertz had no interest, and probably not even the awareness, that radio waves could be used for purposes of communication. But other experimenters did, and during the 1890s radio waves generated by electrical sparks were used for "wireless telegraphy"—Morse code communications between stations difficult or impossible to connect with wires, such as ships at sea. The most successful of the these was Guglielmo Marconi (1874–1937), who at the beginning of the 20th century had created the largest and most far-flung network of radio transmitters and receivers in use.

At this point, radio was nothing more than a means of sending telegraphic messages without wires. Successful transmission of messages more complex than arrays of dots and dashes required a more sophisticated means of sending and receiving radio signals. As a result of efforts by inventors such as Lee deForest (1873–1961), Reginald Fessenden (1866–1932), and Ernst Alexanderson (1878–1975), the spark-based system was supplanted by one based on the use of continuous waves of controlled frequency and length. Continuous waves also made it possible to confine radio signals to a particular part of the electromagnetic spectrum and thereby prevent interference with other radio signals.

Continuous-wave transmission was a key technological step forward, but for many years radio continued to be used primarily as a means of sending messages between two stations. The primary use of radio was for communication between two parties. Amateur radio operators sent out messages with no intended receiver in the hope that someone would pick them up and respond, but this was done as a pastime and not for commercial purposes. Commercial broadcasting, the use of radio with which most of us are most familiar, did not begin in the United States until 1920, when Pittsburgh radio station KDKA began transmitting music, election results, and sports scores. At first, this was not done as a means of making money through the sales of commercial time. KDKA was an affiliate of the Westinghouse Electric Company, and the purpose of the broadcasts was to stimulate the sales of radio equipment manufactured by the parent company.

Commercial broadcasting began to hit its stride in the 1930s, due in part to technical improvements that made the use of radio receiving sets easier to use by people with little technical skill or inclination. Equally important, advertisers began to realize that a huge audience could be addressed through the use of radio broadcasts. Quiz shows, reports of sporting events, dramas, "soap operas," and many other kinds of programming were employed to draw in a listening audience subject to "a word from our sponsor."

During the 1930s, the primary method of using radio waves as carriers for information was "amplitude modulation," or AM. This altered the signal by controlling the height (or amplitude) of each wave. In the 1930s, Edwin H. Armstrong (1890–1954), then an employee of the Radio Corporation of America (RCA), devised the technology of frequency modulation (FM). This allowed for much clearer reception and paved the way for high-fidelity broadcasting in the late 1940s. However, when FM first appeared, RCA, then under the leadership of David Sarnoff, was not eager to scrap much of the investment that was tied up in AM radio broadcasting. Armstrong left RCA, and when the corporation finally began to involve itself with FM, a battle over patent rights with Armstrong ensued. Armstrong's claims were ultimately upheld in court, but by this time the dispirited inventor had committed suicide.

Although hundreds of millions of radio sets are in use in the United States today, radio broadcasting has ceased to be a medium oriented to a mass audience. Today's commercial radio is divided up into many segments, each catering to specialized niches: country and western music, talk shows, all news, contemporary rock, and many other distinct niches. Radio also continues to be of great importance for other forms of communication. For example, it is hard to imagine widespread commercial airline travel without ground-to-air communications based on radio.

See also AMPLITUDE MODULATION (AM); BROADCASTING, COMMERCIAL; FREQUENCY MODULATION (FM); MAXWELL'S EQUATIONS; THERMOIONIC (VACUUM) TUBE

Further Reading: Hugh G. J. Aitken, *Syntony and Spark: The Origins of Radio*, 1985. Hugh G. J. Aitken, *The Continuous Wave: Technology and American Radio, 1900–1932*, 1985.

Radio Telescope—see Telescope, Radio

Radio, Citizens Band (CB)

The citizens band (CB) radio at one time was the fastest-growing electronics communications medium in the United States. In the 1980s, it was displaced by the rise of cellular telephones, but a decade later it had regained some of its popularity. A two-way mobile radio system was introduced by the New Jersey Police Department in 1933, marking the beginning of land mobile radio. The earliest CB radio, proposed by E. K. Jet in 1945, was allocated its own space in the radio spectrum in 1958, when the Federal Communications Commission (FCC) established Class D Citizens Radio Service, operating in 23 public channels within the band frequencies of the 26.965 to 27.255 MHz portion of the high-frequency (HF) spectrum. Although the technology did not involve "pure" communication, in which individuals could speak to each other without the interruption of clearing a channel first (usually by saying, "over"), it did permit two-way communication that was mobile. CB licenses exceeded 49,000 in 1959.

Truck drivers and private motorists quickly saw the value of CB radio. They could talk while driving, sharing information about safety, weather, and (above all) police locations. In 1960, the Highway Emergency Location Plan (HELP) was established; it required truck drivers to have a CB in their cabs, and resulted in the airwaves being filled with "truck driver chatter." Despite their popularity among truckers, CB radios were expensive items until the general consumer market took interest in the technology. By 1976, the FCC attempted to decrease the congestion of the airwaves by increasing CB radio channels from 23 to 40, within the 27.235- to 7.405-MHz range. Popular songs soon immortalized phrases such as "good buddy" and "pedal to the metal" by including the CB conversations over a musical background.

One of the first companies to manufacture CB radios was Dynascan, which was started by Carl Korn in 1954. By 1976, Dynascan sold more than $1 million worth of the radios. Another leading CB manufacturer and retailer, Radio Shack, started marketing its "Realistic" transceiver in 1959; by 1975, the CB had become Radio Shack's fastest growing item of the day, representing 13 percent of the company's overall sales. Units sold for between $60 and $35; a license sold from $4 to $20.

Although Radio Shack stock soared and CB sales rose in the 1970s—growing by 60 percent during some months—the technology remained limited. Any number of users could listen to a conversation, making privacy impossible, but only one person could talk at a time, making it difficult to interrupt a long monologue. The advent of cellular phone technology caused a sudden and substantial dip in the sales of CB radios. Cellular mobile telephones had existed in large numbers since the 1960s, but the poor quality of the mobile phones and high cost of equipment limited the growth of the industry. But cellular phone sales soon accelerated.

The effects of the 1970s oil crises, including long fuel lines and 55-mph speed limits, revived the CB radio, which was commonly used to locate fuel and avoid police. The nation's 16 million licensees owned nearly 45 million CB sets by 1983, and the high number of licenses exceeded the FCC's ability to regulate the medium. As a result, the FCC ceased to license CB radios. Meanwhile, the major limiting factor to cellular phone use was price, and costs started to fall in the 1980s. By 1983, cellular phone sales exceeded $117 million—still far below CB sales, which could surpass that much in 1 month, but the technology had far more potential. Cellular phones provided a much more convenient alternative to hand-operated CBs and allowed a single, private channel to remain open while both sending and receiving, permitting "real time" conversations. Moreover, while CBs had a counterculture appeal, cellular phones represented the technology of the upper classes. CB technology continued to improve: The weight of a CB radio fell from 4.5 kg to .9 kg (10 lb–2 lb), and they continued to offer a cheaper alternative to cellular phones. Equally important for motorists, CBs offered a wide span of frequencies on which to communicate, allowing one to listen in on other people's conversations.—L.S.

See also TELEPHONE, CELLULAR

Further Reading: Marvin Smith, *Radio, TV, & Cable: A Telecommunications Approach*, 1985.

Radio, Transistor

The transistor radio was not the first portable radio, nor was it even the first small portable radio. Crystal receivers, the first commercially successful radios, did not require an external power source, and in this sense they could be considered portable. The more sophisticated radios that succeeded crystal sets used batteries for power; they too were portable in that they did not have to be plugged into a wall socket. But while they were self-contained, these radios were heavy and bulky, making them about as portable as a full suitcase. During the 1930s, improvements in vacuum-tube technology diminished power requirements, allowing the use of smaller batteries and the production of smaller, lighter radios. By the early 1940s, the availability of physically smaller tubes was reflected in the sale of portable radios that weighed around 2.5 kg (5.5 lb) and could fit in a coat pocket. This trend continued after World War II, when the subminiature tubes developed during the war made possible the production of radios small enough to be slipped into a shirt pocket. At the same time, however, these radios were not good sellers; weak reception and a prodigious appetite for batteries limited the appeal of miniature radios.

The Regency, the first commercially marketed transistor radio (courtesy Michael Schiffer).

What made the small portable radio a technical and commercial success was the transistor. With its small size, low power requirements, and high reliability, the transistor eliminated the inherent defects of tube-based radios. The first transistor radio was the Regency TR-1, a collaborative venture of I.D.E.A., Inc., an Indianapolis electronics company, and Texas Instruments, the supplier of the transistors (four for each radio). Introduced in 1954, the Regency sold for $49.95. Hobbled by a high price and poor audio reproduction, sales of the first transistor radios were lukewarm—about 100,000 sets in the first year they were on the market. Audio quality hardly improved in the years that followed, but lower prices drove sales of transistor radios sharply upwards in the late 1950s and the 1960s.

Paradoxically, sales of transistor radios began to take off just when television appeared to pushing radio aside. In reality, radio had not entered a period of decline but one of transformation. While mass-market network radio was rapidly losing its audience, specialized broadcasters were finding their distinctive niches. The most significant new listening audience tuned their radios to the stations that played rock-and-roll music. And a lot of people tuned in. Rock was the music of the vast cohort of baby boomers, the first wave of whom entered adolescence in the late 1950s. This new generation of radio listeners created a huge market for transistor radios. And, in reciprocal fashion, the ubiquitous transistor radio played a large role in popularizing rock.

Many of these radios were produced in Japan, a country that was just beginning to demonstrate its productive capabilities. The first Japanese import was produced by a small firm, the Tokyo Telecommunications Engineering Co. Introduced in 1957, it was so successful that the firm soon took the name of its product: SONY. By the mid-1960s, two-thirds of the portable radios sold in the United States were imports, most of them Japanese in origin. In 1969, the figure had climbed to 94 percent. Yet within a few years, Japanese manufacturers were confronted with rising labor costs that threatened to erode their profits. Just as had happened with American firms, Japanese producers turned to places like China, Malaysia, and Singapore for the low-cost manufacture of transistor radios.

See also CRYSTAL RECEIVER; THERMOIONIC (VACUUM) TUBE; TRANSISTOR

Further Reading: Michael Brian Schiffer, *The Portable Radio in American Life*, 1991.

Radioactivity and Radiation

Radioactivity is a property of certain elements with unstable nuclei. It can be the result of the natural decay of these nuclei, or it can be produced by subjecting atoms to nuclear reactions that add or subtract neutrons. Radioactivity causes the emission of electromagnetic waves and subatomic particles. The emission of these waves and particles is known as *radiation*.

Radioactivity was discovered towards the end of the 19th century. Ironically, this was a time when many scientists believed that the major problems and issues in physics had been resolved. The discovery of radioactivity opened up a whole new world that even today has been only partially explored. In 1896, Henri Becquerel (1852–1908), a French physicist, was intrigued by the recent discovery of X rays. Since his specialty was fluorescence and phosphorescence, he wondered if any fluorescent materials emitted X rays. To find out, he wrapped a photographic plate in black paper and left it out in the sun adjacent to some fluorescent material. When the sunlight struck the material, he reasoned, it might give off X rays that showed up on the photographic plate. This is exactly what happened; the plate was fogged by the X-ray emissions. But then came a string of cloudy days, and Becquerel had to temporarily cease his experimentation. He put the photographic plate and fluorescent material in a drawer in anticipation of sunny weather, which was very slow in coming. A bit bored, Becquerel decided after a few days to develop the photographic plate to see if any residual fluorescence remained. Much to his surprise, the plate was strongly fogged; the fluorescent material apparently did not need light in order to produce X rays. The substance that had produced this effect was potassium uranyl sulfate, a complex compound of potassium, oxygen, sulfur, and uranium. In 1901, Becquerel was able to identify the uranium in the compound as the source of the emanation producing the X rays.

By this time, Marie Sklodowska Curie (1867–1934), a Polish-born physicist working in France with her husband Pierre Curie (1859–1906), had given these emanations the name Becquerel rays. She also coined the name *radioactivity* for the natural process that generated the rays. More important, the Curies, along with Becquerel and Ernest Rutherford (1871–1937), were able to determine that radiation had three basic components: alpha "rays," beta "rays," and gamma "rays."

The Curies also realized that certain substances were even more radioactive than uranium. They concluded that an undiscovered radioactive element in these substances must be causing the high levels of radioactivity. In 1898, they succeeded in refining from pitchblende (a uranium ore) a highly radioactive substance that they named *polonium* after Marie Curie's native Poland. However, the newly discovered element still did not account for the intense radioactivity of uranium ore; something else had to be there. In late 1898, they detected a new element to which they gave the name *radium*. At first, the presence of radium could only be inferred indirectly by spectral analysis and an analysis of the radioactive emanations. In order to isolate the element, the Curies had to refine tons of uranium-rich mine tailings obtained from the St. Joachimsthal mine in Bohemia. By 1902, they had obtained a gram (0.035 oz) of radium from 7,250 kg (16,000 lb) of uranium ore.

In the years that followed, a number a number of new radioactive elements was discovered. One of these was radon, an inert gas discovered in 1900 by the German physicist Friedrich Dorn (1848–1916). In addition to discovering new radioactive elements, physicists also learned that radioactive elements changed into other elements as their nuclei decayed. Radium, for example, was produced by the decay of uranium. In 1905, the American physicist Bertram Boltwood (1870–1927) noted that lead was always present in materials containing uranium. From this he inferred that lead might be the end product of uranium decay (it was later learned that uranium goes through 14 transmutations before ending up as Pb-206 rather than the usual form of lead, Pb-207). Boltwood also suggested that a measurement of the amount of lead in uranium ores, coupled with the previously ascertained rate of uranium decay, could be used to determine the age of the Earth's crust.

The transformation of radioactive elements into other elements occurs according to a predictable schedule. In 1902, Rutherford and his colleague Frederick Soddy (1877–1956) found that radioactive elements lost their radioactivity at a regular exponential rate (i.e., it took the same amount of time for the radioactivity of a substance to fall from 1,000 to 900 units as it did to fall from 100 to 90). The loss of radioactivity is now reckoned in terms of "half-life," the time it takes for a substance to lose half its radioactivity. Uranium loses half its radioactivity in about 4,500 million years, while radium has a half-life of about 1,600 years, and radon has a half-life of only about 3.8 days.

One of the most important scientific consequences of experimentation with radioactive substances was the discovery that atoms were not unitary, indivisible pieces of matter that retained their structure perpetually. Rutherford determined that atoms had an inner core or nucleus, and that its nature could be probed by bombarding it with of one of the products of atomic decay, the alpha particle. Through the use of this technique, in 1919 Rutherford was able to achieve the dream of medieval alchemists, to change one element into another. By shooting alpha particles at nitrogen, Rutherford caused the nucleus of some nitrogen atoms to throw off a proton, turning the nitrogen nucleus into the nucleus of an oxygen atom.

The measurement of radioactivity was facilitated by the invention of the Geiger counter. Another instrument of great value to the study of radioactivity was the cloud chamber. In the early 1920s, Patrick M. S. Blackett (1897–1974) used a cloud chamber to produce a visual indication of the result of the bombardment of nitrogen nuclei by alpha particles, confirming Rutherford's hypothesis that transmutation was taking place.

While physicists were probing the structure of the atom, others were using radiation for medical treatment. Although a considerable amount of quackery surrounded early radiation treatment, it is widely used today for the treatment of certain cancers. Radiation is also used to a limited extent for the preservation of food. On a more troubling note, the study of radioactivity and the process of nuclear decay gave rise to the idea that it might be possible to purposely break apart atomic nuclei and produce a great deal of energy as a result. This expectation began to be realized in the late 1930s with the fissioning of atomic nuclei, and culminated in the detonation of the first atomic bombs in 1945.

The dropping of two atomic bombs on Japan during World War II caused a great amount of physical damage. It also provided graphic demonstrations of the medical consequences of exposure to radioactive substances. It had long been known that radioactive materials could pose a danger to human health: One early victim of radiation was Marie Curie, who died of leukemia that most likely was the result of years of exposure to radioactive materials. In 1986, a massive release of radioactive materials occurred when one of the nuclear reactors at Chernobyl in the Soviet Union exploded. Dozens died within a few weeks of the accident, and thousands of people still face the long-term consequences of exposure to substantial amounts of radiation.

The exact mechanisms through which radioactivity causes damage to the body are not completely understood. It is known that radiation with wavelengths shorter than about 10^{-8} m produces enough energy to dislodge electrons from the atoms of substances absorbing the radiation. Radiation capable of producing ionization is known, logically enough, as *ionizing radiation*. It is the ionization caused by radiation that does the damage to organisms. Stripped of their electrons by ionizing radiation, atoms gain a positive charge. At the same time, the liberated electrons associate themselves with other atoms, giving a negative charge to the latter. These positively or negatively charged atoms are known as *free radicals*. They are strongly reactive, and they can do considerable harm to bodily tissues. For example, the presence of free radicals can cause oxygen to ionize (i.e., gain a negative charge by picking up an electron); the ionized oxygen may then damage or destroy enzymes or DNA.

The effects of radiation on an organism depend to a significant extent on the dosage received. Determining how much radiation has been received is a bit complicated, however, for a number of different units have been used to measure radiation. One fundamental measurement is the rate at which nuclear disintegrations occur. Today, this is usually given in becquerels (Bq); 1 becquerel is one disintegration per second. In the past, this was given in curies; 1 curie equals 37 billion disintegrations per second. In order to determine the effect of radiation on an organism, it is necessary to know how much of it has been absorbed. The absorption of radiation-borne energy by an organism is measured in grays (Gy); 1 gray is 1 joule of energy absorbed by 1 kg of tissue. The gray has largely replaced an older measurement known as a rad, which equals 0.01 Gy. The gray, however, has the drawback of not taking into account the fact that different components of radiation have different consequences for organisms; for example, alpha particles produce more cell-damaging ionization than beta particles. The seivert (Sv) provides a standardized measure that encompasses the type of radiation and the duration of the exposure. It too has replaced an older measure known as the rem, which equals 0.01 Sv.

The determination of a "safe" level of radiation continues to be a controversial matter. There is no mistaking the effects of exposure to high levels of radiation: Burns, nausea, lassitude, and the loss of hair are some of the immediate symptoms. Exposure to smaller amounts of radiation may result in the eventual development of cancer, but many variables affect the extent to which this happens: the type of radiation and duration of exposure, the tissue receiving the most exposure, and the individual characteristics of the people or organisms who have been exposed. Long-term research has shown that survivors of the atomic attack on Hiroshima have suffered from increased incidences of certain kinds of cancer, most notably leukemia, but other types of cancer are not evident to an unusual degree. Moreover, studies of atomic-bomb survivors and others exposed to high levels of radiation have shown that incidences of some kinds of cancer increase in a linear fashion with dosage, while others exhibit a more complicated relationship.

Studies of the survivors of atomic-bomb blasts have provided a great deal of information about the effects of exposure to high levels of radiation. The consequences of low levels of radiation are more difficult to determine. A "low" level of radiation exposure is commonly put at less than or equal to 0.2 Gy. At these levels it is extremely difficult to determine if exposure increases the likelihood of cancer. It has been shown experimentally that even low doses of radiation cause damage to the chromosomes of cells, but the connection between this damage and the initiation of cancer remains unclear. Finally, since radiation is known to cause chromosomal damage, it may be expected that exposure to radiation will damage germ cells, resulting in congenital defects in the next generation. In fact, there has been no statistically significant evidence of inherited damage to the progeny of the survivors of Hiroshima and Nagasaki.

See also ALPHA PARTICLES; ATOMIC BOMB; BETA PARTICLE; CANCER; CHERNOBYL; CHROMOSOME; CLOUD CHAMBER; DNA; ENERGY, MEASURES OF; GAMMA RAYS; GEIGER COUNTER; IRRADIATED FOOD; NEUTRON; NUCLEAR FISSION; X RAYS

Radiocarbon Dating

Until fairly recently, when an archeologist wanted to determine the date of an object, the best that he or she could do was to use geological information to estimate the age of the rock and silt layers in which the object was found. This was a method subject to considerable error. Even more perplexing, many objects were found in places that did not have a stratigraphic record. Many of these problems were solved after the discovery of radiation provided the foundation for a new method of dating ancient objects.

In 1907, Bertram Boltwood (1870–1927), an American chemist, discovered that uranium slowly decays into lead; consequently, rocks containing uranium could be dated by measuring the amount of lead they contained. Boltwood put this idea to good use when he used it to date the age of the Earth. His determination that the Earth had solidified from a liquid mass about 3.7 billion years ago was a vast improvement over the figure of 20 to 400 million years that the eminent scientist William Thomson, Lord Kelvin (1827–1907) had previously put forward.

While Boltwood's method was very useful for geological dating, it only worked when uranium was present, something that was not the case for plants and animals. However, uranium is not the only element that undergoes a process of radioactive decay. Another such substance is carbon-14, a radioactive isotope of ordinary carbon. Carbon-14 is created in the atmosphere by the action of cosmic rays. It is absorbed by plants during the process of photosynthesis, and in turn is taken in by animals who eat the plants, and by other animals who eat the plant-eating animals. In this way, carbon-14 is present in most organisms. Although many plant and animal remains decay rapidly after the death of an organism, some carbon-containing remnants are long lived. Consequently, a variety of organic substances can be tested for carbon-14, notably wood and charcoal, shells, bones, and antlers.

Carbon-14 is not replenished after organisms die, and the existing carbon-14 slowly decays into nitrogen-14. Consequently, the age of an organism's remains can be determined by using a very sensitive Geiger counter to measure the concentration of carbon-14. At some point the amount of remaining carbon-14 is too small too measure, so the method can be used only for subjects that are less than a few tens of thousands of years in age.

Dating with carbon-14 was pioneered by the American chemist Willard Libby (1908–1980) in the late 1940s. In the ensuing years, the technique was used to date the Dead Sea scrolls, to determine the age of the earliest evidence of human beings in North America, and to ascertain the time when the last great Ice Age ended. Libby's technique entailed measuring the concentration of carbon-14 by monitoring its decay into nitrogen-14. In the late 1970s, a technique using mass spectrometry to directly detect carbon-14 atoms was developed. This allowed the use of smaller samples and the dating of older objects. Whereas Libby's initial method was effective for objects that were no more than 50,000 years old, spectrometric techniques can be used to date objects with ages of up to 70,000 years.

Although carbon-14 dating has been of great value to archeologists and other scientists, it does not provide a completely accurate reckoning of an object's age. Part of the problem of accurate dating results from the fact that the production of carbon-14 in the atmosphere is not constant but varies as a result of fluctuations in cosmic-ray activity. Moreover, carbon-14 does not mix rapidly and uniformly in carbon-containing organisms, affecting the ratio between it and ordinary carbon. The ratio may also be affected by the contamination of a sample with carbon from another source. Human activities, most notably the testing of nuclear weapons, have introduced significant quantities of carbon-14 into the atmosphere, making it very difficult to determine the age of recent specimens; most laboratories will not try to ascertain the age of an object that is thought to be less than 100 or 150 years old. Finally, researchers have to take into account the fact that the complex history of an object may lead to misleading results. For example, a wooden post that was excavated at an archeological site may have come from a tree that had been felled hundreds of years before it was used for lumber.

The uncertainties inherent in carbon-14 testing are reflected in the way that dates are presented. Instead of a single date, a range is presented; for example, an object may be given the date 6,800 ± 80 B.P. (before present). This indicates that when the object was tested, its likely age was between 6,880 and 6,720 years old.

See also COSMIC RAYS; GEIGER COUNTER; ISOTOPE; PHOTOSYNTHESIS; RADIOACTIVITY AND RADIATION; SPECTROSCOPY

Railroad

In its most basic form, a railroad is anything that uses equipment with flanged wheels rolling over fixed rails. Primitive railroads (or railways, as they are usually termed in Great Britain) were found in some medieval mines, but the railroad did not emerge as a complete transportation system until the early 19th century. Befitting its status as the first industrial nation, Great Britain pioneered the development of railroads; the Stockton and Darlington Railway, which began operations in 1825, is acknowledged to be the first common-carrier railroad. The Stockton and Darlington used a track gauge (the distance between the rails) of 4 ft, 8½ in. (1.435 m). This is the most common gauge in the world today, although railroads have used—and some continue to use—a profusion of other gauges.

Early railroads experimented with a number of ways of pulling their trains. Stationary engines, horses, mules, and even sails were tried, but it soon became evident that steam power was the best kind of motive power. These locomotives, along with the cars they pulled, the rails they rode on, and the land on which the rails were laid, represented a very considerable fi-

nancial investment. Some of the necessary capital was raised from private sources, but from their inception, the railroads of Europe and the United States received considerable infusions of money from local and national governments. In parts of the world where industrialization came late, the building of railroads was usually undertaken by the government rather than private enterprise. Even in the United States, where an ideology of private enterprise was dominant, railroads benefited from generous government grants of land and money. One example of governmental largesse was an 1850 act of Congress that in the years that followed gave various railroads 3,736,005 acres of land to help them meet their construction costs.

Although early American railroad development had been dependent on technology transfer from Great Britain, it was not long until the United States was a world leader in the building and operation of railroads. In the 1840s, the United States already had about 53,000 km (3,300 mi) of track compared to about 29,000 km (1,800 mi) in all of Europe, and by 1860 the figure for the United States had increased nearly 10-fold. At the same time, however, the rapidity of construction was made possible by adhering to engineering standards that were often well below the ones used in Europe. American railroads were often laid out with steep grades and sharp curves. Bridges were built from wood instead of more durable materials. Tracks were put down on roadbeds that lacked adequate preparation.

What seemed to be nothing more than slovenly construction reflected the need to build, as quickly as possible, a transportation system essential to the settlement of a vast frontier along with the economic integration of a far-flung nation. An expanding network of rails played a major role in uniting the country economically and culturally, but even as this was taking place, the nation faced dissolution over the issue of slavery. The Civil War finally resolved this long-simmering conflict, and in the process it convincingly demonstrated the military importance of railroads. The Civil War has been rightly described as the first railroad war, for it provided many examples of how crucial railroads were for the provision of supplies and the movement of troops.

In 1869, 4 years after the conclusion of the Civil War, the United States completed a railroad line that allowed the passage of people and freight from the East Coast to California. The building of the transcontinental railroad required some heroic engineering and the arduous labors of thousands of workers, the bulk of them Irish or Chinese immigrants. It also entailed substantial financial support by the federal government. The costs of building the first transcontinental railroad were partially defrayed by the sale of land that had been granted to the Central Pacific and Union Pacific railroads. The two railroads had a lot of land to sell, having received 20 sections of land (a section is 1 square mile or 640 acres) for every mile of track laid. In return, the railroads agreed to transport federal troops without charge and to carry mail at low rates. In the decades that followed, the government land grants were more than offset by the low costs of hauling mail, but the fact remains that the transcontinental railroad would not have been built in the absence of these grants.

Government grants were also essential for an equally impressive feat of railroad construction, the building of the Canadian Pacific Railroad. Completed in 1885, the Canadian Pacific had great political importance, for the province of British Columbia had made its joining the Dominion of Canada contingent on the railroad's construction. A more direct approach was taken in czarist Russia, where the trans-Siberian Railroad was built directly by the government, which received substantial technical assistance from citizens of other European countries. It too was motivated by a desire to open up a largely unsettled frontier and to solidify governmental control over a vast territory.

Wherever the railroads were located, their construction, operation, and management presented challenges on a scale never encountered before. As the first example of "big business," railroads of necessity were pioneers of modern management practices. The far-flung nature of railroad operation required a managerial structure that was decentralized and yet kept under firm control at all times. The answer to this seeming paradox was an organizational structure that vested authority with divisional superintendents, but at the same time made extensive use of precise rules and regulations intended to cover every aspect of the business. Many of the principles and procedures developed by the railroads were subsequently adopted by other business enterprises. Of equal importance was the development of communication systems for the running of trains and other operations. These depended on the most modern means available at the time, beginning with the telegraph and culminating with the extensive use of computers by today's railroads.

The railroads also expanded the role of governmental management. As was noted above, governments of nations that came comparatively late to industrialization often acted as railroad entrepreneurs. When railroads began to go into decline in the 20th century, many countries nationalized privately owned railroads in the hope of preventing further deterioration. Even where railroads remained in private hands, government involvement was substantial. An expanded role for the U.S. central government came with the creation of the first federal regulatory agency, the Interstate Commerce Commission. Established in 1887, it had as its purpose the regulation of the rates that railroads charged to shippers and involved itself in many other aspects of the business.

Although the basic technological pattern was set early in the industry's history, railroads subsequently benefited from a plethora of innovations great and small. The airbrake made the operation of trains much safer, while the automatic coupler saved the fingers and hands of many a trainman. Steel rails and hardware greatly increased the operational life of these components.

The railroads also made extensive use of special-purpose cars: Tank cars, hoppers (cars used for hauling coal and similar materials), stock cars, and cars designed to transport nothing but automobiles lowered transportation costs, sometimes dramatically. One of these cars, the refrigerator car, had an immense effect on both individual diets and national commerce by making possible the long-distance transport of fruits, vegetables, dressed meat, and other commodities.

In recent decades, railroads have improved their operations through the use of automatic train controls. These computer-based systems allow trains to be governed from a central facility that may be hundreds of kilometers from most of the traffic. Automatic train control is tied in with signaling systems that keep trains a safe distance apart. Computerized controls also are widely used in classification yards to aid in the assembly of cars into trains, and then their disassembly when they arrive at a terminal. Railroad automation of this sort has substantially improved productivity, but as with many productivity improvements, one result has been the elimination of many jobs, even though railroad unions engaged in a strenuous effort to preserve jobs that technology had rendered obsolete.

A considerable amount of the improvement in railroad productivity has come through hauling longer trains made up of larger cars. For decades these trains were pulled by increasingly powerful steam locomotives, but in the 1930s a revolution in railroad motive power began with the large-scale introduction of diesel-electric locomotives.

By cutting costs and increasing productivity, the diesel locomotive served as a kind of technological fix at a time when railroads were facing increasingly stiff competition from automobiles, trucks, and airplanes. Railroads have been undergoing years of relative decline, as can be seen by comparing their situation today with what existed in the past. When American railroads were at the apex in 1916, they carried 77 percent of intercity freight traffic and 98 percent of the intercity passenger traffic. Today, passenger traffic is largely confined to a few well-traveled corridors, while less than 40 percent of the total goods are transported by rail. Other countries make much more extensive use of trains for passenger travel, but in almost all of these cases their governments provide a significant subsidy for this service. Japan has earned a well-deserved renown for its Shinkansen trains, capable of carrying passengers at speeds in excess of 190 kph (120 mph), but this impressive service was accompanied by financial losses that had to be absorbed by the Japanese government. In the United States, chronic losses put the very existence of passenger trains in doubt until the National Railroad Passenger Corporation (Amtrak) was established by the federal government in 1971.

In the future, passenger travel may undergo a substantial revival in the United States and elsewhere through the application of magnetic levitation technologies. In the meantime, railroads still hold the distinct advantage of being highly energy efficient due to the low friction losses inherent in running flanged wheels on steel rails. Today's railroads earn much of their revenue through the hauling of bulk commodities such as coal and cement, but in recent years, the rise of intermodal transportation has given new life to many freight haulers. It seems certain that a transportation technology with its roots in the early days of industrialization will continue to be essential to the functioning of "postindustrial" economies.

See also AIRBRAKE; AIRPLANE; AUTOMOBILE; COUPLER, AUTOMATIC; ENERGY EFFICIENCY; FEATHERBEDDING; LOCOMOTIVE, DIESEL-ELECTRIC; LOCOMOTIVE, STEAM; MAGNETIC-LEVITATION VEHICLES; REFRIGERATOR CAR; STEAM ENGINE; STEEL IN THE 20TH CENTURY; TECHNOLOGICAL FIX; UNEMPLOYMENT, TECHNOLOGICAL; TELEGRAPHY; TRANSPORTATION, INTERMODAL; TRUCKS

Further Reading: John Armstrong, *The Railroad, 1990.*

Rankine Cycle

The Rankine cycle is a thermodynamic cycle for heat engines. Although ideal, the Rankine cycle can be approximated by actual machinery. It thus provides a more realistic value of a plant's maximum efficiency than the Carnot cycle. William J. M. Rankine (1820–1872) used this cycle to model reciprocating steam engines. He was the Regius Professor of Civil Engineering and Mechanics at the University of Glasgow, a prominent consulting engineer, and one of the leading physical scientists of the Victorian era. Along with William Thomson Lord Kelvin (1824–1907), and Rudolph Clausius (1822–1888), he established the modern science of thermodynamics, his most significant contribution being the publication of *A Manual of the Steam Engine* (1859). This was the first engineering textbook that employed the First and Second Laws of Thermodynamics. It remained the standard thermodynamics text for Anglo-American engineers until the 1910s. There, Rankine employed the Rankine cycle in terms of steam pressure and piston displacement. By the early 20th century, engineers were expressing that cycle in terms of entropy and temperature to model steam plants with turbines.

The Rankine cycle has its origins in the 18th century. Benjamin Robins (1707–1751) presented a diagram of gas pressure versus bullet displacement in *New Principles of Gunnery* (1742).

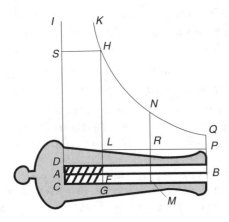

Rankine used this diagram to calculate a projectile's muzzle velocity by assuming Boyle-Mariotte's law (pressure is inversely proportional to displacement) for gas expansion, Newton's Second Law of Motion, and the 39th proposition of Newton's *Principia*. In a manner similar to Daniel Bernoulli's (1700–1782) analysis in *Hydrodynamica* (1738), Robins used integral calculus to find the area underneath the hyperbolic expansion curve, which in turn allowed him to determine the "work" performed by the gas on the projectile. When John Robison published the first theoretical study of a steam engine's performance in 1797, he applied Robins's diagram and analysis:

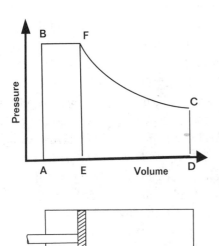

In this case, steam begins entering into the cylinder at point *B* under constant pressure. The steam valve is closed at point *F*, and the steam expands to

point *C*, according to Boyle-Mariotte's law. The exhaust valve to the condenser or to the atmosphere then opens, and the pressure drops to point *D*. Robison showed that work performed by the steam on the piston is the sum of areas *BFEA* and *FCDE*.

Rankine's diagram of an ideal steam engine seems identical to Robison's:

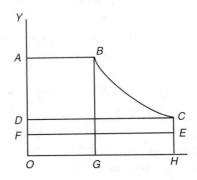

Rankine, however, rejected Boyle-Mariotte's law as a suitable model of steam expansion. He employed either an adiabatic expansion (no heat transfer through the cylinder) or a saturated steam expansion (no condensation or superheating). He also considered the work performed by the cylinder on the steam during the reverse stroke. Steam is thus introduced into the cylinder at constant pressure along line *AB*, and the expansion occurs along *BC*. The efficiency is maximized when the expansion is adiabatic. To calculate the work generated by the steam engine in this ideal situation, Rankine calculated the total work performed by the steam on the piston (area *OABCH*) and subtracted the work performed by the piston on the steam (area *OFEH*). To calculate the engine's efficiency, he divided this work by the heat introduced into the cycle, employing both theoretical considerations and experimental data.

Rankine used this analysis to resolve a controversial engineering issue of that age: whether allowing steam to expand in engines is economical. He recognized that in actual steam engines, the steam starts condensing when it expands adiabatically, leaving moisture on the cylinder's walls. When the pressure drops along line *CE* after the exhaust valve is opened, the moisture reevaporates, drawing heat from the cylinder. Rankine argued that such undesirable heat transfer is minimized when the steam expands along the saturated or dry-steam line. In that case, most of the condensation occurs outside the cylinder after the exhaust valve is opened. Not content with this scientific explanation, Rankine also suggested a technical solution: He recommended heating the cylinder in a steam jacket to ensure that the steam expands along the saturated steam line.

The immediate technical consequence of Rankine's thermodynamic studies during the 1850s was the introduction of compound marine engines by the Scottish shipbuilding firm Randolph and Elder. Rankine was a close friend and engineering consultant of John Elder. By dividing the steam expansion process into a high-pressure and a lower-pressure cylinders, and providing each with a steam jacket, Elder resolved the condensation problem in a commercially feasible manner. According to Rankine's obituary notice of Elder, this innovation was a direct result of Elder's application of thermodynamics. Undoubtedly, he acquired such ability through Rankine's influence.—B.S.

See also CALCULUS, DIFFERENTIAL AND INTEGRAL; CARNOT CYCLE; ENERGY, CONSERVATION OF; ENTROPY; MECHANICS, NEWTONIAN

Rayon

On many occasions a new technology has been developed to overcome the loss or diminution of an important material. Such was the case of rayon, a man-made fiber originally known as "artificial silk." In the mid-1880s, a silk worm disease spread across France, motivating Louis Hilaire Bernigaud de Chardonnet (1839–1924) to devise a substitute. De Chardonnet's process used wood pulp that was converted to nitrocellulose, but by the turn of the century rayon was being made from cellulose acetate following a process developed in 1894 in England by Charles Cross (1855–1935) and Edward Bevan (1856–1921). The cellulose acetate process received a boost after World War I, for that conflict had given rise to a substantial capacity for the production of cellulose acetate used for doping fabric-covered aircraft.

In the 1920s, changing women's fashions created an expanding market for silk and silk substitutes, as rising hemlines led to a need for more attractive stockings. But when used as a substitute for silk, rayon had many deficiencies. A rayon fiber was a single filament, whereas a silk thread consisted of many extremely small fibers twisted together. Consequently, silk threads contained tiny pockets of air that gave silk fabrics their softness, warmth, and capacity to absorb moisture. In contrast, stockings woven from rayon monofilaments tended to sag in a most unfashionable manner.

Although rayon had been initially developed as a substitute for silk, it came into its own only when it began to be used for products to which it was better suited. Although rayon stockings were never an adequate substitute for silk and later nylon ones, many other kinds of garments have been successfully made from rayon. It also has many industrial uses, in the late 1930s it began to be used for tire cords, although it eventually was replaced by polyester.

Chemically, rayon is a polymer, a very large molecule made up of tens of thousands of small molecules (monomers) that are arranged in a long chain. The production of rayon entails converting cellulose, another polymer, to a fluid form while preserving its molecular structure. The fluid is then solidified to form the rayon. Three different manufacturing processes are used to do this. The viscose process chemically converts cellulose into a viscous solution (viscose) that is pumped through very fine (0.05–0.1 mm [.002–.004 in.]) holes known as *spinnerets*. After passing through an acid bath, undergoing a number of treatments, and then being dried, the individual filaments are spun together to make rayon yarn. In the cuprammonium process, the cellulose is treated with copper sulfate and ammonia to form a viscous solution of cellulose cuprammonium. It is then extruded through the spinnerets, and the resulting filaments are drawn out with a stream of water. In the acetate process, the cellulose is converted into a dry acetate that is soluble in acetone. The dissolved acetate is then extruded through spinnerets into warm air.

The oldest of the artificial fibers, rayon is still widely used. It has only fair strength, and fabrics made from it are prone to wrinkling, but its annual monetary value equals that of cotton and surpasses synthetics such as nylon. The commercial appeal of rayon has been enhanced by producing it in the form of staple fibers that can be processed on the same machinery used for cotton and wool fibers.

See also COTTON; NYLON; POLYMERS; TIRE, PNEUMATIC

Razor, Safety

Shaving has long been practiced by men and women; archeological sites more than 7,000 years old have been found containing sharpened flint and horn used for this purpose. In the centuries that followed, beards came in and out of masculine fashion. When they were out of fashion, the device used to remove them was the classic straight-edged razor. These implements could produce a smooth shave, but they required continual sharpening and honing, as well as a deft hand. A moment's inattention or clumsiness could easily produce a nasty gash.

In the late 19th century, King Camp Gillette (1855–1932) was earning a modest living as a salesman for a firm making bottle caps. In 1894, he had published a book entitled *The Human Drift*, in which he outlined his plans for a cooperative megalopolis of 24,000 standardized apartment buildings. Gillette remained an ardent propagandizer of socialist utopias, but his enduring legacy was a new way of shaving. In 1895, he conceived the idea of the safety razor after being prodded by his boss to come up with the commercial equivalent of the bottle cap: something to be used once and then thrown away. His prototype razor contained the two key elements of his system of shaving: a disposable blade that obviated the need for sharpening and a blade holder that prevented the accidental infliction of deep cuts.

As is often the case with successful inventions, the road from brilliant insight to successful product was a long and difficult one. Gillette had no experience in metal working, and metallurgists told him that putting a sharp edge on cheap sheet metal was an impossibility. He was fortunate to have the services of William Nickerson (a singularly inappropriate name for someone connected with shaving), whose efforts eventually provided the machines that hardened steel strips and turned them into sharp blades. In October 1903, 8 years after Gillette's original insight, the Gillette Safety Razor Company sold 51 razors and 168 blades. But only a year later, the company sold 91,000 razors and more than 2 million blades. Within a few years, Gillette was a wealthy man, with sufficient leisure time to promulgate his socialist ideas.

Gillette's safety razor still required some handling

of the blade, a task that was eliminated by Jacob Schick's blade-injector razor. The next logical step was a razor that was simply discarded when the blade lost its sharpness. In the United States today, 60 percent of shaves using blades are done with disposable razors. King Camp Gillette's political ideals were animated by a belief in the superior efficiency of socialism. Were he alive today, he might note with dismay that his razor had evolved into a key artifact of the "throwaway economy."

Reaper

For millennia, grain was manually harvested using nothing more sophisticated than a sickle. During the Middle Ages, the scythe brought some gains in productivity, although it was used more for mowing hay than for harvesting grain. Although still a hand tool, the scythe allowed the user to stand upright and reap a larger area than could be done with a sickle. A further improvement was the cradle scythe, an 18th-century invention that combined several parallel blades, allowing a larger area to be cut with each pass. Although far from ideal, manual harvesting was adequate in places where labor was abundant and the area to be harvested small. But this was not the case in 19th-century America. As settlers moved into frontier areas, vast tracts of land came under cultivation. Relative to the size of the land, the country was underpopulated, making agricultural labor scarce and expensive.

Precursors to the reaper date back to the ancient Gauls, but it was not until the 19th century that inventors seriously took up the challenge of mechanical harvesting. Two Americans stand out, Obed Hussey (1792–1860) and Cyrus McCormick (1809–1884), although the reapers they designed may have been based on a mechanical reaper built in Great Britain by Patrick Bell in 1828. Hussey received a patent for his reaper in 1833, and McCormick received his a year later. Both reapers required considerable modification, and it was not until the mid-1840s that they began to be used in significant numbers. Ultimately, McCormick's was the more successful of the two. Some of the reason for success was technical, due in part to McCormick's willingness to incorporate improvements bought or borrowed from others, whereas Hussey stubbornly depended on his ideas alone. The mechanical reaper was a combination of basic elements: a divider to separate the stalks entering the machine, a reel to push them down, a reciprocating knife to cut them, and a platform to catch them. It was pulled by a horse harnessed in front and to one side. The horse's power also ran the various components through a system of gearing.

Although it was a good design, the McCormick reaper rose to a position of preeminence largely as a result of superior manufacturing and marketing strategies. At first, McCormick licensed his design to other manufacturers, but the poor quality of many of their products led him to fear for the reputation of his own design. In 1847, he built a factory in Chicago; in a dozen years it was producing 12,000 reapers annually, and by the end of the century it had become one of America's largest manufacturing enterprises, employing more than 4,000 workers.

McCormick's firm also took into account its customers' financial circumstances. The reaper was an expensive item, and farmers were understandably reluctant to commit a significant portion of their income to the new device. To counter this reluctance, McCormick advertised heavily in agricultural periodicals, making abundant use of testimonials from satisfied customers. His reapers were exhibited at rural fairs, often in conjunction with field trials that showed their superiority against competitors. Perhaps most importantly, McCormick's agents extended credit to purchasers. In 1849, for example, cash-short farmers could buy a $120 reaper for a $30 down payment, with the final payment due on December 1, after the customer had brought in the harvest and had a supply of ready cash.

Economic considerations were always paramount in the decision to purchase a reaper. It has been argued that the primary consideration in whether or not to buy a reaper was the size of a farmer's acreage. The amount of land available for wheat farming was highly significant, because the purchase of a reaper could only be justified if there was sufficient land on which to use it. Historical trends favored the diffusion of the reaper beginning in the middle of the 19th century, as farm size expanded due to the availability of vast tracts of land opened up by America's westward

An early version of McCormick's reaper (courtesy State Historical Society of Wisconsin; negative number WHi [x3] 26155).

expansion. Sales of the reaper were also stimulated by rising grain prices, a trend that was reinforced by the emergence of a national and even international grain market as expanding railroad and steamship networks allowed the shipment of wheat over great distances.

The mechanical reaper made a major contribution to increased wheat production in the United States during the second half of the 19th century. Due in part to productivity improvements effected by the reaper, the acreage sown to wheat in the United States nearly doubled from 1866 to 1878. At the same time, the expansion of wheat acreage ensured a large market for labor-saving agricultural implements like the reaper. The use of the reaper and the cultivation of new lands formed a reciprocal relationship: The seemingly boundless land of 19th-century America stimulated the diffusion of the reaper, while at the same time the reaper made it possible to turn virgin land into productive acreage.

Recording, High-Fidelity

The term *high fidelity* in the context of music reproduction traditionally referred to equipment that represented an advance over what has been commonly available. Harold Hartley, an innovative British designer of equipment for musical recording and reproduction, claimed to have been the first to use the term in 1926, when he declared the advances of the day to have achieved high fidelity. The achievements in technology that prompted him to coin this term—if he did indeed coin it—were the introduction of electronic amplification and loudspeakers using a voice coil and a magnet. Hartley's phrase gave tribute to record players that could reproduce frequencies from about 60 to about 5,000 Hz (Hz is the unit for cycles of oscillation of a wave per second), with an s/n ratio (signal-to-noise ratio) of perhaps 30 dB and distortion not much better than 10 percent. This quality of reproduction is about what is obtained from telephone reception today.

By these standards, most home stereo equipment is definitely high fidelity, although the term is still used to refer to equipment better than the norm. The modern standard of high-fidelity reproduction for these dimensions is a frequency response of 20 to 20,000 Hz, with a 90-dB signal-to-noise ratio, and distortion under 0.1 percent for the electronics, and around 1 percent for loudspeakers (at moderate sound levels and middle frequencies). These are not the only qualities we now require for high-fidelity reproduction; stereo imaging, dynamic range, and transient response are also important. Before going any further, however, we should examine the meaning of these three traditional basic measures of fidelity of reproduction.

The human auditory system is said to respond to frequencies from 20 to 20,000 Hz. However, this is a very generous and approximate figure; the bottom octave (20–40 Hz) is more felt (through resonances in the lungs and other parts of the body) than heard, and very few adults can hear frequencies above 16,000 Hz. Furthermore, human hearing is not uniform over the frequency domain, with dips and peaks in response throughout, and a roll-off at the frequency extremes such that a sound at 100 or 16,000 Hz can require 100 times greater sound pressure level than a 3,000-Hz tone to equal it in loudness.

Despite these limitations of hearing, listeners do notice improvements as the frequency range of reproduction extends to these limits. Even more, they notice improvements in uniformity of frequency response over the range. Uneven frequency response can create very unnatural sound. The range of frequencies needed for reproduction of musical fundamentals is from 32.7 Hz (the lowest C on an organ) to 4,186 Hz (the high C on a piano). However, the overtones of these notes extend to (and beyond) 20,000 Hz, and they are essential for musical reproduction because it is the overtones, not the fundamentals, that give the instruments their characteristic sounds. Indeed, without the overtones a trumpet and a clarinet would be indistinguishable.

The s/n ratio refers to the difference between the residual noise level of the equipment and the most intense signal it can produce. The noise is electronic hiss, hum, and the noise inherent in the use of the storage medium of records, tapes, or CDs. Obviously, any such noise does not belong in the music, and 90 dB is adequate to make it nearly inaudible.

Often the s/n ratio is confused with dynamic range, which is the difference between the loudest and quietest sounds a music system can produce. Dynamic range in most cases is limited by the loudest sound a system can produce without distortion, not the s/n

ratio. An $^s/_n$ ratio of 90 dB translates into a dynamic range of 130 dB, since the equivalent of 0 dB in the home listening environment is about 40 dB above absolute threshold, even in the quietest of homes. The upper limit of 130 dB, which corresponds to a level 90 dB above this point, is far beyond the capabilities of available audio systems. If a loudspeaker can produce 90 dB of sound pressure with a 1-watt input (a reasonable figure), it would require 10,000 watts to produce 130 dB.

Distortion is a measure of the way the signal is modified, other than in amplitude, by its passage through a system or a component. Harmonic distortion is generated when the shape of the wave is modified. It is termed *harmonic* because changes in the shape produce a host of new frequency components in the signal, all at integral multiples (harmonics) of frequencies already present. Intermodulation distortion is the set of combination and difference frequencies produced by the interaction of frequencies in the signal when they pass through nonlinear amplification stages.

Transient response is important for accurate reproduction. The characteristic sound of instruments depends on their acoustical properties at onset as much as on the overtones, and percussive instruments depend almost entirely on transients. Poor transient response can also show up in the form of "boomyness" or mid-range resonances that color the music in various ways. These resonances are easiest to hear in reproduction of the human voice, to which they give an inappropriate nasal coloration, or a "chesty" sound, or some other coloration.—W.B.

See also AMPLIFIER; COMPACT DISK; PHONOGRAPH; RECORDS, LONG-PLAYING; SIGNAL-TO-NOISE RATIO; STEREOPHONIC SOUND

Further Reading: Technical and practical information can be found in magazines devoted to high fidelity, notably, *Audio*, *Stereophile*, and *Sensible Sound*.

Records, Long-Playing

Although the first phonograph records were cylindrical, by the early 20th century the disk was the most common form. The earliest records were made from wax compounds. During the first decade of the 20th century, phenol resins and celluloid supplanted wax. In addition to being more durable, these substances allowed more grooves to be cut into a record, increasing its playing time. These early plastics also made for clearer and louder recordings, an important advantage in the days before the electrical amplification of sound. Records made from these substances then gave way to so-called shellac records, pressings made from a mixture of shellac, fillers such as slate or limestone, lubricants, and binders.

Records of this sort had a number of deficiencies. They were fragile and plagued by surface noise. Sound reproduction deteriorated with use; a record was only good for 75 to 125 plays before it was effectively worn out by the phonograph needle continuously gouging the surface. These records also had a playing time of only a few minutes. Listeners had to constantly change records, which was especially annoying when a lengthy piece like a symphony or opera was being performed. Record albums were exactly what their name implied: bound collections of individual records, each in its own jacket.

Phonograph records underwent a dramatic improvement in the years immediately following World War II. One important improvement was the use of polyvinyl chloride (PVC) as the material basis of records. In addition to being much more resistant to breakage, PVC could have more grooves cut into it, as many as 260 per inch instead of the 100 grooves per inch that was the maximum for shellac records. The grooves had a bottom width of only .0025 cm (.001 in.), hence the term *microgroove*. They required much less tracking force from the pickup at the end of the tone arm; as a result, they wore more slowly and produced less surface noise.

The first commercially successful long-playing (LP) record was developed at Columbia Phonograph Company by a team led by Peter Goldmark (1906–1977). It rotated at $33\frac{1}{3}$ rpm instead of the traditional 78 rpm, further extending the playing time. After amassing a collection of long-playing recordings and collaborating with Philco in the design of an inexpensive phonograph to play the new disks, Columbia unveiled the LP record in 1948. It was not able to standardize the format of the long-playing record, however, for in the following year RCA began to market its own microgroove recordings.

With a diameter of 17.8 cm (7 in.), RCA's records were smaller than Columbia's 30.5-cm (12-in.) LPs. They were further distinguished by their playing speed of 45 rpm and their large spindle hole.

The resulting "battle of the speeds" (which was further complicated by the introduction of even slower-turning 16-rpm LPs) retarded the adoption of microgroove records, as many consumers delayed purchases of new phonographs and waited to see which would emerge as the standard. When 78-rpm records and their players hung on for a number of years, many recording companies had to produce records of popular songs in both 45- and 78-rpm formats until the late 1950s. New record players were burdened by the complication of multiple-speed turntables to accommodate the variety of records being sold. In the end, $33\frac{1}{3}$- and 45-rpm records settled into distinctive niches, the former being used for long-playing "albums" and the latter for shorter-playing "singles." In the 1980s, the issue became moot as first audiocassettes and then compact disks rendered traditional records obsolete.

See also AMPLIFIER; BAKELITE; CELLULOID; COMPACT DISK; PHONOGRAPH

Further Reading: Andre Millard, *America on Record: A History of Recorded Sound*, 1995.

Rectifier

A rectifier allows current to pass in one direction and inhibits the flow of current in the other direction. Its main application is converting alternating current into direct current. The simplest rectifier is a diode. It has the drawback of only converting a portion of the current that passes through it, that is, a half wave of each cycle of alternating current. A full wave of alternating current can be graphed in this manner:

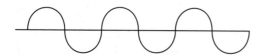

When it passes through a diode, the result is half-wave rectification:

Because of the inefficiency of this form of rectification, most power rectifiers include four diodes, resulting in the following waveform:

The first thermoionic tube, a rectifying diode, was used to receive radio signals early in the 20th century, a time when radio had become the focus of expanding scientific and commercial interest. Vacuum-tube rectifiers were also used in the power-generating industry as a result of Saul Dushman's (1883–1954) invention in 1915 of a vacuum-tube rectifier capable of handling high-voltage electricity.

Prior to the development of vacuum-tube rectifiers, the conversion of alternating current to direct current was done by mercury arc rectifiers. Invented in 1902 by Peter Cooper Hewitt (1861–1921), these devices were spinoffs from a type of arc lamp he had invented a year earlier.

Most vacuum tube and mercury arc rectifiers were rendered obsolescent in the early 1930s, when rectifiers based on polycrystalline metals such as selenium were introduced in Germany. By 1954, monocrystalline semiconductor rectifiers were in commercial production. At first based on germanium, later rectifiers of this type were based on silicon. Another type of rectifier, a mechanical device that employed a motor-driven, rotating switch was less easily dislodged, for it operated at higher efficiencies than existing solid-state rectifiers and could be used for currents of up to 600 volts. But it too was eventually displaced by another solid-state device, the silicon-controlled rectifier (SCR), which had been commercially introduced in 1957.

Cheap, reliable rectifiers are essential components of many products and processes. They provide the direct current necessary for the operation of computers, reversible electric motors, high-voltage electroplating, and automobile electrical systems. Alternating current is more efficiently transmitted than direct current, but direct current is necessary for many applications. Rectifiers make possible the combined use of both of these types of current.

See also ALTERNATING CURRENT; DIODE; DIRECT CURRENT; INVERTER; THERMOIONIC (VACUUM) TUBE

Recycling

The production of large quantities of refuse is one of the hallmarks of a high-income industrial society. All human societies have produced trash, but not in the massive volumes that are commonplace today. Precise figures are elusive, but it estimates of annual solid waste production in the United States run from 5.4 billion metric tons (6 billion tons) to 9 billion metric tons (10 billion tons). This amounts to nearly 60 kg (132 lb) of solid wastes per person each day.

Industry and agriculture account for about 98.5 percent of America's wastes. The remainder, the wastes generated by homes and businesses in the vicinity of urban areas, is known as municipal solid waste. About 73 percent of this waste is dumped in pits and then covered with dirt. This is an adequate solution in most places, at least for now and the near future. Incineration is sometimes used in places where potential landfill is scarce or there are other objections to burying wastes. Incineration presently accounts for 14 percent of the waste disposal in the United States. It is far from being the ideal method, however. An incinerator leaves substantial residues; about 30 percent of the weight of incinerated material remains as ash, and the ash has to be buried in a landfill. Some of the ash consists of toxic materials. Even an incinerator that meets current environmental regulations typically emits 4.5 metric tons (5 tons) of lead, 263 kg (580 lb) of cadmium, and 774 metric tons (853 tons) of sulfur dioxide each year. Incineration is an expensive proposition, costing $400 to $550 per ton of waste. (In comparison, using landfill for waste disposal costs $60 to $270 per ton.)

The remaining 13 percent of refuse is recycled. Strictly speaking, recycling refers to the reprocessing of discarded materials so that they can be made into new products. In this sense, recycling is different from *reusing*, which does not entail the transformation of the refuse; a common example of reusing is the refilling of previously used soft-drink bottles. Recycling sometimes entails the use of discarded materials for their original purpose, as when an aluminum can is melted down and made into another can. Recycling results in a completely new use for the material, for example, grinding up discarded tires and using them as a fuel.

Four major categories of materials are commonly recycled: glass, paper, metals, and plastics. Glass has the advantage of being readily recycled, for it is relatively homogenous and easily processed. In the United States, about 5 billion glass jars and bottles are recycled annually, and 25 to 35 percent of the glass used is derived from recycled materials.

Paper is an excellent candidate for recycling because it is so abundant, comprising about 38 percent of municipal solid waste. Recycling can save some of the 2 million trees that are felled every day for the production of paper products. It also reduces energy consumption as well as the effluents that pollute air, water, and land. Paper is even more extensively recycled than glass; up to 40 percent of all newspapers are recycled. There is a limit to paper recycling, however, as the fibers in the paper are broken by the pulping process. Consequently, paper becomes unusable after being recycled six to eight times.

Many kinds of metal are recycled. The most commonly recycled metal is steel—more than 45 billion kg (100 billion lb) annually. As with other forms of recycling, the recycling of steel confers a number of environmental benefits. Compared to the manufacture of steel from raw minerals, the production of a ton of recycled steel eliminates 90 kg (200 lb) of atmospheric pollutants, 45 kg (100 lb) of water pollutants, 2.7 metric tons (3 tons) of mining wastes, and the use of 22,700 metric tons (25,000 tons) of water. Equally impressive are the environmental gains from the recycling of aluminum. The manufacture of aluminum from bauxite ore is a highly energy-intensive process, and 90 percent of this power is saved when aluminum comes from recycled sources.

In the mid-1990s, 65 percent of aluminum beverage cans were being recycled. In contrast, only about 1 percent of the 6.3 billion kg (14 billion lb) of plastics used annually in the United States is recycled. Plastics are hard to recycle because they come in so many varieties, and mixing them together results in a worthless material. Therefore, successful recycling of plastic materials requires that they be separated according to type. To aid in this, many manufacturers have voluntarily agreed to stamp their products with codes that indicate the type of plastic used. There currently are seven numerical codes for specific categories of plastic: 1 for polyethylene terephthalate (PET); 2, high-density polyethylene; 3, vinyl or polyvinyl chloride (PVC); 4, low-density polyethylene; 5, polypropylene; 6, polystyrene; and 7, all other types of plastics and mixtures of different types of plastic. At present, only the first two categories are recycled to any significant degree. In the future, new processes that reduce plastic polymers to monomers will aid in the recycling of plastics, but these methods require high temperatures and the use of considerable amounts of energy.

Animal and plant remains form a large portion of the wastes requiring disposal. Much of this waste material can be composted and used to improve the soil. Composting is a venerable technology that converts organic wastes into humus that enriches the fertility and texture of soils.

Successful recycling requires the application of the right technologies; no less important is the presence of adequate economic incentives. When a material commands a high price, there will be a strong incentive to recycle it. This can be clearly seen in the case of aluminum cans, which have consistently fetched a price that made their recycling economically worthwhile. In contrast, the demand for wastepaper has fluctuated, so in some years old newspapers have piled up in warehouses, while in other years they have been stolen from curbsides and pickup centers. Finding or creating a market for recycled materials is known as "closing the circle," and it is often a more challenging task than getting people to separate aluminum cans, glass bottles, and garden wastes. On occasion it may be necessary to use tax breaks and other governmental incentives to encourage recycling, but it must be remembered that mining, lumbering, and other raw-materials industries also have benefited from government favors.

Many consumers try to support recycling by favoring products that are supposedly made from recycled materials. Such behavior is laudable, but consumers should be aware that there are no federal standards that guarantee that a product is what its label claims. In some cases, the addition of only a small fraction of recycled material is sufficient as far as the manufacturer is concerned.

Recycling is not the only way of reducing the amount of waste. Also valuable is "precycling," the reduction of packaging materials by manufacturers. In addition to avoiding the use of more packaging than the product requires, precycling can take the form of selling products in a concentrated form. Wastes also can be reduced through "source reduction," i.e., using different materials, changing manufacturing processes, and in general making do with less.

Some strident environmentalists are of the opinion that recycling is at best a halfway measure that just reinforces the "throwaway society." But since most people have no desire to sharply reduce their material standard of living, recycling allows a substantial reduction of the wastes that inevitably accompany high levels of consumption.

See also ALUMINUM; COMPOSTING; PAPER; POLYMERS; TIRE, PNEUMATIC; WASTE DISPOSAL

Further Reading: Michael L. McKinney and Robert M. Schoch, *Environmental Science: Systems and Solutions*, 1996.

Reflecting Telescope—see Telescope, Reflecting

Refracting Telescope—see Telescope, Refracting

Refrigerator

It has long been known that the spoilage of foods could be retarded by keeping them cold. For centuries, the most common way of accomplishing this was through the use of manufactured or naturally occurring ice, although in a number of places evaporating brine was used for this purpose. In the 19th century, the first mechanical devices for keeping things cold were developed, and in the course of the 20th century they became an essential part of everyday life.

Today's refrigerators work on a simple principle: Liquids absorb heat when they vaporize. In a refrigerator, the vaporization of a liquid (known as the *refrigerant*) cools the interior and its contents. The vapor is subsequently turned back into a liquid, producing heat that is released into the surrounding environment. This is done by a motor-driven compressor, which is turned on and off by a thermostatically controlled switch.

In 1834, Jacob Perkins (1766–1849), an American

inventor living in London, built a refrigerator of sorts that used this basic process. Although Perkins's invention had no commercial consequences, the rapidly expanding brewing and meat-packing industries stimulated a great deal of subsequent experimentation. Massive but reasonably effective refrigerators were built in France by Ferdinand Carré (1824–1900) in the 1860s. Carré's refrigerators were built on the absorption principle; they employed a refrigerant (in most cases ammonia) that was vaporized by heat. The ammonia was then absorbed into water, causing it to condense and cool the interior of the refrigerator. The absorption process caused pressure differentials within the system, allowing the refrigerant to circulate without the need for a pump of any sort.

Technically and commercially successful refrigerators were built in Germany from 1873 on by a firm founded by Carl von Linde (1842–1920). Used primarily in breweries and other commercial establishments, these refrigerators worked on the compression principle, using ammonia as the refrigerant.

Ammonia was an effective refrigerant, but it is toxic. A number of other refrigerants were used in the early history of refrigerators: methyl chloride, diethyl ether, and sulfur dioxide, all of which had their drawbacks. The standard refrigerant for many years was a fluorinated derivative of methane that was developed in the late 1920s by Thomas Midgely, Jr. (1889–1944), of the General Motors Research Laboratory (one of the divisions of General Motors was Frigidaire, a leading maker of refrigerators). Although its adverse consequences for the Earth's ozone layer have led to its demise in recent years, this product, usually known by its most common trade name Freon, combined a low boiling point (−29.8°C) with noninflammability and nontoxicity.

Refrigerators also benefited from the development of improved compressors and the electric motors that powered them. Another significant improvement came in the 1920s when air cooling replaced the stream of cold water that had been used to condense the refrigerant. Electrical power demands increased, but this worked to the benefit of General Electric and Westinghouse, key manufacturers of refrigerators who were also interested in advancing the revenues of central generating stations, important customers for the power-generating equipment they also produced.

While sales increased for refrigerators employing the vapor-compression cycle, refrigerators employing the absorption principle were neglected by electrical manufacturers. They were uninterested because these refrigerators, did not need an electric motor; in fact, they had hardly any moving parts. The absorption

The refrigeration apparatus used by a Munich brewery toward the end of the 19th century (from M. Hard, *Machines Are Frozen Spirit*, 1994).

principle was developed for domestic refrigerators in Sweden in the 1920s by Baltzar von Platen and Carl Munzer. Since absorption refrigerators had few moving parts, there was little to go wrong, and the absence of a compressor removed a source of noise. But technical capability was not translated into commercial advantage, due in part to the inability of the firms that made absorption machines to promote them effectively. One American manufacturer continued to make full-size absorption refrigerators until 1956. Today refrigerators of this sort are only used where electricity is not available, as in recreational vehicles.

Although domestic refrigerators were in commercial production by the second decade of the 20th century, it was not until 1930 that the sales of mechanical refrigerators in the United States exceeded those of old-fashioned iceboxes. In the Depression years that followed, many people could not afford expensive items like refrigerators, and many homes were still without electricity. Even so, sales increased steadily. Four million were sold in 1941, and by 1944 nearly 70 percent of American homes had refrigerators.

Features like separate freezing compartments, introduced by General Electric in 1939, helped to make the refrigerator a desired appliance. At the same time, effective marketing solidified its position as a highly sought-after consumer product. Electric utilities promoted refrigerators as a means of selling more electricity, while aggressive price competition by their manufacturers brought them into increasing numbers of homes. In 1920, the average refrigerator was priced at $600, but this declined to $275 in 1930 and to $154 in 1940. Not content to expand the market through offering lower prices, manufacturers and dealers made abundant use of credit plans as well as the shady practice of offering "loss leaders" in order to attract prospective customers into the store so they could be sold on higher-priced models. Styling also became an important marketing tool, with refrigerators undergoing regular styling changes, much like the automobiles of the time. In the mid-1930s, the "monitor top" that housed the compressor was removed, and the machinery was moved to bottom of the refrigerator, giving the refrigerator a more streamlined form.

In recent years, a heightened concern for energy efficiency has resulted in the manufacture of less-wasteful refrigerators, primarily through the use of more-efficient motors and compressors, and the addition of more insulating material. Since 1970, the annual cost to run a refrigerator has declined by $64. Since the refrigerator is typically the largest consumer of electricity in the home, and absorbs 7 percent of total electrical power, even small gains in efficiency can significantly reduce energy use.

See also CHLOROFLUOROCARBONS; ENERGY EFFICIENCY; ICEBOX; MOTOR, ELECTRIC; STREAMLINING

Further Reading: Barry Donaldson and Bernard Nagengast, *Heat and Cold: A Selective History of Heating, Ventilation, Refrigeration, and Air Conditioning*, 1994.

Refrigerator Car

Domestic and commercial iceboxes were widely used for the preservation of food during the 19th century. It therefore did not take a great leap of imagination to put a large icebox on freight car trucks in order to transport perishable items by rail. From 1850 onwards, numerous inventors in the United States devised serviceable refrigerator cars that hauled fruit, vegetables, and meat far from their places of origin. These cars resembled the wooden boxcars of the era, but they were fitted with ice bunkers, usually located at the ends of the car. The walls were insulated with inner layers of felt, hair, or sawdust. Early efforts at insulation were only partially successful, as the wooden cars had many sources of air leaks. Temperatures within the cars were uneven, as areas close to the ice bunkers tended to be colder than areas toward the middle of the car. These were not crippling shortcomings, and early refrigerator cars helped to transform the American diet by making a wide variety of foods available to consumers wherever they lived.

The greatest impetus for the development of early refrigerator cars came from the meat-packing industry. Entrepreneurs like Gustavus Swift (1839–1903) and Philip D. Armour (1832–1901) realized that it was considerably cheaper to ship dressed meat to market than it was to haul livestock to slaughterhouses situated close to consumers. Refrigerator cars allowed the concentration of the meat-packing industry in places like Chicago and Cincinnati. They also facilitated the oligopolistic concentration of the industry as a whole. Most of the

Manually icing a train of refrigerator cars (courtesy Union Pacific Museum Collection).

cars that moved dressed meat were owned by the packing companies themselves, for the railroads were reluctant to acquire refrigerator cars, fearing that the expansion of the dressed-meat industry would undermine their profitable livestock transportation business. After the first decade of the 20th century, the railroads began to perceive that the hauling of meat and produce could contribute to their revenues, and they began to build up their own fleets of refrigerator cars.

Refrigerator car technology changed little during the first half of the 20th century. Cars became larger and better insulated, but cooling was still provided by ice and salt carried in bunkers and loaded through hatches on the roof. This was a labor-intensive process, and it required numerous stops; a train bound from Florida to New York typically required five shops to re-ice its cars. Moreover, conventional refrigerator cars provided no way of regulating temperatures, so when the weather outside got cold, fruits and vegeta-

bles could be damaged by excessively low temperatures.

While the railroads contented themselves with the operation of ice-cooled refrigerator cars, in the 1930s the trucking industry began to operate trucks and trailers cooled by mechanical refrigeration. Offering convenient door-to-door delivery, mechanically cooled trucks began to cut into rail traffic. Manufacturers of railroad cars had built a few dozen mechanical refrigerator cars during the mid-1920s, but the railroads balked at purchasing them, citing the heavy weight of the cars and the inability to use them in any other kind of service. Also, railroads were getting mixed signals from shippers. Distributors of fresh fruits and vegetables had no interest in cars that brought temperatures below what was needed for transporting their produce. At the same time, however, ice cooling was inadequate for the haulage of frozen food, a consumer product that was gaining in popularity during the 1930s.

The refrigeration units used in early mechanical refrigerator cars were powered by the cars' axles as they turned. Power was transmitted directly or it was used to turn an electrical generator. These systems were adequate as long as the car was rolling, but once it stopped, a supplementary source of power was required, adding complication and expense. Efficient and economical mechanical refrigeration was made possible by the development of self-contained refrigeration units powered by gasoline or diesel engines. The first modern mechanical refrigerator car in the United States took to the rails in 1949, and the fleet grew slowly in the years that immediately followed. During this time, the advantages of mechanical refrigeration became evident. Although its initial costs were higher, the operating expenses of a mechanical refrigerator car were lower than those for cars cooled with ice. The mechanical equipment took up less space than the ice bunkers of conventional cars, allowing for a 20 percent larger cargo area. Mechanical refrigerator cars also were lighter than cars filled with ice, which trimmed operating expenses. Finally, mechanical refrigeration eliminated the need for regular icing stops, further cutting costs and improving schedules. As the advantages of mechanical refrigeration were becoming evident, the ice-cooled cars were slowly wearing out, and by the late 1960s they were virtually extinct as far as U.S. railroads were concerned.

See also ENGINE, DIESEL; ENGINE, FOUR-STROKE; FROZEN FOOD; ICEBOX

Further Reading: John A. White, *The Great Yellow Fleet*, 1986.

Regenerative Circuit

In the early years of the 20th century, radio took on growing importance as a practical means of communication and as a source of instructive entertainment for growing numbers of hobbyists. Radio, however, suffered from many imperfections. Two fundamental problems were an inability to transmit strong signals and, in parallel fashion, a difficulty in receiving them.

Reception was improved considerably through the use of early vacuum tubes, first John Ambrose Fleming's (1849–1945) diode and then Lee de Forest's (1873–1961) "audion," the first triode. In 1912, the value of the vacuum tube was greatly enhanced when Edwin Howard Armstrong (1890–1954) invented the regenerative circuit. Armstrong's invention converted the vacuum tube from a mere detector of radio signals into a device that amplified them as well. This was done by creating a circuit that fed oscillating current from the triode's plate back to its grid. It was a very important discovery, but there was more: In addition to amplifying the signal, the regenerative circuit also could be used to make a tube operate as an oscillator for the transmission of radio signals. It generated a much stronger signal than the devices in use at the time, and soon rendered them obsolete.

Armstrong's claim to the invention did not go unchallenged. As sometimes happens in the history of invention, the regenerative circuit, or something very much like it, had been devised independently by several other radio experimenters at about the same time. De Forest himself claimed to have invented the regenerative circuit before Armstrong. He probably did so, although he had no appreciation of its significance at the time. Nor did he understand how it worked; he had to examine Armstrong's patent application to get some idea of the underlying principle. Nonetheless, de Forest was determined to secure the rights to this invention. A battle for patent rights was initiated in federal court in early 1921 and finally resolved in de Forest's favor, albeit on dubious legal and technical grounds, by the U.S. Supreme Court in 1928. Another series of court cases, this time initiated by Armstrong in the name of a radio manufacturer that used regenerative circuits in its products, was not resolved until 1934, again in de Forest's favor. All in all, de Forest and Armstrong's battle had entailed 13 separate court decisions and vast expenditures of time and money. For the great majority of electronic professionals who sided with Armstrong, the legal outcome illustrated the capriciousness of the patent system when priority of invention was left to the courts. The experience was particularly galling to Armstrong, and it affected his later behavior when he tried to defend his proprietary rights to another one of his epochal inventions, frequency modulation (FM).

See also AMPLIFIER; FREQUENCY MODULATION (FM); PATENTS; RADIO; THERMOIONIC (VACUUM) TUBE

Further Reading: Tom Lewis, *Empire of the Air: The Men Who Made Radio*, 1991.

Relativity, General Theory of

General relativity is Albert Einstein's theory of gravitation. The development of this theory, culminating in its publication in 1915, is one of the most creative achievements by a single individual in the history of science. General relativity is not only our current theory of gravity; it is also the framework in which we are coming to understand the large-scale structure of the universe, together with its history and its destiny.

After introducing the special theory of relativity in 1905, Einstein realized that Newton's inverse-square law of gravitation cannot be correct, as it is inconsistent with special relativity. The law implies, for example, that the gravitational influence of the sun is communicated to planets instantaneously, and not at the speed of light or less, as special relativity requires. Einstein therefore set out to find a relativistic theory of gravity. He also wanted a theory in which the fundamental laws of physics could be used by observers in any frame of reference, including accelerating frames, and not only by those in inertial frames, the frames in which special relativity is valid. With the general theory of relativity he accomplished both goals.

An important step was the *principle of equivalence*. This principle maintains that all physical phenomena taking place within a box freely falling in a uniform gravitational field are equivalent to physical phenomena within a similar unaccelerated box sitting at rest, far from any gravitating bodies. Einstein proposed that through no experiment whatever—whether mechanical, optical, biological, etc.—can the experimenter distinguish one box from the other as long as the experiment is done entirely within a box. One can show from the equivalence principle that inertial mass, the mass in Newton's law $F = ma$, must be the same as gravitational mass, the mass in the gravitational-force law $F = mg$. Experiments have confirmed the equivalence of the two kinds of mass to a high degree of accuracy. One can also use the principle to show that high-altitude clocks run fast compared with low-altitude clocks, a prediction confirmed using very sensitive atomic clocks near the Earth's surface.

Einstein thought also about measurements made in uniformly rotating frames. He argued that creatures living on a rotating turntable would find that their length measurements are inconsistent with Euclidean geometry. For example, they would find that the circumference of a circle of radius r about the center of the turntable is greater than $2\pi r$, since the meter sticks they would use to measure circumference would be Lorentz-contracted from the standpoint of outside inertial observers, a consequence of his special theory.

It had already been known for over a half-century that non-Euclidean geometry corresponds to the geometry of curved surfaces. In fact, the 19th-century mathematicians Johann Karl Friedrich Gauss (1777–1855), Georg Friedrich Bernhard Riemann (1826–1866), and others had invented differential geometry, in which the curvature of space differs from one point to another, and the local geometry depends on position as well. The local geometry is characterized by what is called a *metric*, a formula for the infinitesimal distance between two neighboring points, a generalization of the Pythagorean theorem. Geometric properties of the surface can be derived in terms of the metric. For example, the metric is used to find geodesics on a surface, which are the shortest-distance paths between two given points, a generalization of straight lines on a flat surface and great-circle routes on a sphere. The metric is also used to find the local curvature of a surface, as given by quantity called the Riemann curvature tensor. The surface is flat if and only if every component of the Riemann tensor is zero.

It had previously been shown that special relativity is naturally understood in terms of a *four*-dimensional space-time with coordinates x, y, z, and t, a combination of the three spatial dimensions x, y, z, and time. After Einstein had convinced himself that gravity is related to accelerated frames through his principle of equivalence, that non-Euclidean geometries apply in accelerated frames, and that non-Euclidean geometries are appropriate for curved spaces, he attempted a grand synthesis of these ideas. He proposed that gravity is not a force at all; what we observe as gravitational deflections are really the unforced motion of particles in a curved spacetime. A freely falling particle traces a geodesic in spacetime, the "straightest" path possible in the curved spacetime.

The curvature of spacetime is caused by mass and energy in any form, and also by pressures and shear stresses. The relationship between spacetime curvature and the matter and energy sources is governed by the

Einstein field equations $G_{ij} = 8\pi T_{ij}$. Here the subscripts i and j each range over 0, 1, 2, and 3, corresponding to the four dimensions of spacetime. The equation as written is a shorthand for 10 coupled equations, each a second-order differential equation in the metric. The right-hand side of the equation contains information about the matter, energy, etc., in the spacetime, and the left-hand side represents the space-time curvature.

Many exact solutions of these equations have been found. One of the more important is the spacetime surrounding a spherically symmetric, nonrotating massive object. To a good approximation the sun is such an object, so we can compare the predictions of general relativity with observations in the solar system. The theory predicts gravitational red shifts, in which light emitted from the sun is reddened as it escapes from the sun; this phenomenon has been observed from the sun and other stars as well. The theory also predicts that light rays should be bent as they pass close by the sun or other stars; in fact, the bending of a ray passing near our sun should be 1.75 seconds of arc. This prediction has been confirmed from photographs of stars taken during total solar eclipses, and also from radio telescope images of distant radio sources. General relativity also predicts a precession in the orbits of planets; in particular, the perihelion (the point of closest approach to the sun) of the planet Mercury should precess by 43 seconds of arc per century due to relativistic effects. That is, Mercury's elliptical orbit should slowly turn in space, so that perihelion should occur at a slightly different point from one revolution to the next. Such a precession was observed long before the theory was proposed. The successful explanation of these observations was an important early test of general relativity. Other effects are predicted as well; some have already been observed, and others are awaiting experimental verification.

The spherically symmetric solution also predicts the possible existence of black holes, objects from which light cannot escape. Black holes are thought to be formed by the gravitational collapse of stars or collections of stars. Many scientists believe that black holes lurk in certain observed double-star systems and in the nuclei of galaxies. Strong evidence for this has been found by studying the behavior of ordinary stars and other matter in these systems; they appear to be influenced by nearby black holes.

The field equations also predict the existence of gravitational waves, ripples in the space-time curvature that travel at the speed of light. Gravitational waves have been detected indirectly from the inspiraling of a double-star system; it is expected that they also will be detected directly by highly sensitive instruments constructed on the Earth and in space.

The largest spacetime is the universe as a whole. General relativity is one of the theoretical bases of the study of cosmology, our attempt to understand the universe on the large scale, including the Big Bang expansion and the question of whether the universe will expand forever or eventually start falling back again.—T.H.

See also BIG BANG THEORY; BLACK HOLES; DOPPLER EFFECT; GEOMETRY; GRAVITY; MECHANICS, NEWTONIAN; RADIO; RELATIVITY, SPECIAL THEORY OF; TELESCOPE

Further Reading: Albert Einstein, *Relativity: The Special and General Theory*, 1952. Stanley Goldberg, *Understanding Relativity*, 1984.

Relativity, Special Theory of

The special theory of relativity, introduced by Albert Einstein (1879–1955) in 1905, revolutionized our understanding of space and time. And because space and time form the arena within which physical particles and fields move, a changed understanding of space and time affects our understanding of much of the rest of physics as well.

The theory is built on two postulates. The first, called the principle of relativity, asserts that all inertial (i.e., unaccelerated) reference frames are equally valid for the description of fundamental physics. An observer at rest in an inertial frame moving at constant velocity relative to the first. The principle of relativity was not new with Einstein; we have long known, for example, that Newton's second law $F = ma$ is as valid on a constant-velocity train as it is at rest relative to the ground. The second postulate is that the velocity of light is independent of the speed of the source. The velocity of light emitted by a flashlight is the same from our point of view, no matter how fast the flashlight is moving. This second postulate was new.

Combining the two postulates leads to the conclu-

These results, and others as well, are concisely contained in the *Lorentz transformation* between two inertial frames. Consider two such frames, moving with respect to one another at uniform velocity V along their mutual x axes, the primed frame moving to the right relative to the unprimed frame. The transformation is

$$x' = \gamma(x - Vt),\ y' = y,\ z' = z,\ t' = \gamma(t - Vx/c^2)$$

where $\gamma = 1/\sqrt{1 - V^2/c^2}$. That is, if an event takes place at point x, y, z, at time t as observed in the unprimed frame, then that same event takes place at point x', y', z', at time t' as observed in the primed frame. Suppose for example that a clock is placed at rest at the origin of the primed frame, so that $x' = y' = z' = 0$ for all t'. Then from the x' equation, it follows that $x = Vt$, confirming that the clock moves at velocity V in the unprimed frame. Then from the t' equation it follows that $t' = t\ \sqrt{1 - V^2/c^2}$, showing that t', the reading of the moving clock, is less than t, the reading of clocks at rest in the unprimed frame, by the factor $\sqrt{1 - V^2/c^2}$. This is the time-dilation effect:

$$\sqrt{1 - V^2/c^2}$$

Einstein in 1905 (from N. Spielberg and B. D.Anderson, *Seven Ideas that Shook the Universe*, 1987).

sion that the speed of light c is the same in all inertial frames. This remarkable conclusion is only the first in a series of remarkable conclusions. For example, one can show that clocks at rest in one inertial frame run slow from the point of view of observers in another frame, the phenomenon of *time dilation*. This slowing is valid for any clock, whether mechanical, electrical, or biological. Also a moving object is shortened in its direction of motion, the *Lorentz contraction* (named after the Dutch physicist who formulated it, Hendrik Lorentz [1853–1928]). A meter stick is moving past us lengthwise will be less than 1 meter long according to measurements made in our frame. One can furthermore show that simultanity is relative; i.e., if two events occur at the same time in one frame, the events generally do not occur at the same time to observers in another inertial frame. There is no universal time marching forward at the same rate from the point of view of all observers.

The Lorentz transformation states in effect that the arena of physics is not three-dimensional space plus time but is instead four-dimensional spacetime. There is no universal distinction between space and time. What is purely a spatial separation between two events in one frame, for example, (i.e., the events are simultaneous in that frame) is partly spatial and partly timelike in another frame. A number of supposed paradoxes have been proposed to try to show logical inconsistency of the theory, or simply to illustrate the strangeness of relativistic effects. The best known of these is the "twin paradox," in which twin A stays at home while twin B travels to a distant star and then returns. According to A, B's biological clocks have run slow, so B will be younger than A when B returns. The "paradox" is the assertion that one could look at the events from B's point of view, to whom A has been moving, so that A's clocks should run slow from B's point of view. Therefore A should be younger than B when they meet again. Their relative ages cannot, however, be in dispute when they are standing next to one another, so one of their stories must be false, seemingly in contradiction to the relativistic behavior of clocks. The resolution of the paradox is achieved by noting that the results of special relativity apply only to observers at rest in some inertial frame. If A is such an observer, then B must have accelerated for some portion of the trip, and so cannot use the results of special relativity during that period. Therefore A's story is correct; B is younger than A at the end of the journey, and there is no paradox.

Special relativity also changes our understanding of momentum and energy. According to classical mechanics, the momentum of a particle is $p = mv$, the product of the particle's mass and velocity. Both the velocity and momentum of the particle can be arbitrarily large if a force is exerted on the particle indefinitely. According to special relativity, the momentum is instead $p = \gamma mv$, where $\gamma = 1/\sqrt{1 - v^2/c^2}$. Although the relativistic momentum of a massive particle can become arbitrarily large if a force is continuously exerted upon it, the particle's velocity can only approach and never quite reach the speed of light. Only massless particles like photons move at light speed. It is widely believed that no signal can travel faster than light, because serious causality paradoxes could then arise. The device which emits such a hypothetical signal could be switched off before the signal was sent, by some effect of the signal itself.

In special relativity, the energy of a free particle is $E = \gamma mc^2$. If the particle is at rest in some inertial frame, then $\gamma = 1$ in that frame, so the particle's energy is $E = mc^2$, called its *rest energy*. Mass is therefore seen to be a form of energy. As the particle starts to move, its energy increases because γ increases. The energy of motion, the kinetic energy, is $(\gamma - 1)mc^2$, which reduces to the classical result $(1/2)mv^2$ if v is very small compare to c. Total energy is conserved in collisions and in the decay of unstable particles, but a portion of the mass-energy can be converted into kinetic energy, and vice versa. A free neutron is unstable, for example, decaying into a proton, an electron, and a neutrino. The neutron has more mass than the final three particles combined. The loss of mass-energy implies a gain of kinetic energy, shared among the final particles. A neutron incident upon a uranium nucleus can cause the

nucleus to split (or fission) into two pieces, while emitting several neutrons as well. About 0.1% of the mass-energy is converted into kinetic energy in this case, a sufficient fraction so that vast amounts of energy can be released from a small quantity of uranium if there is a chain reaction in which released neutrons induce additional fissions. Mass is also converted into kinetic energy in the form of heat in chemical reactions, including fossil fuel burning. In this case the percentage loss of mass is much smaller than in nuclear reactions.

Innumerable experiments have confirmed the predictions of special relativity, so that the theory has become a well-established and indispensable part of physics.—T.H.

See also LIGHT, SPEED OF; MECHANICS, NEWTONIAN; NUCLEAR FISSION; RELATIVITY, GENERAL THEORY OF

Further Reading: L. Pearce Williams, ed., *Relativity Theory: Its Origins and Impact on Modern Thought*, 1968. David Cassidy, *Einstein and Our World*, 1995. Martin Gardiner, *The Relativity Explosion*, 1976.

Remote Sensing

Remote sensing is used to learn about the Earth or other planets through techniques that do not entail actual contact with the surfaces of these bodies. It is called remote sensing because the data for the image are recorded at a distance from the area being observed. Although photography is one type of remote sensing, other kinds of images may be produced. Many kinds of remote sensing require the use of a computer to construct the images to be analyzed.

The steps involved in a remote-sensing project are data acquisition, data processing, and interpretation. The final product is an interpreted image that answers some question about an area, such as vegetation types and amounts, land use, topography, or surface geology. Photographic images are made through the exposure of light-sensitive chemicals. However, visible light is only one type of electromagnetic radiation, energy that moves with a harmonic wave pattern at the speed of light. Other types of electromagnetic radiation include radio waves, microwaves, infrared, ultraviolet, X rays, and gamma rays. These other types of electromagnetic radiation also can be used to make photographs; medical X rays are a common example. However, the images used for remote sensing are usually made from data that are digitally recorded rather than captured on film. The data are then processed to look like pictures, but they are not photographs. They can be re-processed to present the data in different ways in order to answer different questions.

Images have certain fundamental properties—scale, brightness, tone, texture, contrast, and resolution—that are independent of the type of energy recorded. Scale is the ratio between the distance separating two points on an image and the distance separating them in the real world. Brightness is proportional to the intensity of the electromagnetic energy recorded. Usually, it is displayed in direct proportion to the energy level recorded. Tone is distinguishable variation in intensity, and the variation may be calibrated to a scale useful in the desired interpretation. Texture is the rate at which tone changes. Contrast is the ratio between the most and least bright parts of an image. Contrast is critical to the resolution of the image or the ability to distinguish between two closely spaced objects on an image. These fundamental properties are combined to determine "detectability" or how small an object can be observed. The properties can be manipulated during image processing in order to enhance the image in ways that help with interpretation.

Images are collected from framed or scanned data. A photograph is an example of framed data; it represents an instant in time captured on film because the entire area has been photographed at once. A scanned image is produced through the use of a detector with a narrow field of view that is swept across the area in a pattern of parallel lines. The resulting information is then transferred to a computer that processes it into an image. In general, energy in a particular interval is called a *band*, and the bandwidth is the range of each interval. The multispectral data provided by scanning can be combined in various ways. If three bands are combined and displayed at the same time, with one band represented as the color red, one as blue, and the third as green, a color image is produced. These are easier for people to interpret, because the human eye can distinguish changes in color better than it can distinguish changes in shades of gray.

Common examples are the images produced from infrared data. The wavelengths of infrared radiation are 0.7 to 100 μm (1 μm [1 micrometer] = 0.000001 meter). The radiation is felt as heat, but the wavelengths are too large to be seen by the human eye. Different bandwidths within the infrared region correspond to different characteristics of the Earth, such as amount of energy absorbed by the chlorophyll in plants, the moisture content of soil, and the alteration of minerals in rock by hydrothermal fluids. Images for vegetation and land use studies frequently display combinations of different bands of infrared energy; these allow the identification of different types of vegetation and soils. The image will be in color, but it is not a color photograph. Images of this sort are often referred to as "false-color images" because the colors observed on the image do not correspond to their colors in visible light.

Interpretation and processing are closely related in remote sensing because an image usually is produced to answer specific questions. The interaction of matter and electromagnetic radiation is enhanced by the computer to make the interpretation as unambiguous as possible. Even so, the interpretation should be field checked, if possible, to ensure that the assumptions made in the processing and interpretation are correct.—D.A.B.

See also GAMMA RAYS; INFRARED RADIATION; MICROWAVE COMMUNICATIONS; PHOTOSYNTHESIS; RADIO; ULTRAVIOLET RADIATION; WAVE THEORY OF LIGHT; X RAYS

Further Reading: Floyd F. Sabins, *Remote Sensing*, 3d ed., 1997.

Research and Development

The organized and continuous pursuit of technological change has been one of the hallmarks of the industrial age. In the 19th century, individual inventors and entrepreneurs pioneered the new technologies that played a large role in creating the era we know as the Industrial Revolution. Towards the end of that century, Thomas A. Edison (1847–1931) was establishing his reputation as the most prolific inventor in an age of heroic invention. In addition to his unusual creativity and energy, Edison had another major advantage over his competitors: his well-staffed laboratory. Edison used his laboratories at Menlo Park and, later, West Orange, N.J., to test ideas quickly and build prototypes of new devices such as the lightbulb and phonograph. The research and development laboratory itself would prove to be one of Edison's enduring legacies.

Another early precedent for the modern research and development laboratory was established in the German dyestuffs industry in the 1870s and 1880s. Hundreds of chemists were employed to make and test thousands of chemicals to determine which ones had potential as dyes. By the 1890s, some of these chemicals would be used for other purposes such as pharmaceuticals. Aspirin was one such chemical developed by the Bayer Company. The research and development laboratory soon spread to other industries in Germany and the other countries.

In the second half of the 19th century in the United States, new industries such as railroads, steel, electric lighting, and chemicals were expanding rapidly. The complexity of these technologies and the large investment required to develop them had prompted entrepreneurs to create the modern industrial corporation with its legions of salesmen, accountants, lawyers, clerks, engineers, and general managers. Included in this group were a growing number of professionally trained scientists, usually chemists and physicists, who were employed to improve the properties of products and the efficiency of the processes used to make them. Often the more basic skills such as chemical analysis were the ones that proved to be the most profitable. Much of the work involved solving the myriad problems that production people encountered daily. Not surprisingly, these scientists usually worked in engineering or production departments.

In the first decade of the 20th century, a few large corporations, notably General Electric, DuPont, and AT&T, organized separate research laboratories that had their own directors. Research and Development (often abbreviated as R&D) would now have an independent voice within the corporation. A major reason for making this change was to insulate scientists from day-to-day problems so that they could concentrate on longer-range ones. For example, General Electric established the first corporate R&D laboratory in the United States in 1900 to protect its profitable lightbulb business after the Edison patents expired (General Electric had bought out Edison's interest in electric lighting). GE's new research director, Willis R. Whitney (1868–1958) hired Ph.D. chemists and allowed them to use the latest

scientific theories to understand the operation of lightbulbs. This research led to a new generation of improved lightbulbs that were protected by GE patents. Similar, but less spectacular, research successes at other companies soon convinced many American corporate leaders that R&D did indeed pay.

By the 1920s, the term *research and development* had been joined together to describe the growing industrial research initiative in American industry. The goals of these laboratories continued to be the improvement of existing products and processes but also began to include the development of new products and processes. Invention was still widely believed to be the exclusive domain of the inspired individual and beyond the purview of rational corporate scientists. Industrial chemists, however, were beginning to make important discoveries serendipitously. Most industrial researchers worked to develop technologies that had their beginnings elsewhere. DuPont produced cellophane films; Kodak worked on color film; and General Electric manufactured refrigerators. In addition to the efforts of independent inventors in the United States, Europe continued to be a prolific source of new technologies.

In the 1920s, a few firms, notably DuPont and AT&T, began to create new science instead of just applying old science. In part, these firms believed that academic scientists were not keeping up with the rapidly expanding chemical and electronics technologies. DuPont research in the emerging field of polymer (long-chain) chemistry led to the discovery of nylon in 1934. AT&T research on semiconductor materials led to the invention of the transistor in 1947. Throughout the 1930s and especially during World War II, science became increasingly regarded as the driving force behind invention. Science and scientists claimed much of the credit for the wartime innovations of penicillin, radar, synthetic rubber, and the atomic bomb. The head of the United States wartime R&D effort, Vannevar Bush (1890–1974), argued that the war had stopped the generation of new scientific knowledge on which future technological advance depended. To maintain a steady supply of new basic science, Bush called for federal funding of research in American universities. His ideas led to the establishment of a National Science Foundation in 1950. The programs sponsored by this new agency, however, were dwarfed by massive military R&D spending during the Cold War.

By the early 1960s, federal expenditures accounted for two-thirds of all R&D outlays in the United States. Nearly half of the federal component was defense related. Another third went into the space program. In some industries such as aircraft, federal funding represented about 80 percent of all R&D expenditures. Private funds for R&D also grew dramatically in the postwar era.

In the 1950s, corporate leaders bet heavily on R&D to produce blockbuster new products that would ensure the growth and profitability of their companies. New research centers were built in country club settings where scientists would not be distracted by the more mundane problems of industry. The accepted wisdom of the time was that investment in science today would lead to new products tomorrow. However, by the mid-1960s, when "tomorrow" had not come quickly enough for corporate management, R&D began to lose some of its autonomy, and once again began to be more integrated into the overall functioning of the firm. For example, emphasis shifted from discovering and developing new products to improving old ones.

As American economic performance began to decline, corporate executives began to take a much harder look at R&D budgets. In the 1970s, corporate research funding leveled off at about 1.1 percent of GNP. Management also began to play stronger roles in setting R&D priorities and in managing programs. Companies began to look for new ways of acquiring knowledge and technology. Industry-academia and interfirm partnerships have become much more common approaches to R&D.

Public and private investment in R&D has become a critical component of economic success in the industrial age. Most industrial nations today invest about 2 percent of their GNP in R&D. International competition in science and technology has become intense in part because modern communications and transportation systems have made R&D a global activity. The high cost of development of new products for world markets has led to more international joint ventures. As in the past, R&D has had to adapt to changing circumstances. Although the sources of new technologies remains diverse, the R&D laboratory continues to be essential to the conversion of new technological concepts into useful products.—J.K.S.

See also ASPIRIN; ATOMIC BOMB; DEPARTMENT OF DEFENSE, U.S.; DYES, ANILINE; INDUSTRIAL REVOLUTION; NATIONAL SCIENCE FOUNDATION; NYLON; PENICILLIN; PHONOGRAPH; PHOTOGRAPHY, COLOR; POLYMERS; RADAR; REFRIGERATOR; RUBBER; SEMICONDUCTOR; STEEL, BESSEMER; TRANSISTOR; TUNGSTEN FILAMENT

Resistor

A resistor does exactly what its name implies: It resists the flow of an electric current. Resistors are typically made from a combination of carbon particles and a binder; the fewer carbon particles relative to the binder, the greater the resistance. Another, more accurate form of resistor uses a film of carbon, metal, or metal oxide inside a ceramic cylinder. Resistors can also be constructed using a light-sensitive material such as cadmium sulfide, so that increasing the level of illumination will decrease the resistance. These are also known as *photocells*. In similar fashion, resistors can be made of materials that decrease in resistance as the temperature goes up. An important variety of resistor is the variable resistor, or potentiometer. Here the resistance can be changed in order to adjust the volume of a radio or the brightness of a lamp. When a potentiometer is constructed with two terminals, one of which is connected to a sliding contact, it is sometimes referred to as a *rheostat*.

Resistance is measured in ohms (named after the German physicist George Simon Ohm [1789–1854]). Individual resistors are marked with three colored bands that indicate its value in ohms. When resistors are wired in series, the total resistance is simply the sum of the individual resistances. When two resistors are wired in parallel, the total resistance is the product of the two resistances divided by their sum. For three or more, the total resistance is given by the formula

$$R = \frac{1}{\dfrac{1}{R1} + \dfrac{1}{R2} + \dfrac{1}{R3}}$$

When an electric current passes through a resistor, it produces heat. This can be a useful property. For example, a wire made from a nickel-chrome alloy will get hot and change to a reddish color when electricity is run through it. The heat produced can be used to toast a slice of bread or heat a room. For toasters and hairdryers, this is a reasonable application. It is hard to justify the use of this technology for space heating, however. Electrical resistance heaters are convenient, but they are very wasteful of energy. In fact, this means of producing heat borders on the absurd; heat obtained from burning a fossil fuel is used to power a steam turbine that turns an electric generator, only to have the resultant electricity converted back into heat. Losses occur at every stage of the process, so only a small amount of the energy employed is actually used for its intended purpose.

See also ENERGY EFFICIENCY; FUELS, FOSSIL; TURBINE, STEAM

Resource Depletion

"Resource" is an inherently technological concept, for, like technology itself, it implies some desired purpose or goal. Considered broadly, resources range from the familiar raw materials to intangibles like "know-how." In the sense of resource depletion, however, resources usually refer to "natural" resources used for the production of some value or utility to humanity.

The most generalized understanding of resource is founded in Nicholas Georgescu-Roegen's classic treatment of the relationship between resource value and the Second Law of Thermodynamics. Resource depletion is essentially the increase of entropy, or disorder. In creating value, an economy produces goods and services, while at the same time turning valuable resources (low entropy) into valueless waste (high entropy).

Resources are usually identified in more specific terms, e.g., mineral, timber, or water resources, even recreational, scenic, or aesthetic resources. Regardless of type, resources imply some finitude or limit or scarcity. Resource depletion, then, means the diminishment or exhaustion of some naturally occurring raw material or other environmental amenity that has value to humanity. In simplest terms, depletion of a resource is like the emptying of a tank. This simple analogy applies most obviously to the category of nonrenewable or exhaustible resources, most notably fossil fuels and ores. These exist in nonreplenishable stocks that often

are depleted at accelerated rates. In contrast, renewable resources, like forests, fisheries, and, perhaps, even topsoil, replenish themselves through natural processes over some measurable amount of time. While there exists at any point in time a stock of any renewable resource, that stock can grow as well as shrink, depending on usage and replenishment rates. For nonrenewable resources, depletion is simply a choice of rate, while for renewable resources, depletion is a management decision to consume faster than natural replenishment can occur. If renewable resource usage is set equal to the annual addition to the total stock, depletion is minimized. This is the basis for the promotion of solar energy and for basing materials economies on recycled goods. Maintaining the stock and consuming the annual flow make a *sustainable* resource economy possible. However, entropy has not been vanquished, as recycling can never reach 100%.

Sustainable resource economies have not been a hallmark of industrial societies or a focus of their economic policies. Economic analyses that discount the future (consumption is worth more now than a year from now) treat the resource as an asset that will be expected to offer a return comparable to other types of assets. Accordingly, the rate of discount will influence depletion rates, and economists disagree over whether the appropriate discount rate for society as a whole should be lower than the rate set by the market. The undervaluation of resources by the future-discounting market system results from the fact that future users cannot bid for the resources that current generations are consuming. In order to prevent socially undesirable depletion rates, some economists advocate policies to offset the market's discounting of the future in valuation of nonrenewable resources.

The term *depletion* connotes "using up" or "running out," but resources are seldom used up in their entirety. Instead, the remaining stocks are so small, dispersed, or difficult to use that they become prohibitively costly. For example, overexploited wildlife like whales were not driven to extinction but their resource value for economic production was depleted. In this sense, a resource is depleted before its supply is exhausted. Similarly, resources like oil and ores will never be gone; they will simply get scarcer and more costly, and hence no longer valuable as resources, in that their costs of production exceed the value they might generate.

The problem of quality reduction reflects another key aspect of depletion. Many resources can be *qualitatively* diminished over and above the *quantitative* measure of the depletion of their stock. Typically the richest, most accessible resources are the ones used first. For example, most old-growth timber in the United States has been harvested. The supply of 500-year-old trees has diminished, and even though timber is a renewable resource the quality of second-growth timber cannot offer the same characteristics as the old-growth timber, at least not until the many decades needed to create large old-growth trees have passed. Similar analyses apply to the accessibility and quality of fossil fuel deposits and ores. We have likely pumped and mined and harvested the resources with the lowest cost and the highest yield per unit cost of extraction. With present technologies, oil deposits under the ocean are much more costly to recover than easily extractable oil from the giant Middle Eastern fields. In effect, quantitative measures alone do not provide the full picture of the impact of resource depletion.

One of the most obvious factors influencing the depletion of resources is the politico-economic system. The definition of ownership and the political assignment of profits and royalties can either promote or inhibit depletion. Political choices like the treatment of property and mineral rights, regulation of ownership via zoning and conservation laws, due process rights in eminent domain, and compensation for takings all influence resource depletion. Even such technical matters as the assumptions used in the compiling of national income accounts affect policies regarding depletion. Present accounting methods show Gross Domestic Product (GDP) growing when resources are consumed, but such consumption should be reckoned as asset liquidation. Sustainable timber harvests could constitute income, but mining and oil production can only be asset liquidation. Several adjustments to national income accounts have been suggested to account for resource depletion and measure sustainable national income.

Efficiency of resource use and extraction can influence resource depletion, and such efforts can be either promoted or inhibited by the availability of resources. Efficiency of use not only conserves resources and

maintains welfare by delivering the same goods and services at lower resource costs, it also reduces pollution and waste on the output side due to the overall minimization of throughput. The ready availability of low-priced stocks of resources reduces incentives for efficient use, while limits on resource extraction or advantages for recycling tend to encourage efficiency. Many economists suggest setting resource utilization rates at long-term sustainable levels (either acceptable rates of depletion for nonrenewables or sustainable rates for renewables). These practices have a long and mixed policy history for some resources like forests, but less progress has been made for fisheries, topsoil, and other threatened renewables, although Depression-era conservation programs do demonstrate a long-standing awareness of the problem.

The *interaction* between nonrenewable and renewable resources adds an additional dimension that should not be overlooked. The loss of nonrenewable resource can affect the supply of renewables, as in the effect of wetland loss on wildlife and fishery populations. Urban development obviously shrinks arable land and the renewable return from agriculture. Thus, the interaction of various problems of resource depletion has the potential to be more of a threat to continued human prosperity than the simple aggregate of the individual effects. This problem can arise as a result of the complex interactions of ecological systems (e.g. fertilizer runoff harming fisheries). It can also be reflected in the rising prices of substitutes for the depleted resource (e.g., reduced American oil production leading to larger imports or to greater demand for natural gas).

Responses to the problems of resource depletion range from business-as-usual to dramatic action. Laissez-faire approaches tend to be advocated by interests profiting from the exploitation of resource, as the benefits of resource conservation are spread across the population as well as across time, while the rewards of exploitation accrue directly to those with an immediate economic stake. Not surprisingly, governments, affected industries, and investors with short time horizons tend to be disproportionately represented in resource management decisions. More conservation-oriented analysts recommend depletion quotas (on total annual resource extraction), and taxes (on resources extracted), and even prohibitions or moratoria on resource use (as in whaling or the creation of national parks). "Fee-bates" have also been proposed as revenue-neutral, efficiency-promoting devices, with charges being levied on inefficient resource-intensive processes and the resultant revenues being used to subsidize more resource-efficient users (e.g. taxing cars with low mileage per fuel consumed to reward high-mileage cars).

Analyses of resource depletion note the absence of a long-term perspective in much decisionmaking about natural resources. In addition, the problem of common property resources (resources for which exclusive management or property rights are difficult to define and enforce) can intensify the current generation's short-term perspective by creating zero-sum or negative-sum games where common (unregulated) resources like ocean fisheries may be destroyed due to the inability to prevent overfishing. No fisher wants to see the resource depleted, but each is forced to contribute to the depletion in order to escape being beggared by the other fishers. These "tragedies of the commons" tend to enforce the short-term perspective by penalizing self-regulation. Such situations lead to overexploitation of resources because they promote a gold rush, get-it-before-someone-else-does incentive structure.

As a final point, the limitations—not to say inadequacy—of the resource approach must be acknowledged. The complexity of ecological interdependence and the ethical limits of anthropocentric values suggest that more sophisticated approaches to human relationships with the natural world might offer both improved understanding and a better foundation for decisionmaking than a narrow analysis of "resources." Destruction of species that have no "resource" value would arguably be ecologically shortsighted and ethically benighted. The natural world cannot be seen as just a package of resources.—A.W.

See also CONSERVATION OF ENERGY; ENTROPY; FUELS, FOSSIL; PHOTOVOLTAICS; RECYCLING; SOLAR DESIGN, PASSIVE; SPECIES EXTINCTION

Revolver

For centuries, attempts were made to produce gunpowder weapons that did not have to be reloaded after each shot. One of the earliest examples appeared in the 1580s. This was a weapon with three tubes soldered around a manually controlled axis; each tube contained a bullet that was fired by a spinning wheel that provided a spark. Many examples of multiple-barrel handguns were produced in subsequent decades, but their effectiveness was limited by the low-quality metal from which they were built, the difficulty of shaping parts accurately, and the lack of cartridges containing both bullet and powder. Although the multiple-barrel "pepperbox" pistol was popular as late as the 1830s, like all early revolvers it would on occasion fire more than one bullet simultaneously, often with disastrous consequences for the one who fired it.

In 1815, France's Hyppolyte Lenormand employed the revolver principle for a pistol that fired its bullets with percussion caps. In 1818, H. Collier of England obtained a British patent for a pistol invented in the United States by Artemus Wheeler. Only the cylinder containing the bullets revolved, rather than a combination firing chamber and barrel, as had been the case with earlier guns. It had the drawback of requiring the cylinder to be turned manually after each shot. In 1835, Samuel Colt (1814–1862) of the United States, received a U.S. patent for an essentially modern revolver that automatically turned the cylinder after each shot had been fired.

Colt's revolver fired five shots fairly rapidly, but loading took some time. First the hammer was brought back to half-cock, allowing the cylinder to turn freely. A paper or foil cartridge, or a separate ball and loose powder charge, were placed in one of the cylinders. The cylinder was then rotated to bring the charge under a rammer at the bottom of the gun; when actuated by the loading lever, the bullet and powder were seated in the proper position. A percussion cap was then mounted on a nipple at the rear of the cylinder. This procedure was repeated for the remaining cylinders. Pulling back the hammer all the way back brought a cylinder in line with the firing camber and locked it in place. The Colt revolver was a single-action gun; it had to be manually cocked after each firing.

In subsequent years, other manufacturers produced double-action revolvers through the use of the Beaumont lock: A single pull of the trigger first cocked and then fired the pistol. An equally important innovation was Smith and Wesson's 1857 revolver, which was the first to use metal cartridges. These simplified

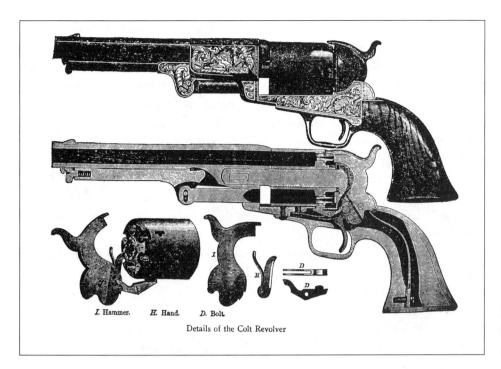

I. Hammer. II. Hand. D. Bolt.

Details of the Colt Revolver

Details of an early Colt revolver (from R. Burlingame, *March of the Iron Men*, 1938).

the loading of the gun and eliminated the danger of multiple discharges. Today, most revolvers are loaded from the back, with the entire cylinder swinging out to facilitate the process.

The revolver was the first repeating personal weapon used on a large scale by military forces. It was, however, not an accurate weapon. Early revolvers could fire a bullet only 100 m (328 ft), and were reasonably accurate at no more than half that distance. Perhaps more important than their effectiveness as weapons was the manufacturing operations they inspired. After its founding in 1848, Colt's armory in Hartford, Conn., developed into one of the prime centers of mechanized production. The factory made extensive use of drop forging for the production of parts, which were then finished through the use of specialized milling machines and turret lathes. The resulting parts were not fully interchangeable, for such a high rate of manufacturing precision was deemed too expensive. Moreover, the heat treating of finished parts often resulted in enough deformation to prevent interchangeability. Even so, Colt's armory was an important example of the enterprises that developed mass production techniques from the middle of the 19th century onward.

See also LATHE; FORGING; SEMIAUTOMATIC WEAPONS

Richter Scale

An earthquake is a shaking of the Earth's crust that in most cases is caused by seismic waves emanating from the Earth's interior. These waves are produced by movements of the tectonic plates on which the Earth's surface rests. Individual earthquakes vary enormously in intensity; some can only be detected by highly sensitive instruments, while others cause vast amounts of death and destruction.

Several scales have been designed to provide indications of an earthquake's severity. Early measurements, such as the Mercalli scale that was widely used after it was devised in 1902, were based on damage caused by the quake and people's reactions to the quake as it occurred. Although scales of this sort provided a way to judge the severity of an earthquake, they had the disadvantage of being affected by factors unrelated to geological conditions, such as population density and types of building construction.

The most familiar measurement of the intensity of earthquakes is the Richter scale, which was devised by Charles Richter (1900–1985) of the California Institute of Technology in 1935, and based on an earlier scale developed by K. Wadati in Japan. The Richter scale rates earthquakes in terms of their magnitude, a number that is proportional to the amplitude of the quake, that is, the height of the maximum wave motion as determined by a seismograph. Since earthquakes vary greatly in size, the amplitudes of one wave can be thousands of times greater than the amplitude of another. To compress the scale, Richter defined a quake's magnitude as the base-10 logarithm of the maximum amplitude. Mathematically, an earthquake's magnitude is given by the formula $\log_{10} E = 11.4 + 1.5M$, where M is the magnitude. Thus a 6.0 earthquake is not twice as powerful as a 3.0 earthquake, but $(10 \sqrt{10}) \times (10 \sqrt{10}) \times (10 \sqrt{10})$ as powerful.

Richter's original intention was to provide a means of distinguishing the large, medium, and small shocks that often occurred in southern California. To make it applicable to other parts of the world, the scale has been modified to take in a number of variables in addition to amplitude, such as the focal depth of the quake and regional peculiarities that affect the quake. Earthquakes of magnitudes given by the Richter scale have the following effects:

Magnitude	Effects
0	Detected only by seismographs
2–4	Can be felt, but causes little or no damage
4.5	Local damage
5	Releases energy equivalent to the detonation of the first atomic bomb
6	Considerable destruction near the epicenter
7	Major earthquake, one causing widespread destruction
8.6	Maximum

Although the Richter scale continues to be widely used, seismologists have made increasing use of an earthquake magnitude scale based on the amount of energy released. This can be rendered as the earthquake's "moment," the average slip multiplied by the slipped fault area multiplied by the Earth's elastic constant. This produces measurements of seismic moment in dyne-centimeters. The largest measured earthquake was the one that struck Chile in 1960. Measured at

8.5 on the Richter scale, it released 10^{30} dyne-centimeters (10^{23} newton-meters) of energy.

An earthquake's magnitude as measured by the Richter scale or any other scale is only a rough guide to the actual damage produced. A great deal depends on the composition of the soil and other geological features, as well as the manner in which structures have been constructed. Due to improvements in construction methods, strong earthquakes in California and Japan have produced less destruction and injury than earthquakes of lesser magnitude that have struck elsewhere.

See also EARTHQUAKES; LOGARITHM; PLATE TECTONICS

Further Reading: Bruce A. Bolt, *Earthquakes: A Primer*, 1988.

Rickshaw

Although an adult human being in reasonably good physical shape has a power output of less than one-fifth of a horsepower, people have long been employed to carry other people. The simplest man-powered vehicle is the sedan chair: a seat, sometimes enclosed, with two poles extending out at each end. In most cases, a bearer is situated at each end, providing the power to carry the passenger at not much more than a walking speed. Human-powered vehicles work much more efficiently when they are equipped with a wheel. For centuries the Chinese used a special wheelbarrow for hauling passengers. This was a fairly heavy contrivance, and not suited for even moderately rapid travel.

In the 1870s, some of the basic elements of bicycle technology (a tubular frame, wire-spoke wheels with pneumatic tires, and ball bearings) were used in Japan to produce the *jinrikisha*, or "rickshaw," as it is usually rendered in English. Its origin is in dispute, some claiming that Jonathan Gable, an American missionary, was the inventor, others arguing for an indigenous origin. Whatever the source, the rickshaw was highly successful. At the end of the 19th century, 200,000 rickshaws were in use in Japan—one for every 200 inhabitants. At that time Japan also exported about 10,000 rickshaws per year, most of them going to other parts of Asia.

Although the rickshaw was an improvement over

A rickshaw with a small passenger in China around 1875 (from L. C. Goodrich and N. Comeron, *The Face of China,* 1978).

other forms of human-powered transport, in some places it came to be a symbol of oppression and human degradation. In China, in the decades before the communist revolution, a number of influential literary works focused on rickshaw pullers and their sufferings. The choice of rickshaw pullers as the embodiment of human misery was appropriate: One Chinese study found that the life expectancy of a rickshaw puller after he went to work was not much more than 18 months.

Rickshaws are seldom seen today except in tourist spots, where they are employed for amusement rather than basic transportation. Human-powered passenger vehicles are by no means extinct, however. In many parts of the world, and particularly in Asia, a standard urban conveyance is the pedicab, a light, pedal-powered vehicle.

See also WHEELBARROW

Rifle

A rifle is a firearm, usually assumed to be a shoulder weapon, that uses helically cut grooves or some analogous shaping of the bore to impart axial spin to the projectile for greater stability and thus greater accuracy. The principle of rifling is an old one, generally considered to derive from the fletching of arrows with offset feathers to produce a similar rotational effect. However, it is important to note that some of the earliest rifling was straight rather than spiral, possibly added in an attempt to reduce the fouling inherent to gunpowder weapons. The earliest rifles date back to around 1500, but this type of weapon was long a relative rarity, as cutting rifling into the bore of a firearm was an arduous and precise—and thus very expensive—process.

The rifle, furthermore, was of limited use on the early modern European battlefield, thus limiting the demand for it even more. In order for the bullet to engage the rifling, a very tight fit between ball and bore was necessary, often requiring the marksman literally to hammer the ball down the barrel—the mallet typically provided by the maker as an accessory—especially as repeated shots fouled the bore. This was, to say the least, inimical to the relatively rapid fire essential to the infantry tactics of the era. In any event, the thick gunpowder smoke that shrouded contemporary battlefields impaired visibility at the ranges where the rifle could be used to advantage. The cheaper, more robust, and quicker-loading smoothbore musket was thus the dominant military firearm of the time; the rifle was mostly used for hunting and recreation, where accuracy was at a premium and rate of fire a secondary consideration. These same considerations meant that the rifle often used wheelock ignition, employing a spring-loaded, serrated steel wheel to strike sparks from a lump of iron pyrite, whereas the musket would use the simpler matchlock, which provided slower ignition but was more equal to the rigors of campaigning.

Despite these disadvantages, the rifle gained a niche in European armies during the last third of the 18th century. The rigorous, linear tactics employed by standing armies worked best in open country and hardly at all in swampy, forested, or mountainous terrain. To this end, most Western armies developed bodies of light infantry for these conditions and for the reconnaissance, outpost, and skirmishing missions for which line infantry was at best indifferently equipped and trained for. The bulk of these forces used variants of the musket, but some used rifles. Such employment was facilitated both by the fact that the new corps were in many cases raised from foresters and game-keepers familiar with the weapon, and also because technological progress had made the rifle more suited to military use. Flintlock ignition had made the weapon more durable, while the adoption of a greased patch wrapped around a slightly smaller ball than those laboriously driven down the bores of earlier rifles facilitated quicker loading, although the rate of fire never approached that of contemporary muskets.

Sturdier, handier rifles, such as the hunter's "Pennsylvania" (miscalled "Kentucky") rifles of the American Revolution and specialized military weapons such as the British Baker rifle of the Napoleonic era made a place for themselves on the battlefield. Nonetheless, the adoption of the rifle as the standard infantry weapon had to wait until the general adoption of the Minié-Delvigny ammunition system in the mid-19th century. This approach relied on a cylindro-conical bullet with a conical hollow in the base. Such a projectile could be made with enough windage (the distance between bullet and bore) to allow loading to be as easy as that of a musket ball, but when the powder charge was ignited, the expansion of the resultant gas into the hollow drove its walls outward to engage the rifling. This innovation resulted in a 5-fold to 10-fold improvement in the effective range of infantry fire, out to 800 to 1,000 meters (2,625–3,281 ft) against extended targets such as massed troops.

Such a range was comparable to that of many of the smoothbore field guns used by contemporary artilleries. As a result, the rifle became by many measures the dominant weapon in land warfare, with the consequence that the close-packed linear formations of the 17th and 18th centuries became fatally inappropriate on the battlefields of the Crimean and American Civil Wars. The dominance of rifle-armed soldiers continued through much of the remainder of the 19th century against a background of constant technological innovation. The muzzle-loading rifle musket was supplanted by faster-firing breech-loading weapons. These rifles themselves progressed through several generations, first with the adoption of metallic cartridge ammunition—an adjunct that made breechloading completely practical for the first time after 4 centuries of experimentation. The capabilities bestowed by breechloading were augmented by the development by repeating, magazine-fed actions, again facilitated by the metallic cartridge, and the replacement of gunpowder by more energetic, nitrocellulose-based, smokeless propellants that allowed the use of smaller-caliber, higher-velocity, more-aerodynamic projectiles. Rifle muskets had fired .52- to .58-caliber (13–15-mm) Minié-pattern bullets weighing around 500 grains (c. 30 g) at over 305 meters (1,000 ft) per second. Early cartridge rifles fired solid lead bullets of approximately .45 caliber (11–12 mm) at a somewhat higher velocity. Even immature smokeless powder technology allowed copper-jacketed bullets of .28 to .32 caliber (7–8 mm) weighing 120 to 160 grains (7–10 g) to be propelled at over 760 meters (2,500 ft) per second. Each improvement brought concomitant gains in performance, until an individual soldier in 1900 armed with a bolt-action magazine rifle could be expected to fire 15 or even 20 aimed shots per minute at a point target hundreds of meters away.

The rifle was displaced from its supremacy in land warfare late in the 19th century (although, again, tactical doctrine lagged behind reality long enough for the discrepancy to contribute to the slaughter of World War I) by the very factors that had made it a mature weapons system. The metallic cartridge and smokeless powder made the machine gun practical, and helped greatly to make quick-firing artillery possible. The rifle nonetheless remained an essential part of the military arsenal and continued to develop technologically. Self-loading, or semiautomatic, rifles had reached some infantry units, especially those of the United States Army, by the outbreak of World War II. Wartime and postwar efforts were made to extend full automatic fire capabilities to semiautomatic designs, but such weapons proved difficult to control and to keep supplied with ammunition. The fully automatic personal infantry weapon, or assault rifle, did not become practical until a conscious decision was made to trade effective range and penetrating power for useful full-automatic fire capability. This was done by using less powerful but lighter ammunition, a radical break in a trend over a century old.—B.H.

See also BREECHLOADING WEAPONS; GUNPOWDER; MACHINE GUN; MUSKET

Further Reading: Howard L. Blackmore, *Guns and Rifles of the World*, 1965.

Risk Evaluation

Risk is the probability of an unwanted outcome. Qualitative concepts of unwanted outcomes have, of course, been with us for a very long time, but the history of efforts to calculate their probability is much shorter. The formal analysis of risk can be traced back to 1713 and the publication of Jacques (or Jacob) Bernoulli's (1654–1705) work *Ars Conjectandi* (*The Art of Conjecturing*). This book, which was published posthumously in 1713, laid the theoretical foundation for calculating probabilities in decision problems, i.e., for calculating risks. Bernoulli's work grew out of a very practical concern of the time: gambling. By the 18th century, gambling had become a national obsession in many European countries, and gamblers sought to gain advantage by paying mathematicians to calculate odds for them. Bernoulli was foremost among mathematicians who wanted to develop rules and theoretical underpinnings for these calculations. Soon other people besides gamblers took a serious interest in Bernoulli's "art of conjecture." Insurance brokers were especially anxious to calculate risk carefully because their repu-

tations depended on it. An insurer was accused of gambling if it was determined that the insurer had assumed a risk that was too great. A risk deemed too low, on the other hand, brought charges of usury. Bernoulli's book thus established probability theory as an important aspect of risk evaluation for virtually all circumstances.

As defined by Bernoulli, the probability or risk of a single event occurring is calculated as the actual number of occurrences of that event divided by the total number of possible occurrences. In the case when one is attempting to calculate the chance of an outcome before any occurrences of the relevant event, the probability must be estimated based on other information, but the general structure of the calculation remains the same. Frequently, it is important to estimate the risk of one among several possible events or a combination of events. The probability that any one of several independent events will occur is calculated as the sum of the probabilities of each individual event. This is called the *addition law*. For example, the probability of rolling either a 1 or a 6 with one toss of a six-sided die is $\frac{1}{6} + \frac{1}{6}$. The chance that several independent events will occur together or in close sequence is determined by using the multiplication law: The probabilities associated with each individual event are multiplied with each other. For example, the probability of rolling two sixes on two consecutive rolls of a six-sided die is $\frac{1}{6} \times \frac{1}{6}$. The probability is the same for rolling a six on each of two dice tossed at the same time.

The accuracy of a prediction based on a calculated probability depends on the number of samples that contributed to the estimate. Bernoulli argued that, if presented with a jar containing a mixture of an unknown number of white and black balls, he could estimate the proportion of white balls simply by taking samples of balls from the jar and comparing the number of white balls to the total number of balls drawn from the jar. The accuracy of his estimate would increase with the number of samples. This dependency is known as the *law of large numbers*.

The method by which samples are collected is also important in determining the accuracy of a calculated risk. Risks of such things as vaccines are often determined in retrospective studies. This type of study looks only backward to ascertain the causes of observed events. The risk of a certain side effect of a vaccine, for example, could be estimated by sampling a population of people who had received the vaccine in order to determine who had experienced that effect. The resulting estimate of risk (number with the effect divided by number sampled) could be compared to the rate of occurrence of that effect in an unvaccinated population to estimate the additional risk contributed by the vaccine. Retrospective studies of risk are frequently less expensive than the major alternative (see below), but hindsight can have a distorting effect on the data collected. Equally significant, other important data, such as predisposing conditions in the case of vaccine side effects, may be not be available. Finally, retrospective studies are effective at establishing correlations, but less useful for determining cause-and-effect relationships between an event and a hazardous outcome.

The major alternative to retrospective studies of risk is a prospective study, whereby a carefully selected group of people is divided into an experimental group that receives some treatment and a control group for comparison. People in both groups are followed closely before, during, and after the treatment period, and differences in number and timing of hazardous outcomes are noted. Ideally, neither the participants nor the people immediately responsible for collecting the data know who belongs to which group. This procedure allows the histories of the subjects to be recorded before the outcome of the study has been observed, and increases the investigator's ability to determine the likelihood of a link between a suspected cause and a particular hazardous effect. While prospective studies usually provide better estimates of risks, they are expensive to do, especially when sample sizes must be large. They also may present ethical dilemmas if a potentially helpful treatment must be withheld from the subjects in the control group in order to learn whether the treatment is actually beneficial.

In many cases, an estimate of risk must include information not only about the probability with which a suspected cause is related to a hazardous outcome but also about how the cause relates to the outcome. The risk of cancer associated with exposure to ionizing radiation, for example, can be estimated by the procedures mentioned above, but an additional question

must be answered. Does exposure to radiation simply add a fixed probability of cancer to the preexisting background risk (additive risk), or does the radiation multiply the risk (relative risk)? The model of risk that pertains, additive or relative, determines the magnitude of the risk estimate. According to the additive risk model, the risk of radiation-induced cancer is the same for people of all ages who experience the same level and type of exposure, but the relative risk model predicts higher risks for older people.

As this brief discussion implies, it can be quite difficult to develop accurate estimate of risks. The level of uncertainty is often quite high. Yet decisions still must be made. Arguments over what policy should follow from a particular calculation of risk are as old as probability theory itself. The situation is further complicated by the fact that public perception of risk often differs sharply from quantitative estimates of risk developed by experts. For example, living near a toxic-waste site will, on average, take 4 days off the lives of the residents, while cigarette smoking causes an average loss of 2,580 days. Yet many people consider the former far more hazardous than the latter. One of the reasons for this anomaly is that people are more likely to fear a hazard over which they have little control or that has a mysterious source than to fear a more substantial hazard that is well understood.

For all of the sophistication of risk analysis, fundamental questions remain. What is safe? What quantitative values should be assigned to a particular hazard? What level of risk is acceptable? What are reasonable tradeoffs? These questions have no easy answers.—N.C.

See also CANCER; IMMUNIZATION; RADIOACTIVITY AND RADIATION; SAMPLING; SCIENTIFIC METHODOLOGY; STATISTICAL INFERENCE

Road Construction

Roads are crucial elements of any civilization, for without them individuals and the communities in which they live are isolated entities that have to make do with whatever the immediate area can provide. Centralized political states also depend on roads for such crucial functions as the collection of taxes, the maintenance of order, and the preservation of territorial integrity. Throughout history, every significant empire has recognized the necessity of constructing networks of roads throughout its territory.

One example comes from South America, where the Inca Empire was held together by 23,000 km (14,300 mi) of roads. This system included the Royal Road, probably the longest road in the ancient world; it stretched 5,700 km (3,540 mi) from what is now Santiago, Chile, to Quito, Ecuador. Extensive road systems also were built by the Assyrians and Persians to consolidate their foreign conquests. One of the most famous roads of the ancient world, the Great Silk Road, was generally used for more pacific purposes. Originating in Xian, then the capital of China, the road connected China and central Asia with the Middle East and ultimately the Mediterranean Sea. Although it was put to good use by the Mongols during their conquest and subsequent rule of China, its main users were merchants who brought such items as jade, porcelain, and, of course, silk out of China. In return, the Chinese imported cotton and ivory, while cultural innovations such as the Buddhist religion also made their way into China by means of the road.

In the Western world, the most famous road builders were the Romans, who tied their far-flung empire together with a system that was unexcelled for centuries. In building their roads, the Romans made abundant use of materials and techniques first developed by other peoples: masonry from the Greeks, cement from the Etruscans, and surveying from the Egyptians. At its peak around 200 C.E., the system encompassed 80,000 km (49,700 mi) of roads that circled the Mediterranean and extended from Asia Minor to Western Europe.

In constructing their roads, the Romans began by digging down to a firm foundation or sinking piles to provide a foundation. Earth also was excavated to form drainage ditches on both sides of the roadway. The earth was then used to fill the roadway to a height of up to a meter (3.3 ft) above ground level. Each side formed a shoulder, while the roadway in the middle was topped with a layer of sand. On top of this was laid a pavement that typically consisted of a layer of flat stones, another layer of smaller stones, and yet another layer of even smaller broken stones, bricks, or similar material. On heavily traveled routes, the road might have a top surface of hexagonal flagstones fitted together. The roads often were 7.6 m (25 ft) wide, but in frontier regions or difficult terrain they could be much narrower.

The Roman road system deteriorated in medieval times, and much of it was abandoned. Consequently, travelers in most parts of Europe had to walk or ride on meandering tracks that often turned into impassable bogs in wet weather. There were several reasons for this decline. A fragmented political order lacked the authority and the resources to maintain the system. In addition, commerce dried up as economies became largely self-contained; as a result, traffic declined, and there was little impetus to keep up the roads. And in dangerous, often chaotic times, many communities preferred to be isolated, so roads were deliberately closed in order to discourage invaders and brigands.

This situation lasted until about the 13th century, when road building began to revive in many parts of Europe. The Crusades created a demand for travel from the interior to seaports, while the growth of towns and the expansion of commerce provided a further impetus for transportation and travel. The Industrial Revolution in the mid-18th century, coupled with the rapid growth of cities that needed connections with the hinterlands, made the need for better roads painfully evident. A growing network of canals and then of railroads met some transportation needs, but in many places there was no substitute for an expanded network of all-weather roads.

Although England was the setting of the Industrial Revolution, the nation that led the way with improved methods of road building was France. In particular, Pierre-Marie Jérôme Trésaguet (1716–1796), an engineer for the Corps des Ponts et Chausees (the government agency for road building that was founded in 1747), can be credited with the development of superior road-building methods. Trésguet worked during the latter years of the 18th century, and beginning in 1831 his ideas received posthumous recognition in a book that had a large readership. Trésguet's method entailed first laying foundation stones about 200 mm (7.9 in.) in size, flat on one side. These stones were arranged to provide a camber that aided drainage. They were followed by a layer of smaller stones hammered into the gaps between the foundation stones. Next, another layer of broken stones was put down and the surface leveled. Finally came a 25-mm-thick (1-in.) layer of broken stones. This technique was a departure from Roman methods in that the pavements were not self-supporting structures. In Trésguet's scheme, the function of the pavement was to keep the natural supporting structure dry, and to absorb the pressures applied by vehicular traffic.

Trésguet's road construction techniques were adopted and refined in Great Britain by Thomas Telford (1757–1834). His roads began with a layer of block-shaped stones. On top of these were placed a layer of hand-packed stones, each one no more than 60 mm (2.4 in.) across. The top surface was a mixture of broken stones and gravel. Unlike most of the road builders of his time, Telford took great pains to ensure adequate drainage and the effective management of the flow of water.

A major step in the building of durable roads came with the work of John Louden McAdam (1756–1836). Providing his services to turnpike companies, often at no charge, McAdam departed from the principles laid down by Trésguet and Telford. His roads had a bottom layer of angular stones no more than 75 mm (3 in.) in width that went to a depth of 200 mm (7.9 in.). For the upper layer, which was about 50 mm (2 in.) thick, stones with a width of no more than 20 mm (.79 in.) were used. These stones were usually produced by men, women, and children who sat by the roadside using hammers to break large rocks into small

ones. In later years, steam-powered crushers lifted this burden and made McAdam's methods easier to employ.

The strength of McAdam's roads came from the interlocking of the individual stones. The surface even improved after some use, as the action of wheels and hooves broke the edges of stones into fine material that settled between the interstices of the larger stones. McAdam also insisted that the drainage be ensured by cambering the surface of the supportive bed and elevating it above the water table. Although others had anticipated these techniques, McAdam did the best job of building and publicizing them, and it is not without reason that roads so built have been given the generic name *macadam*.

Macadam roads suffered from the drawback of being dusty in dry periods and muddy during wet ones, especially when corners were cut during their construction, which was often the case. The answer to these problems was topping the surface of the road with a stable, durable substance. One of the most common of these is asphalt. In precise terms, this is a mixture of bitumen (a heavy fraction of petroleum), sand, and bits of stone. However, the term *asphalt* is often used as a synonym for bitumen. The employment of naturally occurring asphalt as a road material began in mid 19th-century France. Road builders also surfaced roads with a substance similar to bitumen, the tar that came as a by-product of illuminating gas production. In this case, instead of serving as a road material, the tar was used as

A nearly impassable road near Ventura, Calif. (courtesy California Department of Transportation; copy neg. 94-06670-32).

a binder for macadam roads. The resulting road was known as *tarmacadam*, or *tarmac* for short.

In the 19th century, another paving material, concrete, came into widespread use. In 1865, it was employed in Great Britain to bind together the stones of a macadam road, in much the same way that tar already had been used. Concrete was rarely used for road surfaces at first because it was poorly suited to the draft animals that plied the roads of that era. Concrete took on a new importance in the 20th century as automobiles riding on pneumatic tires became the dominant form of transport. But the increased use of concrete did not mean the end of asphalt roadways. Builders of asphalt roads copied macadam roads, using a continuous gradation of stone sizes so that smaller stones fit into the gaps between larger stones. The laying of asphalt roads was further advanced by the development of laying and finishing machines in the 1920s.

Whatever the materials and the methods used, road building is an expensive proposition. Although road construction generally has been a governmental responsibility, in some times and places privately owned toll roads offered an alternative means of financing road construction and maintenance. Where roads have been a responsibility of the government, taxation is required, although it need not take the form of money. Throughout the ages, governments have forced many of their subjects to work on road construction and repair. This labor tax is often known by its French name, *corvée*, and it was one of the irritants that led up to the French Revolution. In the United States, the use of convicts for roadwork has been common, especially in the South.

Roads have often been financed by property and income taxes, as well as by the issuance of bonds. In 1919, the gasoline tax made its appearance in the United States. Introduced in only a few states at first, the gasoline tax was universal within a decade. This tax has the advantage of providing payments that are roughly in accordance with the extent to which a driver uses the roads. Unlike other taxes, gasoline taxes are clearly earmarked for a particular purpose, although in recent years some revenues have been used to finance mass-transportation projects. As a result, gasoline taxes have not been as deeply resented as most other forms of taxation.

See also AUTOMOBILE; CANALS; CONCRETE; INDUSTRIAL REVOLUTION; LIGHTING, GAS; PORCELAIN; RAILROAD; TIRE, PNEUMATIC

Further Reading: M. G. Lay, *Ways of the World: A History of the World's Roads and the Vehicles That Used Them*, 1992.

Rockets, Liquid-Propellant

Unlike their solid-propellant counterparts, liquid-propellant rockets are recent developments. They owe their initial development to three pioneering theoreticians, Konstantin E. Tsiolkovsky (1857–1935) in Russia, Robert H. Goddard (1882–1945) in the United States, and Hermann Oberth (1894–1989) in Germany. Of the three, Goddard alone was engaged significantly in the actual development of rockets. On Mar. 26, 1926, he succeeded in launching what is generally recognized as the world's first liquid-propellant rocket, although it flew to a height of only 12.5 m (41 ft). During the next 20 years, Goddard managed to develop a great many of the technologies later used on large, liquid-propellant rockets. But he never succeeded in combining them in a workable rocket capable of reaching the high altitudes needed for satellite-deploying rockets, ballistic missiles, or space rockets. Moreover, his secrecy prevented him from having a great deal of direct influence on the later development of rocketry.

Consequently, it was Oberth—and, to a lesser extent, Tsiolkovsky—who managed to have a much greater influence on liquid-propellant rocketry. Influenced by Oberth, a group of German engineers and scientists under the technical direction of Wernher von Braun (1912–1977) succeeded in developing the first large, liquid-propellant missile, the A-4 (commonly known as the V-2, for *Vergeltungswaffe* or "retaliation weapon") with which Germany bombarded London, Antwerp, and other allied cities during World War II. With a large staff of technically competent personnel, a huge complex of test stands and other research facilities, and the assistance of German universities, technical institutes, and commercial firms, between 1932 and 1944 von Braun's group succeeded in overcoming many problems in the areas of propulsion, supersonic aerodynamics, and guidance and control to produce a workable missile capable of delivering a warhead of roughly a ton to a distance of about 320 km (200 mi).

This was by far the largest and most important liquid-propellant device developed during the war. Von Braun was captured by the Americans and went with over a hundred of his key engineers and scientists to the United States. There the V-2 became the essential prototype for a great many subsequent U.S. missiles and rockets from the Redstone to the Saturn V "moon rocket" that were developed under von Braun's direction at the Army's Redstone Arsenal and then NASA's Marshall Space Flight Center in Huntsville, Ala.

Meanwhile, the Soviet Union had been active, with some guidance from Tsiolkovsky, in developing liquid-propellant rockets of its own in the 1930s when von Braun had begun work on what became the V-2. The Soviets captured some Germans who had worked for von Braun, but the extent of the Soviet debt to German technology varies with the sources one consults. A Soviet liquid-propellant rocket launched the world's first artificial satellite, Sputnik I, on Oct. 4, 1957, setting off a Cold War race between the Soviets and the United States to see which power could land a human being on the moon first. The United States won this race on July 20, 1969, when Neil Armstrong (1930–) and Edwin "Buzz" Aldrin (1930–) walked on the moon during the Apollo 11 mission. The energy released by the five rockets in the Saturn's lower stage equaled that of a large electrical power plant, and in the process made more noise than any human activity other than the detonation of a nuclear bomb.

Meanwhile, the Cold War had also precipitated a missile race between the two superpowers that lasted until the fall of the Soviet Union in 1991. The initial intercontinental ballistic missiles for both nations used liquid propellants, but the United States converted to solids in the early 1960s. The Soviets were slower to make this conversion, with liquid systems dominating into the late 1970s, although they had solid-propellant missiles by the late 1960s.

Liquid-propellant rockets enjoy two distinct advantages over solids. One is the higher specific impulse (a measure of thrust) of many liquid propellants. The other is the ability to throttle the injection of the propellants into the combustion chamber for thrust variation and even to shut down and then restart the engines. However, to feed the propellants into the combustion chamber re-

quires many complicated features that are unnecessary in solid rockets. These include turbopumps or pressure-feeding systems to force the propellants into the combustion chamber, injectors to provide the proper mixing of propellants (usually a separate fuel and oxidizer), and a cooling system to keep the combustion chamber from melting if it operates for longer than a few seconds.

In the United States, many of the early missiles and rockets owed much to V-2 technology, but there were also numerous independent contributions from non-German sources. For example, a team under Theodore von Kármán (1881–1963) and Frank Malina at what became the Jet Propulsion Laboratory (JPL) in California had developed rockets with self-igniting propellants during and immediately after World War II. Beginning in 1960, the Aerojet General Corporation, originally founded by von Kármán, Malina, and others from JPL, used this basic technology (with different self-igniting propellants) to develop the engines for the Titan II through IV missiles and space launch vehicles. The Titans have the advantage over many other liquid-propellant systems that they use storable propellants. These do not need to be inserted into the rockets' storage tanks just before launch (as do, for example, liquid oxygen and liquid hydrogen).

Another key U.S. innovation was the "steel balloon" developed by Convair Corporation (later General Dynamics) engineers under Karel J. Bossart for the Atlas missile. Instead of the heavy propellant tanks favored by von Braun and his team, Convair pioneered a missile with a thin skin that also served as the wall of the propellant tanks. This reduced the weight of the missile significantly, thereby increasing its range. Such thin skins were later used on the Thor missile and on the upper stages of the Saturn.

Although the Cold War conditioned the United States and the Soviet Union to become the principal centers of missile and rocket development after World War II, France, England, the European Space Agency, Japan, China, India, and Israel have also participated significantly enough in rocket development to have launched satellites into space. Some of their launch vehicles have been powered by solid rather than liquid propellants, but liquid propellants have remained dominant for large space launchers because of their higher performance.—J. D. H.

See also APOLLO PROJECT; GEMINI PROJECT; MERCURY PROJECT; MISSILE, INTERCONTINENTAL BALLISTIC; ROCKETS, SOLID-PROPELLANT; SPUTNIK

Rockets, Solid-Propellant

Solid-propellant rockets have existed for centuries. The Chinese developed primitive ones by burning black powder (saltpeter, charcoal, and sulfur) for use in battle by the first third of the 13th century. Since then, military needs have stimulated most advances in the technology, although solid-propellant rockets have found many civilian uses ranging from life saving at sea to high-altitude research and space launches.

By the beginning of the 20th century, rockets as weapons had temporarily been eclipsed by artillery, but propellants had advanced from black to smokeless powder consisting of nitrocellulose plus a stabilizer, and often nitroglycerine (or a related nitrogenous compound). These propellants were referred to, respectively, as single- and double-base propellants. Several years before World War II, the Russians, Germans, and British began to develop rockets again as weapons, giving rise to discoveries that led to the revolution in postwar solid-propellant rocketry.

From the late 1950s until 1989, the Cold War prompted the development of countless solid-propellant rockets in many countries, with a wide variety of applications: air-to-air, air-to ground, ground-to-air, and surface-to-surface. The history of this technology has yet to be adequately written, and this account must limit itself principally to the development of large, solid, ballistic missiles in the United States. To a significant degree, the history of solid propellants in other countries and for other applications has been similar. But there have been exceptions. For example, the Soviet Union used a composite propellant including a binder made from a resin found only in the Urals. Composite propellants do not contain nitrocellulose; they consist of an oxidizer and other substances dispersed in an elastic binder that also serves as a fuel. Unlike cruise missiles, rockets operate independently of the oxygen in the air.

Two innovations during World War II essentially launched postwar developments in solid rocketry. The most fundamental of these occurred at the rocket research project of the Guggenheim Aeronautical Laboratory, California Institute of Technology (GALCIT). There, a small group of researchers sought a solution to the problem of controlled burning for many seconds of a propellant for jet-assisted takeoff (JATO) units on military aircraft. Previous solid rockets had operated for only 2 or 3 seconds. Frustrated with explosions resulting from cracks in compressed powder after storage, in June 1942 the mostly self-trained explosives expert John W. Parsons came up with a combination of asphalt as a binder and potassium (later changed to ammonium) perchlorate as an oxidizer to form the first castable, composite solid propellant. Although the specific ingredients in later composite propellants and the configuration of the propellant grain (mass) were to evolve significantly, this discovery was the fundamental breakthrough to the sophisticated composite propellants used into the 1990s.

As the GALCIT rocket research project evolved to become the Jet Propulsion Laboratory (JPL), a group of engineers around Charles Bartley advanced the development of composite propellants by replacing the asphalt with a Thiokol Chemical Corporation polysulfide polymer and incorporating the British concept of an internal-burning, star-shaped grain. As a result, three firms with organizational linkages to JPL—Thiokol, Aerojet General Corporation, and Lockheed—came to dominate the production and evolution of composite propellants for large, solid missiles and space boosters.

The other fundamental World War II innovation in rocketry was a process developed by a group around Drs. Henry M. Shuey and John F. Kincaid at the National Defense Research Committee's Explosives Research Laboratory in Bruceton, Pa. It yielded a castable, double-base propellant of comparable specific impulse (a measure of thrust) to the composite propellants made by Aerojet and Thiokol for large missiles, although it was used mainly in upper stages of those missiles. Hercules Powder Company, an offshoot of DuPont, became the principal producer of these rocket motors and others used for tactical missiles.

Other key innovations necessary for the deployment of the first large U.S. solid missiles, Polaris and Minuteman, in the early 1960s included reducing warhead size, improving guidance and control, producing lighter-weight but high-strength materials for the combustion chamber as well as heat-resistant materials for the nozzles, and providing thrust-vector-control and thrust-termination devices. These had little to do with propellants themselves, but one major innovation that did was the addition of aluminum powder to the propellant grain to increase specific impulse sufficiently to power large, solid missiles. Where others had tried and failed, Keith Rumbel and Charles B. Henderson with the Atlantic Research Corp. succeeded in discovering, under contract with the U.S. Navy, that large additions of aluminum (some 15–20 percent of the propellant composition) added enough to the specific impulse to make Polaris feasible.

This permitted the conversion from the more cumbersome liquid-propellant missiles like Atlas and Thor to solid missiles that could be fired from submarines and launched more quickly and easily from land, ranging from Polaris and Minuteman to Trident and MX. In the process, many chemists, such as Karl Klager at Aerojet, contributed to the evolution of ingredients in the propellant grain that permitted increasingly sophisticated solid propellants and to the development of compatible liners separating them from the motor case that also served as a combustion chamber. These propellant ingredients included polyurethane and such unpronounceable as polybutadieneacrylic acid (PBAA), cyclotetramethylenetetranitramine (HMX), polybutadiene acrylonitrile (PBAN), carboxy-terminated polybutadiene (CTPB), and hydroxy-terminated polybutadiene (HTPB).

These and other ingredients in composite propellants, combined with a great variety of internal-burning grain configurations, have yielded not just the missiles of the Cold War but also a variety of strap-on solid rocket boosters used in space launches, such as those on the Titan III and IV, the Space Shuttle, and the European Space Agency's Ariane 3, 4, and 5. Furthermore, while most heavy space launchers employ liquid-propellants because of their higher specific impulses and their ability to stop and start on command, solid propellants are used for perigee and apogee motors, which move satellites from one orbit to another. And some space launch vehicles like the U.S. Pegasus and Taurus use solid motors not just as boosters but in all of their stages.—J.D.H.

See also EXPLOSIVES; GUNPOWDER; MISSILES, CRUISE; MISSILES, GUIDED; MISSILES, SUBMARINE-LAUNCHED; POLYMERS; SPACE SHUTTLE; SUBMARINE, TRIDENT

Rotary Printing Press—see Printing Press, Rotary

RU-486

Conventional birth control pills provide a safe and effective method of contraception. In the late 1970s, a new contraceptive was developed, one that offered an alternative approach to preventing pregnancy. As an outgrowth of his research into steroids, Etienne-Emile Baulieu, a French physician and medical researcher, discovered a chemical that effectively blocked progesterone, a human hormone that helps to maintain pregnancy. This chemical was called RU-486 because it was the 38,486th compound tested at Rousel-Uclaf, a major French chemical firm.

Unlike conventional birth control pills that work by preventing the occurrence of conception, RU-486 goes into action after an ovum has been fertilized by a sperm. Working in conjunction with prostaglandin, a drug administered 2 days later, RU-486 blocks progesterone. This prevents the next stage of pregnancy, the implantation of the fertilized ovum on the wall of the uterus. Because conception has already occurred, RU-486, strictly speaking, causes an abortion.

After successful animal tests, RU-486 was made available in France, where it soon found a large market. By 1991, approximately 100,000 French women had received the drug. In only 3 percent of the cases was the abortion unsuccessful or incomplete, and in only 1 percent of the cases did women experience excessive bleeding. RU-486 seems to meet any reasonable criteria for safety and effectiveness.

RU-486 is widely used in France and other parts of the world because it does what no other means of birth control does: Unlike other methods that require some degree of planning prior to intercourse, RU 486 is taken afterwards. But it is this unique mode of action that has deeply disturbed many people. Some critics charge that because it serves as a "morning after" pill, RU-486 promotes promiscuity. Even more troubling to many is the fact that the drug works by inducing an abortion. The destruction of the fertilized ovum takes place at a very early stage of the embryo's development, but for those who believe that human life begins at the moment of conception, the early abortion is a form of murder. Organized protests forced the withdrawal of RU-486 from the French market in 1988, but counterprotests by gynecologists and other medical practitioners were soon mounted, and the French Minister of Health ordered its manufacturer to put RU-486 back on the market, claiming that it had become "the moral property of women, not just the property of the drug company."

In the United States, opponents of abortion have prevented the use of RU-486 for the termination of pregnancy; any company that attempted to market it would surely face an organized boycott of all of its products. It is, however, used for the treatment of Cushing's disease, for it blocks the receptors for cortisol, a hormone produced in excessive amounts by those afflicted with the disease. Researchers in Europe are also experimenting with RU-486 as a possible treatment for breast cancer and certain kinds of brain tumors.

See also CONTRACEPTIVES, ORAL; STEROIDS, ANABOLIC

Rubber

Natural rubber is the latex of the hevea tree (*Hevea brasiliensus*). Confined initially to the southern portion of the New World, rubber was used for a number of articles by Native Americans. It was introduced to Europe by the French astronomer de la Condamine in 1740, but remained little more than a curiosity for many decades. In the 19th century, rubber found its first substantial market when it was used for a waterproof jacket known as a *mackintosh*, the name having been derived from its manufacturer. Still, rubber remained a troublesome material, crumbly when cold and gummy when warm, until Charles Goodyear discovered the process of vulcanization in 1839.

Vulcanized rubber created a large and growing demand for latex that was extracted from trees growing in a wild or semiwild state in Brazil and the Congo. In

1876, an Englishman named H. A. Wickham managed to smuggle 70,000 rubber tree seeds out of Brazil. The seeds were germinated at the Royal Botanical Gardens at Kew, and then sent to Ceylon (today's Sri Lanka) and Malaya (today's Malaysia), where they formed the basis for a vast expanse of rubber plantations. Although these plantations were well established by the end of the 19th century, the market for rubber was still comparatively modest; world consumption was only 45,900 metric tons (50,500 tons) in 1900.

Demand for rubber exploded in the years immediately following as a result of the rapid growth of automobile ownership. With its seemingly insatiable appetite for tires, the automobile created a market for rubber that had been scarcely imagined up until then; by 1939, world consumption stood at 1,270,000 metric tons (1,400,000 tons). The year 1939 also marked the beginning of World War II, and an even larger demand for rubber. At the same time, however, the Japanese conquest of Malaya in 1942 cut off a major source of natural rubber from the Allied powers. These events were a powerful stimulus to the development of a new industry for the production of synthetic rubber.

The first form of synthetic rubber had been created in 1882 by William Tilden (1842–1926), a British chemist, when he obtained isoprene by cracking turpentine. Even though he was able to vulcanize this substance in 1885, it had no commercial value. During World War I, German chemists used a polymerization process invented by Ivan Kondakov (1857–1931), a Russian chemist, to produce a form of synthetic rubber made from butadiene. The butadiene, in turn, had been obtained from butylene, a petroleum by-product.

The subsequent development of synthetic rubber was thwarted by the low price of natural rubber during the early interwar years. Neoprene rubber, invented by Arnold Collins and Wallace Carothers at DuPont in 1931, found a marketing niche due to its oil-resistant properties, but the major impetus for the development of synthetic rubber was more political than economic or technical. As part of its effort to insulate itself from the capitalist world economy, the Soviet Union was the first country to develop a synthetic-rubber industry. Beginning with a pilot plant in 1930, by 1939 it was producing about 81,000 metric tons (90,000 tons) of synthetic rubber annually. Another country with autarchical ambitions was Nazi Germany. Looking toward the day when Germany might need to be self-sufficient in certain key materials, Hitler asked the German chemical industry to resume research into synthetic rubber. Two German chemical firms, BASF and Bayer, produced Buna-S rubber in 1934, so named because it was produced by the action of sodium (chemical symbol Na) on butadiene. Buna-N, a similar synthetic, began to be made in small quantities by I. G. Farben at about the same time.

The right to use I. G. Farben's technology for both types of synthetic rubber had been acquired by Standard Oil of New Jersey in the 1930s. Standard's ownership of these patents took on great importance after America's entry into World War II. In early 1942, its stockpile of rubber stood at 450,000 metric tons (500,000 tons), a quantity that would meet only two-thirds of a year's need. In response, the federal government initiated a ambitious program to produce synthetic rubber in large quantities. The government already had established an organization known as the Rubber Reserve Company, which requested in 1941, that four major U.S. rubber manufacturers each build a synthetic rubber plant with a capacity of 10,000 tons (9,070 metric tons). It upped the volume to 30,000 tons (27,000 metric tons) shortly after America entered the war.

The government's desire to rapidly increase stocks of rubber was understandable, but the targets were very difficult to meet at first, for the industry faced critical shortages of the styrene and butadiene used as feedstocks. No less important was a lack of technical knowledge. Standard's butadiene patents were the foundation of the program, but as is often the case, holding a patent does not guarantee the possession of sufficient know-how to successfully exploit it.

Hindered by management that was not always up to the task, the U.S. synthetic rubber program was hardly a model for the successful administration of technological advance, but it eventually produced about 770,000 metric tons (850,000 tons) of GR-S (government rubber-styrene) rubber similar to Buna-S. The production of synthetic rubber went into temporary decline after World War II, but output began to increase around 1950. Within 10 years, the production of synthetic rubber had outstripped the production of rubber derived from natural sources. No less important, the development of the synthetic-rubber industry played a large role in the great expansion of polymer science and technology in the postwar era.

See also AUTOMOBILE; POLYMERS; THERMAL CRACKING; TIRE, PNEUMATIC; VULCANIZED RUBBER

Further Reading: Vernon Herbert and Attilio Bisio, *Synthetic Rubber: A Project That Had to Succeed*, 1985.

Rubber, Vulcanized

Most natural rubber is derived from latex produced by the hevea tree (*Hevea brasiliensus*). Unknown in Europe until it was brought back from the Americas, rubber elicited little interest for many decades. Its first significant commercial application came at the beginning of the 19th century, when Charles Macintosh (1766–1843) in Scotland began to manufacture garments made from two layers of cloth with a layer of rubber in between. The success of this venture was dependent on Macintosh's discovery that naphtha is an effective solvent for rubber.

Macintosh's jackets were so successful that *macintosh* became a generic name for this article of clothing. But rubber was used for little else except pencil erasers and crude rubber shoes. Its use was severely hampered by its tendency to become brittle and crumbly in cold weather, and gummy in warm weather. In the 1830s, experimenters in Germany and the United States discovered that the addition of sulfur made rubber less sticky, but the deleterious effects of hot and cold temperatures remained.

At about the same time, Charles Goodyear (1800–1860) in the United States was having some success with rubber treated with nitric acid, but the surface hardening it induced was not permanent. Goodyear persisted in his efforts, and one day in 1839, by fortunate coincidence he knocked over a container filled with latex, sulfur, and white lead. The contents fell on a hot stove, and after things had cooled down, what remained was a glob of rubber that remained stable with no loss of elasticity. We now know that the combination of latex, sulfur, white lead, and heat had caused the latex to polymerize, that is, to form the long molecular chains that confer the essential properties of modern rubber.

Goodyear began to manufacture a number of articles from the new form of rubber. At about the same time, Thomas Hancock (1786–1865) in England was developing techniques for processing rubber. In 1844, he secured a patent on the treatment of latex with sulfur and heat—a year before Goodyear was awarded an American patent on the process. A friend of Hancock gave the process its name, *vulcanization*, after a mythological blacksmith who lived in volcanoes.

After considerable litigation, Goodyear's American patents were sustained, but he failed to capitalize on them. Goodyear set up a number of manufacturing enterprises, but he undercut his position by virtually giving away licenses and manufacturing rights. No stranger to poverty, Goodyear had been thrown into debtors' prison on several occasions before and after his success with rubber. He died at the age of 60, owing $200,000 to his creditors.

Charles Goodyear, who once had hoped to be a minister, had a missionary zeal in promoting the virtues of rubber. Part of his promotional effort consisted of listing hundreds of new uses for rubber. Ironically, rubber tires were nowhere mentioned.

See also POLYMERS; RUBBER; TIRE, PNEUMATIC

S

Saccharin

Saccharin, a widely used sugar substitute, was discovered by accident. One evening in 1879, Constantine Fahlberg (1850–1910), a researcher in Ira Remsen's (1846–1927) chemical laboratory at Johns Hopkins University, noticed an unusually sweet taste in the food he was eating. He correctly guessed that the sweet taste came not from the food but from some chemical that had adhered to his hands. After testing a number of chemicals in his laboratory, he tracked down the sweet taste to the sodium salt of o-sulfobenzimide. He named it *saccharin*, from the Latin word for sugar, *saccharum*. Actually, saccharin is far sweeter than the substance for which it was named, having 300 to 500 times the sweetness of sugar.

Fahlberg received a patent on saccharin in 1885, setting off a bitter dispute with Remsen over the "ownership" of saccharin. Remsen was not interested in the financial rewards (which proved to be substantial), but he wanted recognition as the codiscoverer of the sweetener. In Remsen's eyes, Fahlberg had violated the commonly accepted practice of giving the lab director a share of the credit for a discovery. A large pharmaceutical company offered to financially underwrite a patent suit against Fahlberg, but Remsen declined.

Commercial production of saccharin began in 1900. Financial success was threatened when an official of the Department of Agriculture (which was then charged with overseeing the food and drug industries) tried to have saccharin banned because it was allegedly "poisonous and deleterious to health." President Theodore Roosevelt, himself a user of saccharin, appointed a review committee that subsequently exonerated saccharin. In 1970, saccharin was once again indicted by the federal government when studies showed an increased risk of bladder cancer among rats that had ingested very large quantities of saccharin. The possible dangers of saccharin led to its being banned in Canada. The U.S. Food and Drug Administration refrained from banning saccharin outright, but foods and drugs containing it are required to have a warning label noting that "This product contains saccharin which has been determined to cause cancer in laboratory animals." It is not known if these labels have produced a significant lowering of saccharin consumption. Also open to question is the relative danger of saccharin use when it is compared with the health risks of obesity and other conditions exacerbated by the consumption of sugar.

See also DELANY CLAUSE; FOOD AND DRUG ADMINISTRATION, U.S.; PATENTS; RISK EVALUATION

Safety Lamp

Mining has always been a dangerous occupation. For centuries, one of the things that made mining especially hazardous was underground explosions. All too often a pocket of "fire-damp," a combustible gas largely composed of methane, was touched off by a naked light. The resultant blast would often kill dozens of underground workers. Miners tried to minimize the danger by using phosphorescent materials, such as certain kinds of rotting fish, but this was obviously not a satisfactory solution.

After one explosion killed 92 English miners in 1815, the Society in Sunderland for Preventing Accidents in Mines appealed to the renowned British scientist Humphrey Davy (1778–1829) to devise a safer means of illuminating the mines. Within a few months Davy invented a lamp that was based on what he had learned about the combustion of fire-damp. To combust, the gas required between 6 and 14 times its volume of air. Moreover, it would not explode in a tube with a diameter of less than $\frac{1}{2}$ inch (1.27 cm). Davy devised four different lamps, settling on the simplest version, which worked as well as the others. It consisted of a candle or oil lamp burning inside an enclosure that received air from a set of small tubes. Davy later found that the tubes could be replaced by fine gauze, provided that each hole had a diameter equal to its depth; 740 holes per square inch came to be the standard. This arrangement had the further benefit of causing the light to burn more brightly in the presence of fire-damp, thereby giving the miner a warning of impending danger. Davy did not patent his invention, satisfied with the knowledge that his safety lamp would likely save many lives.

At about the time that Davy was inventing his lamp, a similar device was being created by George Stephenson (1781–1848), who went on to great success as a pioneering builder of railroads and their equipment. The simultaneous invention of the safety lamp thus exemplifies the process whereby technological change may be stimulated by a particular social environment.

The safety lamp also illustrates another aspect of technological change: An apparent boon produced by a technology may be partially negated by its misuse. In the years following the invention of the safety lamp, miners sometimes worked in areas that they would have avoided in the past. This led to lethal explosions being touched off by miners whose safety lamps had given them a false sense of security.

See also MINING, UNDERGROUND

Sailing Ships—see Ships, Sailing

Salt

In chemical terms, a salt is any compound that is formed when one or more hydrogen ions from an acid are replaced by one or more metal ions. In everyday usage, *salt* refers to only one of these, sodium chloride (NaCl). Salt is essential to human survival, for it helps to maintain a proper balance of liquids in the body. Amounts of 6 to 8 grams (.21–.28 oz) of salt must be ingested every day, and even more must be consumed when it is lost through heavy perspiration. Getting enough salt is not a problem for people with normal diets; instead, the problem may be excessive salt consumption, as most diets supply us with more than 15 grams of salt daily. Salt consumption has been linked to high blood pressure and other cardiovascular problems; people suffering from these disorders are often put on low-salt diets.

In addition to being necessary for human survival, salt has played an important role in the preservation of food. Long before there was canning or refrigeration, salt was being used to preserve meat, fish, and vegetables. Ancient civilizations as diverse as Rome and China made heavy use of salted foods. Because of its value, salt also served as a currency. Roman soldiers, for example, were once paid in salt. When soldiers were eventually given money to purchase salt, the funds were known as *salarium*, from which our word *salary* is derived. In the 19th century, salt took on an additional importance as the starting point in the production of sodium carbonate (Na_2CO_3), a chemical compound that was widely employed in the glass, paper, soap, and textile industries.

Salt is obtained by extracting it from underground deposits, and from brine or seawater. In the former case, salt deposits left from the evaporation of ancient seas are mined in underground or surface mines. Underground salt deposits can also be dissolved *in situ* by the injection of water; the resultant brine is then brought to the surface and boiled. Since salt deposits large enough to support mining are not found everywhere, much of the salt used throughout human history has been obtained from the ocean. Seawater contains an average of 30 to 35 grams (1.06–1.23 oz) of various salts per liter. By far the most common of these is sodium chloride, which comprises 70 percent of the salt in seawater, followed by magnesium chloride ($MgCl_2$) at 10.8 percent, magnesium sulfide ($MgSO_4$) at 4.7 percent, calcium sulfide ($CaSO_4$) at 3.6 percent, and potassium sulfide (K_2SO_4) at 3.5 percent, as well as traces of other compounds. For centuries, the extraction of salt from seawater was done through natural evaporation and through boiling. Processes based on evaporation used large, shallow salt pans that held seawater or brine from salt marshes. The water evaporated after a few months, leaving the salt behind. This technique required little energy (other than supplied by the workers, which was substantial), but the product that resulted left something to be desired, as it contained many impurities, including salt-loving microorganisms that caused spoilage when the salt was used for preserving food. Evaporation is still used today, but the salt is continuously removed as it forms at the bottom of the pan, and then goes through a filtration process to remove impurities.

Salt is also produced by boiling saltwater. This was a common technique in China, where natural gas began to be used for this purpose 2,000 years ago. China also pioneered the use of percussion drilling to obtain brine. Today, salt is often extracted from saltwater through a more involved process that entails boiling at low pressures. One widely employed method uses three stages with progressively lower pressures; in the third stage, the low pressure causes the saltwater to boil at only 43°C (110°F). The resultant salt slurry is then filtered, dewatered, and dried at 177°C (350°F) before being screened and packed. In many cases, salt intended for the kitchen and table has iodine added to it. The widespread use of "iodized" salt has prevented many cases of goiter for people whose diet does not include sufficient amounts of iodine.

Since seawater or salt deposits are not found everywhere, there has always been a commercial trade in salt. Throughout history, trading nations and city-states have amassed great fortunes through the salt trade. One of these was Venice, whose salt merchants once stole the body of St. Mark from Alexandria and returned it to their homeland in a cargo of salt.

The importance of salt was readily appreciated by governments throughout the world, which often monopolized the sale of salt or slapped a substantial tax on it. These taxes could produce substantial income for the government. During the early 1900s, China's salt tax accounted for an estimated 7 percent of the central government's revenues. In France, the salt tax known as the *gabelle* was a major irritant that substantially increased the cost of living for common people. Although it filled the government's coffers, the salt tax was one of the prime causes of the French Revolu-

tion. It was abolished in 1790, but in 1804 Napoleon put a tax of 2 centimes on every kilogram of salt, a rate that remained until 1945.

See also ALKALI PRODUCTION; CANNED FOOD AND BEVERAGES; NATURAL GAS; OIL AND GAS EXPLORATION; REFRIGERATOR

Further Reading: Robert Multhauf, *Neptune's Gift: A History of Common Salt*, 1978.

SALT I and SALT II

Beginning in 1945, the destructive power of atomic weapons made it evident that a full-scale nuclear war would be like nothing ever experienced before. In the 1960s, the two superpowers, the United States and the Soviet Union, began a series of talks and agreements aimed at making the outbreak of a nuclear war less likely. The first of these was the 1963 Limited Test Ban Treaty, which prohibited the testing of nuclear weapons above ground. Since its primary goal was to limit radioactive fallout in the atmosphere, it did nothing stop the manufacture and deployment of nuclear weapons. At the same time, however, the treaty served to demonstrate that the two powers could hammer out a workable weapons limitation agreement.

In 1964, President Lyndon Johnson proposed a freeze on the number of strategic nuclear weapons. Since compliance was to be monitored by onsite inspection, the Soviets rejected it out of hand. In 1967, Johnson threatened to initiate an antiballistic missile (ABM) system if arms limitations talks did not get underway. This got the attention of the Soviet leadership, who soon indicated their willingness to begin a dialogue. Negotiations began in 1969 after having been delayed by the Soviet invasion of Czechoslovakia during the previous year.

Both parties had the same basic interests in entering negotiations. Both wanted to prevent a nuclear war that neither could really win, and both hoped to reduce the financial drain of an endless peacetime military buildup. New military technologies also entered into the picture. The United States and the Soviet Union worried about the destabilizing consequences of ABMs, as well as the emergence of nuclear missile warheads known as MIRVs. On a more positive note, technical advances in spy satellites promised accurate yet noninvasive means of monitoring the other side's compliance with any agreements that might be forged.

The first Strategic Arms Limitation Treaty, or SALT I, limited intercontinental ballistic missiles (ICBMs) and submarine-launched ballistic missiles (SLBMs) to the number existing at the time. It also stipulated that the size of their launching silos could only increase by 10 to 15 percent. SALT I also limited both sides to only two ABM sites each. Individual sites were limited to 100 launchers and missiles, and restrictions were also placed on the radar systems they employed. The most significant omission was that no limits were put on the deployment of MIRVs. The treaty was signed on May 26, 1972, by Richard Nixon and Leonid Brezhnev, and was subsequently ratified by the U.S. Senate.

SALT I reflected the political realities of time. The United States was well along with the development of MIRV technology, and was there was a strong unwillingness to abandon such a promising weapon. At the same time, however, the ABM had limited support, having been previously approved by only a one-vote margin in the U.S. Senate. Hence, it was easy to allow its virtual abandonment.

SALT I was intended to be a prelude to more far-reaching arms limitations agreements. The first of these was a 1974 agreement worked out in Vladivostok that froze at 1,400 the total number of strategic weapons (ICBMs, SLBMs, and intercontinental bombers) that could be deployed. According to the agreement, a total of 1,320 ICBMs and SLBMs could be equipped with MIRVs.

SALT II was a considerably more ambitious attempt at limiting strategic arms. The treaty limited the total number of MIRV-equipped missiles, as well as the number of MIRVs that could be used with particular kinds of missiles, such as the U.S. Minuteman and the Soviet SS-18. SALT II also put limitations on the upgrading of missiles and their launchers. The treaty limited the spread of cruise missiles by requiring reductions in the number of MIRVed ICBMs and SLBMs if more than 120 cruise-missile–carrying bombers were deployed. It prohibited interference with each nation's ability to verify compliance, such as the encryption of telemetric data. The treaty also prohibited the use of emerging technologies such as orbital nuclear weapons and long-range, underwater-launched cruise missiles.

SALT II engendered a great deal of political opposition in the United States Some liberals felt that it didn't go nearly far enough, whereas some conservatives were of the opinion that it put the United States at a military disadvantage. In particular, objections were raised concerning the limitation of cruise missiles and a provision that allowed the Soviet Union to deploy 308 heavy SS-18 ICBMs. Concerns were also raised about the largely uncontrolled Soviet Backfire bomber, which some Western military analysts considered to have intercontinental capabilities. Making matters worse, the relationship between the two countries had deteriorated since the early days of detente and SALT I; in particular, the invasion of Afghanistan in 1979 seemed to demonstrate the aggressive intentions of the Soviet Union. The opposition prevailed in the U.S Senate, and the treaty was never ratified, although its provisions were honored until 1987, when the Reagan administration exceeded the limitation on the number of SLBMs. At the same time, the United States indicated its willingness to engage in another round of arms limitations talks, which were given the acronym START, for Strategic Arms Reduction Treaty.

In considering the overall consequences of SALT I and SALT II, it is important to keep in mind that the goal of the treaties was to reduce the threat of nuclear war, not to provide either side with a military advantage. Both the United States and the Soviet Union accepted the need to maintain mutually assured destruction (MAD), the certainty that a first strike would be followed by a devastating retaliatory blow against the aggressor. The goal of the SALT negotiations was to preserve a strategic parity by ensuring that neither side had either an overall superiority in offensive weaponry or the confidence to launch a first strike because it could protect itself from a retaliatory strike.

SALT I was successful because it harmonized with the individual interests of the two countries. With their massive nuclear arsenals, neither side was particularly interested in adding still more weaponry. At the same time, ABM systems were seen by both sides as expensive and technologically dubious. Moreover, they carried the threat of upsetting the nuclear balance. Were they to become effective, ABMs might confer a first-strike capability by giving an aggressor the confidence that a retaliatory strike would be successfully intercepted. In these ways, the SALT agreements reinforced a status quo that wasn't perfect but over the years prevented the Cold War from turning into a hot one.

At the same time, however, the SALT agreements did little to prevent the spread of one of the most potentially destabilizing items in the nuclear arsenal, the MIRV. MIRVs posed a new problem since they held out the prospect of obliterating an adversary's launch facilities and preventing a successful retaliatory strike. But these weapons had gained a great deal of technical and political momentum in the 1970s, and putting sharp limits on their deployment proved impossible.

The difficulty of controlling new weapons like MIRVs points to a general problem with the SALT agreements and with arms-control efforts in general. These agreements have for the most part centered on putting limits on the number of particular weapons that may be deployed. But in the present era the real threat to political and military stability is the emergence of new military technologies, not increases in the number of existing weapons. Arms-control agreements like SALT I and SALT II have often been effective in limiting the spread of weapons that can be easily counted, but they have had a much harder time in dealing with weapons that provide new capabilities altogether.

See also ATOMIC BOMB; MISSILES, CRUISE; MISSILE, INTERCONTINENTAL BALLISTIC (ICBM); MISSILES, MULTIPLE INDEPENDENTLY TARGETED REENTRY VEHICLES (MIRVs); MISSILES, SUBMARINE-LAUNCHED; STRATEGIC ARMS REDUCTION TALKS

Sampling

We are constantly bombarded with the results of various studies based on samples of people. The A. G. Nielsen Company produces ratings of television shows based on a sample of households in the United States. The morning newspaper brings news of the result of a poll of all voters, or Republican voters, or women voters, all of them based on a sample of respondents. Selecting a sample for study and for use in decisionmaking has become commonplace, and yet few of us really know what is involved in sampling.

When we set out to take a sample, we need to be

clear about the nature of that process and what we need to know in order to design a good sampling procedure. To begin with, we have to be clear about the population we wish to sample. For instance, we might want to sample all adults in a community. Alternatively, we might want to survey only those registered voters who voted in the last general elections. The first step, then, is to be clear about the population of interest. Next, we must decide whether we need to take a simple random sample or whether some form of stratification will be best for our purposes.

To assess the state of public opinion on some issue in a community, we simply could go to the largest supermarket in that community and attempt to interview every registered voter who entered the store. We would get a set of answers, but we could not claim that those answers were representative of opinions in the community. Depending on the time of day and the weather, we might get a disproportionately large number of women, or young people, or night workers, or commuters.

What is needed is a principle to guide our sampling that will minimize chances of such obvious bias. The principle can be stated as follows: In order to ensure randomness in any sample we select, we need to devise a sampling scheme that ensures that each person or element in the population has the same chance of being selected into our sample. Following this principle will get us a simple random sample.

In terms of our example above, we might proceed as follows: Obtain from the registrar of voters in our community a list of all registered voters. Assign each voter an identification number. Then, using a random-number table found in any introductory statistics book, select 100 random numbers. Go back to our list of registered voters and pull out each of those voters whose identification number has been randomly selected. In this manner, we can obtain a random sample and one that can be said to be representative of our population. To repeat, the key principle here is to devise a sampling procedure that results in each member of the population having the same chance of being selected into our sample.

For some purposes, a simple random sample might not suit our purposes, whereas a stratified random sample would. Here we would divide the population into appropriate strata and then select a random sample from within each strata. For example, in a study of attitudes toward housing, we might want to divide our population into two groups, homeowners and renters, and select a random sample of each. This would allow us to get at any differences between these two groups in a more efficient fashion than would a simple random sample.

Finally, in some cases where it is not possible to maintain a list of all members of the population (e.g., all adults in the United States), we could resort to multistage sampling in which we first draw a random sample of all counties in the United States, then from each county in the sample draw a random sample of blocks, and then for each block develop a list of adults living on each block and draw a random sample of adults from those randomly selected blocks.

Some issues to be sensitive to when samples are drawn include the following: First, it is likely that the population list in any sample—a simple random sample, a stratified sample, or a multistage sample—is not 100 percent accurate. The Nielsen Company, for example, feels that its list covers about 98 percent of all U.S. households. Some households and some people (e.g., the homeless) will not be represented in their sample.

Second, whenever we select a sample, we do not expect any statistic we calculate, such as a sample mean, to equal the population mean. Rather, we expect to find some differences between sample values and the population value, differences that are referred to as *sampling error*. Part of the goal of statistics is to devise tools that allow us to calculate the chance that our sample was drawn from a population with a specific population mean. When making statistical calculations we need to take into account sampling error. That is, in selecting a sample, we need to be sensitive to the fact that sample statistics will likely differ from population values, and we need to use the tools of statistics to assess the role that sampling error plays in our sampling design.

An important part of designing a sampling procedure involves deciding on the sample size. Such a decision will be influenced by two factors: the amount of money we have to spend in selecting our sample and the amount of error we can tolerate in our statistical analysis. The former is determined by our resources. The latter is based on statistical considerations de-

scribed in most statistics textbooks. We can be guided by the following: the larger our sample, the lower will be our sampling error. However, beyond a certain point there are diminishing reductions in error as we increase sample size. Thus, we do not want to just go out and select a large sample. Rather, we will employ accepted statistical techniques in order to select a sample size with a tolerable error factor.—J.D.S.

See also STATISTICAL INFERENCE; STATISTICS, DESCRIPTIVE

Further Reading: David S. Moore, *The Basic Practice of Statistics*, 1995.

Sanitary Napkin

Until the 20th century, most women used diaperlike garments to absorb the discharge from their monthly periods. Disposable pads made from gauze-covered cotton were produced in the United States by Johnson and Johnson in 1896, but the inability to advertise an "unmentionable" product limited their commercial success, and the pads were eventually withdrawn from the market. During World War I, French nurses began to employ cellulose pads used for bandaging wounds when they had their periods. This was a considerable improvement, for while cloth garments were washed and reused, the pads could simply be thrown away. Kimberly-Clark's Kotex, the first commercially successful disposable sanitary napkin, was put on the market in 1920.

Early sanitary napkins generated respectable incomes for their manufacturers, even though these products were not advertised. Even women's magazines did not run advertisements for sanitary napkins until the 1930s, and for many years the ads said nothing directly about the product itself, but traded in oblique terms like "freshness" and "daintiness." The electronic media were even slower to accept ads for sanitary protection products; radio and television stations that subscribed to the code of the National Association of Broadcasters did not begin to run them until 1972.

Sanitary napkin technology changed only in details until the 1970s, when smaller pads that used a strip of adhesive to fasten them to panties appeared, obviating the need for special belts. By this time, many women were using tampons instead of sanitary napkins, although sanitary napkins still had a large market.

Many factors contributed to the liberation of 20th-century women from confining traditional roles. Without overstating the case, it can be fairly said that beginning in the 1920s, sanitary napkins played a significant role in helping women move through the world with greater freedom and confidence.

See also TAMPON

Further Reading: Janice Delany, Mary Jane Lupton, and Emily Toth, *The Curse: A Cultural History of Menstruation*, 1988.

Sanitary Plumbing Fixtures

In the West, bodily cleanliness has fallen in and out of fashion over the centuries, as reflected in the artifacts of bathing. In the Roman era, citizens enjoyed communal baths in sunken pools filled with warm waters piped in directly from hot springs or heated in underground channels. The precursors to modern hot tubs—large, wooden, communal tubs—gained popularity in the stews, or bathhouses, of European towns during the Crusades, until moralists criticized mixed-gender bathing. Used by royalty since ancient times, single-person metal tubs were included among the bedroom furnishings of 18th-century elites, but individualized tubs equipped with running water and drains were not common in European and American homes until recent times.

If people did not readily submerge themselves in tubs of water, they washed parts of their bodies using a range of vessels, from basins to bidets. Since the ancient era, people cleansed their hair using sets of ewers and pans. In medieval times, diners washed greasy hands at the table in water from shared bowls made from wood, metal, or ceramics. In private chambers, they rinsed teeth, faces, and hands in stone basins, or lavers, sometimes equipped with portable reservoirs that had spouts or taps. Specialized accessories, such as bidets and shaving tables, date from the early modern era, as washing was growing more commonplace, but these forms remained relatively rare. As late as the mid-18th century, makeshift utensils sufficed for most bathers, including English author Horace Walpole

(1717–1797), who washed by simply dunking his head into a bucket of cold water each morning.

The earliest sanitary utensils for the disposal of human waste were probably chipped or battered jars, kept at bedsides for nightly urination. Medieval sleepers relied on vase-shaped containers for waste, while Renaissance slumberers used the wide-mouthed metal or pottery chamber pots, or jerries, which remained the norm for centuries. Chamber pots were the ultimate dual-purpose vessels, for they served as repositories of real and symbolic refuse; potters never hesitated to embellish the interiors of jerries with low-relief sculptures of wart-covered frogs or painted images of crooked politicians. In Western ceramics manufacturing districts, 19th-century potters skilled at making chamber pots eventually utilized their knowledge of clays and glazes to manufacture portable toilet sets for bedroom use. A typical suite included a large water pitcher, a washbasin, a shaving mug, a soap dish, a toothbrush holder, a small water pitcher, a slop pail (for dirty water), and one or two chamber pots.

The modern bathroom equipped with a built-in sink or lavatory, a flush toilet, and a 5-foot bathtub dates from the mid-19th century. Then, physicians, architects, plumbers, manufacturers, and health reformers combined their expertise to improve public and private water and sewage systems, motivated in part by consumers' desires for healthy homes. By midcentury, more and more American middle-class households used piped water and a mixture of portable sanitary devices and built-in fixtures. By the century's end, potters in England and the United States increasingly made a wide variety of ceramic fixtures designed for buildings with running water and sewage disposal systems. Then, waist-high sinks displaced moveable basins and pitchers, while flush toilets replaced chamber pots and outhouses. The footed bathtub, sometimes ornamented with Chippendale-style claw-and-ball feet, was standard issue until the early 20th century, when manufacturers of enamelled metal products embraced new fabricating techniques that enabled the low-cost production of the sunken tub.—R.L.B.

See also SEWERS; TAMPON; TOILET, FLUSH

Satellites, Spy

Most countries look upon reconnaissance flights over their territory as hostile acts. No nation was more concerned about overflights by foreign aircraft, and no nation was more committed to secrecy, than the Soviet Union. It is thus ironic that this country began a process that opened it to prying eyes by launching the first artificial satellite in 1957. Although early Soviet and U.S. satellites did not have reconnaissance capabilities, it wasn't long before spying became a major activity of the two countries' space programs.

In August 1960, the United States concluded its first successful in-space spy mission when a photographic payload was recovered from a *Discoverer* satellite. A major spy satellite program began with the launching of five SAMOS (Satellite and Missile Observation System) and then the orbiting of 145 Corona satellites during a 10-year period. The benefits of the spy satellite program soon became evident. During the Cuban missile crisis of 1962, U.S. satellite surveillance indicated that the Soviet Union was incapable of launching a first-strike nuclear attack; the Kennedy administration could therefore mount a naval blockade with no fear of retaliation. Two years later, satellite photos revealed that China was on the brink of testing its first atom bomb, while satellite photos taken in 1967 revealed the immense military destruction wrought by Israel against its attackers.

Beginning in the early 1970s, the United States launched a succession of Big Bird satellites. These 9,000-kg (20,000-lb) satellites took photographic images that were scanned onboard and sent back to Earth stations. If better photographic quality was required, the film capsules could be jettisoned and then recovered in midair by specially equipped aircraft. The current generation of American KH-11 spy satellites do not use photographic film; they rely entirely on digital imaging, which allows pictures to be sent back in real time. They also orbit at a higher altitude, which extends their operational life to perhaps 2 years or more. In addition to gathering photographic data, some satellites are used to monitor electronic communications. Known as *electronic intelligence* (ELINT) *satellites*—or "ferrets," to use their unofficial name—they can locate transmitters for radio and radar, eavesdrop on radio communica-

tions, and monitor telemetry form missile tests. The United States also is thought to have at least two radar-imaging Lacrosse satellites in orbit.

The quality of the photographic images sent from orbit has to be a matter for speculation, for much of the relevant information is classified. It can be reasonably estimated that objects as small as 30.5 cm (12 in.) or even 5 cm (2 in.) can be identified. Although popular accounts sometimes refer to the ability to read the headline on a newspaper held by a Moscow pedestrian, this is almost certainly an exaggeration. Even with perfect optics, resolution is adversely affected by fluctuations in the Earth's atmosphere caused by temperature differences. To some extent, these distortions are being overcome by adaptive optics technologies used with the latest generation of American spy satellites.

To achieve a high degree of photographic accuracy, it is necessary to select the proper orbital altitude, and to have this orbit angled correctly with respect to the equator, usually between 65 and 115 degrees. Spy satellites orbit at altitudes of 145 to 2,000 km (90–1,240 mi), and have orbits synchronous with the sun. This allows pictures to be taken at the same time every day, so day-to-day comparisons can be made without the need to take changing shadows into account. A low orbit will cause reentry in about a week due to atmospheric drag, but a proper altitude can be maintained by occasionally firing onboard thrusters. The most advanced satellites can use the thrusters to change their orbit so that an area of particular interest can be examined more frequently in the event of a crisis.

For all their technical sophistication, spy satellites suffer from one inescapable limitation: To be of use, the pictures they take have to be inspected by individual human beings. Since a satellite may generate thousands of pictures in a short space of time, this is a significant bottleneck. The need for individual inspectors also means that important military installations might go undetected. A missile site built in a remote area might be overlooked because data handlers on the ground lacked the time or the inclination to examine pictures that had been taken of this area.

Spy satellites have played a dual and even somewhat contradictory role as long as they have been around. On the one hand, during the Cold War they were an integral part of an arms buildup that threatened to destabilize an already tense relationship between the superpowers. On the other hand, spy satellites were, and continue to be, powerful tools for the effort to reduce armaments and lower tensions. By verifying the activities of each party, spy satellites lend confidence to the expectation that neither side is cheating on an agreement.

See also SPUTNIK

Saws, Wood

A saw is a toothed cutting tool made of a sheet or strip of metal which rasps or files a grove called the *kerf*. Typically, saw blades are made of a high-quality tempered steel. The teeth provide the cutting edges. In a particular saw, the teeth are a series of notches all the same size, pitch, bevel, and set. However, among the various types of saws, these characteristics vary considerably depending on the nature of cut to be made. Basically, there are three classes of teeth. For going across grain, the cutting teeth of a crosscut saw are arranged in parallel rows that are offset relative to each other. These teeth cut like little knives. For going with the grain, ripping teeth are set straight across in parallel rows so that they function like a series of chisels that peel away the wood's fibers. Saws with intermediate teeth perform both functions. A variety of saws is available to meet particular needs: rip, crosscut, back, hack, flooring, compass, pattern makers', keyhole, dovetail, bench, buck, stair builders', coping, and fret saws.

Whatever the job, the teeth must be sharp and accurately set for the saw to function properly. On large construction jobs, men used to be hired to sharpen and set saws for the carpenters. Likewise, expert saw filers were highly prized in large woodworking shops. To sharpen a saw, the teeth were hand filed until their extremities were sharp. The angle of the sharpened tooth varied from 60 to 70 degrees for cutting soft wood to 80 to 90 degrees for cutting hard woods. The teeth were then bent or set at an equal distance right and left, so that when a carpenter looked along the edge, the teeth formed two lines that slightly exceeded the thickness of the blade itself. The set allowed the saw dust to be removed as the cut was being made.

Handsaws remain common tools among all woodworkers. They consist of thin, flat blades of steel with a wooden handle on one end. Cutting wood by hand expends considerable energy, so skilled woodworkers maintained their tools and demonstrated an economy of motion when using their saws. A "greenhorn" or "wood butcher" could be identified easily, not only by the manner he used a saw but also by the care he employed in setting it down and picking it up.

Power saws consist of three types. In a reciprocating or fig saw, a blade moves up and down, cutting on each down stroke. Until early in the 19th century, these saws were common in sawmills; they also did the heavier work in furniture, railroad car, and pattern shops. Gangs saws with multiple blades cut large timbers and were capable of producing up to 30 boards at a time. But reciprocating saws shook violently and required a very slow feed, since only half the motion of the saw, the downswing, was a cutting motion. Consequently, circular saws replaced the reciprocating saw. The third type, the jig saw, remained useful in some furniture work, since it was capable of cutting interior patterns inaccessible to other saws.

A circular saw is a toothed steel disk that revolves on a shaft at high rates of speed. Unlike the reciprocating saw, the circular motion of this blade allows for continuous cutting and rapid feeds. In the 19th century, the manufacture of circular blades involved numerous steps, including drilling the center hole, toothing, knocking down (straightening), hardening, tempering, smithing, stamping, grinding, polishing, etching, filing, and careful inspection of the finished product. Blades were made as thin as possible to conserve lumber and power. But thinner blades were subject to wobble; when this happened, they made crooked cuts that offset any potential savings.

In sawmills, large circular saws were powerful and heavy duty. Logs were fed to the saw on carriages with power drives. However, the diameter of a circular saw blade must be twice the diameter of the largest log it cuts. Consequently, for very large timber, two circular blades were used with a smaller top saw positioned above a larger bottom saw. While effective, these saws cut wide kerfs, wasting prodigious amounts of wood. Smaller circular saws were mounted on frames with a table. Table saws functioned like hand saws and were the workhorse of most small shops. They were simple machines adaptable to both large batch and custom order work. Accessories included rolling and tilting tables, fences, spreader blades, and self-feeding mechanisms. Since World War II, the most common circular saw has been the skillsaw, a portable tool used by carpenters and found on every construction site.

Band saw blades are thin strips of tempered steel with teeth. They are formed into endless hoops that can run around two pulleys. Two guides hold the blade in position against the thrust of the wood, and a slot allows the blade to pass through a table. Either the table or the blade itself could be tilted to allow for an angled cut. Cooperage and furniture shops used saws with blades as narrow as 0.3 cm (.125 in.), but the largest bands were 30.5 cm (12 in.) wide and 18 m (60 ft) long. Theoretically, the band saw was superior to the circular saw because the blade was thinner and could be operated at higher speeds. Band saws also cut curves, allowing the saw to create intricate ornaments with greater speed than a reciprocating saw.

In practice, whether large or small, blades were extremely difficult to manufacture and seldom ran true. By the 1850s, French manufacturers achieved success brazing the ends of the blade to form a suitable hoop. However, band saws required a uniform rate of feed, since jarring could fracture the blade or drive it from the pulleys. The tension of the blade also remained problematic, for blades heated after cutting often snapped as they cooled. The solution entailed the use of a spring-loaded top bearing that contracted and expanded to adjust for the tension of the blade. By 1875, improved design and higher-quality steel made band saws common. Blades were double edged to cut in two directions, and some machines included multiple blades for resawing timber into boards. Like all woodworking machines, band saws were redesigned in the 20th century following the introduction of electricity and ball bearings. The redesign eliminated the potentially dangerous exposed belts that drove machines in the 19th century. Manufacturers also provided other safety features, although saws remain extremely dangerous when fed by hand.—J.C.B.

See also BEARINGS; MOTOR, ELECTRIC; STEEL IN THE 20TH CENTURY

Science-Based Technologies

It is often assumed that technology is simply applied science, and that once the scientific principle is discovered, technical applications will naturally flow from them. This is at best a half-truth. The relationship between science and technology is a complex one in which the influence does not always go in one direction. There are many instances where a scientific discovery was made possible by a prior technological advance. There are a plethora of technologies that owe virtually nothing to science, as well as many significant technologies that work, even though *why* they work is not well understood.

A brief review of the histories of science and technology shows that science and technology have usually occupied distinctive realms that only rarely intersected. The ancient Greeks made many contributions to the development of science and mathematics, as they did to mining, metallurgy, and building construction. At the same time, however, Greek technology did not draw upon Greek science to any significant degree. Medieval Europe was the scene of impressive advances in farming, construction, and the use of waterpower, but these achievements owed nothing to the science of the times. In the 18th and 19th centuries, the most advanced science was done in France, yet its scientific inferior England originated the key technologies that helped to launch the Industrial Revolution. Science and technology moved closer together in the 20th century, but few recent scientific discoveries have served as the basis for new technologies. And when scientific knowledge has been essential for the emergence of a particular technology, in most cases the science had been done many years before it found a technological application.

Science and technology differ in the goals pursued by their practitioners. Scientists concern themselves with "Is it true?", whereas engineers and inventors ask "Will it work?" As far as technology is concerned, it is often useful to understand the scientific principles underlying a particular technology, but it is not always essential. For example, people produced steel for centuries despite their being largely ignorant of the basic principles of metallurgy. The absence of scientific understanding has not always resulted in inferior technologies. The great astronomer Johannes Kepler (1571–1630) once used calculus in order to determine the optimum dimensions of beer kegs, only to find out that the coopers who constructed the kegs already were building them to these dimensions.

On occasion, the expected relationship has been inverted, and technology has served as an important source of scientific advance. Obvious examples are devices like electron microscopes and particle accelerators that create new research possibilities. It also happens that a practical technology poses an intellectual problem that cannot be answered by the science of the time. For example, in the 19th century a device known as a steam injector worked very well, but its operation made no sense when heat was conceived to be an actual substance known as *caloric*. If, however, heat was understood to be the result of molecular motion, then the steam injector's operation could be readily comprehended.

Technology also stimulates scientific advance by adding legitimacy to scientific enterprises. Science is valued for the truths that it reveals, but much of its support is derived from the belief that it gives rise to technological applications. Many areas of scientific inquiry have become very expensive, and without the promise of an eventual payback in the form of a new technology, many research projects would die for lack of funding. The withdrawal of financial support for the superconducting supercollider showed what can happen to an expensive scientific project that appears to have few practical applications.

There can be no denying, of course, that scientific discoveries can be the source of new technologies. Put simply, the science "pushes" the technology. For example, Wilhelm Konrad Roentgen's (1845–1923) discovery of X rays in 1895 stimulated the development of a host of technologies. This is not the typical situation, however; most of the science that finds a technological application is "pulled" there. The people doing the pulling are the engineers, technicians, and inventors who draw upon scientific knowledge to gain insight into a specific issue or solve a particular problem. Much of the science involved is hardly epochal; it is comprised of theories, formulas, and data that are deemed relevant to some aspect of the research and development process.

Neither "pull" nor "push" happen automatically. Both are the result of human activities that convert scientific knowledge into technological applications. Since science and technology are separate realms, in many cases the development of a successful science-based technology requires the presence of individuals who participate in both scientific and technological communities. Their presence in these separate realms allows them to serve as "translators" who can mediate between the communities of science and technology. One individual who was able to serve in this capacity was John Ambrose Fleming (1848–1945). A scientifically trained engineer who served as a consultant to Guglielmo Marconi's Wireless Telegraphy Company, Fleming was well situated to translate a scientific discovery into an effective technology. In 1883, Thomas Edison (1847–1931) had detected a flow of current when a metal plate was put between the legs of filament in an incandescent light bulb and connected to the plate's positive leg. This was a scientific discovery of sorts, but it was little more than a curiosity. Fleming's connections with both science and radio technology gave him a unique and important insight, that bulbs modified in this way could serve as detectors for radio waves. Fleming was not an inventor or a full-fledged scientist. Instead, he was a scientifically trained engineer and teacher who had close connections to the electrical industry and with engineering schools. These distinct but interrelated roles put him in an advantageous position to make a connection between science and technology.

Although science and technology are separate realms that often have to be bridged by translators, they do have some important commonalties. Both are based on the gathering of knowledge, and both advance through the cumulative development of that knowledge. Both partake of a generally rational approach to problem solving. Although the practitioners of science and technology may draw on intuitive and other nonrational modes of thought, rationality informs their basic methodology. The methodologies of science and technology include a willingness to challenge traditional intellectual authorities; settling questions through observation, testing, and experimentation; and the development and use of exact methods of measurement. Both use mathematics as a kind of language and as an analytical tool. Scientists and engineers have institutions and media that encourage the diffusion of knowledge: scientific and technical societies, journals, books, and meetings (it should be noted, however, that engineers often treat new information as proprietary knowledge and may be reluctant to see it spread). Finally, at the core of the common culture of science and technology is a belief that the world can be understood and shaped for human betterment. While science and technology build on past successes, they also make good use of their failures, as when a disproved scientific theory paves the way to the formulation of a better one, or a collapsed bridge provides lessons that help to prevent future failures. Science begins with the assumption that the world is knowable, and technology is driven by the belief that things can always be done better. These attitudes contribute to a progressive, essentially optimistic spirit.

As noted above, science and technology have been separate enterprises in most times and places, but during the 20th century they have become more closely linked. This linkage has been a major source of advance for both science and technology. When they are insulated from one another, as has been the historic case, science and technology have developed much more slowly.

See also ARCHES AND VAULTS; CALORIC; DIODE; HEAT, KINETIC THEORY OF; INDUSTRIAL REVOLUTION; INJECTOR, STEAM; MICROSCOPE, ELECTRON; PARTICLE ACCELERATORS; RADIO; RESEARCH AND DEVELOPMENT; SCIENTIFIC METHODOLOGY; SCIENTIFIC PAPER; SCIENTIFIC SOCIETY; SUPERCONDUCTING SUPERCOLLIDER; THERMOIONIC (VACUUM) TUBE; WATERWHEEL; X RAYS

Scientific Management

One of the greatest problems confronting the United States and other industrial countries at the beginning of the 20th century was conflict between labor and capital. This conflict stemmed from the two groups having needs that seemed inherently antagonistic. Many workers held down jobs that provided meager wages in return for putting in long hours amidst dangerous working conditions. Capital and its managers, on the other hand, complained about the difficulty of

getting an adequate output from a poorly motivated labor force. All too often, conflicts between labor and capital became unmanageable, and factories and whole communities were racked by strikes and other labor disputes. Some of these turned into violent encounters that ended only when government troops were brought in to restore order. Many feared that the whole social order was coming apart.

A solution to these problems was proffered by Frederick W. Taylor (1856–1915). What was necessary, according to Taylor, was the application of the principles of science and engineering to the management of labor. This idea came naturally to Taylor, who had racked up many accomplishments as a practicing engineer. Although he had been born to a well-to-do Philadelphia family, Taylor began his working career by apprenticing at a steel company. Through hard work and study, he rose rapidly through the ranks. His engineering triumphs included the development of materials and processes that made high-speed tool steel possible. This accomplishment was a prologue to an even more important task. As Taylor saw it, the scientific approach that allowed obdurate metals to be brought under control also could be used to manage labor, make it more productive, and use that productivity to increase income for management and labor alike.

Taylor called his approach "scientific management." The key to it was scheduling and organizing work activities so that workers had no alternative but to do things in the best manner possible. In Taylor's words:

> The work of every workman is fully planned out by management at least one day in advance, and each man receives in most cases complete written instruction, describing in detail the task which he is to accomplish, as well as the means to be used in doing the work. . . . This task specifies not only what is to be done, but how it is to be done and the exact time allowed for doing it.

The planning and management of tasks was not to be left to foremen and other old-fashioned managers but to a trained cadre of managers who possessed the specialized knowledge necessary for the effective organization of work. This knowledge was to be gained by ". . . gathering together all of the traditional knowledge which in the past has been possessed by the workmen and then of classifying, tabulating, and reducing this knowledge to rules, laws, and formulae."

In addition, trained managers were to analyze and improve on the work being done through the use of time-and-motion studies. By precisely timing the amount of time it took to perform each part of a work operation, scientific managers would be able to eliminate all motions that did not contribute to getting the job done. For the workers, this meant doing exactly as they were told, even down to when and for how long they rested.

Although they had no power over how their work was done, workers were expected to benefit from scientific management because the optimization of work activities would boost workers' incomes by increasing production. Instead of squabbling over how a small pie was to be divided, workers and owners could take home slices cut from a larger pie. Workers would necessarily share in production gains, for they were to be paid according to their output. Piece rates rather than set wages would motivate the workers to cooperate with management and work to the fullest extent of their abilities.

In addition to holding out the prospect of labor peace for the nation's factories, Taylor made the even more grandiose claim that principles of scientific management

> can be applied with equal force to all social activities: to the management of our homes; the management of our farms; the management of the business of our tradesmen large and small; of our churches, our philanthropic institutions, our universities, our government departments.

Taylor's message fell on receptive ears. The enthusiasm for scientific management was not limited to the United States, or even to the capitalist world. According to V. I. Lenin (1870–1924), the first ruler of the Soviet Union,

> The possibility of building socialism will be determined precisely by our success in combining Soviet Government and the Soviet organization of administration with the modern achievements of capitalism. We must organize in Russia the study and teaching of the Taylor system and systematically try it out and adopt it to our purposes.

The actual achievements of scientific management failed to live up to the expectations of Taylor and his acolytes. Some elements, most notably time-and-motion studies, were used in a number of firms, but the complete package was never implemented. Part of the prob-lem, as may be expected, stemmed from the resistance of workers who were unwilling to be turned into flesh-and-blood robots. But managers, and foremen especially, also balked. For them, the full implementation of scientific management would have meant relinquishing many of their powers to engineers and other outsiders engaged in bringing scientific management to their enterprises.

The promoters of scientific management also erred in their assumption that workers are motivated exclusively by the desire to earn more money. Social-psychological studies that began in the 1920s showed that the performance of an individual worker was affected by a host of factors. The desire to earn more money was important, but the informal norms and sanctions of coworkers often affected the speed at which people worked, piecework incentives notwithstanding.

See also TECHNOCRACY

Further Reading: Daniel Nelson, *Frederick W. Taylor and the Rise of Scientific Management*, 1980.

Scientific Methodology

An essential element of the scientific approach is skepticism. Scientists recognize that while both intuition and authority are good sources for ideas, these ideas may be flawed. Accordingly, scientists insist on using scientific methods to validate both descriptions and explanations of phenomena in their discipline. Central to all scientific methods is the requirement that all propositions, or hypotheses, be subjected to empirical testing, and that this research be carried out in such a way that it can be observed, evaluated, and replicated by others. The four goals of scientific research are to (1) describe a phenomenon, (2) predict the occurrence of the phenomenon, (3) determine the origin or the causal event underlying this occurrence, and (4) explain the process creating or producing the phenomenon. The scientific method provides a set of rules for gathering, evaluating, and reporting this information.

Scientific methods allow other scientists to evaluate the validity, generalizability, and replicability of the research. In order to describe a phenomenon, it is important to gather valid and reliable information, or data. That is, the data must, in fact, accurately reflect what you claim you are measuring. *Generalizability* refers to the ability to generalize the results from a given experiment to all instances of the phenomena. *Replicability* refers to the ability to carry out the exact same study under the exact same conditions.

A scientific approach incorporates a variety of methods, and each scientific discipline has its own set of methods best suited to study the questions of the discipline. Broadly speaking, methods may be divided into two broad categories: descriptive methods and experimental methods. Description must precede explanation, and thus descriptive methods serve the first two goals of scientific research. Descriptive methods, which are commonly used in the natural and behavioral sciences (e.g., biology, psychology, or anthropology), as well as astronomy, focus on describing an action or set of actions in a naturalistic setting. The goal of the researcher is to observe a naturally occurring action with little or no experimental interference. The researcher has control over the choice of a sample and setting, but very little control over extraneous factors that might be specific to the given time or location. Thus, descriptive methods do not allow for statements of cause and effect, as it is not possible to isolate specific factors from the influence of other factors.

Experimental methods, in contrast, primarily serve the goals of causal explanation. The use of experimental controls is especially important when attempting to identify and explain causal associations. The researcher, using techniques appropriate to his or her field, attempts to replicate the phenomena in the laboratory, and by varying different factors, attempts to explain the origin and the causal process underlying its occurrence. The experimenter sets up different experimental conditions, in each of which a different factor is isolated and investigated. To ensure that only specifically identified factors are being varied, the experimenter makes sure that all other factors are held constant across the different conditions so that they cannot influence the results. The criteria necessary for establishing cause-and-effect is a controversial issue among both scientists and philosophers. Many argue that cause-and-effect is established only if the cause is both necessary and sufficient for the effect to occur. That is, the cause must *always* produce the effect and must *always* be necessary for the effect to occur. Thus, scientists attempt to determine both the cause *and* the underlying process to provide a complete explanation. The behavioral sciences, which deal with more complex behavior patterns, typically accept causal statements in which either a necessary or a sufficient causal link has been demonstrated.

The first step in all scientific methods is formulating a valid hypothesis. A hypothesis is a proposition, based on prior research and intuition, that may be true and that can be tested. It is a formal statement about how variables are related, and makes a formal prediction about this relationship under specific conditions. Scientific methods define two categories of variables according to their role in the research: dependent variables and independent variables. Independent variables are the variables that the experimenter can control (e.g., the temperature of a solution, the age of a group of people, the angle of the slope of a ramp). Dependent variables are the outcome variables that the experimenter is measuring (e.g., the rate of crystal formation, the rate of learning, the acceleration of an object rolling down the ramp).

Scientific methods attempt to describe and explain the relationship between the dependent variables and independent variables (e.g., the effect of temperature on rate of crystal formation, the effect of age on rate of learning, the effect of angle of slope on acceleration). The research method chosen will then test the relationship between the variables under the conditions specified. If the predicted outcome occurs (the prediction is confirmed), then the hypothesis is supported; if not, it is rejected. A hypothesis can never be proved. This is because scientific methods can only make statements concerning probabilities, but cannot address the issue of definitive proof. Each time a research hypothesis (the hypothesis that the scientist has proposed) is supported, confidence in it is increased. Thus, replicability is an essential part of the scientific method.

Hypothesis testing involves deciding between two hypotheses. The null hypothesis states that no relationship exists between the variables, and this hypothesis is assumed to be true unless rejected by the study's findings. The research or alternative hypothesis states that the predicted relationship exists. When the results indicate that there is a relationship between the variables, the null hypothesis is rejected, and, by simple logic, another hypothesis must be true. However, a scientific method cannot distinguish between the research hypothesis and the possibility of yet another alternative hypothesis that has not yet been tested. Thus the research hypothesis has been supported but not proved to be the true proposition.

Furthermore, within the scientific method, the support and rejection of hypotheses is a statement of probability. That is, scientific methods determine only the probability that the null hypothesis is false. This is because empirical studies always include some component of error that can influence the outcome. Two primary sources of error that must be considered are error in the measurement and error in the sampling. Measurement error refers to the error that is inherent any time a measurement is made. This error may be large or small, depending on the accuracy of the measurement tool. It is also impossible to eliminate, as one can always consider that an even more precise measurement could be made (e.g., if measurements are made with the accuracy of 1/100 of a second, greater precision could be achieved with 1/1,000 of a second). In addition to reporting their findings, scientists also report the confidence level associated with these findings, that is, the probability that their measurements are true and not due to error. The confidence level reflects the likelihood that the same experiment will produce the same outcome (e.g., a scientist may report a finding with a 95 percent confidence level; the probability that this finding is due to error is 5 percent).

Sampling error arises from the fact that research studies cannot examine every possible instance of a given phenomenon, and instead must rely on a representative subset, or sample. For example, a study of the effect of age on learning cannot examine every older adult. Rather, research is always carried out with a sample of older adults, who are selected to represent the population being studied, and thereby permit generalization to the general population. Selecting a representative sample introduces another source of error that must be taken into account when deciding to generalize the outcome to the general population. When the population is simple to characterize or reproduce (e.g., in terms of physical properties), sampling error is generally negligible. However, when the population characteristics are complex, sampling error may be quite large.

Scientists often use statistical procedures to help them decide whether to reject the null hypothesis (for the general population) based on the findings from a given study. Two types of errors may be made in deciding to reject the null hypothesis: (1) The null hypothesis is rejected when it is, in fact, true in the general population (type I error), or (2) the null hypothesis is accepted when it is, in fact, not true in the general population (type II error). Type I error is generally considered to be more serious than type II error. When the null hypothesis is incorrectly rejected, it may determine the wrong course for future research, whereas when the null hypothesis is incorrectly accepted, it remains open to be rejected by a subsequent study. However, when type II errors result in great cost, they are considered to be more serious. In a medical setting, for example, when incorrectly deciding to accept the null hypothesis will result in a patient's death (e.g., incorrectly deciding not to perform an operation), the doctor is likely to decide that is better to incorrectly reject the null hypothesis (in this case, operate) than to incorrectly accept the null hypothesis (not operate).—M.M.

See also EMPIRICISM AND SCIENTIFIC METHOD; EXPERIMENT; HYPOTHESIS; SAMPLING; STATISTICAL INFERENCE

Scientific Paper

In the modern world, almost all new scientific knowledge is formally communicated through the medium of the scientific paper. However, it is becoming increasingly common to circulate preprints of papers to small groups of colleagues through computer networks or by FAX. Although there are alternatives, the scientific paper still serves as the major means of getting new knowledge before the relevant audience. At the same time, it provides a means of quality control, since virtually all journals that publish scientific papers submit proposed manuscripts to referees to check on their originality and quality before accepting them for publication.

Prior to the 1660s, new scientific discoveries were frequently communicated through correspondence networks, such as the ones operated by Marin Mersenne (1588–1648) and Samuel Hartlib (1600–1662), who were in contact with many of the outstanding mathematicians and natural scientists of the early 17th century. A scholar such as Galileo Galilei (1564–1642), Pierre Fermat (1601–1665), or René Descartes (1596–1650) would write to Mersenne, who would in turn either write letters to those whom he thought would be interested, or arrange to publish the material under his own name, giving appropriate credit to his correspondents. This pattern was changed by Henry Oldenburg (c. 1618–1677), who carried on correspondence with many of the members of Hartlib's circle. When Oldenburg became secretary of the Royal Society of London, he broadened his network of correspondents on behalf of the society and began to publish the more interesting correspondence in the *Philosophical Transactions of the Royal Society of London*, having first submitted it for review by selected members of the society. Oldenburg thus set the pattern followed by most journal editors to the present.

An alternative path for the publication of scientific papers was established in 1666 in the French *Journal des Savants*. Papers published in this journal were first presented at meetings of the Paris Academy of Sciences. Then, after being edited to take into account criticisms by the academicians who heard them, they were printed by the academy. This path to publication has continued into the present through the publication of conference proceedings. Most frequently in the case of papers first presented at scholarly conferences, draft papers are initially screened by conference organizers. Those chosen for presentation may then be revised in the light of criticisms from conference attendees.

A special form of scientific paper, the review paper, began to become increasingly important in the early 20th century as specialization increased and the number of papers appearing in many fields became so great that it was difficult for even specialists to keep up with the literature. A number of scientific societies began to publish a new kind of journal, such as *Physical Review Letters*, in which expert scientists were invited to explore the current status of a field, identifying and summarizing those papers that had made important contributions. Such review papers make it possible for scholars to begin working in a new field relatively efficiently.

Almost all scientific papers are now published in highly specialized disciplinary journals, but there are a handful of very prestigious journals with broader coverage, such as the *New England Journal of Medicine, Science,* and *Nature*, in which findings that are likely to have wide-reaching implications are published.

On occasion, a scientist will not disseminate his or her results in articles published in peer-reviewed scientific journals but will seek to go public in the more popular press or the electronic media. There also are persons who pay to have their work published in publications other than journals sponsored by scientific societies. These people and their findings usually are mistrusted because they have circumvented the quality-control mechanisms established to protect the credibility of scientists. To take a recent example, the public announcement of "cold fusion" by two Utah scientists in the late 1980s was quite rightly met with great skepticism, not just because it was surprising in its own right but also because it bypassed the process of publication that scientific papers normally undergo.

The reason given by these two scientists for eschewing the usual publication route was that they did not want to divulge information that could be of great commercial importance. It is not likely that they had achieved what they claimed, but there are many other scientific findings that do have the potential to generate significant profits. Consequently, the possible monetary value of research findings in some fields has restricted the flow of information that is essential to scientific progress. In similar fashion, the sponsorship by the military of a great amount of scientific inquiry has blocked the usual channels through which research findings are disseminated. The U.S. government even publishes scientific journals that have their circulation limited to scientists who have the proper security clearances and a "need to know."—R.O.

See also COLD FUSION; COMPUTER NETWORK; FAX MACHINE; PEER REVIEW; SCIENTIFIC PAPER; SCIENTIFIC SOCIETY

Scientific Society

The first scientific societies emerged as products of Renaissance court life and were often modeled on the Platonic "Academies" that appeared in the courts of numerous European aristocrats during the 15th and 16th centuries. Toward the beginning of the 17th century, groups of mathematicians, mechanics, physicians, and natural philosophers enjoyed the patronage of a number of powerful individuals. Among the most important of these groups was the one assembled by the Emperor Rudolph at Prague, where the astronomers Tycho Brahe (1546–1601) and Johannes Kepler (1571–1630), and the instrumentmaker Joost Burgi worked. Another group collected around the Earl of Northumberland, including the mathematician–natural philosophers Thomas Harriot (1560–1621), Walter Warner, Robert Norton, and Thomas Allen, as well as Sir Walter Raleigh (1552?–1618) and the poet John Donne (1572–1631).

Among the first groups to focus exclusively on natural knowledge, to have a formal constitution, and to expand its scope beyond the purely local was the Accademia dei Lincei, founded by the teenage Duke Federigo Cesi at Rome in 1603. Among early members of the academy was the natural magician Giambattista della Porta. Galileo joined in 1611, and the group soon grew to a membership of 32 people living throughout Europe. The members corresponded with one another, supported one another, and lent one another instruments. The academy published several of its members' books, including Galileo's first major astronomical work, *The Sidereal Messenger*, and the first work containing descriptions of microscopic observations. The Accademia dei Lincei dissolved on Cesi's death in 1630, but it was revived in modern times to become the Italian equivalent of America's National Academy of Sciences.

Between 1657 and 1667, the short-lived but critically important Accademia del Cimento flourished in Florence. Here, a group of court-supported scholars, including Nicolas Steno (1638–1686), Francesco Redi (1626–1697), and Giovanni Alphonso Borelli (1608–1679) joined together to publish one of the first scientific journals, the *Saggi di Naturali Esperienze*, which described experiments performed before and attested to by the group.

In England, several private groups that had been meeting in London and Oxford came together and received a royal charter from Charles II in July 1662 to establish the Royal Society of London for Improving of Natural Knowledge. Within a few years, the secretary of the society initiated publication of the monthly *Philosophical Transactions of the Royal Society*, which printed scientific papers both by members of the society and by nonmembers who submitted papers to the secretary. It is still published today.

Shortly after the founding of the Royal Society of London, French scientists appealed to the country's controller-general of finances, Jean Baptiste Colbert, to establish a French Academy of Sciences. In return for the crown's financial support, the society would provide advice on problems amenable to scientific inquiry. The Paris Academy of Sciences was founded in 1666. Unlike the Royal Society of London which received no crown monies, the French academy had a staff of paid academicians, as well as honorary members, all of whom presented papers that were published in the *Journal des Savants*. Moreover, members of the French academy were asked to serve the government by advising on patent applications, approving scientific books for publication, and advising on a host of government problems.

Throughout the 18th century, scientific societies based on both the Royal Society and the Academy of Sciences proliferated throughout Europe and European colonies. One of the most noteworthy was the Lunar Society of Birmingham, so named because its meetings were scheduled according to the phases of the moon. Its illustrious membership included James Watt (1736–1819), Matthew Boulton (1728–1809), Erasmus Darwin (1731–1802), Josiah Wedgwood (1730–1795), and Joseph Priestley (1733–1804). The society provided a congenial intellectual home for scientifically inclined men, many of them businessmen and engineers. The Lunar Society and institutions like it helped to bring science and technology closer together, and in so doing contributed to the progress of the ongoing Industrial Revolution.

As science became more professionalized and as scientific knowledge became more highly specialized during the early 19th century, general-purpose, closed-membership scientific societies began to be supplemented by two other kinds of institutions. On the one hand, scientists in search of professional status formed more specialized societies, such as Linnaean societies for the study of natural history, geological societies, chemical societies, and physical societies. In part because chemistry was a relatively late science to develop, it was relatively specialized from the beginning; as early as 1792, the Chemical Society of Philadelphia became one of the first specialized scientific societies anywhere. Many of the new specialized societies began to publish their own scientific journals; this continues to be one of the principal functions of scientific societies. The proliferation of specialized scientific societies occurred so rapidly that by the mid-20th century there were nearly 50,000 of them worldwide.

As professionalization drove most scientists into smaller specialized societies, there still remained a need for broader scientific societies to promote public and governmental support for science in general. Accordingly, in the United Kingdom the British Association for the Advancement of Science was formed in 1833, and in the United States, the American Association for the Advancement of Science was founded in 1848.—R.O.

See also INDUSTRIAL REVOLUTION; SCIENTIFIC PAPER

Further Reading: James E. McClellan III, *Science Reorganized: Scientific Societies in the Eighteenth Century,* 1985.

Selden Patent

In the 1880s, several inventors struggled to create powered road vehicles. The most successful of these were Gottfried Daimler (1834–1900) and Karl Benz (1844–1929), who produced the first vehicles powered by internal combustion engines. But unbeknownst to these early inventors, an American had already filed for a patent intended to cover the automobile as a single invention. The man who sought to patent the automobile was George Selden (1846–1922), a patent attorney in Rochester, N.Y. Selden had long been interested in internal combustion engines and their application to vehicles, and had even experimented with some early engines. He never was able to get one to run for any great length of time, and he never installed one in an automobile chassis. But on May 8, 1878, he filed a patent application for an "improved road-engine." At the time of the application, an automobile patent was of little value, for a working automobile had not even been created, much less commercially marketed. Selden accordingly postponed the actual issuance of the patent by adding amendments from time to time and by using other delaying tactics. Consequently, it was not until Nov. 5, 1895, that he received U.S. patent number 549,160.

A patent gives its holder a claim to the exclusive rights pertaining to an invention, but these rights often have to be vigorously pursued, often in courts of law. This in turn can require large amounts of capital, and Selden, although well-to-do, did not have the financial resources necessary to exploit his patent. In 1899, he transferred the license to a recently formed financial syndicate known as the Columbia & Electric Vehicle Company (later the Electric Vehicle Company), although retaining a share of royalties on the patent.

The Electric Vehicle Company had faltered badly in its attempt to build and operate a large fleet of electric cabs to serve a number of U.S. cities. In order to recoup its finances, it began in 1900 to press suits against manufacturers accused of infringing on Selden's patent. But within a few years, it took a different approach. Realizing that legal actions against members of the automobile industry would evince continual hostility and large legal expenses, an executive of the Electric Vehicle Company devised a plan to create a patent pool. Selected manufacturers would be licensed to use the Selden patent in return for a royalty payment, while other manufacturers would be excluded. In this way the licensed manufacturers had a weapon to suppress their competition. Members of the pool were established as the Association of Licensed Automobile Manufacturers (ALAM) in March 1903.

The companies comprising ALAM were for the most part builders of expensive cars marketed to the rich. Henry Ford (1863–1947), on the other hand, was committed to building cheap cars for the masses, but his plans were threatened when ALAM refused to give him a license. The company went on making cars, and in 1903 it was sued by ALAM for patent infringement. A lengthy court case ensued until 1909, when Judge Charles Hough ruled that Ford had infringed on the Selden patent. An appeal proceeding began the following year. In January 1911, the three judges serving the U.S. Circuit Court of Appeals for the Second Circuit overturned the previous decision. They granted that Selden had "invented" an automobile powered by an internal combustion engine, but that the patent covered only automobiles powered by a Brayton engine, a type that had long been extinct.

Although the Selden patent had been due to expire in 1912, Ford's victory brought him a great deal of public acclaim as a foe of "monopolists." Also, the

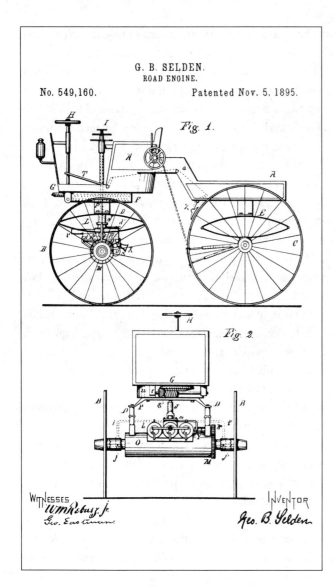

Patent illustration for Selden's "road engine."

Selective Breeding of Animals—see Animal Breeding, Selective

Semiautomatic Weapons

A semiautomatic weapon is a self-loading firearm capable of firing one shot per activation of the triggering mechanism. Semiautomatic pistols and longarms had differing development paths, so it is useful to treat them separately. The first successful semiautomatic pistols (more generally referred to as "automatic pistols") were various types of revolvers. Nonetheless, they had inherent limitations. Even double-action revolvers had limited rates of fire, and the long, hard trigger pull necessary to cock the hammer and rotate the cylinder had an adverse effect on accuracy. Ammunition capacity was generally limited to six rounds, and reloading was fairly slow. The windage between the cylinder and frame let gas out during firing while letting dirt and water in at other times. Finally, the cylinder itself made the weapon intrinsically bulky in relation to its firepower. These problems could be addressed through design refinements, but only in a limited fashion and at a considerable cost in complication.

By the 1890s, however, the combination of higher manufacturing tolerances and the clean-burning character of nitrocellulose- and nitroglycerin-based smokeless powders made another approach possible: using the recoil energy generated by the act of firing to eject and extract a spent cartridge, and then to deliver and chamber a new one. The key problem here was finding a way to hold the breech (the nether end of the barrel chambered to receive the cartridge) and bolt or breechblock (which carried the cartridge and closed the barrel) together while firing. The first semiautomatic pistol to reach production, the Schonberger pistol briefly manufactured by Steyr-Mannlicher in 1892, accomplished this by mechanically delaying the tendency of the breechblock to recoil ("blowback") upon firing. Most later designs, starting with the more successful Borchardt pistol of 1893—which evolved into the more famous Parabellum or "Luger"—used various arrangements to physically lock the barrel and bolt together for firing. Borchardt's design also introduced the standard method of ammunition supply for automatic pistols, feeding rounds from

drawn-out case had starkly revealed some basic flaws in patent litigation, and court procedures were simplified in the years that followed. Finally, the Selden patent demonstrated the problems that occurred when one person or group held a patent vital to a particular industry. In 1915, most American automobile manufacturers joined in a cross-licensing arrangement that made patents held by individual firms available to all.

See also AUTOMOBILE; ENGINE, FOUR-STROKE; PATENTS

Further Reading: William Greenleaf, *Monopoly on Wheels: Henry Ford and the Selden Automobile Patent*, 1961.

a spring-loaded magazine enclosed in the butt. For pistols firing less powerful ammunition, the problem was almost trivial, though by no means intuitive, since the inertia of the breechblock could be relied on to hold the assembly together for the required period, as done with John Moses Browning's (1855–1926) model 1900 pistol. By the outbreak of World War I, the semiautomatic pistol was technically mature and both militarily and commercially successful, and the weapon proved successful in wartime.

After WWI, the major innovations centered on "double-action" (trigger-cocking) mechanisms—which had been invented and produced before the war, but largely ignored—by the German Walther firm, and the high-capacity double-column magazine pioneered by Browning in his last design, and refined and marketed in 1935 by Fabrique National of Belgium. Trigger cocking removed one of the last advantages of the revolver relative to the automatic; the double-column magazine turned a marginal advantage in ammunition capacity into a much greater one.

Over the course of the 20th century, the automatic pistol has largely replaced the revolver in law enforcement and military applications. The last major holdout was the U.S. law enforcement community. Although enamored of the nominal safety and relative ease of training associated with the double-action revolver, most such agencies made the change in the 1980s, spurred by technological advance and the prospect of facing assailants who did not share their priorities. Minor enhancements continue to be made to improve safety, firepower, and accuracy of the pistol, but it has essentially reached a technological plateau, and perhaps a functional dead end. As a military weapon, at least, its function as an auxiliary weapon seems to be eroded by various "personal defense weapons" such as submachine guns and shortened assault rifles. Still, the semiautomatic pistol retains a military niche as a close-quarters weapon and token of rank.

While the pistol has faded into the background as a military weapon, the rifle continues to be the main infantry firearm. The invention and adoption of the machine gun seemed to show that a self-loading infantry weapon was feasible, since it generally used the same ammunition as the bolt-action magazine rifles then being adopted. The logistics of ammunition supply and the difficulties of controlling a handheld rifle-sized machine gun made the fully automatic infantry rifle a poor prospect, but a semiautomatic one seemed within reach. However, producing a serviceable weapon was a prolonged process.

Early semiautomatic rifles were generally unsuitable for field service. Some were too heavy for use as individual weapons, and often evolved into light machine guns, as in the case of the Danish Madsen gun and the U.S. Browning Automatic Rifle. Some, such as the Mexican Mondragon design of 1907, were agreeably light, but were too fragile for the stresses imposed by the cartridges they fired and the rigors of the field. This state of affairs, combined with the considerable capabilities of current rifles, the dislocations of World War I, and enormous war surplus stocks of weapons and ammunition, militated against the adoption of the semiautomatic infantry rifle by even the most advanced armies.

At the outbreak of World War II, only the U.S. Army had standardized such a weapon. This rifle, the gas-operated M1, had been designed by John C. Garand (1888–1974) of the Springfield Armory during the 1920s in a climate of severe financial stringency. The Garand was approved for service in 1932, and finally issued in 1936. A thoughtful initial design combined with long years of refinement made for a highly capable weapon, and the Garand served well in all theaters of the war. By war's end, however, the semiautomatic rifle appeared to be almost as obsolescent as the bolt action. New manufacturing techniques, improved ammunition supply, and the increasing premium on individual firepower brought the fully automatic rifle back into prospect. Indeed, a generation of such weapons, such as the Belgian FN FAL, the Hispano-German CETME, and the Garand-derived M14 were widely adopted for service. However, such weapons proved only marginally controllable in full-automatic fire, and have been largely restricted to semiautomatic use. Practical fully automatic personal weapons required the acceptance of reduced-power ammunition, and arrived in the form of the assault rifle, which has steadily been replacing the earlier types since its introduction by the Germans late in World War II, and enthusiastic adoption by the Soviet Army and its allied forces in the form of the AK-47 and its successive refinements and modifications.—B.H.

See also BREECHLOADING WEAPONS; EXPLOSIVES; MACHINE GUN; REVOLVER; RIFLE

Further Reading: Ian V. Hogg, *Military Small Arms of the 20th Century*, 5th ed., 1985. Edward C. Ezell, *The Great Rifle Controversy*, 1984.

Semiconductor

Most materials either conduct electricity, and are thus known as *conductors*, or they do not conduct electricity, in which case they are termed *insulators*. There are, however, materials that fall into a third category: semiconductors. These are materials that act as conductors under some circumstances and act as insulators under other circumstances. For example, certain semiconductors will conduct electricity at some temperatures and not at others. Some semiconductors take on special electrical qualities when small amounts of other substances are added to them. This makes them especially valuable as the basis for a variety of electronic devices.

Semiconductor properties were first observed in the 1870s by a German physicist, Karl Friedrich Braun (1850–1918). Braun noticed that a sulfide of lead known as *galeria* conducted electrical current in only one direction. This property made it useful as a rectifier, a device that converts alternating current into direct current. Lead sulfide and kindred materials were particularly important to the early development of radio, for they were the "crystals" commonly employed for the reception of radio signals.

Crystals were rendered obsolete by the development of various types of vacuum tubes. But vacuum tubes had a number of defects; they were fragile and bulky, consumed a good deal of electric power, and were prone to burning out. In 1947, the invention of a new semiconductor-based device, the transistor, marked the beginning of the end for the vacuum tube in most of its applications. Within not much more than 10 years, procedures were developed for putting a few transistors and other semiconductor-based components on a chip. Today, a small chip may contain more than a million of them.

The most common basis for semiconductor devices is silicon. Although natural silicon is very common, comprising nearly 30 percent of the Earth's crust, the silicon used in semiconductors is derived from crystals that have been grown under carefully controlled conditions. Added to the silicon are small quantities of impurities such as boron or phosphorous, a process known as *doping*. Silicon doped with phosphorous (P-type silicon) is an aggregation of silicon and phosphorous atoms that share their outer electrons to form a completed valance shell of eight electrons. Each atom of silicon and phosphorous contributes four electrons apiece. But because an atom of phosphorous has five electrons in its outer shell, one electron is free to travel through the crystal. When these electrons travel, an electrical current flows.

Conversely, silicon that has been doped with boron (N-type silicon) has a deficiency of electrons in the outer shell shared by the two atoms. Because the boron atoms have only three electrons in their outer shells, each atom has a vacancy for an electron known as a *hole*. Electrons from nearby atoms can fall into this hole, leaving a new hole in the process, a phenomenon known as *hole flow*. This too allows the movement of an electrical current. Many semiconductor devices are comprised of P-type silicon combined with N-type silicon. For example, one combination of P-type silicon and N-type silicon forms a diode, a device that conducts electricity in only one direction.

See also ALTERNATING CURRENT; CRYSTAL RECEIVER; DIODE; INTEGRATED CIRCUIT; RADIO; RECTIFIER; THERMOIONIC (VACUUM) TUBE; TRANSISTOR

Separate Condenser

The practical use of steam power in the 18th century entailed many improvements to the basic engine invented by Thomas Newcomen (1663–1729). Of these, by far the most important was James Watt's separate condenser. Watt (1736–1819) was employed as a "mathematical instrument maker" at the University of Glasgow when in the winter of 1763–64 he was asked to repair a model Newcomen engine used for classroom demonstrations. Even after he fixed it, the model engine failed to work properly, running out of steam after a few strokes. Watt thought (incorrectly, as it turned out) that the problem lay in the small size of the model, for its large surface-to-volume ratio caused the rapid loss of heat through the walls of the cylinder. Watt tried to counteract this by constructing a cylinder out of wood—a material that did not conduct heat as rapidly as metal—but achieved only limited success.

If Watt had been only a talented repairman, that might have been the end of the matter. But he had a keen interest in science, and enjoyed the acquaintance of accomplished scientists like Joseph Black (1728–1799). Watt's work with the steam engine is sometimes presented as little more than inspired tinkering, but in fact he conducted a number of experiments that were scientific in their methodology and their spirit. He heated water, first with steam and then with boiling water. In so doing he discovered that the steam caused a greater elevation in the temperature of the water, even though it was at the same temperature as the boiling water. Black's theory of latent heat provided the explanation for this phenomenon, and allowed Watt to calculate how much water was necessary to completely condense the steam. But when it used the requisite amount of cooling water, his model engine developed very little power. After conducting more experiments, this time centering on determining the boiling point of water at various pressures, Watt came to the realization that if the walls of the cylinder remained hot, the water remaining in the cylinder would boil in the partial vacuum, with a resultant loss of that vacuum. The efficient running of a steam engine thus depended on the resolution of a paradox: The cylinder had to be kept hot so that heat was not wasted every time that steam entered the cylinder, yet the necessarily high temperature prevented the generation of a good vacuum throughout the condensation phase of the operating cycle.

According to his retrospective account, Watt came upon the answer early in 1765 while strolling across Glasgow Common: He thought that instead of condensing the steam in the cylinder, a separate chamber should be used for this purpose:

> . . . it occurred to me, that if a communication were opened between a cylinder containing steam, and another vessel which was exhausted of air and other fluids, the steam, as an elastic fluid, would immediately rush into the empty vessel, and continue so to do until it had established an equilibrium; and that if that vessel were

kept cool by an injection, or otherwise, more steam would continue to enter until the whole was condensed.

In 1769, Watt patented a steam engine with a separate condenser, and in the same year built an experimental engine with that feature. It was a promising start, but work languished due to the financial reverses of Watt's partner, John Roebuck (1718–1794). In 1773, Matthew Boulton (1728–1809), one of Roebuck's creditors, took over Robuck's share of the patent in return for forgiving Roebuck's debt to him. In 1776, the firm of Boulton and Watt installed the first steam engines with separate condensers, one in a coal mine and the other in a blast furnace. Their value was immediately evident, for they worked at about twice the efficiency of the best Newcomen engines.

See also LATENT HEAT; STEAM ENGINE

Further Reading: R. J. Law, *James Watt and the Separate Condenser: An Account of the Invention,* 1969.

Servomechanism

A servomechanism (or servo) is a motorized device for regulating position. It consists of three main components arranged in a closed loop: an element that senses an "error" by comparing the extent to which actual position diverges from what is desired, an amplifier to send a signal to a servomotor, and the servomotor itself, which acts to correct the positional error. Servomechanisms take a great many forms. The sensor may be thermometer or a gyroscope; the amplifier may be electronic, magnetic, pneumatic, or hydraulic; the servomotor may be electric or hydraulic.

Servomechanisms serve a great many purposes. One of the most common applications pertains to steering vehicles. These devices began to evolve in the latter 19th century, by which time large ships often required several men at the helm during heavy weather, exerting their muscle power against tremendous hydrodynamic forces. In 1853, an American named Frederick Sickles invented a steam-assisted steering gear. This was an *open-loop* system (like power steering in an automobile), not a servomechanism, but at least it introduced the "motor" part of a servomechanism. Steering devices that utilized closed feedback loops appeared in the late 1860s and early 1870s in both France and England, one being employed aboard the *Great Eastern*, the largest ship afloat. Typically, these devices consisted of a chamber containing highly pressurized fluid and a pilot valve to determine on which side of a piston fluid was to be admitted. The piston in turn acted on the rudder, and the rudder was linked back to the pilot valve, stopping the flow of fluid when the rudder held its proper position with respect to the desired heading.

But with such devices, there was still a human element to consider. Sensing an "error"—a deviation from the desired heading—remained with a helmsman reading a compass and turning a wheel. Such systems were rendered fully automatic with the invention of Metal Mike, the gyropilot, by Elmer Sperry (1860–1930) in the early 1920s. Through a system of two feedback loops, the gyropilot was able to compare a ship's desired heading with its compass heading and change its direction accordingly—thereby doing away with the need for a helmsman altogether, except to set a small dial to the proper compass heading. Indeed, Metal Mike represented an improvement over all but the most-skilled helmsman, for when changing course it had a seemingly intuitive capacity to ease off on the rudder at just the right moment and then to put it over the other way to counteract any tendency of the ship's momentum to carry it past the desired heading. Within 10 years, by 1932, Metal Mike had 1,000 "brothers." Through this and many other automatic guidance and control devices, Sperry established himself as a seminal figure in the history of servomechanisms and feedback devices.

The automatic pilot was of course even more dramatically applied to airplanes, with systems of servomechanisms regulating airspeed, altitude, and direction, but such devices were not limited solely to vehicular control. A ponderous optical telescope could be positioned by means of a small handcrank. With both at approximately the same angle, a sensor would read the position of the crank—the "input." Then, through an electrical amplifier, the sensor fed this information back to a motor that moved the telescope itself, bringing both into exactly the same alignment. A similar sequence of events takes place inside a "guided missile." Because they resonated with a growing fascination for extreme accuracy and especially for "robots," servomechanisms

captured the popular imagination as did few other technological devices of the 20th century.—R.C.P.

See also CYBERNETICS; GYROSCOPE; MISSILES, GUIDED

Further Reading: Thomas P. Hughes, *Elmer Sperry: Inventor and Engineer*, 1971.

Sewers

In the United States and most other places, the practice of using sewers to carry kitchen, bathroom, and industrial wastewater to designated points for purification and final disposal is relatively new. Indeed, sewerage was not designated as the preferred means of waste disposal until the 19th century. Before that time, fecal waste was either stored in privy vaults and subsequently carted out of town or permanently buried. Household debris was thrown onto the streets to be consumed by pigs or other roving animals, and industrial wastes were dumped at sites beyond the centers of population. The "out of sight, out of mind" mentality prevailed, and city and town governments seldom interfered with these methods of waste disposal. Only when an individual's waste disposal habits were particularly offensive and complaints had been filed with local authorities would municipal governments interfere with the manner in which individuals disposed of their wastes. When such problems surfaced, the offender was ordered to clean up his mess.

Sewer pipes, or conduits that conveyed stormwater to nearby rivers and lakes had been constructed since colonial days, but they were exclusively laid to move stormwater away from the built environment. Although the wastes left on streets naturally flowed into these sewers, personal and industrial wastes were prohibited from entering these facilities without special authorization. These sewers were generally constructed on an individual basis, and there was little interest in coordinating or integrating them with one another.

Although cities occasionally financed the construction of these facilities to drain a particularly problematic region, generally, individuals paid for conduits to be built to convey floodwaters away from their property to the designated place of outfall. Consequently, a decidedly inequitable pattern of sewers was constructed whereby the wealthiest realized better drainage while the poor were left to wallow in wetness.

The earliest sewers were open channels built either in the center or along the sides of city streets. Early in the 19th century, however, the increased demand for street surfaces as well as new concerns over the dangers of sun-baked water motivated people to sink the conduits below the street surface. Existing open sewers were covered, and bored-out logs were laid to convey the waters. During the second quarter of the 19th century, new materials provided alternatives to this traditional mode of construction. Brick pipes, cemented with quicklime concrete, became available and provided a real advantage, as they did not deteriorate as wood pipes were prone to do. In the 1840s, vitrified clay pipe became available and offered an alternative to the larger egg-shaped or round brick and mortar conduits.

In the middle of the 19th century, traditional patterns of wastewater disposal changed, and many cities endorsed sewerage construction. Initially, wastewater was conveyed through the existing sewers along with floodwater. However, problems mounted as channeling wastewater into the sewers that had been built exclusively for stormwater had serious unexpected consequences. Clogs, breaks, and leaks often undermined the effectiveness of sewers, and the increased volume and new composition of matter entering the facilities aggravated these problems. Problems compounded during the first half of the 19th century as many major cities introduced new water supply systems, thus increasing the volume of water that was to flow through the sewers. In response, lay reform groups, together with many physicians and engineers, lobbied for cities to assume a more active role in municipal sanitation. Brooklyn and Chicago were two of the earliest American cities to reform their sanitary systems. They charged sanitary engineers with constructing new sewers and rationalizing existing sewers into integrated systems. New York City followed suit in the 1860s and 1870s.

After 1870, the technology of sewer construction changed as engineers debated the merits of new building techniques, as well as the advantages and disadvantages of separate or combined systems. Most sewer systems built up to that time were of the combined variety. Sewage together with rainfall and floodwater flowed through a system of conduits to a point of outfall. In the 1870s, sanitarian George Wading began advocating an alternative to this, the separate sewer system. Separate sewerage systems were comprised of a network of pipes that conveyed wastewater exclusively. Stormwater either drained naturally or was conveyed by a independent matrix of pipes. Rudolf Hering challenged Wading's stated preference for separate systems. He argued that the decision to construct separate or combined systems should be made on an individual basis in consideration of local conditions. Whereas the combined system was most suited to large, densely built cities, the separate system could adequately and more economically accommodate the conditions of smaller cities. This precept guided much sewer construction into the 1890s.

Whether combined or separate, both methods assumed that once the wastewater was conveyed to the point of outfall, usually a running body of water, it ceased to pose a hazard. In the 19th century, sanitarians generally believed that only stagnant and putrefying wastes emitted miasmas that caused disease. This principle was effectively challenged by the germ theory of disease, which became generally accepted in the 1890s. For the next 30 years, studies of the methods of filtration and disinfection became the focus of much sanitary engineering research. Most agreed that preliminary screening of suspended solids facilitated purification of the sewage. Secondary treatment then required one of several alternatives that settled and purified the dangerous impurities in the sewage. By the early 20th century, research indicated that oxidation by filtration and disinfection with chlorine freed a significant amount of the bacteria. Through the 1970s, most sewage treatment included both primary and secondary processes.

At present, much of the concern has shifted to the matter of sludge disposal. Sludge refers to the solids removed from sewage during the primary and secondary sewage treatment processes. Sludge had been dumped into the ocean with little afterthought, but since the 1970s an increase in volume has raised concern. As modern computer-assisted detection of sludge heightens our awareness of the amount of impurities that remain after processing, attention continues to focus on the matter of sludge disposal. Composted disposal on land and incineration appear to be the most promising solutions to this problem.—J.A.G.

See also COMPOSTING; GERM THEORY OF DISEASE; SANITARY PLUMBING FIXTURES; WASTE DISPOSAL; WATER SUPPLY AND TREATMENT

Further Reading: Joel A. Tarr, James McCurley III, Francis C. McMichael, and Terry Yosie, "Water and Wastes: A Retrospective Assessment of Wastewater Technology in the United States, 1800–1932," *Technology and Culture*, vol. 25, no. 2 (Apr. 1984): 226–39.

Sewing Machine

Sewing is one of the oldest of human technologies. Stitching things together with needles dates back to Paleolithic times, and needles with eyes opposite their points have been used by people all over the world. By the 19th century, mechanized spinning, weaving, and other technological advances had increased the pace of textile production dramatically. This in turn created a receptive environment for the development of a machine that could rapidly convert cloth into clothing.

Many histories credit Elias Howe, Jr. (1819–1867) with the invention of this machine. In fact, numerous sewing machines had been invented and even put into production before Howe's. The first machine to achieve a modicum of commercial success was patented in France by Barthelemy Thimonnier (1793–1857). Thimonnier's machine produced the kind of stitch that was commonly used in embroidery: A hooked needle passed through the cloth and then caught a thread located underneath the fabric. When the needle moved upwards, it brought a loop of thread above the fabric. A repetition of this cycle produced another loop that linked with the first loop, resulting in a chain stitch. By 1841, 80 of these machines were being used to sew army clothing. Unfortunately, a group of tailors who feared that the machines would take away their work broke into Thimonnier's shop and smashed the machines. Timelier eventually died in poverty.

In its basic operation, Thimonnier's machine replicated the actions of manual sewing. The first machine to depart from this pattern was produced in the 1830s by Walter Hunt (1796–1860) in New York. Hunt's machine produced a lock stitch through the use of two threads, both of them interlocking as one passed through the loop formed by the other. Although it was not the first to use it, Hunt's machine included a highly significant feature: a needle with the eye at its point. As sometimes happens in the history of technology, an important breakthrough was effected by departing from the "natural" way of doing things. But Hunt had little interest in his invention and did not even bother to patent it. It did become important 15 years later, when Hunt's machine was featured in patent litigation over who could rightfully claim to be the inventor of the sewing machine.

A number of other sewing machines were invented in the United States and Europe in the years that followed. Many of them were functional in the sense that they could produce a stitch, and many received patents, but commercial success remained elusive. Many inventors had devised elements of a practical sewing machine, but they did not come together in one machine until Elias Howe, Jr., devised his sewing machine in 1845. Howe's machine used an eye-point needle that produced loops of thread that were engaged by another thread carried by a shuttle located on the underside of the fabric being sewn. After receiving a U.S. patent in 1846, Howe spent three frustrating years trying to interest manufacturers in putting his machine into production. The best he could do was sell the British patent rights for £250.

By this time a number of other inventors had been working on sewing machines of their own, and some of them succeeded in producing and selling a few machines. One of these was Allen B. Wilson, whose machine was subsequently manufactured by the Wheeler and Wilson Manufacturing Company. Among its innovative features was a an automatic cloth feed that is still used in sewing machines today. By far the most successful manufacturer of sewing machines was Isaac Merritt Singer (1811–1875), who in the mid-1850s began to produce a machine built to a design patented in the United States by John Bachelder. Singer could rightfully claim that his sewing machine embodied a number of improvements, but his real genius lay in marketing. Singer's machines were heavily advertised, exhibited in lavish showrooms, and demonstrated by attractive saleswomen. Perhaps most important, Singer pioneered the selling of his machines on the installment plan, making them financially accessible to every middle-class household.

A 19th-century sewing machine at work (from B. Yenne, *100 Inventions*, 1993).

Before achieving a position of dominance, Singer had to contend with Howe's patent. He was not the only one. Most sewing machine manufacturers were paying Howe a royalty of $25 for every machine they produced, while at the same time Howe himself was being sued for infringing on the patents of others. Disputes over patent rights were resolved in 1856 by the formation of the first patent pool, the Sewing Machine Combination. Members of the combination, which included firms headed by Howe, Wilson, Singer, and one other firm, could use one another's patents free of charge. Any other manufacturer who produced a sewing machine using any of the combination's patents had to pay a fee of $15 per machine (reduced to $7 in 1860). The combination lasted until 1877, when the last of the patents expired.

The expiration of the patents held by the combination allowed many new firms to enter the market. Singer remained the dominant company, even though it was slow to use interchangeable parts for its machines. One of the consequences of the proliferation of sewing machines during the latter part of the 19th century was the

expansion of the ready-to-wear clothing industry. People who had been accustomed to wearing clothes made at home or by custom tailors now bought the bulk of their clothing "off the rack." The sewing machine did not create this industry, but it gave it an enormous boost. According to one study conducted in 1861, a shirt that required nearly $14\frac{1}{2}$ hours of hand sewing could be machine sewed in 1 hour, 16 minutes. The sewing machine also stimulated the emergence of new fashions, such as women's hoop skirts and other elaborate garments that lent themselves to machine sewing. Sewing machines had a similar effect on the making of shoes, which passed from being a cobbler's trade to a mass-production industry. By the beginning of the 20th century, sewing machines also were being used for the manufacture of sails, tents, books, flags, luggage, saddles and harnesses, and umbrellas—just about anything that could be stitched together.

By the last quarter of the 19th century, the sewing machine had taken on its essential form and features, changing little in the years that followed. The only major innovation was the introduction of electric power. Electric sewing machines appeared as early as the 1880s, but did not become popular until the second decade of the 20th century. Even then, treadle-operated machines continued to be used by many households, an understandable situation since many homes still were not wired for electricity.

Whether it was used in the home or the factory, the sewing machine brought great benefits to the public by increasing productivity and bringing down the cost of apparel. But all too often, the workers—most of them women—who produced these clothes were poorly paid sweatshop laborers. The sewing machine did not create this segment of the working class; the needle trade had been characterized by low wages and poor working conditions before it came on the scene. But the sewing machine intensified the problem by creating a vast market for semiskilled labor. In many parts of the world, many industrial nations not excepted, sewing machine operators still put in long hours for wages that provide little more than subsistence.

See also MASS PRODUCTION; TECHNOLOGICAL UNEMPLOYMENT

Further Reading: Grace Rogers Cooper, *The Sewing Machine: Its Invention and Development*, 1976.

Shape Memory Alloys

Shape memory alloys (SMAs) are a class of uniform metal combinations displaying a property called the *shape memory effect*. SMAs have a uniform crystal structure that radically changes to a different structure at a distinct temperature. When the memory alloy is below this "transition temperature," it can be stretched and deformed without permanent damage, more so than most metals. After the alloy has been stretched, if it is heated (either electrically or by an external heat source) above its transition temperature, the alloy "recovers" or returns to the unstretched shape and completely undoes the previous deforming.

Shape memory alloys have been made using many different combinations of metal elements. The transition temperature varies for each alloy and can be set during manufacturing by carefully controlling the ratios of the component metals and the presence of other elements. The most common memory alloy, nitinol

(pronounced "night in all"), consists of nearly equal amounts of nickel and titanium atoms. This alloy generated new interest in the SMA field because it was safer, less expensive, and had a better deformation-to-recovery ratio than earlier alloys.

During the 1960s and 1970s, researchers worldwide worked to develop uses for these metals. NASA studied memory alloys for satellite antennas that would unfold and expand when exposed to the heat of the sun. University, corporate, and private researchers explored many uses, including a variety of engines that ran only on hot and cold water, automatic temperature-controlling greenhouse windows, hot-water pipe valves, and automobile fan clutches that engaged when the engine reached a specific temperature. Many early ideas did not pass beyond the research stage. Some were limited by the slow speeds and variable quality of the available alloys, others were limited by their design, and the inventor's lack of experience with SMAs.

Because shape memory alloys can perform useful work with relatively small temperature differences, researchers looked for ways to use heat from the sun, thermal sources like geysers, and waste heat from industrial processes to produce energy. In 1980, McDonnell-Douglas demonstrated a scaled-up nitinol heat engine. The engine produced power in excess of 32 watts (.04 hp), the largest output for a nitinol heat engine demonstrated to that date. This engine demonstrated that scaled-up designs for nitinol-powered thermal engines eventually might be feasible.

At about this time, researchers developed a more complex device that used SMAs to move eight pins in a Braille character display. Their award-winning design permitted a computer to show Braille text for blind persons to read by touch. A 3-line, 60-character display used 480 individual SMA wires and received signals from a standard portable computer. Another application was the Fingerspelling Hand, an anthropomorphic robotic device to serve as a tactile communication aid for deaf and blind individuals, particularly those unable to read Braille. The hand acts as a computer display, presenting one character at a time using a finger-spelling alphabet. The user places his or her hands lightly on the device and "reads" each letter or character by feeling it, one at a time. Data can come from a keyboard, modem, page scanner, or closed-caption TV decoder.

To scale down the size of shape memory actuators, it has been possible to adapt the techniques used to make computer chips to form nitinol shapes on silicon. These thin Ni-Ti films can be used for creating very small mechanisms. Even on a microscopic scale, nitinol has an extremely high strength for its size—much larger than the electrostatic (static electricity) or piezoelectric forces often employed in micromechanical devices.

Other current topics in SMA research include mechanical switching units for optical communication fibers, electrically activated nonexplosive bolts released by an SMA element for sea and space applications, compact fingertip stimulators or "tactors" for virtual reality input/output devices, and thin films of nitinol for creating mini- and micromechanical devices for use in very small robotic systems.

Shape memory alloys can now be found in a wide range of applications and products that take advantage of their thermal recovery and superelastic properties. These include orthodontic arch wires in which preformed structures can be deformed for easier installation, flexible catheters and surgical guidewires that are implanted and anchored before an operation to act as guides to a tumor or other site, and expandable blood clot filters installed inside a vein or artery in a collapsed form and then expanded by the warmth of the body.

New manufacturing methods for SMAs include "spin melt" techniques that form molten metal directly into thin wires. Increased sales volumes, new and improved alloys, and greater awareness of the abilities of SMAs all promise to greatly expand the number and diversity of applications. This will further reduce their cost and increase their attractiveness for use in more products.—R.G.G.

See also AUTOMATION; VIRTUAL REALITY

Further Reading: Roger G. Gilbertson, *Muscle Wires Project Book*, 3d ed., 1994.

Shaping Machines

A shaping or molding machine uses metal cutters to shave the surface of a board or timber to produce a contoured edge. Typically, the work had been done by hand planes that were run with or across the grain of a piece of dressed lumber. The earliest machines employed a cutter fixed into a workbench with a fence for guiding the piece as it was passed over the cutter. However, as early as 1793, Samuel Bentham patented a shaping machine with various contoured cutters that revolved on a vertical shaft above a table. The machine could be cranked by hand or driven by belts as the workpiece was secured to a carriage and passed underneath the cutter. The principle resembled that of planing machines also designed by Bentham.

Molding machines developed more slowly than planing machines, for they produced articles less in demand than the simple floorboards, joists, sheathing, and dressed lumber made by planing machines. In addition, moldings had to be planed to a more precise finish than common building materials. Among early machines, only one end of the axle for the cutter was supported. Consequently, the machines vibrated to a considerable degree, which resulted in an inferior, wavy surface. Also, cutters were difficult to change and sharpen. Consequently, molding produced with a hand plane remained a far superior product until the 1870s. Until then, molding machines produced only cheap, simple goods that could be produced in large batches.

Molding machines improved considerably after 1860. Larger, heavier, cast-iron frames replaced earlier wooden frames, increasing substantially the cost of the machine but reducing vibration. Feed rolls provided a faster and more compact method of bringing the piece to the cutters than the older method of using carriages. Improvements in tool steel improved the quality of the cut and increased the time the machine could operate without the need for sharpening the cutters. Some machines made complex cuts using multiple heads and cutters. Increasingly, factory-produced molding supplied a tremendous demand for door panels, stair stringers, window sash, trim, furniture, and plow handles.

By the 1880s, variety molders were essential machines in most small- to medium-sized woodworking shops. Sometimes called *paneling* and *recessing machines*, they functioned like the modern router and were remarkably versatile shaping tools. Most often, two spindles projected above the surface of a table. When necessary, fences guided the work. The spindles rotated at high speeds in opposite directions, allowing the woodworker to cut with the grain without having to stop in order to reverse the spindle. Cutters were designed that allowed molders to contour the interior and exterior edges of curved pieces like oval picture frames. In the 20th century, electric motors and ball bearings increased speeds and reduced vibration to an even greater degree . These improvements were incorporated into molding machines that could produce dovetails accurately.—J.C.B.

See also MOTOR, ELECTRIC; PLANING MACHINE; STEEL, ALLOYED

Ships, Sailing

The sailing ship was the first means of transportation to make use of nonanimate power. Where or when the first sail was employed on a ship is not known, but its consequences were revolutionary. Vessels equipped with sails could be larger, carry more people and cargo, and travel farther and faster than ships propelled by oars. Sailing ships transformed the seas, which became transportation corridors instead of barriers. Intercontinental trade, conquest, and colonization were all made possible by the sailing ship.

The sailing ship has a long history. As early as 3000 B.C.E. the Egyptians were sailing 500 km (300 mi) to Crete, and by 500 B.C.E. the Carthaginians had sailed down the west coast of Africa. During this time and for many centuries afterwards, sail design remained largely unchanged. The Romans traded and fought throughout the Mediterranean with ships propelled by a single square mainsail that hung from a straight yard (i.e., a spar perpendicular to the mast). One Roman innovation was the use of *brails*, vertical ropes that passed through rings sewn onto the sail. By manipulating the brails the sail could be furled and unfurled as the situation required. Roman ships also employed two triangular topsails that could be raised above the main yard in order to catch light breezes. Roman ships also used a small forward sail known as an *artemon* that facilitated steering by helping to keep the ship in line with the wind's direction.

Roman ships suffered from a problem common to all ships fitted with square sails, the inability to sail into the wind effectively. Since the sail was set at right angles to the ship's axis, the ship was propelled at the most rapid pace when the wind blew directly from astern. It was possible to travel at an angle divergent from the direction of the wind by turning the yard to which the sail was attached. But if the yard was turned to an excessive angle, the ship would be blown sideways and all forward motion would cease.

The solution to this problem was the lateen sail. In Europe this type of sail goes back to at least the 9th century, when first it appears in a manuscript. The name is a corruption of *latine*, the French word for Latin, as Western Europeans first saw this type of sail in the Mediterranean. However, the Mediterranean was not the birthplace of the lateen sail; it may have originated in the East Indies, the Arabian Sea, the Nile, or even China. Whatever its origins, it spread rapidly, and by the 13th century it had become virtually universal in the Mediterranean.

The latten sail is triangular in shape and is attached to a long yard that hangs at an angle from a short mast. Its shape and rigging allows it to be swung around until its yard is almost in line with ship's axis. For this reason it is also known as a *fore-and-aft* sail. It is this kind of sail that most people visualize when they hear or see the word *sailboat*. A fore-and-aft rigged ship works according to the same principles that explain why an airplane's wing provides lift. When sailing into the wind, the ship's sail curves, so moving air has to travel a greater distance over one side of the sail than the other side. This causes the air to accelerate on the side that billows out, while it decelerates on the other. This in turn produces an area of low pressure on the billowing side and high pressure on the concave side. It is this difference in pressure that propels the ship. Many sailboats are fitted with two triangular sails, a mainsail and a smaller forward sail called a *jib*. A slot formed by the space between the two sails improves airflow over the mainsail by increasing the pressure differentials acting on that sail. Fore-and-aft rigging allows a ship to be sailed at least 45 degrees into the wind. It is not possible to sail directly into the wind, but a ship can move in the same direction from which the wind is blowing by "tacking." This is a process of repeatedly moving port-starboard-port in a zigzag fashion that takes the ship in the intended direction.

For all its benefits, the lateen sail is not perfect. It is less efficient than a square sail when the wind comes di-

A contemporary illustration of *The Great Republic,* 1853 (courtesy Smithsonian Institution).

rectly from astern, and shifting to another tack is a cumbersome process that requires a considerable amount of manipulation of the yards and sails. Thus by the 15th century, shipbuilders were making vessels that combined both types of sail: the full-rigged ship. These typically had three masts. The foremast and mainmast each carried two or more large square sails, the mizzenmast supported a lateen sail, and a large spritsail was mounted above the bowsprit (a spar that extends from the ship's bow). Ships rigged in this manner could get up a fair rate of speed and still sail 65 degrees into the wind.

This design endured for nearly 4 centuries, carrying crew, cargos, and passengers to the four corners of the world and profoundly altering the course of history. While the basic principles remained the same, ships became larger in the course of the 17th and 18th centuries, and some of the masts, which had to be constructed from three sections because of their considerable height, carried as many as five sails. In the 19th century, the sailing ship reached its aesthetic apogee in the American clipper ship. Built for speed (up to 21 knots or 39 kph) at the sacrifice of durability and carrying capacity, the clipper had a short life. First taking to the seas in the 1840s, within a few decades these ships found it increasingly difficult to compete with the rapidly developing oceangoing steamship. Also, the completion of the U.S. transcontinental railroad in 1869 diminished the importance of being able to travel from the Atlantic to the Pacific coast via the horn of South America in less than 100 days.

Sailing ships continued to be built in the late 19th and early 20th centuries, but their days as commercial vessels were numbered. The year 1902 marked a high point in one respect, for it was in this year that the largest sailing ship was launched. Built to haul nitrate, from Chile to Germany, the 133.5-m (438-ft) long *Preussen* had 5,580 sq m (60,000 sq ft) of sail and required a crew of 58. Large sailing ships have not disappeared; some of the world's navies use them as training vessels because it is believed that they teach leadership and seamanship better than powered craft. The primary realm of the sailboat today is recreation and competition. The latter activity, which reaches its highest point with the America's Cup competition, entails a high degree of technical sophistication and the expenditure of vast sums of money, even though the energy source, the wind, is available at no cost.

See also BERNOULLI EFFECT; RAILROAD; STEAMBOATS

Sickle-Cell Anemia

Under the microscope normal red blood cells appear to be disks that are concave on both sides. Their role is to transport oxygen from the lungs to all tissues of the body, and then return carrying carbon dioxide. They accomplish this by means of a four-chain globular protein, each chain linked to a ringlike compound containing an iron atom, called a *heme* (from the Greek, for *blood*). The iron reversibly binds four oxygen atoms, conveying them through the capillaries, where they are released in exchange for carbon dioxide. Ordinarily, red blood cells are flexible enough to squeeze through the tiniest blood vessels, but for people who have inherited the so-called *sickle-cell trait* from one of their parents, a small percentage of red blood cells will develop fibers during compression that twist the cell into a sickle shape. At the same time, calcium rapidly builds up. Both the twisting and the calcium influx are reversible, but over time, recurrent sickling causes the cell membrane to harden from calcification, and the deformed shape remains fixed. Those cells can no longer perform their role, but usually they are not numerous enough to cause serious disabilities. Sufferers of "sickle-cell anemia," by comparison, have inherited the recessive trait from both parents and can have as many as half of their red blood cells so crippled, resulting in occlusions, infarctions, and, typically, death around age 40.

In 1910, Chicago physician James B. Herrick described the case of Walter Clement Noel, a 20-year-old dentistry student from the Caribbean island of Grenada, documenting sickle-cell anemia for the first time. At this time, medical researchers had just begun to understand that hereditary disorders conformed to Mendelian principles, and, consequently, sickle-cell anemia became a model disease for the rising field of genetics. Indeed, the intricacies of hemoglobin and heredity lent support to the notion—later proved wrong—that genes were made of protein.

In 1949, Linus Pauling and Harvey Itano published results of their study of normal and sickled hemoglobins, concluding that the difference resided in the globins rather than in the hemes, though they could not precisely identify the cause of the "molecular disease." Eight years later, Vernon Ingram at Cambridge University employed methods that had been used to determine the structure of insulin to show the difference between normal and sickle-cell hemoglobin. Ingram found that in the sixth position of the second and fourth chains, anemic individuals had valine in place of glutamic acid, and he reasoned that a single gene was responsible. Deciphering the genetic code during the 1960s, he revealed that the base sequence of DNA for these two amino acids differed by a position reversal of a single adenine-thymine pair. So simple an error, though, had catastrophic consequences in certain circumstances.

While laboratory researchers figured out the molecular biology, epidemiologists, following a suggestion from Scottish geneticist J. B. S. Haldane (1892–1964), noticed that the sickle-cell trait had a high incidence in African and Mediterranean areas where malaria ranged. This suggested that the hemoglobin abnormality offered some passive selection advantage against the mosquito-borne parasite. British physician A. C. Allison proved that correlation in 1954, explaining that when the parasite infected blood of a person with sickle-cell trait, some red cells distorted and provoked an immune response that protected that individual from malaria's rapid progression. However, because their lives were prolonged into the reproductive years, "carriers" had a much greater chance of giving birth to children with sickle-cell anemia.

Without malaria's pressure on a population, the gene responsible for sickling would disappear over time, for the trait would become a selection disadvantage In the United States, mosquito control virtually eradicated malaria, and the incidence of both the trait and the disease among African-Americans is about one-fourth that of native Africans living in the malaria belt. Consequently, subSaharan Africa, where about 500 million people live with little or no protection against malaria (and the incidence exceeds 200 cases per 1,000 population), continues to be the chief incubator of the sickle-type hemoglobinopathies.

In the United States, sickle-cell anemia affects about 80,000 African-Americans. One ray of hope for this group has been the mid-1990s development of bone marrow transplantation. However, this procedure is limited to patients with genetically matched donors, about 7 percent of those afflicted. A promising alternative, now experimental, is gene therapy. This entails inserting a short chain of artificially made DNA and RNA (ribonucleic acid) into the patient; these bind to the defective gene and initiate a process that repairs the abnormal red blood cells. Future research will focus on genetic therapy aimed at repairing the bone marrow cells that produce the red blood cells.—G.T.S.

See also BLOOD, CIRCULATION OF; DNA; GENE THERAPY; GENETICS, MENDELIAN

Signal-to-Noise Ratio

In an electrical communications channel, the current path usually carries two levels of modulation: the signal that is deliberately modulated to transmit information, and "noise," an undesired result of stray electrical currents or voltages generated from various internal and external sources. Similarly, a wireless radio transmission contains an information signal as well as electrical noise created by natural or man-made sources (electrical storms in the atmosphere, electrical fields around certain types of equipment, etc.). The difference in amplitude between the signal level and the noise level is called the *signal-to-noise ratio*, sometimes abbreviated as the S/N ratio.

In communications systems, the signal must be much stronger (have a higher power amplitude) than the noise level, or the noise will mask or distort the signal so its information content is either lost or disrupted. Noise can be a steady phenomenon, such as a hiss or hum, or random in nature, such as crackling or popping, and is often generically called *background static* or simply *interference*. The noise might be either audible or visible (as "snow" or streaks in a video signal).

The signal-to-noise ratio is expressed with a unit of measurement called a *decibel* (dB). The number of decibels indicates how far the noise level is below the desired signal level. For example, the amplifier in a high-quality stereo receiver might specify the S/N ratio of its output as a maximum of 70 dB, meaning that the noise level is 70 dB below the strongest signal level, or virtually inaudible.

During the 1920s, when telephone engineers were experimenting with radio frequencies as a carrier medium and introduced commercial radio telephony between North America and Great Britain, background noise was a nuisance. Karl G. Jansky (1905– 1950), a young scientist at Bell Telephone Laboratories, published a paper in 1932 classifying three sources of static: local thunderstorms, distant thunderstorms, and an unknown source that produced a steady hiss. A year later, Jansky reported that the hiss noise came from our Milky Way galaxy. Jansky's discovery led to the creation of a new science, radio astronomy, and eventually to the development of the Big Bang theory regarding the birth of the universe.

In telephone systems, the difference between the speech signal and the background noise is hard to establish because the amplitude of a conversation has a wide range of continuously fluctuating energy levels. The louder speech sounds have far greater amplitude than the usual noise level, but when nobody is talking, the background noise might be disturbing. The volume of noise usually depends on the transmission media. A telephone conversation often is transmitted over different wire and cable networks, some of which are invariably noisier than others. Unfortunately, that noise often travels through the entire circuit path, especially in analog systems. A typical long-distance call, for example, must pass through a local network at each end of the calling path, plus the long-distance network, and probably will be carried on a variety of transmission media (twisted-pair wires, coaxial cable, fiber-optic cable, and possibly microwave circuits). The twisted-pair wires are the most common source of undesirable noise, and they are usually located at both ends of the calling path.

Signal-to-noise ratios are commonly of limited use in characterizing telephone voice channels, since the individual talker's signal power fluctuates widely. On the other hand, research has established that users are most annoyed by the background noise heard when nobody is talking. However, if the circuit is *too* quiet during a pause, some users will think the line has "gone dead" and might hang up. Therefore, telephone engineers prefer to specify the maximum idle-circuit noise power allowed in a message channel rather than to use S/N ratios. This is especially true when analog transmission is used. The line amplifiers, or "repeaters," needed at intervals in analog transmission lines to prevent signals from growing too weak, also amplify the background noise together with the voice signals. Digital transmission has the advantage of easier identification of electrical noise artifacts, which can be erased by signal regenerators that are used instead of repeaters.

Various noise spectra sources can influence the transmission quality of electrical communications channels; some sources can be suppressed, but others cannot. The transmission medium most resistant to noise sources is fiber-optic cable, since the conversation is transmitted through special glass fibers using photons rather than electrons, and is digitally encoded as well.

The motion-picture and commercial recording industries are extremely interested in reducing background noise that could degrade the quality of their product, especially musical recordings. Various types of signal-processing techniques, usually based on mathematical algorithms, have been developed to eliminate audible background noise during both recording and playback modes. Perhaps the most frequently used noise reduction (NR) approach in use today was developed in the 1960s by Ray Dolby (1933–) of Dolby Laboratories. Originally designed to control the background hiss in studio recording on magnetic tape, the Dolby method has been expanded into consumer market audio equipment, as well as being used to control sound tracks for premium-quality videotape and film recordings. NR systems function by filtering certain frequencies from the signal, analyzing and suppressing the noise frequencies, then restoring the filtered sound frequencies to the signal. Special microchips are used to accomplish this.

Modern commercial recordings are created with digital technology that diminishes the need for noise reduction techniques, especially if the playback is via compact disks, digital video disks (DVDs), or digital tape systems. However, noise reduction is still employed to control tape hiss during playback of analog audio- or videocassettes. If the audio or video playback equipment lacks appropriate NR decoding circuits, the higher frequencies in the sound will seem too prominent unless tone controls are adjusted to reduce their power level.—R.Q.H.

See also ALGORITHM; BIG BANG THEORY; COMPACT DISK; FIBER-OPTIC COMMUNICATIONS; MICROPHONE; MICROPROCESSOR; TELEPHONE; TELESCOPE, RADIO

Silk

Silk is made from filaments derived from the cocoon of the silkworm, *Bombyx mori*. To make their cocoons, silkworms secrete a protein-based liquid through spinnerets located under their jaws. This is used to spin a filament that is glued together with a substance known as *sericin*, a liquid that hardens as soon as it is exposed to air. The resulting cocoon envelopes the silkworm as it develops into a moth, a process that takes about 2 weeks.

According to legend, silk production began in China around 2600 B.C.E., although it is likely that silk production began in Japan at an even earlier date. Whatever the exact date and place of origin, it is certain that a long-distance trade in silk had been established by the early Han dynasty (206 B.C.E.–220 C.E.). Underscoring the significance of the silk trade is the name by which China was known to the Romans: Serica, "the land of silk." Silkworms were smuggled into Byzantium in the 6th-century C.E., initiating silk production in Constantinople, from where it spread to parts of Europe.

The basic processes for converting cocoons into silk are not complicated, but they require a great deal of care. If left alone, the moths will break through the cocoon, rendering it useless. Consequently, after the cocoon is fully developed, it is necessary to kill the moth by plunging the cocoon into hot water. This is followed by a number of immersions in hot and cold water to soften the sericin. The cocoon is then unwound, yielding a single thread with a length of 300 to 600 m (984–1,968 ft). The unwinding of the cocoon is a process known as *reeling*, which also includes winding together 3 to 10 filaments to form a single thread of the required diameter. Winding can be done by hand, or it can be mechanized. Winding was partially mechanized in China during the Northern Sung dynasty

(960–1126), with a foot treadle supplying the power.

Raw silk is then converted to finished silk thread (also known as silk yarn) by a process known as *throwing*. This too was mechanized at a rather early date, albeit not in China but in 14th-century Italy, where the city of Lucca became a renowned center for silk production. The design of Italian machinery was pirated in the early 18th century by Thomas and John Lombe, who established a silk factory in Derbyshire. At about this time, automatic looms were being developed in France for the weaving of silk cloth with elaborate designs.

Some tentative steps at mechanizing silk production were taken in 18th-century Japan, but these were slow in being adopted, in part due to a fear that machines would replace human labor. In China, efforts to introduce mechanized reeling were subverted by established guilds, which wanted to maintain a monopoly on silk production for their members. This was a self-defeating strategy in the long run. By the end of the 19th century, the Japanese silk industry was more technically advanced than the one in China, and Japan's exports of silk provided most of the foreign currency used for the rapid industrial modernization that characterized the early Meiji period.

While the reeling, throwing, and weaving of silk was gradually mechanized, the cultivation of silkworms and their cocoons remained a small-scale, labor-intensive activity. In many parts of China and Japan, the raising of silkworms was an important sideline activity for farm families, often being the difference between privation and a modicum of prosperity. Whatever wealth was gained did not come easy, however, for the care and feeding of silkworms was an arduous task. Temperatures had to be kept within a narrow range, and often entailed the frequent moving of silkworm beds from one part of the house to another. Great cleanliness was required to prevent the spread of infectious diseases. The silkworms consumed large amounts of mulberry leaves, and these had to be cleaned, chopped, and sieved many times a day prior to being fed to them.

Raising silkworms and processing their cocoons into thread and weaving it into cloth brought needed income into many Chinese and Japanese households, but on many occasions it led to financial disaster. Increased production could cause the price of silk to plummet, leaving many silk producers with a product worth less than the cost of its production. The same thing could happen when the demand for silk dropped, as happened during the worldwide economic depression of the 1930s. Making matters worse, the development of synthetic materials, first rayon and then nylon, seriously undercut the demand for silk and led to the ruination of many who had depended on the silk trade. At the same time, however, there still remains a significant demand for high-quality silk, for no other fabric has all the qualities it possesses.

See also LOOM, "JACQUARD"; NYLON; RAYON; UNEMPLOYMENT, TECHNOLOGICAL

Skates, Roller

Ice skating has a history that stretches back for many centuries, but it wasn't until the 18th century that an anonymous inventor, most likely Dutch, thought of mounting wheels on a small platform that was attached to the shoe. By 1760, roller skates were well enough established for a roller skate shop to open in London. Roller skates seem to have been forgotten in subsequent years, for in 1823 an English periodical noted that:

> A skate has just been invented with the design of rendering this amusement independent of the frost. It is like the common skate, but instead of one iron it has two, with a set of very small brass or iron wheels let in between, which, easily revolving, enable the wearer to run along with great rapidity on any hard level substance, and, indeed to perform, though with less force and nicety, all the evolutions of skating.

By the middle of the 19th century, a performance on roller skates had been incorporated into an opera by the German composer Jakob Meyerbeer (1791–1864), and before this a German beer hall had employed waitresses who made their rounds on skates.

The skates of this era were not totally satisfactory, as their design made it difficult to execute turns. This fault was remedied in 1863, when James L. Plimpton invented the now-familiar four-wheel skate, which had the additional advantage of rubber springing. In 1884, another American, Levant Richardson, increased the efficiency of roller skates by fitting their wheel-and-axle assemblies with ball bearings.

Roller skates retained their essential form for many

years thereafter. In the 1970s, they were improved by the use of urethane wheels, which provided better traction than wheels made from steel. The next major innovation was the inline skate, which as its name implies, has four 72- to 76-mm-diameter wheels arranged in a straight line. These skates, which were invented in 1980 by Scott Olson, a Canadian hockey player, resembled ice skates in the way they handled. Although they are not as suitable for competitive figure and dance skating as conventional roller skates, inline skates are the standard gear for players of roller hockey. The rapid growth of roller hockey in the United States helped to stimulate a substantial increase in the sales of inline skates, from 800,000 pairs in 1990 to 3.9 million in 1992. In that year, sales of inline skates generated revenues of $267 million.

Skylab

With the conclusion of the Apollo program, the American manned spaceflight program found itself at loose ends. It had achieved its major goal, putting a man on the moon, and with no project scheduled for the immediate future, the people and organizations involved in the space effort faced the prospect of permanent unemployment. Some NASA administrators had foreseen this, and well before the first Apollo mission, they began to develop plans for an orbiting space station. The sequence turned out to be an inversion of what many space scientists had formulated, for a space station initially had been regarded as essential to a lunar landing. The various Apollo missions were based on different principles, so a space station had to be given a different mission: serving as an orbiting scientific laboratory.

The basic design of Skylab was settled after the resolution of a technical dispute between the Manned Spaceflight Center in Houston, Tex., and the Marshall Space Flight Center in Huntsville, Ala. The former envisaged the use of an empty third stage of the Saturn rocket being used as the station, while the latter promoted a rendezvous in space between two Saturn rockets, one of which would have its second stage converted into the orbiting laboratory by a crew from the other rocket. After considerable discussion, NASA opted for Houston's version.

Since it was based on a stage of the giant Saturn booster, Skylab was quite large—vast in comparison with the cramped capsules of previous missions. Weighing 91,000 kg (200,000 lb), its two compartments, one the crew's living quarters and the other a work area, were situated in the part of the stage that had contained the hydrogen tank, a space more than 15 m (50 ft) in length with a diameter of a bit more than 6 m (20 ft). The work area contained the control panels for electricity, communications, and environmental regulation. It also contained an airlock for leaving and returning.

The May 14, 1973, launch of the rocket containing the unmanned Skylab produced some unexpected problems. A meteor shield came loose at launch, taking with it one of the solar panels. Eleven days after the launch, Skylab's crew made a successful rendezvous with the damaged space station and performed in-space repairs. They used a rectangular screen to shade the area where the meteor shield had detached and managed to free the stuck solar panel.

During the crew's 28 days in space, they engaged in a number of scientific observations of the Earth and the sun, including the taking of 29,000 pictures of the sun. The crew of a second mission, launched on July 28, 1973, stayed in Skylab for 59 days. The crew of the third mission, launched on Nov. 16, 1973, put in 84 days. All three missions engaged in scientific observations, while at the same time adding to an understanding of how human beings function during prolonged periods of weightlessness.

Skylab remained in orbit for only 6 years. Its orbital height, 443 km (275 mi), exposed it to stray molecules from the atmosphere as well as particles from the sun. As a result, it slowly lost altitude, and in July 1979 it returned to Earth amidst considerable concern that flaming remnants would fall on populated areas. As things turned out, no damage was done—except for the loss of America's first and last space station.

See also APOLLO PROJECT

Further Reading: Leland F. Belew, ed., *Skylab, Our First Space Station*, 1977.

Skylab. Note the improvised cover that replaced a shield that had separated after launch (courtesy NASA).

Skyscrapers

The skyscraper of the late 19th and early 20th centuries is truly an American art form. While the capability to build tall buildings existed in many parts of the world, the will or interest in building skyscrapers did not. However, in both Chicago and New York the peculiar combination of land speculation, urban density, technological capabilities, and hubris created the skyscraper in the last quarter of the 19th century.

Tall buildings existed in Rome during the Imperial Era, with some apartment buildings extending up to 10 stories in height, until building collapses prompted the passage of laws that limited them to 7 stories. However, without an elevator, the trek up and down the stairs must have been arduous at best. The passenger elevator, which had its first commercial application in New York in 1857, made it possible to charge rents on the upper floors of buildings in reasonable proportion to the cost of building them.

The earliest tall buildings in New York featured small towers appended to a larger base. In many cases, these towers had little or no functional space to be occupied; their main purpose was to grant notoriety to the occupants or to look down on adjacent buildings. The development of "Newspaper Row," with rival newspapers building ever-taller buildings in competition with one another, is the clearest example of this stage of development. Chicago developed in a slightly different manner. Most tall buildings were speculative office buildings, not company headquarters. The emphasis was on developing rentable floor plans, not heraldic roof lines. Consequently, Chicago buildings tended to have a consistent floor plan throughout their entire height. Without a tower to "puncture" the sky, their flat roofs "scraped" the sky, hence the term *skyscraper*, which originally distinguished flat-roofed buildings from the so-called "tower buildings" that strove for dominance through recognition and absolute height.

Important developments in skyscraper building occurred in Chicago for a number of reasons. One accidental catalyst was the Chicago Fire of 1871, which illustrated a peculiar fact of American real estate development: Land cleared of all buildings was worth more than when occupied by a building of only modest size. This financial reality drove investors and building owners to tear down and rebuild in increasingly rapid cycles, each time building higher. With no perceived historic patrimony to save (Chicago was only a few decades old, and the fire removed most historic structures anyway), and given the peculiar American frontier mentality, architects and developers set out to conquer and subdue the uncharted heights of the urban landscape.

Chicago was the site of important technical advances. The steel frame and curtain wall were Midwestern innovations that enabled greater building heights and greater profitability. Usually credited to William LeBaron Jenney's (1832–1907) Home Insurance Building of 1884–85, in reality this structure was only a step towards the fully independent frame. Other architects in Chicago, New York, and other midsized American cities quickly capitalized on this innovation, which uses an iron frame to support not only the floor loads but the weight of the enclosing wall as well. This enabled architects to bring more light and air into office spaces and create larger shop windows at ground level. By the mid-1890s architects were stiffening the iron frame to provide independent resistance to wind loads by using portals and/or X bracing. With the publication of simple calculation methods for continuous frames in 1908, engineers and architects pushed building heights as far as economics would allow. These calculations were responsible in part for a string of "tallest" buildings: Metropolitan Life, Woolworth, Chrysler, and Empire State Buildings.

Another technical innovation from Chicago was the development of lightweight terra cotta fireproofing systems that proliferated during the final decades of the 19th century. Terra cotta systems were gradually replaced by concrete fireproofing, especially after the Baltimore fire of 1904 demonstrated concrete's superior performance in actual fires. However, the weight of concrete fireproofing has virtually eliminated its use in very tall buildings. Instead, sprayed-on asbestos and other mineral coatings, as well as paints that blister and insulate steel from high temperatures, reduce dead loads of buildings, an especially important feature in areas of high seismic activity.

One of the most critical factors in skyscraper profitability is vertical circulation. Elevator shafts consume a large proportion of interior space—space that can-

The first skyscraper in Minneapolis is dwarfed by a more recent building (courtesy National Archives).

not be rented out. On the other hand, building occupants who have to wait inordinate amounts of time for elevators will not be willing to pay top dollar for office rents. Early steam-powered or hydraulic elevators had a practical limit of about 12 stories, and they were not particularly fast. The development of the electric traction elevator permitted higher speeds, although early models tended to accelerate with disconcerting jerks as additional windings in the motor were energized. With higher speeds, passengers could be carried in fewer elevators, resulting in more rentable space and higher profitability.

Until World War II, skyscrapers depended on windows for light and air. No office space could be profitably rented that was more than 8.5 m (28 ft) from a window. Consequently, skyscrapers on larger sites had O-, U-, or H-shaped plans to provide windows for offices. Elevators and stairs were typically buried in central cores. However, with the advent of air conditioning and florescent lighting, skyscrapers no longer needed a multiplicity of windows for occupant comfort. Consequently, the floor plate size of recent skyscrapers has increased dramatically, with no relieving indentations into the building mass for light and air.

Zoning has had a dramatic impact on skyscraper development, either through limiting height or shaping form. One of the most important zoning regulations was the 1916 New York zoning law, which required setbacks to allow light and air to reach street level, yet permitted a tower of unlimited height on 25 percent of the site. This law produced a large number of so-called "ziggurat buildings" whose outer profile stepped upward as they followed the legal envelope. Another byproduct of the law was increasing the value of sites large enough so that a 25 percent tower would have a floor plate of sufficient size to accommodate elevators, services, and office space. The Empire State Building is a classic example, in that it was built in midtown Manhattan, some distance from the downtown business district, but occupying one of the largest building sites available.

After World War II, skyscraper design was heavily influenced by new curtain wall systems, most of them using large panels of tempered glass. These glass skins were light, inexpensive, and favored by the modernist aesthetic. Their ubiquitous use in urban America has spawned debate concerning "banal boxes" that all look the same. More recently, "postmodern" skyscrapers have abstracted historical forms in order to avoid the monotony of contemporary urban landscapes, but critics have not always reacted favorably to these buildings.

By far the most exciting development in skyscrapers in recent decades is the advent of computer analysis techniques that have significantly reduced the amount of steel structure required, thereby making tall towers more profitable. These methods of analysis can model the structural arrangements most suitable for resisting wind and seismic loads.

Skyscraper construction in the United States has moderated in the last few decades, slowed by economic rather than technological constraints. In the near future, conditions are likely to be more favorable on the Pacific Rim, and indeed the current world's tallest building is the Petronas Towers in Kuala Lumpur, Malaysia, at 450 m (1,476 ft), although it soon will be eclipsed by buildings in Shanghai or Hong Kong. These skyscrapers will require financing by institutions with enormous capital reserves, large urban sites, and governments willing to bear the cost of providing the necessary infrastructure to support the building.

Although it is technically possible to build buildings of 150 or more stories, it remains to be seen if such construction is desirable. Tall buildings containing tens of thousands of workers put enormous demands on public transportation. Electric, water, and gas utilities also must be able to support what amounts to a small city on a single site. Although there are economies of scale when utilities for a small city are concentrated at a single location, municipalities can't always control where developers decide to build, and it can be a major burden to provide utilities and public transportation in remote locations.

Economics also limits the height of skyscrapers. Taller buildings require larger columns that use more steel and reduce rentable space. They require more elevators that also eat up rentable space. And the water supply and waste piping has to withstand greater pressures, resulting in more costly pipes and valves. A study in 1930 concluded that the optimum economic height of a skyscraper was 63 stories tall, a figure that probably hasn't increased by more than about 20 percent. Not only do construction expenses increase rapidly above this limit, but also the flexibility of tall buildings (required for wind and earthquake resistance) causes them to sway noticeably in the wind at the higher floors, making some inhabitants uncomfortable or even motion sick. Mass dampers have been used to limit sidesway and improve the comfort of inhabitants, but this adds to their cost. Of course, owners can charge higher rents in a building that is a notable landmark or is "the tallest building in the world." But there is a limit to what people will pay for visibility and prestige.

In the 1950s, Frank Lloyd Wright (1867–1959) envisioned a Mile-High Tower, which he assumed would have nuclear powered elevators to whisk inhabitants to their destinations. Despite the naiveté of his proposal, it is still true that the prestige of a building visible for miles, the commanding views, and the tremendous profit that can be earned from a well-located skyscraper ensure that very tall skyscrapers will continue to be built. The only questions are how our technologies will develop to meet the challenge, and how human beings will adapt to such tall structures.—E.C.R.

See also AIR CONDITIONING; ASBESTOS; ELEVATOR; EARTHQUAKE-RESISTANT BUILDING STANDARDS; FIREPROOF/FIRE-RESISTIVE BUILDINGS; GLASS; LIGHTS, FLUORESCENT

Slide Rule

For many decades, a slide rule was a key symbol of science and technology. Used by scientists, engineers, and students of science and engineering, the slide rule seemed the embodiment of esoteric techniques that were beyond the ability of ordinary people. In reality, an ordinary slide rule is a simple device that requires only a bit of instruction before the user is able to manipulate it effectively. The most common application of a slide rule is for division and multiplication. Slide rules may also contain scales for squares, cubes, and trigonometric functions. Specialized slide rules allow the quick solution of equations based on specific formulas, for example Ohm's law, which relates the voltage, current, and resistance of an electric circuit.

A simple slide rule can be constructed to add and subtract numbers by moving one part relative to the other. A regular slide rule does the same thing, but the numerals on

each scale are arranged logarithmically. Consequently, the addition of the logarithms of two numbers produces the logarithm of the numbers' product, which is read on the slide rule's scale as the product itself.

Logarithms were invented in the second decade of the 17th century by a Scottish baron, John Napier (1550–1617). Napier also devised some mechanical aids for multiplication and division, tasks that took much time and effort on the part of astronomers, compilers of navigational tables, and astrologers. Known facetiously as "Napier's bones," these devices consisted of strips imprinted with rows of numbers; the combination of the appropriate strips allowed easy multiplication, division, and the extraction of square and cube roots.

One of the first people to see the value of logarithms was Henry Briggs (1561–1631), professor of geometry at Gresham College in London. Briggs calculated an extensive table of base-10 logarithms, the ones in most common use today. Edmund Gunter (1581–1626), a fellow professor at Gresham College, was intrigued by these tables. His earlier work with calculating devices helped him to realize that the process of adding and subtracting logarithms could be facilitated by engraving a logarithmic scale on a stick, and then using a pair of dividers to add or subtract two numbers. For example, the multiplication of two numbers, X and Y, could be done by opening up the dividers so that the points spanned from 0 to X on the engraved stick. The dividers were then moved so that one point was on Y and the second was to the right of it; the product then could be read as the number on which the second point rested.

An important refinement of Gunter's logarithmic stick was devised by William Oughtred (1574?–1660). It was his idea to add another stick with a logarithmic scale that could be moved relative to the first stick, thus allowing calculations to be done without the use of a pair of dividers. He also conceived the idea of having the scales engraved on concentric disks, thereby creating the first circular slide rule. Oughtred considered his devices to be of little consequence, but when one of his students published a description of the circular slide rule and seemed to take credit for its invention, Oughtred published a book of his own.

The first duplex slide rule with two fixed strips (the "stock" or "body") and a sliding strip (the "slide") in the middle was built in 1657 by Seth Partridge. Despite this refinement, the slide rule seems to have been little used in the ensuing years. In 1850, a 19-year-old French artillery officer named Amedee Mannheim (1831–1906) added a moving runner with a vertical hairline. This made the slide rule easier to use, and in this form it was soon adopted by the French artillery service. The slide rule was much less popular in other places, most notably the United States. In this country, it began to be used by significant numbers of scientists and engineers only after 1890, due in large measure to a series of propagandizing articles in the *Engineering News* written by William Cox.

In recent years, the slide rule has been rendered obsolete by cheap electronic calculators. Calculators have the advantage of being somewhat easier to use and being able to produce more precise results. Yet the apparent precision of operations performed by calculators has often led to the acceptance of erroneous results. Another failing of the electronic calculator is that it does not require the operator to determine the calculated number's order of magnitude (that is, where the decimal point should go), as is necessary with a slide rule. Students who have never used a slide rule may be less able to develop the ability to estimate results and to determine when a numerical result is reasonable or unreasonable. Again, the precise number rendered by a calculator may convey a false confidence in the accuracy of the result.

See also LOGARITHM

Further Reading: Michael R. Williams, *A History of Computing Technology*, 1997.

Smallpox Eradication

Smallpox is an infectious, often-fatal viral disease. It is characterized by skin lesions that leave permanent scars if the afflicted person survives. In the late 18th century, smallpox began to be controlled through the practice of vaccination, but for many decades it continued to kill and disfigure millions of people who had not been vaccinated. As late as the mid-1960s, 10 to 15 million new cases and 1 to 3 million deaths occurred annually as a result of smallpox.

In the years that immediately followed, efforts to vaccinate people where smallpox was endemic were greatly aided by the development of freeze-dried vaccines that remained potent and stable even in tropical climates. In 1966, the World Health Organization, an agency of the United Nations, initiated an ambitious 10-year campaign aimed at nothing less than the complete eradication of smallpox. Despite wars and rebellions in a number of the targeted countries, health workers were able to vaccinate large numbers of people. Although not everyone was vaccinated, public-health workers were able to sharply contain the spread of the disease by targeting areas that were of particular epidemiological importance.

The program was highly successful. In 1977, a

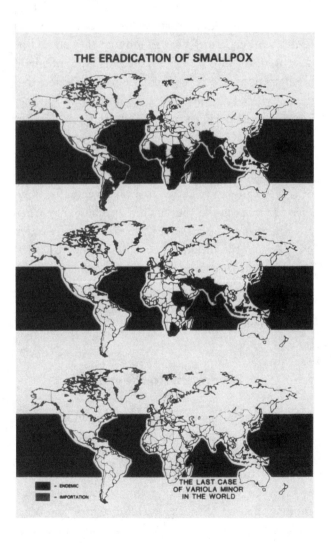

Map showing countries afflicted by smallpox in 1966, 1972, and 1977 (courtesy National Library of Medicine).

cook from the African nation of Somalia gained the dubious distinction of being the last person to contract smallpox through natural transmission. His was not the last case, however. A year later a woman in Birmingham, England, died from smallpox she had contracted when the virus escaped from a research laboratory, traveled through a ventilation duct, and entered a room in which she was working. This was the last recorded new case of smallpox. In May 1980, the World Health Organization declared that smallpox had been eradicated, the first disease to have been eliminated as a result of human intervention.

Although smallpox epidemics are a thing of the past, samples of the smallpox virus still exist in two laboratories, one of them in Atlanta, the other in Moscow. In recent years, there has been lively debate concerning the ultimate fate of these samples. Many health workers are of the opinion that they should be destroyed, foreclosing forever the possibility of smallpox infections. However, there are some who feel that the complete eradication of a form of life, no matter how potentially dangerous it may be to humankind, is an immoral act. On a more practical level, some researchers believe that the samples should be preserved because there is always the possibility that they may be of future value to medical research. Eventually, however, the mapping of the smallpox genome may obviate the need to preserve the virus for research purposes. In the meantime, samples of the virus that causes smallpox continue to reside in their high-security laboratories.

See also EPIDEMICS IN HISTORY; FREEZE DRYING; IMMUNIZATION; VIRUS

Smog, Photochemical

Etymologically, the word *smog* is a combination of *smoke* and *fog*. In some parts of the world, this accurately describes the nature of smog. One extreme example of this kind of smog occurred in December 1952, when 4,000 deaths in London were caused by a choking mixture of fog, soot, sulfur oxides, and other pollutants produced by the burning of bituminous coal. In many other parts of the world, however, severe smog exists in the absence of fog and the production of smoke. This is termed *photochemical smog*, because it is the result of a chemical reaction in the atmosphere that occurs under the influence of sunlight.

The chief raw materials of photochemical smog are the largely invisible products of internal combustion engines. The ignition of an air-fuel mixture leaves some hydrocarbons unburned, while high temperatures and pressures within the engines' combustion chambers convert atmospheric nitrogen into oxides of nitrogen, or NOx. After these exhaust gases are released into the atmosphere, the ultraviolet radiation of sunlight breaks down one of the nitrogen oxides, nitrogen dioxide (NO_2), leading to reactions with oxygen and with the unburned hydrocarbons to produce the other constituents of photochemical smog: ozone and peroxyacyl nitrates (PANs), along with residual NO_2. The combustion process also results in the production of carbon monoxide (CO) and carbon dioxide (CO_2), as well as sulfur dioxide (SO_2) and various particulates. These compounds can cause serious problems, but strictly speaking they are not constituents of photochemical smog.

The likelihood of high concentrations of smog is increased by topographic and meteorological conditions. An area enclosed by mountains with consequent lack of atmospheric circulation will be more vulnerable to smog production, as will an area capped by an inversion layer. In the latter case, a layer of warm air sits on top of a layer of cold air, thereby preventing vertical circulation. These conditions are especially evident in Southern California, and when added to a huge automobile population the result is the worst air quality in the United States.

Photochemical smog damages property and health. Trees and crops are harmed, resulting in substantial monetary and aesthetic losses. Smog further diminishes environmental quality as unreacted nitrogen dioxide blankets an area with a gray-brown haze. PANs (chemicals related to tear gas) irritate the eyes, while ozone makes breathing difficult and even painful. Smog may also increase susceptibility to asthma, allergies, and viral infections. Although precise epidemiological studies are impossible due to continual population shifts, it is possible that living in a smoggy environment increases the risk of lung cancer and cardiopulmonary disease.

Since the 1960s, government regulations have attempted to alleviate the smog problem by diminishing the production of the substances that produce it. Industrial plants are subject to restrictions on the fuels they burn; regulations covering the use of paints and solvents reduce the volatile organic compounds that contribute to smog. It is relatively easy to control these potential sources of smog. Considerably more difficult is the control of automobile exhaust emissions, which in the 1950s were shown to be major contributors to the production of smog. Automobile engines are difficult to control because they run through a wide range of rotational speeds, are run before being completely warmed up, and are often poorly maintained. In the 1960s, California, the state with some of the worst examples of smog pollution, began to implement increasingly strict regulations on automotive exhausts. In 1961, it required the installation of positive crankcase ventilation (PVC) valves on new cars beginning in 1963. In 1966, the California's legislature established the first emission standards for automobiles.

In 1963, the federal government became engaged in air pollution control through the passage of the first of a series of clean-air acts. The federal government's involvement increased substantially with the passage of the 1970 Clean Air Act, which established national air quality standards—a 90 percent reduction in the emission of nitrogen oxide by cars required by 1976—and gave the newly created Environmental Protection Agency (EPA) broad regulatory powers. In 1977, an expanded act mandated a 96 percent reduction of hydocarbons and carbon monoxide. Automobile manufacturers were subsequently given more time to meet these requirements. Even so, at first they could not meet them without sacrificing performance, drivability, and fuel economy. In time, new engine technologies, most significantly catalytic converters, fuel injection, and computerized engine management systems, brought substantial reductions of exhaust-borne pollutants. Despite these advances, air quality in several parts of the country remained below federal standards. In 1990, another clean-air act further tightened emissions standards and required that automotive pollution equipment had to perform acceptably for 10 years. It also mandated that areas not meeting air quality standards had to meet air quality standards according to a specific schedule or face the loss of federal benefits such as highway construction funds.

Parallel governmental efforts in California culminated in an ambitious Air Quality Management Plan aimed at the attainment of federal standards in Southern California by the year 2010. Yet even this distant target could be met only through the imposition of severe restrictions on certain materials (the plan initially called for even the prohibition of charcoal lighter fluid), reductions in driving, and strict emission requirements. These draconian measures met with a great deal of opposition, resulting in several modifications and the deletion of many of the original regulations, notably the California Air Resources Board's requirement that 2 percent of the new cars sold in 1998 had to be zero-emission vehicles.

In recent years, smog-producing emissions have been reduced to a considerable degree. To take one of the most important examples, for cars built in the 1990s, tailpipe emissions of carbon monoxide and hydrocarbons have been reduced by 96 percent compared to cars built in 1960, while nitrogen oxides have been reduced by 76 percent. As a result, air quality has improved substantially. In the Los Angeles region from 1955 to 1992, ozone, a key indicator of smog severity, declined from 680 parts per billion parts of air to 300 parts per billion. At the same time, however, industrial development and an expanding automobile population make it difficult to hold on to the gains that have been achieved. Effecting significant reductions in smog will require continued technological advances and perhaps some changes in the way we live.

See also AUTOMOBILE, ELECTRIC; CATALYTIC CONVERTER; CLIMATE CHANGE; EMISSIONS CONTROLS, AUTOMOTIVE; ENVIRONMENTAL PROTECTION AGENCY; ENGINE, FOUR-STROKE; FUEL INJECTION

Soap

In its most basic form, soap is produced by reacting animal fats or vegetable oils with an alkali. One traditional alkali used in soap manufacture is lye, which can be either Potassium hydroxide (KOH) or sodium hydroxide (NaOH). For many centuries, soap was made with lye that was derived from wood ashes. Soap also can be made from potash (potassium carbonate, K_2CO_3). The reaction of fat and an alkali is done by boiling the two together, a process that in times past made the manufacture of soap a messy and odoriferous procedure.

The reaction of fats and alkalis results in a long carbon-chain molecule, one end of which is attracted to water (hydrophilic), the other end of which is repelled by water (hydrophobic) but attracted to grease. A greasy object is cleaned as the soap molecules form a bridge between grease molecules and water molecules. A little agitation then causes the grease to be lifted off the object and put into suspension with the water.

Soap was first made in Sumeria around 3000 B.C.E., but it was unknown to the ancient Greeks and Romans. Despite their love of the bath, the Romans made do without soap, using various abrasives to clean themselves. Soap was introduced into the late Roman Empire by the "barbarian" Gauls, although it was used more as an ointment than a cleansing agent.

By the 9th century, soap was being used throughout Europe, and cakes of hard soap began to be produced in the 12th century. Even so, the Middle Ages were not a receptive time for the use of soap. Excessive concern with cleanliness was thought to signify a lack of spirituality, and public baths were often associated with prostitution. The situation was no better in early modern times; by the 16th century, public baths had almost disappeared, due in part to growing concern with venereal disease.

A heightened concern for cleanliness began to emerge in the latter part of the 18th century and became an obsession during the 19th century. New attitudes towards personal hygiene reflected of the rise of the bourgeoisie and the ascendancy of "bourgeois tastes." Cleanliness was prized because it bolstered the status of the middle class by distinguishing them from "the great unwashed" working class, as well as the aristocracy, who usually relied on perfumes to mask the consequences of their infrequent bathing. In the United States, the new-found concern with cleanliness was also motivated by the influx of large numbers of immigrants in the late 19th and early 20th centuries; clean clothes and a clean body were taken to be distinguishing characteristics of "real Americans."

Increasing demand for soap was met by new technologies that lowered production costs. Particularly important were improved processes for making alka-

lis. Soap was produced in larger batches as a cottage industry gave way to factory production, but for decades the methods employed largely consisted of scaling up traditional procedures. Fats and alkalis were vigorously boiled in giant kettles and then cooled for several days. Salt was then added to separate the mixture into a layer of soap on top and a layer of alkali at the bottom. Perfumes and other additives were mixed in, and the resulting substance was cast into bars, rolled into flakes, or spray dried into powder. In the late 1930s, continuous process technologies began to be used, speeding up production substantially. Not content with increasing supply, soap manufacturers also made vigorous efforts to inflate demand by means of aggressive advertising that played on people's anxieties about their personal appearance. One soap manufacturer even succeeded in getting the term "B.O." (for body odor) inserted into the national lexicon.

In the 20th century, detergents replaced soap in a variety of uses. Detergents had the significant advantage of not leaving deposits, as soap does when its carboxylate ions react with the calcium and magnesium ions found in hard water. Detergents have largely replaced soaps for laundry and many other cleaning applications, but conventional soap continues to be used for bodily cleansing since it does not have the drying properties typical of detergents.

See also ALKALI PRODUCTION; DETERGENTS

Social Construction of Technology— see Technology, Social Construction of

Sociobiology

Sociobiology is "the systematic study of the biological basis of all social behavior." So said E. O. Wilson in his 1975 volume, *Sociobiology: The New Synthesis.* Still the subject of considerable controversy, sociobiology is increasingly presented as a valid theoretical framework for the study of animal behavior. Although Wilson compiled the best-known book on the subject, sociobiology was shaped by many earlier theorists, such as R. A. Fisher (1890–1962), J. B. S. Haldane (1892–1964), W. D. Hamilton, and R. L. Trivers. The model is also implicit in some of the writings of Charles Darwin (1809–1882).

Sociobiology proposes that, if a given behavioral pattern has a genetic basis, then that pattern is subject to the forces of Darwinian natural selection. Those individuals who behave in ways that improve their reproductive success will, by definition, increase their genetic representation in following generations, and those beneficial behaviors that are influenced by genes should also become more prevalent. Thus, males whose genotype encourages them to perform a particularly spectacular mating dance will enhance their access to mates and will produce more offspring, some of which will be sons who continue to perform the spectacular mating dance and acquire more mates and produce more offspring. Males whose genotype does not induce the proper mating dance will experience lower reproductive success, and those genes will gradually dwindle in number. Over the generations, the mating dance will spread throughout the males of the population.

Following the theory of evolution by natural selection, scientists proposed that individuals should behave in their own self-interest and thereby enhance their genetic contribution to the next generation. However, observations of animal behavior revealed many examples of individuals apparently acting altruistically, sacrificing themselves for the good of others. For example, a prairie dog will often bark at the sight of a predator, alerting its group mates to the danger, but drawing attention to itself and therefore increasing its own chance of capture. Why would this individual behave in this risky manner? Some evolutionary biologists have asserted that these individuals act for the good of the species. However, in a Darwinian world, such a strategy is untenable, particularly if such self-sacrifice has a genetic component: Those individuals with a genetic tendency to sacrifice their own fitness for the good of the group would have fewer offspring in the next generation, and over succeeding generations the tendency for self-sacrifice would die out. In the 1960s, Hamilton solved this conundrum by pointing out that relatives share genetic material, and so individuals who help out kin would actually be enhancing their own fitness, their "inclusive fitness," to use Hamilton's term. An individ-

ual should be selected to help close kin, with whom a high percentage of genetic material is shared, more readily than distant kin. This model is the basis of Fisher's remark that he would lay down his life for two siblings (each related to him by one-half) or eight cousins (each related to him by one-eighth).

While sociobiology as a theory is fairly straightforward, it has met with opposition when it is put forth as an explanatory model. Critics such as Stephen Jay Gould (1941–) and Richard G. Lewontin (1929–) have asserted that sociobiologists are little more than mythologists, telling "Just So" stories about the adaptive value of every behavior observed. Little consideration, they claim, is given to random forces such as genetic drift or the constraints that genes impose on phenotypes. Furthermore, they suggest, there is no evidence that behaviors are controlled by genes, and there are few hard data that behaviors really do enhance reproductive success. In addition, it is argued that too little is known about the genetic structure of populations, such as the degree of relatedness among individuals or the extent of gene flow between groups, to ascertain the influence of factors such as Hamiltonian kin selection. Finally, some have raised the question of how an animal such as a prairie dog is able to assess degrees of relatedness and decide whether or not to help a sibling versus a cousin.

Critiques such as these are valid in many ways, especially for complex and long-lived organisms such as primates. For populations such as these, there are, indeed, few long-term studies that can shed light on the genetic relatedness among individuals, or the extent to which a given behavior enhances lifetime success. However, for social insects, such as paper wasps, data demonstrate that individuals are more likely to help sisters than other, less closely related females. As for the relationship between genes and behavior, the animal kingdom is replete with examples of behaviors that are instinctive, rather than learned, and therefore influenced by genes: egg-laying in marine snails, mating calls in birds, and herding activities in sheepdogs, to name just a few. And despite the claims of those in opposition to the sociobiological model, no sociobiologist would suggest that a single gene would control for a behavior such as a mating dance or an alarm call. Rather, sociobiologists assert that suites of genes may influence behaviors, perhaps through the media of hormones and neurons that put those behaviors into action. As for the prairie dog's ability to evaluate his family ties, this is an unnecessary element of the model. The individual need not be cognizant of its activities or their effect on its fitness. Instead, to put it rather simply, those individuals whose genes encourage them to do the "right" thing will have more offspring who also do the "right" thing. Those whose genes encourage them to do the "wrong" thing will reproduce less successfully, and eventually that "wrong" behavior will wane. Finally, few sociobiologists would attempt to explain all behaviors in terms of adaptation. Clearly some individuals employ maladaptive strategies, behave pathologically, or do something unique and anomalous. These observations are outside the scope of sociobiological explanation. Instead, the paradigm is employed as a theoretical framework to develop hypotheses about patterns of behaviors, about why a population employs one strategy over another. These hypotheses must then be empirically tested in order to determine the extent to which natural selection has played a role in the evolution of that pattern.

For many species, such as primates, learning introduces a great deal of behavioral plasticity; that is to say, the genotype is more tenuously linked to activity patterns. In these cases, it is more difficult to employ the sociobiological model. However, improved data-gathering technologies and additional field research will eventually enhance our ability to evaluate animal ecology, ethology, and population genetics. According to Wilson, "When the same parameters and quantitative theory are used to analyze both termite colonies and troops of rhesus macaques, we will have a unified science of sociobiology."—L.E.M.

See also HORMONE; NATURAL SELECTION; NEURON

Software

Software, the name usually applied to the programs that run on a computer, was a term that did not appear until a decade after the initial development of the modern stored-program computer. The name is of course derived from the fact that the other major portion of a computer system, the electronics, was usually referred to as *hardware*. In modern systems the software usually takes two distinct forms: the programs that control the hardware of the machine—usually called the *operating system programs*—and the application programs that serve some useful function: word processors, spreadsheets, graphics packages, etc. When computers were first developed, all programming had to be done in the machine's own binary language, a very difficult and error-prone activity. After people got used to writing such programs, it became obvious that a sophisticated program, usually referred to as an "assembler" or a "compiler" could ease the programming process by allowing people to write their programs in a language more suited to human use, and having the compiler then translate it into the machine language for execution. These "higher-level" language systems were among the first major software packages produced. Once these were available and in regular operation, it again became obvious that the efficiency of computer use could be improved by having a control program that would take care of a lot of the "housekeeping" jobs—e.g., loading the compiler program into memory, punching out the resultant machine language program, and scheduling the individual jobs to be run—that had previously been done by the human operator. This quickly led to the development of early operating systems programs that not only performed these tasks but also provided facilities that allowed other programs to control the peripheral devices on the computer, items such as the card readers/punches, printers, magnetic-tape units and, eventually, disk drives. Today, these operating systems programs are an integral part of the computer system, and a machine would be worthless without them. They generally take the form of floppy disks and sometimes CD-ROMs that come with the machine. The modern operating systems—DOS, UNIX, and Windows are some of the names given to them—control not only the way in which the application programs are run but also provide the support facilities that enable them to control the monitor screen, read files from the disk drives, perform backups of the information, obtain information from local devices such as printers and keyboards, arrange for information to be packaged so that it can be sent and received over computer networks, and even perform mundane tasks like keeping track of the time and date. While the vast majority of users never has to be concerned with how these very large and complex systems of programs are set up and interact, occasionally users will be faced with a problem that, they are told, involves part of the system with a mysterious name such as TCP/IP and/or a "driver" for a particular device. It is only in these situations that the average user appreciates how complex modern operating systems have become.

Applications software are those programs that run while using the facilities provided by the operating system programs, a situation often referred to as "running on top" of the operating system. These are the only programs that the average user is able to interact with. The creation and marketing of applications programs have been among the largest areas of growth in the computer industry. While the market for computer software of various kinds is almost as old as the computer itself, one particular application gave rise to the vast flood of software for the personal computer. In 1979, an entrepreneur named Don Bricklin produced VisiCalc, the first spreadsheet program. It took the business world by storm; people who had never even considered a computer were now purchasing Apple II computers and peripherals just to run the new concept of doing "what if?" calculations on financial data. In 1981, the Lotus 1-2-3 spreadsheet, which added facilities such as graphs and charts, made the IBM PC an extremely popular choice. Ever since then, computer programmers have been dreaming for finding another "killer application" to make them rich and famous.

Today, while it is still possible to write useful programs in a few hundred lines of code, the major software packages usually are composed of many hundreds of thousands of lines of code, require extensive expertise to produce, and cost many millions of dollars to develop to the point where they are marketable. This is the major reason why the software producers are concerned about the illegal copying of their products and why they often require some proof of purchase (such as typing in a serial number) as part of the process of installing them on a computer.—M.R.W.

See also BINARY DIGIT; CD-ROM; COMPUTER, MAINFRAME; COMPUTER, PERSONAL; COMPUTER LANGUAGES; COMPUTER NETWORK; DISK STORAGE; MEMORY, COMPUTER; PUNCH CARDS; WORD PROCESSING

Solar Design, Active

Active solar energy includes solar power plants, solar water and space heating, solar-thermal concentrating systems, and photovoltaics. The Greeks, Romans, and Chinese undertook the first active application of solar energy, developing curved mirrors that could concentrate solar rays on an object and cause it to ignite. Little more was done for many centuries thereafter, until in the 19th century solar power plants attracted a number of inventors. One of the most prominent was France's Augustin Mouchot. Worried that coal would eventually be used up, he combined the idea of the passive solar heat trap and active solar burning mirror to invent several successful devices, among them a solar oven, solar still, and solar pump. With government support, Mouchot's oven and still found use among French soldiers in colonial North Africa, and he went on to develop a solar motor for the 1878 Paris Exposition that pumped 2,000 liters (500 gal) of water per hour and powered an ice maker. His assistant, Abel Pifre, built several more motors, exhibiting one in Paris that drove a press that printed copies of the *Journal Soleil.*

In the United States, Swedish-American inventor John Ericsson (1803–1889), also deeply concerned about the rapidly expanding use of coal, contended in 1868 that a worldwide fuel crisis could be averted only by the development of solar power. Ericsson constructed a solar engine remarkably like Mouchot's motor, then turned to development of solar hot-air engines, which he hoped to employ in irrigation pumping in Southern California. He finally announced he had perfected the hot-air engine in 1888, but his penchant for secrecy took the details to the grave with him when he died a few months later.

Charles Tellier, a French engineer, advanced solar-motor technology by using pressurized vapor from sulfur dioxide, which boils at –10°C (14°F), in the place of pressurized steam. His work was furthered by H. E. Willsie and John Boyle Jr., who built several low-temperature solar engines in the American Southwest. Philadelphian Frank Shuman achieved even greater success with parabolic-trough reflectors. With investment capital from England, he constructed a solar steam-pumping plant in Meadi, Egypt, which employed seven reflectors that covered 1,200 square meters. Although a success, Shuman's plant was shut down during World War I, and the marketability of his and other solar engines was eclipsed thereafter by abundant and cheap oil and natural gas.

Solar water heating found modest commercial success at about this time. Baltimore inventor Clarence M. Kemp developed the Climax Solar Water Heater in 1891. Consisting of four galvanized iron tanks painted black and placed in a glass-covered pine box insulated with felt paper, the entire device could be placed on a building's roof or attached at an angle to an outside wall; the solar-heated water was then piped inside. In 1895, two entrepreneurs purchased the rights to manufacture and market the Climax in Southern California, and 1,600 were sold by 1900.

This was a start, but it took the efforts of William J. Bailey to make solar water heating truly practical. In 1909, Bailey received a patent for a revolutionary new solar collector. It was a shallow glass-covered box containing copper tubing that coiled back and forth across a copper sheet to which it was soldered. The tubing ran to a storage tank that stood above the collector in the house attic. Bailey's Day and Night Solar Heater Company soon became the world's leading solar heater manufacturer. Whereas the Climax heater cut fuel consumption for water heating in the typical Southern California home by 40 percent, the Day and Night heater reduced it a full 75 percent. By 1920, Bailey had sold some 5,000 heaters in Southern California and Arizona, but intensive marketing of natural gas heating by Southern California gas companies caused the regional solar water heater market to wither.

In Florida, however, solar hot-water heating flourished through the efforts of H. M. Carruthers and his partner Charles Ewald. Carruthers purchased the rights to manufacture and market Bailey's system in 1923, and Ewald improved on it during the 1930s. Without natural gas resources, many Floridians embraced solar hot water heating. More than half of Miami's population used solar water heating in 1941, and its use had spread to the Caribbean. But World War II's moratorium on nonmilitary use of copper killed the industry, and, after the war, high copper prices and cheap electricity transformed the southeastern solar industry into a service business for existing installations.

In the postwar era, while the United States enjoyed cheap and abundant energy supplies, energy shortages elsewhere revitalized solar water heating. Levi Yissar, for example, introduced a solar water heater in Israel in 1953. Rights to his system soon were acquired by Miromit, Israel's largest metal fabricator, which installed nearly 50,000 water heaters in the country between 1957 and 1967, and shipped more to some 60 different nations. Ironically, Israel's victory in the 1967 Six-Day War devastated the industry, as oil from captured fields on the Sinai Peninsula obviated the need for solar water heaters. Meanwhile, on the other side of the world, solar water heating flourished. Government support for solar development helped Australia's Solahart and other solar water heating firms manufacture and sell some 40,000 collectors between 1958 and 1973, and inexpensive soft-vinyl solar heaters and heaters similar to the old Climax heater were introduced in Japan. There, total solar water heater sales between 1958 and 1969 reached 3.7 million.

Active solar space heating also dated to the 1880s, when Edward S. Morse of Salem, Mass., experimented with solar air heaters. Although Morse's work was widely publicized, solar collectors for house heating were ignored until 1938, when researchers at the Massachusetts Institute of Technology began a solar house program. Over the next 20 years, researchers at MIT and the University of Colorado designed and tested solar space heating systems that used flat-plate collectors to heat water, air-heating wall collectors, water walls, and rock bed heat storage. With office buildings, houses, and even a solar-heated school (constructed in Tucson, Ariz.), researchers and designers demonstrated that solar space heating was technically feasible and caused no sacrifice in comfort, convenience, or aesthetics. But in an age of cheap fossil fuels, their work was little appreciated, and their research was terminated in 1962.

A decade later, the 1973 energy crisis gave a renewed impetus to solar water and space heating, although new designers and manufacturers of active solar systems had to rediscover basic fundamentals. But solar technology's historical invisibility was also a source of its strength. It was a technology that operated in a quiet and pollution-free manner that left the landscape untouched. As such, it quickly became part of the 1970s

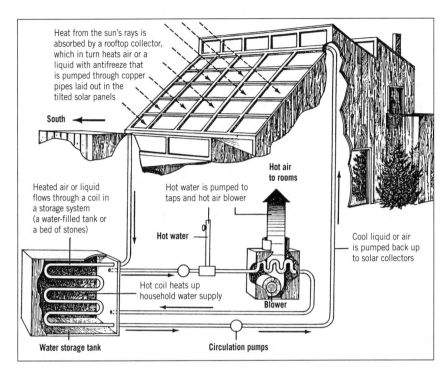

Heat from the sun's rays is absorbed by a rooftop collector, which in turn heats air or a liquid with antifreeze that is pumped through copper pipes laid out in the tilted solar panels

South ◄—

Hot air to rooms

Heated air or liquid flows through a coil in a storage system (a water-filled tank or a bed of stones)

Hot water is pumped to taps and hot air blower

Hot water

Cool liquid or air is pumped back up to solar collectors

Hot coil heats up household water supply

Blower

Water storage tank

Circulation pumps

Diagram illustrating how a house may be warmed by circulating water heated by the sun (from G. Tyler Miller, *Energy and Environment,* 1980; used with permission).

appropriate technology movement, promising a better environment and independence from centralized energy networks, and tapping into traditional American values of self-sufficiency. At the same time, however, commercially available solar collectors remained costly and beyond the reach of most consumers.

Some cynics observed during the 1970s that when big utilities figured out how to meter the sun, solar energy would become a reality. Large-scale solar-thermal concentrating systems, which became the focus of the electric-power industry during this period, in a sense accomplished just that. By using parabolic trough and parabolic dish systems or heliostats aimed at central receivers (in the tradition of burning mirrors), solar energy could be used to generate electricity for the existing utility grid. Stimulated by the Public Utility Regulatory Policy Act (1978), which required regulated public utilities to purchase electricity produced by independent power producers at a price equivalent to the utilities' production costs, LUZ International built nine solar-thermal concentrating plants in the Mojave Desert between 1984 and 1990. The plants' parabolic troughs focus sunlight onto vacuum-enclosed pipes that carry a heat-transfer fluid to a central facility where steam is generated to produce electricity. In 1991, with 345

megawatts (MW) of capacity, LUZ's Solar Electric Generating System facilities produced about 95 percent of the world's solar-thermal electricity.

Among utility-scale thermal electric systems, however, central receivers are considered more promising than parabolic troughs or dishes. In 1982, Southern California Edison opened Solar One (since replaced with Solar Two), the first central receiver plant. Covering 52.6 hectares (ha; 130 acres), its 1,818 mirrors or heliostats—each with a surface area of over 38 square meters (sq m; 400 sq ft)—automatically tracked the sun and concentrated its rays onto a central boiler mounted atop a 90-m (300-ft) tower. Superheated steam then turned a generator below, producing some 10 MW of power for 8 hours a day during the summer and 4 hours during the winter. An additional 7 megawatts of electricity was provided from a thermal storage system for about 4 hours after sunset. Similarly, Pacific Gas and Electric Company, Rockwell International, and ARCO Solar teamed up to build a more advanced 30 MW central receiver plant with 2,000 heliostats, and a 120-m (400-ft) high "power tower" containing liquid sodium and a heat-transfer system.

Although development of solar-thermal concentrating systems thus far suggest that active solar energy may tend to reinforce existing energy distribution patterns, solar water and solar space heating, perhaps in conjunction with photovoltaics, will remain a force for individual initiative in the active solar energy field. In any event, all forms of active solar energy will be utilized increasingly as the world makes its inevitable transition to renewable energy resources.—J.W.

See also APPROPRIATE TECHNOLOGY; DISTILLATION; ELECTRIC POWER BLACKOUTS; FUELS, FOSSIL; NATURAL GAS; PHOTOVOLTAIC CELL; SOLAR DESIGN, PASSIVE

Further Reading: Ken Butti and John Perlin, *A Golden Thread: 2500 Years of Solar Architecture and Technology*, 1980. Pascal De Laquil III, et al., "Solar-Thermal Electric Technology," in *Renewable Energy: Sources for Fuels and Electricity*, 1993.

Solar Design, Passive

Solar energy has been employed by people for centuries, encompassing everything from sun-drying fish for year-round consumption to the passive solar architectural designs of Pueblo Indians of the American Southwest and of the classical Greeks and Romans. The Greeks mastered solar architecture, designing houses to maximize heat from the sun in winter and deflect it during the summer. The Romans improved on such designs, using glass or transparent sheets of mica or selenite to create solar heat traps, and the Chinese developed solar building designs and urban planning much like that in Greece.

Greek adoption of passive solar design probably was in response to diminished supplies of fuelwood as well as to the region's favorable climate. Greek veneration of the sun also encouraged them to embrace solar houses. In these structures, thick walls sheltered the rooms from the north and were open to a south-facing portico through which the winter sun streamed. Earthen floors and adobe walls absorbed much of the sun's radiation during the day and released it in the evening. In summer, the overhanging portico shaded the rooms from the high sun. Solar architecture was applied across class lines in most classical Greek cities.

In Rome, wood consumption was even more wasteful than in Greece. In addition to industrial and construction use, enormous amounts of wood were used for space heating. Wealthy Romans commonly had *hypocausts*, central heating systems in which an underground furnace consumed as much as two cords of wood a day. Not surprisingly, wood shortages eventually turned the Romans to solar architecture. They learned much about it through the writings of Vitruvius (fl. 1st-century B.C.E.), a student of Greek architecture, and from Greek colonists who had settled in southern Italy. By the 1st-century C.E., solar architecture became increasingly common among wealthy Romans. Pliny the Younger (61– c. 113) described the use of solar heat in his villas and noted that one of his favorite rooms was a *heliocaminus* (literally, "solar furnace"). The openings in the southwest-facing room likely were covered with glass or mica, which was increasingly common during the 2d-century C.E.

The use of transparent materials over windows created a solar heat trap, and Romans soon applied this principle to the construction of cold frames for winter plant beds and greenhouses for exotic plants. They also applied the solar heat trap to heat public baths. From the 1st-century C.E. onward, most baths faced toward the winter sunset, and the whole south wall was usually glazed. Access to sunlight became so important, especially for the *heliocaminus* room, that sun-rights for such rooms were recognized and eventually written into the Justinian Code of Law.

After the collapse of the Roman Empire, solar heat traps were almost unknown until the 1500s. Ships returning from Asia, Africa, and America brought exotic plants, and the cold frames and greenhouses were reintroduced by the Dutch and Flemish, with the French and English close behind them. As horticultural greenhouses came of age in Holland and England, they also assumed the more lavish form of conservatories, where plants were displayed rather than cultivated. Architects attached glass gardens onto the south side of living rooms or libraries, and by the 1800s attached conservatories found their way to the northeastern United States and began to filter down in more modest form to the middle class. Likewise, cold frames and greenhouses became widely used.

Unlike the idea of a glazed solar heat trap, solar architecture per se took longer to be rediscovered. Whereas solar building flourished in China and spread to Japan, Europeans virtually ignored it. Northern European builders copying classical Greek and Roman architectural styles during the Renaissance frequently oriented their buildings with the porticos facing away from the sun. Later, as industrial cities appeared in the 18th and 19th centuries, working-class housing typically was built back-to-back, row-after-row, with little access to sunlight. However, like their counterparts in classical Greece, European physicians recognized that sunlight was essential to health and sanitation. By 1900, planned communities with southern exposure for buildings appeared in response to unhealthful urban housing conditions, and sun-rights began winning legal recognition. During the 1920s, Walter Gropius (1883–1969) and other German architects gained international attention with their east-west oriented *Zeilenbau* (row house) plan that permitted maximum sunlight for the entire community. They soon discovered that more solar heat was absorbed by the streets than by the apartments, and a switch to single-story houses with rooms and windows facing south began in the 1930s. The rising Nazi party, however, attacked these solar-oriented worker communities as "communistic."

The United States fell behind Europeans in passive solar design. Not until the 1930s did Chicago architect George Keck launch solar architecture in America. He was amazed by the heat that accumulated in the glass House of Tomorrow that he built for the 1932 Chicago World's Fair, and he began designing homes with glass that trapped solar energy for heat. Within a few years, he understood the importance of southern exposure and overhangs, and he built his first "solar home" for real estate developer Howard Sloan. Sloan was so impressed that he built a small tract of 30 houses called Solar Park in 1941. Shortly afterwards, the rationing of oil and gas during World War II gave solar homes an important boost. In the postwar housing boom, much of the prefabricated home industry and many assembly-line homebuilders adopted solar designs, and heat storage systems improved nighttime heating. Despite these gains, cheap energy and the popularity of mechanical heating and cooling systems buried interest in solar architecture.

The application of passive solar energy resurfaced in the wake of the 1973 Arab oil embargo, as architects and builders oriented buildings with south-facing windows, and made use of heat storage in tile floors, stone fireplaces, and interior brick walls. Additionally, new techniques began to be employed: skylights and evaporating water for cooling, improved thermal glazings to reduce heat loss, electrochromic window films that admit or block sunlight in response to minute electrical currents, and light pipes and holographic films that bring sunlight further into interior spaces. Principles of modern passive solar energy design include "sunshading" for summer periods, "daylighting" to maximize natural light and reduce energy needs for artificial light and cooling, "thermal mass" to heat and cool the interior in winter and summer, and energy control systems such as photocell dimmers and computerized controls.

Passive solar designs, however, have not been widely adopted by the construction industry. Perhaps 250,000 solar homes were built in the 1980s, but architects and builders do not have great incentive to incorporate passive solar features in their buildings since they do not have to pay subsequent heating and cooling bills. Nevertheless, in California some communities, most notably Davis, and some utilities, such as the Sacramento Municipal Utility District, have programs to encourage passive solar design, and these programs are being adopted elsewhere.—J.W.

See also AIR CONDITIONING; CENTRAL HEATING; STOVES, HEATING

Further Reading: J. Douglas Balcomb, ed., *Passive Solar Buildings*, 1992. Simos Yannas, *Solar Energy and Housing Design*, 1994.

Solar System, Heliocentric

Because of the apparent daily motion of the stars, the seasonal motion of the sun, and the longer apparent periods of motion of the planets, virtually every society that has produced any astronomical theorizing has posited the existence of some regular rotational motion about a fixed center. Ancient Chinese and Japanese astronomy assigned to the North Star the role of unmoving center of heavenly motions, including those of the sun. Most premodern Western theories—including those most widely held in Mesopotamia, Egypt, Greco-Roman antiquity, and Islam—held the Earth to be the stationary "center" about which the sun, planets, and stars moved. But a small number of Greek astronomical theorists, including Aristarchus of Samos and the Pythagoreans, held that the Earth and other planets revolved around the sun.

This view was revived in the early 16th century by the Polish astronomer Nicholas Copernicus (1573–1543) in an attempt to improve the predictive accuracy of astronomy. Improving the accuracy of astronomical tables would provide a better foundation for astrology and calendar construction, but Copernicus also intended to provide a better coordination between mathematical and philosophical approaches to astronomy. A heliocentric solar system was made the central presumption of Copernicus's *On the Revolutions of the Heavenly Spheres* of 1543. It is not certain that Copernicus believed that the sun was literally the center of the solar system; his writings largely treat the issue in terms of facilitating astronomical calculations. But largely as a result of the brilliant polemics of Galileo (1564–1642) in his *Dialogue Concerning the Two Great World Systems* of 1632, the heliocentric model became established as the dominant Western astronomical system by the mid-17th century. In retrospect, the heliocentric model of the solar system that began with Copernicus and culminated with Newton was a profound intellectual revolution that displaced human beings from the center of the universe. Subsequent astronomical discoveries demonstrated just how insignificant the Earth is relative to the universe as a whole.

In order to accept Copernican theory, one had to accept the assumption that the diameter of the Earth's orbit around the sun was tiny in comparison with the distance between the sun and the other fixed stars. Moreover, one had to give up the predominant Aristotelian account of heavenly motions, which was difficult for many. Tycho Brahe (1546–1601) offered an alternative system in which the sun and moon orbited the Earth, and the rest of the planets in turn orbited the sun, thus allowing astronomers to avoid the assumption of huge stellar distances. This system appealed for a time to some, but it could not escape the problems inherent in the Aristotelian interpretation. Copernicus's version of heliocentrism, while it raised many philosophical problems for traditional Aristotelians and religious problems for biblical literalists, at least offered a reason for the fact that as bodies got farther from the sun, their orbits had monotonically increasing periods. It also provided a way to establish relative orbital sizes.

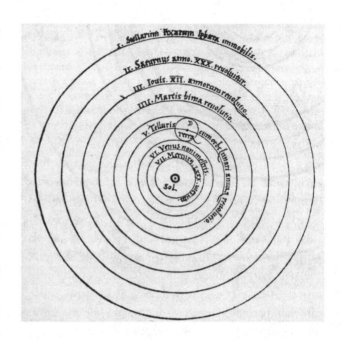

A diagram of the sun-centered universe that appeared in Copernicus's *DeRevolutionibus*.

And it offered a system that did not involve the philosophically suspect mathematical devices known as *equants* that most geocentric theories employed in order to account for the motion of "wandering stars," i.e., the planets.

In order to explain planetary motions, geocentric theories also required the use of "epicycles," circles that were centered on the circumference of other circles. It is sometimes stated that Copernicus's model of the solar system was simpler because it demanded fewer circles in its mathematical exposition. This is incorrect; some modern scholars have even concluded that, if anything, Copernicus's system was more complex than the Ptolemaic system. It was also the case that astronomical tables based on the Copernicus's heliocentric solar system were no more accurate than those based on geocentric systems.

What is often taken to be the "Copernican Revolution" actually began in the early 17th century, when Johannus Kepler (1572–1630) showed that (1) the planets move in slightly elliptical orbits with the sun at one of the foci of the ellipse, (2) equal areas are cut out by the line segment connecting the sun to a planet in equal time periods, and (3) that the periods of the planets orbital motions are proportional to the three-halves power of the average distance between the planet and the sun (the semimajor axis of the ellipse). Isaac Newton (1642–1727) was further able to demonstrate in his *Mathematical Principles of Natural Philosophy* of 1687 that Kepler's observations could be accounted for if each planet was attracted to the sun with a force that was proportional to the inverse square of the distance between the sun and the planet.

Although the term *heliocentric* (centered on the sun) is commonly employed, in fact Copernicus, Kepler, and Newton all had the planets' center of the motion displaced just slightly from the sun. Beginning with Kepler, it was understood that the sun stands at one focus of the planetary ellipse rather than at its center. Still, it has become conventional to designate the model of the solar system that began with Copernicus as heliocentric.—R.O.

See also MECHANICS, NEWTONIAN; PTOLEMAIC SYSTEM

Sonar

Construction of submarines by Irish nationalist John Holland (1840–1914) in the late 19th century convinced the Royal Navy that it had to develop submarine detection equipment. In 1882, Captain C. A. McEvoy referred to his "submarine detector," an induction device with coils thrown over the stern of the ship. In theory, those coils detected disturbances of submerged metal objects. That system was soon replaced by another of McEvoy's devices, a hydrophone that captured underwater sound waves. McEvoy's hydrophone consisted of a microphone enclosed in an open-ended diving bell located on the ocean floor. Air inside the bell picked up sound waves and vibrated a platinum strip that was connected to a control panel ashore. There, a switch completed the circuit. Sounds were picked up on a telephone receiver and displayed by the movement of a galvanometer needle. A refinement caused a bell to ring when the needle moved a sufficient extent.

Picking up sound was different from determining the direction of the sound. Throughout the early years of World War I, British and American scientists working to counteract the U-boat threat improved hydrophones, making them directional and portable. Using baffles for the diaphragms that captured the sound waves, the British made advances in unidirectional reception, while American researchers, whose interest in submarines intensified after America's entry into the war in 1917, developed the multiple hydrophone array that used the ship as a "soundscreen," and employed several hydrophones with a rotating plate device that "scanned" the reception.

True sonar, however, required not only receiving a signal but also sending out a signal and fixing a location based on directional ranging. In 1917, British researchers experimented with sonic and ultrasonic ranging techniques, in which they relied on the work of a Russian émigré, Constantin Chilowsky, and a French physicist, Paul Langevin. Chilowsky had developed an ultrasonic echo detection device using high-frequency electric power, and in 1915 a successful experiment transmitted a signal across the Seine. After working with the British, Langevin made the crucial breakthrough by inventing a quartz transmitter that was durable and seaworthy. The resulting device was referred to as *asdics*, or "ultrasonic detection," and the key test occurred in March 1918 when asdic echoes were obtained from a submarine at a distance of 457 meters. By late summer 1918, trawlers and other ships were fitted with asdics.

Although refinements and improvements to asdics continued over the subsequent 30 years, no watershed technological breakthroughs occurred until after World War II, by which time the U.S. Navy had focused on detection of one submarine by another. The navy had developed omnidirectional scanning based on transducers that were scanned hundreds of times a second for echoes, allowing an image of the submarine's surrounding sea to be presented on a cathode-ray display screen. As antisubmarine warfare (ASW) techniques improved, several varieties of sonar appeared, including "dunking" sonar (utilized by helicopters), towed sonar arrays (TASS), and hull-based sonars in ships and submarines. Modern sonars in the SOSUS (Sound Surveillance System) developed by Bell Telephone and Western Electric, which consist of permanently placed hydrophones in sound channels off the U.S. coast, have exceedingly long range. In addition to range, modern systems have the capacity to identify most known submarine and ship types.

The actual performance characteristics of modern

submarine sonar and of SOSUS—which President Clinton ordered partially dismantled in his first term—remain classified, but authoritative sources have indicated that American operators could track *and identify* Russian submarines from the moment they left their bases. Of course, using evasive maneuvers at sea in deep channels, any sub could escape SOSUS tracking. Nevertheless, modern submarines of the Trident class have exceptional detection capabilities, using two main types of sonar techniques: passive and active. Passive sonars are hydrophone based, and merely receive signals in the water. Active sonar involves sending out an ultrasonic signal—a "ping"—that establishes range to target. Submarine crew only employ the "ping process" to finalize target acquisition, and never randomly send out signals that might give away their position. Finally, advances in dunking or dipping sonar have included sonabuoys, which are dropped by helicopter in an area thought to contain a hostile submarine. The sonabuoys thus cordon off an entire section of the ocean, reducing the search perimeter.—L.S.

See also ELECTROMAGNETIC INDUCTION; SUBMARINE, TRIDENT; SUBMARINES

Further Reading: William Hackmann, *Seek & Strike: Sonar, Anti-Submarine Warfare and the Royal Navy, 1914–1954*, 1984.

Soyuz and Salyut

Long before space travel became feasible, visionaries were devising schemes for orbiting space stations. In the 1970s, these dreams became a reality. The United States orbited Skylab and the Soviet Union began its Salyut program. Part of the Soviets' motivation came from a desire to regain the international prestige that had been attained through their pioneering efforts in space. Toward the end of the 1960s, the United States began to pull ahead in the "space race," most notably by being the first to land a man on the moon. Somewhat disingenuously denying that they had ever been in a race to the moon, the Soviets turned their attention to long-duration stays in space, an endeavor that was to be highly successful.

Initial efforts centered on the linking up of separate spacecraft, an essential procedure for the mainte-nance of a space station and the rotation of its crews. The basic vehicle employed for these efforts was the Soyuz series of spacecraft, descendants of the Vostok and Voshod capsules that sent the first cosmonauts into space. Each Soyuz consisted of three elements: an orbital module (experiment and sleeping quarters), an orbital propulsion module (for maneuvering in space), and a descent module that returned the crew to Earth. Altogether, early Soyuz capsules weighed 6,800 kg (14,990 lb) and were 2.7 m (8.9 ft) in diameter and 7.5 m (24.6 ft) in length, not including two outstretched solar panels that had a span of 8.4 m (27.6 ft). Soyuz enclosed a habitable volume of 10 sq m (350 sq ft).

The first Soyuz mission, launched on Apr. 23, 1967, began successfully, but prior to reentry the craft began to tumble uncontrollably. The parachute harnesses tangled, and V. M. Komorov fell to his death. After an 18-month hiatus, another Soyuz rendezvoused with an unmanned Soyuz that had been launched a few days earlier. On Jan. 15, 1969, the Soviets achieved the first docking of two Soyuz spacecrafts. Two of the three crew members of Soyuz 5 traveled along the outside of the docked pair for 37 minutes before joining the single crewman of Soyuz 4. In October of the same year, three spacecraft, Soyuz 6, 7, and 8, orbited the Earth simultaneously. Dozens of missions followed, many of them flown in conjunction with the operation of the Salyut space station.

The first Soviet spacecraft designed specifically to serve as a space station, Salyut 1, was launched on Apr. 19, 1971 (10 years after the first man was put into space, and hence a "salute" to that accomplishment), and entered an orbit 200 km (125 mi) above the Earth. In April 1971, a Soyuz crew docked with Salyut 1, but did not enter it. On Jun. 7, 1971, the crew from Soyuz 11 moved into Salyut 1. They remained for 3 weeks, but again tragedy struck when an oxygen valve failed on reentry and the three crew members died of pulmonary embolisms. With no immediate prospect of another crew manning it, Salyut 1 was taken out of orbit 175 days after its launch, burning up in the Earth's atmosphere during its return.

Problems continued to hobble the Soviet space program. Salyut 2 was orbited on Apr. 3, 1973, but went out of control 8 days later and was destroyed in the course of its return to Earth at the end of May. The program got on track with the orbiting of Salyut 3 on Jun. 24, 1974. The space station was then inhabited for 2 weeks by cosmonauts from Soyuz 14, which had been launched on July 3. Salyut 3 orbited at a lower distance to the Earth than other space vehicles, 260 km (162 mi), and it is likely that at least part of its mission was military reconnaissance. A series of space station launches followed: Salyut 4 (December 1974) and Salyut 5 (June 1976). Salyut 6, the most successful of the series, was launched on Nov. 29, 1977, and subsequently boarded by two crew members from Soyuz 26, which had been launched on Dec. 10, 1977. It was the first Salyut to have two docking ports; this facilitated docking and made it possible to link up with two other spacecraft at the same time. The first crew stayed for 96 days; they were succeeded by seven long-stay crews and nine visiting crews, eight members of which were from other countries. On Jan. 20, 1978, the first of 12 unmanned supply ships, Progress 1, docked with Salyut 6. These supply ships extended Salyut missions by bringing fuel, food, oxygen, equipment, and scientific apparatus. The supply ships also served as "space tugs," propelling the space stations into higher orbit and preventing their premature return to Earth.

On Apr. 19, 1982, a few months before Salyut 6 returned to Earth (July 29, 1982), the final example of the series, Salyut 7, was launched. This space station had a useful service life, but it required extensive repair work in 1985, done by a crew that had been launched by Soyuz T13. Prior to its return to Earth on Feb. 6, 1991 (after about 50,200 orbits), it had been the home of five long-term crews and four visiting crews, including the second woman in space, who made two separate visits.

In the course of their long stays in space, the Salyut crews performed many experiments and procedures, including crystal growing, soldering metals through the use of electron beams, making astronomical and spectrographic observations, recycling water, observing croplands, transferring liquids in space without the use of pumps, and cultivating plants. Crews also made use of special training suits and exercise techniques designed to prevent bodily debilitation brought on by prolonged weightlessness. The Salyut series provided a great deal of scientific data, along with critical information about living and working in space. The experience gained with Salyut was invaluable when the Soviets began to operate an even more ambitious space station, Mir.

See also APOLLO PROJECT; MIR SPACE STATION; SKYLAB; SPUTNIK; VOSTOK AND VOSHOD

Further Reading: Kenneth Gatland, *The Illustrated Encyclopedia of Space Technology: A Comprehensive History of Space Exploration*, 1981.

Space Probe

A space probe is an unmanned vehicle that uses a variety of instruments to provide information about the moon, the planets, and other bodies within the solar system. Probes receive their power from solar cells or from small nuclear generators, while contact with the Earth is provided by some kind of telemetry system. In most cases, the course of the probe can be altered from the ground by activating onboard vernier thrusters. Some probes have made a single rendezvous with their objective (a "fly-by"), while some have gone into orbit, and others have landed on the surface of the moon or a planet. Since the late 1950s, well over 100 space probes have been launched. Not all of them have been successful, but the majority have provided some useful information. Some of the most important space probes are described below.

The orbiting of an artificial satellite by the Soviet Union in October 1957 can be taken as the beginning of exploration beyond the confines of the Earth. On Jan. 4, 1959, less than 15 months after the orbiting of Sputnik, the Soviet probe Luna 1 achieved the first lunar fly-by when it passed the moon at a distance of 7,500 km (4,660 mi). Luna 2, launched on Aug. 12, 1959, was an unsuccessful attempt to land on the moon, but Luna 3, launched on Oct. 4, 1959, was able to circle the moon before returning to Earth. Its most notable accomplishment was sending back photographs of 70 percent of the hitherto unseen "dark side" of the moon. Luna 9, which had been launched on Jan. 31, 1966, was the first to achieve a soft landing on the moon, from which it sent back a few pictures.

After a shaky start marred by flight mishaps, missed targets, and camera failures, the United States space program met success with Ranger 7. Launched on July 28, 1964, the probe sent back 4,000 pictures of the moon

Model of the Viking space probe that gathered scientific data on the dynamic and physical characteristics of Mars (courtesy NASA).

before it crashed into the lunar surface. Two other successful Ranger missions were followed by seven Surveyor probes launched between May 30, 1966, and Jan. 7, 1968. Some of these achieved their intended soft landing on the lunar surface; others did not. Meanwhile, the United States launched a series of five Orbiter missions from Aug. 10, 1966, to Aug. 1, 1967. All can be counted as successes. Orbiting the moon, they sent back so many photographs that not all of them have been studied to this day.

Parallel efforts to put men on the moon gained the lion's share of public attention, but a great deal of information continued to be provided by unmanned probes, such as the Russian Luna 16 (launched Sep. 12, 1970), Luna 20 (launched Sep. 14, 1972), and Luna 24 (launched Aug. 9, 1976), which landed on the moon, collected rocks, and returned them to the Earth. The Soviets also landed moon crawlers, known as *Lunokhods*, that traveled over the moon's surface, taking pictures and collecting samples. The last of these, Lunar 24, landed on the moon, picked up samples to a depth of 2 m (6.5 ft), and then returned to Earth on Aug. 22, 1976.

While probes were being sent to the moon, efforts were mounted to explore the near planets (Mercury, Venus, and Mars). The first attempt was the Soviet mission to Venus, Venera 1. Launched on Feb. 12, 1961, it

lost communication with the Earth a few weeks after its launch. The U.S. Mariner I mission to Venus was even less successful, for it crashed into ocean after its launch on July 22, 1962, but Mariner 2 (launched on Aug. 26, 1962) achieved a successful fly-by of Venus on Dec. 14, 1962. It made the important discovery that, contrary to some theories, the surface of the planet was very hot, and that its upper atmosphere rotated 60 times more rapidly than the planet itself. On July 22, 1972 the Soviet Venera 8 probe landed on the surface of Venus, and then transmitted data back to Earth for nearly an hour. This was followed by Venera 9 and 10, which landed on the surface of the planet on Oct. 21, 1975, and Oct. 25, 1975, respectively. Each sent back one photograph to the Earth. Subsequent Soviet and American Venus probes provided information that was used to map the surface of the planet, record temperatures, and determine the nature of its atmosphere.

Important information on the planet Mercury was gathered by Mariner 10, which was launched on Nov. 3, 1973, and made its first pass of Mercury on Mar. 29, 1974, about 2 months after it had flown past Venus. At that time, it sent back television pictures that revealed a surface rather similar to that of the moon. It then went into orbit around the sun, and on two occasions it came close enough to Mercury to send back more pictures. Mariner 10 still orbits the sun, but contact with the Earth has been lost for many years, and no subsequent probes to Mercury have been launched.

The Soviets followed their program of Venus probes with rather less successful probes to Mars. The first successful mission to Mars was the U.S. probe, Mariner 4, which was launched on Nov. 28, 1964, and got within 10,000 km (6,200 mi) of the red planet on July 14, 1965. Before flying past Mars and going into permanent solar orbit, Mariner sent back 21 pictures and a great deal of data. One of its prime discoveries was that the surface of Mars showed extensive cratering. Mariner 9, launched on May 30, 1971, went into orbit at 1,370 km (850 mi) around Mars, from where it sent back stunning pictures of Martian volcanoes, one of which, Olympus Mons, was found to be three times the height of Mount Everest. In all, Mariner 9 sent back more than 7,000 photos of the Martian surface. These revealed a landscape of craters, valleys, ridges, and volcanoes, but no evidence of the canals that some astronomers had claimed to have seen. Even more was learned about Mars through the landing of two probes, Viking 1 and Viking 2, which touched down on July 20, 1976 and Sep. 3, 1976, respectively. They found no evidence of life on Mars, but they relayed back to Earth stunning pictures of a rocky, reddish-hued landscape.

The exploration of the outer planets (Jupiter, Saturn, Uranus, Neptune, and Pluto) began with the first Jupiter probe, Pioneer 10. Launched on Mar. 2, 1972, it did not rendezvous with Jupiter until Dec. 13, 1973. In the course of its fly-by, it measured temperatures, analyzed the atmosphere, studied the magnetic field, and examined surface features. It also took pictures of Jupiter's four largest satellites—Io, Europa, Callisto, and Ganymede— and discovered some previously unknown satellites. A follow-up probe, Pioneer 11 gathered more information as it passed by Jupiter beginning on Dec. 2, 1974. The next set of probes were Voyager 1, and Voyager 2, which flew past Jupiter on Mar. 5, 1979, and July 9, 1979, respectively. After a long hiatus, the exploration of Jupiter resumed when Galileo (launched on Oct. 18, 1989) on Dec. 7, 1995, sent a probe through the Jovian atmosphere that gathered data until it crashed into the surface of the planet.

Pioneer 11 had made a brief run past Saturn, but the exploration of Saturn began in earnest with Voyager 1, which began to approach the planet on Nov. 12, 1980. On Aug. 25, 1981, the second outer-planet probe, Voyager 2, passed by Saturn. In addition to providing data on the surface and rings of Saturn, these probes added greatly to our knowledge of Saturn's moons. Voyager 2 then continued to a rendezvous with Uranus. As it approached the planet, the probe discovered previously unknown moons. It then moved on to the vicinity of Neptune, a distance of more than 4.4 billion kilometers (2.75 billion miles), where once again new satellites were discovered. The fly-by also confirmed a previous suspicion, that Neptune was encircled by rings. Both Voyagers continued their journeys, in the course of which they will join Pioneer 10 and Pioneer 11 as the first manmade objects to leave the solar system.

Although the United States could claim the most ambitious and successful program of space probes, funds for space exploration were squeezed in the 1980s. As a result, the United States did not send any

probes to study Halley's comet when it made its return in 1986. However, the Soviet Union launched its Vega 1 probe on Dec. 15, 1984, followed by Vega 2 six days later. Vega 1 passed within 8,890 km (5,525 mi) of the comet on Mar. 6, 1986, followed by Vega 2, which passed within 8,030 km (4,990 mi) 3 days later. The Japanese then launched their first space probes, Sakigake and Suisei on Jan. 8, 1985, and Aug. 18, 1985, respectively. France used a European Space Agency Ariane rocket to launch its Giotto probe to Halley's comet on July 2, 1985, and flew past it at a distance of only 596 km (370 mi) on Mar. 14, 1986.

The Space Shuttle was used as a platform for the launch of several space probes, beginning with the Magellan mission to Venus that was launched on May 4, 1989, and went into orbit around the planet on Aug. 10, 1990. The shuttle was also used for the Oct. 6, 1990, launch of the European Ulysses mission to Jupiter.

In recent years, space probes have been launched at a slower pace, as the collapse of the Soviet Union and budgetary constraints in the United States diminished the ardor of the two main space-faring nations. Even so, major advances have occurred. One space probe in particular, the Pathfinder mission, provided a dramatic example of what unmanned space programs can accomplish. Pathfinder was launched on Dec. 4, 1996. On July 6, 1997, a capsule containing a robotic rover descended through the Martian atmosphere and achieved a soft landing on the surface with the aid of rockets, parachutes, and airbags. Called "Sojourner," the rover was 280 mm (10.9 in.) long and weighed 11.5 kg (25.4 lb). It gathered some data about the Martian surface, but its main mission was to advance the technology of unmanned planetary missions. In this it was an outstanding success. While traveling over the Martian surface, Sojourner relayed back to Earth stunning pictures of the local terrain that millions of people were able to download from the Internet. More missions of this sort will be flown in the future, adding considerably to our knowledge of the worlds beyond Earth, and at a fraction of the cost of manned space missions.

See also APOLLO PROJECT; GEMINI PROJECT; INTERNET; MERCURY PROJECT; SPACE SHUTTLE; SPUTNIK

Further Reading: Patrick Moore, *Mission to the Planets: An Illustrated History of Man's Exploration of the Solar System*, 1990.

Space Shuttle

Although conventional rockets have performed admirably in launching people and hardware into space, the process is inherently wasteful. A large and very expensive booster rocket, usually comprising several stages, can be used only once. A reusable spacecraft offers the prospect of more economical space missions. Moreover, if a permanent space station were to be built, it would require just such a vehicle to resupply the station and ferry crews back and forth.

A reusable spacecraft project began only a few months after the first landing on the moon. A mission to Mars, the next logical venture into interplanetary space, failed to gather sufficient political support. Consequently, the Nixon administration turned to a more popular project, an orbiting space station along with a reusable spacecraft to help in its construction and maintenance. As it was conceived in 1970–71, the spacecraft would have two basic elements: an orbiter and a manned boost vehicle to put the orbiter into space, after which the booster would be piloted back to Earth. A cost estimate of $10 billion by the Bureau of the Budget put an end to this plan. Instead, the orbiter was put into space by less costly means. The Space Shuttle, as the complete package came to be known, was launched by two solid-fuel rockets that were jettisoned after completing their burn and five liquid-fuel rockets supplied by tank that dropped off after the orbiter reached maximum altitude. After completing its mission, the delta-winged orbiter with its crew of up to 10 returned to Earth by landing like a conventional airplane, except that it lacked power. The first landing approach was the only approach, and it had to be right.

Five shuttles have been built: *Enterprise* (used only for testing), *Columbia, Challenger, Atlantis, Discovery* and *Endeavor*. The shuttle's orbiter weighs 91,000 kg (200,000 lb), is 37.25 m (122 ft) long, and has a wingspan of 23.8 m (78.1 ft). Its cargo bay is 18 m (60 ft) long, with a diameter of 4.6 m (15 ft), and can contain a payload of up to 29.5 metric tonnes (32.5 tons).

It is equipped with a 15.2-m (50-ft) robot arm used for the launching of satellites and other payload items. The shuttle is propelled by two small rockets, each producing 2,722 kg (6,000 lb) of thrust (used for achieving final orbit, changing orbit, and initiating return to Earth), and three main rockets, each producing 16,800 kg (370,000 lb) of thrust. Rocket fuel is carried in an external tank 8.2 m (27 ft) in diameter, containing .68 million kg (1.5 million lb) of hydrogen. When the tank drops off, it disintegrates upon reentering the atmosphere, and the remaining fragments fall into the ocean. In addition, there are two solid-fuel rocket boosters containing .45 million kg (1 million lb) of mercury perchlorate, aluminum powder, iron oxide, and a binding polymer. Each produces 1.2 million kg (2.6 million lb) of thrust and burns for the first 2 minutes and 8 seconds of the shuttle's flight (at which point a speed of 4,600 km/hr (2,860 mph) and an altitude of about 50 km (31 mi) has been achieved), after which they are jettisoned. Unlike the liquid rockets' fuel tank, the solid-fuel rocket casings are recovered for reuse.

Once in space, the shuttle is controlled by 44 rocket thrusters in the craft's nose and tail. Onboard power is supplied by fuel cells that produce 12 kW of peak power. To protect it from being incinerated by the nearly 1090°C (2,000°F) heat produced by reentering the atmosphere, the shuttle's outer surface is covered with 31,000 special ceramic tiles. These have required a great deal of hand labor for application; 300 man-years of labor in the case of the shuttle *Columbia*. The complexities of flying the shuttle are accommodated by five onboard computers capable of nearly 2 million operations per second.

After a series of gliding tests conducted with the *Enterprise*, the shuttle *Columbia* made the first powered flight on Apr. 12, 1981, returning to Earth on April 14, after a mission of 53 hours, 5 minutes. A second flight 7 months later indicated that the shuttle was truly a reusable space vehicle, albeit one that required a great deal of maintenance between flights.

Excessive downtime has continued to be a major problem of shuttle operation. When the shuttle fleet was being built, its operators, the National Aeronautics and Space Administration (NASA), believed that 116 shuttle flights would take place during 1981–85. This proved wildly optimistic. Not only was the shuttle more than 2 years behind schedule and $1 billion over budget when *Columbia* made its first flight, but also the fleet as a whole was incapable of the fast turn-around expected. As a result, only 24 flights took place during this period.

The shuttle has not been the often-aloft, relatively inexpensive "space truck" envisaged by its designers. Further limiting its usefulness is its limited maximum altitude of 800 km (500 mi), well below the orbiting distance of most satellites (a geosynchronous communications satellite orbits at more than 56,000 km [35,000 mi]). Accordingly, booster rockets have to be used to put shuttle-borne satellites into final orbit. Economic considerations also have weighed against the shuttle as a satellite launcher. Due to the initial expense of the shuttle ($8.8 billion was spent to develop and build the initial fleet), its intensive maintenance requirements, and the cost of transporting it (on top of a modified 747 airliner) back to Cape Canaveral after the usual landing in California, the cost of putting a satellite into low orbit is approximately $1,400 per pound. In contrast, the cost of doing the same with a European Space Agency Ariane rocket is only $30 million. Despite these drawbacks, the shuttle program advanced the capacity to live, work, and conduct scientific experiments in space. But all of these accomplishments were forgotten on Jan. 28, 1986, the date of the 25th shuttle flight, when *Challenger* exploded 1 minute and 13 seconds after takeoff. This catastrophe was a serious blow to the shuttle program and to the American space program as a whole. The next shuttle flight did not occur until Sep. 29, 1988.

By the first quarter of 1998, space shuttles had flown 89 missions, encompassing 757 days in space and 11,944 orbits of the Earth. By then, shuttle flights included eight dockings with the Russian Mir space station, the initial stage in the construction of an international space station.

See also APOLLO PROJECT; *Challenger* DISASTER; FUEL CELL; MIR SPACE STATION; ROCKET, LIQUID PROPELLANT; ROCKET, SOLID PROPELLANT; SPACE STATION

Space Station

Space stations have been fixtures of both science fiction and actual efforts to explore space. In the 1970s, the Soviet Union orbited a series of Salyut space stations, while the United States followed a number of missions to the moon with the Skylab program. In 1986, the Soviet Union began an even more impressive venture, the orbiting of the Mir space station. The first phase of a much larger and more sophisticated space station is scheduled to be built in the last years of the 20th century. Unlike earlier efforts that were products of the Cold War "space race," this space station is intended to be a joint project involving the contributions of a number of nations, Russia among them.

The International Space Station, or ISS for short, will be built in a number of stages. The first U.S.-built segment, a connecting module known as Node 1, is scheduled to be launched by the space shuttle *Endeavor* in July 1998. Two weeks before this, the first space station component, a power and propulsion unit known as the Functional Cargo Block, will have been launched by a Russian team in Kazakhstan. The subsequent construction of the station will require 73 space missions by three nations over a 55-month period: 44 to bring up components for assembly, 10 to bring crews to the station, and 19 to reboost the station in order to keep it at a proper altitude.

Upon completion, the space station will be by far the largest man-made object in space, weighing 423,000 kg (933,000 lb) and stretching 109.1 m (358 ft) end-to-end. It will also be a very expensive project; by the year 2012, the cumulative bill for launches, construction, and operation is expected to reach $100 billion.

Readers of science fiction and fans of the movie *2001: A Space Odyssey* are likely to be disappointed by the space station's configuration. It will not be an elegant wheel-shaped structure but a seemingly haphazard collection of component parts. The major components will be supplied by the United States, Russia, Canada, Japan, and the European Space Agency. The space station will be built around a backbone, or "primary truss," to use its official name. Fixed to the truss will be large solar arrays for providing power, a robot arm for manipulating large objects, and several modules that will serve as crew quarters and laboratory facilities. The

A computer-generated representation of the International Space Station in its completed form (courtesy NASA).

station will orbit at an altitude of 400 km (250 mi), and will be kept at the proper attitude (its orientation to Earth) by a global positioning system. While orbiting, the space station will be slowly pulled Earthward, and it will be necessary to fire rocket thrusters every 3 months to return the station to its proper orbit.

The first stage of the space station's program began in February 1994, when a Russian cosmonaut served as a crew member on the Space Shuttle. This was followed by a series of missions in which the shuttle docked with the Mir space station, the first of which occurred on June 29, 1995. The joining of U.S. and Russian modules in 1998 will mark the second phase. Three people will be housed in the embryonic space station, which will be enlarged to hold six people by the year 2002. Subsequent projects will include equipping the station with a Canadian-made robot arm to help with the installation of more components, solar arrays, and modules. One of the latter will be the Japan experiment module, which is

to consist of a pressurized laboratory, an open platform for experiments in space, and a logistics unit.

One of the major justifications for the building of the International Space Station is its ability to support experiments that only can be conducted in space. For example, in a zero-gravity environment, it is possible to grow nearly perfect protein crystals that may be of great value for research into the AIDS virus and the causes of cancer. Also, the study of physiological changes that occur in space, such as the gradual loss of bone mass, may be helpful in understanding terrestrial afflictions like osteoporosis.

Other experiments may shed new light on natural phenomena, and the space station may eventually serve as a platform for travel to other parts of the solar system. However, many critics remain unconvinced of the space station's long-term value. Given the financial constraints of the world's governments, the space station's immense cost will be hard to justify. If it is constructed as presently envisaged, the space station's costs will rival those of the U.S. program to put a man on the moon. Today, however, the political climate is less receptive to the funding of costly ventures in space. When the decision to support a permanently manned space station was announced by President Reagan in his State of Union address on Jan. 25, 1984, the United States was deeply entrenched in a rivalry with the Soviet Union that had energized prior space programs. With the collapse of the Soviet Union and the end of the Cold War, the situation has been reversed; Russian technical and financial support is now seen as essential to the program. However, with Russia experiencing considerable political and economic turmoil, its continuing assistance cannot be taken for granted. Given all of the uncertainties of funding and political support, it is likely that some of the plans for the International Space Station will have to be modified in the future.

See also ACQUIRED IMMUNE DEFICIENCY SYNDROME (AIDS); APOLLO PROJECT; CANCER; GLOBAL POSITIONING SYSTEM; MIR SPACE STATION; SKYLAB; SOYUZ AND SALYUT; SPACE SHUTTLE

Further Reading: Regularly updated information about the space station can be obtained on the NASA web page: http://station.nasa.gov/

Spacelab

Largely free from the constraints of gravity and the distortions of Earth's atmosphere, an orbiting space station is able to support scientific observations and experiments that could not be done on Earth. A significant amount of scientific research had been done in Skylab and continues to be done in the Russian Mir space station, but new opportunities arose when the U.S. space shuttle became operational.

Since 1983, the shuttle's cargo bay has on occasion been filled with Spacelab, a self-contained scientific and engineering laboratory that is carried on the shuttle. Spacelab is a collaborative venture between the National Aeronautics and Space Agency (NASA) and the European Space Agency, a consortium of 15 nations that was formed in 1975 and is headquartered in Paris. Spacelab first flew from Nov. 28 to Dec. 8, 1983, aboard the shuttle *Columbia*, the ninth shuttle mission.

Spacelab consists of three basic components. Its crew works in a module that comes in two sizes enclosing 25 or 100 sq m (880 or 3,500 sq ft); it is pressurized, allowing work to be done in normal attire. The second module, an instrument mounting platform or "pallet," is located behind the work module. It is covered by doors in the shuttle's bay that can be opened in order to expose sensors and antennas to a space environment. The third module contains support equipment such as airlocks, viewports, and storage lockers. Some Spacelab missions have included only the manned laboratory or a set of up to three pallets.

Orbiting at an altitude of 185 to 555 km (115–445 mi) for up to 10 days in a microgravity environment, Spacelab offers extraordinary opportunities for scientific research. Many research areas have been explored on Spacelab: biology and medicine, energy, astrophysics, communications, Earth observation and meteorology, navigation, and basic chemical and physical science. Work done aboard Spacelab also has particular significance for industrial processes that eventually may be done in space, such as manufacturing new drugs, growing larger and purer crystals, and making new alloys. Equally important, the international cooperation engendered through the construction and operation of Spacelab provides a model for future, even more ambitious ventures in space such as a large-scale space station.

See also SKYLAB; SOYUZ AND SALYUT; SPACE SHUTTLE; SPACE STATION

Further Reading: *The Cambridge Encyclopedia of Space*, 1990.

Species Extinction

The permanent disappearance of a species of living organism is known as extinction. By most scientific accounts, the world is on the verge of an episode of major species extinction, rivaling five other periods of accelerated extinction over the past half billion years. The most recent of these documented periods occurred 65 million years ago when the dinosaurs disappeared.

Unlike previous die-offs—for which climatic, geologic, and other natural phenomena were the cause—experts say that the current episode is driven by anthropogenic factors, such as the rapid degradation of habitat for human use, the introduction of invasive exotic species, pollution, human-caused climate change, human population growth, deforestation, and other activities that cause harm to ecosystems. While past extinctions occurred over a long period of time—10,000 to 100,000 years—the present rate of extinction is extremely rapid; estimates range from 15 to 150 for the number of species going extinct each day. If present trends continue, our planet could lose anywhere between 20 to 50 percent of its species by the end of this century.

The human role in animal extinctions may extend back to the Stone Age and the Late Pleistocene. Some researchers argue that faunal extinctions correlate with the spread of prehistoric human populations and the development of big-game hunting. Determining the root causes of extinctions, however, is complicated because major cultural shifts in human societies, such as the advent of big-game hunting, may have occurred in response to climatic change. Therefore, Late Pleistocene human beings may have simply finalized the extinction of animal populations already doomed by rapid post-Glacial environmental changes.

When a species goes extinct, there is generally more than one causal factor. Habitat destruction, for example, leaves populations isolated in fragments, where they are increasingly vulnerable to predation and disease. When a species goes extinct, the ecological web and food chain are changed forever, and one extinction may cause other extinctions. A recent analysis of animal extinctions since 1600 found that where the primary cause was known, 39 percent had resulted from species introductions, 36 percent from habitat destruction, and 23 percent from hunting and deliberate extermination. There appears to be a direct correlation between human population growth and the number of species that have become extinct since the mid-17th century.

Habitat loss is considered the greatest present threat to biodiversity, especially in the tropics. Biodiversity is the variation of genes within a species as well as the overall diversity of species, and is critical to viable populations of plants and animals. Even if all the individuals of one species are not killed, the remaining members of a species will go extinct if the genetic diversity of their members is too small. A large gene pool, however, helps species combat diseases and adapt to environmental stresses.

When a local population becomes extinct, the genetic viability of the whole population is decreased. This process, referred to as genetic erosion, is directly related to the loss of biodiversity. Along the west coast of North America, for example, over 100 endemic populations of salmon and steelhead have been lost, largely as a result of the construction of dams that prevent silt and nutrient flow downstream and prevent fish from spawning upstream. As a result, the greater Pacific salmon population is less hardy and more prone to further extinctions.

Most species that are threatened with extinction have not been identified or studied. Estimates of how many species there are on Earth vary significantly. Informed guesses put the total number of species at between 3 and 30 million, of which fewer than 1.8 million have been identified so far. While most birds, mammals, and plants have been classified, little is known about the other orders such as the insects, which make up the overwhelming majority of species described to date.

The number of recorded species extinctions over the past century is small compared to those predicted for the coming decades. This difference is due, in part, to the accelerated rates of habitat loss in recent decades and to the difficulty in documenting extinctions. Since the vast majority of species have not been described, it is almost certain that many will disappear before they are known to science. Moreover, according to internationally recognized criteria, a species is not classified as extinct until 50 years after its last slighting, so figures for recorded extinctions are highly conservative. Finally, since most species must have several thousand individuals to survive over the long term, depleted species may persist for several generations without hope of species recovery.

Extinction is a biological reality, and is part of the process of evolution. One theory, for example, holds that the extinction of dinosaurs occurred after the Earth was struck by a massive asteroid, which created a cloud of dust and smoke that blocked the sun, killing plants and animals. The extinction of dinosaurs made way for the evolution of new and varied life forms.

Presumably, some modern extinctions are also caused by natural events. In any period, including the present, there are species that are naturally doomed to disappear, either through overspecialization and a corresponding inability to adapt, or through the effects of natural cataclysms such as drought, volcanic eruptions, and floods. The scale of current human-induced extinctions, however, renders natural extinction trivial by comparison.—P.F.

See also DEFORESTATION; DINOSAUR

Further Reading: R. Edward Grumbine, *Ghost Bears: Exploring the Biodiversity Crisis*, 1992.

Spectroscopy

When light is passed through a prism it is dispersed. Each frequency of light present in the entering beam is diffracted through a different angle by the prism. The emerging beam can be viewed to reveal the frequencies present in the incoming beam and the intensity of light at each of these frequencies. White light, such as that produced by ordinary lightbulbs, will produce a rainbow band when viewed through a prism. The resulting band, a display of the light's intensity as a function of frequency, is called its *spectrum*, and a device that produces a spectrum is called a *spectroscope*. The study of the spectrum and its interpretation is called *spectroscopy*, which is an important part of atomic physics, astrophysics, and chemistry.

The first step in spectral analysis was taken in 1814, when Joseph von Fraunhofer (1787–1826), a German optician, passed sunlight through a narrow slit and then through a prism of his own manufacture. The resulting spectrum resembled an ordinary spectrum, but interspersed with the colored bands were a number of fine, dark lines. Fraunhofer eventually found more than 700 of these lines, which came to be known as Fraunhofer lines, but the reason for their existence remained a puzzle. By the middle of the century, a number of scientists began to suspect that the lines represented places where light was absorbed by particular elements. This hypothesis was the basis of studies conducted by the German physicist Gustav Kirchhoff (1824–1887), who in the late 1850s discovered that when heated to incandescence, each chemical element emitted light that produced a unique pattern

of spectral lines. In 1859, Kirchhoff found a material with a spectral line different from those of all known elements. In fact, it was new element, to which the name cesium (Latin for sky-blue, the most prominent color in its spectral line) was given.

In addition to identifying new elements on Earth, spectral analysis could be used to determine the chemical composition of extraterrestrial bodies. In 1862, the Swedish astronomer Anders Jonas Ångström (1814–1874) used spectroscopy to discover hydrogen in the sun. In 1868, while the French astronomer Pierre Jules César Janssen (1824–1907) was observing a solar eclipse, he noted the presence of a spectral line unlike that produced by any known element. Norman Lockyer (1836–1920), an English astronomer, decided that it was a new element; he named it *helium*, from the Greek word for the sun. Nearly 3 decades passed before helium was found on Earth by the Scottish chemist William Ramsay (1852–1916).

Modern spectroscopes may still make use of prisms, but more common are diffraction gratings, thin transparent sheets with hundreds of parallel lines etched onto their surface. The grating produces an effect similar to that of a prism when light is either passed through it or reflected from it. The light that passes through a prism or diffraction grating is emitted when a charged particle is accelerated. In most common situations, the charged particle is an electron. To take a common example, in an incandescent lightbulb, an electrical current passing through a thin wire heats the wire and causes the atoms in the wire to move about randomly at high speeds. Many of these atoms collide with each other, and the energy exchanged in the collision kicks some of the electrons orbiting the nuclei of these atoms into higher energy orbits. The electrons then quickly fall back into lower energy orbits, and the excess energy, equal to the difference in energy associated with the two orbits, it radiated as light. The frequency of the light radiated reflects the energy available according to the equation $E = hf$, where f is the frequency, E is the energy, and h is a constant, called Planck's constant, after Max Planck (1858–1947), the founder of quantum theory.

Every element has its own unique spectrum. Light emitted from a neon light is a familiar example; the reddish light we see is due to the particular frequencies of light emitted by the gas inside the glass tube. In an incandescent lightbulb, the atoms or molecules in the thin wire filament are so close together that they interact with each other, and the energy levels of the individual atomic orbits become distorted, so the light we see has many different frequencies and appears as white light to our eyes. The study of the spectrum of different elements has been an important tool for understanding the atomic structure of the elements. The Bohr model of the atom, the first quantum mechanical model, was based on the phenomena revealed by spectroscopy.

The temperature of the material that is emitting the light has an important effect on the light. The higher the temperature, the greater the average speed of the molecules, atoms, or ions, and the higher the frequency, and hence energy, of the average light or photon emitted. The exact relation between the frequency of the emitted light and the temperature of the source is known as the Planck radiation law, and scientists can use this law and the observed spectrum to determine the temperature of the emitting object. This is an important tool for astronomers wishing to determine the temperature of a distant star, or for an engineer who needs to measure the temperature of a blast-furnace interior. All objects radiate light in this manner, but most objects we encounter on a daily basis are cool, and according to the Planck law, most of the light they radiate is in the infrared region of the spectrum, where it cannot be detected by our eyes.

When light is passed through a cool cloud of gas, some of the light will be absorbed by the molecules, atoms, or ions in the gas. The frequencies of the absorbed light must correspond to the energy differences between the orbit of the electrons in the gas and the more energetic orbits allowed for the particular elements present in the gas. The resulting spectrum will show a continuous rainbow-colored background interspersed with Fraunhofer lines that correspond to the frequencies where the light has been absorbed. If the gas is hot rather than cold, it will produce a set of emission lines at the same frequencies; the emission lines are due to excess light at these frequencies. These absorption or emission lines allow a scientist to determine the chemical makeup of the gas the light passed through.

The presence of a particular line merely indicates the presence of a particular element or chemical compound, but careful analysis of the line's strength, or the total amount of light absorbed at that particular frequency, allows an accurate determination of the abundance of the element or compound in the gas. This technique has obvious applications in chemistry and astronomy. Astronomers have determined the chemical makeup of stars and planets this way, beginning with the aforementioned discovery of solar hydrogen. The exact relation between the number of photons (i.e., "packets" of light) absorbed or emitted and their frequency depends on temperature, relative abundance of the element producing the absorption, and the motions of the material, including any possible rotation. Thus, careful analysis of the spectrum can reveal a great deal about the object. For this reason, spectroscopy ranks as perhaps the most valuable tool available to the astrophysicist.

When an object moves either towards or away from an observer, any light it emits experiences a shift in frequency. The shift is to higher frequencies if the object is moving towards the observer, and to lower frequencies if the object is moving away from the observer. The shift itself is proportional to the object's velocity. This phenomenon is known as the Doppler effect. Accordingly, measuring the shift in frequency of the absorption or emission lines in an object's spectrum allows the scientist to determine the object's speed. The measurement is possible because we can accurately measure the exact frequencies for a laboratory source that is not moving and compare the frequencies of the lines seen in the spectrum of the moving object.

The analysis of the spectrum becomes more difficult when molecules rather than atoms are involved. Since molecules are made of several, or perhaps even several thousand, atoms, there are many complicated ways that light can interact with the molecule. In addition to excited electronic states that correspond roughly to the orbital states of atoms, molecules can also posses energy states that involve the rotation of one atom in the molecule about another, as well as vibrational motions of different atoms about their normal "equilibrium" position within the molecule. By studying the electronic, vibrational, and rotational spectra of a molecule, chemists can uncover information about the size and shape of the molecule and about the nature of the bonds that hold it together. A number of specialized techniques, such as Raman spectroscopy, have been developed specifically to aid in the determination of molecular structure. Finally, a somewhat different type of spectroscopy involves the interaction of nuclear energy states and magnetic fields. Nuclear magnetic resonance has become a major tool of the chemist and recently has seen important applications in medical imaging.—S.N.

See also ATOMIC THEORY; DOPPLER EFFECT; INFRARED RADIATION; MAGNETIC RESONANCE IMAGING; NEON; QUANTUM THEORY AND QUANTUM MECHANICS

Further Reading: James W. Rohlf, *Modern Physics from A to Z*, 1994.

Speed of Light—see Light, Speed of

Sphygmomanometer

Although its proper name is not well known outside the medical profession, the sphygmomanometer is one of the most common instruments found in a doctor's office; it is the device that measures blood pressure. The readings produced by the sphygmomanometer are given as two numbers, one high and one low. The high number, *systolic* blood pressure, expresses the pressure of the blood (expressed in millimeters of mercury, or mm Hg) on the walls of blood vessels while the heart is contracting. The low number, *diastolic* blood pressure, measures pressure while the heart is between contractions. High blood pressure, also known as *hypertension*, is generally considered to be indicated by a reading of more than 140/90 mm Hg. It may be dangerous in itself, or it may be an indication of other health problems. People with this condition are at some risk, for high blood pressure has been statistically correlated with a greater-than-average propensity to suffer strokes, kidney failure, and heart disease.

The use of the sphygmomanometer has allowed the quick and painless measurement of blood pressure. The first device to closely resemble the modern sphygmomanometer was created in 1896 by an Italian physician, Scipione Riva-Rocci. By the second decade of the 20th century, its use had become a regular feature of medical practice. More than simply a diagnostic device, the sphygmomanometer was emblematic of the effort to make the practice of medicine less of an art and more of a science. However, this goal was not supported by all medical practitioners, and the sphygmomanometer became the focus of a controversy regarding the nature of medical practice. Physiologists praised the instrument for its capacity to produce objective, quantitative, and precise data. Many practicing doctors were not so enthusiastic. Physicians had long been concerned with the diagnosis of cardiovascular problems, but their methods did not produce objective data. In particular, physicians depended on the subjectively evaluated feel of a patient's pulse to give indications of the presence and nature of cardiovascular abnormalities. Diagnosis made by feeling a patient's pulse required considerable experience and skill, and doctors who were proficient in this practice were loath to see their well-honed technique supplanted by a simple device. They were also concerned that the device would act as a barrier between them and their patients. The sphygmomanometer, as many doctors feared, would substitute a piece of equipment and technical proficiency for skill-based, patient-oriented medical practice. The result, it was feared, would be inferior care and a diminution of the physician's prestige and authority.

In the end, use of the sphygmomanometer became standard practice, and succeeding generations of doctors never even learned the old techniques for determining the state of a patient's cardiovascular system. Medicine had become more precise and "objective," but at the cost of its practitioners losing touch—both literally and figuratively—with their patients.

See also STETHOSCOPE

Further Reading: Hughes Evans, "Losing Touch: The Controversy Over the Introduction of Blood Pressure Instruments into Medicine," *Technology and Culture*, vol. 34, no. 4 (Oct. 1993): 784–807.

Spinning Jenny

An improvement on the single spindle technology of the spinning wheel, the spinning jenny was the first viable multiple-spindle machine for the production of cotton or wool thread. James Hargreaves (1720–1778) is credited with the invention of the jenny in 1764.

Tradition has it that Hargreaves, a poor Lancashire spinner and millwright, came upon the idea for the multiple spinning technology while observing the workings of a spinning wheel accidentally tipped on its side by his daughter Jenny. (In some versions his wife was responsible for the upset wheel. The name may also be a derivation of *engine*.) Noticing that the overturned spinning wheel continued to produce thread for a few moments while the spindle, customarily horizontal, revolved vertically, Hargreaves reasoned that the correct relative positioning of wheel and spindle could eliminate the necessity for the manual twisting of raw carded fibers. Mechanization of the operations of drawing and twisting offered a means to remove a critical technical bottleneck in textile production by enabling a single spinner to produce a number of threads simultaneously. Indeed, the potential threat to employment provided by the jenny was recognized by a group of spinners who, fearing for their livelihood, broke into Hargreaves's workshop and destroyed a number of his machines.

Hargreaves's jenny, patented in 1770, consisted of a rectangular frame that supported a row of vertical spindles driven by a hand-cranked, belted wheel. In operation, rotating spindles drew and twisted loose bands of fiber or "rovings" from large spools. Threads traveled through a movable carriage or tension bar located at the top of the machine and facing the spindles. Placed in the backward position, the carriage drew out the thread to give it fineness; in the forward position the carriage allowed spun thread to be wound around bobbins. While the spinning jenny produced thread of more even quality than could be achieved by distaff or wheel, jenny-spun yarn was rather weak and thus suitable largely for weft or filling-in weaving. In addition, the design of the machine limited the productivity of adult spinners by forcing them to work nearly doubled over. Further breakthroughs in thread quality and labor productivity awaited the development of the spinning "mule" by Samuel Crompton in 1779.—L.S.

See also LUDDISM; MULE SPINNING; UNEMPLOYMENT, TECHNOLOGICAL; WEAVING

Further Reading: Julia de L. Mann, "The Textile Industry: Machinery for Cotton, Flax and Wool, 1760 to 1860," *in* Charles Singer, ed., *A History of Technology*, vol. IV, 1958.

Spinning Wheel

Spinning is the process of taking the filaments of short-staple fibers such as wool, flax, and cotton, and twisting them together to produce a length of thread. Until the late 12th or 13th century, European production of thread relied either on simple hand spinning or on the traditional distaff and drop-spindle method. In the latter technique, the spinner drew out and guided lengths of raw fibers from the distaff, while a spindle, to which a lead length of fiber was attached, was spun and dropped. Disk-shaped weights or whorls attached to the spindle functioned as flywheels to regulate the speed of the spin and, together with the extent of the drop, worked to determine the fineness of the thread produced, while the spinner's skill determined the uniformity of the product. In the spinning wheel, the flywheel principle embodied in the spindle whorl was achieved through the first known application of a belt-driven transmission: The former weighted disk functioned as a pulley linking a large hand-driven wheel and a horizontally mounted spindle. The operator (or "spinster" in its original meaning) twisted the fibers between the fingers and wound the resulting thread around the turning spindle. The turning motion of the spindle in turn helped the operator to draw out more fibers and twist them together.

While the precise birthplace of the spinning wheel technology remains uncertain, it is thought to have originated in India or China and spread throughout Asia well before its appearance in Europe. In China, the spinning wheel was in all likelihood an outgrowth of silk-winding machines, devices that appear in literary texts as early as the 2d-century C.E.. The exact date of the spinning wheel's emergence in Europe remains uncertain, although there is some evidence for the use of the spinning wheel in the Italian woolen industry in the 12th century. But by the next century, the spinning wheel was common enough to provoke regulations that restricted or prohibited the use of wheel-spun yarn, as happened in Speyer in 1280 and Abbeville in 1288.

Early spinning wheels, which required the spinner to turn the wheel with one hand and to twist and guide the yarn with the other, appear to have increased the productivity of spinners by a factor of three over drop-spindle spinning. Further improvements in productivity

Spinning wheel used in a Tennessee home in the 1930s (courtesy National Archives).

were realized with the 15th-century development of the so-called Saxony wheel, which incorporated a U-shaped flyer that was mounted on the spindle and driven by the wheel, allowing spinning and winding to proceed simultaneously. This innovation required the successful coordination of three moving parts—the spindle, flyer, and wheel—all rotating at different speeds. A final improvement, the addition of a crank, connecting rod, and foot-operated treadle, was made in the 16th century.

Spinning was one of the first processes to be industrialized, and by the 19th century the output of machine-spun thread far exceeded that produced by the spinning wheel. But in the 20th century, the spinning wheel took on symbolic importance when the Indian leader Mohandas Gandhi (1868–1948) pushed for an expansion of hand spinning in the hope that it would provide jobs and serve as an antidote to industrialization and westernization, and the perceived evils that accompanied them.—L.S.

See also COTTON; INDUSTRIAL REVOLUTION; SILK; SPINNING JENNY; WOOL

Further Reading: John Munro, "Textile Technology," *in* Joseph Strayer, ed., *The Dictionary of the Middle Ages,* vol. XI (1988).

Spinoffs

In the United States and in many other industrial nations, a vast amount of research and development is oriented towards military needs. On occasion, a technology that has been developed for the military ends up having civilian applications. Technologies of this sort are sometimes known as *spinoffs,* as they have "spun off" from the military to the civilian sector. The term is also used for technologies that originate with space programs.

At many times in history, the military has been a source of technologies that are new and significant. The World War II era is particularly notable for new or much-improved technologies that went on to find significant civilian markets. Some of the most important of these are radar, penicillin, the digital electronic computer, and jet aircraft. Nuclear power, while not an object of wartime development, was a beneficiary of the intense effort to create nuclear weapons. Some new products were the unexpected results of wartime innovations, one of the most remarkable being the microwave oven, a spinoff from radar technology.

The government of the United States continued to pour vast resources into military research and development (R&D) in the decades following World War II. The Cold War between the United States and the Soviet Union was a time of considerable anxiety, and many military and civilian officials were convinced that national security depended on a rapid pace of technological change in weaponry and related systems. It was widely agreed that computers and electronics were key elements of technological modernization, and by the end of the 1950s, an estimated 85 percent of electronics R&D was being funded by the federal government, most of it oriented to military needs and the space program. It was during this time that the inherent drawbacks of using vacuum tubes for computers were overcome by the employment of transistors, the development of which owed a great deal to government-sponsored R&D

Of equal importance was the federal government's establishment of a guaranteed market for transistors and other solid-state components. The military paid a good price for the transistors it purchased, and revenues received by the manufacturers were plowed back into further research. This pattern was repeated when the integrated circuit was invented as a means of reducing separate componentry. Although many of the firms that pioneered integrated circuits deliberately avoided military funding so that they could maintain their ownership of the technologies involved, they forged ahead with the development of new products and manufacturing processes, secure in the knowledge that the military would provide a large, predictable market for their products. Of particular importance was the Air Force's use of integrated circuits for the Minuteman II missile program. In 1965, this program was responsible for no less than one-fifth of all integrated circuit sales. A similar story could be told about the development of magnetic storage devices for computers, where funds provided by the government and the prospect of military contracts gave a strong impetus to the development of disks and disk drives.

Although the collapse of the Soviet Union and the end of the Cold War in the late 1980s removed a major justification for high levels of defense spending, the military remained a major source of R&D funds. Although military spending was reduced from its Cold War levels, in the 1990 fiscal year, nearly 62 percent of federally sponsored R&D was performed in the de-

fense sector, accounting for about 30 percent of all R&D conducted in the United States.

High levels of spending for military R&D are justified primarily in terms of their contribution to national security, but spinoffs are often mentioned as a valuable by-product of that spending. However, critics point out that successful spinoffs are relatively rare, and that expecting defense-oriented R&D to be a major source of nonmilitary technological advance is unrealistic. Overreliance on spinoffs also disregards what economists call "opportunity costs." The funds and human resources used for R&D are necessarily limited, so when these are used for military purposes, they are not available for civilian R&D. It is also the case that military R&D is sharply skewed; in the early 1990s about 70 percent of military R&D was concentrated in aerospace and electronics. Consequently, a great deal of R&D effort has been oriented to technologies that have no obvious civilian application—stealth aircraft, for example.

In addition to the kinds of technology being pursued, successful military spinoffs are infrequent, because defense industries are different from civilian ones in some fundamental ways. Above all, the armed forces demand the highest levels of performance; cost is a distinctly secondary factor when survival may depend on having a superior technology. In similar fashion, military R&D is usually oriented toward effecting radical breakthroughs, and not the continuous, incremental improvements that typify the goals of commercial R&D. Firms oriented towards military or civilian needs also differ in the nature of the market that they supply. Whereas the latter typically supply a market made up of a multitude of customers, firms serving the military have only one buyer, the federal government. To make up for the absence of competition, the government draws up a myriad of regulations, specifications, and standards that contractors have to meet. Adherence to these requirements can hinder a firm's ability to serve the civilian market. In addition to increasing the costs of production, government requirements tend to lock products into patterns that may be hard to change. In contrast, firms producing for the commercial market are accustomed to making continual changes in their products in order to meet the needs of a changing marketplace. Finally, the internal policies and practices of firms producing military goods and firms producing civilian goods usually reflect the very different environments in which they operate. Even when a single firm makes both military and civilian products, these activities are kept in separate departments, with little connection or even communication between the two. These barriers further limit the possibility of advanced technologies moving from the military to the civilian sector.

Many technologies have both military and civilian applications; for this reason they are known as *dual-use* technologies. As noted above, many dual-use technologies originated in the military sector and then made their way into the civilian economy. However, the movement is not only in one direction; technologies originally developed for commercial purposes may end up having important military applications, as has been the case in recent years with computers and navigation systems. This pattern has become increasingly important in the post–Cold War era, because as military R&D budgets shrink, modernization is becoming more dependent on technologies that were initially developed for the civilian economy. To be sure, the military sector will continue to generate many advanced technologies in the future, and some of them will have important civilian applications. But spinoffs from the military cannot be expected to be a primary source of future technological progress.

See also ATOMIC BOMB; DEPARTMENT OF DEFENSE, U.S.; DISK STORAGE; INTEGRATED CIRCUIT; MISSILE, INTERCONTINENTAL BALLISTIC (ICBM); NUCLEAR REACTOR; OVEN, MICROWAVE; PENICILLIN; RADAR; RESEARCH AND DEVELOPMENT; STANDARDIZATION; STEALTH TECHNOLOGY; THERMOIONIC (VACUUM) TUBE; TRANSISTOR; TURBOJET

Further Reading: John A. Alic et al., *Beyond Spinoff: Military and Commercial Technologies in a Changing World,* 1992.

Spontaneous Generation

There is no greater scientific mystery than life; a living being is different in many ways from an inanimate object, but what makes it different is still a matter of intense speculation. In the second half of the 19th century, much of the debate over the nature of life centered on the theory of spontaneous generation. According to the

adherents of this theory, life continuously emerged from nonliving matter. For example, maggots could be produced by decaying matter and molds could emerge from a nutritive medium. Opponents held to the belief that although life on Earth may have arisen from nonliving material, that event occurred billions of years ago; afterwards only life could beget life.

The most important opponent of the theory of spontaneous generation was the great French scientist Louis Pasteur (1822–1895). At the time of the debate, which peaked in the 1860s, Pasteur had made fundamental discoveries in microbiology, many of which had significant practical applications. He was a firm believer in the experimental method, and he approached the issue of spontaneous generation by putting it to a series of experimental tests. In one typical experiment, sterile flasks containing a sterilized nutritive substance were exposed to purified air and then sealed; if Pasteur was correct, no living organism would be observed in the flasks at any time.

Conventional accounts of Pasteur's experiments note that nothing appeared in the flasks, a decisive refutation of the spontaneous-generation theory. In fact, the experiments were not as definitive as might appear at first glance. Supporters of the theory pointed out that there remained the possibility that in the course of sterilization, some nonliving substance essential for spontaneous generation had been destroyed. Conversely, when some of their experiments seemed to indicate the presence of spontaneous generation, opponents argued that there was no way to conclusively ascertain if the air and the nutritive substance had been completely sterilized; either might harbor microorganisms that gave rise to more of their kind. This in fact was the argument that Pasteur used against Felix Pouchet (1800–1872), another French scientist, who had observed apparently spontaneous generation in flasks supplied with oxygen that had been generated by the reduction of mercury oxide.

Other experiments that used presumably sterile air collected at high elevations seemed to confirm spontaneous generation when conducted by those who believed in it, while similar experiments conducted by Pasteur seemed to refute the theory. The issue was eventually referred to the French Academy of Sciences, which created two successive commissions to evaluate the experimental evidence. Both found in Pasteur's favor, not surprising since their membership was drawn from opponents of the spontaneous-generation theory. Pouchet, correctly feeling that the cards were stacked against him, withdrew his experimental results from further consideration. Ironically, his flasks may have lent support to the spontaneous-generation theory, for it was later discovered that the hay infusion they contained harbored a spore that often survived boiling. Since this fact was unknown at the time, the growth of microorganisms in Pouchet's flasks could easily have led an unbiased observer to accept the theory of spontaneous generation. As it turned out, Pasteur was right about spontaneous generation, but in retrospect it is evident that the experimental evidence could not provide a definitive basis for the support or the rejection of the theory.

See also PASTEURIZATION

Further Reading: Harry Collins and Trevor Pinch, *The Golem: What Everyone Should Know About Science,* 1993.

Springs

A spring is an elastic device that is able to return to its former shape after being compressed, stretched, bent, or twisted. Springs are often used to absorb the shocks and vibrations that are harmful to a mechanism or to the person operating it. A spring can also be used as a means of storing energy; a common example is the mainspring of a mechanical clock or watch.

The use of springs goes back to prehistoric times, when early human beings used saplings and other elastic materials for traps. Another very early use of a springy material was the bow, which dates back at least 20,000 years. The use of metals as elastic media began to be considered in the 3d-century B.C.E., when Ktesibius (c. 270 B.C.E.) and his student Philo (or Philon) of Byzantium (c. 250 B.C.E.) proposed the construction of a catapult powered by curved bronze plates. In all likelihood, this was never built, given the difficulties of producing bronze with the requisite amount of elasticity. Philo also realized that air is an elastic medium, and could therefore be used as a spring. A catapult powered by an air spring is mentioned in a book of mechanics he authored, but again, it is unlikely that it ever was built, or that it would have worked if it had been built.

Philo's proposed spring may be classified as a "semi-elliptical" spring because its curved shape gave it the form of half an ellipse. Springs of this general shape have been employed in large numbers for the front and rear suspensions of automobiles. They have the advantage of being cheap to manufacture, and they save on overall costs because they locate an axle in a fore-and-aft as well as a transverse plane, obviating the need for additional linkages. Springs of this sort usually are made of several separate leaves. One advantage of this arrangement is that the spring will continue to function even if several leaves break. Also, when the spring flexes, friction between the leaves causes the spring's motions to be damped so that the motion does not continue for an unacceptably long time. This self-damping quality is not usually sufficient, however; the springs used in automobiles and many other mechanisms are usually complemented by "shock absorbers" that dampen the springs' movements. Since the springs and not the shock absorbers absorb most of the forces, the British term *damper* provides a more accurate description of the function of these components.

Coil or helical springs were known to the Romans, who used them for the fastenings of personal ornaments, but, as best can be determined, they were not used for larger mechanisms. Large springs were beyond the capability of preindustrial societies, for their manufacture required the development of fatigue-resistant steels with just the right amount of elasticity. Making these springs in quantity also necessitated the invention of mechanized spring-winding equipment. For these reasons, heavy coil springs were practically unknown before the second half of the 19th century, when the demands of the growing railroad industry created a large market for them. Modern coil springs may employ progressive winding, in which one end of the spring has the coils placed more closely together than the other end. This provides for the progressive absorption of a shock; as force increases the spring becomes more resistant to it.

A third type of spring, the torsion bar, is essentially an unwound coil spring (alternatively, a coil spring is a torsion bar wound into a coil). One end of a torsion bar is connected to the moving part; the other end is connected to a fixed point. Deflection of the moving part causes the torsion bar to twist, absorbing the force that deflected it.

The action of springs is governed by Hooke's law. Named after the man who formulated it in 1678, Robert Hooke (1635–1703), the law states that the force that restores a spring to its equilibrium position is directly proportional to the distance the spring has been displaced from the equilibrium position. In reality, Hooke's law does not hold in all instances, for the action of springs is not perfectly linear; a spring may behave differently when it is close to the end of its compression (or extension) than it does when it is less compressed (or extended).

See also BOW; CATAPULT, TORSION; CLOCKS AND WATCHES; METAL FATIGUE

Sputnik

During the 1950s, the United States, building on earlier German accomplishments in rocketry, developed the capacity to put an artificial satellite into orbit. Although the satellite was to be launched by a rocket of military origin, the project was to be kept in the civilian realm. The year 1958 had been declared the International Geophysical Year (IGY), and Project Vanguard, the U.S. satellite program, was to be one of the IGY's crowning achievements. Tying Vanguard to an international scientific enterprise was also advantageous to the United States. An orbiting satellite had a considerable potential for military reconnaissance. The United States hoped that its first satellite would pass over other countries without provoking strong objections, especially from the Soviet Union, which had a deep aversion to overflights by other countries.

The assumption that the United States would pioneer the orbiting of artificial satellites was shattered on Oct. 4, 1957, when the Soviet Union became the first country to put the a satellite into orbit. Called Sputnik (loosely translated as "fellow traveler"), the satellite was little more than a 23-inch (58.4-cm) sphere containing a radio transmitter. But its impact was enormous. Although the Soviets had indicated their intention to launch a satellite (including the announcement of the radio frequency it would broadcast on), few in the West had paid much attention. The assumption of American technological superiority was deeply engrained, and the shock of being trumped by the So-

viet Union provoked emotions similar to those felt after the attack on Pearl Harbor in 1941. Less than a month later, on Nov. 3, 1957, the Soviets confirmed their lead with the launch of Sputnik II, a 1,121-pound satellite that contained a dog.

In a state of near panic, the United States rushed its Vanguard rocket into a satellite-launching attempt. On Dec. 6, 1957, in full view of the world's media, the rocket rose 1.2 m (4 ft) off the launching pad and then exploded. American pride was salvaged on Jan. 31, 1958, when Explorer I was launched from Cape Canaveral, Fla. Using the well-tested Army Redstone rocket, the 4.8-kg (10.5-lb) satellite quickly made a number of discoveries, most notably the Van Allen radiation belts circling the Earth.

On May 15, 1958, the Soviets launched Sputnik III, but by then the score had been evened, for the United States successfully orbited Vanguard I on March 17, followed by another Explorer on March 26. Use of the word *score* is unavoidable, for the early days of space exploration were suffused with a contest mentality. What began as a scientific exercise with strong military overtones had turned into a competition between two economic and political systems, a competition that was energized by the dubious assumption that superiority in space flight was an indication of overall technological superiority, and, even more ludicrously, emblematic of superiority in general. Sputnik had been a profound shock to the United States. It was the opening event of the space race, and the desire to never again be second-best was a prime motivator for subsequent U.S. space efforts, culminating with the first lunar landing.

See also APOLLO PROJECT

Further Reading: Walter A. McDougall, *The Heavens and the Earth: A Political History of the Space Age*, 1985.

Stained Glass

From medieval to modern times, brilliant stained-glass windows have been essential components of Western ecclesiastical architecture. Initially used as heuristic devices alongside mosaics and sculptures for conveying scriptural lessons to illiterate parish members, stained-glass windows assumed other functions and meanings over the centuries. Still crucial teaching tools, ecclesiastical windows hallmarked the technical and artistic achievements of glassmakers, designers, and painters by the late Middle Ages. In some instances, prominent church patrons commissioned windows as tributes to favorite saints or loved ones, and the medium became a vehicle for conspicuous display of power and wealth.

The first stained-glass artisans were probably monks, who adorned monastic chapels with colorful holy pictures constructed from small vitreous fragments set into lead channels. In the early 12th century, the monks of Tengersee created the clerestory windows at Augsburg Cathedral, the oldest complete example of a stained-glass installation. Inspiration for these and other brilliant chromatic decorations probably came from various sources, including rich ceramic mosaics in Byzantine churches, the painting style of illuminated manuscripts, and dazzling examples of Venetian enamelwork, introduced to Europe in the 10th century. Cathedral building accelerated in the 12th and 13th centuries, with great structures constructed at Chartres, Canterbury, Cologne, and other cities. At this time, itinerant glass craftsman traveled throughout Europe, setting up workshops in the shadows of building sites, often erecting small glassworking furnaces on the fringes of nearby forests. An expensive material, glass was entirely appropriate for the decoration of houses of worship. Colored glasses—in deep reds, blues, greens, and purples—suggested the richness of precious gems, suitable offerings to a Supreme Being. The state of medieval glassmaking technology shaped the appearance of church windows, for it limited the spectrum of colors and the dimensions of glass bits available to the first glaziers. Indeed, the dim, mysterious light cast by early windows and the jumbled character of their motifs owed less to artistic expression than to relatively primitive glassmaking methods.

At first, the clergymen who designed or directed the design of stained-glass windows were less concerned with realistic representations than with forceful, symbolic depictions of biblical images. All of this began to change by the 13th and 14th centuries, when shifts in ecclesiastical thought, changes in approaches to church construction, and minor advances in glass-

making technology combined to reshape the appearance of windows. Naturalistic depictions replaced cramped, stylized figures in the monumental windows, and carefully selected palettes celebrated daylight rather than mystical darkness. The preference for interior light dictated a shift away from deep blues and reds toward yellows and other warm tones, a development that was facilitated by the introduction of silver nitrate stain. The technique of painting lines or stippling on colored glass with enamels was an innovation that enabled stained-glass artisans to create more delicate, detailed, and accurate scenes based on observations in nature. Artisans learned to scratch, etch, stencil, and otherwise manipulate the surfaces of windows to achieve decorative effects on larger lights, or pieces of glass. Beginning in the 15th century, the stained-glass window was less a medium in its own right than a vehicle for glass painters. The painters' triumph led to the decline in the craft of making stained-glass windows, so that medieval techniques were almost forgotten by the 17th and 18th centuries.

In the 19th century, American and European romantics reinvented the craft of stained glass as they imagined it had been practiced by medieval craftsmen. America's first stained-glass artist, William Jay Bolton, designed and executed window installations in the gothic revival style, fusing medieval and Renaissance design elements. Because glazing is a relatively simple job, Bolton and other craft revivalists easily determined how to construct windows by studying European examples and extracting tidbits of data from secretive glaziers. After making a sketch for patrons, Bolton created a life-size cartoon that served as the window's outline. (Lacking large sheets of paper, medieval artisans had probably drawn their templates on whitewashed tables used for laying out the glass.) Medieval glaziers had cut their glass by drawing a hot iron rod across incised lines. After the glass cracked, they broke off the desired section, trimmed excess material with special tools, and laid the fragments side by side on the table for leading. If the glass were painted or tinted, the leaded window was heated to a relatively low temperature in a kiln (sometimes several times), allowing the enamels or stains to fuse with the colored glass. While medieval artisans made their own glass from sand, potash, and mineral oxides, 19th-century craftsmen like Bolton purchased leading, sheets of colored glass, enamels for painting, and prepared oxides for staining from firms making and selling those products to glaziers and painters. The finished window was placed in an armature—that is, a frame made from wood, iron, or bronze—in preparation for architectural installation.

To be sure, stained glass has not always exclusively been a church art. As glass became inexpensive and widely available in the 19th and early 20th centuries, small stained-glass workshops made windows in various sizes, shapes, and motifs for commercial, domestic, and ecclesiastical buildings. Some stained-glass studios created elaborate works on commission by wealthy patrons, but most workshops created more modest artifacts for mass consumption. From homes to offices, doors and doorways were sometimes ornamented with stained or etched glass, purchased by builders from glass workshops or mail-order catalogues. At the turn of the century, design reformers advocated a stained-glass aesthetic based on the principles of the arts-and-crafts movement. As fitting testimony to the craft's heritage, the gothic style, inspired by the work of medieval and Renaissance glaziers and painters, remained the preferred mode of most stained-glass studios into the late 20th century.—R.L.B.

See also GLASS

Standard Time

For most of human existence, the time of day was determined by local conditions; the sun's highest point marked noon, and all other times were determined from this point. In the 19th century, the needs of an expanding commercial network, scientific community, and railroad system motivated a conversion to a way of reckoning time that transcended particular localities. Although the railroads are usually given credit for the introduction of standard time, in fact it was the scientific community that initially took the lead. Many scientific endeavors, such as astronomical observations that needed to be simultaneously conducted in a number of places, required a universal time. In contrast, as long as railroads traversed short distances there was no need for a standard time; the time of day

at one particular station could simply be taken as the railroad's own "standard time." Although British railroads had begun to adopt standard time in the 1840s, American railroads showed no similar inclination.

For astronomers and other scientists, there was no simple way to cope with the multiplicity of local times, and beginning in the early 1870s, they began to promote the idea of a universal time. A crusade to establish standard time was headed by Cleveland Abbe (1838–1916), an astronomer and meteorologist. Abbe had been annoyed that a recent study of the aurora borealis had been marred by the absence of proper time reckoning, and that other kinds of astronomical observations had been flawed for the same reason. Also, at about this time, the railroads were beginning to encounter difficulties with time reckoning as their systems expanded. The federal government too was beginning to see the advantages of standard time, and in 1882, Congress authorized the President to convene an international conference to regulate time across the globe.

The railroads' slowly emerging realization of the value of standard time was powerfully reinforced by the efforts of William F. Allen, the secretary of the General Time Convention, an organization established by the railroads to coordinate the schedules of individual railroads. Unlike Abbe and other scientists pressing for standard time, Allen was a railroad insider, and hence carried more weight within the industry. The General Time Convention threw its support behind Allen's proposals and moved rapidly towards the implementation of standard time. At noon on Nov. 18, 1883, most railroads made the changeover, and by the end of the following year, all but two small railroads in the Pittsburgh area ran their trains according to a system of standard times in three time zones. For the most part, there was little opposition to the new system. Most complaints were voiced by people living on the boundaries between zones, for they experienced sunrises and sunsets that came earlier or later than had been the case before the adoption of standard time.

The introduction of standard time continued a process that began with the invention of the first clocks, as people submitted to an artificial reckoning of time and its passage. A few towns and cities on the borders of time zones held to their local time, but theirs was a rear-guard action, and by the early 20th century almost every settlement in the United States adhered to standard time. A further step away from "natural" time was taken on Mar. 18, 1918, when daylight savings time was introduced.

See also CLOCKS AND WATCHES; NOT-INVENTED-HERE SYNDROME; STANDARDIZATION

Further Reading: Ian R. Bartky, "The Adoption of Standard Time," *Technology and Culture*, vol. 30, no. 1 (Jan. 1989): 25–56.

Standardization

A key aspect of industrialism has been the drive for uniformity, and consequent efforts to produce standardized parts, processes, and even people. In its ultimate form, this mechanical vision imagined factories that ran without any people at all, a system of production that depended on the ability to replicate anything, exactly, and in any number. Dreams of this sort had their origins in an odd mixture of fervor and calculation. On the one hand, dramatic increases in productivity through the application of machinery promised wealth untold. On the other hand, rationality manifested in standardization held out the prospect of a conflict-free social order.

The development and use of standardized components and procedures is intimately connected to the history of the Industrial Revolution. The emergence of new sources of energy was a major feature of the Industrial Revolution, but no less important were the standardized machines that the water turbines and steam engines powered. Among these are multiple-spindle spinning machines and turret lathes, industrial machines that produced standard products while being driven by a central power source. Also of critical importance to industrialization was the development of jigs, fixtures, and gauges that allowed the mass production of components that could be assembled into final products with little or no filing, shaping, or forcing.

Another key aspect of 19th-century standardization was the creation of screw threads that followed a single pattern. Prior to this time, machinists followed their own idiosyncratic patterns when they made threaded fasteners. This practice began to change in the 1840s,

when Joseph Whitworth (1803–1887), working in conjunction with Great Britain's Institution of Civil Engineers, developed a set of standardized screw threads that ensured that all nuts and bolts of a particular size could be freely interchanged. A similar project was undertaken in the United States, where William Sellers devised a set of standardized screw threads that were an improvement on Whitworth's in that they were easier to manufacture.

In the 20th century, an early, and perhaps the greatest, triumph of standardization was the Model-T Ford. Although the Model-T's design underwent many changes over time, its manufacture was based on the use of parts that freely interchanged and often remained unchanged for several years. The result was a cheap, easy-to-repair car that was made in vast numbers and sold at progressively lower prices. In 1909, its first full year of production, 13,840 Model-Ts were made, and sold for about $1,000 each. In 1916, 585,388 were made, and the cheapest model cost only $360.

Paralleling the standardization of industrial products were efforts to standardize the actions of the workers who made them. The assembly line was one of the key agencies of this form of standardization; the relentless pace of the line ensured that workers had no time to do anything but go through their prescribed motions. Also of considerable importance were efforts to apply the principles of "Scientific Management," F. W. Taylor's system for getting work performed in "the best possible way." When this happened, Taylor and his adherents believed that output would be so great that labor and management would have little reason to squabble over who was to get what.

The impetus to standardize has not been confined to the industrial sector. The most far-reaching example of standardization commenced in late 18th-century France, when the revolutionary government replaced a jumble of weights and measures with a rational system based on a metric system. Even the reckoning of time, a seemingly inviolable natural phenomenon, was altered by the division of the world into standard times zones.

The role of governments in the promotion of standardization increased during the closing decades of the 19th century. The first governmental agency for the setting of standards was established in Germany in 1868. Russia followed suit in 1878, and the government of

Great Britain founded its Standard Department a year later. The United States was a relative laggard, not establishing the National Bureau of Standards until 1901.

Government agencies have done much to promote the development and use of standards, but of even greater importance are trade associations. Virtually every industry has at least one trade association, and these agencies have often played the leading role in the setting of standards for the products of that industry. Coordinating and promoting the standardization efforts of trade associations are umbrella organizations like the American National Standards Institute (ANSI), which encompasses about 900 firms and 200 trade, labor, technical, professional, and consumer organizations. National standards associations like ANSI are in turn members of international agencies, of which the most important are the International Organization of Standardization (ISO) and the International Electrotechnical Commission (IEC).

Standards lower production costs, facilitate repairs, and make products easier to use, but they have some negative consequences as well. The biggest drawback of standardization is that once a standard is set, it tends to lock things in place and obstruct needed changes. To note the most famous example of the locking-in process, the QWERTY keyboard is not the most efficient layout for rapid typing, but once it was adopted as the standard, it became very difficult to induce typists to learn how to use a different arrangement of keys. Another example of the hazards of early standardization is the American television set. Eager to commence quantity production of TV receivers in the late 1940s, American manufacturers adopted a system that used the best technology available at the time, one that rendered the televised image in 525 horizontal lines. In Europe, the slower commercialization of television allowed the use of a more advanced scanning technology that produced 625 of these lines. By this time, so many sets had been sold in the United States that a conversion to the superior European technology was not feasible.

Further complicating the process of standardization is the fact that the setting of standards is often more than a technical exercise. Standards often have economic implications, so government organizations and other independent agents of standardization may have to balance the various interest affected. Of particular importance is preventing the most powerful players from dominating the setting of standards—for example, the common situation whereby local building codes are formulated by the producers of construction materials. The "democratic" setting of standards is particularly difficult given the likelihood that the people with the most knowledge and experience will be in the employ of the most powerful groups.—B.S.

See also ASSEMBLY LINE; AUTOMATION; AUTOMOBILE; GAUGES, INDUSTRIAL; INDUSTRIAL REVOLUTION; MASS PRODUCTION; NATIONAL INSTITUTE OF STANDARDS AND TECHNOLOGY; STANDARD TIME; STEAM ENGINE; SCIENTIFIC MANAGEMENT; TYPEWRITER; WEIGHTS AND MEASURES, INTERNATIONAL SYSTEM OF

Statistical Inference

Imagine that you are given the task of determining whether a new ice cream shop in your city sells, on average, the same number of cones per day as do other, more-established shops. You are told that the average number of cones sold per day in the latter is 167. Given time limitations, you decide to select a random sample of 40 days from the first year of operations. You find that the for the new shop the sample mean is 173 cones sold per day, and the sample standard deviation is 10. The question you must answer now is the following: Could a sample with a mean of 173 cones per day have been drawn from a population of shops whose mean is 167? Put another way, does the new shop sell, on average, the same number of cones as other shops in the area so that the apparent difference is due to sampling error?

This kind of question is the basis for statistical inference. Statistical inference involves selecting a sample from some population and using a sample statistic (e.g., the sample mean) to estimate a population parameter (e.g., the population mean). The following assumes that the reader has already read the entries on descriptive statistics, normal curve, and sampling.

You can recast the problem by noting the following: If you had taken 1,000 samples, the mean of those sample means would have been the population mean, and they would have had a standard deviation called the *standard error of the mean*. The standard error of the mean is obtained by dividing the sample standard deviation by the square root of the number of observations. As you increase the size of each of those 1,000 samples, the distribution of sample means would approach a normal distribution, irrespective of the shape of the population distribution. Statisticians refer to this as the "central limit theorem."

The illustration presents the problem graphically. If the new shop was drawn from a population with a mean 167, the sample means would be distributed normally and would have a mean of 167. The problem now is determining whether the difference between this assumed population mean and the sample mean is significant.

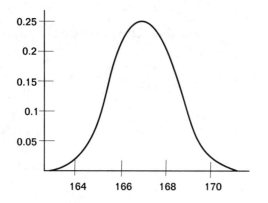

The standard error of the mean here is given by 10 divided by the square root of 40 (the size of the sample) or 1.58. The next step is to calculate a Z score for your sample mean:

$$Z = \frac{171 - 167}{1.58} = 2.53$$

Thus, the sample mean is 2.53 standard deviations above the population mean. To answer our original question, we need to determine the area beyond a Z score of 2.43. This will tell us the likelihood of drawing our sample mean from a population with a mean of 167. This probability equals .006. In effect, this means that the chance that we would have gotten a sample mean of 171 from a population with a mean of 167 is only 0.6 percent, a very small number indeed. Thus, you are led to conclude that the difference between the new ice cream shop and the others is statistically significant and that the new shop sells, on average, more cones than do other shops.

Statisticians call this approach *hypothesis testing*.

There is another approach much preferred by most statisticians: establishing confidence intervals. Briefly, this involves calculating, with a stated degree of confidence, the upper and lower bounds which we expect to bracket the true population mean. The calculation of a confidence interval is more complicated then determining a Z score, but its interpretation is more useful. For our problem, a 95 percent confidence interval is 167.91 to 174.90.

Again, return to our hypothetical 1,000 samples. If we had calculated 95 percent confidence intervals for each of those samples, we would expect that 95 percent of them would bracket the true mean. We only calculate one confidence interval, but we assert that we are 95 percent confident that it brackets the true mean. Note that our confidence interval does not include the population mean of 167 and that its width gives us some idea of the range of possible values for the mean of the new shop.

Two additional points need to be stressed. First, statistical significance refers to the probability of drawing our sample from the whole population; as such it does not make any statement about substantive relevance. At the same time however, a finding of statistical significance should lead us to explore broader questions of relevance. In the illustrative case, we would next ask how the new ice cream shop differs from the other shops. Second, in calculating the standard error of the mean, we divided by the square root of N, the sample size. Accordingly, as sample size increases, the standard error of the mean decreases, and we will find smaller actual differences proving to be statistically significant. This must be kept in mind as tests of hypotheses and confidence intervals are interpreted.—J.D.S.

See also HYPOTHESIS; NORMAL CURVE; SAMPLING; SCIENTIFIC METHODOLOGY; STATISTICS, DESCRIPTIVE

Further Reading: David S. Moore, *The Basic Practice of Statistics*, 1995.

Statistics, Descriptive

One often encounters the term *average* as a way of describing some phenomenon or issue. For example, we have been told that the average number of children in families in the United States is 2.2, that the average cost of a house in Southern California is $175,000, and that the average consumer spends $12.35 going to the movies each month. As we consider such claims, we need to understand that there are a number of "averages" or "measures of central tendency" that can be used to describe a set of observations; one can, for example, briefly describe three: the mode, the median, and the mean. Also consider the standard deviation, the measure of how different the raw scores are from one another.

A goal is to be clear about these three kinds of averages. All can be useful as they are, and some can be used for further statistical analyses. Means, in particular, play an important role in estimating population values from sample values.

The median is that grade above which and below which half the grades fall. In this figure, the median grade is C. The mean is the arithmetical average computed by summing all raw scores and dividing by the number of scores. To determine this, convert the letter grades into numerical grades: an A = 4 grade points, a B = 3 grade points, a C = 2 grade points, a D = 1 grade point, and an F = 0 grade points. In this distribution, the sum of the raw scores is:

$$(2 \times 4) + (3 \times 3) + (4 \times 2) + (2 \times 1) + (1 \times 0)$$

which equals 27. Divide this total by the number of observations, 12, so that the mean is 27/12 = 2.25.

Note that each average (the mode, median, and mean) tells us something different, and that in this case, they are not all the same value. The only time that all averages have the same value is when the data are distributed normally.

Now develop a measure of dispersion: the measure of how different the raw scores are from one other. Start with a simple set of numbers.

Consider the following table in which the mean is subtracted from each raw score and sum those signed differences:

	Score	Mean	Difference	Square
	3	4	-1	1
	5	4	1	1
	2	4	-2	4
	7	4	3	9
	3	4	-1	1
Totals	20	20	0	16

Notice that the sum of the Difference column equals 0. This is because the arithmetical mean is at the center of the distribution, and any deviation below the mean is balanced by one or more deviations above the mean. Thus, the mean can be visualized as the physical balance of any distribution.

Next, calculate the following:

$$\text{Variance} = \frac{16}{5 - 1} = 4.0$$

In other words, this is the sum of squared deviations divided by the number of observations less 1. This provides us with a measure of dispersion, but in terms of squared deviations. Put this number back into the original units by calculating the

$$\text{Standard deviation} = \text{square root } (4) = 2$$

Thus for this set of numbers we can see that the mean, or most typical score, is 4, while the standard deviation, or typical deviation, is 2.

Note that we divided the square root of 4 by $N - 1$ and not N. Remembering that the sum of signed deviations around any mean equals zero, imagine that we have four observations, and present the first three signed deviations as follows: –3, 2, –4. What would be the fourth deviation from the mean? Since the sum of the signed deviations equals 0, that last deviation must be 5. Thus, the last deviation is not free to vary, and we must take this into account in calculating the standard deviation. If we were to divide by N, we would underestimate the standard deviation. $N-1$ gives another important statistical measure known as *degrees of freedom*. In this particular case, there are four observations and 3 degrees of freedom.

In summary, there are a number of averages, and each tells us something different about a distribution. Whenever we are given a mean, we should also ask for a standard deviation, for this allows us to get some sense of the numbers that lie on either side of the mean. The larger the standard deviation, the more widely are the raw scores distributed about the mean.—J.D.S.

See also NORMAL CURVE; STATISTICAL INFERENCE

Stealth Technology

Military forces have long sought to evade detection by an enemy. For most of military history this meant resorting to some kind of camouflage. Camouflage was fairly effective when all that mattered was to avoid being seen, but beginning in the late 1930s it became possible to detect targets at great distance through the use of radar. The radar used during World War II could sometimes be foiled by scattering strips of metal foil known as *chaff* or *window*, but improvements in radar largely negated this possibility. Moreover, new detection measures based on the sensing of an airplane's hot exhaust also began to be employed. It had become evident that the survivability of aircraft hinged on evading detection through the reduction or even elimination of their heat and radar signatures.

In the early 1960s, the United States began to incorporate stealth technologies in new aircraft designs. One example was the Lockheed SR-71; not only was it fast enough to hold the coast-to-coast speed record, it also had a number of stealth features built into it. The 1970s underscored the need for stealth technologies; combat experiences from the Vietnam War and the 1973 Arab-Israeli War made it evident that radar-guided antiaircraft weaponry had reached high levels of effectiveness. At the same time, the lengthy Cold War with the Soviet Union required the maintenance of military prowess and encouraged the development of new capabilities such as stealth technologies. Ironically, however, the theoretical breakthrough that allowed the United States to develop radar-evading technologies came from the Soviet Union. Building on earlier work by James Clerk Maxwell (1831–1879) and Arnold Sommerfeld (1868–1951), a Soviet scientist named Pyotr Ufimstev in the 1960s developed a set of formulas for calculating the radar cross sections of objects. These in turn could be used in computer programs to design an airplane with a minimal radar signature. Beginning in mid-1970s, Lockheed Aircraft employed this technique for the design of the F117A fighter-bomber. The F117A had no large, flat surfaces and was devoid of sharp angles, such as the one where the wing meets the fuselage. This was done by breaking up the airplane's surface into a multitude of small panels or facets. When the F117A first flew in 1981, it was virtually undetectable by existing radar.

At the time the F117A was being designed, computers had insufficient power and memory to allow the design of three-dimensional designs or rounded shapes. The designers of the next stealth airplane, the Northrop B-2, benefited from improved computers that made it possible to dispense with faceting. The B-2, which first flew in July 1989, is a flying-wing design with no separate fuselage or tail. Like the F117A, its flying characteristics are affected by stealth requirements, and it needs computer-based "fly-by-wire" control systems to maintain stability.

Complementing radar-resistant shapes, stealth is also obtained through the use of radar-scattering materials such as composites and ferrite coatings that also scatter radar energy. About two-thirds of an airplane's stealthiness comes from its shape and one-third from radar-scattering materials and coatings. Some elements of these technologies can be used to endow existing aircraft with stealth capabilities. For example, by blending surfaces, installing baffles on jet inlets, and using radar-scattering materials, it was possible to give the B-1B bomber a head-on radar cross section that is approximately 10 percent of the original B-1's cross section.

Although many aspects of the F117A and B-2 remain classified, it can be reasonably expected that stealth is achieved at the cost of lower performance in terms of speed, range, and payload. Stealth aircraft also require considerable preflight preparation. For example, to maintain its ability to evade detection, the F117A has to have all gaps, such as those between access panels and the fuselage, sealed by special materials that have to be cut to fit.

Stealth aircraft are costly; at $43 million apiece, the F117A has a hefty price tag, although it is not vastly more expensive than a conventional fighter. The B-2, however, is an exceptionally expensive aircraft. Designing and building the B-2 required the solution of many novel problems and a consequent expenditure of funds. Costs were further escalated by a policy of concurrency, that is, putting the plane into production before prototypes have been built and tested. It has been estimated that by 1991 the B-2 program had cost $64.7 billion, of which $43.8 billion (reckoned in 1981 dollars) was spent before the first plane took to the air. With 75 B-2s expected to be built, this came to $863 million for each airplane. Cuts in the defense

The Lockheed F-117A stealth fighter (courtesy Lockheed Aircraft).

budget that followed the end of the Cold War resulted in only 20 B-2s being built at the astounding cost of $3.24 billion per plane. If more B-2s are built, their unit cost will go down, but of course more money will have to be spent to achieve this result.

At present, these astronomical costs have been justified by the great advantages conferred by stealth. That these advantages are not just theoretical was demonstrated in the 1991 Gulf War, when F117As destroyed Iraq's air defense operations centers, allowing nonstealth aircraft to perform their missions with little danger of being shot down. The B-2 has never been deployed on a combat mission, and as its prime task is to drop nuclear bombs in an all-out war, it is to be hoped that it never will be. In the years to come, it is possible that new detection technologies will negate stealth capabilities, but until that happens, these aircraft will have a substantial military advantage.

See also COMPOSITE MATERIALS; FLY-BY-WIRE TECHNOLOGY; PRODUCTION CONCURRENCY; RADAR

Steam Engine

The steam engine has made vast contributions to the supply of energy and the development of an industrial civilization, yet it began as little more than a toy. The first recorded steam-powered device was the *aeolipile* invented in the 1st-century C.E. by Hero of Alexandria. The aeolipile was a hollow sphere with two arms terminating in nozzles that were put at right angles to the arms and pointing in opposite directions. When water inside the sphere was heated, the resulting steam was ejected through the nozzles, imparting a rotary motion to the sphere.

Although Hero and his associates designed a num-

Watt's rotative beam engine from 1788. Note the sun-and-planet gearing at the lower right.

ber of pneumatically driven devices, efforts to use the power of steam soon ceased, not to appear again until the 17th century. At that time, an interest in the properties of air led scientists like Galileo and Evangelista Torricelli to calculate the pressure of the atmosphere, while in Germany Otto von Guericke (1602–1686) performed several famous demonstrations that vividly demonstrated the force exerted by air. In 1690, Denis Papin (1647–1714), a French Huguenot refugee living in Germany, built a device that used steam to move a piston inside a cylinder. Water beneath the piston was heated to steam, which pushed the piston upwards until it was arrested by a catch. When the steam cooled and the catch was released, the vacuum below the piston pulled it downward, and lifted a weight that was connected to the piston by means of a cord running over a pulley.

This was an ingenious demonstration, but it had no immediate practical consequences. The first use of steam power to do productive work operated on somewhat different principles. Patented by England's Thomas Savery (1650–1715) in 1698, this engine did not use a piston and cylinder, and its only moving parts were a number of valves. Savery's engine, which he called "the miners friend," used the condensation of steam in a re-

ceiving vessel to produce a vacuum that drew water from flooded mines. Steam pressure was then used to expel the water that had been drawn into the vessel. Engines built to the Savery design developed as much as 20 horsepower, and although they were very wasteful of fuel, they served a real need. The Savery engine continued to be used for a number of decades after its invention; about 150 were in operation around the middle of the 18th century, and sometime after 1765 one was used in a silk mill for a purpose now unknown. This seems to have been the first use of a steam engine in a textile mill, a preview of what was to come.

Although Savery engines continued to be installed through the last decade of the 18th century, the future of steam power lay with piston engines. The pioneering inventor of these engines was Thomas Newcomen (1663–1729), an ironmonger in the southwest of England who had learned firsthand the problems miners encountered in keeping water out of their mines. In 1712, Newcomen installed his first engine in Staffordshire, although it is possible that others had been built earlier in Cornwall. Newcomen's engine was called an *atmospheric engine*, for air pressure is what made it work. The major components of the engine were a piston and cylinder that were situated above a boiler.

Steam was admitted through a valve into the cylinder, causing the piston to rise. When the steam condensed, it created a vacuum in the cylinder, causing the descent of the piston, whereupon the process started again. At first, the condensation took place when the steam came into contact with the cool cylinder wall, but as the cylinder heated up, it failed to condense the steam. Surrounding the cylinder with a water jacket helped somewhat, but it was not a completely adequate fix. The solution to the condensing problem was discovered when a leak in a soldered patch in the cylinder allowed the sudden injection of water that immediately condensed the steam so that air pressure pushed down the piston with great force. A considerable amount of detail work with the valving and water injection was necessary to get the engine to run reliably, but once it did it, the Newcomen engine found a number of applications in mines, waterworks, and factories.

Improvements made to the Newcomen engine, most notably by John Smeaton (1724–1792), brought it up to as much as 72 horsepower, although most Newcomen engines developed much less power. Unfortunately for Newcomen, machinery for "raising water by the impellant force of fire" was covered in a patent awarded to Savery, and he had to enter into some sort of partnership with the owners of the patent. Although it had been rendered obsolescent for many decades, the last Newcomen engine was put into service in a Welsh slate mine in 1906.

In 1763, a young instrument maker named James Watt (1736–1819) was asked to repair the model Newcomen engine used for demonstrations at the University of Glasgow. Watt was puzzled by the poor operation of the model, and after much thought he came to the realization that it would run much better if the condensation of the steam took place in a chamber separate from the cylinder. From that insight and the experimentation that followed, Watt was able to build a more efficient steam engine, one that worked at nearly twice the efficiency of the best Newcomen type engine.

Watt's technological success eventually was paralleled by a commercial one, due in good measure to his partnership with Matthew Boulton (1728–1809), a Birmingham manufacturer. In 1776, the first Watt engine was put into service in a coal mine. At this time, steam engines were capable of only producing a reciprocating (up-and-down) motion; this was fine for pumping water out of mines, but direct rotary motion was necessary if the steam engine was to be used for industrial processes, especially in the rapidly expanding textile industry.

Inventors before Watt had devised steam engines that produced rotary motion, but uneven power strokes prevented smooth operation. In 1782, Watt built his first rotative engine. The problem of uneven power strokes was addressed by making his engine double acting: Steam was injected alternately on both sides of the piston. A more regular motion was also the result of controlling the admission of steam with a governor. As with the Newcomen engine, an overhead walking beam connected the piston rod to a pump rod or gear-driven flywheel. The gearing that translated reciprocal motion into rotary motion was a set of sun-and-planet gears, as the more straightforward method of using a crank had already been patented by another builder of steam engines. The piston rod was connected to the walking beam by a parallel motion linkage (which Watt later described as the invention of which he was most proud), which kept the rod moving in a straight line.

By 1800, 496 engines had been built by Boulton and Watt's company. In the succeeding years, the Watt rotative engine was widely used for running textile machinery, pumping water from mines and marshes, grinding grain, and many other applications. But it was slowly being supplanted by high-pressure (initially 40 psi, but later much higher) steam engines that used only the force of steam to drive the piston and dispensed with the condensing. Although it was evident that engines of this type delivered more power and better fuel economy, Watt was always suspicious of the use of high-pressure steam. His qualms centered on safety, a reasonable concern given the state of metallurgical and engineering knowledge at the time.

High-pressure engines eventually won out due to their greater thermal efficiency. They also lent themselves to using steam expansively (shutting off the flow of steam while the piston was still moving and using the expansive force of the steam to complete the stroke). A further extension of this principle was compounding, using the steam exhausted from one cylinder to move a piston in a second, larger cylinder. In some cases, triple and even quadruple compounding was employed. Although compound engines were generally more efficient than simple ones, compounding never became universal due to the higher maintenance costs it imposed. For many applications, most notably steam locomotives, it was rendered unnecessary by the introduction of superheating. This improved efficiency by heating steam well above the boiling point of water, thereby preventing the loss of heat through the condensation of steam on the cylinder walls. The potential merits of superheating were recognized by the middle of the 19th century, but their realization depended on the introduction later in the century of stronger materials and better lubricants.

The great improvement in the operation of steam engines can be seen in the sharply declining consumption of fuel relative to horsepower produced. Savery's engine required 14 kg (31 lb) of coal to generate 1 horsepower in 1 hour, while Watt's 1768 engine used only 4 kg (8.8 lb). A high-pressure engine built by Oliver Evans (1755–1819) used 3 kg (6.7 lb), while a compound engine built in the latter half of the 19th century used only 1.02 kg (2.25 lb).

Toward the end of the 19th century, steam engines began to be used to run electrical generators. This function put new demands on steam engines for greater power and higher rotational speeds. By the beginning of the 20th century, steam engines for electrical generation were turning at a rate of more than 250 rpm, more than four times the speed of the typical steam engines used in factories and mills. Achieving such speeds required substantial improvements in valving, lubrication, governors, and balancing. The largest reciprocating steam engines ever built were the eight engines designed in 1898 to generate electricity for the New York City subway system. Each engine was a four-cylinder compound with 112-cm (44-in.) high-pressure cylinders and 224-cm (88-in.) low-pressure 88 cylinders. Turning at a stately 75 rpm, each 485,340-kg (535-ton) engine put out 11,000 hp. This marked the apogee of the reciprocating steam engine. Within a few years, large engines of this sort were obsolete, their place taken by steam turbines for large installations, and diesel engines for smaller ones.

In its heyday, the steam engine was the backbone of industrial society. By the middle of the 19th century, the steam engine had become the prime source of energy and provided a dramatic demonstration of the immense power made available through the burning of fossil fuels.

The famous assertion by Karl Marx that "The hand mill gives you the society with the feudal lord; the steam-mill, society with the industrial capitalist" cannot be taken literally. Nevertheless, it contains an important truth, for the use of the steam engine was intimately connected to important changes in the ownership of productive assets and the organization of work. Steam engines were expensive items, not easily obtained by small producers. Hence, the steam engine reinforced an emerging pattern of centralized ownership and production. The expense of steam engines made it uneconomical to run them sporadically. In many factories, the operation of steam engines entailed regimented labor and the imposition of shift work. As with all epochal technologies, the steam engine brought many benefits, but these benefits were not distributed equally.

See also AIR PRESSURE; CONDENSER, SEPARATE; FACTORY SYSTEM; ENGINE, DIESEL; GOVERNOR; TURBINE, HYDRAULIC; TURBINE, STEAM; WATERWHEEL

Further Reading: Richard L. Hills, *Power from Steam: A History of the Stationary Steam Engine*, 1989.

Steamboats

The steam engine was the first technological device capable of converting thermal energy into power. Although this principle had been demonstrated in Alexandria as early as the 1st century, its intensive development dates from the 17th and 18th centuries at the hands of Thomas Savery (1650–1715), Denis Papin (1647–1714), Thomas Newcomen (1663–1729), and James Watt (1736–1819). The earliest practical engines, used for pumping water out of coal mines, were both unwieldy and inefficient because the cylinder was alternately heated and cooled. In 1763, Watt was repairing a Newcomen engine and noted that this problem could be obviated by condensing the steam in a vessel that was external to the cylinder. Thermal efficiency was greatly enhanced, and it now seemed feasible to apply steam power to the propulsion of human conveyances, specifically watercraft. Feasible perhaps, but in much of the world the incentives were slight: Sources of wood for fuel were problematic in England,

and maritime entrepreneurs almost everywhere were inherently conservative, especially when confronted with the prospect of sacrificing cargo capacity in order to accommodate engines, boilers, and bunkers.

Such constraints were less strongly operative in the United States, however, where natural waterways provided the basis for a superior system of internal commerce. By the time of the Revolutionary War, a great variety of watercraft was plying the rivers of the Eastern Seaboard, particularly their broad lower reaches. But upstream voyages were awkward at best, and it was in response to this situation that much of the initial experimentation with steamboats took place. Although credit for the first successful steam voyage goes to France and the Marquis de Joffroy's *Pyroscaphe*, which steamed up the Saone in 1783, the United States was the place where the initial development of the steamboat took place. In the late 1780s, James Rumsey powered a ship with a steam-powered pump that forced a jet of water out the stern. His premature death put an end to his experiments. More successful was John Fitch (1743–1798), who built several steam-powered vessels, one of which put in a few months of regular service on the Delaware River in 1790. These early efforts were overcome by technological, financial, and psychological difficulties. But the first years of the 19th century saw the emergence of steamboats that would decisively liberate mariners from whimsical winds and contrary currents. Shortly the United States would attain—and long retain—top rank in the size of its fleet of steamboats.

Fitch, Rumsey, and lesser-known contemporaries such as Griffen Green and Samuel Morey were of a type that fills the annals of technology. They could conceive an idea full of promise, but they were not sufficiently sensitive to the subtleties of design or to the transcendent import of entrepreneurial skill. A man who stands in stark contrast was Robert Fulton (1765–1815), whom schoolbooks traditionally proclaim to be "the inventor of the steamboat." And in an important sense he *was* the inventor, notwithstanding the operation of steamboats by others when Fulton had scarcely thought about the subject. For what Fulton indisputably did do was make the steamboat a commercial success. With everyone before him, economic viability had been tenuous at best. Even a vessel that is sometimes described as the first completely practical steamship, the British tug-

boat *Charlotte Dundas*, remained in service for only a short time in 1801. The key to Fulton's success was a wealthy, influential, and technically astute patron, his father-in-law, Robert Livingston. In August 1807, a vessel prosaically named *The Steamboat*—40.5 m (133 ft) long and equipped with side paddlewheels and a power plant imported from Boulton and Watt, the world's premier manufacturer of steam engines—made the 240 km (150 mi) from New York City to Albany in 32 hours, including an overnight stop at Livingston's estate, Clermont. At a later time, this vessel came to be known as the *Clermont*. By the end of the second war with England in 1815, 21 steamboats had been built for shipping enterprises in which Fulton was interested.

At the same time, steamboats were also appearing on the Ohio and other rivers west of the Alleghenies, which were generally shallower than those of the Eastern Seaboard. Because of this circumstance, these craft were designed quite differently, with paddlewheels at the stern, very little draft, and high-pressure engines mounted right on deck. The foremost innovator was Henry M. Shreve (1785–1851), who successfully contested an attempt by Fulton and Livingston to monopolize all steamboat operation and also supervised the removal of major impediments to navigation in the West. By the middle of the 19th century, hundreds of shallow-draft sternwheelers were in operation on the "Western Rivers." Races between competing boats were common, leading to overwrought boilers and consequent explosions. So great was the carnage that the normally laissez-faire U.S. Congress began to enact laws governing the operation and maintenance of steam boilers—the first instance of federal regulation.

By the middle of the 19th century, the hegemony of steam-powered riverboats began to be threatened by the railroads, which enjoyed much more direct routes and could keep rolling even when watercourses were frozen, but on the high seas steam power was just coming into its own. The first steam crossing of the Atlantic was accomplished by the USS *Savannah* in 1819, although it was under steam for only about 80 hours of its 28-day voyage. The first ship to traverse the Atlantic entirely under steam was the British *Sirius*, which beat another British vessel, *The Great Eastern*, by only a few hours despite having a headstart of 4 days.

The development of oceangoing steamships was

marked by technological changes following one another in rapid succession after midcentury, notably screw propulsion, steel hulls, and compounding. Compound engines (wherein steam was used first in a high-pressure cylinder then exhausted into a low-pressure cylinder) soon gave way to triple expansion and quadruple expansion engines as liners approached 215 m (700 ft) in length, and by the turn of the century the size of reciprocating engines had reached a practical limit. In the early 20th century, designers turned to steam turbines. In the heyday of the ocean liner, the power produced by these engines was astonishing; ships like the British *Queen Mary* and the French *Normandie* were propelled by turbines delivering 134,000 kW (180,000 hp) and 119,00 kW (160,000 hp), respectively. Although most vessels are now powered by diesel engines, and steam propulsion is increasingly rare except in very specialized sorts of operations, one needs to take only a casual look at the history of technology to realize the tremendous significance of this innovation during the recent past.—R.C.P.

See also ENGINE, DIESEL; PROPELLER, MARINE; SEPARATE CONDENSER; STEAM ENGINE; TURBINE, STEAM

Steel in the 20th Century

It is only in the past 150 years that steel has been produced in large quantities. A small step forward took place in the 1740s, with the development of the so-called *crucible process*. In the 1850s, Henry Bessemer (1813–1898) began producing steel by blowing air through molten pig iron in a tilting pear-shaped furnace, thereby removing most of the carbon from the iron. A decade later, Siemens and Martin developed the open-hearth process, which increased tonnages enormously.

Prior to about 1890, most of the world's iron and steel was produced in Great Britain; indeed, iron and steel, along with coal and textiles, were the very foundations of the British Empire. By the end of the 19th century, the lead had passed to United States, coinciding with the advent of the skyscraper era and a shift in the primary demand from railroad rails and equipment to framework for buildings, and, later, sheet steel for automobiles and appliances like refrigerators and washers. Vast increases in the scale of North American

production were based on exploitation of ore deposits around Lake Superior and the simultaneous development of enormous freighters for transporting ore to steelmaking centers south of the Great Lakes.

World steel output totaled about 25.5 million metric tons (28 million tons) in 1900, and continued to climb until the Depression of the 1930s. This was only a temporary setback, however, as the demands of World War II led to a massive increase in production; U.S. output soared from 28.7 million metric tons (31.5 million tons) in 1938 to almost 82 million metric tons (90 million tons) in 1944. In 1950, American mills produced 190 million metric tons (208 million tons), almost half the world's production. By 1993, world production had risen to 741.3 million metric tons (814.6 million tons), a stunning increase. But this growth was not the only important change, for America's share in world steel production had dropped to about 12 percent. The United States had been the world's primary steel-producing nation for most of a century, but the combined effects of technological and economic changes, along with managerial complacency, began to take their toll in the 1970s.

When the 20th century began, steel was already one of the most capital intensive of industries, for most large American producers adopted the logic of vertical integration, acquiring coal and iron mines, steamship and rail lines to move those materials, and expensive plants to produce coke, smelt ore into pig iron, refine iron into steel, and fabricate the raw steel into wire, plates, bars, structural shapes and beams, and rails. U.S. Steel, the largest corporation in the world after its formation in 1901, exemplified this development with its control of the Mesabi iron range in northern Minnesota. Huge shovels attacked the high-grade iron ore from the surface and loaded it into rail cars for transport to lake freighters, all owned by the company. With the development of enrichment processes that concentrated lower-grade ores into taconite pellets after World War II, Mesabi has remained a leading source of ore, although huge new ore bodies have been opened in Canada, South America, and Australia since 1945.

A crucial theme of the 20th century is the continued process of change in production technologies. In the steel industry, many changes were incremental, such as the adoption of by-product coke ovens that recovered gases, tars, ammonia, and other residues. These coal-based chemicals proved valuable in themselves, and the gas was burned in the ovens to generate the heat that released more gas. The new coke ovens were developed in Germany, but by World War I had been introduced everywhere. Blast furnaces for making iron grew steadily taller and larger, while mechanical handling equipment and huge cranes replaced hand shoveling. And after 1900, the open-hearth process of Siemens/Martin displaced the Bessemer process as the usual method of refining iron into steel. All equipment, including rolling mills and the specialized apparatus for making wire, pipe, and plate, grew larger and heavier, and during the 1920s, electric motors replaced steam engines as motive power.

More radical was the introduction of electric furnaces. Experiments using electric currents to break down iron were made in the 1880s, but success came through the efforts of Paul Héroult (1863–1914) in France in 1900. The electric arc from two huge electrodes melted iron and scrap quickly, and these furnaces proved well suited for making special-purpose alloy steels for tool steels and automobile uses. Another innovation involved mechanization of one of the most arduous jobs in the industry—rolling sheet—in the 1920s. The American Rolling Mill Company (later Armco Steel) pioneered the development of semicontinuous sheet-rolling mills, called *hot-strip mills*, that eventually produced sheet steel in continuous rolls perfectly suited for consumer appliance and automobile-makers.

Two technologies shook the industry after World War II. One change returned to Bessemer's insight and blew oxygen (not air) into a vessel of molten steel, this time from the top. Several groups experimented with this idea in the 1930s, but oxygen steel making was not successfully introduced until 1950 in Linz and Donawitz, Austria. American firms, still building open-hearth furnaces at this time, chose not to play the pioneering role, although oxygen furnaces cut costs while maintaining high quality, and created fewer environmental problems. In 1969, only 1 year after the last American Bessemer converter was closed, oxygen steel surpassed open-hearth production in the United States, but by this time foreign producers were ahead of American firms in the use of this technology.

The other crucial postwar innovation was continu-ous casting, a process that simplified the shaping and rolling of steel. Traditionally, steel was poured from furnaces into ingots and then placed in "soaking pits" (furnaces) for several hours to permit even internal heat distribution in the ingot and formation of proper grain structure in the steel. The ingots were then rolled into billets or slabs that could be shaped or rolled in smaller finishing mills. Continuous casters, on the other hand, transform a stream of molten steel from the furnace directly into billets or slabs ready for the finishing mills. The technical challenges were immense, for the molten metal had to be cooled enough to emerge like toothpaste from a tube, but not cooled too much. A Swiss company, building on German experiments in the 1930s, commercialized continuous casting of small billets in the mid-1950s, but slabs proved harder to cast continuously. In 1963, McLouth Steel near Detroit, Mich., installed the first American slab caster. Larger American firms were very skeptical, despite the elimination of production steps and significant savings in energy and wastage. Nippon Kokan, however, brought the first Japanese unit on line in 1967, and by 1980, 60 percent of Japanese steel was continuously cast, compared to 20 percent of American steel.

These numbers suggest the central theme of the steel industry after 1960: international competitiveness and challenges to American producers. The American decline was exacerbated by the maturity of the American steel industry, which meant that American facilities were much older than the facilities built in other countries after the destruction of World War II. U.S. firms also grew complacent because they faced no real competitors in the late 1940s and early 1950s. By 1960, however, European, Japanese, and other producers were successfully attacking the American market. Import quotas failed to stem a loss of market share and eventual red ink on American balance sheets. By the late 1970s, U.S. plants were closing and jobs were lost. In the 1980s, workers often kept their jobs only by accepting pay cuts and reductions in fringe benefits. Some, but not all, American firms adopted cooperative management styles, while all benefited from governmental assistance in the form of tariffs and retraining funds. In 1986, the steelworkers union struck U.S. Steel, refusing to accept additional cutbacks. Thirty years earlier, steel strikes could cripple the economy. President Harry Truman even attempted to take over the mills in 1952 to avoid a strike. The 1986 work stoppage lasted 184 days, the longest in industry history, but this time almost no one outside the industry noticed.

Yet not every American steelmaker was in trouble. New "mini-mills" combined organizational changes with technical innovations in order to produce steel at lower costs. These enterprises eschewed the massive integrated plants of the giant firms in favor of electric furnaces and continuous casters. They also used nonunion workforces to produce smaller products for consumer markets (e.g., reinforcing rods). They succeeded brilliantly, proving as damaging to the integrated giants as foreign competition.

Clearly, the steel industry had changed fundamentally. From a high of almost 137.5 million metric tons (151 million tons) in 1973, American production dropped to a low of 67.8 million metric tons (74.5 million tons) in 1982; it recovered to 88 million metric tons (97 million tons) in 1993. Nor were the changes limited to the United States. Automakers have reduced weight and shed pounds by shifting to plastics and aluminum. Iron pipe has been replaced by the PVC variety, and newer materials, like ceramics and the carbon filaments replacing the steel shafts on golf clubs, continue to invade steel's market. With these substitutions, the age of heavy industry as the cornerstone of the modern economy seems to be at an end.

It is interesting, however, to observe the cycles of this industry. In the late 1870s, England began to worry about its industrial competitors, the United States and Germany, and it seemed a genuine catastrophe when both countries surpassed Great Britain in iron production in 1890. Americans viewed that achievement as proof of their nation's technological vitality and confirmation of superior cultural, organizational, and technical skills. Not surprisingly, Russia in the 1930s built the world's largest steel mill, copying U.S. Steel Corporation's facility at Gary, Ind. Constructed in Magnitogorsk at enormous human cost, Russia celebrated the plant as proof of the superiority of Russian-style communism. The Soviet Union took enormous pride in surpassing the United States as the largest producer of steel in 1974. In the end, however, it took more than raw steel to prevail in the contest between capitalism and communism.

By the early 1990s, every Western nation had seen the decline and restructuring of its steel industry. Yet in other nations, iron and steel are still the first steps to building industrial economies. In 1993, the United States accounted for about 12 percent, Europe about 18 percent, and Eastern Europe about 27 percent of world production. Total U.S. production was the same as in 1962. Asia, on the other hand, made more than 36 percent of the world's steel. Japan provided a model for many nations when it embraced iron and steel as keys to its economic recovery in the 1950s and 1960s. Japan's enormously efficient mills soon set world standards, and Japanese exports of less-costly, higher-quality steel to the United States caused the first trade complaints between the two countries. Yet in one of the ironies of the current global economy, Japan in 1993 cannot compete with newer steel producers like Korea and Brazil, and has closed and downsized its steel mills. To emerging industrial nations, iron and steel stand as crucial components of a modern economy.—B.E.S.

See also COMPOSITE MATERIALS; ENGINEERING CERAMICS; POLYVINYL CHLORIDE; SKYSCRAPERS; STEEL, ALLOYED; STEEL, BESSEMER

Further Reading: William T. Hogan, S. J., *An Economic History of the Iron and Steel Industry in the United States,* 1971.

Steel, Alloyed

Once metallurgists began to understand processes for making steel by carefully regulating the carbon content, they started devising other alloys by adding measured quantities of a third element, and often a fourth. As was often the case with artificial polymers (plastics), metallurgists sometimes devised new materials in advance of any clear idea of what to do with them. But certain new alloys immediately proved valuable in cutting ordinary steel and so were known as *tool steels.* Crucible steel, invented in the 1740s by an English clockmaker named Benjamin Huntsman (1704–1776), had this cutting capability, but it was extremely costly. In 1868 an English ironmonger, Robert Mushet (1811–1891), found that adding 7 percent tungsten to carbon steel resulted in a material-that could be used to cut ordinary steel with lathe tools. For the next 40 years, most of the major developments with alloy steels would take place in Great Britain, particularly in Sheffield, England. Nevertheless, Mushet's discovery invigorated the metalworking and machine-tool industries in the United States as well as in England. The first important American contribution in the realm of tool steels was not a material but rather a technique for treating tool steels by heating them almost to the melting point and then quenching them rapidly. This process was devised by the originator of "Scientific Management," Frederick Winslow Taylor (1856–1915), working with a Bethlehem Steel metallurgist, Maunsel White. Cutting tools treated by the Taylor-White process remained strong and sharp even when they became red-hot.

At the turn of the century, Robert Hadfield (1858–1940) of Sheffield alloyed carbon steel with a small percentage of manganese, the resulting material being extremely tough and hence useful for railroad rails and, later "burglar-proof" safes. Manganese steel was the first specialty steel produced on a commercial scale, and manganese remained a component of virtually all specialty steels. Others were made by alloying nickel; vanadium, molybdenum, and even uranium, but ultimately the most important alloying agent was chromium, the product being called *stainless steel*. Stainless steel was patented in 1915, and an American chemical company placed the first large commercial order—to make tanks for nitric acid—in 1924. Its use soon spread to petroleum refining, to the manufacture of paper, textiles, and lacquers, and to the processing of food, beverages, dyes, and pharmaceuticals. In the late 1920s, the Budd Company, a maker of automobile bodies in Philadelphia, developed a method for welding stainless steel. By 1930, U.S. production of stainless steel amounted to 54,400 metric tons (60,000 tons) annually, about half the worldwide output. Applications ranged from dinnerware to the internal parts of internal-combustion engines. Stainless steel came to symbolize "machine-made America" as embodied in automobile-trim, building facades, and lightweight, streamlined trains such as the diesel-powered "Zephyr" that was delivered to the Chicago, Burlington & Quincy Railroad by the Budd Co. in 1934.

Adding a third element, molybdenum, to steel and chromium resulted in chrome-moly steel. This was used in high-stress applications such as box wrenches and for making the tough thin-wall tubing used in the aircraft and bicycle industries. Alloy steels proliferated in the mid-20th century. The Society of Automotive Engineers (SAE) classified four chrome-moly steels in the 1920s; by the 1970s there were almost 40. To keep order, the SAE and the American Iron and Steel Institute have grouped steels under about 100 major headings—within each of these, subgroupings specify ingredients for specialty steels with different properties. For example, 4130 chrome-moly calls for certain ranges of carbon, manganese, phosphorus, sulfur, and silicon, in addition to chromium and molybdenum, and 4340 chrome-moly specifies the same amount of phosphorus, sulfur, and silicon, but different quantities of other materials and a relatively large amount of nickel. Each alloy has specific mechanical, thermal, physical, and electrical properties.—R.C.P.

See also LATHE; SCIENTIFIC MANAGEMENT; STEEL IN THE 20TH CENTURY

Steel, Bessemer

Steel, an alloy of iron and carbon, was a precious metal for much of human history, for it could be produced only from high-quality iron ores. Special skill, long experience, and some luck were always required, and the quality varied widely from batch to batch. Until the late 1700s, smiths and ironmasters did not know that the key was the amount of carbon in the metal.

No furnace in antiquity was hot enough to melt iron, so steel typically was made by heating wrought-iron bars containing almost no carbon in wood charcoal fires. This process added enough carbon (0.15 to 1.5 percent) to the outer surface of the iron to produce a thin surface layer of steel. Only the best iron could be "steeled" in this way, and the harder, more-durable metal that now held an edge became the cutting surface of knives, tools, and weapons. No measuring tools or chemical tests were available; only the skill of the craftsman produced this wondrous and expensive material. As the French scientist Buffon (1707–1788) commented in his *History of Minerals* in the 1780s, "What difficulties to conquer! What problems to solve! How many arts piled one on the other are necessary to make this nail or pin which we value so lightly."

China was one of the places where high-grade iron ore permitted quantities of steel to be produced in antiquity, beginning at least by the 2d-century B.C.E. and by the 5th-century C.E. Uniquely, Chinese ironmasters produced steel by placing wrought iron in liquid cast iron, thereby increasing the carbon content of the surface of the wrought iron. In the Hyderabad region of India, small quantities of a special steel called *wootz* were made from about 400 B.C.E. Wrought iron was heated in small clay crucibles with charcoal and plant leaves, producing a small cake of wootz steel. Used in Damascus swords, the crystalline structure of wootz was visible in beautiful wavy lines in the polished metal.

By the Middle Ages, steel was produced in various places, including Spain, France, Germany, and Sweden. The Japanese also produced superior steel swords using selected iron produced in a special *tatara* furnace and carefully forging and reforging the blade. But not until the 1740s, when Benjamin Huntsman (1704–1776) in England introduced cast crucible steel, was uniform steel available at reasonable expense. As with other British innovations in iron production, Huntsman relied on coal fuel, developing a furnace that held crucibles, closed pots made of special clay. He packed high-quality Swedish wrought iron with charcoal inside crucibles: after several hours, the white-hot liquid steel was poured into ingots of homogeneous steel. Britain monopolized this process for almost a century, gaining a crucial advantage over would-be competitors, for Huntsman's method permitted regular production of steel of consistent quality. Britain soon dominated production of edge tools, files, springs, cutlery, and other steel products, and by the mid-1850s Britain was producing 55,000 metric tons (60,000 tons) of steel annually.

Still, steel was expensive until an inventor named Henry Bessemer (1813–1898) found a faster way to make steel. Experimenting by blowing air through a ladle of molten cast iron, he produced a violent reaction as the carbon was burned off. In an 1856 paper to the British Association for the Advancement of Science, Bessemer announced discovery of "The Manufacture of Malleable Iron and Steel with Fuel." Through this process, he refined cast iron into tons of steel in about 15 minutes. Licensees of his patent rushed to build their own plants, only to encounter wretched results, for Bessemer had fortuitously used a low-phosphorous iron. More typical ores with high-phosphorous content produced very brittle steel, and Bessemer was denounced as a charlatan until David Mushet (1811–1891) resolved the problem by adding *spiegel*, an alloy of iron and manganese that eliminated excess oxygen in the molten steel. After Bessemer patented a tilting furnace or converter in 1860, the revised process was widely adopted, although it remained tricky to control. It was never easy to leave exactly the right amount of carbon in the steel, and ironmasters soon burned off all the carbon and added back just the right amount. Formal chemical and metallurgical knowledge were for the first time essential, although many resisted linking science to this traditionally empirical industry.

Bessemer's innovation allowed an enormous expansion in steel output while slicing costs. His process spread widely after Sidney Gilchrist Thomas (1850–1885) installed a converter lining of basic dolomite $(CaMg[CO_3]_2)$ rather than the acid silica brick used by Bessemer, permitting utilization of high-phosphorous iron ores. Production stood at 450,000 metric tons (500,000 tons) in 1870 and 25.5 million metric tons (28 million tons) in 1900. Nowhere was the increase faster than in the United States, where production vaulted from 20,000 metric tons (22,000 tons) of steel in 1867 to 10.4 million metric tons (11.4 million tons) in 1900. The first American converter was built in Wyandotte, Mich., in 1864, but failed. Development was then delayed by a patent contest between American William Kelly (1811–1888) and Bessemer after the U.S. patent office granted Kelly priority. But if Kelly actually made steel before Bessemer, he never developed the process, and Bessemer's British patents were the real key to quantity steel production. American engineer Alexander Holley (1832–1882) was the key technical figure, for he helped build and operate almost every American Bessemer plant through the 1880s, inspiring an operational pattern stressing speed and quantity. "Hard driving," this was labeled, symbolized by such innovations as removable bottoms for converters that could be changed after every 25 "blows" in only 15 minutes.

Holley and engineers built the plants, but Andrew Carnegie (1835–1919) best realized the business potential of the new technology. He entered the iron industry in the 1870s and moved into steel after Holley had solved the main problems. Carnegie's strategy centered on always being the lowest-cost producer. A ruthless businessman, he forced this style on the industry, a style that matched the drive of the technicians who made evolutionary improvements at his mills. Like other American Bessemer steel makers, Carnegie made rails, paying little attention to other uses until the 1890s. Quantity, not quality, mattered, and his companies dominated the American scene until they were made part of the United States Steel Corporation in 1901, then the largest business firm in the world.

This pattern led to the appearance of other enduring elements, especially a sense of workers as interchangeable cogs in a giant machine. American steel makers deliberately segregated different ethnic groups by job, disrupting communications that might lead to unions. And while the most physically demanding jobs were soon mechanized, the hours labored by steel workers rose while wages fell throughout the late 19th century. The 12-hour day and the long turn (a 24-hour shift) every other week, when workers moved from days to nights, remained a fixture in the American industry until the industry was shamed into ending it in the mid-1920s. Still, the comment of British crucible steel maker Harry Brearley should never be forgotten:

> That there are workmen, in any steelmaking district in the world, who command astonishingly high wages, may be taken as an indication that, in spite of every available academically devised aid, there are jobs depending on the exceptional judgment of the doer. . . . The importance of making good steel, for its own sake, must be in the mind of the workmen, because it is the only reliable guarantee the steel has been well made.

He also noted: "I hope it will not be taken amiss if I say that workmen are often much wiser than their masters."

Bessemer production dominated, but there were other ways of making steel by the end of the 19th century. First, cast-crucible-steel output climbed in England and elsewhere, for this method produced the highest quality alloy steels for tools and special purposes. But crucible steel was eventually displaced by another new development, the Siemens/Martin or open-hearth process. Much less dramatic than Bessemer's technique, iron ore and scrap metal were melted in furnaces that made steel in hours, not minutes. Tight control of temperatures allowed better control of the quality and character of the resulting steel. By 1910, open-hearth steel had surpassed Bessemer steel in total output.

Steel making was always hard and dirty, and workers suffered enormously. But the violent, opposition of American firms generally kept unions out until the 1930s. Moreover, the environmental consequences of steel making were always evident in dirty mill towns; open-hearth furnaces in particular left a coating of red dust everywhere. Yet steel mills were among the most important facilities in every country, be they those of Krupp in Germany, Schneiders in France, Coalbrookdale in England, Carnegie or Bethlehem in the United States. Moreover, this industry held a special fascination for many people, a fascination that highlights the importance of enthusiasm for the growth of technology. It is captivating to watch steel being made, especially at night. The hustle and bustle of American society, the cultivation of size for its own sake, and the enthusiastic exploitation of machinery as almost an end in itself can all be found here. Men—and the steel industry has been a bastion of the male workplace until quite recently—took pride in wresting steel from nature. As well as anyone, Stewart Holbrook captured what this was all about in *Iron Brew* (1939):

> One of the converters was in blow as we entered the shed. Tilted almost but not quite straight up, the mouth of it belched flame like a cannon built for the gods. It was a terrifying sight, and hypnotic. . . . The roar was literally deafening; and little wonder, for here was a cyclone attacking a furnace in its brief but titanic struggle. . . . The roaring continued. The red fire changed to violet, indescribably beautiful, then to orange, to yellow and finally to white, when it soon faded. . . . I saw the great vessel rock uneasily on its rack. . . . A locomotive pushed a car close under. The hellish brew was done.
>
> Slowly the converter tilted over, and from its maw came a flow of seething liquid metal—Bessemer steel. A Niagara of fire spilled out, pouring into the waiting ladle, and sixty feet away the heat was too much for comfort. A cascade of sparks rolled out over, a sort of spray for this cataract, and it seemed everything in the shed danced with light. . . . It is the most gorgeous, the most startling show that any industry can muster, a spectacle to make old Vulcan's heart beat faster, enough to awe a mortal. No camera has ever caught a Bessemer's full grim majesty, and no poet has yet sung its splendor.

No one has witnessed this spectacle in America since 1968, when the last Bessemer converter was retired. But it was a pillar on which industrial society was built. Both the excitement and the patterns of production and organization created by the development of steel exemplified the development of large-scale technology at the end of the 19th century.—B.S.

See also CAST IRON; IRON; IRON, WROUGHT; STEEL IN THE 20TH CENTURY; TECHNOLOGICAL ENTHUSIASM

Further Reading: Theodore A. Wertime, *The Coming of the Age of Steel*, 1961. K. C. Barraclagh, *Steelmaking, 1850–1890*, 1990.

Steel Frame Building Construction— see Building Construction, Steel Frame

Stereophonic Sound

The prefix *stereo* comes from the Greek *stereos*, which means solid, or three dimensional. And indeed, even though stereophonic sound has come to imply the use of two loudspeakers (as opposed to one loudspeaker for monophonic sound, four for quadraphonic, and five or more for surround sound), the original intent of stereophonic sound was to recreate the impression of a three-dimensional acoustical space in front of the listener, complete with depth and height of auditory image, and a replication of the placement of instruments along a stage from left to right. The original experiments with stereophonic sound at Bell Laboratories that began in 1931 used three channels; stereo sound in movie theaters had four channels. But when stereo sound became commercially available in the 1950s, only two channels were used. The chief reason for reduction to two channels was that the LP-record format could not easily accommodate more than two channels. Two channels were, however, enough to afford a qualitative improvement in realism, and stereo quickly became the standard format for commercial recordings.

Normal auditory space perception is based largely on the differences in time of arrival and intensity of a sound when it arrives at the listener's ears. A sound coming from a point directly in front of a listener will arrive at the two ears simultaneously and at the same intensity. Consequently, any sound striking both ears at the same time and at the same intensity can be localized directly in front of the listener. If the sound arrives at one ear sooner than at the other, it will be localized in the direction of the ear where it is first heard. Likewise, if it is louder at one ear than at the other, it will be heard on the side of that ear, because the sound that has traveled around the head to reach the other ear will be somewhat attenuated when it reaches that ear.

Accordingly, the auditory system uses intensity and time cues to localize sounds in space. However, localization requires additional cues, and the processing of that information is quite complex. For example, a sound arriving at the same intensity and time at both ears is not only consistent with a source directly in front of the listener but also with any other source that could be located in the saggital plane (i.e., in the plane passing vertically through the center of the body and dividing the left and right sides). Additional cues therefore are needed to determine whether the sound comes from the front, the back, or above the listener. These cues are most likely echoes and room reflections, as well as the acoustical signals that are subtly modified as they enter the ear from different directions.

Stereo sound effects derived from loudspeakers can simulate the intensity differences between the two ears, but they can only with the greatest difficulty recreate time differences, and only then for a person whose head is carefully positioned in an exact point in the sound field created by the speakers. Even the intensity differences will be correct at only one point in space.

Stereo recordings can be made with a single pair of microphones, or with many microphones whose output is mixed into the two channels of stereo. How to place the microphones and combine their output is as much art as science. If the placement and combination have been done properly, the illusion created by a pair of speakers can be rather convincing.—W.B.

Further Reading: Ian R. Sinclair, ed., *Audio & Hi-Fi Handbook*, 1995.

Sterilization

All temporary methods of preventing conception have their drawbacks. Most require some advance planning and preparation, many carry some health risks, and all are less than 100 percent effective. This situation will not change soon, for there are no immediate prospects for a breakthrough in birth control technology. Worried about lawsuits that would likely accompany the marketing of a new product, most pharmaceutical firms have abandoned contraceptive research. In the absence of a perfect means of avoiding pregnancy at a given time, many men and women have opted for a permanent means of avoiding pregnancy: sterilization.

Every year, approximately 700,000 women in the United States undergo a tubal ligation, an operation that renders them permanently incapable of conceiving. Requiring only a half-hour of surgery and often performed under a local anesthetic. a tubal ligation entails closing off the fallopian tubes, the ducts through which ova pass from the ovaries to the uterus. Many of these operations make use of a relatively new surgical technique known as a *minilaparotomy*. In this procedure, an instrument is inserted through the cervix so that the uterus can be moved into a position that makes the fallopian tubes more accessible. The tubes are then sealed through the use of heat or an electric current, or they are tied off with sutures, rings, or clips. A surgeon also may also gain access to the fallopian tubes by inserting an instrument through a specially equipped laparoscope, a device that allows viewing the abdominal cavity. Alternatively, this procedure may require making a separate incision for the insertion of the instrument.

Although sterilization is properly viewed as permanent, in some cases it is possible to use reversal procedures to restore fertility. Sterilizations that have been done with rings and clips are the easiest to reverse; those done with an electrical current the most difficult. Reversal procedures are effective only about half the time and are expensive. The decision to undergo sterilization is therefore a serious one, and the operation should be performed only on women who are certain that they do not want to bear children in the future. Most of the women choosing to be sterilized are over 30 and have completed their families, but increasing numbers of younger women are now undergoing sterilization. It remains to be seen if the number of reversal operations will increase in the years to come.

The sterilization procedure used for men is known as a *vasectomy*. It entails making a small incision in the scrotum, severing the vas deferens (the ducts that transport sperm), then closing it off. The closure blocks the passage of the sperm, which are eventually absorbed into the lymphatic system. As with women undergoing tubal ligations, vasectomies are most often performed on men who are at an age when more children are not desired. About 15 percent of American men over the age of 40 have undergone vasectomies. Although vasectomies are simpler operations than tubal ligations, the latter outnumber the former by up to 300,000 a year.

Neither male nor female sterilization should have any effect on sexual desire or performance. Nor has either procedure conclusively been shown to produce serious health consequences. Some studies have indicated a possible higher risk of prostate cancer for men who have undergone vasectomies, but on the other hand there is some evidence that men who have had vasectomies enjoy a slightly lower likelihood of developing coronary artery disease. For women, having undergone a tubal ligation actually decreases the risk of contracting ovarian cancer.

Although both male and female sterilizations are done only on willing patients today, in the not-too-distant past they were often performed on unaware or unwilling men and women who were deemed genetically inferior. Sterilization was viewed as a major weapon in the effort to improve the genetic stock, and it was employed widely and often indiscriminately.

See also CONTRACEPTIVES, ORAL; EUGENICS; INTRAUTERINE DEVICE; RU-486

Steroids, Anabolic

As a general category, steroids comprise a wide variety of fat-soluble organic compounds. Male and female hormones, adrenaline, cholesterol, and vitamin D are all steroids. First synthesized in 1953, anabolic steroids are used for the treatment of a variety of diseases, notably rheumatoid arthritis, skin and eye disorders, and allergies. They also are injected or taken orally in the hope that they will increase muscle mass and strength.

Anabolic steroids are synthetic versions of the male hormone testosterone. *Anabolic* refers to a constructive bodily process whereby simple substances are converted into complex products like muscle tissue. Natural testosterone combines anabolic effects with androgenic effects, i.e., the development and maintenance of male sexual characteristics. Synthetic steroids maximize the anabolic effects and minimize the androgenic ones.

By themselves, high dosages of steroids do not stimulate muscular growth. They do serve to increase the protein synthesis that occurs as a natural result of physical training. To be effective, therefore, steroids have to be used in conjunction with a vigorous workout program and a high-protein diet. While anabolic steroids increase protein synthesis, much of their effect may be the result of the body's increasing its production of creatine phosphate, which allows the muscles to work harder during training. Consequently, gains in strength and bulk may come largely from more vigorous workouts, and only secondarily from the physiological actions they generate.

Whatever their source, the effects of the intake of steroids tend to sharply diminish after a month or two of use unless dosages are continually increased. Even this may not be enough, as the body's increased production of serum cortisol may counter the anabolic effects of the steroids. A cessation of steroid use may be accompanied by a diminished production of naturally occurring hormones, and a consequent loss of strength and muscle mass.

While steroids contribute to the growth of muscle mass and strength, there is no clear evidence that they significantly contribute to speed, endurance, and overall athletic performance. At the same time, the use of steroids has been associated with a number of physiological and psychological problems. A person taking steroids may suffer from depression and mood swings, and may exhibit the aggressive behavior patterns known colloquially as "roid rages." Steroid use also may do permanent damage to the liver and kidneys, and may contribute to atherosclerosis (the thickening and hardening of the aorta). Men may suffer testicular shrinkage, a severe drop in sperm count, and loss of scalp hair. Women who use steroids may undergo clitoral enlargement as well as a permanent deepening of

the voice and growth of facial hair. Steroid use is particularly hazardous for teenagers and pre-teens, as it may cause a premature closing of the ends of the bones, resulting in a permanent stunting of growth.

The negative consequences of steroids have led to their being banned by the International Olympic Committee and by many amateur and professional sports associations. In order to enforce antisteroid regulations, many of these organizations require that athletes submit to random urine tests. In 1988 the U.S. federal government enacted a law that made it illegal to distribute anabolic steroids for nontherapeutic purposes.

These restrictions have not put an end to steroid use, for many steroid users are more interested in physical appearance than athletic performance. Young people in particular have been major consumers of steroids; half of the million estimated steroid users in the United States are teenagers. One survey conducted in 1991 found that 1.4 percent of high school seniors and 2.4 percent of the male seniors in the United States admitted to having used steroids during the previous year. It has been estimated that more than half of the steroid users began before they were 16 years of age.

The abuse of anabolic steroids for physical enhancement has led some medical professionals to view it as "reverse anorexia." While some people have starved themselves in an effort to achieve an extreme state of slenderness, others have turned to steroids in order to acquire a physique that, in its own way, is no less abnormal. Both groups put themselves at considerable risk in their pursuit of body types that appear more often in the mass media than they do in the real world.

See also CHOLESTEROL; HORMONES; VITAMINS

Stethoscope

Since at least the time of Hippocrates (5th-century B.C.E.) physicians have known that sounds within a patient's chest provide important clues concerning the functioning of the heart, lungs, and other bodily organs. Until the 19th century, however, most physicians paid little attention to the sounds generated in the thoracic region. Even those physicians with an interest in these sounds were hampered by the difficulty of hearing them. To do so required that the physician put his ear firmly on the patient's chest, a procedure that was inconvenient, uncomfortable, and embarrassing to doctor and patient alike.

In 1816, the French physician René Théophile Hyacinthe Laënnec (1781–1826) surmounted these difficulties through the simple expedient of rolling some sheets of paper into a cylinder, and then placing one end of it on a patient's chest while he put his ear to the other. Laënnec subsequently made an improved model that consisted of a 1-foot-long (.3-m) piece of wood with a longitudinal hole in its center. He called it the *stethoscope*, derived from the Greek words for *chest* and *I view*. Laënnec also wrote a lengthy textbook that familiarized many physicians with the stethoscope and its use.

For all their advantages, early wooden stetho-

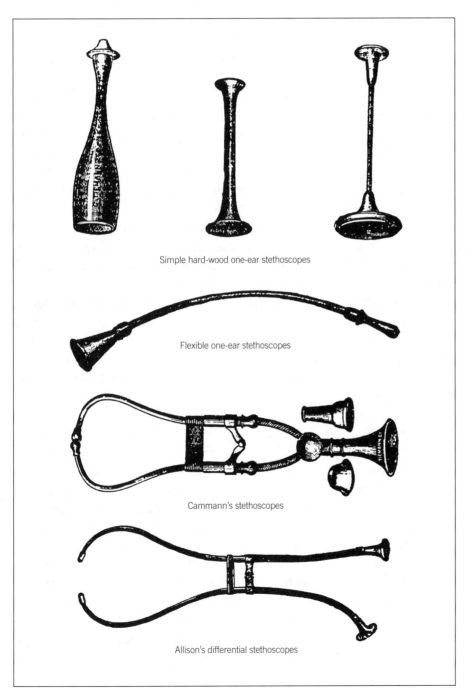

Simple hard-wood one-ear stethoscopes

Flexible one-ear stethoscopes

Cammann's stethoscopes

Allison's differential stethoscopes

Stethoscopes from the late 19th century.

scopes were not the easiest instruments to use. The situation was improved in 1829 when a stethoscope with a jointed middle section was invented, followed by the invention of fully flexible versions in the 1830s and 1840s. In 1851, Arthur Leared devised the first practical binaural stethoscope, which helped the listener shut out extraneous noise, while at the same time aiding in the hearing of faint sounds in the chest.

The stethoscope and the diagnostic techniques associated with it were not rapidly accepted. Many physicians resisted use of the stethoscope because it required the acquisition of new skills and changes in long-established ways of practicing medicine. Some doctors objected to the use of any sort of instrument, fearing that use of devices like the stethoscope would make them appear to be mere artisans. In those days, surgeons—practitioners with lower status than physicians—were the ones who wielded instruments. Traditionally minded physicians continued to base their diagnoses on direct observation and patients' description of their symptoms. Diagnoses done in this manner were subject to many errors, for different diseases could manifest the same symptoms, and the same disease might produce different symptoms in different people. In contrast, the sounds transmitted through the stethoscope could be directly apprehended and connected with particular illnesses. In the years following the invention of the stethoscope, the validity of this method of diagnosis was repeatedly confirmed by autopsies conducted after the patients had died.

Opposition to the stethoscope was a rear-guard action, and its use came to be a fundamental part of medical practice. But as with many other medical technologies that promised objective results, the use of the stethoscope led to diagnostic techniques that discounted the importance of other symptoms, especially those that were less objective in character. At the same time, the stethoscope did not produce uniformly objective results, for in actual practice the effective use of the stethoscope required some subjective interpretation of the sounds heard through it. Although the stethoscope is still a mainstay of medical diagnosis, there is some irony in the fact that its importance has been diminished by the appearance of new medical instruments (e.g., X-ray machines) that promise even more objective data.

See also SPHYGMOMANOMETER; X RAYS

Further Reading: Stanley Joel Reiser, *Medicine and the Reign of Technology*, 1978.

Stirrup

Some inventions are not as obvious as they seem to be. People have ridden horses for thousands of years, yet for much of that time they relied on nothing more than a saddle and reins to keep them mounted. The idea of using the legs and feet to hold on to a horse seems to have originated in India, where by the 2d-century-B.C.E., riders were steadied by loops through which they slipped their big toes. True stirrups were being used in China by the early 4th-century C.E. From there they diffused westwards, arriving in Europe by the 9th century.

The primary use of early stirrups was to aid riders as they climbed into the saddle. It was eventually realized, however, that stirrups also gave mounted warriors a substantial military advantage. A securely mounted rider could do more than shoot arrows or hurl spears. Equipped with a lance or spear, he could charge an enemy and deliver a blow that had the full power of the horse behind it.

The vastly improved capability of the mounted soldier had ramifications that went beyond the military realm. According to a controversial hypothesis by Lynn White, a historian of medieval technology (*Medieval Technology and Social Change*), mounted combat did more than change the nature of warfare; it also altered the structure of medieval society. According to White, the substantial costs associated with knightly combat—horses, weapons, armor, and prolonged training—split medieval society into two distinct groups: a small warrior class, and the large majority of the population who were primarily engaged in farming. The resulting economic and political order is known as *feudalism*.

Since its formulation in the early 1960s, White's thesis has been challenged on a number of fronts. Critics have pointed out that mounted shock combat was not the primary mode of medieval combat. The military campaigns of Charles Martel (688–741), the Frankish king who is credited with initiating the feudal order, were primarily centered on laying siege and conducting raids. Even mounted troops may not have engaged in massed charges. The Bayeaux tapestry (c. 1080) that celebrates the victory of William the Conqueror (1028–1087) at the Battle of Hastings depicts knights hurling their spears and lances rather than engaging in mounted charges. Critics have also argued that origins of feudalism were complex and represented much more than an adaptation to a new mode of combat. It has been pointed out that a feudalistic division of society between nobles and serfs existed long before the introduction of the stirrup; some elements can be traced back to the *latifundia* of Roman times.

The stirrup undoubtedly changed some aspects of military conflict, and its influence certainly was felt beyond the battlefield. However, to invest it with the ability to restructure a whole civilization is to push technological determinism beyond reasonable limits. But in fairness, it must also be pointed out that White never claimed that the stirrup in and of itself caused the creation of the feudal order.

Further Reading: Bernard S. Bachrach, "Charles Martel, Mounted Shock Combat, the Stirrup, and Feudal Origins," in *Studies in Medieval and Renaissance History*, vol. 7 (1970): 47–76.

Stoneware

Stoneware is one of the many varieties of ceramics. Developed in Asia and northern Europe, stoneware is best described as a hybrid. While it often resembles pottery, stoneware has many of the characteristics of porcelain, for it is made of refractory clays that vitrify at approximately 1,200° to 1,400°C (2,200°– 2,550°F). Whereas pottery is often porous, stoneware and porcelain are both resistant to liquids, including most acids. Fired at low temperatures, pottery (or earthenware) is fragile and will break when dropped, while wares made from stoneware or porcelain clay bodies are serviceable and sturdy. Together, pottery, stoneware, and porcelain constitute a ceramics trinity; before the late 20th century, most ceramics were made from one of these three classes of materials.

In the East, Chinese potters developed a flourishing stoneware industry during the Tang Dynasty (618–907 C.E.), producing the world's first celadon glazes on fine gray-white wares for various ceremonies. In Japan,

stoneware potters eventually catered to the tastes of their countrymen for asymmetrical, naturalistic designs by capitalizing on stoneware's distinctive characteristics. Whether in Korea or in Japan, items of stoneware incised with fish and lotus, coated with dark, iron-rich glazes, or painted in the earthy Oribe palette of green, brown, and white, were counted among the prized possessions in elite households. Stoneware's resistance to liquids and its durability (the results of firing at high temperature and glazing with wood ashes) also made it an ideal material for storage containers, and, in this form, the material saw widespread use.

In northern Europe, German potters, who had been making unglazed stoneware since the Middle Ages, developed methods for glazing their high-fired ceramics with salt in the late 14th century. Vessels crafted from highly refractive clays were glazed when the kiln men threw salt into the ovens; in the intense heat, this material volatilized to cover the wares with a glassy film. The appearance of German stoneware varied from region to region, and much depended on the vessel's intended use. Some wares, such as the Jakobakanne wine jugs made in 15th-century Siegburg, are undecorated, while other products, including the Bellarmine jugs of Cologne, are embellished with people's faces and detailed naturalistic moldings. Stoneware could be made into very refined products, as demonstrated by Johann Friedrich Böttger's experiments with red and black stoneware at Meissen in 1715, which, however, reached a deadend following his rediscovery of porcelain.

Immigrant craftsmen carried the Germanic saltglazed stoneware tradition to England in the 17th century and to the United States in the 18th century. In America, artisans used German methods and techniques to make a wide array of products for commercial and domestic use, from tiny ink pots to enormous butter churns. As the nation urbanized and industrialized in the late 19th century, managers in stoneware potteries added sewer pipes and chemical wares to their factories' product lines. Today, curious passersby at urban construction sites might look down and see dark brown shards, the fragments of old sewage systems—one of the many legacies of America's stoneware industry.—R.L.B.

See also PORCELAIN; POTTERY

Stoves, Cooking

The first stoves, made of cast iron, appeared in Europe as early as the second half of the 15th century. However, for another 4 centuries most families continued to use open, wood-burning fireplaces and attached ovens for heating and cooking. In America, the reason for the reluctant move from fireplaces to cast-iron stoves was primarily the prohibitive cost of the stoves, due to the expensive high-grade iron required for durable stoveplates and the transportation costs of those stoveplates in the sparsely populated new nation.

This situation began to change in the 1830s when Jordan Mott of New York perfected a process for casting stoveplates from remelted low-grade pig iron, which turned out to be more durable than iron made from higher-grade ores. Mott was also the first to manufacture complete stoves rather than just assemble parts from different iron producers. These two innovations lowered the cost of cast-iron stoves and made them more widely available. At first, though, some people were suspicious of using stoves in their homes. They were concerned that stoves emitted unhealthy hot air, their cooking temperatures seemed difficult to monitor, they required daily cleaning and frequent polishing to prevent rust, and their use required more training than the use of fireplaces. However, the advantages of the cast-iron stove in terms of fuel economy and heating ability, more convenient cooking height, and relative safety and portability slowly won over the critics. By the end of the Civil War, cast-iron stoves were common features in American homes.

The move from open fireplaces to cast-iron stoves was not the only development in the 19th century. Starting in the 1830s, the single stove, which served both cooking and heating functions, evolved into two separate entities. The cast-iron cooking stove became larger and gained rings specifically for pots, while the heating stove became smaller and was designed for better, more economical heating. The other important development was the growing variety of fuel sources for both cooking and heating stoves. Coal increasingly replaced wood as the principal fuel source because it became easier to acquire and cheaper to transport, while wood became scarcer and more expensive. Gas, which was already being used for lighting and indus-

trial applications, also emerged as a fuel source. Used as early as 1802 for cooking at home, gas was a significant improvement over wood and coal because it was clean burning, did not require carrying and hauling, and ignited instantly.

Gas began to overtake coal as the primary cooking fuel around 1920. Intensifying competition between gas companies and rising electric companies helped fuel the change. To remain competitive in the utilities market, where they were rapidly losing lighting revenue, gas companies began lowering the cost of gas while sponsoring improvements in the quality and design of cooking stoves. Gas (or combination gas and coal) stoves were made more efficient and easier to clean, and the broilers were improved to emit less smoke. Also, gas stoves were designed to look more like elegant pieces of furniture, gaining slender cabriole legs and becoming lighter and more compact. During the 1930s and 1940s, manufacturers began producing cooking ranges with more burners, ovens located below the cooking surface, thermostats, and porcelain surfaces. By the 1950s, gas ranges were streamlined to match the height of the other kitchen work surfaces, creating the now familiar tabletop design.

Electric companies, attempting to keep up with the gas companies, developed and designed similar cooking ranges. Some restaurants had electric stoves by 1889, and 2 years later the first domestic electric stove was sold. However, electric stoves did not compete successfully with gas or coal stoves at first because few Americans could afford or even gain access to electric power. By 1907, only 8 percent of homes in the United States had electricity, although that figure increased dramatically over the next few decades. Not until about World War II did electric ranges compete effectively due to decreases in electricity rates and the development of reliable thermostats that allowed precise regulation of heat.

Since the 1930s, improvements of both gas and electric ranges—such as oven lights, safety locks, convenience compartments, and better surface units for faster and more even heat—have developed simultaneously. One of the more recent innovations in gas ranges is the adaptation of an electric ignition instead of a standing pilot flame to make the ranges safer and more energy efficient. As long as the kitchen remains the central focus of the home, gas and electric manufacturers will continue to compete by introducing innovations in kitchen cooking ranges and kitchen appliances in general.—M.S.

See also CAST IRON; COAL; ELECTRICAL POWER SYSTEMS; LIGHTING, GAS; PORCELAIN; STOVES, HEATING; STREAMLINING; THERMOSTAT

Further Reading: Ruth Schwartz Cowan, *More Work for Mother: The Ironies of Household Technology from the Open Hearth to the Microwave*, 1983. Susan Strasser, *Never Done: A History of American Housework*, 1982.

Stoves, Heating

Heating, as applied to human comfort, has been a concern as long as people have existed. Archeological evidence indicates that ancient human beings warmed themselves with campfires as long ago as 1.5 million years. Warming by an open fire persisted for millennia, and more sophisticated methods did not appear until after 2000 B.C.E. By at least 1200 B.C.E., a rudimentary form of central heating was being used.

Fireplaces as well as chimneys came into use in Europe and England during the 14th century. Within another century, fireplace grates and even the use of outside air for better combustion were introduced. Individual room heating by stoves began even earlier, dating to the 9th-century C.E. By the year 1000, stoves made of clay or tile were being used in Germany and Switzerland. Over the next 600 years, their use had spread to much of Europe, the stoves having evolved into large, elaborately decorated works of art. Many of these so-called Dutch, Russian, or Swedish stoves exist today, valued as museum pieces.

Early heating devices were crude, there being little understanding of the scientific principles underlying their functioning. Eventually, however, heating devices became the subject of scientific investigation. Greater sophistication in construction and use evolved. For example, a round metal heating stove was introduced by Kesslar in Germany in 1618. His stove featured controls for inlet air, a chimney damper, and improved heat transfer by use of a zigzag passageway for flue gases. The first text devoted to heating was possibly *La Mecanique du Feu* written under the assumed name of Gauger by Cardinal Polignac in 1713.

One of the most famous improvements made to stoves was Benjamin Franklin's (1706–1790) Pennsylvania fireplace, or Franklin stove, as it came to be known. Invented in 1742, it used a baffled airbox. Smoke moved in the airbox from the front of the baffle, spilled over the top of the baffle and then moved downwards past the rear of the box. It continued in a downwards direction past the rear of the baffle, and finally entered the chimney flue at ground level. This arrangement allowed the retention of heat that would otherwise have gone straight up the chimney, while keeping warm, fresh air flowing continually into the room being heated. The stove also had the advantages of producing little smoke and consuming modest amounts of fuel. The Franklin stove is also famous for its inventor's unwillingness to patent it. As Franklin explained: ". . . as we enjoy great advantages from the inventions of others, we should be glad of an opportunity to serve others by any invention of ours, and this we should do freely and generously."

During the 18th century, improvements in foundry technique led to more complicated stove construction and an ability to make the finished product more ornate. Like the earlier tile stoves of northern Europe, cast-iron stoves evolved into real works of art. The late 19th century was the zenith of ornate cast-iron stove manufacturing, with hundreds of manufacturers competing with elaborate and unusual designs. Original examples are avidly collected today, and some are still being produced today, particularly in Europe. Although stoves do not provide the all-around comfort of central heating, their simplicity and low cost ensure their continued widespread use. However, anyone using a heating stove must be sure that it is adequately vented; otherwise, an accumulation of carbon monoxide could be fatal to those in the room being heated.—B.N.

See also CENTRAL HEATING; PATENTS; THERMOSTAT

Strategic Arms Reduction Talks (START)

The nuclear arms control agreements that were negotiated in the 1970s limited the arsenals of the United States and the Soviet Union, but they did not win universal acclaim. In the United States, some conservatives believed that the treaties needlessly and dangerously weakened the United States and surrendered the initiative in world affairs to the Soviet Union. In some Republican circles, adherence to these treaties was portrayed as a Western sellout of morality and strategic advantage, and was even compared to the betrayal by Great Britain and France of Czechoslovakia with Hitler at Munich in 1939. Out of this world view emerged organizations such as the Committee on the Present Danger that tried to unite opposition to the Carter administration by presenting apocalyptic visions (later quietly abandoned) that the Soviet Union could at any moment attack U.S. nuclear forces without fearing an effective American response.

President Reagan began his administration on a strident note by calling the Soviet Union an "evil empire," while Secretary of Defense Caspar Weinberger spoke publicly of "winning" a nuclear war. The administration's arms control positions were equally unyielding: zero intermediate nuclear weapons for both sides ("zero-zero") and a 50 percent reduction in strategic nuclear forces. The leaders of the Soviet Union—Leonid Brezhnev and his two successors, Yuri Andropov and Konstantin Chernenko—refused to negotiate on such terms. This suited the hard-liners in the administration, but the lack of progress did not suit the President, who felt a deep aversion to nuclear weapons, or Secretary of State George Shultz, who hoped to achieve a better working relationship with the Soviets.

When the administration persisted in its tough line, it lost control of public opinion in the West: Anti-American demonstrations erupted in Europe, the House of Representatives voted for a nuclear freeze, and numerous town meetings around the country declared themselves in favor of nuclear-free zones. The negative impression was offset somewhat by the President's offer to the Soviet Union to share the technology of the Strategic Defense Initiative (SDI), or Star Wars, a plan for the use of lasers for a space-based, nationwide antimissile defense.

SDI offers a glimpse of the maneuvering on both sides that characterized START. To the President, SDI offered a way of avoiding Armageddon. To critics, such as Secretary Shultz and almost every American scientist in a relevant field, the idea of space-based

missile defense was a dangerous, costly absurdity that would accelerate the arms race. But to the American National Security Advisor Robert McFarlane, the threat that the United States could build a nationwide defense against strategic missiles would force the Soviet Union to accept a "grand compromise": In return for an American decision to scrap or share the technology, the Soviet Union would reduce its strategic nuclear forces to levels low enough to eliminate any advantage they might have, thereby making a defensive system unnecessary.

Reagan's faith in SDI worked to complicate and ultimately deny him the chance to reduce strategic nuclear weapons during his term of office. The controversy over SDI that pitted the executive branch against the U.S. Senate, coupled with Soviet suspicion and opposition, stalemated negotiations until the second Reagan administration came to an end. Only Mikhail Gorbachev's rise to power and his reformer's need for good relations with the United States allowed agreement to be reached in the very last weeks of the Reagan administration for 50 percent cuts in the strategic forces of both sides. This agreement was only "in principle," and it fell to the administrations of George Bush and Bill Clinton to actually carry out reductions in strategic nuclear weapons.

START I (signed on July 31, 1991) called for both sides to reduce their arsenals in three phases over 7 years to no more than 1,600 strategic nuclear delivery vehicles (SNDV) and 6,000 "accountable" warheads, of which no more than 4,900 may be on ballistic missiles. These amount to cuts from 30 to 50 percent in launchers and warheads. The United States renounced all heavy intercontinental ballistic missiles (ICBMs), of which it had none, and the Soviet Union agreed to a limit of 1,540 warheads on no more than 154 heavy ICBMs. Both sides agreed to a limit of 1,100 warheads on mobile ICBMs and to equal ceilings on ballistic missile throw weight (essentially, the warheads carried).

In line with the American view of strategic nuclear stability, the "counting rules" of START favor bombers carrying short-range missiles and bombs or longer-range, air-launched cruise missiles (ALCMs). The bombs and short-range missiles count as only 1 warhead regardless of the actual number on board, while 2 ALCMs count as 1: The first 150 U.S. bombers may carry up to

20 ALCMs, to be counted as 1 SNDV and 10 warheads. The first 180 Soviet bombers may carry up to 16 ALCMs, to be counted as 1 SNDV and 8 warheads. Thus, the 50 percent reductions agreed to "in principle" during the Reagan administration actually amounted to about 33 percent in practice. Other START I provisions allowed for reducing the number of warheads carried by certain existing types of missiles (downloading), limits on mobile missiles, and elaborate verification procedures, including numerous on-site inspections.

Gorbachev was nearly overthrown in an attempted coup d'état a few days after signing START I, and his subsequent downfall and the disintegration of the Soviet Union complicated the implementation of that treaty and delayed ratification of START II, which was concluded on Jan. 3, 1993. After 3 years of difficult negotiations, all the successor states of the former Soviet Union accepted the terms of START I.

START II, a much more ambitious undertaking than its immediate predecessor, calls for the reduction of strategic forces to 3,000 to 3,500 warheads within a decade, the elimination of all ICBMs with multiple warheads, the limitation of warheads on submarine-launched ballistic missiles (SLBMs) to 1,750 on each side, and the counting of all warheads one for one, regardless of their delivery system. While the treaty meets the test of nuclear stability by removing weapons that might be used in a first strike, its bias against ICBMs obliges Russia to undertake costly military modernization at a time when the country faces severe economic problems. Political pressures on both sides also impede early ratification. Although the U.S. Senate passed START II in January 1996, at this writing (early 1997) START II remains unratified.—P.E.H.

See also MISSILE, INTERCONTINENTAL BALLISTIC (ICBM); MISSILES, CRUISE; MISSILES, MULTIPLE INDEPENDENTLY TARGETED VEHICLES (MIRVs); MISSILES, SUBMARINE-LAUNCHED; SALT I AND SALT II; STRATEGIC DEFENSE INITIATIVE

Further Reading: The journal *Arms Control Today* frequently has articles on START, including START Supplement (Nov. 1991), START II Supplement (Jan./Feb. 1993), and START II Resolution of Ratification (Feb. 1996).

Strategic Bombing—see Bombing, Strategic

Strategic Defense Initiative

Intercontinental ballistic missiles (ICBMs) armed with nuclear warheads are weapons of immense destructive power. Each missile can travel thousands of kilometers in less than an hour and then destroy a substantial portion of a large city. With no means of stopping a catastrophic missile attack, civilian and military leaders developed the strategy of "mutually assured destruction" (MAD) in the 1960s. MAD meant that each nuclear power would launch an immensely destructive retaliatory strike against a country that had struck first. This situation has been described as a "balance of terror." On a more positive note, beginning in the 1970s, the United States and the Soviet Union attempted to diminish the nuclear threat by agreeing to certain limitations on the size and quality of their nuclear arsenals.

Treaties and MAD diminished the threat of a nuclear Armageddon, but it was an uneasy peace. Some military and civilian leaders hoped that the development of an effective antimissile system would eventually provide more protection than mutually assured destruction or negotiated limitations on nuclear weapons. Although both the United States and the Soviet Union had given up the attempt to build antimissile systems in the 1970s, the election of Ronald Reagan gave new life to the hope of destroying ICBMs before they reached their targets.

The Reagan administration came to power in 1981 with a more confrontational attitude towards the Soviet Union than had been the case with past administrations. Many Americans were unsettled by the deteriorating relationship between the two superpowers and feared that the danger of a nuclear war was rising. One response to the deteriorating international climate was the growing support in the United States for a freeze on the production, testing, and deployment of nuclear weapons.

With antinuclear sentiments on the increase, President Reagan came forward with a plan that promised to eliminate a major nuclear threat by destroying ICBMs before they reached their targets. On Mar. 23, 1983, the President used a televised speech to outline his plan for making offensive missiles "impotent and obsolete." The Strategic Defense Initiative (SDI), as he called the program, would use advanced technology to shoot down incoming missiles. The speech came as a surprise even to high-ranking members of his own administration, for the President announced the SDI program without prior consultation with the Joint Chiefs of Staff, the Secretary of State, or the Secretary of Defense.

Although the technical details of the proposed program were sketchy, SDI—or "Star Wars," as its opponents called it—was subsequently approved by the U.S. Congress. Research began in 1984, and the system was approved for acquisition in 1987. In 1988, Congress approved phase I of the system. Had it been built to completion, phase I would have been very expensive, costing an estimated $69.6 to $87 billion, plus an additional $10.1 billion for operations and support prior to the completion of deployment. However, phase I, even if it had worked as designed, would not have provided complete protection against a massive missile attack; its main purpose was to demonstrate the feasibility of the technologies involved.

Many critics of SDI were unconvinced that the system would perform as advertised. In 1990, these concerns were reinforced by the federal government's General Accounting Office's (GAO) review of the SDI program. In its report, the GAO concluded that deployment of phase I by the 1993 target year would be "premature and fraught with high risk." By this time, the technological underpinnings of SDI had come under intense scrutiny. In particular, the efficacy of key SDI components known as "directed-energy weapons" was very much in doubt. Directed-energy weapons were intended to shoot down attacking missiles by firing intense beams of particles like hydrogen ions from the ground or from orbiting satellites. In 1990, Congress eliminated funding for this component of the SDI system. This decision was somewhat moot, for SDI proponents had shifted their attention to a new program known as "brilliant pebbles," thousands of tiny, independent weapons that would permanently orbit the Earth until they were called upon to repel a missile attack.

Although the United States is still a long way off from actually deploying an anti-ICBM system, SDI has made significant claims on the defense budget, absorbing about $36 billion from 1984 to 1996. The election of Bill Clinton in 1992 meant that SDI no longer had an advocate in the White House, but by this time the program had accumulated many supporters in the armed forces and the defense industry. The collapse of the Soviet Union removed the major rationale for the program's development, but SDI supporters now claim that an antiballistic missile system is needed to protect the United States from attack by a "rogue nation." This justification for an anti-ICBM system has failed to convince critics of SDI, for even in today's tumultuous world there exists no nation with nuclear weapons, the means to deliver them, and a leadership foolish enough to risk a massive counterattack by the United States.

In 1993, the Strategic Defense Initiative was renamed the Ballistic Missile Defense Organization. Spending on the program, while slightly scaled down from Reagan-era budgets, is still substantial. The Clinton administration plans to provide $3 billion in annual funding for antimissile research until the end of the decade. There are, however, no plans for actual deployment.

Support for an antimissile program continues despite serious questions about its workability. To be effective, a missile defense system has to be able to detect incoming missiles and then distinguish them from the many decoys that an attacker is likely to put into the air. This task is made particularly difficult by the existence of warheads that contain several independently targeted "reentry vehicles." After detection and discrimination, an anti-ICBM system has to track its targets and destroy them. This all has to happen in 35 minutes or less, and the system has to be close to 100 percent effective, for any missile or warhead that gets through is capable of causing immense damage on the ground.

At the heart of any effective antimissile system is a network of computers. This network is controlled by computer programs containing 40 to 100 million lines of code. It is not likely that such an extensive set of programs could be completely error-free, and there is no way to test the system in its entirety. It would have to work flawlessly the first time it is tried—that is, in the event of a missile attack on the United States. The performance of many recently developed high-tech weapons gives little cause for encouragement on this score. It also has to be kept in mind that even a perfectly functioning system would have no effect on other forms of nuclear attack. Low-flying cruise missiles would not be affected, and an antiballistic missile defense system would be of no help if a terrorist organization succeeded in smuggling a nuclear bomb into the center of a large city.

See also MISSILE, INTERCONTINENTAL BALLISTIC (ICBM); MISSILES, CRUISE; MISSILES, MULTIPLE INDEPENDENTLY TARGETED REENTRY VEHICLES (MIRVs); MISSILES, SURFACE-TO-AIR; STRATEGIC ARMS REDUCTION TALKS

Streamlining

Streamlines are the lines that sweep around an object as it moves through a fluid. The practice of streamlining refers to the shaping of a body to reduce the resistance of the air or water through which it moves. A streamlined body is characterized by laminar flow, i.e., one in which the fluid moves in orderly, parallel layers around the body. The absence of laminar flow is manifested by the generation of wakes, vortices, and eddies as the fluid separates from the body's surface. Streamlining has been of particular importance for the development of aviation, but its influence has extended well beyond the design of aircraft. More than a set of principles for achieving efficient movement, streamlining has been employed as the basis of good design, used as a marketing tool, and invoked as an emblem of modernity.

The first experiments in streamlining were conducted in the 18th century in an effort to improve the design of ships' hulls, but these had little or no influence over contemporary practice. The theory of streamlining began to be developed in the early 19th century, before it had much applicability to the design of actual objects. In 1809, George Cayley (1773–1857), an Englishman who pioneered the systematic study of heavier-than-air flight, measured the girth of a trout at regular intervals. These measurements were then converted into a set of mean diameters that were used to form a wooden cylinder. When split lengthwise, the form could be used to model the optimal shape for the hull of a ship. Cayley also made the important observation that shape of the trailing end of an object was at least as important as the shape of its front, for turbulence at the rear produces a region of low pressure that impedes the object's forward motion.

Empirical inquiries into streamlining were greatly aided by the invention of the wind tunnel in the 1870s. By injecting smoke or some other visible substance into the air stream, it was possible to observe the flow of air around various objects. In this way, actual streamlines could be seen or photographed. Wind tunnels were important tools in the development of the airplane, and not coincidentally the airplane was one of the first objects to be consciously streamlined. Beginning in the late 1920s, the reduction of air resistance went hand in hand with key innovations like monoplane design, all-metal construction, and the development of more powerful engines to effect substantial improvements in aircraft performance.

Other means of transportation were quick to imitate the streamlined airplane. Efforts to build wind-cheating trains went back to the late 19th century, but it was not until the 1930s that the railroads, feeling the pressure of airplane and car competition, began to run streamlined trains. Some of these trains were headed by new diesel-electric locomotives, but many others were pulled by conventional steam locomotives that had been streamlined by the addition of specially designed shrouds. These efforts reduced air resistance considerably, but contributed only modestly to the speed and fuel economy of the trains. Even so, many of the locomotives and the cars they pulled were beautiful expressions of the art of industrial design, and they conveyed a sense of modernity to a public that had viewed railroads as distinctly old-fashioned.

Production automobiles began to be cautiously streamlined in the 1930s, as rectangular forms were smoothed and rounded. The most adventurous was the Chrysler Airflow, the final design of which was the result of extensive wind tunnel testing. The Airflow was not a commercial success, however, due in part to its being ahead of its time, but also because its shape did not resemble the teardrop form that the public had come to identify with streamlining. A far more successful streamlined automobile was designed at about the same time by Ferdinand Porsche (1875–1951) in Germany: the Volkswagen.

Although there were good reasons to apply the principles of streamlining to transportation equipment, in the 1930s streamlining also became a design principle for fixed objects like buildings and appliances. "Streamlined moderne" emerged as a recognized architectural

A Boeing Model 200 Mono-mail, an example of a 1930s streamlined form (courtesy Boeing).

style, and in some cases inspired buildings that are aesthetically pleasing today. Other applications of streamlining, such as pencil sharpeners that looked like aircraft engine nacelles, bordered on the ridiculous. In many cases, streamlining was nothing more than a marketing gimmick, a way to convince potential customers that a refrigerator or washing machine was a great advance over older models, even though its actual performance represented scant improvement.

When streamlining emerged as a design principle in the 1930s, it also seemed to offer a vision of a better world that lay in the future. While people were grappling with the hardships of the Great Depression, streamlining held out the promise of prosperity through efficiency and productivity. On a more mundane level, by producing and marketing consumer goods that had been restyled in accordance with an aesthetic of streamlining, manufacturers hoped to win customers, boost spending, and defeat the Depression. Finally, streamlining also took on a symbolic meaning. In an era marked by frustration and blocked opportunities, streamlining presented an image of a world with much less friction and resistance. Streamlining in this sense entailed using rational, scientific principles to reform organizations and even whole societies. This vision of a streamlined society was a natural com-plement to technocratic approaches to government.

See also LOCOMOTIVE, DIESEL-ELECTRIC; LOCOMO-TIVE, STEAM; TECHNOCRACY; WIND TUNNEL

Further Reading: Donald J. Bush, *The Streamlined Decade*, 1975.

Streetcar (Trolley)

Urban conveyances that operate along tracks laid flush with the roadway are called streetcars or trolleys in North America, and tramcars or just trams in most other parts of the world. Power is ordinarily supplied via electricity from an overhead wire (with one rail completing the circuit), but occasionally from a wire in an underground conduit between the rails. Street railways remain commonplace in Eastern Europe and Asia, although they are relatively few in North America, and at one point in the 1970s were all but extinct. Yet historically they played a major role in the development of nearly all major North American cities.

Since colonial times, there had been public transit via "omnibus" (literally "for everyone") in various cities on the Eastern Seaboard. But streets were often unpaved, or paved with rough cobblestones at best, and there were obvious advantages to the combination of

A PCC trolley rounds a corner in a Los Angeles residential neighborhood (from The Trolley Calendar).

flanged iron wheels on iron rails, the essence of railroad technology. The first horse-drawn street railway vehicles appeared in 1831, on Broadway in New York City. The use of so-called *horsecars* (although often the beasts were in fact mules) extended to other cities gradually in antebellum times, and rapidly after the 1870s. By 1890, there were 90,000 km (56,000 mi) of track in the United States alone, and 28,000 cars using about four times that many horses or mules. In many cities, street railways had been part and parcel of significant spatial expansion, the development of "streetcar suburbs." But animal power was unsatisfactory for numerous reasons, and an active search was underway for improved modes of propulsion. Power from small steam engines was one proven possibility, as was propulsion via moving underground cables. Following its introduction in San Francisco in 1873, cable railway technology had spread rapidly, with cable cars operating along 500 miles of track in some three dozen cities by the latter 1880s. But cable railways were tremendously expensive to build and operate, and they were rapidly phased out after the practicality of electric propulsion was demonstrated, most notably by Frank J. Sprague (1857–1934) in the city of Richmond, Va., in 1888.

The population of the United States grew nearly 50 percent between 1890 and 1910, with most of this growth taking place in urban areas. These same 2 decades were the heyday of the trolley car as well. Cities spread out radially as street railway systems were expanded, and systems then expanded even further to accommodate a desire for residence away from the downtown congestion. Through the construction of a network of streetcar lines, it was hoped, urban problems could be literally left behind by workers commuting from their suburban residences.

But developing almost simultaneously with the technology of the electric streetcar was the technology of the automobile, and even by World War I the growing popularity of autos foretold the decline of "surface" public transit, that is, transit operating on city streets. Subways and elevated railroads operating on separate rights-of-way were not affected to the same degree. Many streetcar systems were in financial trouble by the 1920s, and this fate befell most of the rest during the Depression decade of the 1930s. If and when a system was expanded, it was with buses, which cost less and did not entail the high fixed overhead of trolley lines. Many smaller cities replaced trolley cars with buses altogether during the 1930s, and this trend spread to even the largest cities after World War II. The development of a relatively high-tech standardized streetcar design, the PCC car, by a consortium of transit company officials (PCC stood for President's Conference Committee) slowed but did not stay the attrition of street railways, which were abandoned wholesale in the 1950s. Although accusations have been made that this was the result of a conspiracy involving General Motors, the Standard Oil Company, and the Firestone Tire and Rubber Company, in reality it was largely due to seductive economic reasoning: The fixed costs of track and overhead-wire maintenance was a tremendous drain on finances, and the deferred state of routine maintenance had left many streetcar systems in sad physical condition. Even with bus substitution, transit companies could rarely show a profit, and in the third quarter of the 20th century virtually all urban transit systems were taken over by governmental agencies that were not compelled to keep the books balanced.

As for streetcars, by the 1970s they survived in only a handful of North American cities, and where they did survive they were mostly regarded as charming (or annoying) anachronisms. But technology is often driven by fad, even whimsy, and by the 1980s electric, railborne transit was being regarded in a different light. What had been seen as hopelessly old-fashioned a generation previously (a prime rationale for abandoning the last streetcars in Washington, D.C., was that no other major capital in the Western world still had such vehicles) was now perceived as something new. So-called light-rail vehicles—their technology little different from a PCC streetcar (and little different from tramcar designs commonplace in Europe)—began to appear in one North American city after another, and a perception grew that any city *without* such transit was old-fashioned. Hence there are now electric railways once again in cities such as Los Angeles, San Diego, Baltimore, Buffalo, Portland, St. Louis, Guadalajara, and Vancouver, where for decades they had been missing altogether.—R.C.P.

See also CABLE CAR; RAILROAD; SUBWAY

Submarine Cable

In the 19th century, the revolutionary capability of the telegraph to speed up communication gave rise to the desire to connect places separated by large bodies of water. The first such installation was the work of a professor of chemistry at Calcutta Medical College who had never even seen a telegraph. In 1838, he strung 22 kilometers of wire on bamboo poles and laid 3 kilometers of insulated wire under the Hooghly River. It worked, but because India was far from centers of telegraphy it remained largely unknown. In 1850, an underwater cable was used to span the English Channel, linking Dover in England with Calais in France. It was the first to use a superior insulating material known as gutta-percha, a substance made from the latex of a tropical tree. Unfortunately, within a few hours it was broken when fishermen accidentally brought it to the surface. The following year, a more substantial cable consisting of four wires, a gutta-percha insulating layer, and an iron rope sheath was laid across the English Channel. It remained in service for 37 years.

During the same decade, moderately successful efforts were being made to lay cables across the Mediterranean, motivated in large measure by France's desire to forge better communication links with her developing colonial empire in North Africa, and Britain's need to keep in touch with its Mediterranean naval bases. The most exciting project, however, was the Atlantic cable. An American, Cyrus Field (1819–1892), along with Britain's John Brett and Charles Bright and several financiers formed the Atlantic Telegraph Company in 1856 to lay a cable connecting Newfoundland with Ireland. After two failed attempts, the cable was successfully laid, allowing the exchange of greetings between Queen Victoria and President James

C.S. *Long Lines,* used by American Telephone and Telegraph to lay the longest portion of the first transoceanic fibre-optic cable (courtesy Lucent Technologies).

Buchanan on Aug. 13, 1858. But the signals soon weakened, and little more than 2 months later they stopped altogether. Part of the problem was that the cable was too light and not well insulated, resulting in its rapid deterioration. Also, a few years later, William Thomson (later knighted as Lord Kelvin [1824–1907]) discovered that long cables acted as capacitors, thereby smoothing out the distinct electrical pulses. A short reverse pulse sent immediately after the main pulse solved the problem. In 1865, a new cable was unreeled from the hold of the world's largest ship, the *Great Eastern,* and stretched across the bottom of the Atlantic. In the course of being laid it broke, but was reconnected the following year. Another cable had been laid 6 weeks prior to the successful splice, so Europe and North America were now connected by two cables.

The laying of submarine cables did not require massive engineering breakthroughs. What was needed were cables with conductors thick enough to prevent the attenuation of signals and insulation strong enough to protect the conductor. During the 19th century, the conducting wires contained anywhere from 98 to 162 kilograms of copper per kilometer. The insulation and shielding made the cables much heavier; depending on the application they weighed as little as 1 metric ton or as much as 13 metric tons per kilometer. Cable-laying technology also had to be developed. In 1873, Great Britain launched a cable-laying ship second in size only to the *Great Eastern.* It was equipped with special equipment to unwind the cable at the proper speed, all the while maintaining a true course no matter what the weather or ocean conditions. Special repair ships were also built to locate breaks in a cables, haul up the broken ends, and splice them together.

Submarine cables also motivated the development of a faster substitute for the Morse code, using positive and negative pulses in place of dots and dashes. Duplexing, which allowed for the simultaneous transmission of messages from both ends of the cable, was developed between 1875 and 1879. An automatic transmitter using punched tape allowed the sending of messages at a faster rates. By the 1890s, telegraph messages from Paris to London could be sent through the transatlantic cable via New York faster than they could be sent directly from one of the cities to the other. By the end of the 19th century, the Atlantic was spanned by 12 working cables, only one of which failed in the succeeding 50 years.

Many submarine cables primarily served commercial needs. At the same time, the colonial expansion of European countries was an important motivation for laying cables beneath the oceans. Rapid telegraphic communication facilitated centralized administration of far-flung empires. While a crisis such as a military conflict might temporarily increase the autonomy of local administrators, the overall trend was toward bureaucratic controls administered via the telegraph. However, while the power of the administrators was augmented, administration by remote control did not always produce good results.

A key element of communications in the 19th century, the submarine cable was augmented and partially eclipsed by new communications technologies in the 20th century. Long-distance radio communications emerged as a serious rival, especially in the 1920s when the potentialities of short-wave transmission began to be explored. Radio was cheaper than long-distance telegraphy, and it had the additional advantage of being able to carry voice transmission. The first transatlantic submarine cable capable of carrying voices (which required the installation of amplifiers, known as *repeaters*) did not go into service until 1956. A few years later, satellite communications links began to be developed, but submarine cables were not rendered obsolete. Although satellite-based systems play an important role in international communication, the dozens of cables lying on the floors of the world's oceans provide a reliable and less-expensive means of conveying messages.

See also CAPACITOR; COMMUNICATIONS SATELLITES; MULTIPLEXING; RADIO; TELEGRAPHY

Further Reading: Daniel R. Headrick, *The Invisible Weapon: Telecommunications and International Politics, 1851–1945,* 1991.

Submarine Communications

Long considered one of the most vulnerable areas of undersea operations, submarine communications have undergone extensive improvements since the 1970s. Even by standards applicable to the communications available to other weapons in wartime, submarines have attained reasonably reliable and effective communications.

World War I– and World War II–era submarines were extremely limited in their ability to send and receive information. Traditional Ultra High Frequency (UHF) or Very High Frequency (VHF) radio signals could not penetrate the ocean, forcing submarines to rise to the surface or to extend an antennae from just below the surface to receive or send messages. During that time, the submarine—lacking antiaircraft guns or radar—was vulnerable to attack by aircraft or even the guns of nearby surface ships. Under such conditions, transmission or reception of information was limited to the short time the vessel dared remain exposed.

With the advent of strategic, or ballistic-missile, nuclear submarines (SSBNs), the problems of "C3"— command, control, and communications—were exacerbated. Nations had to have assurance that SSBNs would not launch unauthorized attacks against enemy targets, and, conversely, had be confident that orders to launch such an attack would be received accurately. As the stakes rose to keep SSBNs "survivable," or hidden from enemy antisubmarine forces, traditional surface communications by UHF no longer were viable, although they were highly accurate and secure through the use of encoding equipment. Instead, navies installed extremely long antennae (in the case of the Trident SSBNs, over 1,500 ft) to trail behind the submarine and receive Very Low Frequency (VLF) signals from an aircraft circling overhead. These aircraft are known by the acronym TACAMO, for TAke Charge And Move Out. The TACAMO aircraft put into service by the U.S. Navy in 1964 were C-130 Hercules, which have undergone several updates and improvements since their introduction. TACAMOs generated a 200-kilowatt signal fed into the antennae of the aircraft, which dangled vertically as the airplane circled. But VLF communications only penetrated 30 to 50 feet of water, making it necessary for the subs to extend their own antennae near the surface. Both the aircraft and the antennae were vulnerable to various types of visual and radar detection.

To improve communications, the U.S. Navy as early as 1958 had conceived a system called the Extremely Low Frequency (ELF) transmitter, which was buried in the ground and generated a low, and very slow, pulse. ELF could not deliver complicated or rapid messages because it broadcast at only about 15 to 30 seconds per character, compared to the VLF antennae systems that permitted teletype communications. Nevertheless, ELF communications were secure from interruptions and jamming, and were all but impervious to any but a direct missile attack on the broadcast facility—itself a "message" that war had started. ELF facilities, however, faced resistance from environmental groups, and early attempts to establish an ELF facility in Clam Lake, Wis., and in Texas failed due to environmentalist pressures. However, Michigan agreed to serve as an ELF site, and in the mid-1980s the facility started to broadcast. At best, ELF remained a "bell ringer" to notify subs that an important message was awaiting on one of the other channels.

In a problem related to that of radio/voice communication, submarines traditionally have had problems establishing their exact position. Once again, position fixes required surfacing in order to make time-tested stellar navigational plots, and boats that surfaced for such fixes suffered from the same vulnerabilities as subs that exposed themselves for radio contact. Satellites brought new solutions to those problems, however. Originally called NAVSTAR Global Positioning System, the Rockwell-built system is today called simply GPS. It consists of a space-based radio positioning and navigational satellite constellation providing three-dimensional position data accurate within 10 m (32.8 ft), velocity information, and time to which military users can gain access. In 1988, the last of the 18 satellites in the constellation went into orbit at 20,100 km (12,500 mi) in altitude. The satellites broadcast information constantly to monitoring stations that correct the data for differences in clocks. Submarines with access to GPS can fix their position with great accuracy in a matter of minutes, proving especially useful to SSBNs that needed the accurate positioning for target data.

Recent experiments by the navies of the United States and the former Soviet Union focused on using blue-green lasers to transmit communications. Retaining the integrity of the laser beam remained a problem into the 1990s. In more exotic efforts, the Soviets even attempted to use "paranormal" or "extrasensory perception" transmission, apparently without success. Achieving clear and effective communications with underwater vessels remains a challenge for any nation that has a submarine force.—L.S.

See also GLOBAL POSITIONING SYSTEM; MISSILES, SUBMARINE-LAUNCHED; POWER LINES AND ELECTROMAGNETIC FIELDS; SUBMARINE, TRIDENT

Further Reading: Tom Clancy, *Submarine: A Guided Tour Inside a Nuclear Warship*, 1993.

Submarine, Trident

The Trident submarine program originated in 1966 as a result of a Department of Defense (DOD) study called STRAT-X, that, among other things, examined replacement systems for the Polaris/Poseidon submarines. The study investigated more than 125 different missile-basing systems for their ability to survive enemy missile attacks, and concluded that the new submarine would need a "brand new, 6,000-mile [9,656-km] missile." A submarine concept called the Underwater Long Range Missile System (UMLS), which incorporated the missiles into a new, if rather conservative, design specifically tailored to stealth and survivability, provided the first outline of the Trident submarine system.

Through 1970 and 1971, the U.S. Navy examined design possibilities, but ultimately, as Admiral Isaac Kidd, Jr., stated, "the missile sized the submarine." Consequently, early designs included a larger hull diameter (12.8 m [42 ft]) as well as additional length, although the navy conducted studies that compared the cost effectiveness of different hull lengths with the optimality of different numbers of missiles. Utilizing a new, larger nuclear reactor, the submarine could effectively carry 24 Trident I (C-4) missiles. The design contract, authorized in 1972, reflected the large size, with dimensions of 170.7 m (560 ft) in length and 17,100 tonnes (18,750 tons) of displacement.

Like most American submarines, the *Ohio*—the first of the Tridents—had a single screw and power train. Powered by a 90,000-horsepower nuclear-fueled

A Trident missile emerges from an undersea launch (courtesy U.S. Naval Institute).

GE S8G reactor, the *Ohio* had an officially stated speed of 46+ kph (25+ knots), but analysts estimate that it can attain up to 74 kph (40 knots) for short periods. But neither the speed nor the diving capacity of the Tridents constituted their main strength: instead, their capability for stealth was unmatched in the world. Using a combination of anechoic coatings to foil sonar, extremely quiet machinery, and advanced defensive sensors, the Tridents proved more difficult to locate and track than even the U.S. Navy had expected.

These improvements allowed the navy to change the requirements for the missile system in the 1980s. Originally planned to carry the Lockheed-built Trident II (D-5) missile with a range of 11,100 km (6,000 n.m. [nautical miles]) as an upgrade to the C-4 (7,400 km [4,000 n.m.] range), the D-5 provided greater operating range, adding to the submarine's ability to hide. Trident I C-4 missiles carried either seven or eight Mk-4 reentry vehicles, depending on the combination of range and warheads needed. The Trident II D-5 was originally planned to carry 8-10 Mk-12A reentry vehicles.

Construction on the *Ohio* started in 1974, and, after a number of labor difficulties and design problems, Electric Boat Company, the shipbuilder, delivered the first Trident to the navy in 1981. The navy anticipated procuring more than 20 of the submarines.

Electric Boat invested more than $450 million on a revolutionary ship welding system, and the Tridents were the first submarines built with the "continuous welding/circular welding" process.

During the construction process, a controversy erupted over a number of welds the navy quality-control supervisors found "defective," as well as other work they deemed unacceptable. Electric Boat responded by maintaining that the navy had foisted on it additional change items not included in the original contract, and the additional work had come at the expense of quality control. Eventually the Secretary of the Navy, John Lehman, resolved the controversy by threatening to withhold the contract for the 10th Trident submarine. Electric Boat agreed to pay half the overrun costs, and the navy paid half.

Two new submarine bases were established for the Tridents. The first, at Bangor, Wash., included a training center and a missile-handing facility specially tailored to the larger Trident submarines. Bangor provided Pacific Ocean access for the Tridents; a second base established at Kings Bay, Ga., permitted Atlantic access.

The Tridents were highly successful. Some analysts have credited the appearance of the Tridents with forcing the Soviets to abandon any notions of a "first strike" against the United States in the early 1980s. The exceptional undetectability of the submarines helped shift the strategic balance back to the United States in the 1980s—so successfully, in fact, that by the end of the decade the Soviet Union had started to unravel, in large part due to its unsustainable levels of military spending. That, in turn, allowed the navy to cap production (as of present) of the Tridents at 18. The end of the Cold War also allowed the navy to reduce the range requirements of the D-5 missile and to abandon plans to retrofit four of the Tridents (originally fitted with C-4s) with D-5 missiles.—D.D.D. and L.S.

See also MISSILES, SUBMARINE-LAUNCHED

Further Reading: D. Douglas Dalgleish and Larry Schweikart, *Trident*, 1984.

Submarine-Launched Missiles—see Missiles, Submarine-Launched

Submarines

Attempts to create underwater vessels began several centuries ago. As early as 1620, a Dutch inventor worked on a navigable submarine, and English and Spanish inventors sought to perfect diving bells during the next century. But not until 1775 was a submarine used for martial purposes. The craft was the *Turtle*, an egg-shaped underwater vessel created by an American, David Bushnell (1742–1824). During the Revolutionary War, it made an unsuccessful attempt to blast a hole in the HMS *Eagle* by placing a charge on its bow. Robert Fulton (1765–1815) proposed a submarine design to France, Great Britain, and the United States at the turn of the century. Fulton's sub, called the *Nautilus* after the submarine in Jules Verne's novel *20,000 Leagues Under the Sea*, impressed Napoleon, who provided some financing for it. Fulton never built his "sub" (as such vessels eventually were called) for Napoleon, and failed to sell the concept to the British or American governments. He had, however, established a principle and a goal, and had included some of the basics on modern submarines, such as a ballast system, horizontal and vertical fins for steering, and a propeller to push it underwater.

Those concepts found their way into the *Hunley*, a submarine constructed by the Confederate navy during the Civil War. The *Hunley*, despite unsuccessful trial runs in which it had sunk several times, conducted an attack on the Union ship *Housatonic*; it went down one last time with all its crew, but it took the *Housatonic* with it.

The goal of sinking surface ships by underwater attack still captivated naval designers, none more than an Irish rebel, John Holland (1840–1914). Holland, whose Irish background left him hostile to Great Britain, had made a series of sketches and calculations for a submarine when he immigrated to America in 1873. After he sent his designs to the Secretary of the Navy, Holland patiently waited for the bureaucracy to see their value. Eventually, a group of Irish-American rebels, the Fenians, funded his project—essentially envisioning it as a terrorist weapon. Holland built a model and, amid much publicity, it sank. Undeterred, Holland produced a second boat, the *Fenian Ram*, which verified the principles of submarine operation. Still, the U.S. Navy was not interested until 1883, when Holland built another sub, the *Zalinski Boat*.

Another decade passed until the navy awarded a contract for the USS *Holland*, at which time Holland himself had incorporated the Holland Torpedo Boat Company. The *Holland IV* made its first successful dive in 1898. A year later, Isaac Rice, president of Electro-Dynamic Company, the firm that had made the electric dynamo for the *Holland*, incorporated the Electric Boat Company, which soon acquired the assets of Holland's firm. Rice soon mated electric battery power to Holland's design, thereby allowing it to cruise under water for long periods of time.

When World War I erupted, the European powers all had submarines of varying sizes and designs. The German Navy had 29 *Unterseeboots*, or U-boats, and had come to the fateful conclusion that it could use subs effectively against British surface shipping. Powered by diesel engines, the U-boats early in the war fought on the surface, mostly with deck guns, and used the unreliable torpedoes only to sink a defeated vessel. Changes in surface defenses—especially "Q-boat" decoy ships—soon forced the U-boats to attack ships from below the surface, by stealth, using the torpedo as their primary weapon. This policy resulted in U-boat attacks on neutrals or passenger ships, as occurred with the sinking of the *Lusitania* in 1915 with the loss

The nuclear-powered missile-carrying submarine *Maine* (courtesy U.S. Navy).

of 128 American lives. "Unrestricted submarine warfare" became the most important cause of war when the United States joined the conflict in April 1917. American entry into the war may have sealed Germany's fate, but by the time the war ended, U-boats had accounted for the sinking of at least 11 million tons of Allied shipping.

Typical U-boats of 225 to 590 metric tons (250–650 tons) could make more than 14.8 kilometers per hour (8 knots) below the surface and more than 27.8 (15 knots) above. They could remain submerged for approximately 10 hours. When surfaced, they were vulnerable to air attacks or attacks from better-armed warships. Submerged, however, submarines proved difficult to detect and kill. No major alterations in submarine design occurred after the war, however, although numerous improvements in operations, speed, and lethality occurred.

By World War II, the navies of the United States and

Germany had determined that submarines had to play a key role in military operations. Germany lacked a significant surface fleet, while the United States in the Pacific had its major surface warships destroyed at Pearl Harbor in 1941. Both, therefore, made extensive use of submarines as commerce raiders. In the Atlantic, German U-boats sank 295,000 metric tons (325,000 tons) of shipping in April 1941 alone, seriously threatening the supply lines from the United States to Great Britain. During the 9 months prior to United States entry into war, Great Britain lost .9 million metric tons (1 million tons) of shipping to submarines. However, American Admiral Ernest King organized a new convoy protection plan, and British advances in sonar and depth-charge devices all but eliminated the German undersea threat by 1943. Japan, meanwhile, had made few preparations to use submarines, and quickly found its sea lanes vulnerable to American subs raiding in Japanese waters. By the time the atomic bomb was dropped in 1945, American submarines had eliminated all seaborne supply to the main islands of Japan.

In the 1950s, the Cold War between the United States and the Soviet Union brought a new era of submarine design in which nuclear reactors powered the vessels. In 1954, the United States launched the *Nautilus*; several classes of nuclear submarines followed. Nuclear power enabled subs to remain submerged for weeks and eventually months, making them even more undetectable than ever. In 1960, the USS *George Washington* fired the first ballistic missile from below the ocean's surface, bringing all enemy territory "under the guns" of the opponent's navy. By 1970, a full-fledged arms race was underway in submarine design and construction.

By this time, two major types of submarines had evolved. The first type, the attack submarine (in the U.S. Navy designated "SSN"), hunted and attacked surface ships and other submarines. A second type of vessel, the ballistic missile submarine ("SSBN"), was a strategic weapons platform intended to deter enemy attack by maintaining a survivable force at sea at all times. Eventually, cruise missiles were mated to submarines for a third category, the SSGN or cruise missile submarine.

Submarines grew quieter, less detectable, and, with the addition of better and more ballistic missiles, much more lethal. When it went into service in 1981, the lead vessel of the *Ohio* SSBN-726 class carried 24 multiple warhead ballistic missiles. A missile attack launched from single submarine could vaporize the major population centers of any nation on earth. At least four other nations have constructed SSBNs: France, Great Britain, China, and the Soviet Union, and most analysts agree that India also has the capability to put an SSBN to sea.

Although attack submarines could be used in most naval warfare circumstances, the ballistic missile subs' usefulness declined after the demise of the Soviet Union in the early 1990s. Consequently, planned production of Trident (*Ohio* class) subs was capped at 18 (with future modifications to reduce the number of SSBN *Ohio*s to 14), and no new ballistic missile boats were designed. Meanwhile, attack submarines, outfitted with cruise missiles, proved stable and effective platforms from which to attack Iraq in the Gulf War of 1990-91. Still, the future of expensive boats such as those remains in doubt; and as of 1995, the U.S. Con-

gress has struggled with the need to fund a *Seawolf* submarine, which even the navy has admitted is not necessary, in order to maintain the production capabilities to build new submarines in a future emergency.

Meanwhile, a total of 48 nations have submarine forces, including countries such as Albania and Syria. The total number of subs operating in 1993 was 705, of which 294 were nuclear powered.—L.S.

See also ENGINE, DIESEL; MISSILES, CRUISE; MISSILES, SUBMARINE-LAUNCHED; PROPELLER, MARINE; SONAR; SUBMARINE, TRIDENT

Further Reading: Dan Ven Der Vat, *Stealth at Sea: The History of the Submarine*, 1995. Antony Preson, *Submarines*, 1982.

Subway

A subway is a subterranean thoroughfare, most often a railway, for vehicles transporting passengers and, rarely, freight. Subway cars may emerge to run on the surface, or even on structures elevated above the surface, and typically they are stored in open-air yards. But a subway is distinguished from a tunnel because it comprises an entire underground system, with branching and intersecting lines as well as stations that must be reached by stairways, escalators, or elevators.

Certain subways, notably those dating from turn-of-the-20th century Boston, New York, and Philadelphia, were built largely by digging trenches and then roofing them over, a process called *cut and cover*. Others were built with the use of boring machines working completely beneath the surface, sometimes at considerable depth. This is the nature of much of the London system ("the tube," as it is known to Londoners), which in some places is 76 m (250 ft) below grade. It is likewise the case with many other systems built later in the century, such as those in Chicago and Moscow, as well as substantial portions of the system in Washington, D.C., where stations are reached by awesomely long escalators.

Whether or not the cut-and-cover technique is employed in construction depends a great deal on local geology. For example, Manhattan Island is rocky, making deep boring very difficult, while parts of Washington have clay substrates that are easily penetrated. The mode of construction also may be determined by the density of the subsurface utility network. Often there are so many pipes, electrical conduits, passageways, gas mains, sewers, and drains lying underground that it is more practical to leave them undisturbed and bore beneath them. Cut-and-cover methods are also enormously disruptive of the ordinary business of a city.

Construction in areas of groundwater incursion or beneath waterways may entail the use of airlocks to maintain pressures that are higher than normal atmospheric. The lining of subways may be in the form of an elliptical reinforced-concrete arch, sometimes boxed steel or concrete, often cellular. Stations, particularly on older systems, may seem claustrophobic, but those on newer lines are often capacious, even grandiose, especially when built with funds from the public purse and with a calculated effort to create showplaces.

As for propulsion, there were experiments in the mid-19th century with pneumatic power (similar in principle to the systems designed for conveying currency and receipts inside department stores). The first underground lines powered by steam locomotives opened for business in London in 1863. These used engines with condensing apparatus that deposited smoke in tanks pulled along behind. But any significant spread of subway systems was contingent on the practicality of powering trains by means of electric motors underneath the cars themselves, and of rendering each of those motors controllable by an operator stationed at the head end. In the 1890s, a workable form of multiple-unit (MU) control was devised by Frank J. Sprague (1857–1934), who is also credited with making the electric streetcar commercially feasible; indeed, the MU system may well be the more significant attainment. MU was first put into regular operation in 1897 on the South Side Elevated in Chicago. Elevated railways, running above normal traffic on surface streets, were another form of what became known generically as *rapid transit*. These lines appeared in New York as early as 1870 and became a dominant vista in several of the largest U.S. cities for much of the 20th century. They were eventually deemed a nuisance, and many have now been demolished, although they may still be found in Chicago, which has recently rebuilt portions of its storied "Loop," as well in a handful of other cities.

The first successful underground electric railway was the City and South London, running from King William Street beneath the Thames to Stockwell, a distance of about 4.8 km (3 mi). It opened in 1890. The first line in the United States, which ran beneath Tremont Street in Boston, opened 7 years later. As anticipated, it greatly alleviated surface congestion. Previously, Tremont had often seen the passage of no less than 400 electric streetcars hourly. The subway line simply routed streetcars underground, so that they were freed from mutual interference with other traffic. But the next line that opened in Boston was designed for MU trains that loaded from platforms at the same level as the floors of the cars, a much more expeditious way of getting large numbers of passengers on and off. "High level" loading became one of the primary attributes used in distinguishing "true" rapid transit from other forms of urban railway, including what is now called "light rail transit."

Boston was the site of the first American subway, but New York City is where subways burgeoned during the early 20th century into a network that remains the world's largest, with stretches of four-track line and thousands of trains transporting millions of passengers daily between outlying boroughs and Manhattan. The subways built in Boston and New York City at the turn of the century, as well as those in Philadelphia, were a matter of private enterprise; for a very brief period of time, there actually were profits to be made in operating mass-transit systems. But that situation was short-lived. Beset by competition from the automobile, public transit of all kinds went into a financial tailspin after World War I, one from which it never recovered. During the second half of the century, this economic weakness resulted in the delivery of virtually all urban transit into the hands of public agencies.

Construction of new lines in the United States all but ceased after World War I, even as cities became terribly congested with automobile traffic, the only notable exception being a line built beneath State Street in Chicago during World War II. Beginning in the 1950s, however, new wellsprings of public funding brought rapid transit to cities that had never before had such systems. Toronto pushed construction of new subway lines beginning in the 1950s, followed by San Francisco in the 1960s and Washington in the 1970s.

As the end of the 20th century drew near, a majority of the world's major cities had at least one subway line, and many had networks. Therein lays an irony. While any great metropolis was *expected* to have subways, the costs had become astronomical. This was especially true for cities that got going very late in the game. For example, when construction of a subway in Los Angeles finally started in late 1980s—after nearly a century of stymied or aborted plans—it was at a price of well over $1 billion a mile. Understandably, there were ongoing protests that this was a appalling misappropriation of public funds. And yet the existence of a subway was so intertwined with urban self-image and notions of civic prestige that such projects were difficult to rein in; they seemed to have a momentum of their own. Looking to the future, however, it appears that new lines such as the one in Los Angeles will remain only fragmentary reflections of the great systems built nearly a century ago in cities like New York and London, when the costs could reasonably be borne by private capital. In most places, new railborne transit will likely be limited to the less-expensive light variety, if indeed even that.—R.C.P.

See also AUTOMOBILE; CONCRETE; LOCOMOTIVE, STEAM; MOTOR, ELECTRIC; STREETCAR; TUNNELS

Sugar

In chemical terms, a sugar is any group of sweet-tasting, water-soluble crystalline carbohydrates. Among these are lactose (milk sugar), fructose, maltose, and glucose (also known as dextrose). In common usage, sugar usually refers to one particular kind of sugar, sucrose (also known as saccharose). Sucrose has the chemical formula $C_{12}H_{22}O_{11}$. When dissolved in water, it splits into two simple sugars: fructose and glucose. Sucrose naturally occurs in honey, fruits, seeds, flowers, and all green plants. Commercially produced sucrose is obtained from sugarcane and beets. These two sources account for 56 and 44 percent, respectively, of the world's commercial sucrose supply.

The cultivation of sugarcane began in New Guinea some time between 15,000 and 8000 B.C.E. By about 6000 B.C.E., it had appeared in India, where it received the Sanskrit name *sarkara*, which was eventually latinized to *saccharum*. It then made its way from India to China and the Middle East. Sugar may have been known to the ancient Greeks and Romans, but there is no definitive proof of this. A sweet substance that may have been sugar is occasionally mentioned in ancient texts, but even if it was sugar, its rarity meant that it was used for medicinal purposes rather than as a food or condiment.

A major impetus for the widespread use of sugar was the expansion of the Moslem empire. Sugar was produced in several parts of the Moslem world. One of these parts was the island of Crete, which bore the Arab name for crystallized sugar, *Qandi*, and is the source of the word *candy*. The Crusades introduced many Europeans to sugar, and during the Middle Ages sugar was a prized commodity, literally worth its weight in silver. The desire to obtain sugar motivated many of the voyages of discovery and exploration that began in the 15th century.

Some of the lands that had been "discovered" by the Europeans were subsequently used for the cultivation of sugarcane. Sugar plantations in the Caribbean, South America, and parts of the southern United States yielded vast amounts of wealth for their owners and for the economies of European nations. But with the expansion of sugar production came a great amount of misery for the slave labor force of African origin that toiled in the cane fields and sugar refineries. In other parts of the world where sugar was produced, such as Hawaii and Southeast Asia, the labor force was nominally free but often worked under conditions that were scarcely better. Today, many of these places continue to depend on sugar as a mainstay of their economies. Sugar production provides income and jobs, but often at the cost of a deformed economy and a damaged environment.

The expansion of sugar production vastly broadened the market for sugar, and by the mid-17th century it was no longer reserved for medicinal uses or consumption by the very rich. Sugar was used in a variety of foods, and the enthusiasm for coffee, tea, and chocolate further expanded its consumption. Then, as now, sugar appeared as a solid of varying color or as molasses (treacle). Substantial amounts of sugar were also used for rum, an alcoholic beverage that was first produced in the 14th century and grew in popularity as sugar production increased.

At this time, all sugar was produced from cane, but as early as 1575 it was known that certain kinds of beets had a very high sugar content. In 1745, a German chemist succeeded in extracting sugar from beets, and by 1786 a beet sugar refinery was in operation in Silesia. The cultivation of sugar beets was given a substantial boost by Napoleon, who ordered their cultivation in order to counteract the British blockade that had cut France off from her sugar-producing colonies. By the middle of the 19th century, beet sugar was cutting into the market for cane sugar, and by the end of that century almost half of the world's sugar was being obtained from beets.

Today, sugar continues to be consumed in large quantities. In the United States, for example, more than 45 kg (100 lb) of sucrose per capita is consumed annually. In many instances, consumers are not even aware that they are ingesting significant quantities of sugar with the processed foods they eat. The consumption of large quantities of sugar has been a source of concern. Sugar has been attacked as a nutritionally deficient source of "empty calories" that provide nothing for the body except some quick energy. Additionally, sucrose intake has been blamed for a number of ailments, including diabetes, obesity, and heart disease. However, there has been no definitive proof that sugar poses a serious danger to health. According to a 1976 report by the U.S. Food and Drug Administration, the only medical problem clearly linked to sugar consumption is dental caries (cavities).

Sugar's position as the preeminent sweetener has been threatened by artificial sweeteners, first by saccharin, and then by cyclamate and aspartame. In recent years, the emergence of corn sweeteners has further undercut the market for cane and beet sugar.

See also ASPARTAME; CYCLAMATE; DISTILLATION; MONOCULTURE; SACCHARIN

Further Reading: Sidney W. Mintz, *Sweetness and Power: The Place of Sugar in Modern History*, 1985.

Sulfa Drugs

In the second half of the 19th century, it was discovered that a large number of diseases and infections were caused by various microorganisms. One immediate benefit of this discovery was the practice of antiseptic surgery, which greatly reduced the number of deaths following surgical operations. In a parallel development, a growing number of diseases were being prevented or cured through vaccination or the administration of antitoxins and serums. At the same time, however, many bacterial diseases such as meningococcal meningitis, puerperal fever, and gonorrhea had no known cure.

The first drugs to address these and many other infectious diseases were the sulfonamides, or as they came to be popularly known, sulfa drugs. The first hint of the value of sulfonamides was provided by the observation that certain dyes applied to the skin prior to making surgical incisions tended to prevent surface infections. In Germany, the I. G. Farben Industrie was one of the leading manufacturers of both dyes and pharmaceuticals. One of its researchers, Gerhardt Domagk (1895–1964), began a program of testing the antibacterial qualities of the firm's azo dyes (compounds with two nitrogen atoms linked by a double bond) because of their strong chemical affinity for protein-based materials like silk and wool. In 1932, he discovered that one of these dyes cured mice that had been injected with hemolytic streptococci in quantities sufficient to kill them. Nearly 3 years later, a French research team discovered that this dye split into several compounds once it was inside an organism. One of the compounds, a substance synthesized in 1908 and later named sulfanilamide, was responsible for killing the bacteria. It is possible that Domagk had been aware of this, and that he did not immediately publicize the therapeutic value of sulfanilamide so that his team could search for a similar but hitherto unknown compound that could be patented. If so, this is one of many examples of the decidedly mixed benefits of the patent system.

Domagk won the 1939 Nobel Prize for medicine for his role in the discovery of the first sulfa drug, although the Nazis forbade his acceptance of the prize (he finally received it in 1947). There is no doubt that Domagk deserved a share of the prize, for he had played a crucial role in the discovery of sulfa drugs. But others, most notably the chemists Joseph Klarer and Fritz Mietzsch, had done equally important work. As with many modern research programs, the discovery of the first sulfa drugs was a team effort that required the collaboration of medical and chemical researchers, a fair amount of staff support, and the administrative skills of Heinrich Hörlein, a research manager at I. G. Farben. As a number of critics have pointed out, the awarding of a Nobel Prize often ignores the fact that an important scientific discovery may not always be easily traced to the efforts of one person, or even of a small group.

Clinical trials soon demonstrated the value of sulfanilamide and derivative compounds, although the actual mechanism through which they worked was not understood. It wasn't until the early 1960s that two British researchers discovered that sulfa drugs worked by attaching themselves to bacteria in place of a chemically similar compound necessary for the bacteria's metabolism. The principles underlying the effectiveness of sulfa drugs may have been unknown, but there was no doubt of their efficacy. World War II created perfect conditions for the spread of bacterial infections, and many lives were saved by the administration of sulfa drugs. At the same time, however, a new problem emerged. Although sulfa drugs could forestall and cure a wide range of infections, when they were administered as a preventative measure, an inherent defect of chemotherapy emerged; new drug-resistant strains of streptococcus bacteria began to appear.

The administration of sulfa drugs was also marked by a tendency to use them indiscriminately and with little in the way of careful clinical trials for specific applications. One drug firm in the South sold a solution of sulfanilamide in diethylene glycol, an industrial solvent. More than 100 people died after taking this "Elixir of Sulfanilamide," a tragedy that led directly to the passage of the Food, Drug, and Cosmetic Act in 1938. Even medical professionals had a tendency to misuse these "wonder drugs." Their apparent potency and ease of administration led many doctors to prescribe them before making an adequate diagnosis; on many occasions sulfa drugs were taken for diseases for which they were totally ineffectual. Important as they were, the curative powers of sulfa drugs were only part of the story. Sulfa drugs had shown that many bacterial infections could be subdued, and in so doing they helped to energize a search for even more powerful drugs.

See also ANTIBIOTICS, RESISTANCE TO; FOOD AND DRUG ADMINISTRATION, U.S.; NOBEL PRIZE; PATENTS; PENICILLIN

Further Reading: Harry F. Dowling, *Fighting Infection: Conquests of the Twentieth Century,* 1977.

Supercharger

The power developed by internal combustion is closely related to how much fuel and air can be packed into the combustion chamber. In a normally aspirated engine, the air-fuel mixture is drawn into the combustion chamber by the suction that is produced as the piston moves away from the combustion chamber. Superchargers are compressors that force a greater volume of air to enter the combustion chamber, resulting in a denser charge of air and fuel. A supercharger can be directly powered by the engine or it can use the engine's exhaust as a source of power. Since the latter is covered in a separate entry, this entry is limited to mechanically driven superchargers.

Superchargers have a number of mechanical configurations. The centrifugal supercharger is built around a rotating plate with curved axial vanes. The rotation of the plate causes air to be thrown from the inlet to the periphery of a spiral-shaped housing. The shape of the housing causes the air to slow down so that velocity is converted to pressure. The pressurized air is then delivered to the engine's inlet ports.

The second major type is the eccentric or vane supercharger. It uses an off-center rotating drum equipped with a number of vanes that slide radially as the drum rotates, thereby maintaining contact between the tip of the vanes and the interior of the supercharger's housing. As the drum rotates, the vanes trap air between the drum, the housing, and one another. Because the drum is eccentrically located, its further rotation causes a contraction of the space between it and the housing, thereby compressing the air. Superchargers of this type effectively compress the air, but friction between the tips of the vanes and the interior walls of the housing (which gets worse as speeds increase and the vanes are pushed outwards by centrifugal force) limits the speed at which they can operate efficiently.

Friction is not a problem for the Roots-type supercharger, a design that goes back to the 19th century when devices of this sort were used to remove dust from grain elevators. A Roots supercharger uses two long rotors, each with two or three lobes along the rotor. Each lobe occupies a space between two lobes of the other rotor without quite touching them. When the two rotors rotate in opposite directions, they collect air at the intake, and then carry it around the diminishing space between the rotors and the housing, compressing the air in the process.

The history of the supercharger is almost as old as that of the internal combustion engine itself. Piston air compressors were used before 1900 to help with the distribution of fuel-air mixtures and the scavenging of exhaust gases from large diesel and gas engines. The demands of military aircraft during World War I hastened the development of supercharging, although few supercharged engines actually went into service. A considerable amount of work was being done with supercharged aircraft engines in the 1930s and 1940s, culminating in the widespread use of superchargers during World War II. Supercharged diesel engines began to be produced in significant numbers during the 1930s, but other automotive applications lagged. Supercharged engines were primarily used for racing and high performance cars, such as the Mercedes SSK (the "K" stood for *Kompressor*, the German word for supercharger). At present, a few passenger cars are fitted with superchargers, but for most automotive applications turbochargers are preferable. Since a supercharger is mechanically driven by the engine, it absorbs power even at low speeds unless the engine is fitted with a clutch that engages and disengages the supercharger in accordance with the demands placed on the engine. Superchargers also increase the temperature of the air-fuel mixture, heightening the danger of detonation unless compression ratios are reduced or higher octane gasoline is used.

See also ENGINE, DIESEL; ENGINE, FOUR-STROKE; OCTANE NUMBER; TURBOCHARGER

Superconducting Supercollider

After the development of the atomic bomb, the federal government devoted an increasingly large share of its scientific funds to advanced research in physics. Part of the funding went to the construction of particle accelerators that could speed up protons in order to learn more about the nature and behavior of energy and matter. Specifically, scientists assumed the existence of the Higgs Boson, a particle that was essential to certain aspects of the so-called "standard model" of the universe that has been generally accepted in the 20th century. A step toward proving the reality of the Higgs Boson occurred in 1983, when European researchers discovered other particles predicted by the unification theory of the standard model. However, proof of the particle's existence would require energy beyond that provided by current accelerators. In 1984, Congress allocated funds to the Department of Energy (DOE) for the production of the superconducting magnets needed to propel the particles inside an ambitious project called the *superconducting supercollider* (SSC).

The SSC that was proposed by the DOE consisted of two gigantic racetrack-shaped rings 85 km (53 mi) in circumference, each composed of a vacuum chamber surrounded by up to 5,000 superconducting magnets and utilizing a proton beam of 20 trillion electron volts. Prior to the SSC proposal, the most powerful facility in the United States, the Fermi National Accelerator Laboratory in Batavia, Ill., could produce only 0.9 trillion volts per beam. As proposed, the SSC would produce protons and send them through three accelerator rings that would sequentially increase their energy. When they reached their point of maximum energy, the protons would be injected in opposite directions into the two rings of superconducting magnets. The rings, constructed in underground tunnels, would be mounted on top of each other and intersect at four interaction chambers, allowing the accelerated protons to be steered into collision paths. The collisions could last more than a day before the number of protons was depleted and the accelerator had to be reloaded. Specialized detectors would capture the collisions and measure the energy and trajectory of the interactions, with the evidence from the patterns confirming or disproving the presence of the particles that had only been theorized.

When proposed, the DOE projected that the SSC would cost $5.3 billion in 1988 dollars. It promised to be an economic windfall for the state that received the contract. Seven states (Arizona, Colorado, Illinois, Michigan, North Carolina, Tennessee, and Texas) were finalists; each lobbied Congress with impressive delegations touting the local advantages and scientific base. Texas eventually was awarded the site, and after winning the award, the state spent $500 million of its own funds. In 1990, Congress authorized construction of the SSC, but already expenditures on expensive projects such as the National Aerospace Plane and the SSC had come under strong attack due to continuing federal deficits. If the program had stayed within budget it most likely would have survived; but mismanagement and cost growth doomed it. A 1993 study for Congress showed that cost overruns already had afflicted the program, and its price tag had moved up to $11 billion. More important, it became difficult to convince the public of the value of spending more than $5 billion simply to accelerate an invisible particle. In 1994, Congress canceled a number of high-cost science projects, including the National Aerospace Plane and the SSC. Congress allocated $640 million to shut down the program, which already had spent $142 million.

Particle accelerator research was expected to be continued at the Large Hadron Collider, a next-generation European collider project in which the United States was invited to become a participant. Research also was to be done using a slightly different type of collider design, the Stanford Linear Collider. However, these projects were estimated to come on line only 15 to 20 years after the SSC, and in the late 1980s, the Stanford collider was shut down due to technical problems.—L.S.

See also ATOMIC BOMB; HYPERSONIC FLIGHT; PARTICLE ACCELERATORS; PROTON; SUPERCONDUCTIVITY

Further Reading: Most of the material on the SSC is in government documents, with the most helpful being the Hearings before the Committee on Science, Space, and Technology, U.S. House of Representatives, 1988–1994.

Superconductivity

Under normal conditions, the passage of an electrical current is hindered by the resistance within the conductor, the material through which the current passes. Many materials, however, exhibit a phenomenon known as *superconductivity*; at very low temperatures, the resistance vanishes. Thus, after it has been initiated, a current can move through the material for an indefinite period. This phenomenon was first observed by Heike Kamerlingh-Onnes (1853–1926), a Dutch physicist. In 1908, Kamerlingh-Onnes had succeeded in liquefying helium, the last gas to be brought to a liquid state. Three years later, he discovered that the electrical resistance of mercury vanished at a temperature of 4.2 Kelvins (i.e., 4.2°C above absolute zero; absolute zero is defined as –273°C). This temperature is known as the *transition temperature*, for it marks the initiation of superconductivity. Subsequent research by Kamerlingh-Onnes showed that the presence of a magnetic field caused a drop in the transition temperature. Moreover, since an electrical current generates its own magnetic field, an increase in the current also caused a drop in the transition temperature. In other words, a higher current requires a lower temperature in the material to maintain superconductivity.

The exact reason why superconductivity occurs is still poorly understood, even though much theoretical and experimental effort has been expended since the initial discovery by Kamerlingh-Onnes. It is known that the cause of electrical resistance is the action of free electrons colliding with a lattice of immobile positively charged ions. (These collisions are also responsible for another phenomenon that accompanies the flow of current: the generation of heat.) It was therefore hypothesized that below the transition temperature electrons became organized in such a way that their motion is not affected by collisions with the positive-ion lattice.

In 1933, Walter Meissner (1882–1974) and his colleague H. Ochsenfeld discovered that a superconducting material had no magnetic field in its interior. This phenomenon, which is known as the Meissner effect, helped Fritz and Heinz London to determine that electron flow in superconductors was confined to a very thin outer layer of the superconducting material. In 1957, the key theory of superconductor behavior appeared, the BCS theory, named after its creators: John Bardeen (1908–1991), Leon Cooper (1930–), and J. Robert Schrieffer (1931–). According to this theory, superconductivity resulted from the formation of electron pairs (known as Cooper pairs). These pairs are created at very low temperatures when an electron exerts a force on a positively charged ion that causes the latter to leave its lattice and move toward the electron. The positive charge in the region attracts other electrons, creating an environment that causes two electrons to overcome their usual repulsion. The resulting Cooper pairs flow in the same direction, producing a flow of current that meets with no resistance.

In the mid-1960s, theoretical advances along with a great deal of cut-and-try experimentation resulted in the discovery of a group of superconducting materials that were produced by combining niobium with tin, gallium, or germanium. In 1973, the latter compound superconducted at 23.2 K, a record that stood for 13 years. These superconducting materials were brittle and generally hard to work with, but they found application as the basis for small but very powerful magnets that were used in an important medical diagnostic device known as the nuclear magnetic resonance (NMR) imager.

Up to this point, all superconducting materials were metals. In the 1980s, superconductor research moved forward at a rapid pace as experimenters began to develop ceramic superconductors. In 1986, while working in an IBM research laboratory in Zurich, Johannes Bednorz and Karl Mueller pushed the transition temperature up to about 35 K with a barium-lanthanum-copper oxide superconductor. An even more significant step was taken in 1987 when Paul Ching-wu Chu of the University of Houston and Mau-kuen Wu of the University of Alabama used an oxide of barium, copper, and yttrium to produce a superconductor with a transition temperature of 95 K. This took superconducting beyond an important threshold. Up to that point, the low temperatures required for superconductivity could only be achieved through the use of liquid helium, the coldest substance on earth. Relatively higher temperatures allowed the use of liquid nitrogen (which boils at 77 K), thereby lowering the costs of material and equipment. Moreover, the new superconducting material (known as *123 Compound* because of its relative atomic proportions of yttrium, barium, and copper) is not particularly difficult to make.

The discovery did have a major downside, however. It could not be explained by the regnant BCS theory, for the electron-lattice coupling is not sufficiently strong to explain the high transition temperatures characteristic of 123 Compound. Physicists come up with a number of hypotheses to explain electron pairing at relatively high temperatures. Some of these are based on the idea that electrons attract because they spin in opposite directions. Another hypothesis upsets the conventional belief that electrons are elementary particles. Instead, they are seen as consisting of two separate particles, one that carries the electron's charge and another that carries its spin.

While theoretical physicists have speculated, experimenters have added to the list of superconducting substances. There are now more than 6,000 super- conducting materials. One of them—a compound of thallium, barium, calcium, copper, and oxygen—has exhibited a transition temperature of 125 K. At the same time, however, important achievements in the laboratory have not given rise to practical applications of commensurate significance. The components for high-speed computers with superconducting circuits have been tested in laboratories, but commercial application may be a long way off. Superconducting materials are an integral part of research into fusion energy, but there is little expectation that fusion will be an important source of power anytime soon. The slow emergence of superconductor-based technologies is due in part to the considerable complexity and expense engendered by the use of liquid helium or nitrogen. The difficulty of shaping superconducting materials is another problem, although not an insurmountable one.

One unrealized use of superconducting materials was the superconducting supercollider. Intended to answer fundamental questions about the nature of matter and energy, its great projected cost resulting in its being killed by the U.S. Congress in 1993. Meanwhile, research into superconducting quantum interference devices (SQUID) has proceeded apace. SQUIDs can detect extremely small voltages and magnetic fields. They have considerable promise in fields as disparate as mine detection and magnetoencephalography. In the latter application, they may be used to pinpoint specific brain cells responsible for such ailments as Parkinson's disease and Alzheimer's disease.

In the future, the practical application of superconductivity may lie more in applications like pinpointing specific brain cells, and not in the enhancement of existing products and processes. To take one important example, although the long-distance transmission of electricity with no resistance losses is an appealing prospect, it is not likely that the existing infrastructure of power lines with its immense sunk costs would be supplanted by a technology based on superconductivity. It is much more likely that superconductivity will be used for technologies that are still in a developmental state, such as trains that use magnetic repulsion for power and support. Superconductivity may eventually become the basis of a set of significant technologies, but its ultimate applications are only dimly perceived at the present time.

See also FUSION ENERGY; HEAT, KINETIC THEORY OF; MAGNETIC-LEVITATION VEHICLES; MAGNETIC RESONANCE IMAGING; RESISTOR; SUPERCONDUCTING SUPERCOLLIDER

Superheterodyne Circuit

Practical radio reception requires the amplification of radio signals. A major improvement in amplification was achieved in 1917, when Edwin H. Armstrong (1890–1954) invented the superheterodyne receiver. Armstrong was serving in France with the U.S. Army Signal Corps, and was engaged in an effort to intercept German radio communications. This was a very difficult exercise, for the Germans were using high-frequency (500,000 to 3 million cycles per second), short-wavelength signals. When the intercepted signals were amplified, the result was unsatisfactory, as feedback caused the receiving tubes to act as oscillators that generated a great deal of noise.

Armstrong solved the problem by building on the heterodyne principle first used by Reginald Fessenden (1866–1932) at the beginning of the century. *Heterodyne* came from the Greek, meaning a mixture of forces—in this case radio waves of different frequencies. The principle in turn was derived from a basic phenomenon of acoustics. If two musical notes are played simultaneously, they produce a third note with a frequency equal to the difference between the two original notes. For example, a combination of middle C played on a piano (256 cycles per second) and B (240 cycles per second) generates a third sound at 16 cycles per second. When applied to radio reception, the transmitted signal provided one frequency, while an oscillator within the receiver produced the other. The result was the generation of a frequency intermediate between the other two. This was useful because it allowed a radio frequency to be converted into a much lower frequency, where it could drive the diaphragm of an earphone or speaker, thereby producing sounds within the range of human hearing.

At the time he discovered it, Fessenden's heterodyne principle was difficult to translate into practice because there was no way of generating a precise frequency in the receiving apparatus. Within a few years, however, the invention of the vacuum tube allowed the use of a reliable oscillator for the generation of radio waves. Armstrong's improved radio receiver employed vacuum tubes in conjunction with one of his earlier inventions, the regenerative circuit. His *superheterodyne* circuit used a tube oscillator to produce the heterodyned wave, which was then amplified. This was in turn detected by his regenerative circuit, converted to direct current, amplified again, and then sent to a speaker.

The superheterodyne principle made possible the practical use of high-frequency radio waves, which were to become increasingly important as radio developed in the 1920s. The fixed frequency of heterodyned radio waves made them stable and easily amplified. The stability of the heterodyned waves in turn allowed the production of receivers that could be precisely tuned to individual stations whose signals were unaffected by those of other stations. Although Lucien Levy in France and others had some claim to the invention of the superheterodyne principle, in 1920 Armstrong was able to sell the patent rights of his regenerative and superheterodyne circuits to the Westinghouse Electric Company for $335,000. Much of it went for legal expenses accumulated in the course of past patent suits, as well as those that were to come.

See also AMPLIFIER; PATENTS; REGENERATIVE CIRCUIT; THERMOIONIC (VACUUM) TUBE

Supersonic Flight

Sound travels at a rate that depends on the temperature of the medium through which it moves. When sound travels through air, its speed is proportional the square root of the air's absolute temperature. The temperature of the air drops with altitude, so at higher altitudes the speed of sound is consequently reduced. At sea level, the speed of sound is 1,223 kph (760 mph); at 10,973 m

(36,000 ft) it is only 1,062 kph (660 mph). Early inquiries into the speed of sound were conducted by Ernst Mach (1838–1916), an Austrian mathematician and physicist. In recognition of his pioneering work, high speeds are given as "Mach numbers," with Mach 1 representing the speed of sound, Mach 2 twice the speed of sound and so forth.

Beginning in the 1930s, the effects of traveling near the speed of sound were of considerable interest to aerodynamists and aeronautical engineers, for wind tunnel tests indicated that strange aerodynamic effects began to occur at transonic (approaching the speed of sound) velocities. In 1935, reporters took note of the comments of W. F. Hilton, a British aerodynamist, that a wing encountered resistance that "shoots up like a barrier" as the speed of sound was approached. Rather inaccurately, his statement was presented as though there was a physical "sound barrier" that blocked efforts to fly at and beyond the speed of sound.

No physical barrier existed, but serious problems did arise when aircraft approached the speed of sound. The main difficulty was the "compressibility" that was caused by a shock wave traveling ahead of the airplane that compressed the air, resulting in diminished lift. And with a loss of lift came a loss of control and stability. This became evident to pilots of high-performance fighter aircraft during World War II who experienced sometimes-fatal difficulties in keeping their aircraft under control after approaching the speed of sound in dives.

During World War II, rapid improvements in the performance of combat aircraft naturally inclined engineers and scientists to consider the possibility of flying at Mach 1 and beyond. In the course of the war, Germany was in the lead due to its development of high-speed rocket and jet-powered aircraft, but afterwards the United States became the center of research into supersonic flight. The Army Air Force (in 1947 it became a separate military arm, the U.S. Air Force) and the National Advisory Council on Aeronautics engaged in a not-always-harmonious effort to build and operate an airplane capable of flying faster than the speed of sound. The airplane, called the XS-1 (for experimental sonic, number 1), was built by the Bell Aircraft Corporation. A small airplane a bit under 9.5 m (31 ft) in length, the XS-1 was powered by a rocket engine fueled by ethanol and liquid oxygen capable of developing

2,722 kg (6,000 lb) of thrust. The XS-1 was designed to be dropped from the underside of a B-29 bomber from a height of 6,100 m (20,000 ft); from this point it climbed to more than double this height after the rocket engines were started. After numerous glide and subsonic tests, the XR-1 "broke the sound barrier," achieving a speed of Mach 1.06 on Oct. 14, 1947.

Within 10 years of the first supersonic flight, a number of supersonic fighters were in service throughout the world. The last in the line of high-speed experimental aircraft, the rocket-powered North American X-15, achieved a speed of 7,297 kph (4,534 mph) or Mach 6.72 on Oct. 3, 1967. In the 1960s, the fastest operational aircraft, the Lockheed SR-71 and the Soviet MiG-25, attained speeds up to Mach 3. On July 27, 1976, the SR-71 set the current world record for jet-peered airplanes when it attained a speed of 3,531 kph (2,194 mph) at 25,935 km (85,089 ft). Since that time there has been no increase in flying speeds, although design studies for hypersonic (more than Mach 5) aircraft are in progress.

Commercial flights at the speed of sound were inaugurated with the Anglo-French Concorde and the Soviet Tu-144. The latter had a short service life, but the Concorde continues to carry passengers today, a technical success, if not commercial one.

See also HYPERSONIC FLIGHT; SUPERSONIC TRANSPORT; TURBOJET

Supersonic Transport (SST)

A new era in aviation was initiated in 1947 with the first supersonic flight. Within a few years, military aircraft with supersonic capabilities were entering service in the world's air forces, while experimental aircraft were being flown at several times the speed of sound at the edge of space. In the commercial realm, however, progress was slower. Jet-powered commercial aircraft began to carry passengers in the 1950s, and by the 1960s jet travel had become commonplace. Jet airliners represented a significant improvement in speed over propeller-driven craft, but their top cruising speed of 885 to 965 kph (550–600 mph) remained well below the speed of sound.

Designing and building passenger aircraft capable

of traveling faster than the speed of sound posed formidable technical challenges. Supersonic flight produces high temperatures on the surface of the aircraft, and protecting the airplane and the people in it often requires the use of exotic materials and production techniques. From an aerodynamic standpoint, it is difficult to design an airplane that maintains good flying characteristics at speeds that range from takeoff to beyond the speed of sound. Another major difficulty is the generation of pressure waves or "sonic booms" that can cause psychological stress and physical damage on the ground.

From the outset, efforts to build an SST were strongly influenced by international politics and economics. According to the conventional wisdom of the 1950s and 60s, a successful SST would bring money and prestige to the country that created it. In the 1950s, the Soviet Union had demonstrated its technological virtuosity through the launching of the first artificial satellite, and it hoped to score another technical coup with a supersonic airliner. In this it met with initial success; the first SST to take to the air was the Tupolev TU-144, which flew for the first time on Dec. 31, 1968. Powered by four jet engines, each one capable of producing up to 20,000 kg (44,092 lb) of thrust,

the TU-144 was also the first jetliner to fly at twice the speed of sound, which was accomplished on June 5, 1969. But these were the airplane's last triumphs. When the second prototype TU-144 appeared at the Paris Air Show in 1973, a flight demonstration took a tragic turn when the plane crashed and the flight crew was killed. The plane was finally put into regular service between Moscow and Alma Ata in the Mongolian People's Republic in late 1975, but only as a carrier of mail and freight. It began to transport passengers in November 1977, but the fatal crash of a test flight in 1978 ended this service in June of that year. The TU-144 went back into freight service for a while, but it was permanently withdrawn in the early 1980s.

None of the countries of Europe had the single-minded determination to build an SST that had been exhibited by the Soviet Union. If an SST were to be built, a collaborative venture that transcended national boundaries would be required. At the same time, even a cooperative venture between a number of very large firms would not provide adequate funding, and from the outset, the development of a European SST was predicated on substantial governmental support. The SST that was given the name Concorde was the product of collaboration between the British Air-

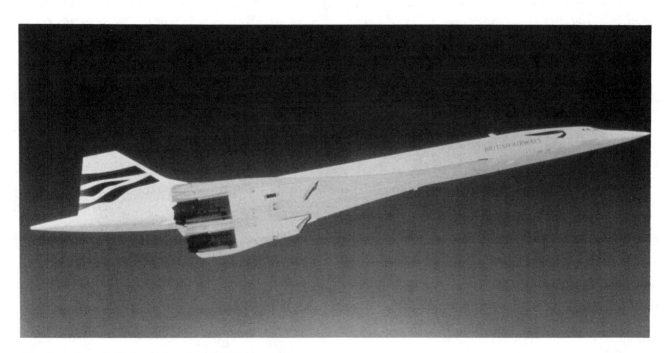

The Anglo-French *Concorde* (courtesy British Airways).

craft Company and the French firm Aerospatiale. Powered by four turbojets that each developed up to 17,260 kg (38,050 lb) of thrust, the Concorde first flew on Mar. 2, 1969, and exceeded Mach 2 (twice the speed of sound) on Nov. 4, 1970. Passenger service began in January 1976. Air France flew a Paris-Rio de Janeiro route; British Airways flew the route between London and Bahrain. Flights between London and New York and Washington soon followed, but the Concorde never made a profit for its operators. Beset by high operating costs (about twice those of subsonic airliners) and small payloads, the Concorde needed continual government subsidies to keep it flying. Production of the Concorde terminated in 1979 after 16 planes had been built.

The United States had given up on supersonic airliners before the Concorde went into service. But before this happened, American aviation firms and the U.S. government spent millions of dollars in an effort to build an SST. The program began with the allocation of $11 million in the 1962 fiscal year and $20 million in the 1963 fiscal year for technical and economic feasibility studies. In 1967, the Federal Aviation Administration selected Boeing and General Electric to produce the SST's airframe and engine, respectively. The first flight of the SST prototype was expected by early 1973 at the latest. The original American SST design was even more ambitious than the Soviet and Anglo-French designs. One of its key features was a variable-sweep wing that would optimize performance at both low and high speeds. Technical problems proved insurmountable however, and the design ended with a fixed delta wing similar to the ones found on the TU-144 and the Concorde.

Even with substantial design modifications, the American SST failed to meet technical and cost-benefit requirements. By the late 1960s, government studies indicated that the SST should be canceled or suspended, but the Nixon administration tried to keep the program alive. At this point, environmental issues were added to the list of technical and economic obstacles to the SST's development. In addition to concern over the inevitable sonic booms, there was a fear that exhaust gases vented in the upper atmosphere would damage the Earth's ozone layer. Faced with mounting opposition, the U.S. Congress terminated financial support for the SST pro-

gram in 1971. By this time $920 million of government funds had been spent.

See also OZONE LAYER; SUPERSONIC FLIGHT; TURBOJET

Surgery, Antiseptic

In the 19th century, as in all preceding centuries, a person undergoing a surgical operation faced an uncertain future. Beginning in the 1840s, the use of anesthesia diminished some of the pain that inevitably accompanied surgery, but even a relatively painless and successful surgery posed a serious danger to the patient. Every surgical wound was subject to infection, and many surgeries resulted in the death of the patient. Surgeons took no particular pains to provide a clean environment. Standard practice was to treat the wounds of an infected patient, and then to move on to one who was not infected, frequently resulting in the transmission of the microorganisms that caused infection. But, as the doctors of the time reasoned, what was the point of cleaning one's hands and instruments when almost every wound was foully infected? Further limiting the desire of doctors to maintain a sterile environment was the belief that infection was a good thing. As the Greek physician Galen (c. 130–c. 200) had stated, the fact that a wound suppurated showed that it was healing, hence the many references to "laudable pus" in the case notes of physicians.

One doctor who had some different notions regarding infection was the Hungarian physician Ignaz Semmelweis (1818–1865). While working as an assistant at a hospital in Vienna, Austria, he observed a dramatic difference in the death rates from puerperal (or childbed) fever. In one obstetric ward, nearly 10 percent of newly delivered mothers died, while in the other the rate was less than 4 percent. At that time, infection was generally thought to be the result of vapors known as *miasmas*, but this hardly explained why two places in the same hospital exhibited such striking variation.

The difference between the two wards, as Semmelweis noted, was that the one with the lower death rate was run by midwives, who generally kept their domain clean. The other ward was the province of medical students, who thought nothing of attending at a birth immediately after performing an autopsy. They didn't bother to wash their hands, and they wore the same blood-spattered coat all the while. When a friend died of an illness similar to puerperal fever after having cut himself while dissecting a corpse, Semmelweis's suspicions were confirmed. As he saw it, unsanitary medical practices were responsible for the transfer of "putrid particles" into the wounds of previously healthy people.

Putting his idea into practice, in 1847 Semmelweis required that medical students wash their hands with a disinfecting solution of lime chloride before entering a ward. As a result, in 2 years the death rate from puerperal fever in the hospital dropped to 1.27 percent. But Semmelweis's colleagues remained unimpressed by the spectacular improvement. Many of them refused to follow his orders, and when Hungary revolted against Austria, the hospital's administrators had all the reason they needed to rid themselves of the troublesome doctor. Returning to Budapest, Semmelweis built up a successful practice in gynecology and obstetrics. In the course of an operation, he cut his hand, and infection set in. By this time he had become clinically depressed, and at the age of 47 he died of streptococcal blood poisoning, 12 days after having been admitted to an insane asylum.

Semmelweis was not the only one to connect infec-

tion with unhygenic practices. In 1843, Oliver Wendell Holmes (1809–1894), an American obstetrician/gynecologist (and accomplished essayist), presented a paper at the Boston Society of Medical Improvement that noted how infections seemed to be propagated by medical staff who moved from infected patients to newly delivered mothers. But Holmes met with the same reception as Semmelweis; most doctors—"gentlemen all," as one eminent professor of medicine called them—refused to heed Holmes's suggestion that they regularly wash their hands. As Holmes noted, "Medical logic does not appear to be taught or practiced in our schools."

Both Holmes and Semmelweis had noted only a correlation between unhygenic practices and the spread of infections; the exact mechanisms through which infections were spread remained unknown. In the 1850s, Louis Pasteur (1822–1895) began his research into such phenomena as the spoilage of wine, and concluded that hitherto unknown microorganisms were the cause of a variety of disorders. The English physician Joseph Lister (1827–1912) took inspiration from what he called "the beautiful researches of Monsieur Pasteur," and in 1865 (the year of Semmelweis's death), he performed two antiseptic operations using carbolic acid (C_6H_5OH) as a disinfectant. Neither was successful, owing to "improper management," as he later put it. But a few months later he successfully set a compound fracture,

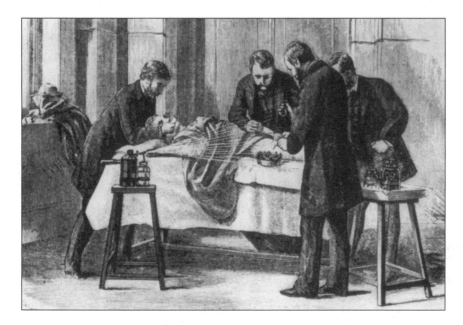

A surgical operation conducted in the late 19th century. Note the carbolic acid spray in the left foreground (courtesy National Library of Medicine).

an injury that usually required amputation of the limb and often ended in the death of the patient. Lister documented his antiseptic procedures in the British medical journal *The Lancet* in 1867, and subsequently published a more lengthy treatise, *On the Antiseptic Principle in the Practice of Surgery.*

As had happened before, the medical establishment was slow to embrace his methods. His carbolic acid spray, which he used as a general disinfectant, came in for particular scorn. Lister assumed that the spray was necessary to kill airborne germs in the operating room, but most doctors and nurses disliked the spray, which irritated the eyes, nose, and throat. The spray gained a measure of acceptance when Lister used it while successfully draining an abscess from the left armpit of Queen Victoria, but some surgeons persisted with their criticisms. In 1890, Lister admitted that the air was not a significant source of infectious microorganisms and that the spray was of little value.

A more important component of antiseptic surgery first appeared in late 1889. William Stewart Halstead (1852–1922), the surgeon-in-chief of the Johns Hopkins University Hospital, made it a practice to use mercuric chloride as a sterilizing agent. One of his nurses complained that the solution produced dermatitis on her hands and arms. In response, Halstead asked the Goodyear Rubber Company to provide his staff with thin rubber gloves. As it turned out, not only did the gloves prevent dermatitis, they also provided levels of sterility not possible with bare hands, no matter how well or often they were washed. The use of rubber gloves was an important element in the transition from antiseptic surgery (killing microorganisms that are already present) to aseptic surgery (keeping the environment free from microorganisms from the outset).

Although great strides have been made in aseptic surgery during the 20th century, surgical procedures still carry some risk of infection. Surgery is a major source of nosocomial infections (infections contracted in the course of a hospital stay). Although surgical operations account for only about 40 percent of hospital admissions, operations are responsible for approximately 70 percent of infections acquired while in hospital). Common infectious agents like *Staphylococcus aureus* and *E. coli* can be successfully treated with antibiotics most of the time, but paradoxically, the wide-

spread use of antibiotics in hospitals has encouraged the growth of antibiotic-resistant bacteria that are very difficult to eradicate.

See also ANESTHESIA; ANTIBIOTICS, RESISTANCE TO; *E. coli*; GERM THEORY OF DISEASE

Suspension Bridge—see Bridge, Suspension

Systems Analysis

Many of the entries in this encyclopedia are devoted to a particular kind of system: the circulatory system, an electrical distribution system, an airplane's flight management system, and so on. In these instances, the term *system* is used to indicate the presence of a multiplicity of parts that interact in a regular fashion in order to get something accomplished. In reference to the first example, the circulatory system consists of the heart, arteries, veins, capillaries, blood, lymph, and lymphatic vessels that transport materials through the body, thereby helping to keep it alive. In the most basic terms, systems analysis is used to identify the components of a system and how they interact, often with an eye towards improving the performance of the system.

Systems analysis begins by distinguishing what is in the system from what is on the outside. The latter will include parts that comprise the system's "environment." The distinction between system and its environment may be an arbitrary one. For example, should the membership of a school system be confined to its teachers, administrators, and maintenance staff? Are students and their parents therefore part of the school system's environment, or should they be considered part of the system itself? However the lines may be drawn, it is still the case that many systems interact with an environment or set of environments, and coming to an understanding of these interactions can be an important part of systems analysis.

A major stage of systems analysis is the description and comprehension of how the different components of a system interact. The interaction of these components can be mathematically examined through a set of techniques known as *operations research.*

Mathematical methods also are used when the attributes of a system are abstracted so that the system can be mathematically modeled. Although the model is not presented as an exact representation of reality, it allows the intensive study of key relationships within the system. Modeling techniques allow the analyst to vary key components of the system in order to determine how these changes affect the system as a whole. Mathematical modeling is particularly valuable when a new system is being designed or an existing one is being modified, as it allows the consideration of alternative means of achieving a given end; in this way it can serve as a powerful tool for informed decisionmaking. Of course, the conclusions arrived at will reflect the prior selection of particular criteria for what constitutes success or failure. The tools of systems analysis can always be used to design a system that works in a highly effective manner while achieving what some would consider to be the wrong outcome.

Systems analysis often incorporates the principle of feedback into its methodology. Many systems do not operate in a linear fashion, with inputs moving in one direction through the system as they are transformed into outputs. Instead, the components of the system may be changed as the system operates, and this will alter the future operation of the system. For example, the introduction of a new kind of fish into a lake may lead to the buildup of waste products that provide nutrients for an insect that had been present in small numbers. The proliferating insects provide a food source for a kind of fish indigenous to the lake, and their population increases substantially. These fish then drive the newly introduced fish out of the places in which they customarily hide from their predators. As a result, large numbers get eaten, leaving far fewer fish than had been anticipated when they were introduced.

Systems analysis originated as a means of coping with the management problems presented by large, complex weapons systems. It achieved significant results in this realm, but efforts to transfer systems analysis to other kinds of systems have not always been successful. Systems that include human actors are particularly difficult to analyze, for people do not always behave with the regularity of mechanical or electronic components. Systems analysis also assumes that the goals and purposes of the system are unambiguous, but this is not always the case with social, political, and economic systems.

See also CYBERNETICS; ECOLOGY; MATHEMATICAL MODELING; OPERATIONS RESEARCH

Tampon

An alternative to sanitary napkins, tampons are small cylinders of absorbent materials, usually cotton and rayon fibers, that are inserted in the vagina for the absorption of menstrual flow. The basic idea is a very old one; women in various parts of the world have used tampons made from moss, paper, small sponges, wool, and vegetable fibers. The first tampon to be commercially produced was Tampax, which was put on the market in 1933. At that time, some concerns were raised that these products might be unsuitable for girls and women who had never experienced sexual intercourse, but this notion has no physiological basis.

This is not to say that tampons are completely harmless. In 1978, a previously unknown illness that was given the name *toxic shock syndrome* (TSS) appeared in the United States. Its symptoms included fever, nausea, vomiting, diarrhea, low blood pressure, muscle tenderness, and a reddish rash that was followed by peeling of the skin. Between 1979 and 1985, 2,814 cases of TSS were reported in the United States, of which 122 were fatal. The use of tampons was soon implicated in the syndrome. In 1985, researchers announced that TSS most commonly occurred in young women who had used superabsorbent tampons. These tampons used clumps of polyacrylate fibers and polyester foam instead of the usual cotton and rayon fibers. It was hypothesized that the extended use of individual tampons allowed the buildup of toxins released by some strains of the *Staphylococcus aureus* bacteria. However, more is involved than the use of superabsorbent tampons, for TSS has afflicted women using regular tampons. Moreover, men, children, and postmenopausal women also have been stricken with TSS. Although TSS is not always associated with the use of tampons, it is still wise to follow some basic precautions. Users of tampons are advised to avoid continuous use, to change them frequently, to avoid tampons made from synthetic materials, and to use smaller sizes as menstrual flow decreases.

Due in large measure to the toxic shock episode, tampons have come under closer scrutiny. In 1980, the U.S. Food and Drug Administration classified tampons as class II medical devices, which means that they have to meet certain performance standards. Moreover, since 1982, packages of tampons have had to carry warnings about the danger of TSS and its symptoms. Since the risk of TSS is related to the absorbency of tampons, packages also have to indicate how absorbent the tampons are. However, since manufacturers do not use a uniform rating system, comparing the absorbency of different brands may be difficult.

See also COTTON; FOOD AND DRUG ADMINISTRATION, U.S.; RAYON; SANITARY NAPKIN

Further Reading: Janice Delany, Mary Jane Lupton, and Emily Toth, *The Curse: A Cultural History of Menstruation*, 1988.

Tankers

The first significant extraction of subsurface petroleum took place during the 1860s in the Appalachian region of the United States, particularly in Pennsylvania. As late as 1899, this region was the source of 93 percent of the nation's total crude oil. But a huge surge of production in the midcontinent area—in Texas and Louisiana, and particularly in Kern and Ventura counties in Southern California—reduced this to 10 percent by the end of World War I. More significantly, these new fields produced more oil than could be absorbed domestically. Even though millions of automobiles were taking to the road, oil was rapidly supplanting coal for heating, the western railroads were turning to fuel oil, and the navy and merchant marine were consuming 50 million barrels of oil annually, by 1919 the domestic market could consume only 85 percent of U.S. production, which was then approaching 300 million barrels. And then, in the 1920s, the exploitation of newly discovered reserves—gigantic oilfields in Oklahoma and in Los Angeles County—resulted in levels of production that altogether outdistanced domestic consumption. A burgeoning export market beckoned, as most of Europe had no apparent oil reserves, and production was just getting underway in regions that would ultimately dominate world markets: the Middle East and South America.

The attractions of a vast export market led to the practice of sending crude oil across the oceans in great floating tanks. Shipping oil in barrels had been fairly common by the turn of the century, particularly out of Texas ports. The *Thomas W. Lawson*—the only seven-masted schooner ever built—had a cargo of oil aboard when she capsized off the Scilly Isles in 1907, thereby creating the first significant oil spill, and at a spot not far from one of the most famous, the *Torrey Canyon* disaster 60 years later. By the time the *Lawson* went down, however, specialized vessels with integral tanks had already come into existence, the first being the 2,720-metric-ton (3,000-ton) *Glukauf*, built in 1886 in Britain for German owners whose traffic was with American oilmen. Among the advantages of such vessels for shippers was their greatly expanded capacity and their capability to take on their liquid cargo from offshore via flexible pipes.

In 1900, there were only three tankers under the U.S. flag, with a total gross tonnage of 7,611. By the 1950s, the number of vessels had topped 500 and the tonnage 5 million—and that was only about a quarter of the world total. In the 1960s, U.S. tonnage remained about the same, but dropped to less than 10 percent of the world total as foreign fleets grew larger, and as individual ships began reaching enormous dimensions.

By the 1960s, the United States was on its way to becoming a net importer as continental reserves dwin-

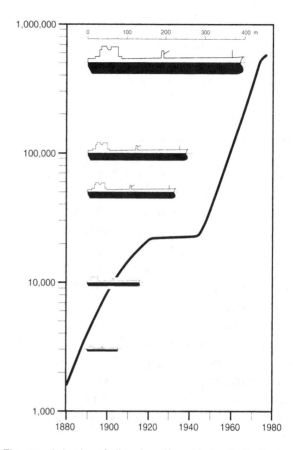

The growth in size of oil tankers (from Vaclav Smil, *Energy in World Civilization*, 1994; used with permission).

dled. Exports had peaked during World War II (at 1.7 billion barrels in 1944), and by 1969 imports were eight times greater than exports. In a spectacular 1969 experiment, the *Manhattan*, the largest American merchant ship afloat, was converted into an icebreaker and managed to batter her way through 1,050 km (650 mi) of ice to reach Alaska's North Slope, thereby becoming the first merchant vessel to traverse the fabled Northwest Passage. While this demonstrated the technical feasibility of transporting Alaskan oil to the East Coast via the Arctic, the cost proved prohibitive. Shortly thereafter, as exports from the South America and particularly the Middle East came to dominate world markets, the 305-m (1,000-ft) *Manhattan* had been eclipsed by supertankers of quadruple her tonnage—more than $\frac{1}{2}$-million tons—none of them built in the United States, and only rarely flying the stars and stripes.

Tankers have been designed to transport other

commodities besides liquid petroleum. One called the *Tropicana* was used for shipping orange juice, 2,460,500 liters (650,000 gal) at a time. Tankers designed to carry liquefied natural gas (LNG) date to the 1940s, the first being the *Natalie O. Warren*, which was converted from a standard tanker of a type that had initially been developed by the U.S. Maritime Commission in the late 1930s. LNG is transported at temperatures below –108°C (–162°F) and thus occupies only 1/600 of its volume at normal temperatures. But most tankers are designed for liquid petroleum, and it was America's mounting dependence on foreign oil that resulted in the fantastic magnification in the size of these vessels. Today's typical tanker is a huge vessel, displacing 181,900 metric tons (200,500 tons) and requiring engines producing 22,400 MW (30,000 hp) to move it and its cargo over the seas.

The development of tanker fleets is a reflection of the world's economic geography. The regions that now produce most of the world's oil actually consume very little of it themselves, while those that use most of the oil produce less and less. So it is that tankers, a distinct novelty only a century ago, are among the most commonplace vessels to be seen on the world's sea-lanes.—R.C.P.

See also AUTOMOBILE; FUELS, FOSSIL; OIL SPILLS; RESOURCE DEPLETION; STOVES, HEATING

Tanks

Within a few months after the outbreak World War I, the Western Front was gripped in a stalemate. Machine guns, artillery, and barbed wire prevented any significant advance, and opposing sides secured their defensive lines by digging trenches that extended from the Swiss border to the North Sea. The weapon devised to break through this impasse was the tank, a tracked vehicle capable of traveling over difficult terrain while protecting its crew from machine gun fire. The idea of an armored vehicle running on tracks went back to the early 20th century, but credit for conceiving the tank and how its should be deployed is usually given to Major General Ernest Swinton, who drafted a paper that eventually caught the attention of the British military. Swinton's ideas were realized in the Mark I tank. This was a 25,400-kg (28-ton) tracked

vehicle with a top speed of 6.5 kph (4 mph). On board was a crew of eight, four of whom were employed with driving and gear shifting; the remainder attended to two small cannons and four machine guns.

These tanks first appeared in battle on Sep. 15, 1915, but it was not a successful debut. All 20 were put out of commission by mechanical problems and a muddy, shell-pocked terrain that was impassable even to tracked vehicles. A more successful foray took place on Nov. 20, 1917, when at the Battle of Cambrai a total of 314 tanks were deployed, and many succeeded in breaking through the German line. But poor battlefield leadership and miscommunication, along with an impassable canal, stopped the advance. A week later, a German counterattack pushed the British back to the original point of advance. The offensive power of the tank had been demonstrated, but much remained to be done if its potential was to be realized.

The years following World War I saw substantial technical refinements, but military thinking often lagged behind the tank's expanding capabilities. Field Marshall Douglas Haig (1861–1928), whose tactics sent tens of thousands of men to their deaths in hopeless attacks during World War I, voiced a common belief of traditional infantry officers when he stated that "I am all for using aeroplanes and tanks, but they are only auxiliaries to the man and the horse." Reflecting this kind of thinking was a British military budget that in 1935 spent twice as much money on cavalry as on tanks. The French high command was equally lacking in vision; tanks were designed to move at the pace of a foot soldier, clearing the way for the infantry but not assuming an independent combat role.

In contrast, Germany, despite having been forbidden by the Treaty of Versailles from having any tanks at all, was able to cultivate a more forward-looking group of officers. Defeat in 1918 had discredited the old leadership, and a new generation of officers gained positions of authority while the Nazis were aggressively pushing German rearmament. The result was the creation in the 1930s of panzer divisions that made tanks and other armored vehicles key components of blitzkrieg, "lightning war." In 1939, these weapons were used to good effect when the German army subdued Poland in a few weeks. In July 1943, the greatest tank battle in history pitted 1,500 German and Soviet

World War II infantrymen and a Sherman medium tank (courtesy National Archives).

tanks against each other near the Russian town of Kursk. Neither side scored a decisive victory, but the German offensive had been blunted.

During World War II, tanks were partially checked by the development of effective antitank weapons like land mines and the bazooka. Today, wire-guided weapons pose a far more formidable threat. Even so, tanks continue to be the key elements of land warfare. At the most basic level, they are still the armored tracked vehicles that first appeared during World War I, but their effectiveness has been vastly improved by better mechanical components, laser sighting devices, and electronic aiming computers. Even the armor has reached a high level of sophistication. Instead of a thick slab of steel, many tanks are protected by Chobham armor, layers of metal, plastic, and ceramics that absorb and break up the projectiles fired into it. Some tanks are equipped with "reactive" or "active" armor that uses an explosive charge on the outer surface to counteract the force of a projectile.

The typical battle tank weighs about 50,000 kg (55 tons), is propelled by a 1,000-hp engine, and carries a 125-mm (4.9-in.) cannon. A key improvement has been the installation of mechanical gun loaders, which eliminates the need for one crewman and allows the tank to present a smaller target. Modern tanks have a top speed of from 45 to 70 kilometers per hour (28–43.5 mph), but they are optimized to travel at a speed of 14.5 to 29 kilometers per hour (9–18 mph) on cross-country runs.

For all the improvements that have been made, tank design still entails striking compromises between the conflicting requirements of protection, firepower, and mobility. Today's tanks are more potent weapons than their predecessors, but with technological advances have come high costs: A single U.S. Army M1 battle tank costs about $2.5 million in the 1980s, seven times as much (in constant dollars) as the World War II–era Sherman tank. It also has a voracious appetite for fuel and requires a lot of maintenance time. Tanks continue to be plagued by less-than-perfect reliability. A division with 300 tanks can be expected to suffer 100 breakdowns in the course of a road march covering no more than 100 km (60.2 mi).

See also BAZOOKA; LASER; PRECISION-GUIDED MUNITIONS; TRACKED VEHICLES

Taxonomy

Taxonomy is the theory and practice of classifying living and extinct organisms. It is important not just because it satisfies the human urge to bring order out of apparent chaos; taxonomy also has played an important role in the progress of biology by stimulating the effort to find similarities and differences within and between groups of organisms. In the course of classifying different groups of organisms, 18th-century biologists began to consider how separate but apparently related groups might have shared common ancestry. The insights that came from these investigations were of critical importance for the development of evolutionary theory. It is no coincidence that Charles Darwin's epochal book on evolution was entitled *On the Origin of Species*.

Although taxonomy is one of the foundations of biology, it is necessary to admit that classificatory systems contain much that is arbitrary. This is inevitable, given the vast number of species extant and the immense complexity of the processes that produced them and their predecessors. New groupings do not always emerge in ways that clearly demarcate them from their ancestors, or from other groups that also have evolved from these ancestors. The vagueness of the demarcations between one group of organisms and another means that the botanists and zoologists may make judgments colored by their own prejudices. To take a notable example, when attempting to classify a particular set of organisms, biologists often have to decide if the group should be split off from a set of organisms with similar features or lumped together with them. Their decision may reflect little more than the personal predilection of the person making the classification, whether he or she is a "lumper" or a "splitter."

Efforts at classification are at least as old as the Greek philosopher and scientist Aristotle (384–322 B.C.E.). Inspired by his teacher Plato's (429–347 B.C.E.) belief in the existence of ideal forms, Aristotle created categories that, in his belief, reflected the essence of particular plants and animals. In so doing, Aristotle formulated two classifications, *genos* and *edos*. *Genos* referred to broad categories of animals, reptiles or birds for example, while *eidos* signified the specific kinds of animals within these groups, like snakes and sparrows.

Aristotle's taxonomy was intentionally hierarchical. The animals that we now call *mammals* were placed at the top of the hierarchy because they were born alive and complete. Reptiles and birds were of a lower order because they hatched from eggs, while creatures like flies and mosquitoes were put at the bottom because Aristotle thought that they arose through spontaneous generation.

Aristotle's biological principles held sway for many centuries thereafter, and it was not until the 17th century that serious efforts were made to devise and apply better taxonomic principles. To a significant extent, this endeavor was sparked by European exploration and colonization, which greatly expanded the number of plants and animals known to the West. About 6,000 species of plants were known to Europeans at the beginning of the 17th century; within a century that number had tripled. The desire to create classificatory systems also reflected a belief that the natural world was orderly because it followed a coherent plan that had been ordained by the Creator. For those less inclined to invoke the Deity, there was the example of physics and Isaac Newton's discovery of the orderly nature of the universe; the same presumably could be done with the living world.

One of the most comprehensive zoology books of the Renaissance era, Konrad von Gesner's (1516–1565) *Historia Animalium*, relied on Aristotle's taxonomy. The following century saw the introduction of new terminology, in particular the word *species*, which was first used by the English naturalist John Ray (1627–1705). Although the definition of the term is a matter of considerable controversy even today, Ray defined a species as a group of organisms capable of interbreeding and producing fertile offspring. Ray's schema was strongly resisted by many other naturalists, who took issue with the idea that a single characteristic could be used to determine membership in a particular group.

The greatest name in the history of taxonomy is Carl von Linné (1707–1778), the Swedish botanist who is usually known by his Latinized name, Linnaeus. As had been the case with Ray, Linnaeus's taxonomic efforts were guided by religious beliefs. Naturalists were charged with completing a task originally given to Adam, naming the plants and animals, and contemplating God's goodness as it was manifested in His many creations. An appropriate system of classification would reveal the essence of particular groups of organisms, and in so doing provide important insights into God's plans and purposes.

Linnaeus's taxonomy was presented as *Systema Natura, or the Three Kingdoms of Nature Systematically Proposed in Classes, Orders, Genera, and Species*, which was in published in 1735. As with all classification systems, Linnaeus had to decide which aspect of a plant or animal best captured its "essence." Like Aristotle and other naturalists that had come before him, Linnaeus used methods of reproduction as the basis for his classificatory system of plants. Accordingly, the number and character of stamens (the male reproductive part) determined a plant's class, while the number and character of pistils (the female reproductive part) determined its order.

In 1753, Linnaeus introduced the now-familiar "binomial" nomenclature that places all organisms in a specific genus and species. First used for plants and later for animals, it uses Latin names for the genus (plural *genera*) and species. According to the conventions of this system, the name of the genus and species are underscored or italicized, and the first letter of the genus is capitalized, while that of the species is lower case. The assigned names of individual genera and species are used by biologists the world over, eliminating the possibility of two scientists calling the same organism by a different name or vice versa. The universalization of taxonomic nomenclature began to be solidified in the early 20th century with the publication of *International Rules of Zoological Nomenclature*.

The taxonomic system used by Linnaeus and successive generations of biologists is hierarchical, with lower categories nested into higher categories. Species are grouped into genera, and genera are grouped into a more inclusive grouping known as a family. Families are grouped into orders, and orders are grouped into classes, which in turn are grouped into phyla (the plural of *phylum*). In many cases, a more complex set of categories has to be employed, and groupings like subphylum and superorder make their appearance.

To illustrate, the taxonomic classification of a house cat, *Felis domestica*, proceeds in the following order:

Kingdom: Animalia
Phylum: Chordata
Class: Mammalia
Order: Carnivora
Familia: Felidae
Genus: *Felis*
Species: *domestica*

As with Ray's, Linnaeus's schema was opposed by naturalists who resisted the idea than a single characteristic defined a group, or even that groups existed at all. In the words of the influential French naturalist Georges Louis Leclerc, Comte de Buffon (1707–1788):

The more one increases the number of divisions in natural things, the closer one will approach the truth, since there actually exist in nature only individuals. . . . The Genera, Orders, and Classes exist only in our imagination.

Buffon, however, did accept Ray's supposition that two individuals belong to the same species if they are able to produce fertile offspring.

The basic nomenclature used by Linnaeus is still employed today, although a number of taxonomic systems are now used, either singly or in conjunction with other systems. One of these systems, which is known as *phenetics*, begins by taking note of all of the characteristics that the classifier deems significant. It thus differs from Linnaeus's approach in that it uses the maximum amount of information rather than a single feature like Linnaeus's stamens and pistils. The presence or absence of these characteristics are then used as the basis for deciding where the organism belongs. Since many characteristics may be taken into account, the practice of phenetics often requires computer analysis. In fact, this system received its great impetus in the 1950s when computer algorithms began to be applied to taxonomy.

Phenetics does not take into account the evolutionary history of the organisms under consideration, although it does not preclude it. Evolution, however, is central to the taxonomic system known as *orthodox classification* or *evolutionary taxonomy*. In this system, groups of organisms are arranged into a hierarchy that reflects what is assumed to be their evolutionary history. In general, each group is supposed to derive from a single origin, although groups may be classified differently from their ancestors if they are distinctive in some important way, or if the group is quite large. As with other taxonomic systems, it is hierarchical; that is, the evolutionary relationship of different groups of organisms gets closer together as we move toward the species level.

A third means of classification goes under the name *cladistics* (sometimes called *phylogenetic systematics*). Originated by W. Hennig (1913–1976), a German entomologist, this system puts organisms into groups (*clades*) on the basis of common features (known as *homologies*) that are believed to indicate common ancestry. These groupings are usually depicted on a branching diagram known as a *cladogram*, in which each branching point represents a divergence from a common ancestor. A cladogram that includes many organisms may have branching points at a number of different levels. Because new species are assumed to branch off suddenly, cladistics is particularly well suited to the theory of punctuated evolution.

In recent years, the theory and practice of taxonomy has been enhanced by the use of new biochemical discoveries and techniques. One of the latter is electrophoresis, which is used to separate organic substances for purposes of analysis. Through the use of electrophoresis, comparative studies of enzymes provide important clues about relationships between groups of organisms. Similarly, DNA analysis can reveal a great deal about the taxonomic proximity of different groups of organisms.

See also ALGORITHM; DNA; ENZYMES; EVOLUTION; EVOLUTION, PUNCTUATIONAL MODEL OF; NATURAL SELECTION; SPONTANEOUS GENERATION

Technocracy

The general term for a science-based mode of government is *technocracy*, rule by those with scientific and technical expertise. The idea of selecting a governing class on the basis of their wisdom and expertise goes back at least to Plato's advocacy of a reign of philosopher-kings. In the 17th century, Francis Bacon made the famous assertion that "knowledge is power," although in his time political authority was more often legitimized by appeals to religious beliefs, as with the doctrine of the "divine right of kings." In the 19th cen-

tury, the great intellectual and practical achievements of science and technology seemed to offer a new foundation for political rule. One of the most influential champions of technocratic rule was the French political philosopher Henri Comte de Saint-Simon (1760–1825), who anticipated the emergence of a society in which the old religious-political governing class would be supplanted by a new elite of scientists and industrialists.

The awe-inspiring progress of 20th-century science and technology made this idea all the more appealing, especially when the achievements of scientists and engineers were contrasted with the workings of governments run by often corrupt officials. One of the intellectual leaders of the early 20th-century technocratic movement was the American economist-sociologist Thorstein Veblen (1857–1929). Veblen's analysis of the American economy and society centered on a conflict between rationality and the search for efficiency on the one hand and a profit-oriented business culture on the other. The former was the domain of the engineer, while the latter was the realm of a "leisure class" whose prime interest was conspicuous consumption. As Veblen saw it, the existing business elite only impeded economic and social progress; future economic and social progress rested on the political ascendancy of engineers and technicians. Veblen hoped to see the realization of a technocratic government in a "soviet of technicians," a vaguely defined administrative group that would come to power as a result of a general strike of workers and technical personnel.

At about the same time that Veblen was dissecting American society, a group of engineers and their supporters were organizing a movement to establish a technocracy. In 1919, an organization known as the Technical Alliance recruited a few scientists and engineers to its cause, including some with impressive credentials like the legendary mathematician and engineer Charles Steinmetz (1865–1923). As a prelude to a technocratic government, the alliance began to conduct an ambitious survey of the U.S. industrial economy and its energy and resource requirements, but it was never brought to completion. Technocracy went into hibernation as a political movement, but the Depression of the 1930s brought new urgency to the need for social and political transformation. At that time, a technocracy movement centered at Columbia University's Department of Industrial Engineering gained a brief notoriety. But the movement soon foundered due to incompetent leadership, internal disputes, the inability to form a coherent plan of action, and an inability and unwillingness to organize a mass movement.

Engineers, supposedly the key players in the new technocratic order, showed little interest in technocracy as a philosophy or political plan of action. The U.S. technocracy movement was comprised of doctors, lawyers, middle managers, and small businessmen; engineers were conspicuous by their absence. The fact that the movement had little appeal to engineers indicated that their ambitions lay elsewhere. Many engineers aspired to be more than technical minions, but most of them attempted to realize their ambitions by working within the existing system. They sought to enhance their authority by acquiring managerial positions within their firms, not through a quixotic attempt to seize political power. In this venture, another doctrine derived from engineering practice, Scientific Management, proved to have greater appeal.

In the United States and elsewhere, organized technocracy movements have been historical failures. Nonetheless, elements of technocratic thinking can still be found in a variety of governmental structures and practices. To a certain extent, technocratic approaches have enhanced the functioning of government by bringing expertise to bear on specific problems. At the same time, however, an excessive enthusiasm for technocratic modes of governance can be harmful. The reasons for this lie in the basic precepts of technocracy.

An essential assumption of supporters of technocratic governance is that the problems confronted by government can be solved by the same methods used to solve technical problems. As one electrical engineer claimed in 1917, "It matters not whether the problems before him are political, sociological, industrial, or technical, I believe that an engineering type of mind is best fitted to undertake them." In fact, the notion that there is no essential difference between technical and sociopolitical issues is the major fallacy of technocracy as a form of government. Social and political problems are inherently more difficult to solve than scientific and technical ones, if for no other reason than that they entail conflicts of interest between groups and individuals. Under these circumstances, there is usually

no "right" solution; either a compromise is hammered out or one particular group succeeds in imposing its will on everyone else.

Technocratic modes of governance attempt to substitute administrative procedures for the workings of politics. In its proper place, administration is an essential part of governance, for it uses rationally derived rules to achieve given goals. The selection of goals can be a matter of dispute, however, and many of them have to be determined through a conflict-laden political process. Moreover, technology and science cannot be the source of the values that inform our choice of goals. In particular, by themselves science and technology cannot resolve a fundamental issue confronted by societies and their governments: how to distribute limited goods, opportunities, and power among the citizenry. Addressing these matters always entails differences of opinion and a fair amount of conflict.

The differences between political and technical problems can be illustrated by considering one of the greatest triumphs of science and technology, landing men on the moon. The moon landing was an extraordinarily difficult technical problem, but it was easier to accomplish than many other objectives of government. There were few philosophical or ethical objections to landing on the moon. The goal was clear-cut, and there were no organized pressure groups seeking to divert the space program to a landing somewhere else in the solar system. There was initial disagreement about the plan of attack, but it was resolved largely through a consideration of objective data. And the moon did not try to prevent a landing on its surface. In short, once the political decision to go to the moon was made, most issues were resolved on technical criteria alone. Most other issues of governance are not so straightforward. If they were, technocracy would be the preferred form of government.

See also SCIENTIFIC MANAGEMENT

Technological Enthusiasm

A persistent debate among historians and sociologists of technology concerns the question of why human beings invent new things. Sometimes a distinction is made between various "utilitarian" motives and other motives that have more to do with curiosity, self-expression, kinesthetic pleasure, a sense of play, or the exercise of ingenuity for its own sake. Taken in these terms, the debate may be cast as a simplified but nonetheless instructive dichotomy: inventors pursue novelties they believe will make a genuine difference to their own world, or they invent what they believe may be possible to invent, aside from any substantial concern about social consequences. The impulse behind inventions calculated to affect society directly may be pecuniary or altruistic or something else altogether; it may even be a product of antisocial or destructive urges. At first thought, motivations for the other sort of invention—when social receptivity seems wanting or feasibility problematic—may seem more elusive. Actually they are not so difficult to understand. They may be subsumed under the rubric of "technological enthusiasm."

Henry Ford (1863–1947) once observed that the success of an invention is contingent on an affirmative answer to three questions: Is it needed? Is it practical? Is it commercial? He was expressing a viewpoint that seems logical and yet, on consideration, raises more questions than it answers. Inventions in certain realms, notably military technologies, are "commercial" only in the sense that military contractors may stand to profit from government contracts. "Practicality" is always a slippery notion, and "necessity" even more so. Many inventions, from the telephone to automobiles and airplanes to new materials such as celluloid and aluminum, met only marginal needs at the outset. Needs with substantial economic significance had to be *contrived*; turning an old adage on its head, invention became the mother of necessity. Indeed, one of the most intriguing perceptions deriving from studies in the history and sociology of technology is that so many inventions seem to have derived from the question, Is it possible?

One can stir up a lively debate by asserting that "If something can be invented, it will be," for this skirts dangerously close to the discredited realm of determinism. But unless one concedes that inventions may often bear little connection to need, practicality, or commercial potential, many technological artifacts are impossible to explain. The realm of transportation technology is rife with examples. Supersonic airliners seem positively mundane compared to magnetic levitation or very-high-speed transit (VHST), passenger-carrying "gondolas" floating on magnetic fields moving at speeds of 16,000 kph (10,000 mph) in underground tunnels devoid of air. Partisans of VHST insist that the idea is both technically feasible and economically sound. Since it proved possible to send a man to the moon, it is hard to dismiss the claim to feasibility out of hand. Strong doubts about the economics surely seem warranted, and yet this does not mean that we will never see VHST materialize, provided that, somehow, a persuasive rationalization can be constructed for its utility. Utility is the one justification for technological novelty that our urban-industrial civilization (though not all past civilizations) will almost invariably accept as valid. Indeed, in the 19th century, the term *useful arts* was often used interchangeably with *technology*, and the Industrial Revolution in essence represented the discovery that technology could be an instrument for profit. Calvinism identified the useful with the profitable, and both with the Godly.

And yet capitalist ideology has served to mask powerful and persistent nonutilitarian motives for novelty, foremost among them being sheer enthusiasm. In reality, enthusiasm has driven many technologies that ultimately proved to be needed, practical, and commercial. It has been noted, for example, that powerful steam locomotives were a product of a great human urge to invent everything that developing means permit; perhaps the personal computer stemmed from this same urge. But enthusiasm has also driven inventions that never had any chance of gaining an affirmative answer to Ford's three questions. Rather, we may consider these inventions as responses to deep-seated psychological urges that may fairly be termed irrational—examples including those 16,000-kph gondolas or a plant designed "to run without men" and turn out automobile frames at the rate of 1 every 10 seconds, even when there was no such demand for frames nor any scarcity of labor.

The explanation for such curious technological novelties may be found to a significant extent in sociological studies indicating that "love of inventing" constitutes a prime motivation for more inventors than does the prospect of financial gain. Some enthusiasts may remain satisfied simply to see that love fulfilled, nothing more. But with extraordinarily costly contrivances such as VHST, or superconducting supercolliders or manned space stations, which have no commercial possibilities attractive to potential investors, there are strong inducements for enthusiasts to *invent* social benefits in order to gain funding from the public purse. Governmental purse strings tend to open most readily when the benefits are portrayed in terms of military advantage, international prestige, "competitiveness," or a particular vision of cosmic destiny. Whether or not inventors attempt to construct some such grand rationale, and whether or not they succeed in dipping into the public purse, it remains clear that the impulse behind technological novelty may often be traced less readily to any chance of an affirmative answer to Ford's questions than to the psychological makeup of people who find novelty seductive and find invention emotionally rewarding in and of itself—that is, to the phenomenon of technological enthusiasm.—R.C.P.

See also COMPUTER, PERSONAL; INDUSTRIAL REVOLUTION; MAGNETIC-LEVITATION VEHICLES; SPACE STATION; SUPERCONDUCTING SUPERCOLLIDER; SUPERSONIC TRANSPORT (SST)

Technological Fix

All too often technology seems to advance at a faster rate than many other aspects of human life. The last 30 years have seen millions of transistors packed onto a fingernail-sized chip, space probes that have left the solar system, and computer networks that literally have wired the world together. But alongside these triumphs can be found the same stubborn problems, some of which may even be intensifying: poverty, ethnic strife, and violent crime, to name only some of the most prominent. This contrast has led to a desire to use technology to solve not just technical problems, but social ones as well. A technology that is used for the latter purpose is known as a *technological fix.*

Many examples of technological fixes may be cited. Drugs such as Methadone and Antabuse are used to counter heroin and alcohol addiction, respectively. Special paints have been developed to allow the easy removal of graffiti. Airbags protect victims of car accidents. Technological fixes, diverse as they may appear, share a number of commonalties. In the first place, the technology that is applied may make some headway with the immediate problem, but it does not

attack its source. The administration of Methadone and Antabuse does not address the social, psychological, and economic causes of drug and alcohol addiction. Special paints may allow easy removal of graffiti, but they do nothing to rectify the attitudes and social conditions that lead some people to deface every available space. Airbags save lives and reduce injuries, but they are often deployed because a drunk or otherwise incompetent driver was incapable of driving safely.

Technological fixes can be effective, but in many cases their value is limited because social problems are fundamentally different from technical problems. One of the major differences between social and technical problems is that technical problems are usually far less ambiguous than social ones. Putting a man on the moon was a stupendously difficult task, but at least the goal was absolutely clear. This made it very different from goals such as reducing crime. "Crime" encompasses a wide variety of transgressions: drug dealing, burglary, spouse abuse, and embezzling, to list only a few. No single technological fix can cover all of these equally well. Moreover, some fixes may serve only to move the problems from one place to another. The installation of a sophisticated burglar alarm system may prevent one house from being robbed, but only because it convinced a burglar that the house next door was a better target.

To make matters even more difficult, social problems are inextricably bound up with human motivations and behaviors. The physical entities that technology customarily deals with are far simpler; the components of the most complex electronic circuits are models of regularity and predictability when compared to individual human beings. Trying to understand human actions is an exceedingly difficult task; changing these actions is even harder. The moon didn't fight back when Apollo 11 landed on it, but people will usually resist efforts to change their behaviors.

Effective technological solutions are generally achieved only within closed systems, that is, collections of interacting components that are isolated from external influences. This means that when problems arise, only a small number of factors need to be considered. For example, when a refrigerator starts making a strange noise, there are a limited number of parts and subsystems that may have malfunctioned, and all of them lie within the refrigerator itself. In contrast, when an educational system produces large numbers of illiterates, the sources of the problem are likely to extend well beyond the classroom.

Finally, no problem, technical or otherwise, is ever really solved; no mechanism is perfect, nor is any social system perfect. The persistence of unsolved problems is of course one of the major sources of technological advance, for they motivate efforts to at least come closer to a solution. The same may be said of social problems, but again, the nature of the new problems may be fundamentally different. On many occasions the new problem that crops up may prove even more intractable than the old one, or at least less amenable to the application of another technological fix. The outstanding example of this is medicine, where new technologies have saved lives and eased pain, but at the same time have inflated costs and generated a host of ethical dilemmas that are nowhere near solution today.

The inherent difficulty of applying technological fixes to nontechnical problems has not prevented occasional attempts to take the politics out of governance by treating it as a technical exercise. To be sure, some issues confronted by governments are largely technical, such as determining the best material for surfacing a road. But the important questions, such as where the road should be built—or if it should be built at all—are by their very nature nontechnical.

In sum, technological fixes may successfully treat certain kinds of problems, but not others. In contemplating the use of a particular technological fix, it is necessary to keep in mind the nature of the problem, and how this is likely to determine the effectiveness of the solution. It is also to important to understand that technological fixes usually cannot serve as substitutes for economic, social, or political reforms.

See also TECHNOCRACY

Technological Innovations, Diffusion of

New technologies are constantly being created. Most of them, however, never amount to much, and few inventors realize their dreams of wealth and glory. To be successful, a technology has to spread from its origin to factories, offices, homes, or farms; this process is known as the *diffusion of technology*. Through diffusion, an *invention* becomes an *innovation*, something that has made its way into the economy and society.

One form of diffusion occurs when a technology that has proved its worth is copied somewhere else. Historically, this has been one of the major avenues of technological advance. Although Western nations have been at the forefront of technological change in recent centuries, in times past many crucial inventions originated outside the orbit of Western civilization: the spinning wheel, the magnetic compass, and an effective horsecollar, to name a few.

Today, a substantial amount of technology diffuses through both authorized and unauthorized copying. The availability of an international stock of technology gives technologically backward places a chance to catch up through what has been labeled "reverse engineering." Even so, copying an artifact or process is not always a simple, straightforward procedure. Possessing a microchip is of no value if the equipment for making microchips is not available. In general, successful reverse engineering requires the presence of complementary inputs. Equipment and materials have to be present, as do the needed skills—often the rarest input of all.

Occasionally, the importation or copying of foreign technologies will make it possible for a country to leapfrog several stages of development and gain the most advanced technology available. But the human and material inputs have to be in place, and past experiences with earlier technologies often provide the firmest bases for their development.

Indigenous abilities may be essential because many technologies have to be substantially modified as they diffuse from one place to another. For example, the steam locomotive was invented in England, from which it quickly made its way to the United States. There it encountered a geographical and economic environment that was markedly different from that of England. At this time, the United States was rapidly populating its frontier regions, and the quick construction of railroads was a key part of this expansion. Under these circumstances, the planning and construction of railroads often was a hurried process. Tracks were laid down with as little excavation as possible, so many lines had numerous sharp curves on their routes. English locomotives were ill suited to trackwork like this, so from any early date, it was necessary to equip American locomotives with pilot wheels that helped them ease into the abrupt curves. And this innovation was the work of American engineers.

Some important examples of technological diffusion do not involve the direct transfer of a technology, only the general idea underlying that technology. This process has been labeled *stimulus diffusion*. An example of stimulus diffusion is the construction of windmills in medieval Europe. The first windmills were built in the Middle East, where they were duly noted by Crusaders and other travelers. They then carried back to Europe the notion that wind could be harnessed as a source of power, although the windmills eventually constructed in Europe followed an entirely different design than the ones that had been built by Arabs and Persians.

Diffusion often entails substantial changes to a technology as it moves from one setting to another. Many technologies have to be altered just so they work at an acceptable level in a new environment. The technology also may have to be modified so that it is within the purchasing range of prospective users. It is one thing to make something work in a laboratory situation; it is quite another to manufacture it at a price that customers are willing to pay. And, after having been bought, the product has to work with a modicum of reliability even though its users may subject it to a great deal of intentional and unintentional abuse. This can be a challenging task; it has been observed that nothing can be made completely foolproof, for fools are infinitely ingenious.

Ideally, all the bugs are ironed out while the technology is still under development, but in many cases, development has to proceed even after the product has been put on the market. It is often the case that customer needs will necessitate a substantial redesign of a device or process. Often, it will be the customers themselves who take an active role in the modification or extension of a technology. For example, a great deal of the technical progress of digital computers was the result of their customers finding new uses for them. These applications often stretched the technical capability of the computers and provided a major stimulus

The leading truck of the *John Bull* allowed this English locomotive to be used on the more sharply curved track of an American railroad (from Brooke Hindle and Steven Lubar, *Engines of Change*, 1986).

for the development of improved hardware and software.

The diffusion of a technology means the acceptance of something new, and novelty usually presents some potential hazards. Any prudent adopter of a new technology needs to make a careful assessment of a new technology and its likely effects. Individuals have to weigh the costs and benefits, and organizations have to take into account how well a new or substantially modified technology can be accommodated to existing infrastructures and organizational cultures. The potential rewards of using a new technology may be substantial, but so can be the risks.

Some of these risks center on the technology itself and how well it functions, but even greater risks often attend the marketing of a new technology. Indeed, the probability of technical success is usually easier to determine than the probability of marketing success. "Will it fly?" is easier to ascertain than "Will anybody pay good money to fly in it?" Some technologically advanced products such as videophones and the Concorde supersonic airliner absorbed vast amounts of money but never found a market of adequate size.

Real or perceived, the riskiness of a new technology can be a major impediment to its diffusion. In organizations, qualms about adopting a new technology may be reduced when it has a "champion" within the organization, someone with a personal stake in the successful adoption of the new technology. Champions are particularly likely to be present when the technology has been developed inside the organization, and its advocate has been involved in some aspect of its development. In contrast, a technology that comes from the outside lacks this built-in support. This makes it likely to fall victim to the "not-invented-here syndrome," the passive or active resistance to an invention that has come from elsewhere.

Internal resistance is not the only thing blocking the diffusion of technologies. Sometimes, deliberate attempts are made that prevent the transfer of a particular technology. Individuals, firms, and whole countries have on occasion sought to block the movement of technologies in order to maintain their monopoly control. This sometimes has entailed the use of harsh measures, as when the city-state of 16th-century Venice sent assassins to poison expatriates who had carried the secrets of Venetian glassmaking to other lands.

Today, technologically advanced nations sometimes use export licenses in order to prevent the export of technologies deemed vital to national security. These efforts are not always successful because the most crucial piece of information simply may be that a technology exists. During the late 1940s, the explosion of the first atomic bomb by the Soviet Union was the occasion of great anguish in the West, and it was assumed that Soviet success was the result of its having obtained secret information from the United States. But the Soviets had long understood that the fissioning of an atomic nucleus unleashed immense amounts of energy, and knowing this, they poured vast amounts of money and expertise into their nuclear program, assured that a fission explosion was possible. In general, knowing that a problem is capable of solution is sometimes the most important step in its solution.

Governments also restrict the diffusion of certain technologies through the enactment and enforcement of patent regulations. The issuance of patents is justified because they are believed to promote inventive activity by giving the inventors exclusive rights to the commercial use of their inventions In this way, patents retard the diffusion of some technologies, but in other ways patents may stimulate technological diffusion. Of particular importance is the fact that the filing of a patent application makes an invention public. Although direct copying is forbidden, access to the public record may give a clearer sense of how a particular technical problem has been addressed and may provide insights into how alternative solutions may be found. Alternatively, firms can obtain licenses to patents by paying a fee to the patent holder. This arrangement may promote the more rapid diffusion of a technology because important supplementary material is often included as part of a licensing agreement.

Finally, the diffusion of technology may be affected by more than economic considerations; a new technology may be rejected because it conflicts with a people's habits and preferences. In the late 1940s, agricultural extension agents succeeded in getting many farmers in northern New Mexico to adopt hybrid corn. Their yields immediately doubled, yet within 2 years the farmers had reverted to the planting of traditional strains. Hybrid corn was rejected because it could not be made into tortillas with acceptable taste and texture. Given existing culinary preferences, hybrid corn was not an appropriate technology.

The issue of a technology's appropriateness is particularly salient today, given the fact that the vast majority of new technologies are developed in the industrialized world. Under these circumstances, most instances of diffusion consist of technologies moving from the more-developed to the less-developed parts of the world. However, technologies that have been created in countries with abundant supplies of capital and human skills may not be suitable for countries lacking these attributes. In particular, the diffusion of capital-intensive, labor-saving technologies may be counter-productive for countries with shortages of capital and an abundance of low-skill labor. For these reasons, there has been a strong interest in the development of appropriate technologies for these countries.

See also APPROPRIATE TECHNOLOGY; ATOMIC BOMB; COMPASS, MAGNETIC; CORN, HYBRID; HORSECOLLAR; NOT-INVENTED-HERE SYNDROME; PATENTS; RAILROAD; RESEARCH AND DEVELOPMENT; RISK EVALUATION; SPIN-NING WHEEL; LOCOMOTIVE, STEAM; SUPERSONIC TRANS-PORT (SST); VIDEOPHONE; WINDMILLS

Further Reading: Everett Rogers, *The Diffusion of Innovations*, 3d ed., 1983.

Technological Unemployment—see Unemployment, Technological

Technology Assessment

Technology assessment is a governmental process providing public disclosure of anticipated effects of the introduction of new technologies in the environment, as well as in economic, social, and political life. Information provided by technology assessments is used by legislators, administrators, and decisionmakers to shape public policy, and to evaluate alternatives for governing the use and diffusion of new technologies in society. Technology assessments may also identify potential political or foreign-policy issues that arise from the application of new technology.

Technology assessment focuses on anticipating future results of new technologies. For the most part, it has not been applied to analyzing the consequences that followed the introduction of technologies in the past, although scholars have occasionally produced "retrospective technology assessments" as an intellectual exercise.

Until recently, technology assessment was conducted for the U.S. Congress by the Office of Technology Assessment. State legislators use technology assessment to inform local development and tax policy, as well as assess risks to public health and environmental quality of new technologies. The techniques of technology assessment are employed by industry to evaluate benefits and costs of new product development, and to forecast long-term market trends.

Planning techniques used in technology assessment such as forecasting, econometric modeling, organizational analysis, decision theory, and cost-benefit analysis date from the 1930s when New Deal national planning addressed the economic and geographic dislocations of the Great Depression. Significant innovations in these techniques also emerged as part of the military effort of World War II and the onset of the Cold War, as well as the regional planning and development activities of the Tennessee Valley Authority.

Interest in technology assessment resulted from intense social and political concerns about technology arising in the late 1950s and early 1960s. The Port Huron Statement of 1962, authored collectively by Students for a Democratic Society, observed that "automation, the process of machines replacing men in performing sensory, motoric, and complex logical tasks, is transforming society in ways that are scarcely comprehensible."

Another early statement of concern about technological change came from the Ad Hoc Committee on the Triple Revolution. The Ad Hoc Committee wrote to President Lyndon Johnson in 1964 urging a "reexamination of values and institutions," and citing a "cybernation revolution" in which "the combination of the computer and the automated self-regulating machine" creates a system of "almost unlimited productive capacity which requires progressively less human labor." The Ad Hoc Committee stated that it was necessary to adopt "just policies for coping with cybernation and for extending rights to all Americans" in order that public discourse about "the supreme issue, peace, can be reasonably debated and resolved."

During the 1960s, technology assumed a prominent and problematic place in the lives of Americans. The National Aeronautics and Space Administration made good on President John F. Kennedy's promise to land Americans on the moon in July 1969; the integrated circuit developed rapidly; some drugs such as DES and thalidomide were found to cause cancer and birth defects, respectively; efficiently destructive chemical weapons such as napalm and Agent Orange were used extensively in Vietnam; and environmental disasters such as the diffusion of DDT and the Santa Barbara oil spill highlighted the unfortunate consequences of technology. And all the while these technological stories were being transmitted almost instantly through the medium of television.

The accelerating pace of change led to calls by leaders in science, technology, and government to use the methods of science and technology to shape the course of technological advance. The National Academy of Sciences stated in 1969:

> . . . as the number of people who share in the material benefits of technology has grown, advances in science and technology have brought advances in our ability to anticipate the secondary and tertiary consequences of contemplated technological developments and to select those technological paths best suited to the achievement of broad combinations of objectives.

In the ensuing years, the process of technology assessment developed a general methodology, with structures and practices similar to those of environmental assessment. Assessment teams are typically interdisciplinary, reflecting the belief that problems raised by new technologies cross traditional intellectual and professional boundaries. Technology assessments begin with a determination of the range of issue areas to be studied. A description of the proposed technology is developed, including the breadth and nature of its application, along with a description of the existing social and/or environmental conditions. The cause-and-effect interactions of the new technology and society and/or the environment are then analyzed, including potential productivity gains, health risks, economic costs, and so on. Alternatives are developed by the analysts, and their effects are compared with the effects of the proposed technology, including whether the alternatives could achieve the same desired effects.

The technology assessments performed by the U.S. Congress's Office of Technology Assessment (OTA) spanned a vast range of topics in the 23 years of its existence. For example, OTA authored reports on healthcare, genetic engineering, water resources, agricultural policy, the effects of nuclear war, computerized manufacturing, early warning systems for plant closures in industrial communities, high technology implications for U.S. competitiveness in the global economy, the need for new statistical systems and new economic concepts for interpreting the changing U.S. economy, and the relationship of multinational corporations to "the national interest."

Over the years, practitioners of technology assessment borrowed forecasting techniques from demography, epidemiology, probability theory, and other mathematical techniques to address the problem of risk in the application of new technologies. These studies usually are referred to as *risk assessments*, and th∏ese forecasting techniques are often used to estimate the potential increase in illnesses and deaths that might be reasonably attributable to the adoption of a new technology, such as a drug, a nuclear-power plant, radioactive and hazardous waste disposal sites and management systems, application of pesticides, and genetically engineered microorganisms.

As economic production assumed a more global character during the 1980s, technology assessment enabled policymakers to grapple with issues of trade policy, technological change, employment, and access to international markets. In the uncertain environment of global markets, corporations and government integrated technology assessment into their decisionmaking processes pertaining to trade, capital investment, and employment and training issues.

Technology assessments clearly increased the volume and availability of information on new technologies and new policy initiatives. However, information about new technologies has not catalyzed actions hoped for by the Ad Hoc Committee on the Triple Revolution. Nonetheless, at key moments, such as publication of the Office of Technology Assessment's "The Effects of Nuclear War" report in 1982, technology assessments have influenced policy debates, as was originally envisioned by its champions.—T.S.

See also AGENT ORANGE; APOLLO PROJECT; CYBERNETICS; DDT; DES; INTEGRATED CIRCUIT; NAPALM;

National Academy of Science; Office of Technology Assessment; oil spills; risk evaluation; statistical inference; Tennessee Valley Authority; Thalidomide

Technology, Social Construction of

The key idea underlying "the social construction of technology" is that social arrangements create, shape, and determine technologies and how they are used. From this perspective, technological change is not a simple matter of "necessity being the mother of invention," i.e., technologies emerge because there is a need for them. Nor do technologies advance through their own internal processes, e.g., technical deficiencies become evident and are duly overcome. Rather, technological change is a process that is strongly affected by social and political arrangements, economic interests, and cultural patterns. By taking into account nontechnical factors; the social construction of technology has much in common with the "externalist" approach to the history of science, from which it had received much inspiration.

Scholars who describe and analyze the social construction of technology do not see society as being homogenous or in a state of peaceful equilibrium. From their perspective, society is composed of a variety of actors, both individual and organizational, with interests and agendas that affect the development and choice of particular technologies. Consequently, technologies succeed or fail not because of their intrinsic virtues (or lack of them) but because individuals, groups, and organizations have an interest in a particular outcome. Discussions of technological change that involve social construction often take into account how the different actors are connected and the strategies and tactics that they employ.

The influence exerted by individuals and organizational actors over technological change can be of paramount importance because the uses and value of a particular technology are not always obvious. From the standpoint of a social construction of technology, the actors involved with particular technology may have quite divergent notions regarding the significance of a single device or an entire system. This characteristic is known as *interpretive flexibility*, and it has an important implication for how a technology is evaluated. Whether or not a technology "works" is not an intrinsic quality, nor is it determined by technical factors. What matters are the needs, expectations, and beliefs of the people who interact with it. For example, the high-wheeled bicycle of the late 19th century was difficult to ride and more than a bit dangerous. If a bicycle is defined primarily as an instrument of transportation, the high wheeler has to be deemed an unsuccessful technology, for it doesn't work very well in this capacity. However, if a bicycle is seen as a sporting instrument that gives its riders opportunities to demonstrate their bravery and skill, the high wheeler is a successful technology. Conversely, the "safety bicycle" with its equal-sized wheels and lower seat height was a better transportation device, but it lacked the inherent drama of the high wheeler. Thus, there is no universal standard for determining which type of bicycle constitutes the more effective technology; it all depends on who is in a position to determine a bicycle's "proper" use.

When the relevant social groups have formed a consensus regarding the purposes and suitability of a particular technology, a state of "closure" is said to have occurred. The technology is no longer subject to interpretive flexibility; it has become a "black box" to its users, a taken-for-granted technology that seems almost to be part of the natural environment. A technology that has reached closure is so imbedded that people find it difficult to think of alternative ways of doing things. Closely related to closure is "stabilization." Technologies reach a point of stabilization because the social roles and statuses connected to them have themselves stabilized. Stabilization sometimes occurs because an individual or group is in a position to exert power, and it is in their interest to see to it that a particular technology succeeds. Alternatively, a technology may stabilize because compromises have been made between significant individuals and groups.

Because technologies are socially constructed, an "invention" does not take place at a particular moment. Instead, invention is a social process that involves a number of actors over what may be a long period of time. In most cases, only after a technology had reached closure will it even be recognized as an invention. When this happens, the assumption is made that the invention is superior to other ways of doing things, and that it was destined to achieve this position. Adherents of the perspective of social construction of technology have pointed out that narratives of how inventions happen are similar to histories of scientific discovery. The social processes that shape the pathways to scientific knowledge are ignored, and the scientific discovery is seen as successful because it accurately reflects the natural world.

The approach of social construction of technology has provided new insights into the nature of technological change, but it has not met with universal approval. In particular, it has been criticized for paying insufficient attention to purely technical considerations. According to critics, technologies are not solely the product of social shaping: The manner in which a technology develops has to reflect inescapable physical realities. Some technologies fail to take root because they cannot escape the constraints of the real world, no matter how strong their social support. This criticism aside, it is safe to say that valuable insights into the nature of technology and technological change can result from scholarship that takes into account the intersection of technical and social factors.

See also BICYCLE; HISTORY OF SCIENCE, INTERNAL AND EXTERNAL

Further Reading: Wiebe E. Bijker, *Of Bicycles, Bakelite, and Bulbs: Toward a Theory of Sociotechnical Change*, 1995.

Teflon®

Teflon is a perfect example of a serendipitous discovery—something of value that has been found while looking for something else. In 1938, Roy Plunkett of the DuPont Chemical Company was trying to develop a new refrigerant when he found that a bottle thought to contain tetrafluorine gas instead contained a slippery white powder. Unbeknownst to Plunkett at the time, the tetrafluorine had polymerized into a kind of plastic that was resistant to all known acids and solvents, withstood a great range of temperatures, and acted as an electrical insulator. The substance was polytetrafluoroethylene resin, or Teflon, as it was named by DuPont (the name is a combination of *tef*, an informal

name for tetrafluorine, and *lon*, a suffix that DuPont used for a number of its products, most notably nylon).

Subsequent research determined that Teflon gained its unique properties from its molecular arrangement, a long carbon chain in which individual carbon atoms are surrounded and hence shielded by fluorine atoms. The unique properties of Teflon made it a valuable substance during World War II, when the program to build the first atomic bombs used Teflon as a gasket material because it was not affected by uranium hexafluoride. Kept secret throughout World War II, Teflon appeared on the commercial scene in 1948, although it found few uses at first. In late 1954, two Frenchmen, Louis Hartman and Marc Grégoire, discovered that cooked food would not stick to a pan coated with Teflon. In 1956, a French company introduced a Teflon-coated nonstick frying pan. Unfortunately, these early utensils suffered from a serious drawback: The Teflon coating was not tightly bonded to the metal and so would easily come off when mishandled. Techniques eventually were developed to bond the Teflon tightly to the metal, making nonstick pots and pans truly practical.

Teflon is also widely employed as an insulating coating for electrical wires and in a number of other electrical applications. Teflon has also made artificial knees and hips possible because parts made from Teflon are not rejected by the body, as is the case for most other foreign substances. This quality has also made possible the implantation of artificial arteries and veins made from Teflon.

See also ATOMIC BOMB; HIP REPLACEMENT

Telegraphy

The ability to transmit messages across long distances had been sought for centuries. Fire, smoke, pigeons, and people have all been used for this purpose. The use of mechanical devices goes back to the early 19th century when the French constructed a chain of semaphores. Through the use of a standard code, these could send messages more than 1,000 miles as they relayed a message from one station to another. Although meeting with some success, the semaphore system was expensive to build and staff, and was useless at night or when weather conditions impeded vision.

Practical long-distance signaling became a possibility as people began to harness the mysterious force of electricity. In 1774, a Swiss inventor named Louis Le Sage devised a communication system that used two pith balls suspended at the end of a wire. When the other end was electrically charged, the two balls repelled each other. With a sufficient number of wires, each letter of the alphabet could be indicated, and messages could be sent by charging the appropriate ones. In the 1820s and 1830s, more-sophisticated applications of electricity employed early electrical meters known as *galvanometers*. Their needles would deflect in accordance with the strength of the current; thus, specific letters and numbers could be indicated by the extent of the deflection of the meter's needle. In 1833, the eminent mathematician Karl Friedrich Gauss (1777–1855) and his physicist colleague Eduard Weber (1804–1891) constructed such a system to link the University of Gottingen's observatory with its physics laboratory, a distance of $1\frac{1}{2}$ miles.

A few years later Charles Wheatstone (1802–1875) and William Cooke (1806–1879) of Great Britain collaborated on the development of a system of telegraphy that employed an important principle, first observed by Hans Christian Oersted (1777–1851), that an electric current will deflect a magnetized needle one way or another, depending on the direction of current flow. Using five needles that pointed to individual letters arranged on a grid, they were able to spell out words and sentences. Their system was employed by two British railroads and achieved fame when it was employed to apprehend a murderer who sought to escape by boarding a train in London. When he debarked at Slough, more than 32 km (20 mi) away, a policeman who had been telegraphically alerted was waiting for him. By 1880, 5,000 Cooke-Wheatstone telegraphs were in service, transmitting the equivalent of 15 million columns of *The Times* of London annually.

For all its success, the Cooke-Wheatstone system, as with the others before it, suffered from an excess of complexity. A simpler system, one that used a code for individual letters, made telegraphy a more-practical proposition. We associate this system with Samuel F. B. Morse (1791–1872), but in fact the derivation of Morse's telegraph and its associated code is a complicated story. Morse had no scientific training; his early career was as a portrait painter, writer, and pro-

fessor at the University of the City of New York (later New York University). He also had the dubious distinction of heading a proslavery organization during the Civil War and involving himself with anti-immigrant groups. In devising his telegraph, Morse made heavy (and unacknowledged) use of ideas of Joseph Henry (1797–1878), who from 1846 to his death served as the head of the Smithsonian Institution. Henry was one of the first to experiment with electromagnets. In 1831, he constructed a circuit comprising a battery, electromagnet, and switch. The electromagnet's core attracted one end of a bell's magnetized clapper. When a current was passed through the electromagnet's coil, an opposite magnetism was induced in the core, repelling the clapper and causing it to strike the bell. Morse employed this principle in a considerably more complicated apparatus. His apparatus consisted of a pendulum equipped with a piece of iron situated near an electromagnet, along with a pencil that made contact with a moving paper strip. Current flowing through the electromagnet attracted the iron, pushing the pendulum at a right angle to the moving paper. A brief passage of current produced a V-shaped mark on the paper as the pendulum quickly moved back and forth. A longer passage of current produced a V with a flat bottom as the pendulum moved, paused, and then returned to its initial position.

The opening and closing of the circuit was governed by lead plates that were grooved to form sharp- and flat-bottomed V's. A message would be translated into the appropriate series of protuberances, which contacted an arm of the switch that controlled the flow of current. In principle, a message encoded on lead plates could be sent over wires to the receiving apparatus some distance away. In reality, the system was like many emerging technologies in that it was needlessly complex.

A spectator at a demonstration of Morse's telegraph, a recent graduate of New York University named Alfred Vail (1807–1859), was intrigued by the device, so much so that he volunteered his time and money for the project. Before long he had replaced the cumbersome plate-and-pendulum system with a simple key for transmitting messages and a speaker for receiving them. The differently shaped Vs were replaced by dots and dashes, still known as the "Morse code," even though it is possible that it was actually invented by Vail. In time, skilled telegraphers were able to convert a sequence of dots and dashes into English at a rate of 60 words per minute, making telegraphy a very rapid way of transmitting messages.

In 1838, Morse, Vail, and two other partners journeyed to Washington in the hope of convincing Congress to financially underwrite the construction of an experimental line to demonstrate the telegraph's potential. Congress was receptive to the proposal and referred it to the House Commerce Committee. The Committee's chairman, Francis Smith of Maine, saw in the system an advantage for the nation—and for himself. He asked to be made a partner in the enterprise, and this being granted, he saw to it that the House appropriated $30,000 for the construction of a line between Washington and Baltimore. Even by the loose ethical standards of the time, Smith's involvement represented a flagrant conflict of interest, but nothing was done about it.

Supervising the project was Ezra Cornell (1807–1874). The initial plan was to bury the transmitting wires underground. But after a few miles had been laid, it was apparent that the wires were likely to short out. Cornell came up with the idea of attaching the individual wires to glass insulators mounted on the crossbars of a pole, a method still employed today. Although messages were sent while construction of the line was in progress, the conventional date for the "invention" of the telegraph is given as May 24, 1844, when Morse's famous query "What hath God wrought?" was sent from Washington to Baltimore.

In the ensuing years, the telegraph became an essential means of communication. The first transcontinental telegraph line was strung up in 1861, and within 5 years the U.S. telegraph network encompassed 100,000 route miles (160,000 km). The telegraph came to be a crucial medium for the operation of railroads, the sending and receiving of business and personal correspondence, and the transmission of news stories. In these ways, the telegraph extended the orbit of everyday life and quickened its pace. It vastly increased the amount of available information and made it readily available—everything from stock quotations to news of wars halfway around the world. This expansion of telegraphically transmitted information was not always beneficial. As critics have argued, the telegraph was the first stage in the creation of an environment that surrounds us with unconnected bits of information that have no direct relevance to our lives.

It can be argued that the telegraph marked a more revolutionary break with established methods of communication than any of the media that ultimately succeeded it. It was the first to employ electricity, and the first to send messages at the speed of light (the velocity of electrical current). It was also the first to use a binary system (the dots and dashes of the Morse code) for communication. So dominant was the telegraph that when radio first began to be employed for communication it was seen as nothing more than a wireless telegraph.

As it turned out, radio and other electronic means of communication severely undercut the supreme position of the telegraph. After World War I, the federal government, concerned about foreign control over radio technology, encouraged the formation of the Radio Corporation of America (RCA). Although it had a reasonable claim to be a participant in the new company, America's largest telegraph firm, Western Union, was not invited to join. Radio, originally used as a supplement to wired telegraphy, in time came to be the dominant medium.. Telegraph companies went into a slow, irreversible decline. A technology that had once been revolutionary was eclipsed by subsequent technological changes.

See also RADIO

Telephone

A telephone is a two-way communication device connected to transmission lines and switches that form a network serving many telephones and other devices. Some observers consider the American telephone network a giant computer, in which millions of telephone sets are terminals linked together. Telephones are designed mainly for voice communications, but other sounds may be transmitted and received as well. Special telephones are capable of sending and receiving video images together with speech.

Scottish-born Alexander Graham Bell (1847– 1922)

patented the first telephone in Boston on Feb. 14, 1876. An American-born inventor, Elisha Gray (1835–1910), tried to file a caveat for a similar device in the same patent office 2 hours later. There then ensued a series of unsuccessful lawsuits brought by Gray against Bell. This was not the only connection between Bell and Gray. At an earlier date Gray had been one of the founders of the Western Electric Company, which in 1882 became the manufacturing unit of the Bell Company. Over the next 100 years, it made most of the telephones and other apparatus for the American telephone system.

Bell had an ambitious set of long-range design goals: The instruments should provide a usable quality of speech, be efficient enough to transmit over a useful distance, be simple enough to use without special training, be stable enough to avoid excessive maintenance or adjustment, and, lastly, be produced by the thousands at a low unit cost. With the involvement of many other people, all these goals were achieved within 15 years.

Today, many different forms of telephones are available for purchase or lease. The most popular are desktop models used with a single line. Others include wall-mounted phones; cordless phones; cellular phones; telephone/answering-machine combinations; call director phones connected to multiple lines serving a number of different phones; extension phones connected to private branch exchanges (PBXs) used in hotels, hospitals, etc.; and public coin telephones (some of which can be activated by inserting a special credit card). Most telephones use analog transmission technology, but telephones based on digital transmission are available for commercial, government, and residential use.

One of the most controversial new capabilities for telephones is encryption, the extremely secure digital encoding of voice or data signals. Commercial customers can now buy a nonclassified version of a telephone set that government and military users know as the Secure Telephone Unit III (STU III). When used at both ends of a voice, fax, or data circuit, the commercial version encrypts transmission by simply pressing a button. Or it can be used in what telephone engineers call the Plain Old Telephone Service (POTS) mode. Even greater convenience is offered by a telephone security device, an easily portable unit that connects to conventional phone sets to give them encryption capability.

In the United States, some 150 million telephone lines currently include both public and private switched and point-to-point networks. The vast majority (almost 96 million lines) are analog public switched lines serving residences. The number of telephones connected to these lines is unknown, since a single commercial number, for example, might serve thousands of extension phones.

A telephone has three basic sections: the handset, the base, and the dial (which may be part of either the handset or the base). In the handset are a tiny microphone and a miniature loudspeaker. The microphone, or "transmitter," changes the sound waves of speech into electrical signals. The sound waves cause a flexible diaphragm to vibrate, activating other components that generate a fluctuating current. During the first 50 years of telephony, some 100 different transmitters were designed, though most were not used commercially. The carbon-granule microphone design was the basis for almost all telephones for more than 80 years. It has been largely replaced by the foil electret microphone.

The electrical signals are transmitted across the wires, optical fibers, and microwave links of the telephone network to another telephone, where the small loudspeaker, or "receiver," in another handset converts the electrical signals back into sound waves that can be heard by the person being called. Both the caller and the person being called can converse simultaneously because the telephones are connected by a two-way, or "duplex," circuit. All telephone conversations through the network will pass through copper wires as electrical signals at some point in the transmission, but at other points the signals also may be transmitted as radio waves or as light pulses.

The miniature loudspeaker, or receiver, converts the incoming electrical signals into sound waves so that speech becomes audible. The receiver has three basic parts: a diaphragm, a permanent magnet, and an electromagnet. The magnets shake the diaphragm, producing air vibrations that reproduce the caller's voice within the limits of the frequency spectrum generated by the transmitter. Normal human hearing has an acoustical frequency range of about 16,000 cycles per second or 16,000 Hertz (1 Hz equals 1 cycle per second). Telephone systems use only a narrow part of this frequency spectrum, a total range of 4,000 Hz. Of this, about 3,600 Hz are the audio portion; the other 400 Hz are used to carry dial tones, busy signals, and other operating requirements). Most people can instantly recognize the typical "telephone voice" sound, often used in radio programs, television, movies, and stage productions. To telephone engineers, the 4,000-Hz (or 4-kilohertz) spectrum is known as a *voice channel.*

Electricity for the telephone circuit is provided from a switching location called the "central office" or "central exchange." The power consists of 48 volts of direct current, supplied from a large rectifier/battery system that is kept constantly charged. This enables the telephone system to keep working if the local electric utility has a power failure. Central batteries were introduced in 1894, to replace the earlier system in which each telephone had its own onsite battery. Many of those phones also used a *magneto* or small hand-turned electric generator to send a ring signal to the operator.

Early telephone systems, especially in rural areas, often were set up with "party lines," in which the telephones of a dozen or more users (or "subscribers") were interconnected on a shared line. Special rings identified which party was being called, but anybody else on the line could listen as well. Business executives and government leaders wanted more privacy, so private lines were set up between the switch and their individual telephones. Eventually, most telephones in major cities were installed on private lines or at least minimum-number party lines.

The first telephones did not have a dial; a call was placed by a human operator at the central office who used plugs to connect the caller's line to the line being called. Human operators were gradually replaced by electromechanical switches, which evolved through several generations to the present-day digital electronic switching system for central offices, introduced around 1980. The early switches automated local calls only. Instead of giving a number to an operator, the caller used a dial to control how the switch directed the call through the network. Most local calls now require dialing seven digits. Direct dialing of long-distance calls was inaugurated in 1951, using 11 digits. In 1967, the first direct-dialed calls to London and Paris were made from New York City. Today, as many as 15 digits may be needed for an international dialed call.

By the end of World War I (1918), dial service was coming into use in significant numbers. The telephone sets were mostly of the desktop "candlestick" design, consisting of a vertical tube with a microphone on its top and a separate receiver hanging from a hook on the side of the tube. A rotary dial was mounted on its base. Rotary dials also were used on the so-called "French" telephones introduced in Europe at first and then the United States in the late 1920s. This design combined the transmitter and receiver in a handset that rested in a cradle atop the base. Many telephones still use the rotary dialing system, but most are now equipped with "touch-tone" dialing that employs pushbuttons. During the first 100 years or so, telephones were leased from the phone company, with the costs included in the monthly bills. Individual ownership of telephones was prohibited by the Bell System's 22 telephone companies (and many of the some 1,400 non-Bell or "independent" telephone companies as well) until October 1977, when the U.S. Supreme Court upheld a decision by the Federal Communications Commission, which enabled Americans to own their telephones if they so chose.

When the telephone handset is lifted from its cradle, an on-off switch called a *hook* is released. When a telephone is in use, it is "off-hook"; if the set is not in use, it is "on-hook." The switch in the central exchange detects the hook's status and either provides a steady humming sound, called a *dial tone,* or disconnects the switch from the line when the handset is replaced on the hook. If a dial tone is not heard when the handset is lifted, something is wrong with the line, and it cannot be used until the dial tone returns. If a number is dialed and a series of buzzing sounds is heard at the rate of about 60 interruptions a minute, the line being called is in use, or "busy." If the buzzing rate is twice as fast, the network is overloaded and all circuits are busy in that calling area.

The cords that connect the handset to the base unit, and the base unit to the telephone wall plug, are modular and can be replaced by depressing small plastic levers on the connection plugs. Various lengths of cords are available, so a telephone can be located at some distance from the wall plug that is linked to the telephone line. Cordless telephones are similar to the basic corded set, except that the handset is a combination receiver/transmitter/dial unit equipped with a small internal battery and a tiny radio receiver/transmitter system, including a small antenna pole. Cellular telephones, sometimes called *mobile phones*, are special radio sets designed for use in vehicles or in locations served by cellular networks.

Sociologically, the use of the telephone has been associated with radical changes in human society over the past century. The rapid spread of telephones enabled people to communicate with each other by voice instantaneously, even though they might be separated by many miles. The telephone has made possible an extended network of relationships, but not at the expense of local ties. People today often are ambivalent about the omnipresence of the telephone: The device is considered an absolute necessity, but when some callers abuse its availability, it is considered an inconvenience. The ease with which people can reach others by telephone means that written correspondence has decreased. Yet telephone calls can keep people in touch with each other on a far more immediate basis, relieving anxieties and promoting relationships.

Economically, the telephone has had immeasurable impact since its introduction. The scale and pace of commerce were increased beyond the comprehension of business in the late 19th century through the use of what computer experts call "real-time" decisions. Today's commerce could not function without the immediacy of communication via telephone line—including voice, data, fax, and video functions on all continents. National and global conferences are held by means of telephone networks that carry both voice and video. Door-to-door selling of products has been largely replaced by "telemarketing," the soliciting of sales by telephone. Catalog sales increased enormously with the introduction of "toll-free 800" numbers, in which the catalog company pays for incoming long-distance calls from customers. But most important of all, telephone networks freed commerce from the old limitations in which operations of most business firms—manufacturing, administration, warehousing, etc.—were clustered within a single region. Eventually, the telephone enabled companies to become true global enterprises. In fact, the world's telephone companies are themselves among the major employers in virtually every national economy.—R.Q.H.

See also AREA CODES; COMMUNICATION SATELLITES; DIRECT CURRENT; FAX MACHINE; FIBER-OPTICS COMMUNICATIONS; MICROPHONE; MICROWAVE COMMUNICA-TIONS; TELEPHONE ANSWERING MACHINE; TELEPHONE RECEIVERS; TELEPHONE SYSTEMS, CELLULAR; VIDEO-PHONE

Telephone Answering Machine

A telephone answering machine is a recording device attached to a single line, usually a phone in a residence or very small business office. It automatically goes off-hook (answers the incoming call), plays a recorded voice greeting, and offers the caller the opportunity to leave a brief recorded message, after which the machine automatically goes on-hook (disconnects the call). Its message storage capacity is limited to the capacity of a single small tape cassette or a digital microchip—usually no more than 30 minutes, and frequently less. Stored messages can be played back or erased by anybody with access to the machine, unless the unit requires a password to enter its stored contents.

In 1952, the Bell System introduced the first commercial telephone answering machine, which had been developed at Bell Telephone Laboratories. The recording medium was a magnetic recording drum. Individual subscribers could not purchase the equipment. At that time, all telephones and auxiliary equipment such as answering machines and PBX systems were available only on a lease basis; if any unrepairable malfunction occurred, the telephone company simply replaced the offending equipment with a matching item. The Bell System maintained that the phone networks were too sensitive to allow any "foreign" (i.e., made by anyone but its manufacturing subsidiary, Western Electric) hardware to be connected to it.

This restriction led to a lawsuit in 1966, involving AT&T and Carter Electronics of Dallas. The small company had developed an acoustical coupling device, known as a Carterfone, which could interconnect private mobile radio systems with the nationwide long-distance network. Although no "hard-wired" connections existed between the private and public networks, Bell System engineers contended that the device could cause disruptions in service and possibly bring down the long-distance network. The Federal Communications Commission disagreed and in 1968 struck down interstate telephone tariffs that had prohibited attachment or connection to the public telephone system of devices not supplied by the telephone companies. Known as the Carterfone Decision, this ruling opened the door for U.S. telephone subscribers to purchase their own telephone sets, PBX switches, answering machines, and, eventually, fax machines and personal computers—any communications device that can be connected to the telephone system.

The recording medium in answering machines gradually evolved from magnetic drums to magnetic tapes, and then to microchips capable of storing digitized voice messages. A blinking light identifies how many messages are stored. Messages can be played back, one by one, and saved or erased. By the mid-1990s, telephone answering machines often were integrated into sophisticated desk telephone sets also equipped with touchtone dial pads, speakerphones that enabled the owner to screen incoming calls, caller identification systems that recorded the subscriber phone number of the caller, storage of frequently called numbers for speed dialing, and other features.—R.Q.H.

See also MICROPROCESSOR; FEDERAL COMMUNICATIONS COMMISSION

Telephone Receivers

When early telephone users wanted to place a call, they picked up the earpiece and waited for an operator to ask what number they wished to call ("number please?"). They then gave the number by speaking into the mouthpiece of the telephone. This was not always viewed as burdensome. Many telephone users enjoyed the human contact with the operators, and in some places operators happily served as messengers and providers of information. But not everyone liked this arrangement. One displeased customer was a Kansas City undertaker named Almon B. Strowger, who believed that local operators were being bribed to divert potential customers to his competitors. This motivated him to invent an automatic switching system that obviated the need for human operators.

Strowger's automatic switching system was installed in a small town in Indiana in 1892. In the years that immediately followed, the use of telephones with rotary dials made telephoning easier and more conve-

nient. A rotary-dial telephone was patented in 1896 by A. E. Keith, J. Erickson, and C. J. Erickson, and by the first decade of the 20th century, telephones with rotary dials were being manufactured in significant numbers. Still in use today, the standard rotary dial telephone has a rotating finger wheel with 10 holes. By inserting a fingertip into a given hole, the wheel can be turned until the finger stop is met. When the fingertip is removed, the wheel automatically returns to its original position. Under the wheel is a number plate with the numerals 1 through 9 and 0, spaced so that each numeral is seen under a finger hole. Each number also has three alphabetical letter groups (i.e., 2-ABC, 3-DEF, 4-GHI, etc.) that were part of telephone numbers until they were replaced by all-digit numbers; for example, OLdfield 6-1010 became 656-1010. Each set of number and letters has a distinct number of pulses, ranging from 1 to 10, assigned to it. These pulses trigger the switching apparatus that sends the call to its proper destination. About half of the United States

The classic "candlestick" telephone that was in use until the late 1950s (courtesy Lucent Technologies).

had dial service by the early 1940s, rural and remote areas usually being the last to be converted.

Rotary dial phones served adequately for many years, but they were slow and prone to errors made by their users. In 1963, AT&T effected a significant improvement when it introduced the 10-button touch telephone. Touchtone dials use a different electrical signaling arrangement called *dual-tone multifrequency*. Each numeral key that is pressed emits two tones selected from each of two mutually exclusive groups of four frequencies.

A touchtone dial consists of a pushbutton pad equipped with 12 keys. In addition to the numerals 1 through 9 and 0, the buttons include one marked * (star) and another marked # (pound). These additional buttons are used to activate special services in switches or when calling computer-controlled operations with "menu" directions. More sophisticated models may include additional buttons for muting the microphone temporarily, automatically redialing the last number called, entering frequently called numbers into an internal memory bank, activating an internal call-answering system, etc., and may even have a small liquid-crystal display (LCD) panel showing the date, time, the number being dialed—and even the number of an incoming caller.

Compared to a rotary phone, a caller equipped with a touchtone telephone can dial a call in half the time and make fewer errors. This improvement in speed and accuracy was in part the result of the placement of the individual keys, which was the result of a great deal of ergonomic research. Most people using touchtone telephones have no difficulty with them, but accountants and other people who often use 10-key calculators or numerical keypads on computer keyboards may have problems, as the telephone puts its highest numbers in the bottom row, the opposite of the arrangement found on keypads.

In addition to adding speed and accuracy to the dialing telephone numbers, the touchtone telephone has allowed the telephone to serve as an interface with a variety of computer-based systems. Touchtone telephones have made possible such diverse services as voice mail and credit-card calls. Indeed, just about anything that requires the caller to punch in a string of numbers can be accommodated to touchtone telephones.

See also CREDIT CARD; ERGONOMICS; LIQUID CRYSTAL DISPLAY; TELEPHONE; VOICE MAIL

Telephone Systems, Cellular

Cellular-telephone technology is based on three primary components: a digital central controlling switch, a honeycomb-patterned network of low-power radio transmitter/receiver antennas connected to the switch, and battery-powered wireless communications devices that interact with the antennas and the switch.

A cellular-phone system consists of a service area subdivided into a number of relatively small areas called *cells*, which are linked to one other in a network. Each cell contains a low-power radio antenna that interacts with radio signals to and from mobile cellular telephones. A cell can range from less than 2 km (1.2 mi) in diameter to more than 32 km (20 mi) wide, depending on the capacity needs of the system and the type of terrain.

The network of cellular antennas is controlled by a central switch, which in turn is linked to the conventional telephone networks. The switch can forward calls from a "cellphone" to other cellphones, or connect the cellphone's wireless calls to conventional wired telephones in other networks. Conversely, the cellular switch can locate a cellphone in its service area and direct any incoming calls to it. The connections from the switch to each antenna are usually either fiber-optic cable or coaxial cable, or sometimes microwave radio channels. Although the switch employs digital technology, it can serve either analog or digital cellular antennas.

Cellular telephones are special battery-powered, two-way wireless mobile radios that enable voice or data calls to be made to and from vehicles or pedestrians. Other wireless equipment used in cellular networks include portable laptop computers, fax machines, and pagers. The cellphones and other devices use radio links to one or another of the various cellular antennas nearby. The portable devices are capable of automatically changing radio frequency settings to match the different frequencies used by each cellular antenna as the wireless phone moves from one to another. As the mobile phone set moves away from an antenna during a call, the signal strength weakens. A complex interaction among the antennas, the cellphone, and the switch transfers the call to the next appropriate antenna in a procedure called a *handoff*. Neighboring antennas usually have different frequency groups, so the handoff also requires the cellphone to change its frequency setting. However, the frequency group assigned to any one antenna is duplicated in one or more other antennas as well in other parts of the network. This approach greatly increases the system's overall capacity by enabling the same channel frequency to be used simultaneously by many callers in different areas of the network, provided there is no signal overlap that would cause interference.

The first commercial mobile phone systems had only one high-powered antenna per city. Introduced by AT&T in 1946, they used analog technology and operated at a relatively low radio frequency, 150 megahertz; this was raised to 450 megahertz in 1969. Operating frequencies are assigned in the United States by the Federal Communications Commission, or FCC, serving as the referee among the television and radio broadcasters, local government agencies, transportation companies, and other competitors fighting for the same wireless frequency bands.

Cellular telephone systems were introduced commercially in 1983 to a few thousand users of car telephones in Chicago, Ill., and Washington, D.C. By 1996, they were serving well over 100 million mobile wireless communications customers worldwide, and were continuing to grow in the United States alone by more than 1,000 new subscribers per hour. In 1997, there were about 40 million cell phones in operation in the United States.

Like the first mobile telephones, early cellphones were installed only in vehicles. But many subscribers also wanted to carry portable phones with them in trains, on buses, and on foot. Advances in electronic technology gradually shrank the wireless transmitter/receiver's size and weight. Within a few years portable cellular phones—as well as other wireless communications devices—became ubiquitous, used for all kinds of social and business calls as well as emergencies.

The popularity of wireless mobile telephone service is based mainly on the communications-oriented culture of the industrial world, plus the desire by many people to be free of the tether imposed by the cord of a conventional wired telephone. Even more significant, a study conducted by the Electronic Industries Association found that 89 percent of cellular service subscribers in 1995 cited "ability to communicate in emergency" as the most important reason to own a wireless mobile phone.

By the end of 1995, the Cellular Telecommunications Industry Association reported that the United States had 1,627 cellular systems, operating with 22,663 cell sites. The industry directly employed more than 68,000 people, or a total of 275,000 people when related industries were included.

Each cellular network (the FCC has allocated two competing networks per "market") has an assigned frequency bandwidth of 25 megahertz, allocated from within the 800- to 900-megahertz region of the electromagnetic spectrum. Each antenna within a network uses a small portion of those frequencies to provide about 100 analog or 300 or more digital two-way radio channels simultaneously within a cell. A single two-way analog voice call occupies a channel of about 30,000 cycles (30 kilohertz) in a radio frequency system, but the same bandwidth can handle at least three digital two-way voice channels using time-division multiplexing to interleave the pulse-code modulated (PCM) signals in a single stream.

From 1983 to 1994, all U.S. cellular phone systems used analog transmission technology. Commercial digital transmission over wireless channels was developed during the 1980s, but competition between different digital standards delayed its introduction in cellular telephony. There are two radically different versions of digital signals used in cellular telephone systems. In the United States, the most widespread in the mid-1990s was the Time-Division Multiple Access (TDMA) or Interim-Standard 136 technology, introduced commercially in 1994 by AT&T Wireless Services. The second, which began commercial deployment about 2 years later, is the Code-Division Multiple Access (CDMA) system, based on earlier military technology known as *spread spectrum* and now known as the IS-95 standard, which is promoted by Qualcomm, Inc., and others. Internationally, a digital system known as the Global Standard for Mobile (GSM) communications uses time-division techniques that are somewhat different from TDMA in the United States. TDMA and CDMA systems are not compatible with each other, so a TDMA-equipped digital cellphone would revert to analog operation in a CDMA cellular system, and vice versa, provided the phone is a dual-mode (digital/analog) model.

In 1996, a multiantenna digital mobile radio telephone system known as Personal Communications

Services (PCS) was introduced in the United States. Basically designed for use by pedestrians and somewhat similar to cellular systems, it uses a much higher radio frequency.

Perhaps the most expensive cellular communications device is the mobile satellite telephone unit, which cost several thousand dollars in 1996. Often used by emergency workers in catastrophe areas, or reporters in remote areas, this briefcase-sized unit is capable of tuning in directly to special communications satellites, either those in geostationary orbit or others that are in low polar orbits. Its calls can be relayed to and from any place in the wired networks.—R.Q.H.

See also COMMUNICATIONS SATELLITES; FAX MACHINE; FEDERAL COMMUNICATIONS COMMISSION; MULTIPLEXING

Further Reading: Additional information can be obtained from the Cellular Telecommunications Industry Association, Washington, DC 20036.

Telescope, Radio

For centuries, a great deal of valuable astronomical observation was done with the naked eye. The invention of the optical telescope in the 17th century vastly enhanced human ability to discern what was occupying the universe. In the 1930s, there appeared a new observational technology that proved to be as significant as the invention of the optical telescope.

Optical telescopes collect and magnify the portion of the electromagnetic spectrum that we know as visible light. This spectrum also includes gamma rays, X rays, ultraviolet and infrared light, microwaves, and radio waves. The latter have proved to be of particular importance for astronomy, for they allow much to be learned about objects in space that produce little or no visible light and hence cannot be seen. The detection of radio waves from space began in 1928 when Karl Jansky (1905–1950) was asked by his employer, Bell Telephone Laboratories, to determine the source of the static that was interfering with trans-Atlantic radio-telephone communications. In 1931, Jansky used two rather crude dish antennas to locate three sources of radio interference. Two of them were the result of electrical storms, but the third seemed to be of extraterres-

trial origin, coming from the vicinity of the constellation Sagittarius in the southern portion of our Milky Way galaxy. Jansky was fortunate in that he conducted his observations at a time when, unbeknownst to him, the sun was in a quiescent phase as far as the production of radio waves was concerned. Had it been otherwise, all other far-off sources of radio waves would have been drowned out.

Jansky's employer saw no reason to follow up this discovery, and professional astronomers seemed equally uninterested. The next step was therefore taken by an amateur astronomer named Grote Reber (1911–), who in 1937 built a 9.5-m (31-ft) receiving dish it in his backyard. Through the use of this dish, the first radio telescope, Reber was able to produce maps that indicated the locations of many objects in space that were the source of radio waves.

Perhaps because their country was particularly well endowed with optical telescopes, American astronomers did not immediately take to radio telescopes. The bulk of the early development of radio astronomy was done in Britain, a country that had also been a center of wartime research into radar. Under the leadership of Bernard Lovell (1913–) a 66-m (216-ft) parabolic receiver went into operation at Jodrell Bank in 1947. Ten years later, that site became home to a 76-m (250-ft) diameter disk, the largest telescope built up to that time. Observations at the Jodrell Bank observatory determined that intense radio waves were being produced by whole galaxies and by immense stellar explosions known as *supernova*. Since then, even larger radio telescopes have been put into operation. The largest steerable radio telescope is the 100-m (330-ft) parabolic receiver operated by the Max Planck Institut für Radioastronomie at Effelsberg, Germany. The largest single individual receiver is the 305-m (1,000-ft) nonsteerable dish in Arecibo, Puerto Rico, that was built in 1963.

In 1955, Martin Ryle (1918–1984), another British astronomer, devised the radio interferometer. This consisted of 12 individual radio telescopes that were trained on the same object. The data they received were sent to a receiver where they were synchronized and analyzed by a computer. In 1955, this instrument found the first galaxy to be discovered solely through radio astronomy. In 1977, a considerably larger interferometer consisting of 27 radio telescopes, the Very Large Array, went into operation in Socorro, N.Mex.

Radio telescopes have provided information that has substantially changed our view of the universe and its constituents, such as the 1967 discovery of pulsars, extremely dense neutron stars that have been created by supernova. Radio telescopes also have been essential to the discovery of some of the most fascinating yet perplexing objects in the universe, quasi-stellar radio sources or quasars for short.

One of the most significant discoveries made through the use of a radio telescope came in 1964. In that year, two scientists employed by the Bell Laboratories, Arno Penzias (1933–) and Robert W. Wilson (1936–), measured radio sources from a part of the Milky Way galaxy through the use of a 6-m (20-ft) radio antenna that originally had been used to receive signals from one of the first communications satellites. One of their discoveries was a considerable amount of microwave noise at a wavelength of 7.35 cm. Unlike other forms of radio noise, these waves had no identifiable source, but were coming from all directions. After further inquiry, it became apparent that the radiation detected by Penzias and Wilson was the remains of the Big Bang that had initiated the creation of the universe billions of years ago.

See also Big Bang theory; gamma rays; microwave communications; quasars; radar; radio; telescope, reflecting; telescope, refracting; X rays

Telescope, Reflecting

Beginning with Galileo, the telescope revolutionized our view of the universe. All of these early instruments were refracting telescopes. Telescopes of this sort continue to be used for astronomical observations even today, but in the 17th century astronomers had to cope with what was thought to be an inherent optical fault: chromatic aberration, i.e., fringes of color around the images. As Isaac Newton (1642–1727) had demonstrated, different colors do not refract equally, so a focal length that brought red into focus would leave the other colors slightly out of focus, hence the colored rings around the magnified object.

Newton concluded (incorrectly, as it turned out) that no shape or arrangement of lenses could prevent chromatic aberration. He therefore invented a telescope based on a different principle, one that magnified through the use of a curved mirror rather than a set of lenses. Newton was not the first to conceive of a reflecting telescope, but he was the first to actually build one. In 1668, he constructed a telescope with a 2.5-cm (1-in.) reflector that had been fashioned from an alloy of copper, tin, and arsenic. Although it was only 15 cm (6 in.) in length, it had the magnifying power of refracting telescopes that were as much as 2 m (6.6 ft) in length.

In Newton's telescope, light from the reflector traveled to a flat mirror set at a 45-degree angle, from which it then made its way to the eyepiece, which also magnified the object. An alternative arrangement was created in 1672 by Casegrain (c. 1650–1700). In a Casegrain telescope, light from the reflector travels to a convex secondary mirror instead of a flat mirror. This reduces the magnification, but it also cuts down on spherical aberration, a distortion of the image that is caused by a mirror with a spherical cross section.

Although the reflecting telescope was theoretically superior to the refractor, its advantages were hard to realize. Early reflecting mirrors were fashioned from metal, which was difficult to grind to the proper shape. These mirrors also failed to reflect much of the light that fell upon them, a situation that got progressively worse as the metal inevitably tarnished. At the same time, however, reflectors were easier to construct than refractors, since casting a large mirror blank was easier than casting an equivalent refractor lens.

In the late 18th century, the capability of the reflecting telescope was dramatically demonstrated by William Herschel (1738–1822), a German by birth who did all of his astronomical work in England. In 1781, Herschel used a telescope with a 15-cm (6-in.) reflector that he had ground himself to discover a new planet, which came to be called Uranus. Enjoying the patronage of King George III, he went on to build a reflecting telescope with a mirror 122 cm (48 in.) in diameter. While using it, Herschel discovered two new moons of Saturn, and with it and other reflectors he was the first to discern binary stars and to see the individual stars comprising certain nebulae. He also was able to calculate the speed at which the sun and its planets were moving through the galaxy. This observation disturbingly implied that the sun was not the center of the universe. Herschel also was able to confirm the theory of Thomas Wright (1711–1786) that our solar system resided within a lens-shaped collection of stars, or *galaxy* as it came to be known.

Herschel's discoveries marked a temporary plateau in the use of the reflecting telescope. In 1842, a massive telescope with a 184-cm (72-in.) reflector was built under the direction of William Parsons, the Third Earl of Rosse (1800–1867). Although it was used to discover the first spiral nebula in 1845, difficulties of manipulating the huge telescope, coupled with poor viewing conditions at Rosse's estate in Ireland, limited its usefulness. Although Rosse's telescope was a bit of a white elephant, it was still true that the large reflecting telescope had a number of potential advantages over large refracting telescopes. It was very difficult to cast big, defect-free lenses for a refractor, and a large rim-mounted lens would inevitably sag, with consequent distortion of the image being viewed. In contrast, a glass reflector would be relatively easy to cast, grind, and mount; all it lacked was a reflecting surface. This situation began to be addressed in 1856 when the German chemist Justus von Liebig (1803–1873) worked out a way to precipitate silver on the surface of glass. This technique was soon applied to telescope reflectors by Carl August von Steinheil (1801–1870) in Germany and Jean B. L. Foucault (1819–1868) in France. Foucault also invented a means of checking the contour of the reflector that allowed the detection of irregularities as small as a hundred-thousandth of a centimeter.

By the early 20th century, the use of glass-based reflectors allowed the construction of telescopes with superior light-gathering and optical qualities. In 1904, the excellent viewing conditions on Mt. Wilson above Pasadena, Calif., were put to good use when a 153-cm (60-in.) reflector was installed under the direction of George Ellery Hale (1868–1938). Hale then went on to spearhead the construction of the 254-cm (100-in.) Hooker reflecting telescope that went into service in 1918 at the same location. Through the use of this telescope, astronomers were able to gain a much better understanding of the size of our galaxy as well as the number of stars it contains. The Hooker telescope also made possible the discovery that many nebulae were not clouds of dust but whole galaxies, a process that

began in 1919 when Edwin Hubble (1889–1953) used the telescope to take photographs of the Andromeda nebula. Hubble went on to determine that this galaxy was on the order of 800,000 light-years from us (subsequent measurements put it 2,300,000 light-years away). Measurements such as these became the basis for the revolutionary idea that the universe was not static but was in a state of constant expansion. All in all, these 20th-century discoveries were as significant as the ones made during the early 17th century, when the telescope provided evidence that the Earth was not the center of the universe.

In 1934, work got underway on the mirror used for the telescope that was the largest in the world for many decades, the 508-cm (200-in.) Hale telescope on Mt. Palomar near San Diego, Calif. This telescope made

The Hale Telescope at Mt. Palomar, Calif. For many years it was the world's largest reflecting telescope (courtesy California Institute of Technology).

use of a new kind of glass for its reflector, a product of the Corning Glass Works that bore the trade name Pyrex. Grinding the reflector to shape took many years and the use of 31 tons of abrasive that removed 4,500 kg (5 tons) of glass from the surface. Dedicated in 1948, the telescope was able to detect galaxies that were hundreds of millions of light-years from the Earth.

For many years, the Hale telescope seemed to represent the upper limit on the size of reflecting telescopes. The Soviet Union built a telescope with a mirror 1 m (3.28 ft) larger than that of the Hale telescope, but it failed to live up to expectations. Beginning in the 1980s, however, a new generation of large reflecting telescopes began to be built. These took advantage of new technologies such as the use of fused quartz for the mirror blanks and the casting of these blanks in spinning ovens that imparted a dish shape, cutting down dramatically on the time needed for grinding their surfaces. The greatest breakthroughs were made possible by computerized controls that allowed a reflector to be made from a number of mirror segments. Today, the most powerful telescope is located at the Keck Observatory on Mauna Kea in Hawaii. Its 10-m (32.8-ft) mirror is composed of 36 computer-controlled hexagonal mirrors, each about 2 m (6.6. ft) across. Computers are also the basis of adaptive optics, a technology that uses measurements of laser-generated artificial stars to adjust an instrument's optics in accordance with changing atmospheric conditions. In 1990, one of the most ambitious scientific projects of all time, the Hubble telescope, put a reflecting telescope into space, where it could operate free from the distortions caused by the Earth's atmosphere.

See also EXPANDING UNIVERSE; GLASS, HIGH-TEMPERATURE; HUBBLE SPACE TELESCOPE; TELESCOPE, REFRACTING

Further Reading: Isaac Asimov, *Eyes on the Universe: A History of the Telescope*, 1975.

Telescope, Refracting

Many civilizations of the past developed sophisticated astronomical systems that rested on nothing more than observations made with the unaided eye. In the early 17th century, vast new possibilities were created by the invention of the telescope, an instrument that brought into view sights that had never been seen before. Observations with the telescope, in turn, became the basis for a radically different understanding of the solar system and everything that lay beyond it.

The invention of the telescope was an outgrowth of the manufacture of eyeglasses. It was in one of the centers of eyeglass making, the Netherlands, that the first telescope was constructed. Although there are other claimants, credit for the first telescope is usually awarded to Hans Lippershey (c. 1570–c. 1619) of Middelburg in the Dutch Province of Zeeland. According to an oft-told story, one of Lippershey's apprentices discovered that two lenses held a short distance apart made far-off objects appear to be much closer. Lippershey encased the two lenses in a metal tube, and in so doing made the first telescope.

Lippershey then offered the instrument to the Dutch Navy, which quickly recognized its military value, and rewarded the inventor with 900 florins and a contract to produce more instruments. But the real influence of the telescope lay in its contributions to astronomy. One of the first to turn a telescope towards the heavens was Galileo Galilei (1564–1642). By using plano-convex lenses (flat on one side, curved outward on the other) for the objective, and plano-concave lenses (flat on one side, curved inward on the other) for the eyepiece, Galileo produced telescopes that eventually were capable of magnifying objects to 33 times their diameter. Although they were weaker than a modern binocular, these instruments allowed Galileo to mount a profound challenge to the accepted view of the heavens and the Earth. When Galileo first used his telescope in 1609, the dominant view was that the objects in the sky were luminous, perfect spheres, and that they revolved around the Earth. But when Galileo looked at the moon through a telescope, he observed mountains, craters, and smooth expanses that he called "seas." By noting the moon's imperfections, Galileo called into question the belief that the heavens were populated with perfect bodies. An even stronger challenge to the accepted view came in 1610, when Galileo trained his telescope on the planet Jupiter. What he found were three spots of light in the vicinity of the planet. A few days later he found a fourth. It was soon apparent that these bodies were orbiting

Jupiter, much as the moon orbits the Earth. In the established scheme of things, all heavenly bodies circled the Earth, so the discovery of four bodies orbiting Jupiter was anomalous at best. Galileo then discovered that Venus, one of the "wandering stars" (i.e., planets) went through periodic phases that were similar to those exhibited by the moon. This could not be explained if it was assumed that Venus orbited the Earth, but it made perfect sense if Venus orbited the sun. Finally, through the use of the telescope Galileo was able to discern spots of the sun, the movement of which seemed to indicate that the sun itself rotated. And if the sun, the brightest and largest object in the heavens rotated, perhaps it was not unreasonable to suppose that the Earth also rotated. Taken together, Galileo's observations with the telescope severely undercut the picture of the universe that had been accepted for centuries and provided support for the one first outlined by Copernicus a few decades earlier.

In addition to having little magnifying power, the telescopes used for these epochal observations also presented distorted images due to chromatic aberration and spherical aberration. Chromatic aberration manifests itself as fringes of color surrounding the objects being observed, while spherical aberration results in a fuzzy image. During the early years of the telescope, spherical aberration was mitigated by the development of new shapes for lenses. Shortly after Galileo's observations, Johannes Kepler (1571–1630) discovered that spherical aberration occurs when the light refracted through a telescope's lens does not converge at a single point. This problem could largely be eliminated through the use of lenses that had parabolic rather than spherical sections, but grinding such lenses was difficult. Alternatively, it was possible to greatly reduce spherical aberration by using a lens with a gently curving surface, for such a surface caused the light to refract by only a small amount. This in turn necessitated making the telescope quite long, so that the light rays passing through the lens could converge. By the middle of the 17th century, astronomers were using refracting telescopes of considerable length, such as the 3.6-m (12-ft) refractor that Christian Huygens (1629–1695) used to discover a satellite orbiting the planet Saturn. In being the first to see the rings around that planet, he used a telescope that was no less than 37 m (123 ft) in length. Some as-

tronomers went even further, and built telescopes so long that it was virtually impossible to keep them in alignment for any reasonable length of time.

The problem of chromatic aberration took longer to solve. In 1672, Isaac Newton (1642–1727) presented to Britain's Royal Society a reflecting telescope that eliminated chromatic aberration altogether. It soon proved highly useful, but it did not completely supplant the refracting telescope for a very long time. Newton believed that chromatic aberration was an inescapable accompaniment of refracting telescopes, and so great was his prestige in scientific circles that this belief went largely unchallenged. One of the few who disagreed with Newton in this regard was Chester Moor Hall (1703–1771), an English lawyer and mathematician who also had a keen interest in optics. Hall discovered that chromatic aberration could be overcome by using two different kinds of glass for the objective lens. This could be done because the two types of glass "stretched" the spectrum (a property known as *dispersion*) by different amounts. The lens had two components: a convex lens made from ordinary window glass (known as *crown glass*) and a concave lens made from flint glass (a dense, highly transparent glass containing lead compounds). The two were then fitted together to make a single biconvex lens that converged light to a focus without causing a separation of colors. For this reason, a lens of this type came to be known as an *achromatic lens*.

Refracting telescopes still suffered from the inability of lensmakers to produce lenses with diameters greater than about 10 cm (4 in.). But early reflecting telescopes had their own problem, the difficulty of keeping the metal mirrors polished. Refractors therefore still had a place in astronomy if their lenses could be made larger. Early in the 19th century, Pierre Louis Guinand (1748–1824) of Switzerland joined up with Joseph von Fraunhofer (1787–1826), a German optician, to produce lenses of unprecedented size. By developing techniques to make the glass homogenous, they were able to produce a 24-cm (9.5-in.) lens with excellent optical properties. The telescope that used the lens was much easier to handle than a reflecting telescope, and it could be fitted with a clockwork drive to track a single object throughout the night.

By the end of the 19th century, the largest and

best refractor lenses were being made in Cambridge, Mass., by Alvan Clark (1804–1887) and his son Alvan Graham Clark (1832–1897). Their efforts culminated in a 101-cm (40-in.) lens that was used for the telescope of the Yerkes Observatory in Lake Geneva, Wis. This, however, was the swan song of the large refracting telescope. Difficulties in producing lenses free of imperfections, coupled with the difficulty of mounting large, heavy lenses, spelled the end of the refractor as an instrument of advanced astronomical research. Still, the basic arrangement discovered in the early 17th century has not been abandoned, for refractors are still widely used as binoculars and small telescopes.

See also EYEGLASSES; PTOLEMAIC SYSTEM; SOLAR SYSTEM, HELIOCENTRIC; TELESCOPE, REFLECTING

Further Reading: Isaac Asimov, *Eyes on the Universe: A History of the Telescope*, 1975.

Teletypewriter

Teletypewriter systems—also known as Teletype, TTY, TWX, or Telex systems—are used to transmit text messages as digitally coded electrical signals over telegraph lines between electromechanical teletypewriters. Functioning as transmitters and receivers, teletypewriters use either manual keying or perforated ("punched") paper tape to enter the message for transmission, which can be either instantaneous ("real time") or saved for bulk ("batch") transmission. The teletypewriter is also known as a *page printer* when operated in the receiving mode.

Teletypewriters were the electromechanical equivalent of today's computer monitor, in that they converted a digitally encoded electrical signal into a text display, but at a much slower speed. Although the ability to transmit the printed word electrically goes back to the 19th century, teletypewriters only went into common use when the American Telephone & Telegraph Co. (AT&T) began leasing them (along with private circuits) to press associations in 1916 and to commercial customers in 1917. The Teletype Corporation was established in 1923, when the Morkrum Co. of Chicago (a manufacturer of teletypewriters) merged with the Kleinschmidt Electric Company. Teletype was acquired by AT&T in 1930.

Teletypewriter services over circuits manually switched by AT&T operators were introduced in 1931, connecting about 900 stations, and soon were favored by news organizations (Associated Press, United Press, etc.) and various commercial services such as railroad, bus, marine and airline operations, stockbrokers, and hotel chains. Within 7 years, the switched Teletype system—by then known as TWX (pronounced "twix") for Teletype-Writer-Exchange—served 11,000 stations in 160 cities, transmitting asynchronous digital pulses representing the characters. AT&T's operators serving the system communicated with users by means of teletypewriters instead of their usual transmitters and headsets. Customer dial-up capability over automatic switches was implemented in 1962, by which time the U.S. network was serving 60,000 stations. Because their operation usually was quite noisy, teletypewriters were often located at one side of newsrooms and other facilities, but their activation produced a distinctive clatter that compelled attention.

Early versions of Teletype machines used a typewriter ribbon for inking, a moving basket of type, and a paper roll inside the machine cover (which muffled but did not silence the noisy clatter). The various letters, numerals, and control signals of the teletypewriter language originally were transmitted in the Murray form of the five-unit Baudot code, with the machines reaching reliable transmission speeds of 60 words per minute by 1930. By 1962, new machines transmitted 100 words per minute. The transmission channels assigned for teletypewriter use were classed as subvoice grade and were operated at speeds ranging from 50 to 150 baud. Unlike telephone transmission, the teletypewriter's output does not modulate the signal carrier. Instead, a serial bit pattern is transmitted by opening and closing a switch contact (much like a telegraph key), using a revolving shaft, arm, and distributor.

Although teletypewriters have been used primarily by industrial, commercial, and government organizations, one application continues to involve certain consumers. People with speech or hearing impairments can use the telephone networks via a system known as TTY, in which portable teletypewriters provide them with the means to communicate with a special operator, who in turn can dial up a conventional telephone. The operator serves as a relay between a speaker and

A teletype machine from the 1940s (courtesy Lucent Technologies).

the TTY user. In recent years, most TTY devices have been replaced by portable TTDs (Telecommunication Devices for the Deaf), which display text on small electronic screens.

The TWX system served the domestic U.S. markets; the Telex system continues to serve the domestic U.S. and international teletype markets. Shipping and financial markets still use Telex services extensively (it operates at the digital transmission rate of 50 baud), but the customer base was eroding rapidly by the mid-1990s. The electromechanical transmission of text messages has been largely replaced by fax machines and electronic mail and other text reports exchanged among computer terminals over telephone lines, using the packet-switched Internet.—R.Q.H.

See also COMPUTER NETWORK; ELECTRONIC MAIL; FAX MACHINES; INTERNET; TELEPHONE

Television

The electrical transmission of stationary visual images was first accomplished in the 1860s. Efforts to transmit moving images began in the 1880s, although many decades were to pass before workable systems were created. During the 1920s and early 1930s, modestly successful television systems emerged, although television was still in an experimental phase and far from being established as a mass medium. At that time, the development of television was proceeding along two different technological paths, one mechanical, the other electronic. Mechanical systems were built around spinning discs that were used to transmit and receive pictures. The most successful application of this principle was the mechanical television system developed in Great Britain by John Logie Baird (1888–1946). Baird was able to transmit and receive pictures for the first time in 1924; 3 years later he succeeded in sending pictures from London to Glasgow. In 1929, the British Broadcasting Company (BBC) began to transmit television programs using Baird's technology. But despite a promising start, the BBC terminated Baird's mechanical system in 1936 because it was incapable of putting enough lines on the screen to provide good picture definition. The inertia of the spinning disc also made it hard to keep rotational speeds within close limits. Another problem of mechanical television was the very high levels of illumination necessary when scenes were being shot; the consequent high temperatures caused great discomfort to people in the TV studio.

Electronic television also required high illumination and had a number of other drawbacks, but it had a greater potential for further development. The basic principles of electronic television had been outlined as far back as 1908 by Alan Campbell-Swinton in a letter published in the British scientific journal *Nature*. Swinton never tried to translate his ideas into a working system, for he knew that a vast amount of development work would be required to make electronic television a practical reality. A continent away in 1922, a Utah schoolboy named Philo Farnsworth (1906–1971) outlined his own ideas for an electronic television system to one of his teachers, and unlike Campbell-Swinton he commenced to turn his ideas into reality. By 1927 he succeeded in transmitting 60-line televised images, and in 1930 he was awarded a U.S. patent for a television camera tube he called an *image dissector*.

At about this time, Vladimir Zworykin (1889–1982), a Russian emigré in the employ of Westinghouse, had been working on another electronic television system which, like Farnsworth's, was grounded in many of the principles proposed by Campbell-Swinton. In 1930, Zworykin moved to the Radio Corporation of America (RCA), whose chief executive, David Sarnoff (1891–1971), put RCA's weight behind the development of television. At the core of Zworykin's system was a camera called the *iconoscope* (from the Greek words for *image* and *to watch*). The iconoscope was based on the phenomenon of certain materials inducing an electrical current when light falls on them. In Zworykin's iconoscope, light from the image to be transmitted struck a screen of photoelectric cells at one end of a cathode tube, thereby inducing an electrical current. When the cells were scanned by an electron beam, they gave off an electrical current corresponding to the intensity of the light that had struck the individual cells. After being transmitted, the image then could be reconstructed through a reverse process. The video signal controlled electrical currents that caused a phosphorescent screen to glow in accordance with the intensity of the currents.

After demonstrating the iconoscope, Zworykin informed Sarnoff that $100,000 might be required to fully develop it. As it turned out, between 1930 and 1939 RCA spent $7 million developing TV, and another $2 million on costs associated with patent litigation. Some of the money was very grudgingly paid as royalties to Farnsworth because the iconoscope shared a number of technical features with Farnsworth's image dissector.

During the 1930s, experimental television broadcasts were made from a number of American cities, and in 1939 RCA put on regular broadcasts in conjunction with the New York World's Fair. But only a few hundred TV sets (priced at $625 each) were sold. At this point, the United States was well behind Great Britain, where electronic television broadcasting began in 1936, and 20,000 television sets were in use by 1939.

Television broadcasts in the United States at first produced a picture composed of 343 lines. In 1942, a standard of 525 lines was adopted (European television produces a higher-quality picture since it uses 625

lines). Further development of television was arrested by the demands of World War II. However, the development of radar, one of the greatest technological accomplishments of World War II, generated knowledge and equipment that was directly applied to television. In the late 1940s, sales of television sets began to take off, and by the late 1950s television was embedded in modern culture. At about this time a massive program of research and development began to bear fruit with the commercial introduction of color television. Yet for all of the technical triumphs underlying it, few technologies have been as disquieting as television. Psychologists, social critics, and ordinary people continue to debate the role of television in shaping behavior. Although firm conclusions are elusive, there is little question that not all of television's effects have been beneficial.

See also CABLE AND SATELLITE TELEVISION; FAX MACHINE; NIPKOW DISC; TELEVISION, COLOR; TELEVISION AND POLITICS; TELEVISION AND VIOLENCE

Further Reading: David E. Fisher and Marshall Jon Fisher, *Tube: The Invention of Television*, 1996.

Television and Politics

From the beginning, television has been closely intertwined with government and politics. Many governments have used television to advance their policies and solidify their power. At the same time, however, television has affected the conduct of government in a number of ways, and it also has been a powerful influence on the electoral process.

Even when television broadcasting is a private enterprise, government has an important power: the ability to award rights to use specific portions of the frequency spectrum. In the course of awarding broadcasting rights, the government also can dictate hours of broadcast and transmitter power, and it can hold broadcasters to certain standards. In some countries, government influence goes much deeper; television broadcasting is a government monopoly and the medium is freely used as a means of advancing the government's interests. This use of television has been most blatant in communist countries and in many Third World nations, but it also has been a feature of broadcasting in more democratic nations. In France, television programming was strongly affected by government oversight until the early 1980s. As a result, when that nation began to be convulsed by the conflict over Algerian independence, television broadcasters did not offer a single program on Algeria from 1956 to 1959. And in 1961, when a coup by the French Army in Algeria posed a major challenge to the authority of President Charles de Gaulle, the president used his monopolization of television coverage to mobilize popular opinion against the insurgents.

Most political uses of television are far less dramatic. In everyday politics, politicians and their advisors are engaged in a continual event to put a "spin" on events, that is, to get television and other media to report events in a manner favorable to their cause. Political advisors also make every effort to develop a good media image of a candidate or official. What matters here is creating a favorable, easily understood idea of the individual, one that is not cluttered up with complex policy issues.

In striving to construct and present good images for themselves (and bad ones for their opponents), politicians can make effective use of one of the chief characteristics of television. TV is not a medium well suited for the presentation of complex, nuanced issues; rather, it works best when strong visual images and simple concepts are being presented. In this setting, "sound bites"—striking phrases that substitute for lengthy analysis—become the highlights of political programming. A simplified political rhetoric is well suited to television because much of the programming that appears on television conveys the notion that problems can be quickly resolved—in less than 30 minutes in the case of many shows, and in 30 seconds where commercials are concerned. When carried over into the realm of politics, this notion favors candidates who offer simple, quick, and largely painless solutions to problems. These "solutions" are neatly packaged for television; the typical political commercial consists of a minute of visual images and a slogan or two.

Further simplifying the televised presentation of politics is the tendency of broadcasters to treat political campaigns like sporting events. In their coverage of elections, for example, the electronic media tend to stress the "horse race" aspect of campaigns and to de-

vote considerable attention to conflicts and personal attacks. Again, what often gets left out is a discussion of policy differences and issues.

Television also has been indicted for distorting democratic processes by increasing campaign costs. Television advertising has become an essential part of campaigning, and it is expensive. In the 1996 election, the Clinton and Dole campaigns spent $73.5 and $76.7 million, respectively, on their campaigns. Of these sums, a very substantial portion was spent on televised appeals. As television has increased the cost of campaigns, the importance of campaign contributions has increased accordingly. The result, as some political analysts have argued, are federal, state, and even local governments staffed by officials beholden in some way to the individuals and organizations that have made significant campaign contributions.

This having been said, it must also be noted that televised advertisements do not motivate most voters to vote for a particular candidate, although they do help to crystallize choices and solidify existing beliefs. Still, there are substantial numbers of voters who do not make their decisions until late in a campaign. For example, one study found that in the 1980 presidential campaign (one that offered voters a very clear-cut choice of ideologies and policies) 10 percent of the electorate did not make up their minds until election day. Voters like these tend to be apolitical people who have not followed the campaign closely. Their interest emerges only during the final days of a campaign, and for this reason most candidates reserve a sizable share of their television advertising for the last days of a campaign.

Voters who are not likely to be swayed by a last-minute barrage of televised political commercials usually have a strong sense of party identification. But party loyalties have been eroding in recent years, and although it is not the only reason for the decline of party loyalty, television has been a significant factor. Instead of hitching their fortunes to their party, many candidates now appeal directly to the electorate through the use of television. These candidates may be the products of state primaries, where televised appearances and political commercials have taken the place of selection by party bosses in smoke-filled rooms.

Although candidates and government officials attempt to manipulate television to serve their purposes, the medium cannot always be molded to meet their needs. On occasion, the televised images have been so vivid that they have been a significant factor in bringing about profound political change. In the 1960s, the need for civil rights legislation to protect the rights of minorities became apparent in the 1960s when television showed civil rights demonstrators being beaten by the police. In similar fashion, public support for the war in Vietnam was seriously undercut by televised scenes of death and destruction in a land halfway around the world. As one of the 20th century's most significant technologies, television has been a powerful tool in the hands of those who have been able to use it for their own purposes, but in some important ways the medium has been beyond the reach of political control.

See also COMMERCIAL RADIO BROADCASTING

Further Reading: Anthony Smith, ed., *Television: An International History*, 1995.

Television and Violence

Although television can be considered one of the greatest technological accomplishments of the 20th century, not all of its consequences have been beneficial. One of the most problematic aspects of television has been its possible contribution to violent behavior. There is little question that violence is a staple item for television. In 1996, a major research project, the industry-funded National Television Violence Study, reported that 57 percent of the programs surveyed (which excluded sports and newscasts) contained violent acts. The cumulative tally of televised violence is vast; by the time he or she is 18 years old, the average young person has seen several thousand televised killings. Worse still, much of the violence that is depicted on TV gives little sense of the consequences of violence. According to the National Television Violence Study, 47 percent of televised violence showed no harm to the victims, and 58 gave no indication that pain was involved. Moreover, in 73 percent of the cases, the perpetrators of violent acts received no punishment.

All of the mayhem appearing on television has led critics to indict television as a significant cause of the rampant violence that plagues American society. Particular attention has been given to the possible connection between television viewing and violent behavior by children and adolescents. However, determining the extent and nature of the relationship has been problematic. Some of the research has made use of laboratory experiments. In a typical experiment, one group of children is exposed to a film depicting violence, while a control group sees a nonviolent film. Their subsequent behaviors are then compared to determine if the former group is more inclined to engage in aggressive or violent acts. Most of the studies of this sort have concluded that such a connection does exist. However, these studies have been criticized for their artificiality. As with all laboratory studies, while attempting to eliminate or at least reduce extraneous influences, experimenters have been unable to avoid creating an unnatural environment. A film clip depicting violent behavior is not the same thing as a television program, where violence may appear in a complicated context. Experiments also have the defect of dealing only with immediate, short-term relationships, and do not take the long-term, cumulative effects of televised violence into account. Finally, in these experiments many of the subjects may have thought that aggressive behavior was permitted or even expected. For all these reasons, the real-world applicability of controlled experiments is questionable.

Other researchers have tried to avoid the inherent limitations of laboratory studies by conducting studies that are more closely related to actual situations. In studies of this sort, one group of children views regular TV programming, while another group does not. Their subsequent behaviors are then observed or are reported by teachers or parents, or both. Taken as a whole, these studies have been inconclusive; some studies have shown a positive correlation between viewing violent TV programming and subsequent violent behavior, while others have not. Some studies have even found a diminution of violent and aggressive behavior among those watching the violent programming, raising the possibility that at least for some children watching violent programming may serve as a kind of safety valve.

A third type of research examines actual viewing habits to see if there is a correlation between viewing violent television shows and aggressive or violent behavior, as inferred from self-reports or interviews with the subjects. Most of these studies have found a positive correlation between watching violent programming and aggressive and even violent behavior. However, this conclusion must be tempered by the realization that correlation is not the same thing as causation. Youngsters with violent proclivities might be particularly inclined to watch violent programming. It is also possible that a third factor, such as poverty or parental abuse, may be the cause of both viewing preferences and patterns of behavior.

Televised violence may not influence the majority of children to commit violent acts, but it may have an effect on some. Violent or excessively aggressive behavior is the product of a complex set of motivations. Most people have moments when they want to commit a violent act; however, intention usually does not turn into action for a number of reasons. People know that a violent act may be met with retaliation. They also know that violence rarely solves the problem at hand. Additionally, most people have internalized a code of behavior that inhibits the commission of violent acts. Acting in concert, all of these prevent most of us from giving vent to our violent urges.

Televised violence may mitigate these inhibiting factors. As was noted above, the majority of violent acts depicted on television are not punished, or if they are, it occurs at the end of the show, when the connection with the violent act is less evident To make matters worse, the second inhibiting factor, the belief that violence is a poor way of solving problems, is not always supported by television programming, where violent acts are sometimes used to attain goals that are altogether appropriate. Finally, there is the tendency of television to desensitize children to violence, thereby attenuating the development of a code of ethics that discourages such acts.

It cannot be said with complete certainty that televised violence affects substantial numbers of viewers in these ways. As powerful as television is, it does not always override the influence of family, friends, church, and school. Research programs that try to ascertain the precise effects of television face the formidable challenge of separating the influence of television from all of the other things that shape behavior. It may never be possible to say with certainty that television has been a di-

rect cause of violent behavior, but much of the research into the subject seems to indicate that it has at least been a contributing factor for some individuals.

See also TELEVISION; V-CHIP

Television, Color

Color television is almost as old as television itself. In 1923, the British television pioneer John Logie Baird (1888–1946) demonstrated an electromechanical color television, and a purely electronic system was demonstrated by Herbert Ives of the Bell Laboratories in 1929. Neither provided an adequate picture; Baird's color television had a resolution of only 15 lines, while the Bell version provided a picture of 50 lines and was about the size of a postage stamp. Color television was little more than an interesting diversion during the 1930s, while the first half of the 1940s saw all work on television shouldered aside by the demands of World War II. But when television began to capture a large audience in the late 1940s and early 1950s, color television seemed to be the next logical step.

By the mid-1950s, the essential technological features of color TV were in place. Then, as now, color television was based on the principle that all colors can be made by combining three primary colors: red, green, and blue. Note that these primary colors differ from the primary colors used for inks and paints; they form a color spectrum through a subtractive rather than an additive process.

A black-and-white television uses an electron gun to energize a fluorescent material on the screen. A color television works according to the same principle, except that it uses three electron guns and a screen coated with tiny dots of different fluorescent materials. These glow with red, green, or blue hues when they are excited by a corresponding electron gun. Interposed between the screen and the electron guns is a plate with thousands of perforations; the plate is arranged in such a way that when the picture is scanned, beams from the electron guns pass through the appropriate holes and do not energize the adjacent fluorescent dots. Although each dot glows in only one color when it is energized by an electron beam, the eye cannot resolve the separate colors and instead sees a single color. The overall form and color of the picture on the screen is determined by the combination of energized dots that appear at a given time.

Much of the developmental work for color television was done by the Radio Corporation of America (RCA), which also owned the National Broadcasting Corporation (NBC). RCA began to market its first color sets in 1954. These early sets met with considerable consumer resistance; by late 1961, fewer than 800,000 of them were in operation. Part of the problem lay in high prices and poor reliability. Also, only the NBC network offered much in the way of color programming. A number of years were to pass before the industry as a whole began to produce and broadcast significant numbers of color programs.

The dominant position of NBC and RCA in early color television could be traced back to decisions made by the Federal Communications Commission, the agency of the federal government charged with regulating radio and television broadcasting. In the late 1940s and early 1950s, NBC's major rival, the Columbia Broadcasting System (CBS), also had also been involved in the development of color television. Under the leadership of Peter Goldmark (1906–1977), CBS had brought out an electromechanical form of color broadcasting by 1946, but it was not "compatible color"; a black-and-white TV set could not pick up one of these color broadcasts. Moreover, the CBS system was based on the ultrahigh frequency (UHF) portion of the electromagnetic spectrum, whereas contemporary TV broadcasting was done using the very high frequency (VHF) part of the spectrum. Other broadcasters and manufacturers understandably wanted to protect their investment in the VHF portion of the spectrum, and lacking any commercial interest in color TV, they were quite content to see the emerging industry begin with black-and-white television.

Even so, the CBS system offered the prospect of a leap directly into color television, and in September 1950 the FCC accepted the CBS system as the standard for color television in the United States. It did not remain so for long. Under pressure from RCA and other manufacturers of television equipment, the FCC reversed itself in December 1953 and threw its support behind RCA's compatible-color technology.

Color television began to grow rapidly in the 1960s as the price of sets dropped and their reliability improved substantially. This helped to turn a vicious circle into a virtuous circle. In the late 1950s and early 1960s, relatively few sets were sold due in part to the lack of color programming. At the same time, broadcasters other than RCA had little incentive to produce color shows when there were few color sets in operation. When both the number of sets and the number of color broadcasts began to grow, each reinforced the growth of the other. The result has been a consumer product that is virtually universal. In 1995, American homes contained 94 million color TVs, about one set for each home.

Although the United States led the world in the development of color television, by the mid-1990s no American firm was producing color sets. Color TVs were still being made in the United States, but the parent firms were all foreign. As often happens with a product that once represented the cutting edge of technology, color television eventually became a standardized commodity. Product innovation became less important than process innovation, and the manufacturers with the lowest production costs took over the market. High-definition television (HDTV), a product that represents a major improvement over conventional television, may offer the opportunity for a resurgence of American television manufacturing. It remains to be seen, however, if HDTV will be able to stimulate the growth of a large market in the same way that color TV did when it began to replace black-and-white TV.

See also FEDERAL COMMUNICATIONS COMMISSION; TELEVISION; TELEVISION, HIGH-DEFINITION

Television, High-Definition

Television technology has evolved from crude, tiny boxes in the early 1950s to runny color sets contained in large consoles in the early 1960s to magnified screens with stereo sound in the 1990s. The next generation of television, according to some analysts, will be "high-definition color television" (HDTV), which represents a substantial advance in the image presented to the viewer. The detail, aspect ratio (the ratio of picture width to height), and overall picture quality of the HDTV is superior to even the best existing televisions. HDTV promises to provide twice as many lines of resolutions as conventional television, with an aspect ratio 25 percent higher than the conventional image (i.e., 4:3 in conventional televisions, 16:9 in HDTV). In addition to improvements in picture, the HDTV offers improved multilevel sound.

HDTV development was promoted by the National Cooperative Research Act (NCRA) of 1984, a politically charged piece of legislation that was passed at a time when the debate about the need for a national industrial policy was raging. The NCRA diminished the threat of antitrust litigation for selected cooperative research activities, namely those the government deemed useful. Under the NCRA, 36 corporations merged to create 7 HDTV alliances. Some critics maintained that Japan, already enjoying a lead in electronics technology, would solidify its dominance in television technology if it developed HDTV first. Naturally, American electronics manufacturers wanted any government assistance they could obtain.

Proponents of HDTV subsidies argued that the new technology would create spin-off effects throughout the video industry. For example, the consortia created specialized memory chips, known as VRAM or *video random access memory chips*, which by themselves constituted a potentially huge market. One analyst envisioned VRAMs making up 30 percent of the $160 billion global multimedia market by the year 2000.

With such high stakes, both the United States and Japanese governments have played active roles in encouraging HDTV technology, although Japan's government poured more than $5 billion into the Japanese consortium with little success. More important, the largest potential market in the world for HDTV, the United States, held the trump card for the actual introduction of the technology. The federal government retained final approval for the HDTV signal architecture under the authority of the Federal Communications Commission. In 1991, "single" and "wide channel" HDTV systems were tested by the Advanced Television Test Center, but as of 1996 no American systems were in operation. American companies expected to achieve the necessary technical requirements and receive approval, but their architecture promised to be substantially different than that of the Japanese. Meanwhile, Japan has a satellite-transmitted HDTV

signal that has been experimentally broadcasting on a daily basis since 1988 and started full-time service in the early 1990s.

HDTV manufacturers not only have to be able to deliver higher-quality pictures, they also have to provide the product at a reasonable cost. HDTV receivers for the home will have an initial selling price of at least $2,000, but the consumer cost only represents a small fraction of the industry expense to convert to HDTV. Televisions stations would have to modify their broadcasting systems from conventional signals to HDTV signals, replace most, if not all, of their present equipment, and find a way to service the millions of consumers who did not switch to HDTV. Stations will need new cameras, tape equipment, converters, and perhaps even new transmitters and towers. One estimate placed the cost of switching by local stations and other parts of the broadcast web to as much as $20 billion.

Even more unsettling to both American and Japanese HDTV developers is the continued evolution of the personal computer, which, when linked through fiber-optic lines, can process digital information, essentially giving it television capability. According to prophets of the new electronic era, the real breakthrough will involve interactive entertainment, in which a viewer watching a football game can zoom in and out or follow a certain player, or while watching a murder mystery a viewer is able to "vote" for one of several alternative endings. Recent telecommunications deregulation has opened the door for telephone companies to enter the market of computer technology, and thus the marriage of computer and television may be sooner rather than later. The proliferation of computer games and personal computers in homes underscores the notion that HDTV may be dead on arrival. Indeed, surveys of youth taken in 1996 have revealed that modern teenagers spend less time than before watching television, but spend more time on the computer (including the Internet), while reading time has not changed. Simply providing a higher-resolution image may not justify a substantial national commitment to HDTV with such trends coming into focus.—L.S.

See also COMPUTER, PERSONAL; FIBER-OPTIC COMMUNICATIONS; INTERNET; TELEVISION; TELEVISION, COLOR; VIDEO (COMPUTER) GAMES

Further Reading: George Gilder, *Life After Television*, 1990.

Tennessee Valley Authority

Hydroelectric power has been eagerly pursued as a source of energy and a stimulus to economic development. In many countries, its development has been the responsibility of state-owned power companies, but in the United States the early development of hydroelectricity was the work of privately held electric utilities. However, many political "progressives" pressed for the public ownership of hydroelectric power; this was realized with the creation of the Tennessee Valley Authority (TVA) in 1933.

The need for electrical power in the region to be served by the TVA was evident. Many of the 2 million inhabitants of the seven Appalachian states drained by the Tennessee River suffered from chronic flooding, erosion, and poverty. Few had electric service, and most private utilities assumed that indigent farmers would never purchase enough power to make the business profitable. The TVA project aimed to facilitate navigation, provide flood control, produce fertilizer, encourage soil conservation and reforestation, promote economic growth, relieve unemployment, and, partly to facilitate these goals, to generate electricity. The TVA also constituted a massive attempt at social engineering, and New Deal Democrats lauded its programs as triumphs of grass-roots democracy. While no other region replicated the TVA experiment, the initiative succeeded admirably in electrifying the valley. In 1933, the TVA's residential rates ranged from 0.4 to 3.0 cents per kilowatt-hour, compared to a national average of 5.5 cents. In addition, the project convinced private utilities throughout the nation that—stimulated by a bountiful power supply, promotional campaigns, and clever marketing of electric appliances—rural Americans could consume just as much electricity as their city cousins.

TVA has been a thoroughgoing success in other ways as well. The 32 dams it owns or controls have brought many benefits. The danger of flooding has been vastly reduced. The Tennessee River is now navigable for a distance of more than 965 km (600 mi), allowing the shipment of 22,000 million kg (24 million tons) of cargo on the river each year. In other respects, however, the TVA has failed to fulfill the intention and spirit of its original charter. Although a stated goal of TVA was to involve the people of the region in decisionmaking, TVA soon found it expedient to accommodate itself to existing power structures. For many years, one unfortunate aspect of this accommodation was its lending at least tacit support to racial discrimination. TVA's energy policies have also been questionable. A major reason for the creation of TVA was the exploitation of hydroelectric resources, but in the years that followed, resources were deemed inadequate, and coal-fired and nuclear-generating plants came to supply more than 80 percent of TVA's electricity. The installation of thermal plants had particularly unfortunate consequences for the region, as it gave rise to the strip mining of coal, laying waste to large areas in the process. Finally, in its zeal to effect modernity through electrification, TVA pushed highly inefficient uses of electricity, such as space heating.

TVA was a child of the New Deal, and as such it has been buffeted by shifting political winds. After some strident debates about the need for government-owned electrical utilities in the 1950s, a general agreement on the role of TVA emerged; supporters of privately owned power accepted TVA's legitimacy, while TVA halted all efforts to expand beyond the region it was serving.—A.S. and R.V.

See also ELECTRIC-POWER SYSTEMS; HYDROELECTRIC POWER; MINING, SURFACE; STOVES, HEATING

Further Reading: David E. Lilienthal, *TVA: Democracy on the March*, 1944.

Tennis Rackets

Tennis enthusiasts in the 16th century were critical of a new piece of equipment—the racket. Playing tennis with the hand was considered excellent exercise, but the racket minimized effort. Even more important than efficient play to the Renaissance elite was maintaining the grace and elegance of the game. For this reason, many players chose to "leave the nets to fishermen" and play the far more aesthetic game of tennis without a racket.

The racket had emerged slowly over the centuries as tennis left behind its origins as a game of hand ball, *jeu de paume*. At first, players used a glove to protect their hands, which gave way to a rope woven around the hand, and then to a wooden bat. At the beginning of the 17th century, the game was becoming a middle-class pastime, and the racket strung with sheep intestines came into prominence. From the middle of the 16th century to the middle of the 20th century, the tennis racket changed very little. Racket lengths grew, and enthusiasts experimented with a variety of asymmetric shapes and stringing patterns, but the nature of the materials set strict limits on racket design. Although the International Tennis Federation (ITF) began to regulate the size, weight, compression, and performance of tennis balls in the late 19th century, it completely ignored rackets. Until 1978, a player could use *anything* to strike the ball.

The use of revolutionary new materials transformed rackets and the game of tennis. A steel racket head had been tested in 1922, but wooden rackets continued to dominate the courts. Inspired by steel-shafted golf clubs, Rene Lacoste (1904–1996), the French champion, patented a stainless steel racket frame in 1965. This racket was significantly lighter and more aerodynamically efficient than wooden rackets. The open handle and decreased weight allowed the player to increase racket velocity and, thus, shot power. Wilson marketed Lacoste's racket as the Wilson 2000 and put it into the hands of professional players like Jimmy Conners. In 1968, the same year that major tennis tournaments were opened to professionals, bringing an influx of money, Spalding marketed the aluminum racket. New metals provided more power, essential for the recreational player, but many elite players preferred the feel of wood. In the 1970s, composite materials such as graphite and wood increased the power of rackets without compromising their feel.

These new light and powerful rackets brought new speed to the hackers' and women's games. In addition, the composite rackets allowed experimentation with racket shapes that the nature of wood had previously restricted. In 1885 a large-head racket had appeared on the court, but the wood frame could not withstand the tension of a tightly strung racket. Not until the 1970s would the large-head racket revolutionize tennis.

In 1976, a recreational tennis player, frustrated by his off-center hits, blamed his racket for the difficulty of making good contact. As he had 20 years earlier, when frustration on the ski slopes led to a revolution in ski materials and manufacturing, Howard Head used his engineering ingenuity to design a racket that would not twist when hit off-center. By increasing the width of the racket face, Head increased its resistance to angular motion. By-products of the new shape were an increase in the "sweet spot" by four times over the traditional racket and a reduction in the vibrations that caused tennis elbow.

While recreational players snuck their oversized Prince rackets onto the courts for their weekend games, the racket did not immediately offer an advantage to the skilled player. Professionals did not need the larger sweet spot, for they hit off-center less frequently than weekend hackers. More problematic, however, was the flexibility of the frames, which along with too-low stringing tensions produced a "trampoline effect." It took the introduction of a stiffer frame and a 20 percent increase in stringing tension to make the new racket truly revolutionary. By 1982, the hottest items on the Wimbledon court were oversized racquets. Although many professionals were slow in adopting the racket, it increased the power, playing years, and skill of legions of junior, senior, and recreational players.

Despite the dramatic effects of new materials and large-head rackets on the game, the International Tennis Federation continued its tradition of allowing any and all racket innovations. This policy changed in 1978 in response to a new stringing system, called "spaghetti strings," which added dramatic top spin to mediocre shots. In 1978, the ITF limited head size and overall length, and initiated stringing regulations. Even so, the new rackets in the 1990s are transforming the sport from a complex one of skill, technique, spin, touch, and power into little more than a serving contest. The ITF is justifiably concerned that racket technology is usurping the challenge and complexity of the sport. It is faced with the dilemma of setting limits that preserve the integrity of the game while allowing the use of equipment that improves the game for recreational, junior, senior, and female players.—J.N.G.

See also COMPOSITE MATERIALS

Thalidomide

Thalidomide is the trade name for α(N-phthalimido) glutarimide ($C_{13}H_{10}N_2O_4$). In the early 1960s, Thalidomide was prescribed in Canada and Europe as a sedative and hypnotic for use as a sleep aid. Most people who took Thalidomide experienced no difficulties, but for pregnant women it led to catastrophe, for Thalidomide interfered with the induction of the cells forming the arms and legs of a developing fetus. The result was the birth of babies with amelia (the absence of limbs) or phocomelia (hands and feet attached to the trunk of the body by a small, irregularly shaped bone). A considerable number of babies had these deformities; until the problem became evident, as many as 12,000 babies had been afflicted. Half died in infancy.

Except for the babies of a few mothers who had been given the drug on an experimental basis, the Thalidomide catastrophe did not occur in the United States. Dr. Frances O. Kelsey of the U.S. Food and Drug Administration, the federal agency that certifies new drugs for commercial release, noticed that the drug did not make rats sleepy, as it was supposed to do for human beings. Suspecting that Thalidomide affected rats and human beings in different ways, she requested that more data be submitted, while withholding certification of the drug for use in the United States. By early 1962, the consequences of the drug had become evident, and a warning was sounded by Helen Taussig (1898–1986) in the May 25, 1962, issue of *Science*.

The near-disaster of Thalidomide in the United States alerted government officials and the public to the potential dangers posed by new drugs that had not been adequately tested. The Thalidomide tragedy led directly to a strengthening of federal food and drug laws through the Kefauver-Harris amendments to the 1938 Food and Drug Act. Signed into law by President Kennedy in 1962, the amendments gave the FDA more control over experimental drugs prior to their approval, as well as the power to withdraw approval from drugs already on the market if there were specific reasons for doing so. Along with addressing safety, the amendments also took up the issue of effectiveness by requiring that drugs be demonstrated to be effective on the basis of "substantial evidence" obtained through "adequate and well-controlled investigations."

While its use produced tragic consequences in the early 1960s, more recently Thalidomide has been successfully used for the treatment of leprosy. It has also been shown to be of considerable value in alleviating certain conditions related to AIDS, such as mouth ulcers and severe weight loss. Thalidomide also has shown considerable promise for the treatment of lupus, glaucoma, and the rejection of bone marrow transplants. Of course, future medical uses of Thalidomide will have to be carefully monitored to ensure that the drug is never taken by pregnant women.

See also ACQUIRED IMMUNE DEFICIENCY SYNDROME (AIDS); ANIMAL RESEARCH AND TESTING; FOOD AND DRUG ADMINISTRATION, U.S.

Theodolite

A theodolite is a surveying instrument that measures the angles formed by horizontal and vertical lines. In the United States, the term *theodolite* is generally reserved for instruments capable of measuring to 1 second of arc or less. This distinguishes it from a less-accurate device known as a *transit*. In other places, the term *transit* is not used, and theodolite is used for both devices, irrespective of their accuracy.

The angles measured by a theodolite are used to determine horizontal and vertical distances through a geometric procedure known as *triangulation*. The basic principles used in surveying with a theodolite were embodied in an earlier device known as a *dioptra*, the invention of which is attributed to Hero of Alexandria (c. 20 C.E.– ?). A theodolite measures angles in much the same way as did the dioptra, but its use is facilitated by equipping it with a small, low-power refractor telescope. The theodolite's tripod mount is equipped with a spirit level that indicates when the instrument is level, while a plumb line is used to ensure that it is vertical.

A theodolite's telescope sight has cross hairs that allow the instrument to be precisely trained on a target. Through an appropriate mechanism, the telescope can be rotated about either its vertical or horizontal axis. Both of these axes have concentric graduated circles that indicate the angles defined by the telescope as it is fixed on a target. The angles read from these circles are viewed through another eyepiece. Some theodolites are accurate to 1 minute of arc (1 minute = 1/60 of a degree), while more precise examples are accurate to a few hundredths of a second (1 second = 1/60 of a minute).

To determine a horizontal angle, the telescope is

put at the intersection of the two lines forming the angle, and in line with one of the legs of the angle. It is then rotated until it is in line with the angle's other leg. The size of the angle in degrees can then be read from the graduated circle. The same basic process is used to find vertical angles. These procedures can be done in reverse in order to define an angle that has an existing line as one of its legs. The measurement of vertical angles to determine elevations is done by training the theodolite on a graduated rod held at the point being measured. This procedure, known as *differential leveling*, is covered in any book about surveying.

See also TELESCOPE, REFRACTING

Theory

In the most basic sense, a theory is a mental creation, rather like a story that one might think through and write down. A theory introduces characters and things, gives them properties and relations, suggests ways in which these characters interact, and tells us how we can observe them or their effects. Unlike a fictional story, however, a scientific theory is created with a specific, technical intention. Theoretical scientists seek to model the world that we experience. A theory is usually created because we are curious about something that has been observed, and we want to understand it better. The theory is created as a possible explanation of what happened. But a theory can also be created because we want to anticipate future experience—we want to predict what will happen.

There has been extensive discussion about the nature and purposes of scientific theories, and virtually all of this discussion has centered around the problems of understanding exactly what we mean by saying that a theory is intended to model what we experience. Commentators have expanded the notion of model to cover both explanation and prediction of experience. They have also suggested important issues regarding what it is that we actually experience. Do our models merely attempt to replicate what we identify directly through sensation? Or do our models attempt to describe a world that we believe to stand behind what we sense, a "real world?" Finally, no matter what intention we pur-

sue within a given theory, we must ask when and how we know that we have succeeded.

Most of the contemporary discussion of scientific theories has been in the "empiricist tradition"; that is, commentators have assumed that theories are intended, at the very least, to model what we experience directly through our senses. From this perspective, the successfulness of a theory is measured, at least in part, on the basis of its relation to whatever we experience in sensation. The philosopher of science, Bas van Fraassen, has called this *empirical adequacy*.

A simple example of a theory with empirical intent is embodied in the solution of a crime and its ultimate trial in a court of law. Suppose that a burglary has occurred. No one has witnessed the break-in, but some valuable jewels have been taken from a safe. Investigating detectives interview various people and collect evidence. On the basis of evidence, hints, leads, and pure intuition and imagination, they create a theory of what happened and who did it. The theory suggests an account (explanation) of the burglary, but it also may suggest when, how, and where the supposed burglar will strike again (prediction). The detectives will probably work in both directions at once, trying to perfect their account of the burglary that has already happened but also lying in wait for other burglaries that they suspect may happen. Presuming that they never catch the burglar in the act but that they do perfect their account to their satisfaction, they will transfer the case to the district attorney's office for prosecution.

The trial in a court of law is a good analogy to what actually happens in the scientific community when a scientist presents his or her theoretical ideas in one or more technical papers. The theory itself is presented in the prosecution's opening statements; this tells exactly what the prosecution thinks happened and how the accused individual played the critical role in executing the burglary. At this point, the prosecution's case is pure story line, pure speculation. There is no reason why the jury should believe the accusations. The trial itself centers around the presentation of evidence.

How does the prosecution's theory make connections with the world that we can actually experience? Have the jewels, in this case, been found and can people connect the jewels with the accused in any way? Does the accused have the expertise to open a locked safe? Did the accused leave anything at the crime scene: fingerprints, personal effects, etc.? Was the accused in the position, by other people's testimony, to commit the burglary at the particular time and place that it happened? In short, what can people tell us about facts of experience, and how can these be connected with the prosecution's general theory of how this burglary took place?

Eventually, the case goes to the jury, and they decide whether or not the defendant is guilty or innocent of the charges. In effect, they decide whether the evidence presented has convinced them to subscribe to the theory. When a jury decides in favor of the prosecution, this does not make the prosecution's theory true; it simply means that the jury agrees, on the basis of evidence, that acting as though the theory were true is the best way to proceed. The defendant will be sentenced and, in that respect, some very important decisions about his future will be made on the basis of the prosecution's theory.

In a similar way, scientists who have considered the evidence for a scientific theory must decide whether it has enough merit to apply it in future situations or to continue to study it under other conditions. In both cases, people are asked to consider the doubts they may have about the empirical adequacy of the theory as presented and defended. Are there, indeed, reasonable doubts, and how may they be overcome? Is a particular theory a sound basis for a variety of future scientific activities: planning research programs, spending money on relevant equipment, perhaps even risking a person's safety by using the theory to guide a specific procedure?

There are certain structural elements to any theory, including the district attorney's case, as described above. A theory must name certain objects and give them properties. It must involve a certain logic of activity and object relations so that we can reason out what these objects will do in their theoretical world. But the theory must also be connectable to the world we experience. A fingerprint is a good example of this connectability. The print itself is something that we can see and capture in the sensed world. The claim that a fingerprint is "attached" to an individual is a connecting link in criminological theories, a standard kind of evidence for the relationship between a theoretical object and a crime scene. We take the print as

indicative of an individual's presence in a place even though we are not actually witnesses of that presence. In a scientific theory, a spot on a photographic plate may act, in a similar way, as an indicator for the presence of a high-energy particle beam.

Theories almost always suggest objects, properties, and environments that are quite different from those of the world that we experience in sensation. Their empirical adequacy is, as we have seen, tested on the basis of how theories can be directly connected with experience. But if a theory is very successful in this way, can we go further and suggest that it actually models a "real world" that stands behind and causes the world that we experience? In effect, are theories merely good instruments for predicting experience, or are they descriptions of real worlds that we cannot experience directly. This issue was debated in antiquity, and no convincing resolution has yet been discovered.

Theories and hypotheses share a great deal in common. A hypothesis, however, need not develop characters, objects, or properties that cannot be directly experienced. Also, a hypothesis is usually rather narrow in scope and can be confirmed or denied with a small number of experiments. Theories usually have a wide scope and require many diverse evidential experiments before they are worthy of support or belief.—T.B.

See also EMPIRICISM AND THE SCIENTIFIC METHOD; EXPERIMENT; HYPOTHESIS; PARADIGM; SCIENTIFIC METHODOLOGY

Thermal Cracking

In the early 20th century, traditional distillation procedures were used to convert petroleum or "crude oil" into useful products. At that time the most important of these was kerosene, which was widely used for illumination. Gasoline, hitherto a substance of limited value, became increasingly important as automobiles began to be produced in large numbers. Since only 15 percent of the petroleum could be converted into gasoline through distillation, there was a distinct possibility that the growth in the numbers of automobiles would be sharply limited by fuel shortages.

This obstacle began to be surmounted when William Burton (1865–1954), an employee of the Standard Oil

Company and one of the industry's first research chemists, began to investigate the breaking of hydrocarbon molecules into smaller ones, a process known as *cracking*. The cracking of petroleum molecules was discovered by Benjamin Silliman (1779–1864) in the mid-19th century, but since it had no commercial value at the time, the idea was not pursued. Burton, working in conjunction with Robert Humphreys, developed the equipment and techniques that cracked petroleum through the use of high temperatures and pressures. Although a pilot operation was successful, his employers refused to approve the construction of a full-scale plant. However, shortly afterwards, Standard Oil was broken up by government decree, and Burton's new employer, Standard Oil of Indiana, was more receptive to the venture. In January 1913, the first thermal cracking plant went into operation.

Although the amount of gasoline produced was doubled, it wasn't exactly equivalent to straight-run gasoline produced through distillation; due to sulfur contamination it was yellow in color and had a bad smell. Even though it was sold at a lower price, it found few buyers until it was chemically treated to remove the sulfur. The production process was also improved through the use of Edgar Clark's technique of cracking the petroleum in a tubular still, the availability of stronger metal sheets and tubes, and better ways of collecting and disposing of carbon residues (known as *coke*). Collectively, these enhancements lowered production costs by as much as the original Burton process did in comparison with distillation. Taken as a whole, the cracking process represented an impressive payoff for research and development; expenditures of $236,000 generated $150 million in profit and royalties.

After developing rapidly, the Burton process reached its limits, and other cracking technologies took on increasing importance. One was developed by C. P. (for Carbon Petroleum) Dubbs, who started with a process invented by his father, Jesse Dubbs, in 1909. The original intent was to remove seawater that was mixed with the petroleum in an emulsified state. Dubbs's process successfully broke the emulsion, and it also cracked some of the petroleum, although the inventor was not aware of it at the time. Working for Standard Oil of New Jersey, C. P. Dubbs invented a "clean circulation" process, which, unlike the Burton process,

cracked the petroleum through a continuous process rather than one batch at a time. It also had the advantage of using straight petroleum for a feedstock, whereas the Burton process required that the petroleum be put through an initial distillation. The Dubbs's process also benefited from a number of auxiliary improvements, most notably the use of a hot-oil pump invented by engineers from Shell Oil, which increased output from 500 to 700 barrels per day.

With Standard of Indiana owning the rights to the Burton process and the Universal Oil Products Company owning the rights to the Dubbs process, oil companies found themselves in the uncomfortable position of paying sizable royalties for their use. Many set about developing their own thermal cracking processes, and by the early 1920s, petroleum was being refined through the Cross process, the Holmes-Manley process, and the tube-and-tank process. All of them were based on the use of heat and pressure for the cracking of petroleum, and all of them were covered by various patents, many of which overlapped with one another—the Patent Office had been quite liberal with the issuance of patents. As a result, the refining industry faced the prospect of endless suits and countersuits by its constituent firms, draining resources and leaving a great amount of uncertainty.

In order to resolve the problem of lawsuits, a number of firms gathered together in 1923, and after some negotiation agreed to enter into patent pooling agreements. With a major uncertainty removed, oil refiners installed more cracking facilities, nearly doubling the amount of cracked gasoline within 2 years. The Department of Justice challenged these agreements, and the issue made its way through the court system until 1931, when the Supreme Court found them to be legal.

Not everyone benefited from the widespread adoption of thermal cracking. In the early years, the Universal Oil Products Company helped smaller refiners to acquire and operate the necessary equipment for using the Dubbs process, thereby enhancing these firms' prospects for survival. But ongoing improvements in thermal cracking were directed at taking advantage of economies of scale. Smaller refiners could not make the necessary investments, and many of them either went out of business or merged with larger firms.

The development of thermal cracking was essential to the great expansion of automobile use that took place in the 1920s. Yet the demand for gasoline still outstripped the ability of refiners to produce it. Meeting the expanding needs of an automobile-based transportation system required the invention of a new means of refining petroleum, catalytic cracking.

See also CATALYTIC CRACKING; OIL REFINING; PATENTS

Further Reading: John Lawrence Enos, *Petroleum Progress and Profits: A History of Process Innovation*, 1962.

Thermoionic (Vacuum) Tube

A vacuum tube uses heat to liberate electrons from its negatively charged cathode. These electrons migrate to the tube's positively charged anode, in the course of which they are controlled by other parts of the tube. It is called a *vacuum tube* because its interior has been evacuated. In Great Britain, a vacuum tube is sometimes known as a "valve," for just as a hydraulic valve regulates the flow of water, a valve of this sort regulates the flow of electrons. This regulated flow of electrons is the basis of many technologies: radio, television, computers, stereos, and radar.

The principles underlying the thermoionic tube began to be explored in the 18th century. At this time, several experimenters looked into the relationship between electrical charge and the temperature of the object holding the charge. Of particular significance, in 1739 Abbe Nollet found that hot metal rapidly lost the electrical charge it previously had been given. The invention of the Leyden jar allowed more elaborate experiments. In 1777, it was noted that a Leyden jar lost its charge when its electrode was brought close to a red-hot iron. In 1838, Michael Faraday (1791–1867) discovered that the interior of a tube containing a rarefied gas glowed when a current was passed through it. This was the foundation of many important discoveries, such as Edmond Becquerel's demonstration in 1853 that the conductivity of gases increased at high temperatures. The transmission of an electric current through space was discovered in 1884 by Johann Wilhelm Hittorf (1824–1914), whose experiments showed that considerable current flowed when the negative pole of a battery was connected to the cathode of an evacuated tube and the battery's positive pole was connected to the anode.

These phenomena were quite puzzling, for they seemed to indicate the flow of current through empty space, a phenomenon that contemporary physical theory could not account for.

Scientific inquiries of this sort were complemented by experiments conducted by Thomas Edison (1847–1931), a man with little interest in what we would today call "pure science." In the early 1880s, Edison was trying to understand why the interior surface of his recently invented incandescent electric light was being coated by a dark deposit. In 1883, he put a metal plate between the two legs of the filament and discovered that if the plate was connected to the filament's positive leg, there would be a flow of electric current, but nothing would happen if the plate was connected to the negative leg. He also noted that the current was proportional to the incandescence of the filament. Once again, a puzzling phenomenon was observed, and the best that could be done at the time was to give it the name Edison effect. Edison himself patented the modified lightbulb, hoping that it might have commercial application as a device for measuring the flow of current, but nothing came of it.

The great potential of the vacuum tube moved towards realization in 1904, when John Ambrose Fleming (1849–1945) used it for the detection of radio waves. The vacuum tube became even more useful when in 1906 Lee De Forest (1873–1961) created the triode by inserting a control grid between the anode and cathode. Again, the application was for the detection and amplification of radio waves, and it proved to be of great importance for this purpose.

Despite these successful applications, the reason for the flow of current was not apparent. Many practitioners thought that functioning vacuum tubes had not been completely evacuated, and that minute amounts of ionized gases were actually conducting the current. The basis for an accurate understanding was provided by J. J. Thomson (1856–1940), who in 1897 used a vacuum tube–based apparatus to demonstrate the existence of negatively charged atomic particles that came to be known as *electrons*. Still, a number of years passed before engineering practice caught up with scientific theory.

The most important early use for vacuum tubes was for the generation, detection, and amplification of radio waves. In succeeding years, vacuum tubes be-

John Ambrose Fleming (courtesy Institute of Electrical Engineers).

to heat the filament, and high-voltage direct current was needed to impart an adequate positive charge to the control grid or grids. Much of the energy supplied was dissipated as heat. Filaments lost their ability to produce electrons and had to be replaced at frequent intervals. The solution to these problems lay in the creation of solid-state devices, beginning with the transistor. In recent years, however, the vacuum tube has undergone a small renascence. Some audiophiles have built their sound systems around vacuum-tube amplifiers because, in their estimation, these produce sounds that are warmer and more natural than those produced by solid-state devices. The reasons for this are not clear, but some audio engineers have speculated that while all amplifiers cause some distortion, tube-based amplifiers do it in a more gradual manner, making for a more pleasant sound.

came more complex through the addition of more controlling elements, resulting in tetrodes, pentodes, hexodes, and so on. The value of vacuum tubes was also greatly enhanced by the development of improved means of evacuating the tube, down to one-billionth of atmospheric pressure. Improved means of effecting good seals between glass and metal components made them more reliable, as did new materials for the tube's filaments. In 1927, a significant advance came with the introduction of the AC (alternating current) tube. Its filament could be heated by house current that had run through a transformer, obviating the need for bulky batteries that required frequent recharging. In 1933, mixer or converter tubes were devised to amplify incoming signals while generating their own signal; this allowed one-tube heterodyne reception.

Specialized vacuum tubes were essential components of many technologies, but they had their drawbacks. A considerable amount of power was required

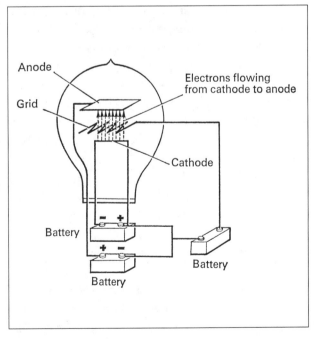

The operation of a triode: The application of a small current to the grid allows the control of the much greater flow of electrons passing from the cathode to the anode (from S. Handel, *The Electronic Revolution*, 1967).

See also ALTERNATING CURRENT; CAPACITOR; DIODE; ELECTRON; RADIO; SEMICONDUCTOR; SUPERHETERODYNE CIRCUIT; TRANSISTOR; TRIODE; TUNGSTEN FILAMENT

Further Reading: Gerald F. J. Tyne, *Saga of the Vacuum Tube*, 1977.

Thermometer

It is often incorrectly assumed that temperature is the measure of heat in a body. A basic understanding of the nature of heat makes it evident why this is not so. Heat is produced by molecular motion; the more molecules there are and the greater their motion, the more heat there is. A lake with its many molecules of water contains more heat than a cup of water, even when the former is just above freezing and the latter is at the boiling point. Temperature is a measure of the speed of molecular motion and is not directly related to the number of molecules. The cup of very hot water contains fewer molecules than the lake, but since these molecules move at a greater speed, the water registers a higher temperature than the lake.

For centuries, people considered temperatures in a subjective fashion. Heat, according to Aristotle, was a quality like smell or color, something that by its very nature could not be measured. There could be common agreement about the general meanings of *hot, cold, cool,* and *lukewarm,* but these terms were severely lacking in precision. Early scientists discovered some of the properties of heat but built no instruments for the measurement of temperature. In ancient Greece, Ctesibius (fl. c. 270 B.C.E.) constructed a device that indicated temperature changes by connecting two liquid-filled vessels with a U-shaped tube; when one of the vessels was heated, the expansion of the liquid caused some of it to pass into the tube. This device illustrated an important consequence of temperature change, but it gave no numerical indication of the extent of these changes.

Ctesibius's device appears to have been reinvented by Galileo (1564–1642), and one of his colleagues subsequently endowed it with a temperature scale. The temperature of melting snow was used for one temperature extreme, and water boiling over a candle flame for the other. Between these extremes were 110 gradations; hence a single gradation could be read as one degree. It

is significant that this attempt at quantification occurred when it did. One of the fundamental requirements of science is accurate measurement. In the course of the accelerated development of science that began in the 16th century, great scientists like Galileo, Robert Boyle (1627–1691), Christian Huygens (1629–1695), and Isaac Newton (1642–1727) naturally turned to thermometers and temperature scales. The most successful scale turned out to be the one created in 1714 by Gabriel Daniel Fahrenheit (1686–1736), an instrument-maker who was born in Danzig (present-day Gdansk) and worked as an instrument-maker in England and the Netherlands.

Although temperature is a measure of the speed of molecular motion, there is no direct way of determining the speed of individual molecules. Rather, speed is inferred from the behavior of a material made up of these molecules. Specifically, most substances (water is a notable exception) expand as their temperature goes up, and contract when it goes down. Fahrenheit's thermometers were thin glass tubes filled with alcohol or mercury that behaved in just this way. For one base point, Fahrenheit used the temperature of the coldest substance he was able to produce, a mixture of water, ice, and sal ammoniac (ammonium chloride or NH_4Cl). Another fixed point was body temperature, which Fahrenheit defined as 96 degrees. It is not certain why he chose this number; perhaps he did so because it could be easily divided by 2, 3, and 4. In any event, subsequent research determined that normal body temperature is 98.6 degrees. With these points fixed, Fahrenheit divided his tubes into 100 equal gradations; 32 gradations from zero happened to be the freezing point for water. Extending the scale upward gave 212 degrees for water's boiling point.

All of this assumed that mercury and alcohol expanded in a linear matter. That is, increases or decreases in molecular motion produced the same changes in the length of a column of these liquids, irrespective of the starting point. Fortunately for Fahrenheit and the scientists that used his thermometers, this in fact is true for the temperature ranges covered by his scale. If it had not been the case, the emerging science of thermodynamics would have made much slower progress in the years that followed.

Fahrenheit's contributions should not be overstated. The idea underlying the thermometers he constructed had been around for some time, and his numerical system was only one among many others. The Fahrenheit scale gained ascendancy not because it was somehow more rational than all the others but because Fahrenheit's thermometers were more accurate than most others; a temperature given by one of his thermometers would likely be the same as the temperature measured by another of his thermometers.

Thermometers graduated with the Fahrenheit scale are still in widespread use in the United States. However, for scientific and technical work, thermometers are usually graduated according to the Celsius or centigrade scale. This scale is also used almost everywhere outside the United States for measuring indoor or outdoor air temperatures, as well as body temperatures.

Although Fahrenheit's scale has been supplanted in many places and for many purposes, the basic instrument he used, a liquid free to expand or contract inside a glass tube, is still in widespread use. For specialized purposes, however, it may be necessary to use thermometers based on different principles. For example, an outdoor display thermometer that indicates the temperature with a moving pointer operates through the differential expansion of two dissimilar metals that are bonded together and coiled into a spiral. Changes in temperature cause the spiral to tighten or loosen, which in turn moves the pointer connected to it. Other thermometers make use of the changes in vapor pressures or electrical resistances brought on by temperature variations. Very accurate readings of small temperature changes are possible through the use of thermometers built around thermistors, solid-state devices that are more sensitive to temperature changes than metallic resistors.

See also ASPIRIN; WEIGHTS AND MEASURES, INTERNATIONAL SYSTEM OF

Thermostat

A thermostat is a device that automatically regulates the operation of a heater, air conditioner, or both in order to maintain a constant temperature within a room or building. Because of its mode of operation, a thermostat is often used as an example of a cybernetic device. As with all cybernetic systems, the operation of a thermostat forms part of a feedback loop. The thermostat receives information (the temperature of a space) and acts on the basis of this information. When the temperature of a space gets below a certain limit, the thermostat turns on the furnace. When the temperature rises above that limit, the thermostat activates the air conditioner. This is an example of negative feedback, for the response goes in a direction opposite to the existing state; that is, when the space gets warm, actions are taken to cool it, and when it gets cool, actions are taken to warm it.

Thermostatic control devices go back to the 17th century, when Cornelius Drebbel (1572–1634) invented a pressure-actuated damper that controlled the temperature of a furnace by regulating the flow of air. It is worth noting that Drebbel's invention was motivated by his research in alchemy, which is today viewed as a pseudoscience. Drebbel's thermostat was not the first and certainly not the last practical technology associated with erroneous scientific principles and endeavors.

Many of today's thermostats are based on the differential expansion of two different metals (brass and iron, for example), a principle discovered in the 18th century. Differential expansion causes an assemblage of the metals to change shape when heated; this change of shape can then be used to actuate a mechanical linkage. The first use of a thermostat based on the differential expansion of two metals is attributed to a Parisian poultry breeder named Bonnemin, who in 1777 used a bimetallic device to regulate the temperature of an incubator. A thermostat based on bimetallic strips was patented in 1831 by Andrew Ure (1778– 1857), who also coined the word *thermostat* in his widely read *Dictionary of Arts, Manufactures, and Mines*, published in 1839. Ure's patent was based on the use of bimetallic strips that had been riveted or soldered together, so that when heated the strip flexed to one side or another. Thermostats built to Ure's design were never commercially manufactured, which is just as well; they probably would not have worked well, for riveting or soldering is an unsatisfactory way to make bimetallic strips. Modern bimetallic strips are made by using heat to fuse the two strips together. Other thermostat actuators use the differential expansion rates of two metals, but in a rod-and-tube arrangement instead of bimetallic strips. Actuators also make use of liquid expansion, the saturation pressure of a liquid-vapor system, and temperature-sensitive electrical resistors.

One commonly used thermostat uses electricity working in conjunction with compressed air. Patented by Warren S. Johnson in 1885, this device was first used to control radiators by means of an electrically actuated compressed-air valve that in turn operated the radiators' diaphragm valves. By the end of the 19th century, Johnson's thermostat and those of his competitors were widely advertised as essential components of a modern heating system.

See also ALCHEMY; CYBERNETICS; STOVES, COOKING; STOVES, HEATING

Further Reading: Barry Donaldson and Bernard Nagengast, *Heat and Cold: Mastering the Great Indoors*, 1994.

Three Mile Island

The most serious nuclear accident in the history of the United States occurred on Mar. 28, 1979. On that day, the core of reactor unit number 2 at the Three Mile Island nuclear power station near Harrisburg, Pa., was seriously damaged. In the succeeding hours, many feared that the reactor would experience a meltdown and spew dangerous radiation throughout the area. As it turned out, this did not happen, and the local populace was not exposed to high levels of radiation. Even so, the accident demonstrated critical flaws in the design and operation of nuclear-power plants and seriously tarnished the reputation of nuclear power.

The damaged reactor at Three Mile Island was a pressurized water reactor, the most common type in the United States. In this design, circulating water plays a double role. It cools the reactor core and, having gained heat in the process, it powers the steam turbines that run the electrical generators. However, the cooling water does not go to the turbines; rather, it is routed through a heat exchanger that heats up water in a secondary cooling system, generating the steam that goes to the turbines. The performance of the cooling systems and the pumps that run them is crucial to the safe operation of the reactor.

The sequence of events that led up to the accident encompassed a complicated series of human and me-

chanical failures. Problems began when a maintenance crew attempted to clean out filter tanks in the secondary cooling system. The workers used high-pressure air to dislodge accumulated sludge, but a faulty valve in one of the tanks allowed water to leak into the air line. The water then got into another line that controlled the valves to the other filters. The entrance of the water caused a sudden change in pressure that automatically closed the valves that controlled the flow of water through the filtering tanks. As a result, the circulation of water in the secondary cooling system was blocked.

The designers of the reactor had anticipated a failure of this sort and had provided a number of backup safety measures. Among these were auxiliary pumps that fed water to the blocked system. Unfortunately, the valves controlling these pumps were not operating. Two panel lights were supposed to indicate the operation of the valves, but one of them was obscured by a maintenance tag, and the other simply was not seen. With no heat-absorbing water circulating through the secondary system, heat and pressure began to build up in the reactor's core.

At this point, a valve on an auxiliary tank connected to the primary cooling system was supposed to allow the venting of steam and water into a drainage tank. The valve opened as it was designed to do, but it then stuck in the open position. With the valve remaining open, water began to drain from the primary cooling system, causing serious overheating of the reactor's core. To lower the temperature, high-pressure water was automatically pumped into the primary cooling system. This might have saved the reactor had not a pressure-measuring device indicated dangerously high pressures in primary cooling system. This was a misleading signal; the entrance of steam, not excess water, had caused the buildup of pressure. Misled by this reading, an operator shut off one pump and reduced the flow of water in the other, thereby diminishing the flow of cooling water to the core. In short, where initially there had been excess pressure in the cooling system, there was now dangerously low water pressure. This important change was not perceived by the operators.

After about an hour, the accumulation of steam in the primary cooling system produced bubbles that caused the cooling pumps to vibrate dangerously. The pumps were therefore shut off while, unbeknownst to the operators, the core was still losing water. As temperatures climbed, the main pipe got so hot that some of the water decomposed into hydrogen and oxygen. This led to the formation of a large hydrogen bubble that threatened to completely block the flow of cooling water and initiate a meltdown of the reactor core. This catastrophe did not occur, and the reactor's containment structure prevented the release of radioactive material into the surrounding area. Some radiation did escape, and people within 80 km (50 mi) of the plant received an average dose of less than 20 μSv (2 mrem). This is not considered to be a harmful level of exposure; however, many local residents suffered substantial psychological trauma as a result of the crisis.

In the years that followed, the accident at Three Mile Island deepened concerns about the safety of nuclear reactors. In 1986, a far more serious accident in the Soviet Union further undermined nuclear power, although the design and operating procedures of the Chernobyl plant were quite different from nuclear plants in the West. More generally, the accident at Three Mile Island demonstrated the vulnerability of large, complex systems. The series of events that culminated in a near meltdown provided dramatic indication of how human beings and mechanisms can fail in an unpredictable manner. The lessons learned from Three Mile Island have been used to improve the safety of nuclear reactors, but some analysts are convinced that occasional failures of nuclear reactors and other complex technological systems are inevitable.

See also CHERNOBYL; "NORMAL ACCIDENTS"; NUCLEAR REACTOR; RADIOACTIVITY AND RADIATION

Further Reading: The President's Commission, *The Accident at Three Mile Island*, 1979.

Tidal Power

Driven by the gravitational fields of the sun and moon, the tides of the world's oceans contain a great amount of kinetic energy as they ebb and flow. On occasion, some of that energy has been captured for human use. In the European Middle Ages, a number of places featured tidal mills that provided energy for various tasks. The most famous of these was part of London Bridge. Since the Thames is a tidal river, its regular ebbing and flowing could be used to turn a turn a waterwheel that operated below the bridge.

The first modern large-scale tidal power unit was installed in 1968 at the mouth of the River Rance, near St. Malo in northwestern France. With an installed capacity of 2.4×10^5 kW, its 24 turbines supply about 5×10^8 kilowatt-hours of electricity per year. In 1969, a much smaller (400 kW) tidal power station was built near Murmansk in the Soviet Union. Although successful, these installations did not initiate a worldwide tidal power program. There are no commercial tidal power units in the United States, although it has been estimated that Alaska and Maine, the states with the greatest variation in the level of high and low tides, could be the sites of tidal power stations with a total capacity of 4.5×10^6 kW.

Tidal power has a number of advantages over conventional means of generating electrical power. It does not pollute the air and it does not consume nonrenewable resources. On the other hand, a large-scale tidal power installation requires the building of dams, dikes, sluiceways, and reservoirs, and these can disrupt the local ecosystem. Also, because high tides occur at intervals of about 12 hours and 25 minutes, peak generating capacity may not coincide with periods of high power demand. The unevenness of power delivery can be smoothed out using some of the power generated to pump water into reservoirs for later release, but this adds to the cost and diminishes the overall efficiency of the project. Although tidal power's operational costs are low, the expenses of building facilities in deep water with swift currents are high. As a result, electricity delivered by tidal power ends up being two to three times more expensive than electricity provided by conventional sources. Barring a substantial rise in the costs of the latter, tidal power is unlikely to become a significant source of energy, even in places where conditions are favorable.

See also RESOURCE DEPLETION; TURBINE, HYDRAULIC; WATERWHEEL

Tiles

For thousands of years, tiles have comprised a major category of decorative and utilitarian building materials. Colorful tiles ornamented with various designs were used to embellish interiors in ancient Egypt, Assyria, and Babylonia, but the world's first significant tile industry flourished in the Islamic nations of the Middle East and North Africa from the 14th through the 19th centuries. By the 11th century, Islamic ceramists learned to impart lustrous sheens to brilliant pictorial wall tiles designed for royal palaces, shrines, and mosques, and by the 13th century, they perfected techniques for making white tiles from a stone paste, a composite of ground quartz, frit, and white clay. These stone-paste wall tiles were embellished with religious inscriptions and iconography in brilliant colors, including cobalt and turquoise, and covered with an alkaline-based glaze, sometimes rendered white by opacifiers. The 13th-century introduction of stone-paste methods was a reinvention of faience techniques used by potters in ancient Egypt and pre-Islamic Iran to make beads, but the impetus for its revival stemmed from desires to emulate blue-and-white porcelain from China. Influenced by Islamic arts, ceramists in Renaissance-era Europe, especially Spain and Italy, created thriving faience industries, devoted to the production of tin-glazed pottery and tiles.

Italian ceramists figured prominently in the creation of the Dutch tile industry, which constituted the first golden moment in European tile making. In the late 15th and early 16th centuries, fire regulations dictated the use of nonflammable construction materials, and a large-scale pottery industry, centered in cities such as Delft, provided the requisite bricks, roof tiles, and lead-glazed floor tiles. As the Netherlands became one of Europe's wealthiest and most urban nations in the wake of political and religious upheaval, faience wall tiles became essential tools in the middle-class housekeeper's battle against fire, dampness, and dirt, and even modest tile installations signified a burgher's participation in urban culture. Paintings by the Dutch masters provide glimpses of sparse interiors made colorful and comfortable by additions of walls, baseboards, and mantels covered with polychrome or blue-and-white tiles. By the 18th century, the domestic market for faience tiles shifted to the countryside, and a new international trade in Delft tiles catered to the demands of consumers throughout Europe and the Americas.

Following the ascension of William of Orange (1650–1702), Dutch craftsmanship gained popularity in England, and enclaves of Dutch ceramists built regional centers of faience production in London and Liverpool. But Great Britain's major tile industry dates from the 19th century, when Staffordshire potters picked up the gauntlet for the Gothic Revival promoted by A. N. W. Pugin (1812–1852) and other architectural reformers who were refurbishing old churches and cathedrals and constructing medieval-style structures, such as the Houses of Parliament. To create lasting floor tiles emulative of inlaid earthenware tiles in medieval buildings, Staffordshire potters developed methods for making two-colored encaustic tiles from clay dust, which were subsequently fired in conventional updraft kilns. Dry-pressing was a major addition to the English tilemaker's repertoire, which also included transfer printing, molding, and other methods dating from the era of industrialization.

By the late 19th century, architects, builders, and middle-class consumers readily embellished public and private spaces with tiles. No floor and wall space, large or small, escaped tile mania. Shops, pubs, and hotels boasted pictorial facades. Hospitals, dairies, eateries, and dormitories had tiled rooms for the sake of cleanliness. And houses and cottages featured tiled, individualized porchways, hallways, fireplaces, kitchens, and bathrooms. In a fitting testimony to the achievements and the international scope of the British tile industry, architects of the United States Centennial International Exhibition of 1876 in Philadelphia covered the floors of the show's major buildings with Staffordshire encaustic tiles.

In the United States, tile manufacturers never rivaled the output of their British counterparts in terms of aesthetics, but they operated some of the world's largest tile factories in the 20th century. Small batch-production factories in Massachusetts, New Jersey, Ohio, and California catered to the fashion for faience, while enormous quantity-production firms such as the American-Encaustic Tiling Company (Ohio and New York) captured markets for more standardized products, from plain white tiles used in bathrooms to those lining the Holland Tunnel.

In the post–World War II era, the growing popularity of other materials, including linoleum and asbestos, eroded the competitive position of the ceramic tile industry in Europe and North America. Today, interiors decorated with imported faience tiles are enjoying a minor comeback among upper-class consumers and middle-class handymen, but the enthusiasm of contemporary revivalists pales in comparison to the zealousness of Dutch and English consumers in earlier eras.—R.L.B.

See also KILNS; PORCELAIN; POTTERY

Tire, Pneumatic

The pneumatic tire is an invention that emerged, died without a trace, and then reemerged once a market for it was established. The invention of vulcanized rubber made it possible to use rubber for a broad range of products. One of these was an air-filled rubber tire, patented in 1845 by Robert W. Thompson and intended for use with horse-drawn carriages. It was successful to the extent that one carriage went 1,600 km (1,000 mi) before the tires needed replacement. Yet Thompson's invention was largely forgotten until first the bicycle and then the automobile created a market for pneumatic tires. Unlike Thompson, John B. Dunlop (1840–1921), a Scottish veterinarian living in Northern Ireland, was in the right place at the right time when he reinvented the pneumatic tire in 1888. Dunlop's tires were given a strong boost when they were successfully used by bicycle racers in the year following their invention. Even more significantly, in 1895 one of the entrants in the Paris-Bordeaux-Paris automobile race was a car driven by the Michelin brothers, scions of a family whose firm was the leading producer of pneumatic bicycle tires. This car was the first to use pneumatic tires, and although it was the last car across the finishing line, it did complete the race, something that 13 of the starting field of 22 had failed to do. Even though the brothers had to patch their tires more than 20 times in the course of the race, the improved speed and comfort conferred by the pneumatic tire were becoming evident. By the beginning of the 20th century, pneumatic tires had become standard equipment on most automobiles, although many trucks continued to use solid rubber tires well into the 1930s.

Pneumatic tires were advantageous for cars and drivers, as well as the roads they traveled on. Compressed air acts as a spring, so pneumatic tires absorbed some of the shock generated by travel on rough roads. The roads themselves deteriorated less rapidly because the tires dis-tributed the vehicles' weight over a larger surface area. The manufacturing of the first pneumatic tires began with pressing rubber into cotton fabric, combining several layers (plies), and then using a mold to form them into a tire. Tires made in this manner lasted for only about 6,500 km (4,000 mi) because the fabric moved around as the tires rotated, causing eventual breakage and separation of the plies. Around 1915, the fabric began to be replaced by separate cotton cords; tires became stronger and longer wearing because these did not rub against one another. The durability of tires also was improved by the addition of carbon black to the rubber, a practice that began in 1912. First used to hide the scuff marks inevitably suffered by tires, carbon black had the unexpected quality of strengthening the rubber. Pneumatic tires were further improved in the mid-1920s when "balloon" tires came into use. Mounted on smaller-diameter wheels, balloon tires had thinner sidewalls and a wider cross section, and carried inflation pressures of about 2.1 kg per cm^2 (30 lb per in.^{2} or 30 psi), half the pressure of earlier tires.

Basic tire design remained largely unchanged in the decades that followed the introduction of the balloon tire. The casing of the tire consisted of a rim of bead wires, to which was attached several layers of plies, each consisting of a layer of diagonal cords that had been fused into two layers of rubber. Each ply had cords running in the direction opposite to the plies above and below it, producing a crisscross pattern overall. The fabric was originally made from cotton, but this fiber was replaced by synthetics such as rayon in the late 1930s, nylon in the late 1940s, and polyester in the early 1960s.

The introduction of new cord materials made for better performance and longer wear, but the basic technology of tire construction remained largely unchanged: Rubber was molded around a casing to form the tread and sidewalls. The cross-ply tires that were made in this way were cheap to manufacture and gave adequate service, but they suffered from rapid wear and left something to be desired as far as braking, cornering, and overall handling was concerned. Beginning in the late 1940s, a new type of tire construction began to be used, one that used radial plies in the place of conventional crossplies. The result was improved performance and longer wear.

Although radial tires have greatly improved tire life, vast numbers of tires reach the end of their service life every day, creating an enormous disposal problem. From 2 to 3 billion discarded tires sit in dumps all over the United States, and more than 250 million more are added every year. Many are buried in landfills, where they tend to float back to the surface in a few years. Their rubber cannot be reclaimed, because the process of manufacturing a tire permanently changes the rubber. However, some recycling is possible; worn-out tires are now being ground up and combined with asphalt or concrete as a surfacing for roads. A road surfaced with this material improves traction and is almost as durable as a road made from conventional materials. It is also possible to obtain low-grade oil, gas, and carbon from tires by subjecting them to pyrolysis—bringing them to a high temperature in the absence of oxygen. Although technically feasible, using pyrolysis to obtain raw materials from tires is at present too expensive to be economically viable. Given current technologies the most practical use of old tires is as a fuel source, a practice that grew rapidly in the 1990s. By the middle of the decade, more than half the tires discarded annually were being consumed in cement kilns, paper mills, electrical plants, and industrial boilers.

See also AUTOMOBILE; NYLON; RAYON; RUBBER, VULCANIZED; TIRE, RADIAL; TRUCK

Tire, Radial

For decades, car and truck tires were reinforced with diagonal fabric plies. In 1947, the French Michelin Company introduced a significantly different tire design. Instead of being stiffened by diagonal plies, this tire used radial plies; that is, the cords of the plies ran at right angles to the tire's rims. Between the plies and the tire's tread were steel reinforcing belts. Radial construction allowed tires to have very flexible sidewalls, which improved cornering and handling, and helped to dissipate heat. Better heat dissipation in turn gave radial tires a considerably longer service life than conventional bias-ply tires.

Many European tire companies introduced their own radial designs in the years following the introduction of the Michelin's radial. American manufacturers, however, were slow to adopt the new design. Tire manufacturing in the United States was done on a much larger scale than anywhere else in the world, and tire firms were reluctant to scrap their existing plants to make way for the new production methods that radial tires required. Then too, American drivers generally drove more sedately than Europeans and did not demand the more precise handling and shorter stopping distances made possible by radial tires.

By the 1970s, however, the superiority of radial tires had become evident to large numbers of drivers. American tire manufacturers tooled up to make radials, while at the same time introducing new materials and production techniques that reduced costs. It was not a painless changeover, however. Longer-wearing radials did not have to be replaced as often as bias-ply tires, and reduced tire sales inevitably led to a smaller tire industry with resultant job losses at all levels.

See also TIRE, PNEUMATIC

Toilet, Flush

In agricultural societies, human wastes were often carefully collected and used to improve the fertility of the soil. Excrement was a commodity of some value; in preindustrial Japan, jokes were told about guests who hurried home to relieve themselves lest their host's fields be the beneficiaries of their visit to the privy. As societies urbanized, industrialized, and grew in size, human wastes lost much of their value, becoming instead a massive problem for residents. Outdoor privies and open sewers were a visual and olfactory blight, and they played a large role in the transmission of epidemic diseases. The collection of wastes from privies and chamber pots was a disagreeable task that had to be carried out regularly.

More pleasant and sanitary ways of disposing and collecting wastes occasionally appeared in the ancient world; for example, more than 1,000 years before the beginning of the Christian era, the palace of Knossos was served by toilets flushed with running water. Citizens of Rome often benefited from ample supplies of running water, but by medieval times little thought was given to the disposal of human and other wastes. An indication of a renewed concern with sanitation was the publication in 1596 of a tract by John Haring-ton that described the construction of a primitive flush toilet. Employing an overhead cistern of water that was used to flush a basin located directly below, a toilet of this sort was installed in Queen Elizabeth's palace at Richmond. Although the queen was said to be delighted with it, Harington's design could not be widely used because ordinary dwellings rarely had an adequate supply of water.

Towards the end of the 18th century, those fortunate enough to have ready supplies of water could install the ancestor of the modern flush toilet. The valve apparatus that made it possible was patented in England in 1778 by Joseph Bramah (1748–1814), a man who was also responsible for many significant improvements in machine tools. Toilets using Bramah's valve enjoyed a modest initial success; 6,000 of them were sold

A 19th-century flush toilet (from H. Petroski, *The Origin of Useful Things*, 1992).

by 1797. No less important than the valve was the water seal, invented in 1782 by John Gaillait, which prevented sewer gases from entering the house through the bowl. Early toilets were constructed largely out of cast iron, but in the mid-19th century pottery bowls began to be used. The final major improvement to the toilet came around 1915, when the elevated water reservoir began to be replaced by one located behind the bowl. Subsequent improvements have made flush toilets quieter and more reliable, but there has been no substantial change in their basic technology.

Modern toilet systems still leave much to be desired. For one thing, sitting on a toilet induces or exacerbates hemorrhoids in many people; the Asian custom of squatting over a low basin is much better in this regard. Flush toilets are also major consumers of water; about 40 percent of the water brought into American homes is used to flush the toilets. Put another way, the body wastes of one person are flushed away by nearly 50,000 liters (13,208 gal) of water in the course of a single year. A considerable savings in water can be effected by installing toilets that use less water to dispose of liquids than they do for solids. Composting toilets completely eliminate the use of water for waste disposal, but installation costs are high and their long-term safety is still a matter of debate.

The convenience, cleanliness, and privacy afforded by flush toilets have been made possible not just by the toilets themselves but also by elaborate waste-disposal systems. Well into the 19th century, toilets were connected directly to sewer pipes that simply carried the wastes to a nearby river. This simply transferred the wastes and the pathogens they carried from one place to another. The effective disposal of the wastes deposited in toilets required the long-term development of many sewage-treatment technologies. For places far from sewer connections, the septic tank, a late 19th-century invention, was a vast improvement over the traditional cesspool. All in all, the familiar flush toilet and its associated disposal systems have made massive contributions to our health and comfort. They have also allowed us to exhibit a squeamishness about bodily functions that earlier generations would doubtless have found amusing.

See also COMPOSTING; WATER SUPPLY AND TREATMENT

Tomato Harvester, Mechanical

The mechanical tomato harvester, transforming hand power to mechanical power, represented a major change in American agriculture. To put it simply, a machine was invented to harvest a crop that had been developed for harvesting by machine.

The traditional tomatoes most in demand by consumers were round and red in color, with a rather thin skin. However, these standards were less important in tomatoes for processing into products like ketchup and tomato paste. Tomatoes usually had a number of fruits on the same plant that ripened at an uneven rate. An effective harvester needed a plant on which the tomatoes were preferably round, would ripen at the same time, and had a skin capable of withstanding movement within a mechanical system. The machine would go down the row only once, cutting the plants close to the ground in order to save as many tomatoes as possible while picking up a minimum of dirt.

Although a number of earlier attempts were made, success in developing a truly useful mechanical tomato harvester came only when scientists at the University of California at Davis joined in what might be called a package of technology or a "systems" approach to the problem. A plant scientist, G. C. Hanna, and an agricultural engineer, Coby Lorenzen, developed the concept. In 1943, Hanna began a search of breeding stocks to find a firm-fruited type that would withstand machine handling, a key to the success of a harvester. In addition, nearly all the fruit had to mature at the same time, be resistant to bruising, and, it was hoped, yield at least 30 tons to the acre. Since it takes 10 generations to fix a characteristic in a tomato, Hanna grew crops in Mexico and Puerto Rico, and worked with growers in California. Meanwhile, Lorenzen and others of the agricultural engineering staff experimented with various devices for doing each part of the harvesting. They worked on devices to cut the plants, to lift the plants onto the machine, to separate the tomatoes from the vines, to sort and clean the tomatoes, and to convey the fruit to a suitable container. The machine that resulted was field tested in 1959 and patented by the University of California, which licensed the Blackwelder Manufacturing Company to undertake its commercial manufacture. The company built 15 machines in 1960. In 1996, two Cali-

fornia companies were producing mechanical tomato harvesters.

Other universities and commercial firms also experimented with mechanical harvesters, with Michigan State University taking the lead. The Michigan team began work in 1958, and by 1960 a model was being tested in Michigan and Florida. Some half-dozen firms had at least experimental models in the fields by 1962. By then, the machines, all operating on somewhat the same principles, were on the way to being a technical and commercial success.

Successful mechanization depended on the development of tomatoes that would meet the standards previously mentioned. By the late 1960s, Hanna had bred varieties that met these goals. Within a few years virtually all tomatoes picked for processing, along with a few picked green for fresh consumption, were being harvested by machine. Many growers were forced to adopt machines because they cut the cost of harvesting; as a result, prices declined, and growers had to increase their production in order to maintain income levels.

Many crops are harvested by machine. The tomato experience was different in that a package of practices was used to develop tomatoes that could be picked by a machine that was built to harvest tomatoes developed for this purpose. Since then, virtually all agricultural production problems have been attacked using a package of practices. As a result, American agricultural productivity has increased substantially and likely will continue to do so in the future.—W.R.

See also AGRICULTURAL REVOLUTIONS IN INDUSTRIAL AMERICA; COTTON HARVESTER; REAPER

Topology

Topology is a branch of mathematics that studies those properties of space or geometric objects that are unchanged by continuous transformations. Whereas the ability to measure distance is central to geometry, this feature is completely absent from topology. For if we allow ourselves to continuously deform an object, twisting and stretching it as we please (but not tearing it), then clearly its size and shape become irrelevant. Nevertheless, properties besides geometric ones, such as dimension or connectedness, remain viable even in the presence of continuous transformations.

Consider, for example, a tetrahedron, a pyramid with three triangular sides and a triangular base. The tetrahedron can be continuously deformed into a sphere. To see this, imagine pumping air "into" the tetrahedron. Its faces will bulge outward, expanding into triangular regions of a sphere. Thus to a topologist, the tetrahedron and the sphere are indistinguishable, as are a cube and a sphere or, for that matter, any simple polyhedron and the sphere.

What properties, if any, are preserved as one polyhedron is continuously deformed into another? One of the earliest results in topology asserts that the *Euler number* of the polyhedron—which is given by $v - e + f$, where v, e, and f are the number of vertices, edges, and faces respectively, of the polyhedron—in fact is invariant. For a tetrahedron we have 4 vertices, 6 edges, and 4 faces, while a cube is different, having 8 vertices, 12 edges, and 6 faces. But the Euler number is the same in both cases: $4 - 6 + 4 = 2 = 8 - 12 + 6$. Although discovered in 1752 by Leonhard Euler (1701–1783), this invariant was actually known to René Descartes (1596–1650) in 1640.

If we consider a torus, or the surface of a bagel, then it seems obvious that this surface cannot be continuously deformed into a sphere, and thus is topologically different. We can prove this using the Euler number. Since all polyhedra that are equivalent to a torus must have the same Euler number, by examining just one such polyhedron, the reader can check that the Euler number is 0, not 2 as it is for the sphere.

Other properties of spheres and tori are also preserved as they are continuously deformed into other shapes. For example, their dimension remains equal to 2. No matter how these surfaces are stretched or distorted, a creature "living" in them would still enjoy "two degrees of freedom" to travel about the surface. Note, however, that by comparing the dimension of the sphere and the torus we cannot tell them apart. Something stronger, like the Euler number, is needed to distinguish them.

Another important topological property of a space is "orientability." In his 1847 book *Vorstudien zur Topologie* (where the word *topology* was first used in print), J. B. Listing described a surface that cannot be coherently oriented. Now known as the *Möbius band*,

after the 19th-century astronomer August Ferdinand Möbius (1790–1868), this surface can be modeled by a long strip of paper whose ends have been joined together with a half-twist. The surface has only one edge, a (topological) circle that travels twice around the band, rather than two edges which would be present if the half-twist were absent. Moreover, the surface cannot be oriented. Suppose a small circle is drawn somewhere on the surface and oriented in one of the two possible directions by labeling it with an arrowhead. By continuously moving the circle once around the band, it can be brought back to its original position, but with its orientation reversed! This phenomenon cannot occur on a surface like a sphere or a torus, which are therefore called *orientable surfaces*. (It is interesting to consider a nonorientable three-dimensional space: a right-handed glove could be moved around the space until it returned as a left-handed glove!)

A large class of objects (which includes the sphere and the torus) are *manifolds*. These are spaces that are *locally* indistinguishable from ordinary Euclidean space. That is, if a small piece is cut from a manifold, then it can be continuously deformed into a small piece of a Euclidean space of some dimension. The Euclidean space of dimension n is the "ordinary" space where each point is identified with n coordinates of real numbers. Thus, 1-dimensional Euclidean space is the line, 2-dimensional Euclidean space is the plane, and so on. An early, and important, result of topology asserts that the dimension of a manifold is well defined. In other words, a small piece of a manifold cannot be equivalent to both n-dimensional and m-dimensional Euclidean space, where $n \neq m$.

One of the goals of topology is to classify all n-dimensional manifolds, or n-manifolds. This has been done for $n = 0$, 1, and 2. The only 0-manifolds are points, and the only 1-manifolds are lines and circles. There are infinitely many 2-manifolds or surfaces, but all the possibilities are known. In fact, the sphere, torus, and Möbius band are the basic building blocks. By cutting pieces from (an unlimited supply of) these three surfaces, and then gluing the pieces together along their edges, we can build every possible surface. Moreover, any two surfaces are equivalent if and only if they are either both orientable or not, and they have the same Euler number.

At the turn of the century, Henri Poincaré (1854–

1912), building on the ideas of Georg Friedrich Bernhard Riemann (1826–1866), figured prominently in the development of algebraic topology, wherein algebraic objects such as groups are associated with topological spaces. This led to deep connections between the theory of functions, the topology of surfaces, and non-Euclidean geometry, namely, the theory of the group of motions of the hyperbolic plane. Algebraic topology grew rapidly in the first half of the 20th century and led to important advances in our understanding of topological spaces, especially manifolds.

Although manifolds are not equipped *a priori* with a geometry, it may still be possible to impose a geometry on a manifold. In the 1970s and 1980s, W. Thurston made dramatic progress in the understanding of 3-manifolds by utilizing this point of view, especially using non-Euclidean geometry. We now seem tantalizingly close to a classification of 3-manifolds: Thurston conjectures that every 3-manifold can be cut into a finite number of pieces, with each piece capable of supporting one of eight possible kinds of geometries.

While understanding 3- and 4-manifolds has obvious applications to cosmology, topology also has important applications to other fields. Know theory, a branch of topology, can be used in chemistry to help analyze knotted synthetic molecules or to study the actions of various enzymes in DNA. Graph theory, another area of topology, has applications in computer science, coding theory, electronics, and engineering. Topology is also extremely useful in the study of fractals and other interesting sets that arise in the study of dynamical systems.—J.H.

See also DNA; ENZYMES; FRACTAL; GEOMETRY, EUCLIDEAN AND NON-EUCLIDEAN; KNOT THEORY

Trace Italienne

The *trace italienne* style of fortification was the principal military-engineering innovation of the Italian Renaissance. It was designed to resist the powerful artillery that emerged in the mid-15th century and obliterated the thin stone walls of previously impenetrable fortresses. Although Renaissance artist-engineers experimented with new fortification techniques, the Italian city states were helpless when the French invaded with their new artillery in 1494. The devastating Hapsburg-Valois Wars (1494–1559) that followed stimulated the innovation of the trace italienne. During the 1530s, it emerged as the dominant style in Western military architecture because it solved the paradox of simultaneously resisting artillery fire with thick and low-lying walls, and massed infantry assaults with elevated vertical walls.

Italian engineers responded to Charles VIII (1470–1498) of France's invasion by lowering medieval walls and augmenting them with inner earthen embankments. This created new problems for the active defense, however. The troops above could not place the wall below under gunfire when stationed behind the thick parapet that protected them from artillery fire. Medieval fortifications suggested a solution: Extend the wall at regular intervals so that the soldiers could safely defend them with a flanking fire. Military engineers subsequently built arrowhead-shaped bastions so that guns placed on their flanks covered the walls. When combined with the fire from the adjacent bastion, a crossfire resulted. By laying out the fortress walls as a polygon and placing bastions at each corner, engineers brought all approaches to the fortress walls under first a uniform crossfire and then a flanking fire. This quickly silenced exposed artillery batteries and made frontal infantry assaults virtually suicidal.

The next step was to augment the fortification's active defense with a passive one. As the Pisans accidentally discovered in 1500, an earthen ramp or *glacis* absorbed virtually unlimited artillery shot, unlike a thick masonry wall. A glacis extending from a fortresses wall hardly inconvenienced an infantry assault, however. To resolve this dilemma, engineers eventually separated the glacis and the thick fortress walls with a deep trench or ditch. During the Dutch revolt of late 16th century, the engineers of the northern provinces began employing flooded ditches. These innovations significantly increased the trace italienne's power.

Although the trace italienne failed to protect the Italian city states from Spanish domination, it did constrain subsequent military campaigns. Large infantry armies could rarely afford to leave them unreduced because of the threat they posed to their lines of communications. Moreover, battles became increasingly irrelevant near trace italienne fortresses because they enabled even a defeated army to maintain effective resistance. Consequently, when large battles did occur they often involved the relief of a fortress. Siege warfare became paramount during the Dutch revolt against Spain and the wars of Louis XIV (1638–1715), where infantry combat involved little else.

The principal impediment to the trace italienne's application was economic. Its construction involved enormous building and maintenance expenses compared to medieval fortifications protecting the same area. By the mid-16th century, it had spread to the Netherlands, the Huguenot strongholds of southwestern France, and to many ports of the Spanish Empire—locations that combined intense military conflict with substantial economic resources. Elsewhere, medieval walls with earthen supports continued to be useful, as demonstrated by many German towns during the Thirty Years' War (1618–1648). By the late-17th century, however, most strategic locations in Europe employed the trace italienne.

The trace italienne offered significant economic advantages in the era of the "Military Revolution." Unlike a field army, the payments for fortress construction could be spread out. Princes, both absolute and constitutional, usually squeezed much of these costs from their subjects. During a siege, a fortress's military gov-

ernor could also rely on additional civilian support, not to mention the dedication of otherwise undisciplined militias. In an era when wars were fought until one side went bankrupt, a properly defended trace italienne ensured that the reduction costs outweighed the construction and garrison costs. More critically, a lengthy siege bought the defending state more time to mobilize both militarily and diplomatically. Thus during the War of the Spanish Succession (1702–1714), Louis XIV's fortresses stopped the Allied invasion attempt, just as the Hapsburg fortresses blocked the numerous Turkish offenses.

The trace italienne was a central feature of the Military Revolution, yet it also had significance for the "Scientific Revolution." By the late 16th century, fortification engineers were routinely applying Euclidean geometry and trigonometry; by the early 17th century, fortification design became a branch of mechanics. Numerous leaders of the Scientific Revolution—Descartes, von Guericke, and Boyle—thus received their initial exposure to mathematical science by studying fortification theory in their youths. Others augmented their incomes by teaching or applying such theory, including Tartaglia, Stevin, and Galileo. The inspiration that fortification design gave to the creation of scientific knowledge peaked during the 18th and early 19th centuries. This

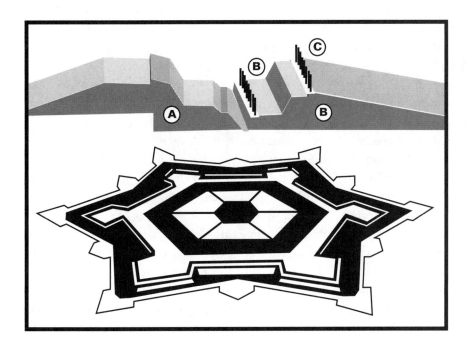

Upper view: A cross section of a trace italienne with the shot-absorbing glacis (B) and the rampart (A) where the defending artillery is mounted.

Lower view: A symmetrical fortress layout with quadrilateral bastions (from Allesandro Capra, *La nuova architettura civile e militare*, 1717).

included the structural and soil mechanics theories of Belidor and Coulomb, Monge's descriptive and differential geometry, Meusnier's defilading analysis, and Cauchy's theory of elasticity.—B.S.

See also GEOMETRY, EUCLIDEAN AND NON-EUCLIDEAN; MILITARY REVOLUTION IN EARLY MODERN EUROPE

Tracers, Radioactive

Radioactive tracers are minuscule amounts of radioactive material that are put into the body orally or through injection. Radioactive tracers are chemically similar to elements normally found in the body, so the tracers will be involved in the body's usual biochemical reactions. Because they are radioactive, these tracers can be monitored in order to determine what happens to them as they are processed by the body.

A commonly used radioactive tracer is iodine-131, an isotope of iodine. The thyroid gland accumulates iodine, but if it is overactive (hyperthyroidism) it will take up more iodine than a normally functioning gland. Conversely, when the thyroid is underactive (hypothyroidism), it will not accumulate as much iodine. The thyroid processes iodine-131 exactly as it does ordinary iodine. But unlike ordinary iodine, iodine-131 emits gamma rays, which allows its detection by a scanner or scintillation camera. These devices convert the gamma rays into images displayed on a video screen. The patient's amount of radiation exposure is minimal, equal to the cosmic-ray radiation received while flying in a commercial jet airplane.

Some of the tracers now in use emit positrons rather than gamma rays. A positron is the antimatter equivalent of an electron. A positron and an electron annihilate each other when they come into close proximity, producing a pair of gamma rays in the process. The detection of these gamma-ray pairs allows the location of electron-emitting atoms with great precision. Positron-emitting atoms are produced by cyclotrons and other particle accelerators. Positron emitters in use today include carbon-11, nitrogen-13, and oxygen-15, all of them isotopes of atoms commonly found in the body. These isotopes are the basis of detailed studies of physiological processes like the flow of blood in the brain and oxygen metabolism in the body.

The use of positron emission in conjunction with computerized imaging is known as *positron emission tomography*, or a PET scan for short.

Although radioactive tracers have been primarily used for medical diagnosis and physiological research, they have not been limited to these applications. Radioactive tracers have been used to find leaks in underground pipes, to measure minute wear patterns in machinery, and to determine the flow of industrial wastes. Radioactive tracers even have been used to study the migration of cockroaches in a sewer system.

See also ANTIMATTER; COSMIC RAYS; CYCLOTRON; ELECTRON; GAMMA RAYS; ISOTOPE; RADIOACTIVITY AND RADIATION; PARTICLE ACCELERATORS

Tracked Vehicles

A tracked vehicle rides on two sets of tracks, metal links that have been pinned together. The tracks each engage sprockets at the forward end that are powered by the vehicle's engine, while the weight of the vehicle is borne by a number of wheels or rollers located inside the tracks. This arrangement spreads the vehicle's weight over a large surface. In effect, the vehicle is continually laying its own road where it travels. The vehicle is turned by changing the relative speeds of the two tracks; for example, running the tracks on the right side faster than the ones on the left side will cause the vehicle to turn left. Tracked vehicles are commonly employed when the ground is too soft or rough for the effective operation of wheeled vehicles. Tracks are used for road-building equipment, some agricultural tractors, and military vehicles.

The idea of using tracks instead of wheels goes back centuries. In 1770, Richard Edgeworth took out a British patent for a vehicle that ran on slats of wood connected to an endless chain: "a portable railway, or artificial road, to move along with any carriage to which it is applied." Other designs followed over the decades, such as Boydell's steam tractor of 1854 that featured flat plates linked to the rims of its wheels. The development of the railroad may have undercut the market for tracked vehicles for a while, but the growing popularity of steam tractors in the late 19th century helped to stimulate the demand for vehicles that did not require a relatively smooth and firm surface.

Credit for the invention of the first practical tracked vehicle is usually given to Benjamin Holt (1849–1920), a New Hampshire native who moved to California in 1883. There he and his brother managed a wagon-wheel business that expanded into a farm implement firm. In 1891, Holt invented a new kind of combine that allowed the harvesting of grain that grew on hillsides. At this time, steam-powered tractors were beginning to be used in increasing numbers to pull farm implements. Holt himself had built a 20,400-kg (45,000-lb) steam tractor in 1890. Like most of the tractors of the time, it worked reasonably well on firm ground, but its great weight caused it to sink into soft earth. These were the conditions found in California's Sacramento Delta, an area of potentially rich farmland that had caused steam tractors to sink up to their axles. Holt and his mechanics set about producing a tractor that would allow the cultivation of this region.

Holt's first tracked vehicle used wooden slats measuring 7.6 cm (3 in.) by 10 cm (4 in.) that had been pinned together to form a set of tracks 2.7 m (9 ft) long and 1.1 m (3.5 ft) wide. It was successfully field tested in late 1904 and then returned to Holt's factory. While seeing it in motion, a company photographer remarked that "It crawls like a caterpillar." The name stuck, and in 1910 the Caterpillar Tractor was registered by the company.

World War I created a whole new market for tracked vehicles as armies bogged down in the mud of northern Europe. At first used for hauling large guns, the tracked vehicle soon became the basis of a whole family of weapons, most notably armored tanks. By this time, gasoline engines had replaced steam. Today's tracked vehicles are generally powered by diesel engines.

See also COMBINE HARVESTER; ENGINE, DIESEL; ENGINE, FOUR-STROKE; FARM TRACTOR; TANKS

Traffic Signals

Signals to control the movement of trains date to the beginning of the railroad era. In 1868, a railroad signaling engineer named J. P. Knight installed a signal for a different purpose, to control the movement of pedestrians outside the British Houses of Parliament at the intersection of George Street and Bridge Street. It consisted of semaphore arms along with red and green gas lamps. Unfortunately, a few months after the signal was installed, a gas explosion in the signal killed a nearby policeman.

As with many inventions, it is difficult to determine the first use of a signal for controlling vehicular traffic. According to one account, red-and-green lights were first installed in 1912 by Lester Wire, a Salt Lake City, Utah, policeman. The documentary evidence for this is thin, however, and Cleveland, Ohio, is generally credited with being the site of the first traffic light, which was installed in 1914. In 1920, another policeman, William Potts of Detroit, may have been the first to install the now-standard three-light signal.

The three-light system is taken for granted by most drivers, but it does pose a problem to those who are color blind. Traffic lights are now standardized to have the red light on top and the green at the bottom so that color-blind drivers can make distinctions on the basis of position. To further help color-blind drivers, the red in the light includes some orange, and the green some blue.

Early traffic signals were manually controlled. In the mid-1920s, the first automatic signals made their appearance, followed by vehicle-actuated signals a few years later. The majority of today's vehicle-actuated signals are based on inductive loop detectors; when a vehicle passes over the loop buried below the surface of the road, the difference in inductance causes the signal to change. However, as bicyclists and motorcyclists will attest, their vehicles often fail to affect the loop, leaving them stuck at the light. Many of today's traffic signals are computer controlled to optimize traffic flow. In the years to come, an increasing portion of a vehicle's operation will likely be governed by traffic control technologies that go well beyond traffic signals.

Controlling traffic through the use of signals is an expensive proposition; the installation of signals at just one intersection costs between $70,000 and $140,000. And a modern city requires many of them. Los Angeles has more than 4,000 intersections governed by traffic signals. The indispensability of these traffic control systems are amply demonstrated when a natural disaster or power failure renders them inoperative.

The price of reasonably orderly traffic is having

one's behavior controlled by a mechanical object. This is an ironic consequence of the automobile age. The automobile, which originally promised freedom and unrestricted mobility, soon created so much chaos on the roads that it had to be tamed by traffic signals and other instruments of control. Acceptance of the traffic light's authority entails a loss of personal time as well as personal freedom. According to one estimate, over the course of his or her lifetime, the average driver cumulatively spends several months waiting for the light to change.

See also INTELLIGENT TRANSPORTATION SYSTEMS; LIGHTING, GAS

Transformer

A transformer changes the voltage and amperage of alternating electrical current. A step-up transformer raises the voltage and lowers the amperage; a step-down transformer does the opposite. A step-down transformer contains a large input coil (the primary) and a small output coil (the secondary). Conversely, a step-up transformer has a small input coil and a large output coil. The ratio of the primary voltage to the secondary voltage is known as the *voltage ratio*.

Step-down transformers are used when it is desirable to reduce the current to safe levels or when the components of an electrical device require low voltage for their operation. On the other hand, television sets require step-up transformers to deliver 20,000 or more volts to the picture tube. Transformers are also essential for the distribution of electrical power, for they make possible its efficient transmission over long distances. In a modern electrical transmission system, after being generated at around 4,000 volts, electrical current is increased by a step-up transformer to 400,000 volts or more. It is then transmitted to a power substation, where the voltage is reduced to about 440 volts by a step-down transformer. It is reduced again to 240 and 120 volts by another transformer mounted on a utility pole close to the home or business using the electricity.

The basic principle underlying the operation of a transformer was discovered by Michael Faraday (1791–1867) in 1831. He found that the interruption of a current flowing through wire wrapped around an iron rod would produce current in an adjacent wire that was also wrapped around an iron rod. For Faraday, the transfer of current from one wire to another (a process known as *induction*) was the significant thing. Only later was it realized that the number of wire windings affected the voltage of the current being transferred.

This principle was first used in 1848 in induction coils employed by electrical experimenters. The first commercial application came in the late 1870s with the powering of arc lamps by an induction coil. This apparatus delivered current to the lamps at higher voltages than the supply current; that is, it served as a step-up transformer. The step-down transformer was first used in 1883 as a key element of an electric lighting system devised by Lucien Gaulard, a Frenchman working in England in partnership with John D. Gibbs. Their patents were overturned in legal proceedings, however, and Gaulard died a broken man in a Parisian hospital for the insane in 1888. Other inventors continued to develop the transformer. Important work was done in Hungary by Otto Titus Blathy (1860–1939), Charles Zipernowski (1853–1942), and Max Deri (1854–1938), and in England by Sebastian Ziani de Ferranti (1864–1930). In the United States, William Stanley (1858–1916) did pioneering work in the application of alternating current. In 1886, working in a loose partnership with the Westinghouse Electric Company, Stanley set up an alternating-current system in Great Barrington, Mass., that successfully served a number of local businesses. Before long, Westinghouse Electric was producing components based on Stanley's system for a large number of installations in the United States, including a transformer design patented by Stanley. By 1890, there were 300 hundred Westinghouse central stations in operation, and Westinghouse's alternating-current system was engaged in a "battle of the currents" with Thomas Edison's direct-current system, a battle that was eventually won because the transformer had made possible the economical and efficient transmission of electrical current.

See also ALTERNATING CURRENT; DIRECT CURRENT; ILLUMINATION, ELECTRICAL; INVERTER

Transgenic Organisms

Transgenic plants and animals are those that have been manipulated so that they carry new genes from other organisms. Just as cloned DNA sequences have "transformed" bacteria, so can recombinant DNA technology change the genetic makeup of plants and animals. In contrast to traditional breeding techniques that are the mainstay of agriculture, transgenic technology allow the exchange of genetic material between species that cannot interbreed.

The procedure is carried out by first constructing a "transgene," a construction that includes the gene to be transferred and some regulatory DNA that will allow the gene to be expressed in a new host animal or plant. The actual transfer process can be carried out in a variety of ways. Transfer to plant cells has been accomplished using biological vectors such as plant viruses, or by physical methods such as a gene gun. In animal cells, the transgene typically is injected into the nucleus.

Crops for which transgenic methods have been developed include corn, rice, cotton, oilseed rape, tomatoes, potatoes, cabbage, and lettuce. Transgenic plants have been developed in order to improve the quality of produce available to the consumer. The Flavr-Savr tomato, for example, is a transgenic tomato that can remain on the vine longer to ripen, yet does not soften during shipment. Transgenic plants that are naturally resistant to crop pests also are in the development stage, and these plants should help reduce the use of chemical pesticides.

Transgenic animals are very valuable in biomedical research. In an early experiment that demonstrated the feasibility of the technique, an unusually large transgenic mouse was produced after it incorporated a rat growth hormone gene. Another transgenic mouse has made a normally resistant strain of mice susceptible to the HIV virus, providing an experimental animal for efforts to develop treatments for AIDS. In agriculture, there are many possible applications of the technology. Other efforts are developing transgenic pigs with leaner meat and transgenic sheep that will secrete pharmaceuticals in their milk.

In February 1997, a transgenic animal experiment stunned the world when the Scottish scientist Ian Wilmut reported the successful cloning of an adult sheep. Contrary to the prevailing scientific dogma that differentiated cells have gone through an irreversible developmental pathway, Wilmut reported that when he transplanted the nucleus from an adult sheep into an enucleated sheep's egg, that transgenic egg went through complete embryonic development in a surrogate ewe. The cloned lamb was genetically identical to the sheep that had donated the cell nucleus. This development opened the possibility of a means to sustain prize breeds of stock animals, but also raised important questions about responsible use of the technology, especially if it were to be attempted in human beings. Creating identical organisms runs contrary to Darwin's studies of animal evolution in the mid-19th century, work that clearly demonstrated the survival value of genetic diversity.—C.I.

See also ACQUIRED IMMUNE DEFICIENCY SYNDROME (AIDS); CLONING; INSECTICIDES; NATURAL SELECTION; SELECTIVE BREEDING OF ANIMALS

Transistor

Like the vacuum tubes that they have largely replaced, transistors use a small flow of electrical current to control a larger current. This property allows them to be used for the detection, switching, amplification, or rectification of electrical currents. Whereas a vacuum tube works through the migration of electrons across an evacuated space, in a transistor the electrons move through a solid. For this reason, a transistor is known as a "solid-state" device.

During the first half of the 20th century, vacuum tubes had become essential components for a wide variety of electrical devices: radios, televisions, medical instruments, computers, and many others. But vacuum tubes had their drawbacks. They were bulky, fragile, and had short operating lives. They consumed a lot of electrical power, and they got hot. The vacuum tube's deficiencies meant that a new device free of these problems would find a ready market. In this sense, the transistor's development was driven by the requirements of the market. But this is not the whole story. Many things that would be eagerly snapped up do not exist because there is no way of making them. The transistor emerged only after advances in science and engineering made them possible. Finally, the transistor did not just simply serve as a substitute for vacuum tubes; it made possible the creation of new products, new processes, and new industries.

The invention of the transistor was the result of the advances in physics that occurred on several fronts. One crucial contribution was the development of quantum mechanics, which provided crucial insights into the behavior of electrons. One of the most important of these was Albert Einstein's use of quantum physics to explain the photoelectric effect: the emission of electrons from certain materials when they were struck by light. This in turn stimulated research into the behavior of certain materials. In particular, physicists turned their attention to semiconductors, substances that allowed an internal flow of electrons only when heat or light were applied to them, or when they were treated with minute amounts of certain impurities.

Scientific knowledge was essential to the development of the transistor, but it had to be used in an appropriate setting. Far from being the product of inspired tinkering by a solitary inventor working in a basement, the transistor was the result of the collective efforts of many people who received adequate financial and organizational support. The development of radar and the atomic bomb during World War II had demonstrated what could be done when large teams of specialists pooled their talents. Research into radar also had a direct applicability to electronics; radar technology was built around crystal detectors, which gave rise to research into germanium and silicon semiconductors. Scientists at research universities like the Massachusetts Institute of Technology and Purdue University paved the way for the transistor through their work on the purification of semiconductors, and learned more about their unique properties. Indeed, a research team at Purdue came very close to inventing the transistor, although they did not know it at the time.

While university researchers were free to pursue inquiries into "pure" science, most industrial research laboratories concerned themselves with the solution of immediate technical problems. A great exception was the Bell Labs, the research arm of American Telephone and Telegraph (AT&T). Few universities could match the research capabilities of the Bell Labs, which in the late 1940s had 5,700 employees, 2,000 of them scien-

tists, engineers, and technicians. Many of its research staff were free to pursue basic research that had no necessary commercial application. At the same time, however, the management of AT&T was keenly aware that the expansion of the telephone system would eventually require something better than the mechanical relays that were used to switch telephone calls. Consequently, a strong research effort was directed at alternative switching devices.

This effort culminated with the invention of the point-contact transistor, which was demonstrated on Dec. 23, 1947, and announced to the world on June 30, 1948. Although it was destined to be one of the most important inventions of the 20th century, the transistor aroused little interest at first. Mention of the transistor appeared in *The New York Times* on page 46, where a brief description occupied 4.5 column inches (11.4 cm) at the bottom of the radio section. The inventors of the transistor are customarily identified as John Bardeen (1908–1991), Walter Brattain (1902–1989), and William Shockley (1910–1989), all of whom received the 1956 Nobel Prize in physics for their contribution. In actual fact, Bardeen and Brattain were the co-inventors of the first transistor, while Shockley can be credited with the 1951 invention of the junction transistor, a much more practical and reliable device than the point-contact transistor.

Transistors began to be used in the Bell System for oscillators in late 1952 and for the automatic routing of telephone calls in 1953. At about the same time, they were being installed in hearing aids, their first consumer application. But the use of transistors was not easy or automatic. Early transistors suffered from a number of defects: They could not handle large currents and high frequencies, and they generated unwanted noise. Manufacturing was also a problem; costs were high, and the quality was variable. Many of the device's shortcomings were solved by the junction transistor, which entered production in 1952, and the silicon-based transistor (early transistors used germanium), which was introduced by Texas Instruments in 1954. Still, manufacturing tended to be hit-or-miss, and firms sometimes determined the use and value of an individual transistor only after testing it.

Throughout the 1950s and much of the 1960s, the major improvements to the transistor centered not on the device itself but on the way it was made. A major production improvement was effected by the Fairchild Semiconductor Company when they developed the planar process in 1958 and used it for production a year later. The planar process entailed oxidizing a silicon wafer to form a mask over its surface. "Windows" were then opened up on the oxide layer through a photo-resist process. Impurities were then diffused onto the exposed silicon. More generally, transistor manufacturers moved up a learning curve in the late 1950s as they solved a host of major and minor manufacturing problems. As a result of this learning process, production skyrocketed and prices plummeted. In 1957, the average price of a silicon transistor was nearly $18; by 1965 it had fallen to less than $1.

A great deal of learning also had to go into the development of applications for transistors; they could not simply be substituted for vacuum tubes in existing circuits. Many electronic devices benefited from being redesigned to use transistors instead of vacuum tubes, but the transistor's full potential became evident only when it began to be used for applications for which vacuum tubes were ill-suited. One such case was the electronic computer. First-generation computers used thousands of vacuum tubes that consumed vast amounts of electrical power and burned out with depressing regularity. In 1955 IBM introduced a second-generation computer that used 2,200 transistors instead of 1,250 vacuum tubes, with consequent improvements in reliability and computing speed.

The military had not been involved with the invention of the transistor, but it was a major customer as soon as the device went into quantity production. By the late 1950s, the U.S. military accounted for half the value of semiconductor sales. In addition to providing transistor manufacturers with a major portion of their income, the military provided an important stimulus for the development of transistors and transistorized equipment. Cold War tensions and consequent large military budgets meant that the services were generally unconcerned about costs; this allowed the development of sophisticated transistor-based devices too expensive for the commercial market. Many of these devices and the production techniques used to make them eventually made their way into the civilian realm. In addition to guaranteeing a market, the armed forces pushed the development of transistors by providing hundreds of millions of dollars for solid-state R&D.

By the 1960s, transistors had become essential com-

John Bardeen, William Shockley (seated), and Walter Brattain, the inventors of the transistor (courtesy Lucent Technologies).

The first transistor. A point-contact device, it was almost immediately superseded by the field-effect transistor (courtesy Lucent Technologies).

ponents of a wide range of products. The federal government continued to be a major player; the U.S. Army, Navy, and Air Force relied on transistors for a new generation of weaponry, and the fledgling space program would literally not have gotten off the ground without transistorized circuits. In business, transistors made possible many vital tools of commerce such as electric typewriters and photocopying machines. For many consumers, "transistor" was virtually synonymous with "radio," as a flock of miniaturized radios followed the first example, the Regency, which had been introduced in October 1954. As a result, radio, which seemed to have been superseded by television, gained a new lease on life. Among its other effects, the transistor radio helped to change musical tastes by creating a large audience for rock-and-roll, a musical genre that, like "rhythm and blues," had been previously restricted to black listeners.

The transistor was only the beginning of the semiconductor revolution. A little more than 20 years after the invention of the transistor, the integrated circuit, a no-less revolutionary invention, made its appearance.

Like the transistor, it represented and improvement of what had come before, but also like the transistor, it created technological possibilities that had scarcely been conceived at the time.

See also ATOMIC BOMB; COMPUTER, MAINFRAME; INTEGRATED CIRCUIT; QUANTUM THEORY AND QUANTUM MECHANICS; RADAR; RADIO, TRANSISTOR; RESEARCH AND DEVELOPMENT; SEMICONDUCTOR; SPINOFFS; THERMOIONIC (VACUUM) TUBE; TRIODE

Further Reading: Ernest Braun and Stuart MacDonald, *Revolution in Miniature: The History and Impact of Semiconductor Electronics*, 1978.

Transportation, Intermodal

Intermodal transportation allows the shipment of freight by rail, truck, and ship without the need to unload and reload every time the cargo is transferred to another type of carrier. The basis of intermodal transportation is simplicity itself: a large container. Containers are big boxes with doors at one end. They have a sturdy steel internal frame and an outer skin made from aluminum, steel, or fiberglass. Each container has castings at all eight corners that allow the container to be hoisted by special cranes or side loaders. In order to facilitate the worldwide use of containers, the dimensions and shape of these castings conform to standards that have been set by the International Standards Association. A container can be put on a truck trailer, a railroad flatcar, and the deck or hold of a ship. By the mid-1990s, 4 million box and tank containers were in use worldwide.

One commonly used container has dimensions of 20 ft by 8 ft by 8 ft (6 m by 2.5 m by 2.5 m). This container provides a standard measurement of a container ship's capacity, the TFEU—an abbreviation of "twenty-foot equivalent unit." Another standard container has the same width and height, but is 40 ft (12 m) in length. Containers of this size weigh 2,720 kg (6,000 lb) and can hold cargo weighing up to 27,200 kg (60,000 lb) Much larger containers are used for transportation within North America; the longest of these stretch 53 ft (16.1 m). However, containers used for truck travel in the United States are limited to loads of 21,800 kg (48,000 lb).

A key advantage of containerization is that it obviates the need to "break bulk" when the mode of transportation has to be changed; this produces considerable savings in time and labor. In addition to greatly facilitating the transfer of freight, containers provide a watertight and damage-resistant repository for freight. Containers also reduce the pilferage that often occurs when goods are out in the open.

The basic idea of enclosing goods in a container is of course very old; boxes and barrels may be reckoned to be the first freight containers. In the 1920s, a few railroads made use of special containers and cars for shipments that were too small to completely occupy a boxcar. Shipments of this sort were known as LCL for "less than carload lot." Although successful, the Interstate Commerce Commission ruled in 1931 that these operations violated established rate regulations. The need to make complicated rate calculations coupled with high labor costs soured the railroads on LCL and stunted the development of containerization. A few electric interurban railroads used containers that were transferred onto trucks, but containerization did not become a significant part of the railroad scene until the 1960s.

Containerized freight was revived in 1956 through the efforts of Malcolm McLean, the owner of Sea Land Service, a steamship line that plied the Eastern and Gulf coasts of the United States. Two years later, the Matson Navigation Co. began to use containers to ship freight between the West Coast and Hawaii. Motivated by a desire to cut expenses in a declining industry, the use of containers was successful from the outset. Within a few years, many railroads and trucking companies were hauling large numbers of containers, many of them containing freight that had spent some time at sea.

As often happens with an emerging technology, intermodal transportation became more efficient as it expanded in scale. Although many containers were no larger than the ones first used, the ships and railroad cars that hauled them grew in size. The first container ship, Sea Land's *Ideal X*, was a converted tanker that had room on its deck for 58 containers. About 20 years later, the world's oceans were being navigated by specially designed container ships with capacities of 1,200 to 1,600 T FEUs. On land, the movement of containers by rail has been made more efficient by the development of special flatcars that allow containers to be carried in stacks of two.

The widespread use of containers is both cause and effect of the large increase in international trade that has occurred over the last 2 decades. On the one hand, the expansion of trade created a need for more efficient

A container train on the Santa Fe Ry. The first five cars are an articulated set, in which two cars share a common truck (from S. Glischinski, *Santa Fe Railway*, 1997, p. 61).

transportation systems. On the other hand, by cutting shipping costs, intermodal transportation reduced the cost of imported goods. Whatever the exact mix might be, the shipment of freight by container has been a key element in an international economic system that puts Korean television sets in American living rooms and German machine tools in Brazilian factories.

The use of containers for ocean transport also gave a strong boost to the overland haulage of containers by truck and rail. From 1980 to 1994, rail intermodal grew by an annual average of 7 percent. So efficient is intermodal operation that many containers traveling from Japan to Europe (or vice versa) never pass through the Panama Canal, for it is cheaper and faster to transport them by rail between Pacific and Atlantic ports. In general, intermodal traffic has played an important role in the revitalization of American railroads. Although the railroads began to augment their revenues in the 1950s through the use of piggyback service, intermodal transportation has diminished the importance of trailer-on-flatcar transportation. On the railroads, container loadings now exceed trailer loadings. Together, the transportation of containers and truck trailers has transformed railroad traffic. In 1957, U.S. railroads operated 750,000 boxcars; by the early 1990s the number had shrunk to 200,000. In similar fashion, the number of stevedore jobs has been drastically cut by the advance of containerization. As sometimes happens with technological change, an advance in economic efficiency may come at the cost of substantial local disruptions.

See also PIGGYBACK TRANSPORTATION; RAILROAD; STANDARDIZATION; UNEMPLOYMENT, TECHNOLOGICAL

Further Reading: David J. DeBoer, *Piggyback and Containers: A History of Rail Intermodal on America's Steel Highway*, 1992.

Trebuchet

While the Romans used their torsion catapults against the "barbarians" with some success, once the barbarian tribes had overrun the Roman Empire, it is not certain that they were able to acquire this technology from their former conquerors. Some historians have contended that the barbarians were unable either to use or to construct Roman-style catapults. They argue that although there is some evidence of early use of artillery at the siege of Thessaloniki by the Goths in 269 and at Tours by the Alemanni or Franks a century later, in both instances the defenders disabled the catapults by hurling blazing missiles at them. Moreover, by the 6th century there is no further mention of them, and catapult technology seems to have passed into obscurity.

Other historians argue that the real reason for the abandonment of the torsion catapult was the adoption of an alternative missile launcher, the trebuchet. It is well established that trebuchets originated in China between the 5th- and 3d-centuries B.C.E., and from there diffused westward to the Islamic lands by the end of the 7th-century C.E., where they continued to be used until the 15th century. The earliest of these artillery pieces were quite large and were designed with a rotating beam placed on a fulcrum that was supported by a wooden tower and base. The beam was positioned unevenly on the fulcrum (at a ratio of 5–6:1 for a light trebuchet and 2–3:1 for a heavy trebuchet), with the largest end of the rotating beam holding a sling in which projectiles, generally large stones, were placed. On the opposite, small end of the beam were 40 to 125 ropes that were pulled by a team of men estimated to number between 40 and 250. By pulling in unison, the team generated a strong force that hurled a projectile weighing between 1 and 59 kg (2.2–130 lb) in a relatively flat arc for a distance of between 85 and 133 m (279–436 ft). It is this source of power that gives the artillery piece its modern name, the *traction trebuchet*.

None of these points is questioned by any of the historians mentioned above. But whereas the first group maintains that the traction trebuchet was not introduced to Western Europe until it was seen by the Crusaders when they attacked the Muslims on the First Crusade, the second group contends that trebuchets were known and used by Western Europeans as early as the 6th century. As evidence, they point to an eyewitness account of the siege of Thessaloniki by the Avaro-Slavs in 597, an account written by John, the archbishop of Thessaloniki. It can be further established that the technology for these weapons was transferred to the Avaro-Slavs by a captured Byzantine soldier named Bousas a decade before the siege of Thessaloniki. Other references to siege machines appear frequently among the chronicles of the early Middle Ages, indicating perhaps a continual use of the trebuchet, although sources in western Europe are scarce until the 12th century.

In 1147, two traction trebuchets were reportedly used by the Crusaders to capture Muslim Lisbon. They were operated by crews organized in shifts of 100 pullers, firing 5,000 stones in 10 hours (250 shots per hour or 1 shot every 14.42 seconds). After this, traction trebuchets appeared at many sieges throughout Western Europe. Still, despite their frequent usage, the accuracy and power of the discharged missiles were often inconsistent, largely because the accuracy and force of the missile were dependent on the strength and unity of a team of pullers. When a team was well trained in the firing of these catapults, their efficiency must have been excellent. However, when a team of pullers was poorly trained or had suffered losses in numbers, accuracy and power of discharge diminished. Consequently, there was a search for an alternative power source, ultimately leading to the creation of the *counterweight trebuchet.*

The counterweight trebuchet differed little from its predecessor. The only significant change was the substitution of the pulling ropes with a fixed counterweight, usually a box filled with stones, sand, or some other heavy body, which provided the power to discharge the missile. Not only did the counterweight allow for a more balanced discharge force, but weighing an estimated 4,500 to 13,600 kg (9,920–30,000 lb), it could propel rather heavy projectiles (45–90 kg [100–200 lb]) an estimated 300 m (984 ft).

The counterweight trebuchet appeared as early as the mid-12th century in the Mediterranean area and then spread into northern Europe, the Middle East, and North Africa. It may have a Byzantine provenance, as the earliest recorded use of the counterweight trebuchet was at the Byzantine siege of Zevgminon in 1165. The Byzantines may have especially favored this weapon, and it appeared often in the Holy Land, where it was used frequently against the Christians. The weapon appeared as far north as Flanders, England, and Scotland. It was also of great interest to many technical writers and draftsmen of the late Middle Ages and the Renaissance, with detailed descriptions and drawings of the mechanism by such eminent technical authors as Villard de Honnecourt

and Giles of Rome in the 13th century; Conrad Kyeser, Marino Taccola, Roberto Valturio, and the "Anonymous of the Hussite Wars" in the 15th century; and Leonardo da Vinci and Agostino Ramelli in the 16th century.

Both traction and counterweight trebuchets seem to have been effective siege weapons. Although they were rarely used in large numbers, sometimes they brought a more rapid conclusion to a siege by breaching the fortification's walls. Trebuchets were also used often to intimidate and destroy the morale of the besieged. At these times stone missiles were replaced by incendiaries, the carcasses of putrefying and diseased animals, or even the bodies or body parts of compatriots of the besieged.

For all its success, the trebuchet was short-lived. By the mid-14th century, it was apparent that gunpowder weapons were making the catapult extinct as a siege weapon. Still, because of the novelty of gunpowder technology, it was not until the 1380s that the trebuchet went into decline as a siege weapon. Almost all of the early sieges of the Hundred Years War included them, and sometimes they were even used alongside early gunpowder weapons in both attacking and defending fortifications.—K.D.

See also CATAPULT, TORSION; GUNPOWDER

Further Reading: Donald R. Hill, "Trebuchets," *Viator,* vol. 4 (1973): 99–114. Kelly DeVries, *Medieval Military Technology,* 1992.

Triode

A triode consists of three basic elements: a cathode that supplies free electrons, a negatively charged anode to which they migrate, and a grid located between the two that controls their flow. The grid is so constructed that electrons are able to pass through it and move towards the anode; the charge applied to the grid determines the rate at which this occurs. A small electrical charge applied to the grid will control a much stronger current that has been supplied by a battery or other source of electricity. A triode can act as an amplifier, a detector of radio signals, and an oscillator that produces these signals.

Although control grids had been used to control the

flow of electrons in early cathode-ray tubes, the first functioning triode was invented in 1906 by Lee De Forest (1873–1961). He was seeking an amplifier for radio signals, so he called his device an *audion*, for it allowed the audio reproduction of radio signals through the movement of ionized gas in the tube. Except for the crucial addition of the control grid, De Forest's invention was similar to the vacuum-tube diode invented by John Ambrose Fleming (1849–1945) a few years earlier. De Forest always denied that he had known about Fleming's diode prior to the invention of the audion, but this is unlikely.

De Forest was correct in his assumption that the movement of ionized gas was the basis of the audion's operation, but other triodes operated with much greater efficiency when their tubes were evacuated of as much gas as possible. When the triode's elements were surrounded by a near-vacuum, free electrons were directly produced by heating the cathode, not by the ionization of gas. Still, several years were to pass before there was a full understanding of the triode's operation and the key role of electron flow. Here technology lagged behind science, for in 1897 J. J. Thomson (1856–1940) had conducted a series of experiments that demonstrated the existence of electrons. In 1900, Owen Richardson (1879–1959) had hypothesized that a metal heated to sufficiently high temperature would throw off electrons, and in 1904 he provided a mathematical proof for this hypothesis. The utility of producing electrons directly from the cathode instead of through the production of ions was demonstrated in 1913 by Irving Langmuir (1881–1957) of the General Electric Company. The tube containing the triode was fully evacuated and as a result the triode performed much better than De Forest's audion.

The audion was one of the most important inventions of the 20th century, but several years passed before its potential began to be realized. As events were to show, the development of the triode required the resources of a large corporation, the American Telephone and Telegraph Company (AT&T). AT&T was trying to set up a cross-country telephone service, and the audion had great potential value as an amplifier of telephone signals. In 1913, De Forest sold the patent rights to the audion to AT&T for all uses except wireless telephony and telegraphy. A year later the company bought a nonexclusive license for wireless telegraphy as well, and in 1917 it bought all the remaining rights, although De Forest was free to manufacture his invention for certain purposes. One of the immediate benefits of the acquisition was the improvement of the triode through the efforts of Harold D. Arnold (1883–1933), an AT&T scientist. Arnold used the German-invented and recently marketed Gaede Molecular Pump to evacuate tubes down to .0001 millimeters of mercury. German technology was also drawn on when AT&T manufactured tubes with the oxide-coated filaments invented by Artur Wehnelt in 1903.

The final application of the triode was as an oscillator for the production of radio signals. De Forest had observed self-oscillation in the earliest audions, but at the time he considered it nothing but a problem, for the audion's oscillation produced audible howls, whistles, and shrieks. However, the same process that produced oscillations in an audio frequency could produce them in radio frequencies. By the time De Forest realized the value of an oscillating audion, other experimenters had discovered the same phenomenon. The ensuing conflict over patent rights absorbed 19 years of litigation before it was resolved in De Forest's favor.

The triode that had first emerged as De Forest's audion was an essential part of generations of radio transmitters and receivers, as well as many other electronic devices. Triodes continue to be important devices today, but they are constructed as solid-state devices and not vacuum tubes.

See also CATHODE-RAY TUBE; DIODE; ELECTRON; PATENTS; SEMICONDUCTOR; THERMOIONIC (VACUUM) TUBE; TRANSISTOR

Further Reading: Hugh G. J. Aitken, *The Continuous Wave: Technology and American Radio, 1900–1932,* 1985.

Trolleybus

The trolleybus (or, sometimes, trolleycoach or trackless trolley) merged the technology of the electric streetcar with that of the motorbus. Like a streetcar, it utilizes electric power, drawing current from overhead wires; like a bus, it runs on rubber tires and can be steered by the driver. Such vehicles were in commercial operation as early as 1910 in Hollywood, Calif., and had been

A modern trolleybus used in Budapest, Hungary (courtesy Robert C. Post).

tried out even before the turn of the century in Europe. But it was not until the 1930s, when urban transit systems found themselves in dire financial straits—especially after legislation designed to break up utility holding companies set many firms adrift from their parents—that they were first utilized on any appreciable scale. They offered several advantages to operating companies. Management was not saddled with extensive track maintenance, nor were vehicles subject to delays when tracks were blocked, yet they could still use the generating and distribution systems developed for street railways. A trolleybus accelerated more quickly than a motorbus—a significant advantage when there were frequent stops—and it made no more noise than the hiss of pickup shoes sliding along the overhead wires. It still lacked the flexibility of a motorbus, however, and the unit cost was higher. There still remained the expense of maintaining the wires, whose complexity was multiplied by the fact that each vehicle had to have *two* trolley poles extending to paired wires in order to complete the electric circuit.

Eventually some 10,000 trolleybuses operated in more than 70 North American cities, carrying a major portion of the transit load in San Francisco, Seattle, Boston, and elsewhere, as well as in smaller cities such as Dayton, Providence, and Shreveport. Indeed, for a time Shreveport's transit service was nearly 100 percent trol-

leybus. As the vehicles wore out in the 1950s and 60s, however, trolleybus lines were abandoned in city after city, and by the early 70s they remained in only a dozen: Boston, Calgary, Dayton, Edmonton, Hamilton, Mexico City, Philadelphia, San Francisco, Saskatoon, Seattle, Toronto, and Vancouver. And then came a partial turnabout, as the rise of the environmental movement enhanced concerns about the depletion of nonrenewable energy sources, particularly petroleum-based fuels such as diesel oil, and about the air pollution created by the exhaust from diesel buses. In recent years, a handful of municipal systems have ordered new trolleybuses, and brand-new lines have even been opened, notably in San Francisco and Toronto, while urban transit in Dayton is still provided primarily by trolleybus lines. These hybrid vehicles, it appears, will retain a small but significant niche in the mosaic of urban transit.—R.C.P.

See also STREETCAR

Trucks

As the automobile became a reasonably practical proposition at the end of the 19th century, the possibility arose of using automotive technology to transport freight rather than people. The motor truck (or *lorry*, as it is often called in Great Britain) had a bright future, for it did not suffer from the limitations inherent to the existing modes of land transportation. Railroads were well suited to long hauls, but they were inflexible and often could not provide door-to-door delivery. Horse-drawn wagons sufficed for local transport, but their capacity was limited, and teams of horses were good for only about 30 km (18.6 mi) a day.

In 1891, the German automotive pioneer Gottlieb Daimler (1834–1900) built the first truck by mounting an engine under the floor of a wagon, and in 1896 his compatriot Karl Benz (1844–1929) built a truck capable of carrying up to 5 tons of cargo. Both were powered by internal combustion engines. As with some contemporary passenger cars, early trucks also were powered by steam or electricity. These failed to achieve a large following, although small numbers of steam trucks continued to be built in Great Britain up to 1950.

The miserable state of the roads kept early trucks largely within the boundaries of large cities. Even so, their potential was becoming evident. One study conducted in 1911 found that a 5-ton truck was nearly twice as expensive to operate as a three-horse team pulling a load of the same weight, but its greater speed made its ton-mile cost one-half that of a horse-drawn wagon. The armed services were quick to see the value of trucks, and in the years immediately prior to World War I the governments of Germany, Great Britain, and France provided subsidies for owners of trucks that could be pressed into military service. World War I showed that this interest in trucks was not misplaced, as trucks transported vast quantities of men and materiel under very difficult conditions. American trucks were exported to Europe in large numbers; many of them were laden with freight and then driven long distances to their ports of embarkation, thereby demonstrating the feasibility of long-haul trucking.

When the war ended in 1918, the United States had 525,000 trucks in operation, up dramatically from only 700 in 1904. Many of these had been built for the U.S. Army, and their subsequent sale to civilians gave a strong boost to the infant trucking industry. In the years that immediately followed, the technical specifications of trucks improved considerably. Many trucks rode on solid-rubber tires (this was an improvement over the solid-steel tires used before

about 1910), but the use of pneumatic tires made life much easier for the driver, the truck, and the pavement. The size of tires increased as trucks got larger, but there was an upper limit to how far this could be taken. Accordingly, in the early 1920s, manufacturers equipped some of their trucks with tandem axles at the rear. Since each axle carried four wheels and tires (two on each side), much of the truck's load was spread over eight tires.

By the late 1920s, big trucks were being equipped with large-displacement six-cylinder engines, giving them the ability to reach speeds of more than 80 kph (50 mph). Higher speeds put greater demands on braking systems. Fortunately, Westinghouse had begun to supply airbrakes to the truck industry in 1921. Chain-drive (see photo) was in the process of being replaced by driveshafts, although some manufacturers contin-

Mack trucks from the late 1920s. Notice the chain drive transmission to the rear wheels.

ued to produce chain-drive trucks through the early 1950s.

One of the most important advances was the adoption of the diesel engine as a source of power. Diesels offered a number of advantages, especially low fuel consumption. In 1932, Kenworth offered as an option the first diesel truck engine, a four-cylinder, 75-kW (100-hp) Cummins. The same year saw the introduction of the first truck with the cab mounted over the engine. Cab-over design (or "forward control" as it is known in Great Britain) became increasingly common in subsequent years, as it allowed more load space in a truck of a given length, an important consideration in places where trucks were subject to length limitations.

By this time many trucks were tractor-trailer combinations. Building the tractor (i.e., the cab, engine, transmission, and front wheels) separate from the trailer meant that the latter could be parked and loaded while the tractor was out hauling another trailer. Tractor-trailer rigs also were more space-efficient and easier to maneuver than the all-in-one "straight truck." Trailers are divided into full trailers, with wheels at both ends, and semitrailers, with wheels only at the back end.

The prolonged depression of the 1930s subjected the U.S. economy to severe stresses, but it stimulated the growth of the trucking industry, as business firms sought to keep afloat by cutting transportation costs. But in desperate economic times, many truckers were shoestring operators who engaged in cutthroat competition that reduced rates to a bare minimum. Moreover, financially strapped owner-drivers often cut corners as far as safety was concerned and few of them had adequate insurance coverage. The response to "excessive" competition in the trucking industry was an expansion of governmental regulation. Prior to 1935, the U.S. federal government had little to do with trucking, but in that year the Motor Carrier Act gave the Interstate Commerce Commission the power to set rates, enforce safety standards, and regulate the number of hours that a driver could be on the road. It was not until the early 1980s that the enthusiasm for regulation waned, and trucking became a more competitive enterprise.

The United States had nearly 4.6 million trucks on the road when it entered World War II. As had happened during World War I, the truck manufacturing industry enjoyed a period of great expansion, as the United States and Canada truck factories turned out more than 2.6 million military trucks and nearly 530,000 trailers. The industry continued to expand through the next decade, and long-haul trucking received an enormous boost from the construction of the Interstate Highway System.

Today, trucking is a significant sector of the U.S. economy. In 1994, 2.9 million men and women worked as truck drivers, operating everything from small delivery vans to huge tractor-trailer combinations. In the United States, trucks are classified according to their gross vehicle weight (gvw), the combined weight of the vehicle and its load. This ranges from class 1 (up to 2,700 kg [6,000 lb]) to class 8 (over 14,850 kg [33,001 lb]). The federal government does not set size and weight limitations, as this is deemed a responsibility of the individual states. However, tractor-trailer combinations weighing no more than 36,300 kg (80,000 lb) and trailer lengths of no more than 14.6 m (48 ft) are permissible anywhere in the United States.

From the 1920s onwards, the rise of the trucking industry was closely paralleled by a decline of the railroad industry. In recent years, however, trucking has been integrated with rail transportation through the development of intermodal transportation, a process that began with the hauling of trailers on railroad flatcars. The truck remains an indispensable part of a modern transportation system, and to an increasing degree it is being integrated with railroad, ship, and barge transportation so that each does what it is best suited to do.

See also AIRBRAKE; AUTOMOBILE; AUTOMOBILE, ELECTRIC; AUTOMOBILES, STEAM-POWERED; CHAIN DRIVE; ENGINE, DIESEL; ENGINE, FOUR-STROKE; INTERSTATE HIGHWAY SYSTEM; PIGGYBACK TRANSPORTATION; RAILROAD; ROAD CONSTRUCTION; TIRE, PNEUMATIC; TRANSPORTATION, INTERMODAL

Further Reading: Niels Jansen, *Pictorial History of American Trucks*, 1995.

Truss Bridge—see Bridge, Truss

Tungsten Filament

An incandescent lightbulb gives off light when a current is run through its filament. A number of things were necessary before this source of light could become a practical reality, most notably a source of current and a means of removing air from the bulb so that the filament did not immediately burn out. When Thomas Edison (1847–1931) and Joseph Swan (1828–1914) began experimenting with incandescent illumination, the most vexing part was finding a filament that gave off sufficient light for a reasonable amount of time. To this end, Edison is said to have tried out 6,000 items, everything from fish line to a whisker plucked from the beard of one of his assistants. The winner proved to be a charred cotton thread, which in 1879 glowed for 45 hours before it finally burned out.

Although Edison's incandescent lamp brought illumination to many homes, shops, and offices, the filament's abbreviated lifespan limited its usefulness. The first step toward the solution of this problem was taken in 1900 when, despite managerial resistance, one of the world's first industrial research laboratories was established within the General Electric Company. Under the leadership of its director, Willis Whitney (1868–1958), the laboratory first subjected conventional filaments to very high temperatures. This transformed their structure, resulting in a 17 percent gain in efficiency. The real breakthrough came when William Coolidge (1873–1975) came to the laboratory intent on exploring the use of tungsten for the filament. Tungsten has the highest melting point of any metal, so it seemed well suited to this purpose. A sintered form of tungsten had already been used with some success by Austrian researchers, but it suffered from a short life. What was needed was a ductile form of the metal, that is, tungsten capable of being drawn into a wire.

Coolidge first tried alloying tungsten with a combination of other metals. This produced a substance that could be successfully drawn into wire, but when electric current was passed through it, the tungsten separated from the other metals. The tungsten was still not ductile, but it was in the purest form ever achieved. Coolidge found that it could be worked at lower temperatures than previously. He then devised some new machines for pressing, rolling, and swaging (in this case, hitting with small hammers) the tungsten, thereby forming it into rods about 1 millimeter in diameter. These were in turn heated to redness and drawn through a series of dies until they emerged as wires with a diameter of .01 millimeter. Finally, they were put into a vacuum, and current was passed through them. The result was a ductile tungsten filament that increased the life of bulbs nearly 30-fold. By 1914, 4 years after Coolidge began his investigations, 85 percent of incandescent lightbulbs were equipped with tungsten filaments.

Producing a practical tungsten filament required individual genius and determination, but it was also a team effort, one that required the services of 20 research chemists and many technical assistants. The laboratory that produced the tungsten filament subsequently supported the research of Irving Langmuir (1881–1957), whose inquiries resulted in the use of an inert gas for the bulb's interior, thereby increasing its life even more. The work done at the GE laboratory made incandescent lighting a more practical proposition. No less important, it demonstrated the commercial value of the industrial research laboratory.

See also ILLUMINATION, ELECTRICAL; RESEARCH AND DEVELOPMENT

Further Reading: Elting W. Morison, *From Know-How to Nowhere*, 1974.

Tunnels

Many ancient civilizations built tunnels, primarily for water supply or sewage disposal. According to one ancient chronicler, the Babylonians were able to build a 900-m (2,950-ft) walkway beneath the Euphrates River around the year 2150 B.C.E. No archeological remains have been found of this tunnel, which supposedly connected a royal palace and a temple. If it had been built, the tunnel would have soon flooded, for the Babylonians did not have the technology to pump out the water that would have rapidly intruded.

Much more firmly grounded in historical fact is the tunnel constructed around the year 522 B.C.E. on the Greek island of Samos. Built under the direction of Eupalinos of Megara (born c. 600 B.C.E.), the tunnel was used to convey water from a spring to a town located on the other side of Mount Castro. The tunnel, which still exists, is 1,006 m (3,300 ft) in length, and is about 1.7 m (5.5 ft) in both width and height. The tunnel was chiseled through solid limestone from both ends, and when the two bores joined, they were off by about 6 m (20 ft) horizontally and 0.9 m (3 ft) vertically. The methods through which Eupalinos achieved this degree of accuracy are still a matter of some dispute.

The Romans adopted the Assyrian practice of digging tunnels to subterranean aquifers in order to obtain irrigation water. In 40 B.C.E., a Roman engineer named Lucius Cocceius Auctus built two tunnels to open traffic bottlenecks in Naples. According to the Roman rhetorician Seneca (c. 4 B.C.E.–65 C.E.), passage through the dark and dusty tunnels was most unpleasant. The Romans also built a number of tunnels to serve as aqueducts. Very few tunnels were built in the Middle Ages or early modern times, although military engineers would occasionally resort to tunneling in order to undermine a fortification or the walls of a city.

The building of extensive canal systems during the 18th and 19th centuries provided a major impetus for tunneling. In 1789, the 3.2-km (2-mi) Sapperton Tunnel was built to link the Stroudwater Canal with the River Thames. The Blisworth Tunnel, the longest canal bore in England, was built in 1805 and extended 930 m (3,056 ft). The first major tunnel in the United States was the 135-m (450-ft) bore that was built from 1818 to 1821 for the Schuylkill and Susquehanna Canals in Pennsylvania. The subsequent development of railroads provided an even more powerful impetus for the building of tunnels.

In 1827, the British engineer Marc Brunel (1769–1849) initiated a radically new construction process when he began work on a 366-m (1,200-ft) tunnel 4.3 m (14 ft) beneath the Thames in London. The tunnel had to bore through porous gravel, and this presented a constant danger of cave-ins. Inspired by the actions of the teredo shipworm, which had shell plates that allowed it to bore through wood and push the sawdust behind, Brunel designed a 11.6-by-6.7-m (38-by-22-ft) cast-iron shield that was moved forward by screw jacks as construction progressed. The shield was equipped with a number of slots, through which up to 36 men could attack the earth with picks and shovels. The tunnel was opened to traffic in 1843, and eventually was incorporated into the London underground railway (subway) system.

Within a few years of the completion of the Thames tunnel, the construction of underground passages was greatly facilitated by the use of pneumatic rock drills and high explosives. The use of gunpowder for construction projects goes back to the early 17th century, but its meager explosive force limited its value. The invention of nitroglycerin, followed by the invention of dynamite, made blasting much more effective. Although blasting was always a dangerous part of construction, it was made safer by the use of safety fuses, which had been invented in Cornwall in 1831.

Even with the assistance of pneumatic drills and explosives, tunneling was a hard, hazardous occupation. Miners often worked in near-darkness, soaked by muddy water while they breathed stale air. Cave-ins were a constant threat, and many lives were lost to blasting that was improperly done. One of the greatest engineering feats of the 19th century, the 14.9-km (9.25-mi) St. Gotthard Tunnel through the Swiss Alps, was completed in 1882 at the cost of 310 men killed and 877 seriously incapacitated. In many places, construction workers survived the hazards of tunneling, only to succumb eventually to silicosis, a lung disease caused by prolonged exposure to dust.

Beginning with an effort to tunnel under the Hudson River in 1879, excavators began to use pressurization in order to bore deeply beneath bodies of water. Inspired by the success of the pneumatic caisson for the construction of bridge piers, the technique entailed pumping compressed air into the tunnel as it was excavated. An airlock at the open end allowed the passage of equipment, materials, and workers. Working in a pressurized bore was a risky occupation. Air leaks could develop, leading to the rapid influx of water and mud. Workers were also subjected to a crippling illness known as caisson's disease or "the bends," an intensely painful accumulation of nitrogen bubbles in the blood and tissues caused by too-rapid depressurization when workers returned to the surface.

In recent decades, the mechanization of tunnel construction has made it a faster, less-dangerous process. In the early 1960s, tunnel-boring machines began to be used on an extensive scale. Capable of boring tunnels with diameters of 10.7 m (35 ft) or more, the machines' rotating cutter heads work like giant drills as they cut through earth, clay, sandstone, and shale. They are less effective on hard rock or very soft soil. These "mechanical moles" can cut very accurate bores when controlled by laser guidance systems.

The most widely publicized tunnel project in many decades is the Channel Tunnel ("Chunnel") that connects Folkestone, England, and Calais, France. The notion of tunneling under the English Channel goes back to the 18th century, and some construction was done in the late 19th century, but political obstacles—notably Great Britain's fear of military invasion from the Continent—prevented its realization until much later. Construction of the Chunnel began in December 1987. In 1990, French and British construction teams met at a point 22.2 km (13.9 mi) from the English coast and 14.5 km (9.7 mi) from the French coast. The tunnel was officially inaugurated by Great Britain's Queen Elizabeth and French President Francois Mitterand on May 6, 1994, although full commercial service did not begin until more than a year later. The tunnel allows railroad trains as well as trucks and automobiles carried on special rail cars to cross the Channel in less than 30 minutes. Its construction costs have been estimated at $15 to $16 billion.

For all the attention it has received, the Chunnel is not the world's longest tunnel; that honor belongs to the Japan's Seikan tunnel. Completed in 1988, it stretches 53.9 km (33.5 mi) between the islands of Honshu and Hokkaido. The longest highway tunnel is the new St. Gotthard tunnel in Switzerland, which burrows 16.3 km (10.1 mi) beneath the Alps.

See also AQUEDUCT; CAISSON; CANALS; DRILLS, PNEUMATIC; EXPLOSIVES; LASER; PUMPS; RAILROAD

Turbine, Gas

A gas turbine is any rotary device operated by an expansible medium, such as steam or the products of internal combustion. In contemporary usage, virtually all gas turbines share two characteristics: They are "internal combustion gas turbines"—that is, they do not depend on an external source of working medium, such as a steam boiler or a heat exchanger—and they employ a continuous combustion cycle (fuel is ignited and burned continuously) rather than an intermittent cycle (as in reciprocating engines). Thus gas turbines have, minimally, a rotary compressor, a combustion chamber where fuel is burned continuously, and a turbine that extracts the energy to power the compressor. In a turbojet, excess energy is expanded in a nozzle to produce thrust directly; in a turboprop, the turbine also extracts mechanical energy in rotary form to turn a conventional propeller. Gas turbines used in marine, electrical, automotive, or industrial applications provide mechanical power analogous to a turboprop. In all cases, gas turbines achieve their highest thermal efficiency only very near their optimal or design rotational speeds and ordinarily are not efficient at off-peak speeds. This virtually universal characteristic of gas turbines sharply limits the uses to which they can reasonably be applied.

Internal combustion gas turbines were proposed contemporaneously with the first steam turbines in the 19th century. An internal combustion turbine had appealing hypothetical technological and economic advantages: Dispensing with boiler, piping, and condenser would reduce size, weight, complexity, manpower requirements, and cost. Charles A. Parsons (1854–1931), the inventor of the Parsons steam turbine, experimented, unsuccessfully, with an axial flow, multistage air compressor (essentially one of his steam turbines run backwards), apparently intending to use it in an internal combustion gas turbine. At the turn of the century, René Armengaud and Charles Lemale in France, and, independently, Sanford A. Moss in the United States (at Cornell, and later at General Electric) developed internal combustion gas turbines. None was successful. The best-reported thermal efficiency was less than 3 percent.

Unbeknownst to them, these experimenters shared a common fallacy. All turbines prior to the 1920s were designed using classical hydrodynamic assumptions derived from water turbine practice: that the medium was incompressible, that flow between turbine blades was laminar (or at least not turbulent) and that no flow separation from the turbine blades occurred, and that energy was neither added to nor withdrawn from the flow. These assumptions were not so deleterious for steam turbines, but in axial-flow air compressors, the increasing pressure gradient promoted flow separation, which resulted in catastrophic inefficiency.

Design of efficient gas turbines depended on theoretical and empirical investigations carried out in the 1920s, again independently, by Alan Arnold Griffith (1893–1963) at the Royal Aeronautical Establishment in Great Britain, by Jacob Ackeret at the Zurich Polytechnic University in Switzerland, and by Adoph Betz and W. Encke at Göttingen University's Aeronautical Research Institute in Germany. These researchers applied mature subsonic aerodynamic theory to the problem of flow in cascades of aerofoils and in turbines and compressors. Aerodynamic lift and drag calculations permitted accurate portrayal of energy transfer and of the conditions for laminar, nonseparated flow along the turbine or compressor blades. Griffith used his experiments as the foundation for design of a very complex contrarotating, contraflow turboprop, development of which was overtaken by the turbojet revolution. Ackeret's research fed into Brown Boveri's development of stationary, gas-turbine–powered electric generators just before World War II, while Betz and Enke's results undergirded German development of turbosuperchargers and turbojets.

Despite the promise of research in the 1920s, the first truly successful gas turbines were the turbojets developed during the 1930s and early 1940s. Turbojet development, in turn, not only laid the design foundation for all succeeding turbine systems but also stimulated development of essential technology in manufacturing and, especially, high-temperature materials. Heirs to the turbojet include all other gas turbine aircraft engines, turboprops, and turbofans, as well as all post–Second World War industrial, marine, and automotive turbines.

Aside from their near dominance of airplane and helicopter applications, gas turbine prime movers have been widely adopted for use only in relatively specialized niches in which their light weight, small size, and very high power-to-weight ratios more than offset their continuing disadvantages in fuel efficiency and, sometimes, initial cost. For industrial applications, the balance of advantages and disadvantages is highly contingent and context specific. For example, a very large (2,013 kW [2,700 hp]), 12-cylinder reciprocating, natural-gas-powered compressor system used offshore to compress natural gas before delivery to collection pipelines costs approximately 50 percent more per installed horsepower than a comparable gas turbine system, weighs several hundreds of tons, and requires a building as big as a good-sized barn to house it. In contrast, a gas turbine gas compressor system not only costs less initially but also weighs less than 1,800 kg (2 tons) and can fit in the back of a pickup truck. Yet the gas turbine typically burns about 70 percent more BTUs per horsepower-hour than a reciprocating system. Especially in offshore applications, where space, weight, and, often, time are critical, gas turbine systems thus offer major advantages, although these depend on contingent fuel costs and the expected length of time over which the system will be amortized, as well as applicable capital discount rates and expectations about the future value of natural gas. In general, industrial gas turbine systems have been adopted for use in remote or environmentally challenging locations (off-shore oil platforms, remote pipeline compressor stations, Middle Eastern deserts, the Arctic), where energy costs are low or not critical, or where small size and light weight are decisive.

Marine and automotive adoption of gas turbines is governed by the same general factors. Beginning in the early 1960s, gas turbine engines derived from contemporary turbojets were used for high-speed propulsion (over 40 knots [74 kph]) in fast patrol craft, as well as in later hydrofoils. These boats typically carried auxiliary diesel engines to provide low-speed, longer-range patrol endurance. These hybrid systems thus exploited the gas turbines' very high power-to-weight ratios and packaging advantages, while retaining the diesels' superior fuel efficiency at low speeds. The U.S. Navy began to use gas turbine engines for its major surface ships in the early 1970s, beginning with the *Spruance*-class destroyers. The problem of the gas turbines' off-peak inefficiency is ameliorated by the simple expedient of treating them as "units," and only operating as many of the turbines as are necessary to attain the desired ship performance, thus running each engine near its peak efficiency. All subsequent U.S. Navy surface combatants, with the exception of the very large aircraft carrier, have been powered by gas turbines. In contrast, recent fleet support and transport ships, which do not require very high speeds, remain diesel powered.

In automotive applications, despite considerable experimentation, especially by Chrysler Corporation in the late 1960s, gas turbines' superior power-to-weight ratios have in most instances not been sufficient to overcome their disadvantages in initial costs and fuel efficiency. Moreover, gas turbines' operational limitations—particularly their poor performance at partial power and the high temperature of their exhausts—continue to make them not well suited to road-going uses, where continual acceleration and deceleration, limited opportunity to exploit very high continuous power, and dense traffic are the norm. In very large off-road mining and earth-moving equipment, where weight is less critical, the perceived robustness and partial-load efficiency of diesels has precluded adoption of gas turbines. For similar reasons, gas turbines have not been widely adopted for railroad locomotives.

Likely the most notable current automotive application of the gas turbine is the power plant used in the M1 family of American Main Battle Tanks. In this application, the gas turbine's compact size and very high power-to-weight ratio reduced the necessary volume of the armored envelope and, more importantly, the proportion of the vehicle's total weight that had to be devoted to its power plant. In addition, the turbine's inferior part-throttle efficiency was judged to be more than counterbalanced by the extraordinary full-throttle acceleration and maximum cross-country speed it offered, both of which were critical to the tank's intended modes of tactical employment. The turbine's manufacturer also claimed that, although the gas turbine is less fuel efficient than a comparable diesel, its total POL (petrol-oil-lubricants) requirements are about the same, since the diesel requires considerably more lubricating oil, as well as more frequent oil changes. Thus, although poor fuel efficiency at idle and a significant thermal signature remain problems, the gas turbine in the M1 family is widely considered to be remarkably successful.

Outside aircraft and related applications, the gas turbine remains very much a specialized, niche prime mover. For those circumstances that play to its intrinsic strengths, it has been intensively developed and widely adopted. But the gas turbine has not yet become the dominant prime mover its enthusiasts once envisioned.

The gas turbine is a maturing technology, and further incursion into new markets and applications most likely will result from intensive development, especially of components: carbon fiber compressor blades, ceramic or single-crystal alloy turbine blades, better combustor design, improved alloys, and better digital sensors and controls. Yet the elegance and simplicity that Charles Parsons and his successors saw in the gas turbine still whisper their seductive siren song.—E.W.C.

See also COMPOSITE MATERIALS; ENGINE, DIESEL; ENGINEERING CERAMICS; TANKS; TURBINE, STEAM; TURBOCHARGER; TURBOFAN; TURBOJET; TURBOPROP

Turbine, Hydraulic

The harnessing of moving water as a source of energy began with the invention of the waterwheel. These devices did a great deal of useful work, but they left much to be desired in terms of speed, durability, and energy efficiency. In the early 19th century, engineers began to apply scientific principles to the development of water power. One of them, Benôit Fourneyron (1802–1867), completed a novel waterwheel at Pont-sur-l'Ognon in 1827, which he called a *hydraulic turbine*. To Fourneyron and his American admirers, *turbine* meant a scientifically designed water motor with a theoretical limit of efficiency of 100 percent. It seemed like a visitor from a radiant future. In particular, Fourneyron's third turbine, which was built at Saint-Blaise in the Black Forest, was a dazzling success. It used a relatively small stream of water at an unprecedented fall of 114 m (374 ft), generating some 44.7 kW (60 hp) at the startling speed of 2,300 revolutions per minute.

These turbines were based on the prior work of Fourneyron's former teacher, Jean-Charles de Borda (1733–1799). Borda had published in 1767 the general criteria that had to be met in order to design an ideal water motor: entrance without shock (a sudden change of velocity) and exit with minimal velocity. A third criterion was implicit in Borda's theory, that frictional losses be as small as possible.

Borda's ideas had been rejected by his peers, due in large measure to French engineering career lines stressing mathematical theory that, it was hoped, would carry the successful engineer to membership in the Academy of Sciences. This had the effect of subordinating engineers to scientific censorship. Borda's ideas were founded on his rejection of the doctrine of the conservation of *vis viva* or "living force," a cherished foundation of the prevailing model of physical science. Scientific disapproval caused the neglect of Borda's work until 1818, when early glimmerings of the conservation of energy made the loss of mechanical energy an acceptable concept.

Fourneyron showed that entrance without shock could be achieved by having the water enter the turbine parallel (tangent) to the vane. Exit with minimal velocity could be achieved if the water was discharged in the direction opposite to the motion of the rotor, and in a direction as nearly tangent to the wheel as possible so that the two velocities came close to canceling out.

In Fourneyron's turbine, the water entered vertically through a cylindrical penstock. At the bottom of the penstock were fixed guide vanes that turned the water through a 90-degree angle and guided it horizontally into the rapidly spinning rotor. The path of the water then widened as it moved through a runner, leading to its spreading and eddying, and hence to some loss of efficiency. Though Fourneyron's configuration was not very widely used, his turbine theory was the exemplar for dozens of turbines designed in Europe and America in the next generation or so. Fourneyron's simple trigonometric theory could be adapted for just about any turbine configuration.

A second aspect of scientifically designed turbines was the replacement of the rule-of-thumb measurements of millwrights by exact data measured by precision instruments. Baron Rich de Prony developed a friction dynamometer that Fourneyron used to measure the power and efficiency of his turbines. Typical of scientifically designed machines was a continuing interaction between them and the development of measuring devices. Fourneyron's tests were not terribly accurate at first, but Captain Arthur-Jules Morin did careful comparative tests in which he developed the Prony dynamometer as an instrument of precision.

The Fourneyron turbine diffused to the textile manufacturing city of Lowell, where Uriah A. Boyden, assisted by James B. Francis (1815–1892), made dramatic improvements in the Fourneyron turbine. Boy-

den addressed the neglected third condition of efficient operation: reduction of friction. To this end, he developed a novel suspension of the turbine from above by means of a low-friction bearing that was easy to lubricate, being well above the level of the water level, along with many other improvements that raised efficiency by reducing friction.

More fundamentally, Boyden and Francis translated the Fourneyron's mathematical theory into graphical theory. Francis published the method whereby the path of the water moving through the runner vane was represented by a visual plot. By calculating the motion of a particle of water as it moved through the runner vanes at successive short time intervals, Boyden and Francis produced a graphical plot which, in principle, made the mathematical derivation of the three angles unnecessary. The angle of tangent entry, for example, could be determined without mathematics simply by laying a straight edge on the plot of the curved path of the water so that the last element of the guide vanes and the first element of the runner vane were on one straight line.

Boyden's graphical method of design led directly to the Francis-type turbine: that is, a plot of the path of the water in an outward-flow turbine emphasized the spreading of the water at its exit. Reversing the direction of flow eliminated spreading, but progressively narrower passages accelerated the water too much and compromised efficiency. Francis was probably first to realize that by varying the depth of the channels, so that the water moved downward as well as inward, the engineer could control the cross-sectional area of a stream of water, and hence its velocity and pressure. However, it was Boyden who realized that the difference in velocity on the opposite sides of a runner vane created a pressure of the water against the vane, in much the same manner that the Bernoulli theorem explained the lift in an airfoil. In his early Fourneyron turbines, Boyden calculated the pressures and designed the passages so that the water was uniformly accelerated as it moved through the turbine, producing a constant pressure on the vanes.

Boyden and Francis opened up the design of turbines to self-educated farm boys (such as Boyden himself). This led to counterrevolution in turbine technology in America. Craftsmen seized the leadership in turbine design in

America, beginning with the turbine of the millwright and patternmaker Asa M. Swain, who patented his turbine in 1860. The counterrevolution gained momentum with the remarkably compact and cheap "New Departure" Francis turbines based on the designs of John B. McCormick. The reality of this counterrevolution was driven home when in 1875 Francis saw his only two full-scale mixed-flow "Francis" turbines at the Boott mill at Lowell ripped out and replaced by Francis turbines that had been designed by Swain. Francis tested them and found them much superior to his own turbines, particularly in the cost per installed horsepower.

The craft counterrevolution had important social and intellectual consequences. Boyden and Francis were content to see the benefits of the turbine go to large textile corporations, whereas millwrights such as James Leffel and John Temple (designers of the two best-selling turbines in America) successfully spread the benefits of efficient, cheap water power to country mills, artisans' shops, and small factories.

Neither Francis nor Boyden anticipated this counterrevolution. There was still much in turbine design that went beyond the traditional abilities of millwrights. It turned out that a few millwrights, such as Swain, were able to learn the theory of the turbine and use it creatively, while other millwrights copied these exemplars. Even so, the millwrights had insights the engineers lacked. Perhaps the most notable (but certainly not the only one of these) was their ability to get more out of turbine testing than the engineers. Unknown to Boyden and Francis, an American millwright, Zebulon Parker, had developed a superior testing tool in 1846, a glass-walled testing flume. This device was then used by James Leffel and other millwright designers. By means of small particles in the water, designers could trace the flow patterns of water passing through their turbines and redesign sources of instability and turbulence. Another craft improvement in testing came from James Emerson, a former sea captain, who patented a version of the Prony dynamometer that permitted quick and cheap, albeit less accurate, turbine tests. Emerson not only tested turbines, he also helped improve them. By repeated tests and tinkering, craft designers could improve the performance of their turbines even though their theoretical knowledge was deficient.

Emerson also propagated a radical democratic and antiscientific ideology that gathered a following among American millwrights. He frankly espoused the independent American mechanic or craftsman as the best possible technologist. Believing that technical creativity was inborn and could be developed only by experience, he held that college training for engineers was worthless. Emerson even encouraged a number of millwright supporters to reject Newton's laws of motion, and to see the key to improvement in the public testing of turbines, rather than in science or education.

The combination of scientifically derived principles and seat-of-the-pants empiricism resulted in the widespread use of hydraulic turbines as industrial power sources. In 1875, hydraulic turbines accounted for 80 percent of the installed power in Massachusetts, a prime industrial state. But this success was short-lived, as technical improvements and inexpensive coal combined to make the steam engine a cheaper source of power. By the beginning of the 20th century, most of the water turbines in operation did not provide direct industrial power; instead, they had found a new niche as sources of power for electrical generators. Today, water turbines running at 95 percent efficiency with capacities of up to 900,000 kW are the basis of hydroelectric power installations throughout the world.—E.L.

See also BERNOULLI EFFECT; CONSERVATION OF ENERGY; DAMS; ELECTRICAL-POWER SYSTEMS; ENERGY EFFICIENCY; STEAM ENGINE; WATERWHEEL

Turbine, Steam

The invention of the steam engine vastly increased the amount of power at human disposal. Conventional steam engines, however, did not convert fuel into energy with a high degree of efficiency. Moreover, their reciprocating motion had to be converted to rotary motion through crankshafts, connecting rods, and transmission belts that consumed power and required regular maintenance. A steam engine that directly produced rotary power would obviously have widespread applicability. During the 1880s, such a device began to emerge in the form of the steam turbine.

Predecessors of the steam turbine can be found in the distant past. The first steam engine, the *aeolipile* created in the 1st-century B.C.E. by Hero of Alexandria, was a rotating engine, although it was little more than a toy. In 1629, Giovanni Branca (1571–1640) hoped to obtain rotary motion by directing a jet of steam at a set of vanes arranged on the circumference of a wheel, but his idea was never translated into actual machinery. On a more practical level, in 1831 William Avery in the United States actually sold 50 devices based on Hero's rotating engine. The engines were used to power sawmills, but their lack of power, difficulty of operation, noise, and unreliability soon consigned them to oblivion.

The first successful use of a steam turbine was in a cream separator. Patented in 1883 by Sweden's Carl Gustaf de Laval (1845–1913), it was a reaction turbine built around an S-shaped tube. When steam rushed out at each end, the reaction made the tube rotate at speeds of up to 43,000 rpm. De Laval's turbine made inefficient use of the steam that powered it, and was eventually abandoned by its inventor in favor of an impulse turbine. In an impulse turbine, steam is directed at a set of blades to produce rotary motion. For his turbine, de Laval used a steam nozzle with an interior space that converged and then diverged. Steam passing through a nozzle of this shape expands, loses pressure, and increases in velocity. De Laval's impulse turbines produced up to 500 horsepower (hp), operating at considerably higher levels of efficiency than conventional reciprocating steam engines. High rotational speed was the key to this and subsequent turbines, but achieving high speeds required effective axles and bearings, and a great deal of work to keep everything in balance.

In 1884, the year after de Laval patented his reaction turbine, Charles Parsons (1854–1931) in England began to develop an impulse turbine for the generation of electricity. Parsons's steam turbine was similar in principle to existing water turbines, except that the operating fluid was steam rather than water. In Parsons's turbine, steam traveled through the turbine's axle, where it lost pressure and gained velocity prior to impinging on the blades. Whereas de Laval's turbine used its steam in a single impulse, Parsons's turbine used steam in a series of stages. Since the steam lost pressure as it traveled through the turbine, the blades that received lower-pressure steam were larger and of dif-

ferent shapes than those that received the first blast. The turbine also made use of the reaction principle: Fixed vanes in the turbine accelerated the steam as it passed through, so when it struck the blades it generated a force similar to the aerodynamic lift that keeps an airplane aloft.

Although Parsons's axial flow design was quite successful, internal disputes caused Parsons to leave the turbine company he founded. Since this firm held rights to his patents, Parsons had to develop a turbine based on a different principle. The result was the radial-flow turbine. In this design, steam entered through the axle and then expanded at right angles to impinge on a series of concentric blades that were alternately attached to one of two rotating disks. As a result of this arrangement, the blades of one rotor also acted as guide vanes for the set of blades on the rotor behind it. As with the axial-flow turbine, the size and shape of blades differed in accordance with their placement in the turbine and the pressure of the steam that acted on them.

Parsons's first turbine developed 10 hp at a speed of 18,000 rpm. By 1890, power had increased sufficiently for a Parsons turbine to power an electrical generating station, the first application of this sort. Ten years later, turbines of similar basic design were being built with a rating of 1,000 kW, and in 1923, a turbine rated at

The pioneering steam-turbine powered *Turbinia* (courtesy National Maritime Museum Greenwich, Conn.).

The low-pressure steam turbine used by the steamship *Mauretania*. The man at the lower right gives an idea of its size (from E. Constant, *Origins of the Turbojet Revolution*, 1980).

50,000 kW was in operation. Parsons's turbines were also used for marine propulsion. The potential of the steam turbine as a ship power plant was demonstrated by Parsons's own ship, *Turbinia*, which achieved the astounding speed of 34.5 knots (65.7 kph) in 1897. By the second decade of the 20th century, the steam turbine had become the standard power plant for large ships, both civilian and naval.

In 1895, Charles Gordon Curtis (1860–1953) patented a steam turbine that combined axial and radial steam flow. Curtis's turbine employed a principle known as *velocity staging*; steam remained at the same velocity as it moved one row of turbine blades to the next. His design was adopted by General Electric as part of that firm's effort to build a vertically integrated system of electrical generation, distribution, and utilization. In France, another multistage turbine was patented in 1896 by Auguste Rateau (1863–1930) and 2 years later a turbine was built to his design. In Rateau's turbine, steam expanded as it passed through a number of stages, with a consequent increase in volume and decrease in pressure at each stage. Along with its other advantages, Rateau's design made for a slower-moving turbine, allowing the direct connection of an electrical generator and the elimination of conventional gearing or belting.

By the first decade of the 20th century, steam tur-

bines had emerged as the logical choice for driving electrical generators. Today, steam turbines are responsible for more than 95 percent of the world's electricity. They are the most powerful single unit prime movers by far, capable of putting out more than 800 MW of power in thermal plants, and 1,300 MW in nuclear plants. This enormous power output is due in part to the sheer size of these engines. A modern steam turbine in a thermal power plant can be as much as 30 m (98 ft) in length and weigh more than 9,000 kg (19,842 lb). Turbines used in nuclear-power applications are even larger. At the same time, their high power output is the result of 35 to 40 percent thermal efficiency, considerably better than most internal combustion engines. A major reason for efficient operation is the high steam temperatures allowed by turbine design, commonly 540°C (1,000°F). Nonnuclear turbine-generating plants generally use fuel oil, but many can be switched to natural gas should air quality regulations require the use of this fuel at certain times of the year.

Efforts to adopt steam turbines for uses other than the generation of electricity and the propulsion of large ships have been less successful. Turbines are occasionally used as power plants for individual factories, and they have powered a few experimental locomotives, but on the whole these applications have been of limited consequence. Turbines are most effective when they are very large in size, making them ideal for electrical generating stations. At the same time, the extensive electrical grids that steam turbines make possible have eliminated the need for individual power sources, except for a few specialized examples.

See also BEARINGS; BERNOULLI EFFECT; ELECTRICAL-POWER SYSTEMS; ENERGY EFFICIENCY; ENGINE, FOUR-STROKE; STEAM ENGINE; TURBINE, HYDRAULIC

Turbocharger

The power developed by any piston engine depends on the rate of energy release, a function of the amount of fuel burned in its cylinders. Fuel-burn in turn depends on the amount of air (which provides oxygen) inducted and compressed in the cylinders. As air density declines with altitude, the power developed in any piston engine therefore also declines. This characteristic of piston engines first became a serious practical problem during World War I, as fighter pilots seeking advantage in combat by climbing above their opponents quickly encountered the limitations of conventionally aspirated piston engines.

One solution to this difficulty is to compress the air before it is inducted into the cylinders, a process usually called *supercharging*, which can maintain sea-level power to much higher altitudes. Also, by forcing more air into the cylinders, supercharging can be used to increase engine power at sea level, albeit at the cost of greater fuel consumption. There is an immense variety of possible supercharger, or compressor, configurations; the most common are the Roots displacement type and the centrifugal rotary type. Both types are gear driven, so a fair amount of engine power has to be used to run the supercharger itself.

Turbocharging (or *turbosupercharging*) is an alternative method of supercharging that obviates this problem by using engine exhaust to run the compressor. Piston engines typically reject about one-third of the energy released in their cylinders as exhaust heat. By using this otherwise-wasted energy to power a gas turbine, which in turn drives a centrifugal compressor, better performance can be had nearly "for free" in terms of fuel costs. However, the power gains provided by turbochargers come at the expense of greater mechanical complexity and the need to use expensive materials to cope with high temperatures and rotational speeds of tens of thousands of revolutions per minute.

Auguste Rateau in France, and Sanford A. Moss at General Electric in the United States, began development of turbochargers for aircraft engines during World War I. Moss's group at GE, with support from the U.S. Army Air Corps, continued uninterrupted development during the interwar years. The resulting turbosuperchargers were used during World War II on such well-known aircraft as the Lockheed P-38, the Republic P-47, the Boeing B-17, the Consolidated-Vultee B-24, and the Boeing B-29. Postwar applications included the Convair B-36 long-range bomber, as well as the Douglas DC-7, Boeing 377 Stratocruiser, and Lockheed Constellation passenger aircraft. Late models (mid-1950s) of the DC-7 and Constellation used an advanced version of turbosupercharging, turbocompounding, in which excess power developed in the gas turbine of the turbocharger was geared directly to the crankshaft of the piston engine.

The superior speed and reliability of turbojet aircraft soon rendered turbocharged piston aircraft obsolete for mainline commercial service. But less-expensive turbocharged piston engines persist in smaller corporate and private aircraft. In addition, the oil crises of the 1970s stimulated renewed interest in turbocharging for automobile engines: Sea-level turbocharging permitted relatively economical, small-displacement engines of four or six cylinders to offer power and performance comparable to much larger and less economical V-8s. Early turbocharged cars were sometimes criticized for their characteristic "turbo lag," the delay between pressing down on the accelerator and getting a response from the engine. Although this problem has been largely solved, turbocharged automobiles are not currently in favor, as improvements in normally aspirated engines have substantially narrowed the performance gap between them and turbocharged engines. Turbocharged engines are still popular in some classes of racing cars, even though their greater power is partially offset by requirements that they be smaller in size (i.e., have less cylinder displacement) than normally aspirated engines.—E.W.C.

See also SUPERCHARGER; TURBOJET

Turbofan

In a turbojet engine, propulsive efficiency depends on the relationship between an aircraft's velocity and that of the effluent providing the propelling reactive force: The closer the two velocities, the greater the efficiency. In the late 1950s, as the gas turbines at the core of turbojets operated at higher and higher temperatures and pressures, and thus attained ever greater thermal efficiency, designers of some turbojets began to take some compressed air from the air compressor and bypass it around the combustion chambers and turbines. This stratagem increased total air mass flow and reduced exhaust velocity, thus enhancing propulsive efficiency.

From low bypass engines (in which bypassed and direct air flows were about equal, or in a ratio of 1:1), it was but a short step to full-fledged turbofans. Vastly enlarged compressor first stages, or ducted fans, at the front of the gas turbine engine provide very high air mass flow at relatively low velocity. This bypassed air typically does not enter the gas turbine core at all, traveling instead between the core and the engine's outer concentric casing. Contemporary large turbofans have bypass ratios above 8:1. These very large turbofan engines are now approaching 45,000-kg (100,000-lb) thrust, fully two orders of magnitude greater than the original aircraft turbojets. Specific fuel consumption (s.f.c., reckoned in kilograms or pounds of fuel burned per thrust-pound hour) has fallen from upwards of .9 kg (2 lb) to .30 kg (.65 lb). This design progress has permitted unparalleled operating efficiencies as well as the use of fewer engines on even very large aircraft; some transatlantic routes are now flown using twin, rather the three- or four-engined aircraft. Virtually all contemporary turbojet engines of any size are of the turbofan variety.—E.W.C. See also TURBOJET

Turbojet

A turbojet consists of a gas turbine used to generate a high-temperature, high-velocity exhaust stream that propels an aircraft (or other vehicle) by reaction: thrust equal in magnitude and opposite in direction to that of the exhaust. All successful turbojets comprise the standard gas turbine elements (compressor, continuous combustion system, and turbine). They also have an exhaust nozzle designed to maximize thrust and an inlet that has been designed to minimize drag and turbulence and to maximize ram effect (the air compression resulting from the forward motion of the aircraft).

The physical conditions for reasonably efficient turbojet operation are stringent. The propulsive (or Froude) efficiency of any vehicle (ship or aircraft) moved by action on a fluid medium is a function of the relative velocities of the vehicle and its propelling exhaust stream: The more nearly equal the two, the higher the efficiency. Because the exhaust effluent of a pure turbojet has a relatively high velocity (typically above 1,200 kph [746 mph]), no pure turbojet can be propulsively efficient at low speeds (perhaps 725 kph [450 mph]). Moreover, both air density and temperature fall with altitude. Lower density translates to lower aircraft drag and higher speed for a given thrust. Lower ambient temperature translates into higher thermal efficiency in the gas turbine (or in any heat engine, for that matter). Consequently, pure turbojets are well suited only to high-speed (near sonic), high-altitude flight.

The turbojet is the multiple, independent invention of at least four men: Frank Whittle (1907–1996), England (1929); Hans von Ohain (1911–1998), Germany (1934); Herbert Wagner, Germany (1936); and Helmut Schelp, Germany (1936). These four men shared two critically important social characteristics: None worked or had ever worked for a manufacturer of conventional aircraft piston engines, and all either worked in or were exposed to leading-edge aerodynamic research. Moreover, there is little evidence that any of them were motivated by traditional economic incentives, as opposed to "pure engineering" enthusiasm or perfectionism. Whittle and von Ohain just wanted to go high and fast; Whittle simply made up the idea of a turbojet-powered transatlantic mail plane as an ex post facto rationalization for his invention. Wagner, who had done his doctoral thesis on aerodynamic lift and drag, headed airframe development for Junkers Aircraft. He was profoundly unhappy with the conservatism of the engineers at the sister firm, Junkers Engine. Wagner's search for a radical alternative propulsion system led him quickly to the turbojet. Helmut Schelp headed an office in the Reichsluftfahrt

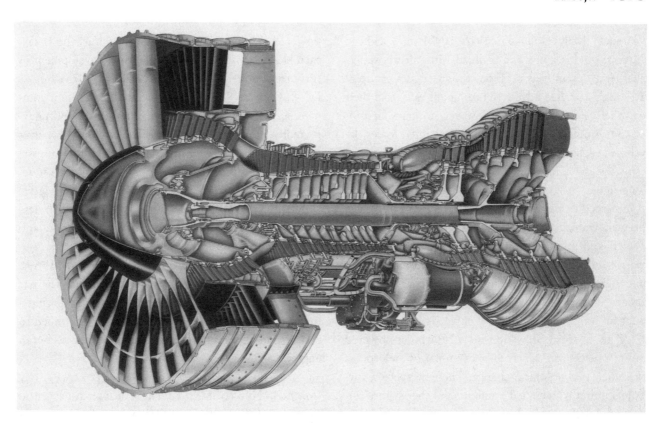

Cutaway view of a turbofan engine (courtesy Pratt and Whitney).

Ministerium (RLM) devoted explicitly to exploring radical aircraft propulsion possibilities; his path to the turbojet lay through Betz and Encke's work on the application of airfoil theory to turbine components.

All four of the original inventors of the turbojet made essentially similar assumptions deduced directly from then-recent advances in aerodynamic science. First, based on research into aircraft drag and sonic phenomena, they believed that conventional, well-streamlined aircraft could fly very close to the speed of sound—upwards of 800 kph (497 mph). This was a time when even fast airplanes might manage 480 kph (298 mph), although the absolute world air speed record did stand somewhat higher. Second, given the rapid increase in drag at near-sonic speeds, it was apparent that conventional propellers were unlikely to function at such speeds, and that in any event some power source considerably lighter and more powerful than conventional piston engines also would be required. Third, they each ultimately recognized that application of airfoil theory to gas turbine components ought to result in efficiencies theretofore thought unattainable.

Reaction propulsion using a gas turbine as its core promised a lightweight, powerful, and reasonably efficient system, at least for very high-speed, high-altitude flight. Elimination of the propeller and its associated gearing reduced weight and avoided the problem of propeller efficiency at near-sonic speeds. By using the exhaust stream of the gas turbine to generate thrust directly, only that proportion of the total energy necessary to operate the air compressor had to be extracted as mechanical power through the turbine. This arrangement reduced the importance of turbine efficiency in pure turbojet designs compared to alternative turboprop proposals, in which nearly all energy would have had to be converted in the turbine into rotary mechanical power.

Development projects in Great Britain and in Germany resulted in production of serviceable turbojets before the end of World War II. Whittle's design and expertise, at the behest of chief of the U.S. Army Air Corps, General Hap Arnold, were shared with the turbosupercharger division of General Electric, and led to the first American turbojets. Several Americans, notably

Clarence (Kelly) Johnson (1910–1990) at Lockheed Aircraft, and engineers in Sanford Moss's turbosupercharger group at General Electric, as well as a group at Westinghouse, had advanced turbojet proposals of their own before the war. Again, however, no manufacturer of conventional piston engines in the United States developed such ideas. In general, the indigenous American efforts were hamstrung by an overriding fixation on long-range fuel efficiency, which reflected America's commercially oriented aeronautical culture as well as its prewar strategic situation and doctrine. Indeed, World War II and immediate postwar turbojet aircraft types were confined to high-altitude, high-speed interceptors and light and medium bombers with relatively short ranges. Large piston-engined aircraft remained dominant for long ranges and heavy payloads.

The postwar development of turbojet-powered aircraft serves as a nearly ideal exemplar of the process of radical technological change. The new species, turbojets, initially occupied a unique and theretofore unoccupied niche: high-altitude, near-sonic flight. In that domain, the turbojet's high fuel consumption and limited life mattered little. As the turbojet further evolved in that niche, it acquired the capacity to diffuse outward into adjacent niches, and ultimately to compete successfully against conventional piston engines in almost all applications. In its original role, the turbojet coevolved with airframes: The combination of swept wings, area-rule designed fuselages, and afterburning (spraying fuel into the exhaust of the turbojet to increase thrust, albeit inefficiently) permitted sonic, then trans-sonic flight. The challenges of trans-sonic flight in turn stimulated a whole cascade of innovations: variable-geometry inlets and nozzles, variable-pitch compressor stators, and careful blending of airframe and turbojet geometries to reduce drag, maximize ram effect, and avoid airflow disruptions into the engines across the extreme range of flight speeds and during violent maneuvers. Intensive development of materials and internal cooling arrangements permitted higher cycle temperatures, greater fuel efficiency, and, most importantly, much greater reliability. These developments in turn led to two-spool designs, with separate low- and high-pressure compressors driven by separate high- and low-pressure turbines.

Meanwhile, successful development of turbojet-powered, medium-range bombers (such as the B-47), then heavy bombers (B-52), promoted development of turbojet commercial transports (Boeing 707, Douglas DC-8, Convair 880). Even these first-generation commercial turbojets proved to be remarkably, and unexpectedly, economically efficient. The aircraft themselves were 50 percent faster than the piston-engined airplanes they replaced, and could thus generate revenue-miles much faster. Moreover, because of their relative simplicity, the turbojet engines required much less maintenance than high-powered piston engines, and thus provided aircraft utilization rates more than double those of the piston airplanes. By the early 1960s, however, the basic turbojet was obsolescent. The gas turbine technology at the core of the turbojet speciated into turboprops, while the turbojet itself continued to evolve and differentiate into contemporary turbofan engines.—E.W.C.

See also AIRPLANE; JET AIRCRAFT, COMMERCIAL; MECHANICS, NEWTONIAN; SUPERCHARGER; SUPERSONIC FLIGHT; TECHNOLOGICAL ENTHUSIASM; TURBINE, GAS; TURBOCHARGER; TURBOFAN; TURBOPROP

Turboprop

A turboprop uses the output of a gas turbine to turn a conventional aircraft propeller through a geared drive. The earliest proposals for gas turbine aircraft engines, which were made in the 1920s, were for turboprops. As it turned out, no turboprop engine was developed before the advent of the turbojet revolution, and almost all the first-generation turboprops that emerged a few years after World War II were in fact based on turbojet gas turbine cores. The first commercial airliner to use a turboprop engine was the British Vickers Viscount in 1948. It was followed by several large commercial turboprop-powered aircraft that appeared in the early 1950s, including the Lockheed Electra and the Bristol Britannia, but they were not successful in competition with faster turbojet and later turbofan aircraft. In general, turboprops are not competitive with turbojets or turbofans for high-speed, high-altitude, long-range flight. But for flight at moderate speeds, on the order of 320 to 480 kph (200–300 mph), and altitudes of 4,500 to 7,500 m (15,000–25,000 ft), the turboprop's superior propulsive efficiency under those conditions gives them the advantage. Turboprops also have a number of advantages when compared with equally powerful piston aircraft engines, as they are lighter, less complex, and considerably more reliable.

For these reasons, turboprop aircraft reign supreme in smaller commercial and commuter airline service, especially over shorter routes. Moreover, where special operational characteristics are critical, such as short takeoff and landing performance, turboprops persist. The turboprop Lockheed C-130 Hercules, for example, has been in continuous production since 1952. Likewise, superior power-to-weight ratios and reliability have made suitably modified versions of turboprop engines dominant since the early 1960s in all but the smallest and least expensive helicopters.—E.W.C.

See also TURBINE, GAS; TURBOFAN; TURBOJET

Turing Test

The Turing test is named after Alan M. Turing (1912–1954), a British mathematician involved in early computer work. Turing was convinced that computers could, some day, simulate the working of the human mind and thus show intelligence. In attempting to explain this concept to people, he had to describe what he meant by "intelligence": Was it simply the ability of a computer to play a game such as chess, or did it mean more than that? Turing was well aware that computers could play games, and that the computer was thought to be exhibiting "intelligent behavior," until the public realized that the computer was simply blindly following a set of instructions programmed into it by a human being. Despite the fact that these instructions were very complex and set up so that the computer could keep track of its results and actually learn not to repeat mistakes in future games, they were still simply a set of instructions, and the computer was still a machine.

In an effort to establish exactly what he meant by "intelligent behavior," Turing postulated that a human being could be communicating via a typewriter-like device with two different "entities"—one a human being, the other a computer. He could ask each of them questions and could bring up any topic he wished. If, after a suitable period of time, he could not distinguish the human being from the computer simply by the answers that appeared on his terminal, then Turing claimed you must concede the fact that the computer is exhibiting intelligent behavior.

While Turing himself didn't live long enough to participate in any real example of such a test, tests of this sort have been held under varying conditions at several places. The most famous are a set of Turing tests that have been conducted by the Computer Museum in Boston. In these contests, the most interesting software programs are pared up with human experts, and a number of people are asked to try to decide which are which. While the topics of discussion are severely limited in these trials (for example to the works of Shakespeare or to "whimsical discussion"), several computer programs have been incorrectly identified as human beings (and vice versa) by the participants. It is, however, still usually the case that the majority of the participants can tell the difference between a human being and a computer.—M.W.

Tuskegee Study

Syphilis is a degenerative disease transmitted by sexual contact. Its first symptoms are painless lesions at the site of infection. After about 2 months, the lesions disappear and are replaced by other symptoms, such as malaise, the appearance of a rash, and enlargement of the lymph nodes. This is a followed by a latency period in which all symptoms disappear. This situation may prevail for the remainder of an individual's life, but in many cases new symptoms appear within 2 to 40 years of the initial infection. If left untreated, the end stage of the disease can cause insanity, heart failure, and the growth of tumors.

In 1932, the United States Public Health Service (USPHS) began to study the long-term course of syphilis through a study of black men in Macon County, Ala. Because the town of Tuskegee is located in Macon County, the program has come to be known as the Tuskegee study. The study centered on 400 men infected with syphilis and a control group of 200 uninfected men. The participants did not know that they were being used as subjects for a long-term medical study; rather, they were told that they were receiving a free treatment for "bad blood," a euphemistic term for syphilis commonly used in the South. In fact, the USPHS researchers had no intention of treating the afflicted group; the purpose of the study was to find more about the long-term course of syphilis, up to and including the subjects' autopsies. Not only were the men untreated by the USPHS, the researchers also enjoined local doctors to withhold treatment, lest the study be "corrupted."

When the study was initiated, there was no real cure for syphilis, although some treatments could be of value. In the 1940s, penicillin was introduced. It provided an effective treatment for syphilis, but this did not alter the terms of the study, as the subjects still received no treatment. This situation continued until July 1972, when accounts of the Tuskegee experiment began to appear in the national press. In the same year, the U.S. Department of Health, Education, and Welfare formed an advisory panel in response to all the negative publicity that was being generated. The panel's report, which was issued in June 1973, duly noted that the study was "ethically unjustified" because the men's participation in the study had not been based on informed consent. In fact, the committee understated the extent of the deception, for the subjects participated in the program because they thought that they were being treated for their disease. Nor did the committee come to grips with an equally noxious aspect of the study: There can be little doubt that racism facilitated the clinical detachment of the researchers and their willingness to withhold treatment. As far as the white researchers conducting the study were concerned, their poor black subjects were less than human. Clearly, their lives were worth very little, and they were deemed much less important than the advancement of science and medicine.

See also PENICILLIN

Further Reading: Allan M. Brandt, "Racism and Research: The Case of the Tuskegee Syphilis Study," in *The Hastings Center Report, vol. 54* (Dec. 1978): 21–29.

Typewriter

Efforts to create a mechanical writing device go back to at least 1714, when Henry Mill was granted an English patent for an artificial machine or method for the impressing or transcribing of letters singly or progressively one after another, as in writing, whereby all writings whatsoever may be engrossed on paper of parchment so neat and exact as not to be distinguished from print. . . .

No model or drawing of Mill's device has survived, and his machine may have been a stencil of some sort rather than a typewriter. In subsequent years, numerous inventors tried to create a workable writing machine. Some of them succeeded in making machines that printed letters reasonably well, but they did so at a slower pace than writing by hand (at best, 30 words per minute, and usually much less). Accordingly, there was little point in acquiring a writing machine.

Credit for inventing the first commercially successful typewriter is usually given to Christopher Latham Sholes (1819–1890), a Milwaukee man who at various times served as a newspaper editor, customs official, and member of the Wisconsin legislature. Working with the assistance of Mathias Schwalbach, Samuel Soule, and Carlos Glidden, Sholes patented in 1868 a machine that embodied the essential features of the modern typewriter: individual keys connected to type bars whose characters struck an inked ribbon when their key was depressed. Sholes and his associates had the right general idea for how a typewriter should be constructed, and equally important they had access to a number of technologies essential to a functioning typewriter, such as accurately machined parts, as well as the example of the piano, which used a set of keys and levers to strike individual notes.

Sholes and his partners were fortunate in receiving financial assistance from James Dunsmore, whose unflagging efforts to promote the typewriter were more important than his small financial contribution. The group was able to get their typewriter put into production by E. Remington and Sons, a firearms manufacturer in need of a new product line in the years immediately following the Civil War. In 1874, Remington's Model 1 became the first typewriter to be commercially marketed, although not with any great success. Only 400 of a production run of 1,000 were sold, one of them going to Mark Twain, who used it to prepare the manuscript of one of his novels. Not only was the typewriter expensive at $125, it had a serious defect: Because the type struck the paper at the underside of the roller, the text being typed could not be seen. Not until 1890 did a typewriter produce a visible text.

Sholes's typewriter had another annoying feature, one that has persisted to this day: the QWERTY keyboard, named after the first six characters of the upper-left alphabet portion of the keyboard. This arrangement makes no sense from an efficiency standpoint, for the position of the keys is not related to the anatomy of the human hand (for example, a touch typist has to use the little finger to strike one of the most frequently used keys, the letter *a*). It is often said that this pattern was created as a deliberate attempt to slow the typist down, for early typewriters were highly susceptible to jamming when the keys were struck too rapidly. In fact, there is no evidence that this was Sholes's intention; the reason for the placement of the keys remains obscure. Whatever its origin, the QWERTY keyboard illustrates a problem that occurs from time to time with developing technologies: premature standardization. Even though there may be a better way of doing things—like a more logically arranged keyboard—people have become so accustomed to the existing order that change comes with great difficulty.

Technical shortcomings were not the only reason that the Remington typewriter was a slow seller at first. Sluggish sales were also the result of aiming the typewriter at the domestic rather than the commercial market. For this it was not well suited, as many people resented receiving typed personal letters. When businesses became the target of aggressive salesmanship, sales began to take off; in 1886, 50,000 typewriters were sold in the United States This expansion of typewriter sales occurred in conjunction with the rapid increase in the size and complexity of business and a corresponding need to communicate all sorts of information. By the end of the 19th century, the typewriter had become an indispensable instrument for the production of letters, reports, invoices, receipts, memoranda, and all of the other pieces of paper essential to the operation of a modern business enterprise.

The widespread adoption of the typewriter also corresponded with an even more far-reaching social revolution, the entry of large numbers of women into office work. Although in the United States the employment of women office workers had begun in earnest during the Civil War, the introduction of the office typewriter cre-

ated a new and important niche for women workers. Women were already being employed in significant numbers as copyists, and the use of the typewriter could be seen as a logical extension of this kind of work. Moreover, women typists (they were originally called "typewriters," just like the machines themselves) had the presumed attributes of nimbleness, neatness, and fidelity, and most importantly, they could be hired at lower wages than men. As early as the mid-1880s, typing was being characterized as "women's work." Unfortunately, like most work so labeled, it was devalued. In contrast to the days when men served as secretaries, women office workers generally were less skilled and were given fewer responsibilities; for example, instead of drafting letters themselves, they simply typed what their boss had written or dictated.

Many of these workers had learned their trade at typing and shorthand schools. A proprietor of one of these schools, Mrs. L. V. Longley, may have been the first person to promote the use of all 10 fingers while typing. Some typists went beyond this and began to

Christopher Sholes's daughter at the keyboard of an early typewriter (courtesy Hagley Museum and Library).

practice "touch-typing." By the late 1880s, touch-typing had become common practice, resulting in considerably higher typing speeds. In the early 20th century, typing actually became a competitive sport, as contestants vied with one another to determine who could type the fastest. Even though 10 words were deducted for each mistake, winning typists typically were credited with around 120 words per minute. One typing champion managed 264 words in a minute—22 finger strokes each second!

The speed of average typists began to improve after electrically operated typewriters came on the market in the mid-1930s. Speed was not the only advantage electric typewriters had over manual ones; they also delivered keystrokes powerful enough to make clear impressions on multiple copies sandwiched between sheets of carbon paper. A further improvement was the IBM Selectric typewriter, which was introduced in 1961. Instead of conventional type bars, its characters were arranged on the surface of a ball that rapidly rotated and pivoted in response to the typist's keystrokes. It had the further advantage of allowing easy changes of typefaces through the substitution of one ball for another. In the late 1970s, advances in microelectronics allowed the production of memory typewriters that were capable of storing and retrieving text. Today's typists are more likely to use computers with word-processing programs, although many still prefer the tactile sensation offered by conventional typewriters.

See also WORD PROCESSING

U

UFO

UFOs (unidentified flying objects) are unusual aerial phenomena. Most of them have prosaic explanations, but some sightings remain mysterious. Among the prosaic explanations are weather balloons, the planets Jupiter and Venus, and meteors. The most popular explanation for those that remain, denied by most scientists, has been spaceships piloted by intelligent extraterrestrials. Sightings have been reported worldwide, especially in the second half of the 20th century.

Although sightings of mysterious aerial phenomena have been sporadically reported throughout history, the first substantial wave of "mystery airships" came in 1896–97, after Percival Lowell's well-publicized theory of artificial canals on Mars but before the Wright brothers achieved heavier-than-air flight in 1903. Although the favored explanation for this first wave was experimental aircraft controlled by individuals or the government, the extraterrestrial hypothesis of piloted spaceships was already mentioned. Despite the appearance of H. G. Wells's *War of the Worlds* in serial form in 1897, however, the extraterrestrial hypothesis was not sustained, nor in fact were there large numbers of reports of mysterious aerial phenomena for the next 50 years.

The modern era of UFOs began on June 24, 1947, when Kenneth Arnold, a reputable businessman flying his private plane near Mount Ranier in Washington state, reported nine disk-shaped objects flying in formation at speeds he estimated to exceed 1,600 kph (1,000 mph). His description of these objects as flying "like a saucer if you skipped it across the water" led the newspapers to coin the term *flying saucer*. Arnold's report precipitated more than 850 additional sightings

during that year. Most of the public ascribed the sightings to illusions, hoaxes, secret weapons, or other earthly phenomena, but Arnold and the media exploited the idea of extraterrestrials. With the publication of books such as Arnold's *The Coming of the Saucers* (1952), coauthored by science fiction publisher Ray Palmer; Donald Keyhoe's *Flying Saucers from Outer Space* (1953) and its sequels; and influential articles in the mass media, the UFO debate was off and running. This time it would be a sustained debate through the rest of the century, further stimulated by the rising popularity of science fiction, which had depicted saucerlike spacecraft at least since the 1920s.

Meanwhile, the U.S. Air Force, charged with the security of the skies over the United States, took charge of UFO investigations, beginning with projects Sign and Grudge (1948–1952), followed by Project Blue Book (1952–1969). It was the U.S. Air Force that applied to flying saucers the more general term *unidentified flying objects* in 1952. The extraterrestrial hypothesis had already been broached by air force investigations in 1948. By that time, astronomers considered intelligent life on Mars and Venus unlikely, and although the concept of abundant planetary systems was coming back into favor, the stars were so distant that the early air force reports concluded that interstellar travel, and thus the extraterrestrial hypothesis, was improbable.

This conclusion was reinforced by J. Allen Hynek, an astronomer hired by the U.S. Air Force as a consultant. Along with Harvard astronomer Donald Menzel (also a skeptic), Hynek was one of the few scientists to enter the debate in the 1950s. It was therefore neither the air force nor scientists, but the media and individual authors, that spread the extraterrestrial hypothesis. The dawn of the space age in 1957 also spawned greater interest in the heavens; it is remarkable that the month after the launching of Sputnik, some 361 reports were received by the air force, compared to a normal average of 50 per month at that time. With little scientific interest or guidance, private organizations such as the Aerial Phenomena Research Organization (APRO, founded in 1952) and the National Investigations Committee on Aerial Phenomena (NICAP, founded in 1956) were left to fill the vacuum scientists refused to occupy.

This situation changed during the latter half of the 1960s, when the extraterrestrial hypothesis of UFOs was the subject of more attention from mainstream science than any time before or since. Whereas for 17 years since 1947, the air force had been the primary investigator of the UFO phenomenon and the media the primary purveyor of the extraterrestrial hypothesis, from 1965 to 1969, at least some scientists turned their attention to the subject, in part because of a new wave of sightings—887 in 1965 alone. Congressional hearings in 1966 led to a study commissioned by the U.S. Air force and led by well-known physicist Edward U. Condon. The so-called Condon study, which employed scientists from a variety of backgrounds, concluded that while not all cases could be explained, further large-scale investigation was unwarranted. Although the National Academy of Sciences gave its stamp of approval to the report, the evidence leading to this controversial conclusion did not satisfy all scientists. But in December 1969, a year after the publication of Condon's *Final Report of the Scientific Study of Unidentified Flying Objects*, the U.S. Air Force dropped Project Blue Book, and since that time no government institution officially has investigated the UFO phenomenon. With a symposium on UFOs supported by the distinguished American Association for the Advancement of Science at the end of 1969, scientific interest in UFOs reached a peak and began a precipitous decline within mainstream science.

The reasons for the decline of interest in the subject within mainstream science (though not among the public and a limited number of maverick scientists) are multifaceted. Although the conclusions of the Condon report were controversial, even its skeptics realized that no clear evidence existed for the extraterrestrial hypothesis, and that to push this hypothesis in the ab-

sence of such evidence was a detriment to further study of other possible explanations. Second, New Wave theories of UFOs as representing a "higher dimension" of reality—adopted by Hynek and others in the 1970s and 1980s—confirmed that this was a subject to be avoided by mainstream scientists. Finally, many claims (the existence of ancient astronauts; an alleged spaceship crash at Roswell, N.M., in 1947; and abduction stories) were based on evidence that most scientists could not accept: statements made while under hypnosis, for example.

The UFO debate is one of the most remarkable and unexpected controversies of the 20th century. It is of great interest not only because it focuses on a phenomenon potentially outside the realm of modern scientific understanding but also because it raises questions of scientific method and the nature of evidence. Moreover, it highlights the interactions between different cultures of scientists, and among scientists and the public, on an issue of potentially great significance. Like the debate about extraterrestrial life, of which it is a part, the UFO phenomenon is one of the myths of the modern age, not in the sense that we know it to be false but as a reflection of the deepest hopes and fears of humankind in the wake of global wars and the dawning of the age of space.–S.J.D.

See also EXTRATERRESTRIAL LIFE; NATIONAL ACADEMY OF SCIENCES; SPUTNIK

Further Reading: Curtis Peebles, *Watch the Skies! Chronicle of the Flying Saucer Myth*, 1994. Steven J. Dick, *The Biological Universe: The Twentieth Century Extraterrestrial Life Debate and the Limits of Science*, 1996.

Ultrasound

Ultrasound is the name given to very-high-frequency sound waves (18,000–20,000 Hz). Their effect on biological tissues was first mentioned by the French physicist Paul Langevin (1872–1946) when he noted the painful effect of ultrasound on his own hands (and the fish it killed) during his work with underwater sound in 1917. Between then and the late 1970s, when diagnostic ultrasound became a regular part of clinical medicine, the technology developed along two separate pathways, with overenthusiastic therapeutic applications setting an adverse atmosphere for efforts to develop diagnostic ultrasound.

In the late 1920s, several Americans—R. W. Wood, A. L. Loomis, and E. Newton Harvey—conducted the first systematic research on the biological aspects of ultrasound. Their work formed the basis for subsequent applications of ultrasound in sterilizing pharmaceuticals, preparing vaccines, and deep heating tissues for physical therapy. By the 1940s, clinicians in Europe and Japan were proclaiming therapeutic benefits for ultrasound in treating everything from asthma to bedwetting to cancer. In 1949, these extravagant claims for ultrasonic therapy led a group of German scientists to call for a moratorium, largely ignored, on the use of clinical ultrasound until scientists could further study the effect of ultrasound on biological tissues.

During the 1950s, physicists such as Theodor Hueter and William Fry in the United States studied the physical action of ultrasound, *in vitro* and *in vivo*, at various frequencies, intensities, and foci. Fry especially worked strenuously for the improvement of research standards on biomedical ultrasound and for the establishment of dosimetry standards, in an effort to counteract the reputation that ultrasound had acquired during the previous decade.

It was in this atmosphere of clinical skepticism and disputed control over research methodology that the idea of diagnostic ultrasound was introduced. Diagnostic ultrasound was not developed directly from therapeutic equipment but from the principle of echo-ranging, extensively used during World War II to detect flaws in metal castings, as well as in sonar and radar. Yet the association with therapeutic ultrasound made it difficult for those investigating diagnostic applications to gain professional credibility, institutional support, research funding, or access to patients. However, many of them did find in the postwar period that they could borrow surplus military and industrial flaw-detecting equipment from equipment manufacturers who were interested in having civilian applications developed for their instruments.

Between 1949 and 1957, European, American, and Japanese researchers in surgery, radiology, cardiology, neurosurgery, and obstetrics/gynecology moved industrial equipment into the clinic. The clinicians who did much of the initial work used different frequencies, intensities, foci, and display methods to extract very different information about the body. With the wide variety of instruments that were developed during the 1950s, it was possible to measure dislocation within the brain, to map tissue surfaces, to characterize tissues as malignant or benign, and to track motion.

With no one discipline or professional group leading the development of diagnostic ultrasound, the progress on instrumentation was slow and erratic, fraught with tension and disputes over theory, methodology, priority, and clinical utility. Yet they all agreed that ultrasound was potentially a very powerful diagnostic tool in medicine. Moreover, it was safer than existing diagnostic methods such as exploratory surgery.

The minute pulsed doses used in diagnosis were minuscule compared to exposures used in therapy. Several clinicians did conduct safety tests on small animals, with histological tissue examination after varied exposures. But most investigators were sufficiently confident in the safety of ultrasound that with little hesitation they applied their instruments to themselves and to their patients.

Despite the multitude of efforts to develop diagnostic ultrasound, it was slow to gain acceptance as a regular part of clinical practice until the late 1970s, when a number of factors converged to make ultrasound clinically viable. In 1969, George Kossoff, an Australian, incorporated gray scale in the image display, allowing the capture of far more information in the image. Research by American, English, and Russian scientists on dosimetry allowed the comparison of clinical results from different research centers, considerably enhancing the scientific credibility for diagnostic ultrasound. As images improved and scientific evidence for the efficacy of ultrasonic diagnosis accumulated, health insurance companies and national health plans began to pay for ultrasonic exams, enhancing the clinical acceptance of the technique.

At the same time, however, the vast proliferation of ultrasonic diagnosis, especially in obstetrics, has raised again the question of safety. The ratio of risk to potential benefit changes considerably when a technique is applied to healthy patients. As ultrasound was increasingly used on healthy pregnant women in the 1960s and 1970s, some people questioned long-term effects, perhaps chromosomal, on the fetus. On the other hand, ultrasonic scans remain the only noninvasive check for fetal abnormalities.

Medical economists have been far more concerned about the economics of ultrasound, as they have attempted to limit the expenditures on a technology that continues to spread most rapidly in obstetrics, with very little measurable effect on birth outcomes or the health of women. With every pregnant woman eligible for scanning, ultrasound has consumed increasingly large parts of every healthcare budget. Still, there are social pressures favoring the use of ultrasound in obstetrics. Patients often request scans, and physicians in such litigious countries as the United States often comply to prevent later malpractice suits, despite widely publicized policy statements encouraging the limiting of ultrasonic scans to only those women with clinical signs of being high-risk patients. In other countries, multiple ultrasonic scans are specifically encouraged for *every* pregnant woman as a means of alleviating anxiety for the mother and increasing compliance with prenatal-care programs.

A much more contentious use of ultrasonic imaging is in gender selection. In cultures where males are highly valued, or only one child is allowed, ultrasound is often used to check the sex of the fetus early in a pregnancy. The role of ultrasound in gender-selective abortion is often cited in ethics debates as an example of a "legitimate" technology run amok.

It is highly ironic that diagnostic ultrasound, originally developed for safe detection of lethal diseases such as cancer, has become instead a technology most frequently used for depicting normality in pregnancy.—E.K.

See also RADAR; SONAR

Ultraviolet Radiation

Ultraviolet radiation is the part of the electromagnetic spectrum occupying wavelengths from approximately 4×10^{-9} to $\times 10^{-7}$ m (some classifications restrict it to the range from 15×10^{-9} to 3.9×10^{-7} m). Ultraviolet radiation was discovered in 1801 by Johann Wilhelm Ritter (1776–1810). In the course of conducting experiments with silver chloride, a substance that darkens when exposed to light, Ritter found that the violet end of the spectrum was more effective than the red

end in effecting this change. He then discovered that the region of the spectrum beyond violet was even better, even though it could not be detected by the eye. Due to its effect on silver chloride, ultraviolet radiations were sometimes referred to as "chemical rays," a term that is no longer used today.

A great amount of ultraviolet radiation is given off by the sun, but much of it is absorbed by the Earth's atmosphere. This blockage of ultraviolet radiation by the atmosphere makes the observation of ultraviolet light from extraterrestrial sources difficult or impossible. This is unfortunate, as the detection and analysis of ultraviolet radiation can provide important information about the composition of stars and other objects in the universe. As with infrared astronomy, putting instrumentation above the atmosphere, first with rockets and then with orbiting satellites, was essential. Beginning in 1968, a number of astronomical satellites capable of detecting ultraviolet light have been launched. The most important of these has been the International Ultraviolet Explorer (IUE), which went into geosynchronous orbit on Jan. 26, 1978. A collaborative effort of American and European space agencies until it went out of service in 1996, IUE captured the ultraviolet spectra of tens of thousands of objects: comets, planets, stars within our own galaxy, other galaxies, and quasars. The orbiting observatory provided information about the upper atmospheres of the planets, discovered haloes of hydrogen around comets, and provided new insights about the still-mysterious energetic sources in the cores of galaxies. Further discoveries made by the ultraviolet-detecting instrumentation aboard the Hubble space telescope may be expected.

On the Earth's surface, the most palpable consequence of ultraviolet radiation is its effect on the skin. Prolonged exposure to ultraviolet radiation with wavelengths shorter than 3.2×10^{-7} m can cause erythema, a reddening of the skin popularly known as *sunburn*. This condition is caused by injury to the epidermis, the outer layer of the skin. This in turn causes the release of substances that migrate to the lower layers of the skin, causing the enlargement of the small blood vessels. A more severe case is characterized by blistering, which results from the accumulation of serum and white blood cells near the skin's surface.

Exposure to ultraviolet light over long periods of time may induce skin cancer in some people. In recent years, there has been a growing concern that depletion of the ozone layer is allowing the passage of excessive amounts of ultraviolet light to the Earth's surface, increasing the risk of skin cancer and other ailments.

Of course, some people deliberately expose themselves to ultraviolet light in order to get a tan. This can be done by sunbathing or through the use of an ultraviolet lamp. In both cases, it is important to avoid overexposure and consequent sunburn. Also, when using a sunlamp, care should be taken to avoid exposing the eyes to ultraviolet light, which can cause temporary blindness and considerable pain. Special glasses should be worn; looking away from the source of the radiation is not sufficient, as it will be reflected by surrounding surfaces.

In addition to producing both tan and sunburn, ultraviolet light triggers a biological process that produces vitamin D. Children who do not receive adequate amounts of sunlight may therefore be prone to rickets, a defective development of the bones. Ultraviolet radiation may also be used for sterilization, a discovery that dates back to 1877. Today, radiation at a wavelength of 2.65×10^{-7} is often used to sterilize food products and to kill airborne bacteria.

Although it is invisible to the eye, ultraviolet radiation is used for the production of visible light in fluorescent tubes. Ultraviolet radiation caused by the electrical excitation of mercury gas hits a phosphorescent coating on the inner surface of the tube, which in turn causes the emission of visible light. The ultraviolet radiation itself is not emitted because it cannot penetrate glass. Specialized fluorescent tubes, popularly known as *black-light lamps*, emit ultraviolet radiation at about 3.6×10^{-7} m. When this radiation falls on fluorescent pigments, it produces a characteristic glow. This can be used to obtain dramatic lighting effects, and it also is employed as a means of finding flaws in machined parts.

See also CHLOROFLUOROCARBONS; COMMUNICATIONS SATELLITES; HUBBLE SPACE TELESCOPE; INFRARED RADIATION; LIGHTS, FLUORESCENT; PASTEURIZARTION; QUASARS; VITAMINS

Unemployment, Technological

Many production technologies are created for the specific purpose of raising productivity. Productivity is usually measured in terms of output per worker, so when productivity increases, fewer workers are needed to produce the same output. It seems only logical, therefore, that productivity-enhancing technological advances will necessarily lead to increased levels of unemployment. But closer inspection reveals that the situation is considerably more complicated. When history and economic theory are taken into consideration, it becomes apparent that there is no connection between improvements in productive technology and increased unemployment. If anything, technological advance increases employment.

The fear that machines will replace human workers has a long history. When an inventor presented the Roman emperor Vespasian with a scheme to haul some huge columns to the Capitol by means of a mechanical device, the emperor declined the offer, adding: "I must always ensure that the working classes earn enough money to buy themselves food." In 1638, the government of England banned the use of "engines for working of tape, lace, ribbon, and such, wherein one man doth more amongst them than seven English men can doe." An even more extreme (although possibly apocryphal) example comes from the Polish city of Danzig, where in 1661 the municipal authorities destroyed a mechanical ribbon loom and drowned its inventor, fearing that the device would put handloom weavers out of work. During the Great Depression of the 1930s, politicians, social commentators, and ordinary people frequently voiced the concern that the steady advance of mechanization was a major cause of the mass unemployment that had racked the capitalist world.

These concerns were understandable, but the underlying reasoning is flawed. Concerns about technological unemployment are valid, but it is essential to differentiate between the loss of *jobs* and the loss of *work*. There is no question that the adoption of some technologies has resulted in the obliteration of certain jobs. In the age of the automobile, there is not much demand for livery stable attendants, nor is there much call for assemblers of vacuum tubes when most electronic devices use transistors and integrated circuits. But the loss of individual jobs, no matter how widespread, does not necessarily add up to a loss of work as a whole. Instead of dwelling on particular industries and occupations, it is essential to look at the larger picture.

One thing seems certain: There will always be work to do in a modern society. If people decided to embrace a subsistence-level lifestyle, leisure might be abundant and little time would be devoted to work. But few people seem to be willing to subject themselves to material deprivation. Although it may not be a particularly noble trait, the desire to acquire more and more material possessions is a very common one, and continuous productive growth does not seem to get us any closer to satiation. And then, of course, there is the great majority of the world's population, for whom attaining the living standards of industrialized countries is a distant dream. Meeting the needs of people that are frequently ill-fed, ill-housed, and ill-clothed is at least a potential source of employment for millions of workers.

Technologies do not only serve existing needs. They also help to create new needs, and in so doing they stimulate the creation of many new jobs. This pattern can be clearly seen in the field of medical technology. Some technologies such as antibiotic drugs, have reduced labor needs by offering a "quick fix" for a number of maladies. But most medical technologies are not like antibiotics. Instead of supplanting existing therapies or diagnostic procedures, they open up entirely new areas of treatment. For example, prior to the invention of the dialysis machine, people with serious kidney problems had a short life expectancy. Today, dialysis offers the prospect of a normal lifespan to people suffering from kidney failure, while at the same time it has created new employment opportunities for medical personnel. More generally, new medical technologies have been the basis of the work performed by great numbers of physicians, nurses, technicians, manufacturing workers, and administrators.

Medical technologies have created these jobs because they stimulate the development of new markets and activities. The same cannot be said of well-established products and the firms that make them. In these cases, it might be thought that productivity improvements will lead to job reductions, as the same level of output can be achieved with fewer workers. Whether

or not this actually happens depends on the relationship between the price of a product and the amount bought, a quality known as *price elasticity* by economists. There are numerous cases of products being consumed in greater numbers as productivity improvements drove their prices down. The most famous example is the Ford Model-T. Through the use of innovations like the assembly line, the Ford Motor Company was able to effect dramatic reductions in the price of their cars. This occurred at a time when relatively few people owned automobiles, and significant reductions in the price of the Model-T allowed millions to buy a car for the first time. And as more cars were produced, more people found work in the automobile industry.

There are, of course, products that have a very low price elasticity, so even vigorous price cutting will not lead to significantly greater purchases. Under this circumstance, it seems inevitable that technology-based productivity improvements will result in a loss of employment. It is true that jobs will be lost in the particular industry, but what holds for the part does not necessarily apply to the whole. When productivity improvements drive the price of a product down, the consumers buying that product have more money left over to spend on other products, increasing job opportunities in the industries that make these products. If they save this money instead of spending it, businesses can draw on the investment funds that allow them to expand, often resulting in the hiring of more workers. Alternatively, a firm may use its increased productivity to augment its profits. These profits may be redistributed to partners, stockholders, or employees, who then have more money to invest or spend, thus creating the conditions that lead to increased employment.

A new productive technology may also create jobs by stimulating the development of other industries. To take the most obvious example, not only has the automobile industry stimulated the growth of jobs for auto workers, mechanics, and sales personnel, it has also indirectly provided jobs for workers in industries as disparate as oil refining and fast food. To cite a more recent example, the airline industry has provided relatively few jobs directly, but it has played a leading role in a vast expansion of travel, which in turn has put many people to work in hotels, restaurants, travel agencies, tourist attractions, and the like.

The travel industry is particularly salient in a discussion of occupational trends, for it is a service industry. Service industries—which encompass such sectors as education, medicine, entertainment, and law—now contribute more than two-thirds the U.S. gross national product. They are the prime source of employment opportunities today. Between 1948 and 1990, the manufacturing sector gained 3.6 million workers, while the service sector added 57.2 million workers. Part of the reason for this massive expansion in employment lies in the difficulty of substituting machines for workers; the services differ markedly from manufacturing and farming in this respect. Also, consumers of services often expect a personal element; few patients would be satisfied by a completely automated diagnostic procedure, no matter how accurate, that did not entail some consultation with a physician.

It is difficult to accept the fact that the majority of today's workers produce little or nothing that is tangible, but in fact the growth of the service sector is simply a continuation of a long-standing trend. Only a few generations ago, about three-quarters of all workers were farmers or agricultural laborers. Today, industrialized countries need only 3 percent of the workforce to be engaged in food production. Technological advance has drastically lowered the number of people needed to raise crops, yet massive unemployment has been a long-term consequence of people leaving the farm in large numbers.

The above discussion has centered on employment in the aggregate, and has argued that technological advance does not destroy work. But as was noted earlier, technological change can destroy specific jobs. In some cases, employees have been able to limit job loss by mandating that they occupy a position even though there is no real work to do. In other cases, workers have received severance payments in partial compensation for the loss of a job. But most workers who have been displaced by technological advance have not fared so well. Technological advance has opened up new job opportunities in many fields, but a 50-year-old worker rarely has the skills needed to take advantage of such opportunities. The inability to respond to new employment opportunities is not always a matter of age, for many young entrants to the labor force are also lacking in requisite skills. For example, one test given

in the mid-1980s to Americans in their 20s revealed that 20 percent could not read at an 8th-grade level, and that nearly 40 percent were below the 11th-grade level. Workers with limited communication and mathematical skills are especially likely to be victimized by technological change, and it will be a major challenge to help them develop the skills needed to benefit from a technologically dynamic economy.

All in all, the fear that technological advance will lead to widespread unemployment is unwarranted. Still, there is no getting around the fact that technological change is often a disruptive, painful process that creates pathetic losers along with spectacular winners. It would be irresponsible and cruel to ignore the needs of workers displaced by technological change. At the same time, however, it would be foolish to attempt to stifle new technologies due to fears of massive job losses. Far from being the product of technological advance, unemployment is much more likely to be a consequence of technological stagnation.

See also ASSEMBLY LINE; FEATHERBEDDING; KIDNEY DIALYSIS; LUDDISM; PENICILLIN

Unit Operations

By the early 20th century, the production of chemicals had become a major industrial enterprise. Paralleling the rise of the chemical industry was the emergence of a new occupation, chemical engineer. More than most other industries, the chemical industry was grounded in scientific knowledge and its application; consequently, well-trained chemical engineers were an essential part of the industry's operation. Yet for early chemical engineers, much of their knowledge base was shared with chemists and mechanical engineers. What was distinctive was too limited. Practicing chemical engineers might know a great deal about specific chemicals and chemical reactions, but they lacked a body of general knowledge unique to their occupation.

In 1915, Arthur D. Little (1863–1935) proposed the concept of "unit operations" as the foundation of chemical engineering. All chemical processes, Little suggested, were brought about through the combination of a few basic operations: drying, grinding, distilling, crystallizing, heating, cooling, heat transfer, filtration, and so on.

There is some irony in the fact that these are physical rather than chemical processes. Even so, a knowledge of these processes gave chemical engineers definite advantages in the planning, design, and management of chemical manufacture. As a result, the concept of unit operations helped chemical engineers to distinguish themselves from chemists and mechanical engineers who worked in the chemical industry, and strengthened their claim to having unique qualifications.

The concept of unit operations came to be the centerpiece of chemical engineering education. In turn, the special knowledge possessed by chemical engineers was used by them to expand their managerial roles, often at the expense of shop-educated managers who lacked formal training. While the concept of unit operations provided a useful intellectual foundation for the development of the chemical industry, it also was used by chemical engineers to defend their turf against the encroachment of managers and workers who lacked the credentials of bona fide practitioners.

Further Reading: Terry S. Reynolds, "Defining Professional Boundaries: Chemical Engineering in the Early 20th Century," *Technology and Culture*, vol. 27, no. 4 (Oct. 1986): 694–716.

Universal Joint

A universal joint allows two rotating shafts to be connected even though the shafts may be at varying angles to each other. A shaft equipped with a universal joint is sometimes known as a *cardan shaft*, named after the 16th-century mathematician Girolamo Cardano (1501–1576). Cardano did not invent it, nor did he claim to do so; the actual invention goes back to the early 9th century. Cardano's name became associated with this kind of a shaft because Cardano described it in a popular book he wrote. The joint described by Cardan is also known, especially in English speaking countries, as a *Hooke joint*. Robert Hooke (1635–1703) was a 17th-century scientist with many other accomplishments to his credit. He used the universal joint for manipulating an instrument that employed a number of reflecting glasses for the safe observation of the sun. He did not attempt to popularize the coupling or apply it to any other mechanism.

Although the names of two Europeans are associated with the universal joint, this kind of coupling was actually invented in China long before it made its appearance in Europe. One poem written in the 2d-century B.C.E. mentions an incense burner that was mounted on metal rings in a manner that kept it upright even though the place to which it was affixed moved about. This was accomplished by mounting the burner inside a series of rings that were connected by pivoted joints to their adjacent rings. The device seems to have been forgotten until it was reinvented in the late 3d-century C.E. From that time on, it was used to hold oil lamps, including those mounted inside balls that could be freely rolled around while the lamp remained upright. By the 18th century, the Chinese were

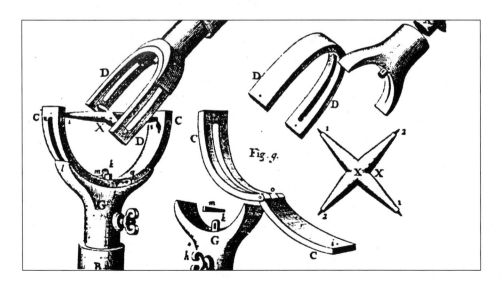

Robert Hooke's diagram of his universal joint, 1676, (from Bill Yenne, *100 Inventions*, 1993).

using this kind of mounting for compasses so that they remained independent of the motion of the ship.

In China and preindustrial Europe, devices of this sort were useful but not of any great significance. Universal joints took on a growing importance as a result of the need to transmit power. They are crucial elements of motor vehicles, in which they are used to connect a driveshaft to the transmission at one end and the differential at the other. They also may be used to transmit power from the differential to the half-shafts that turn the rear wheels on a car with independent rear suspension.

There are a number of variations on the basic principle of the universal joint. All of them are adequate for the transmission of power through a drive shaft or half-shafts. Their use becomes problematic, however, when power has to be transmitted through sharp angles. The few degrees of movement between a drive shaft and a transmission output shaft that are connected by a conventional universal joint are not much of a problem, but in a front-wheel-drive arrangement, the turning of the front wheels may subject the front half-shafts to steering angles of more than 40 degrees. When this happens, the two connected shafts will cyclically accelerate and decelerate, resulting in the two shafts rotating at different speeds. Both will make a complete rotation in the same time, but in the course of this rotation, the speed of one shaft will be greater or less than that of the other. Under these circumstances it is necessary to connect the two shafts with a constant velocity joint.

See also CONSTANT-VELOCITY JOINT; DIFFERENTIAL; FRONT-WHEEL DRIVE

Universal Product Codes

A universal product code (UPC) is the familiar set of parallel lines of different thicknesses that appears on a great variety of products or their packages. UPCs carry information about the product, such as its price, that is "read" by a scanner, freeing a cashier from the need to key in the price. Data extracted from the UPC (or *bar code*, as it is also known) can also be used to determine and record the number of products sold in a given period, the size of the remaining inventory, and other

kinds of useful information. Shoppers encounter UPCs every day, and they are also widely used by libraries for generating checkout records, by factories for keeping track of component parts, and even by the organizers of foot races for recording the order of finish.

Retail establishments had long desired to have the capabilities conferred by UPCs, but the technical capability was lacking. An early effort at producing something like a UPC was made by Norman J. Woodland, who along with Robert Silver devised an arrangement of narrow and wide lines that stood for the dots and dashes of Morse code. These were read by shining a bright light at them; reflected light activated a detector that converted the light into electrical signals carrying the information. Woodland later used a series of concentric circles in place of the parallel lines; this allowed the scanning of the code from any direction. A prototype reader was constructed in 1952 using a 500-watt lightbulb as the source of illumination, while an off-the-shelf photomultiplier tube connected to an oscilloscope served as the detector. Although the system was awarded a patent in that year, it was far from being commercially feasible. There was no way to convert oscilloscope signals into usable data, and the system was very bulky. Nonetheless, the idea had obvious merit, and in 1962, Philco bought the patent from Silver and Woodland. A few years later Philco sold the patent rights to RCA.

In the meantime, a firm known as Computer Identics developed and then marketed a color-based identification code for keeping track of railroad freight cars. It was not a commercial success, but it pioneered the use of lasers for reading the symbols. In 1969, the firm installed two laser bar code systems, one in an automobile factory, the other in a wholesale distribution facility. These were very simple systems that were capable of reading and recording only two digits of information, but they showed that bar codes were a practical proposition. The timing was perfect, for the retail grocery business eagerly awaited anything that promised to cut or eliminate the cost of affixing price labels to individual products, reduce the time required for checkout, and keep track of inventories. Sensing an imminent technological breakthrough, in 1970 a committee representing a consortium of grocery chains established guidelines for bar code development and set up a committee charged with the promotion of standardized codes.

RCA promoted its version of the code, a series of concentric rings used by Woodland for one of his earliest scanning systems. An 18-month test that began in July 1972 showed that the "bulls-eye" code suffered from a serious defect: In the course of printing the symbol, ink was sometimes smeared in the direction of the labels' travel, causing errors in scanning. RCA therefore turned to the Universal Product Code that had been developed by Woodland while he was employed at IBM. On Apr. 3, 1973, it was adopted as an industrywide standard, and on June 26, 1974, a pack of gum sold by a supermarket in Ohio became the first product to be rung up by a UPC-reading scanner.

Interindustry cooperation and standardization were in fact the keys to the bar code's success. Although every manufacturer or retailer was in competition with other manufacturers or retailers, every firm realized that it had to mount a joint effort with other firms if bar codes were to work. Methods of identifying products that were unique to each firm had to be abandoned, and every firm had to identify its products with bar codes that were registered with a Uniform Code Council. The savings offered by the use of UPCs overcame any reluctance on the part of individual firms to surrender some of their independence. Some of the success of the UPC systems can be attributed to technological advances: lasers and computers in particular. But an equally important reason for the rapid spread of UPC-based systems was a long-standing demand for the services that these systems eventually delivered.

See also LASER; OSCILLOSCOPE

Further Reading: Tony Seideman, "Bar Codes Sweep the World," *American Heritage of Invention and Technology*, vol. 8, no. 4 (Spring 1993).

Urea

Urea is a colorless, crystalline compound with the chemical formula CH_4N_2O. Urea is a constituent of human and animal urine. The average person produces about 50 g (1.8 oz) of urea daily. When urea was first iso-lated by the French chemist Guillaume-François Roelle (1703–1770), it was assumed that it could only be produced by living things; consequently, it was considered to be an "organic" substance. The idea that certain substances could be produced only by living beings was part of a larger intellectual framework known as *vitalism*, which postulated that organisms possessed a "vital force" that distinguished them from inorganic entities.

This idea was given a jolt when the German chemist Friedrich Wöhler (1800–1882) was able to produce urea from nonorganic substances. In 1828, Wöhler was attempting to prepare ammonium cyanate by reacting potassium cyanate with ammonium sulfate. But instead of ammonium cyanate, he got urea. As Wöhler noted, it was a striking "example of the artificial production of an organic, and so-called animal, substance from inorganic substances."

In addition to synthesizing urea, Wöhler had discovered *isomerism*, i.e., the existence of two or more dissimilar compounds that had the same elements in the same proportion. Each molecule of ammonium cyanate and each molecule of urea had a single carbon atom, four hydrogen atoms, two nitrogen atoms, and one oxygen atom. But these atoms were arranged differently in the two compounds; hence they were isomers of each other.

Overly romanticized versions of Wöhler's discovery make it appear as though the synthesis of urea sounded the death knell of vitalism. In fact, many supporters of vitalism were not so easily overcome. They pointed out that the potassium cyanate and ammonium sulfate had been derived from organic substances like horns and blood, and they may have contained a vital force that allowed the production of urea. But in 1845, vitalism suffered another setback when another German chemist (and a former student of Wöhler), Herman Kolbe (1818–1884), was able to prepare another organic substance, acetic acid (CH_3COOH), from elemental carbon, hydrogen, and oxygen. Vitalism was in retreat, and organic chemistry came to be regarded simply as the chemistry of carbon compounds.

See also VITALISM

V

V-Chip

The V-chip, or antiviolence chip, was the brainchild of professor Tim Collings of Simon Fraser University. In collaboration with Shaw Communication Inc., Collings developed a computer chip that could be incorporated into television sets and operated by onscreen menu controls. The chip reads a rating of the extent of sex and violence in a program that has been encoded onto the broadcast band used for closed-caption signals for the hearing impaired. Just as an operator can program a videocassette recorder to record a television show based on a numerical code, a parent could program a television equipped with a V-chip to block certain programming. However, the programs must be coded first, and the assignment of a numerical rating has proved to be extremely problematic in that the television industry has neither the capability nor perhaps the willingness to review all its programming for violence and sex.

Pressure for the V-chip had come from family groups and from antiviolence and gun-control groups that wanted to screen children from certain television programming. Lobbying groups such as Action for Children's Television also played a role. In 1996, the Telecommunications Reform Act included a provision mandating that every television with a screen 33 cm (13 in.) or larger be equipped with a rating system–based "blocking device" by January 1997. At this point, a four-category rating system had been defined: TV-G (suitable for all audiences), TV-PG (parental guidance advised), TV-14 (suitable for ages 14 and up), and TV-M (suitable only for "mature" audiences). In addition, these labels sometimes include the letters S, V, L, or D, which stand for sex, violence, coarse language, and suggestive dialogue, respectively.

Despite at least grudging acquiescence by the television industry, significant personnel turnover within the Federal Communications Commission resulted in a lengthy delay of the implementation of V-chip technology. Meanwhile, Canada's more restrictive broadcast system enabled that nation to take the lead in the application of V-chip programming.

In the United States, the widespread use of cable television makes introduction of a V-chip–based system difficult. More than 100 channels now broadcast 24 hours a day, resulting in 2,400 hours of programming every day that would have to be screened, rated, and programmed. Moreover, there would be difficulties with break-in news programs and live violence. Nevertheless, General Instruments has offered a technology based on a movie rating system on its set-top boxes, although the circuit inside was not the V-chip but a competitor. This technology is available only if a consumer subscribes to the premium cable service.

The private sector already has introduced controllers with a edit-programming capability. In 1995, the Zilog Corporation of California introduced a television controller that contains "V-chip-type" technology. And many people have argued that the technology has existed since the creation of the television itself: the set's On/Off switch.—L.S.

See also FEDERAL COMMUNICATIONS COMMISSION; VIDEOCASSETTE RECORDER

Further Reading: "A Public Policy Perspective on Televised Violence and Youth," *Harvard Educational Review* (Summer 1995): 282–91.

Vaccines, Polio

Poliomyelitis (known in the past as *infantile paralysis*) is a potentially crippling disease caused by a virus. The virus enters through the mouth and multiplies in the throat and intestine. It may then attack the motor neurons of the spinal cord and part of the brain, resulting in paralysis of the legs. The virus may also attack the motor neurons that control the diaphragm, and respiratory paralysis may ensue. In this case, the patient may be unable to breathe, and he or she may have to be confined permanently to a respirator. Poliomyelitis (or polio for short) was a major health problem until fairly recently. In the United States, 57,000 cases were reported in 1952, the year of maximum incidence in that country. Of these, half resulted in paralysis.

In 1909, Karl Landsteiner (1868–1943), an Austrian medical researcher, suggested that polio might be caused by a virus. More than 2 decades later, Australian researchers discovered that there were two strains of polio-causing virus. Considerable knowledge was gained about the causes of polio, but attempts to produce an antipolio vaccine were so unsuccessful that many medical researchers believed that a vaccine for polio was an impossibility. Polio, however, was a disease that elicited strong emotions; it was no respecter of social class or ethnic group, and it disproportionately affected children. Polio had also afflicted a President of the United States, Franklin D. Roosevelt. For these reasons, it was easy to mobilize public sentiment and to generate substantial sums of money for polio research.

An important result of concerted research came in 1949 with the successful effort of John F. Enders (1897–1985), Thomas H. Weller (1915–), and Frederick C. Robbins (1916–) to grow a strain of polio virus in cultures made from tissues other than nerve cells. Until this time, most researchers firmly believed that only nerve cells—spinal cords from monkeys were most commonly employed—could be used to cultivate the polio virus. What made the use of other tissues possible was the availability of penicillin and other antibiotics to prevent the growth of bacteria that had ruined previous efforts to grow viruses. The use of cultures based on nonnervous tissues allowed the culti-

vation of massive quantities of polio viruses for experimentation.

Two other important discoveries were made at about this time; a third strain of polio virus was discovered, and polio virus was discovered in the bloodstream. The latter discovery meant that it might be possible to fight polio by injecting antibodies that killed the virus before it invaded the nervous system. This was accomplished by Jonas Salk (1914–1995), who through a great deal of experimentation was able to determine the rate at which to kill the polio virus (using formaldehyde), so that when injected into the bloodstream it stimulated the production of antibodies without posing a threat to the patient. On Apr. 12, 1955, Salk used a tumultuous news conference to announce that extensive field trials indicated that the killed-virus vaccine was 80 to 90 percent successful in preventing polio, and that an even better performance was likely in the near future. The U.S. Public Health Service licensed the vaccine for use on the same day, and except for one tragic instance of 260 cases of polio being caused by improperly treated vaccine, it was a great success. By 1961, the number of polio cases in the United States had dwindled to 1,312.

The Salk vaccine was not the only successful approach to the eradication of polio. Albert Sabin (1906–1993), another American researcher, had long advocated the use of live polio virus for conferring immunity. By 1957, he was able to produce live vaccines for each strain of polio. These were soon administered widely in Eastern Europe and began to be used in the United States in 1961. Live viruses could be used because they were of a type that was too weak to bring on the disease. The Sabin vaccine had the advantage of conferring immunity for a longer period than the killed-virus vaccine, and it could be taken orally. It is the most commonly used vaccine today.

By building on the accomplishments of past researchers, both Sabin and Salk were able to develop vaccines that effectively removed the scourge of polio from much of the world. Unfortunately, a bitter rivalry between the two men clouded their accomplishments. Both were convinced of the correctness of their approach and the inherent error of the other's, and they were not reluctant to make their feelings public. Still, their long-running squabble should not be allowed to diminish their legacy, the near-eradication of polio in the industrialized nations of the world and the promise of its eventual elimination worldwide.

See also PENICILLIN; RESPIRATOR; VIRUS

Vacuum Cleaner

Until fairly recently, keeping carpets clean required regular wielding of a broom or perhaps a manually operated carpet sweeper with rotating cylindrical brushes. These simple implements had limited effectiveness; to get out all of the ground-in dirt, it was necessary to take the carpets outside and beat them vigorously with a stick or carpet beater. In the late 19th century, a number of devices were invented to make carpet cleaning less of a chore. These sucked up dirt by using a bellows to create a vacuum. Since arms or legs supplied the power to the bellows, there was little if any saving of energy over the use of a broom. One such device, a "pneumatic dusting machine" that appeared in 1893, required two people, one to operate the bellows and the other to manipulate the cleaning head. Combining the functions so that one person could handle them was a challenge. One French inventor received a patent for a cleaner equipped with bellows mounted to the operator's feet so that a suction was created as the operator walked through the rooms being cleaned.

An effective vacuum cleaner needed a source of power other than what its operator was able to provide. Steam was the dominant source of industrial power in the late 19th century, and in 1871 it was used to power an "aspirator" invented in the United States by Ives W. McGaffey. Although it was heavy and bulky, these were not impossible drawbacks for a machine that was intended for industrial use.

Another large vacuum cleaner was invented by Hubert Booth in 1901. Mounted on a horse-drawn cart, it used a suction pump powered by a 3.7-kW (5-hp) gasoline engine and connected to a 243-m (800-ft) hose. Booth's vacuum cleaner proved its mettle by cleaning up debris left from the rehearsal of the coronation of Edward VII, and two were subsequently bought by the British royal family, one for Windsor Castle, the other for Buckingham Palace. While Booth's machine was nominally portable, many other vacuum cleaners were

permanently installed in the basements of apartments and other large buildings. Ducts connected the cleaner to outlets in the walls, and a hose was inserted in an outlet when it was time to clean.

Truly portable vacuum cleaners were made possible by the development of small electric motors. One of the first electrically powered vacuum cleaners was invented by Murray Spangler, who subsequently sold the patent rights to a harnessmaker named William H. Hoover (1849–1932). Selling for $75 when it was first marketed in 1908, it was so successful that even today in some parts of the world "hoovering" is synonymous with vacuuming.

The tank vacuum cleaner, a type still in widespread use today, was devised in Sweden by Axel Wenner-Gren (1881–1961) and put on the market by the Lux company (later Electrolux) in 1913. Weighing only 6 kg (13 lb), it used a fully enclosed fan to create a powerful suction. It also had the advantage of being reversible; by changing hose connections the machine could be made to blow, allowing it to be used for spraying and similar applications.

During the 1920s and 1930s, the vacuum cleaner was one of a number of electrically powered domestic appliances that were transforming everyday life. Before long, vacuum cleaners were considered to be necessities rather than luxuries. In 1941, 47 percent of the homes in America had them, even though 20 percent of American homes still had no electrical service.

Although using a vacuum cleaner required less physical effort than older means of cleaning carpets, the overall effect was not simply the saving of labor. Vacuum cleaners actually may have increased household labor, for with their use came higher standards of cleanliness; a weekly vacuuming instead of a yearly spring cleaning became the expectation. The vacuum cleaner also altered gender roles in the performance of household tasks. Unlike the days when men often had the chore of taking up and beating carpets, carpet cleaning had become an indoor household chore, and as such it tended to be defined as "women's work."

See also ENGINE, FOUR-STROKE; MOTOR, ELECTRIC; REFRIGERATOR; STEAM ENGINE; STOVE, COOKING

Further Reading: Ruth Schwartz Cowan, *More Work for Mother: The Ironies of Household Technology from the Open Hearth to the Microwave*, 1983.

Velcro®

In 1948, a Swiss engineer named Georges de Mestral returned from a walk in the woods to find that his clothes had a number of burdock seeds sticking to them. Curious as to why they stuck so tenaciously, he put one under a microscope. He found that it had tiny hooks on its surface that engaged the fabric of his clothes. This was the inspiration for hook-and-loop fasteners, or Velcro, a derivation of the French *velours* (meaning *velvet*) and *crochet* (meaning *hook*). As often happens with successful technologies, a great deal of development work was necessary to transform an inspiration into a marketable product. Eight years passed before Velcro was produced in commercial quantities.

A Velcro fastener consists of two pieces, one with tiny loops, the other with even smaller hooks that engage the loops when the pieces are brought together. Manufacturing the loops was simple, but the hooks presented a problem. This was eventually solved by first forming loops in the supporting fabric and then cutting them to form hooks. Velcro is usually made from a nylon-polyester blend. Two pieces of Velcro made from this material require a force of 10 to 15 pounds per square inch to separate them. Stronger forms of Velcro are made from other synthetic fibers and even stainless steel.

Velcro has many applications today, for it can be used just about anywhere that something has to be held fairly tightly yet be easily removable. Velcro has been particularly useful to the space program, for it keeps things in place that would otherwise drift around in a zero-gravity environment.

See also NYLON

Vending Machine

The original vending machine dates back to the 1st-century C.E.. It was a coin-operated water dispenser invented by Heron (or Hero) of Alexandria (c. 50–c. 120). This device was placed in a temple, where a small amount of water for mutual washing was dispensed after the deposit of a 5-drachma coin. Heron's machine was built around a beam-actuated valve. When a coin struck a pan at one end of the beam, the beam's upward movement opened the valve. The coin then slid off the pan into a receptacle, allowing the other end of the beam to swing down and close the valve.

Many centuries passed before vending machines made another appearance. In the 1600s, some English taverns had coin-operated boxes that sold tobacco and snuff, and more than a century later, Richard Carlile, an English bookseller, attempted to circumvent English restrictions on the press by creating a machine that could sell books to customers on an anonymous basis. A coin-operated stamp dispenser was introduced in London in 1852, but it was vulnerable to the use of slugs (fake coins). Also in England, Percival Everitt patented a coin-operated vending machine in 1882, and in 1887 R. W. Brownhill received a patent for a coin-operated gas meter. The basic principles had been demonstrated, but it took many years of development before vending machine mechanisms reached an acceptable level of reliability.

Credit for the foundation of the American vending machine industry goes to Thomas Adams, the owner of a chewing gum business. In 1888, Adams had machines installed on New York City's elevated train platforms to help sell his Tutti-Frutti gum. The machines used a lever that accepted a coin and then allowed a hand crank to dispense one piece of gum. As vending machines became more popular, the deposit of slugs posed a serious problem. The defeat of slugs required the invention of a number of mechanisms to check the diameter, thickness, weight, alloy composition, and magnetic properties of the coins put into the machines.

While mechanical innovations made the modern vending machine possible, no less important were changed attitudes about buying and selling. Beginning in the 1920s and 1930s, vending machines were one component of a commercial revolution. This was a time in which self-service was becoming the dominant mode of retailing; in a growing number of establishments, customers were expected to help themselves instead of being waited on by a clerk. The widespread use of vending machines presupposed the presence of a buying public that was accustomed to impersonal retailing and was willing to entrust its money to a machine. This culture also created a receptive environment for jukeboxes and, much later, automated teller machines.

World War II caused the government to halt production of vending machines, but sales revived after the war when factories placed thousands of the machines in cafeterias or near the shop floors. In 1945, sales of vending machines rose from $500,000 to $1.5 million by 1953 (in constant dollars), with more than 300,000 machines installed throughout the United States. The invention in 1950 of a new slug rejecter that used a series of magnets made the machines even more popular.

By that time, a clear distinction had developed between manual and automatic machines. The manual machines used a crank turned by the customer that dispensed gum or candy. Automated machines used electronic devices to certify that the money was genuine, to activate the main control circuit board with an electronic pulse, and to allow the customer to make selections based on the deposit of cash. As computer and electronic technology improved, vending machine sales grew more sophisticated, allowing the dispensing of a variety of products, accepting $1 and $5 bills, and even making change. Vending machines have proved especially useful for dispensing tobacco, which is forbidden to minors, who still have little trouble purchasing tobacco products from unmonitored vending machines.

A typical vending machine costs $1,200 to $6,000, and a vending machine route can service 50 to 100 machines. The requirement of having a high number of machines on a route has posed a barrier to entry for the industry, even though the individual machine cost is relatively low. Consequently, while there were an estimated 1.75 million vending machines in the United States in 1994, they were owned and serviced by only about 7,000 companies. The amount of money taken in by these machines is impressive. In the mid-1990s in the United States, more than $23 billion went into the slots of vending machines that dispensed everything from computer software to condoms.—L.S. (with P.J.)

See also AUTOMATED TELLER MACHINES; JUKEBOX

Further Reading: G. R. Schreiber, *Concise History of Vending in the U.S.A.*, 1961.

Video (Computer) Games

The first videogame, a simulation of tennis, was invented in 1958 by Willy Higinbotham, a physicist at the Brookhaven National Laboratory. Although the game aroused some local interest, no attempt was made to commercialize it, and Higenbotham didn't even bother to take out a patent. The history of the commercial videogame industry begins in 1972, when a small American company named Atari created "Pong." After being installed in a California bar, the machine took in up to 10 times as much money as the pinball machine standing next to it. Pinball manufacturer Bally's Midway bought the rights to the game and distributed it, and in 1974 the companies produced more than 100,000 Pong games (although Atari manufactured only 10,000 of them). Atari then attempted to move into the home videogame market with Pong, using Sears as the outlet. Television manufacturer Magnavox also entered the home videogame market in the early 1970s.

The computer revolution brought a sharp drop in the price of microprocessors and computer chips, leading to a new generation of arcade games. By 1980, Americans were spending up to $6 billion a year on arcade games, despite numerous city ordinances limiting the hours of operation or restricting the play of teenagers at them. Home videogames, led by Atari's new 2600 series and gamemaker Mattel, brought in $1 billion a year by 1981. Atari sold 20 million units of the 2600, which had 1,500 different games available. It manufactured an astounding 6 million copies of a new game, "ET: The Extraterrestrial," at the very time that the industry seemingly had saturated the market.

By 1983, however, videogame sales had trickled to $100 million a year. Although several factors contributed to the collapse, including Atari's inability and unwillingness to develop new technology and new generations of game machines, the main problem was an excess supply of games. Prices plummeted, and virtually overnight Atari was out of business. Mattel nearly went bankrupt; Coleco survived only because it tapped into a new fad, "Cabbage Patch Kids." Atari sold its hardware divisions and abandoned the market, leaving a bad taste in the mouths of American videogame consumers.

As Atari and its imitators collapsed, a new giant had risen in Japan, Nintendo Corporation. Nintendo, which originated as a greeting card company, had moved into electronic gun games and had negotiated a license to manufacture and sell a videogame system by Magnavox, which played the popular Pong game. In 1977, Nintendo unveiled its own game of video tennis, and it experimented with the miniaturized computer technology used in pocket calculators. By the early 1980s, more than 20,000 game units a year were sold in Japan, but a huge American market still loomed. Nintendo developed a new system called Famicom, which sold more than a million, yet still did not make substantial inroads into the United States until it released a set of games called "Donkey Kong" and "Super Mario Brothers" in the mid-1980s.

Nintendo introduced its 8-bit system in 1985. Contained in a small gray box, the system allowed the player to compete against the computer at varying levels of difficulty or against a human opponent on a second controller. Consumer demand for the system soared between 1988 and 1990, and as ownership of game machines rose, the demand for games increased apace. By 1990, more than 70 licensees sold millions of copies of games manufactured by Nintendo. The Super Mario Brothers series had become wildly popular, and Nintendo carefully avoided the mistakes of Atari by undersupplying orders for new games, ensuring against the glut that doomed its predecessors.

Nintendo's success spawned a powerful competitor, SEGA, another Japanese company that introduced a 16-bit machine in the early 1990s. Although Nintendo responded with its own 16-bit machine, SEGA gained the upper hand among teenagers with its sports games and its bloody "Mortal Kombat," the biggest selling videogame in history. Electronic game hardware sales exceeded $1.5 billion in 1994, a year that found 30 million game machines in American homes. Yet there was no indication that the market had peaked. In 1995, SEGA introduced a new 32-bit compact-disk system, Saturn, and later that year another competitor, Sony, entered the competition with its own 32-bit system, Playstation. Nintendo, once the industry leader, was the only major company without a 32-bit system on the market.

Videogame technology, however, has vast room for further improvement. New virtual-reality systems with three-dimensional perspective loomed on the horizon of the $5 billion industry, and the personal computer (PC) gaming business had developed as a massive industry of its own. As if to satirize itself, game manufacturer Acclaim put a hidden game of Pong within its new, all-time, best-selling game Mortal Kombat II.

The effects of videogames, especially on young people, continue to be debated. Videogames may reinforce gender differences, for the vast majority of game players are male, and the design of most games seems to ensure that they will have little appeal to girls and women. Playing videogames may sharpen hand-eye coordination, but some critics feel that the more violent games may encourage real-world violence. Such assertions are difficult to prove, however; even after decades of research, there is still no consensus regarding the effects of televised violence.—L.S.

See also COMPUTER, PERSONAL; MICROPROCESSOR; TELEVISION AND VIOLENCE

Further Reading: David Sheff, *Game Over: How Nintendo Conquered the World*, 1994.

Videocassette Recorder (VCR)

Since the appearance of the first motion pictures for a mass public in theaters, there existed a natural progression of attempts to put moving images on film played by equipment available to individual consumers. One breakthrough came with the development of recording tape, which had evolved through a number of stages in Germany and was appropriated as "spoils of war" by the Allies. Several GIs returned from Europe with the technical expertise and went into the tape recorder business.

In theory, there was little difference in an audio recording and a video recording, except that recording a television image was much more complicated (4 million cycles per second for video compared to 20,000 cycles per second for audio). The available technology was similar to that found today. Recording tape consisted of a base made of plastic, a metallic powder, and a coating that sealed the coating to the base. The tape was pulled through a recording electromagnetic head that produced varying fluctuations of intensity or polarity in response to the electrical input from a microphone. The recording process left a trail of magnetism on the tape; when the process was reversed, the magnetized particles on the tape played back to the head, which then sent the resulting signal to the amplifier and then to the speaker.

Early attempts to adapt audiotape to videotape involved speeding up the audiotape process, a method that was used at the Bing Crosby studios in 1951. Subsequent efforts continued to focus on speeding up the audiotape process, resulting in a 1953 demonstration in which 1.6 km (1 mi) of tape was needed to record a 4-minute recording, and requiring the demonstration engineer to wear heavy leather gloves to brake the searingly fast tape reels. Engineers at Ampex Corporation used a different strategy that used a head that moved (as well as the tape), which allowed the designers to slow the tape speed. Further refinements used two or three heads, which solved the speed problem but created new difficulties because the signal had to be shifted from one head to another hundreds of times a second. Eventually the engineers met the challenge by moving the heads to different positions, but the "tracking" problem continued to afflict VCRs throughout their early history.

Ampex introduced the first operational videorecorder in 1956—the first machines were called VTRs for "video tape recorders"—but it would be years before the cumbersome, expensive devices were turned into true consumer items through the application of transistors. In 1960, Sony signed an agreement with Ampex to supply transistorized circuits in return for the right to make VCRs for nonbroadcast customers. Shortly thereafter, Ampex entered into a joint venture with Toshiba. By the early 1970s, the unwieldy 7.6-cm (3-in.) magnetic tape had been replaced with 1.3-cm (.5-in.) tape that made possible consumer VCRs. Manufacturers originally thought that the major use of a VCR would be in playing Hollywood movies that had been transcribed onto tape. Home-recording features were considered only a sidelight, but consumers quickly surprised the producers by using VCRs to "time shift," i.e., to record a program for viewing at a later time or date. Thus, the machine that offered the longest recording time would have a significant advantage.

Two competitive formats arose. Sony introduced the Betamax in 1975, while Japan Victor Corporation (JVC) introduced the Video Home System or VHS format. While the machines had similar technical histories, they used different sizes of cassette tapes, incompatible coding schemes, and substantially different internal mechanisms. The cassette tapes were not interchangeable, generating a "format war." The chief distinction was in the length of recording time, where VHS was almost double that of Sony. Recording capabilities steadily increased the amount of time available on a tape, from 2 hours to 6 hours, but with each improvement made by Sony, VHS was able to exceed the recording time available. Even though critics agreed that the Betamax image was clearer, the time differential made the VHS more popular. In 1978, sales of the Sony product fell behind sales of VHS-format machines. In the years that followed, Betamax was caught in a vicious circle; because the number of machines was relatively small, video rental shops stocked few prerecorded tapes for that format, further limiting its market appeal. Betamax disappeared in the late 1980s.

Even before a single standard was in place, video rental stores had reduced the likelihood that consumers would buy Hollywood movies, especially at the astronomical prices ($100 for *Star Wars* in the early 1980s) demanded by the producers. Instead, the video rental business itself took off, further enhancing the sales of VCRs. As consumers demanded more movie rentals, and as producers received royalties from rentals, an entire secondary market for first-run movies appeared, with profits calculated not only by what a picture would bring in the theaters but in the home rental market as well. With video rental stores, especially the large Blockbuster Video chain, carrying so many movies, the purchase price of movies fell to as little as $10, while "previously viewed" movies sold for as little as $6. By the mid-1990s, video rentals accounted for about 54 percent of the movie studios' revenues. Through rentals, Hollywood reaped substantial profits through a remarkably circuitous route—one that the industry had resisted at first.

VCR technology continued to improve, adding clearer slow-motion, fast-forward, fast-reverse, scan, and still-picture capabilities. VCR sound technology also improved by incorporating stereo hookups. In the late 1980s, an Arizona company developed a dual-cassette deck that allowed direct taping from one VCR to another in the same unit. After a series of copyright lawsuits, the company was successful in offering the machine for recordings for "home use only," not for resale. Standard VCR prices fell to as low as $200, and by the mid-1990s 80 percent of American households possessed a VCR.—L.S.

See also AMPLIFIER; MICROPHONE; MOTION PICTURES, EARLY; STEREOPHONIC SOUND; TRANSISTOR

Further Reading: James Lardner, *Fast Forward: Hollywood, the Japanese, and the VCR Wars,* 1988.

Videodisc Player

In 1965, RCA (Radio Corporation of America), flush with its success in color television, embarked on a new round of technical development with its foray into the videodisc. At that time, color television was a $3 billion industry that only recently had achieved commercial success after its technical development in the 1950s. RCA management sought to sustain the enthusiasm for the color television with a new device that made use of television technology; one result was the videodisc player. Videoplayer systems utilized forms of audio systems, namely magnetic tape or records, both of which were available in the 1950s. Unlike audio, however, video image storage and retrieval required the processing of far more information than that stored on a long-play (LP) record.

Other companies moved into videotape technology, and RCA experimented with tape, particularly in Beta and VHS formats. RCA was not even the first to utilize a videodisc approach, but it was the first to use the information storage and retrieval technology. More important, RCA was capable of putting out an extensive catalogue of programs, a shortage of which plagued other producers. Called Selectavision, RCA's videodisc utilized a 9.1-kg (20-lb) fully enclosed electronic player as the "hardware." The "software" consisted of the videodisc itself, a 30.5-cm (12-in.) vinyl disc that the player spun at 450 rpm, while an electrode on the end of a stylus that rested in the disc groove sent the disc's signal through solid-state circuitry to the player. There, the signal was decoded and

displayed on the television monitor. The user initiated the sequence by inserting the disc in its plastic cassette or "caddie," and these were then put into the disc slot in the front of the player. The disc was extracted from the caddie once it was inside the player.

Each of those technologies embodied new research and engineering, some radical and some well known but pushed to their extremes. RCA hoped that the result would be a picture of superior resolution to any sent over the airwaves, and superior even to magnetic-tape pictures. Most important, the system offered the consumer, for the first time, the opportunity to purchase Hollywood movies and play them at home, free of commercials or network scheduling. But the company knew it had to sell the system for $500 or less, or the product would die. The system had other constraints as well: Most discs at the time could hold only 2 hours worth of material, and some movies ran longer than that.

Research had passed through several phases, including a 1969 demonstration of Holotape, which had shown the technology to still be lacking in crucial areas, and the crucial renewed emphasis on needle-and-disc technology following the European formation of a joint venture to make videodisc players in 1970. At the same time, the company increased its efforts to acquire and record movies, while at the same time making plans to produce its own, "made for Selectavision" movies. In 1975, RCA demonstrated a working videodisc system to the public, and almost immediately other competitors announced their own new disc and tape systems. Philips, in particular, had introduced an optical laser system that technically was superior to Selectavision but was much more expensive than RCA, which itself had to get disc production costs much lower ($2 for a $14 retail-priced disc).

RCA introduced Selectavision in 1981 amid great hoopla, with internal projections showing that the system could become a $7 billion business by 1990. Along with achieving success in this particular technology, the videodisc promised to restore U.S. competitiveness in consumer electronics. But quickly the weaknesses of the videodisc became obvious, especially its inability to record and store programs for future viewing. Indeed, the single most utilized function of the new videocassette technology was in "time shifting," wherein viewers taped programs for viewing at their convenience. Moreover, the purchase of movies for home viewing had not become as popular as it would a decade later, further eroding the usefulness of the videodisc. Finally, their lower costs relative to VCRs were outweighed by the versatility of the latter, whose sales quickly shot past those of videodisc systems. Total sales of videodisc systems totaled 550,000 after 3 years; in comparison, the installed base of VCRs exceeded 20 million by 1985. Although the original videodisc systems disappeared, new digital disc players appeared in the early 1990s with good consumer response.—L.S.

See also RECORDS, LONG-PLAYING; TELEVISION, COLOR; VIDEOCASSETTE RECORDER

Further Reading: Margaret B. Graham, *RCA and the VideoDisc: The Business of Research*, 1986.

Videophone

A videophone is a communications device in which a telephone, a videocamera, and a videoscreen are combined to enable two-way (duplex) audio and video communication via the existing telephone network of transmission lines and circuit switches. It requires a second video telephone at the other end of the circuit in order to exchange video images, but it can be used as a conventional telephone as well.

The first video telephone connected to a telephone line was demonstrated in 1927 in New York City. Invented by Dr. Herbert Ives at Bell Laboratories, an experimental video imaging system in Washington, D.C., transmitted relatively crude images (50 lines transmitted at a rate of 17.7 frames per second) of Herbert Hoover (then Secretary of Commerce) to Walter S. Gifford, president of AT&T. Although voice conversation was two-way, the only video display was in New York.

Experiments with a "picture phone" began at Bell Telephone Laboratories in 1956, using a small videoscreen about 5 by 7.6 cm (2 by 3 in.) on each device. Eight years later (Apr. 20, 1964), the first transcontinental Picturephone call was made between the Bell System exhibit at the 1964 New York World's Fair and Disneyland in California. On June 24, Picturephone service was inaugurated among three telephone network switching centers in New York, Chicago, and Washington, D.C. The first 3 minutes of a call cost between $16 (New York–Washington, D.C.) and $27 (New York–Chicago). A year later, the charges were cut in half.

In 1970, AT&T offered Picturephone service to commercial and industrial customers at a charge of $160 per month for the lease of the equipment, plus a separate charge for each call. The first citywide system was installed in Pittsburgh, Pa., in conjunction with the Westinghouse Corporation. After 3 years of marketing, the service was terminated due to a lack of sufficient public interest.

A Picturephone installation required a touchtone telephone and a special integrated videocamera/video display receiver equipped with a microphone, a loudspeaker, and a cathode-ray tube (CRT) having an image face of 12.7 by 14 cm (5 by 5.5 in.). A special wiring link—a six-wire copper feeder line—was needed between the Picturephone set and "selected" central office switches. Voice-only calls could be made to nonvideophones, but the calls were charged as if for Picturephone service. The video image used 250 interlaced lines. Typed text was described as having "marginal legibility."

The first motion-and-full-color video telephone used over ordinary two-wire phone lines was the AT&T VideoPhone 2500, announced in 1992 and now sold in more than 30 countries. However, the public has been slow to accept it, although the videophone is easy to use. It plugs into a conventional phone jack and requires no special installation. In addition, it does not incur any special usage charges; the video and audio signals are transmitted simultaneously along the usual voice channel path. However, image quality is blurrier than that of broadcast TV, and the number of frames per second is fewer than that of video conferencing, so movement is noticeably jerky. Some industry analysts contend that widespread public acceptance of videophones will come only if the image quality is comparable to commercial television, which requires the capacity of 1,400 phone channels.

The Videophone 2500 is equipped with a liquid-crystal display (LCD) screen having a diagonal of 8.4 cm (3.3 in.). Although the set is used on ordinary analog telephone lines, it is technically a digital device. Digitized video data, audio data, and supervision information are combined into a multiplexed digital signal that is prepared for analog transmission by a built-in high-speed modem that supports data rates of 19.2 and 16.8 kilobits per second.

Because tests with potential customers indicated significant concern about privacy, the camera lens is covered with a shutter that is moved aside by the user when he or she wishes to transmit an image. An additional provision for privacy is a button labeled VIDEO, which must be pressed before the user's image can be transmitted. The content of the video image being sent can be checked by pressing SELF VIEW without interrupting the transmission, so the user can see the same picture as the other party to the conversation.

Other types of video telephones for digital signals carrying both video and audio are designed for use with the Integrated Services Digital Network (ISDN) basic rate interface (BRI) lines offered at extra charge by many telephone companies. These digital transmission lines provide a total capacity of 144 kilobits per second, allowing transmission of better image and audio content but at higher monthly charges.—R.Q.H.

See also LIQUID-CRYSTAL DISPLAY; MULTIPLEXING; TELEPHONE; TELEVISION

Virtual Reality

Virtual reality (VR) is a computer-based technology that situates the user in a simulated environment where he or she perceives and interacts with objects that do not really exist. These objects are seen and felt through the use of special goggles, gloves, and other apparatus that produce sensations similar to those provided by the objects being represented. The term *virtual reality*, which was coined by Jaron Lanier (1954–) in the late 1980s, is not universal; some researchers prefer *virtual environments* or *synthetic environments*.

Virtual reality entered public consciousness in the early 1990s, but some aspects of VR have been around for many decades. An early example of a virtual-reality technology is a flight simulator known as the Link Trainer. At first primarily employed as an amusement park ride, the Link Trainer was widely used to train pilots during World War II. While the pilot trainee was seated in a space resembling an airplane's cockpit, the trainer mimicked certain experiences of flight; as the "pilot" moved the stick and rudder pedals, the trainer yawed, rolled, and pitched in much the same way that an actual airplane would behave. Advanced versions of the trainer also included instruments that gave trainees the experience of "flying blind." Sophisticated descendants of the Link Trainer are widely used today; they constitute a major application of VR technology, although the term is not generally used in this particular context.

Virtual reality has the ability to draw the viewer into a simulated environment because the virtual image appears to have three dimensions, and the relative positions of the objects change as the user moves his or her head. This occurs because the viewer is presented with two slightly different images, one for each eye, that are processed into a single three-dimensional image. In the 1950s, this application of stereoscopic vision enjoyed a brief vogue in the form of 3-D movies, which also might be considered precursors of virtual reality. Enhanced visual experiences were the goal of Morton Heilig, who in 1960 received a patent for a "stereoscopic television apparatus for individual use," a device that eventually became a key component of VR technology.

A seminal event in the history of VR was Ivan Sutherland's 1965 publication of an article entitled "The Ultimate Display," which he followed in 1968 with "A Head-Mounted Three Dimensional Display." These articles described the research conducted at MIT's Draper Laboratory and provided a foundation for subsequent research into VR. A few years later, Sutherland and his associates, now at the University of Utah, created the first head-mounted VR displays. A considerable portion of early VR research was funded by the Department of Defense's Defense Advanced Projects Agency, the Office of Naval Research, and the Bell Labs. The military, and especially the U.S. Air Force, continue to be major sponsors of VR research. Military-sponsored VR technology was used for the training of pilots and tank crews in preparation for Gulf War combat missions.

Today's VR systems include many of the elements initially described by Sutherland. Visual images appear on small television screens directly in front of each eye. The display may be created by a cathode-ray tube or a liquid crystal display. Tactile sensations are transmitted by means of a glove containing bladders and rods that press against the hands of the wearer. The glove is also equipped with a tracking mechanism, usually optical fibers that detect movements of the fingers, while other position sensors attached to the wrist detect movements of the hand as a whole. The glove may appear as a picture on the screen, and this image moves as the wearer's real hand moves. Data supplied by the tracking mechanisms are fed into a computer, which then generates the requisite images and sensations. In similar fashion, a tracking sensor is attached to the wearer's helmet, and as the wearer moves his or her head, the appearance of the simulated objects will change. Perspectives are altered as the viewing angle changes, and the relative size of objects changes as the viewer "approaches" and "recedes" from the scene.

The rapid advance of VR technology in the 1990s has been stimulated by massive increases in computing power, improvements in computer programming, and the development of high-resolution visual displays. By the early 1990s, a number of commercial VR systems were being marketed. However, the realism of virtual reality is still somewhat limited. A highly realistic rendition of a changing landscape would require the generation more than 30,000 polygon shapes per second, and these would have to be updated at a rate of six

Helmet, goggles, and gloves used in a virtual-reality system (from B. Yenne, *100 Inventions*, 1993).

times per second. This is beyond the capability of the most powerful supercomputers. There are also problems with time lags between the users' movements and their representation in virtual reality.

Despite the imperfections of current technologies, VR has already found a number of applications. Entertainment continues to be a major application, and a number of commercial arcade games are enlivened by virtual reality. VR technology also is used for a number of more serious uses. Some Japanese department stores now offer virtual tours of home interiors that allow prospective customers to experience different kitchen layouts so that they can select a kitchen plan and fixtures that will best meet their needs. Similarly, VR can be used in architectural planning so that the interior space of a building is well suited to the building's intended purpose. VR also has been employed in the design of airplane instrument panels so that pilots can try out different arrangements of instruments and controls before anything is actually built.

Virtual reality has a number of actual and potential applications in education. For instance, students can perform virtual science experiments that go well beyond what would be possible in a high school or even university laboratory. Virtual reality also makes possible educational enhancements like virtual tours of art museums and historic places.

Medicine is becoming another important user of VR technology. Through VR, medical students can gain experience by performing virtual operations before they do surgery on real patients. Operations such as tumor removals may be made safer and more effective by having a surgeon first perform an "operation" on a virtual image that replicates a patient's organs and the tumor to be removed.

For all of its potential and actual contributions, some critics have been bothered by the artificial nature of virtual reality. From their standpoint, VR is the latest in a series of technologies that have separated individuals from nature and from other human beings. Virtual technology is therefore condemned for the role it may play in blurring the distinction between what is real and what is not, and condemned because it may divert people from the pursuit of authentic life experiences. Similar charges have been leveled at other technologies, and it remains to be seen if virtual reality will be a continuation of existing trends, or if it marks the beginning of a new era in human life, work, and perception.

See also DEPARTMENT OF DEFENSE, U.S.; ERGONOMICS; INSTRUMENT FLYING; LIQUID CRYSTAL DISPLAY; MOVIES, 3-D

Further Reading: Ralph Schroeder, *Possible Worlds: The Social Dynamic of Virtual Reality Technology*, 1996.

Virus

Like someone charmed by a set of Russian nesting dolls, medical researchers have discovered ever-simpler lifeforms among disease-causing organisms: parasites, bacteria, viruses, viroids, prions, and so-called "naked DNA." From the bacteria, on up the evolutionary ladder, pathogens possess elements for their own replication; in short, they are alive. From the viruses, on the lower rungs of the ladder, pathogens are unable to reproduce without aid from a living cell. Consequently, viruses occupy a peculiar boundary between the living and the nonliving—not dead, but only potentially alive.

During the 23 centuries when "humoral theory" dominated medical concepts of disease, *virus* meant any poison, or literally, a "slimy liquid." Its relevance lay in harmfully altering the vital balance of blood, phlegm, black bile, and yellow bile that maintained health. When the germ theory of disease first challenged classical thought in the late 19th century, research physicians still used *virus* as a generic word for all pathogenic organisms, until Pasteur's colleague Charles Chamberland noted in 1884 that some disease-causing agents could pass through a porcelain filter that retained all known microorganisms. These invisible pathogens he termed "filterable viruses." Then, almost simultaneously, in 1892, Dimitri Ivanovski (1864–1920) in Russia and Martinus Beijerinck in Holland reported that a filterable virus transmitted mosaic disease in tobacco. Further, Beijerinck observed that the pathogen would not grow in a cell-free medium; it required living cells for its reproduction. Tobacco mosaic virus thereafter became a model organism for the independent discipline of virology.

With the discovery of the first virus in plants, investigators pursued the possibility that such pathogens existed among animals too. Friedrich Loeffler (1852–1915) and Paul Frosch confirmed that suspicion in 1898, finding that a filterable virus caused foot-and-mouth disease in

cattle, and probably also explained certain diseases in man, such as smallpox, that had close variants among livestock: the pox diseases of birds, cattle, swine, sheep, and goats. Indeed, without realizing it, Loeffler and Frosch touched on the source of all human viruses: mutations that permit a pathogen to transit from one species to another. In 1915, British bacteriologist Frederick William Twort (1877–1950) discovered that viruses also can infect bacteria—a class that Felix-Hubert D'Herelle (1873–1949) soon named *bacteriophage* (literally, "bacteria eater") or simply, "phage." More than 50 years later, researchers learned that a phage of *Corynebacterium diphtheriae* was capable of activating a "tox+" gene that caused diphtheria instead of the less-serious disease, croup, the bacterium more commonly inflicted on human beings

Throughout history, viruses have caused epidemics among people. Populations, either by their expansion into new territories or by their immunologic naiveté, provide opportunities for virus to spread. The Spanish introduction of smallpox into the New World, for example, caused the decimation of native Americans who had no immunity to Old World virus. Similarly, measles— a paramyxovirus—devastated armies on both sides of the American Civil War because soldiers (who as children on farms had not been exposed to the pathogen when their tolerance of its ravages was greater) succumbed when exposed as adults in crowded camps. Perhaps no disease better represents the crowding principle than the viral disease of hog cholera, which first appeared in the Ohio Valley in about 1833. Without an Old World incidence or an alternative host in the New World, hog cholera arose, most likely, from a mild or even asymptomatic swine virus of European origin that developed new mutations for greater virulence as it passed through herds amassed for the huge packing plants around Cincinnati (a.k.a. Porkopolis). Thereafter, hog cholera took an enormous toll on American farms until it was eradicated from the United States in 1978. Luckily, hog cholera never "jumped" species, to human beings, as did swine influenza, which broke out at the end of World War I and claimed about 25 million lives worldwide.

In the history of viral diseases, two momentous events occurred in 1977: the last natural case of smallpox, and the first case of AIDS. The World Health Organization realized that if it could find every individual on Earth with smallpox and vaccinate all people in contact with those cases, the disease could be eliminated, and so started the WHO Smallpox Eradication Program in 1966. Public-health officials were able to whittle down the case load, until on Oct. 26, 1977, they identified Ali Maolin in Somalia as the last reservoir for one of mankind's oldest plagues. In that very moment of triumph, WHO officials did not know that two homosexual men in New York City, diagnosed with rare Kaposi's sarcoma, probably were the first recorded cases of a developing pandemic, Acquired Immune Deficiency Syndrome. By 1981, AIDS had been identified as a disease, and 2 years later, researchers in France and the United States isolated the probable pathogen, HIV-1, a retrovirus that inserts its genes into the chromosomes of certain human lymphocytes that ordinarily amplify immune response to infection. Because HIV-1 appears to be related to lymphotropic retrovirus of African green monkeys, some epidemiologists speculate that the simian virus simply developed a mutation that opened the human reservoir.

Viral-caused diseases range in seriousness from the common cold and warts to infectious mononucleosis and mumps, to aseptic meningitis and adult leukemia, to the horrible hemorrhagic fevers called Lassa, Marburg, and Ebola. However, for more than 200 years, medicine has employed viruses for therapy, arguably beginning with Jenner's cowpox vaccination against smallpox in 1796, and continuing with killed, attenuated, or subunit virus vaccines for poliomyelitis, influenza, rabies, measles, mumps, yellow fever and hepatitis B. During the 1980s, molecular biologists conceived the idea of harnessing the ability of viruses to enter cells as a means for delivering new genes, either to correct inherited deficiencies or to give cells new characteristics. Research evolved, leading to the first human gene therapy experiment in 1990, in which clinicians at the National Institutes of Health used the Maloney Murine Leukemia Virus, first disabled and then reengineered, to convey three foreign genes (including one for the human enzyme adenosine deaminase) into the lymphocytes of a 4-year-old girl suffering from Severe Combined Immune Deficiency (due to hereditary adenosine deaminase deficiency). Since then, over 100 gene therapy experiments have proceeded using recombinant viral vectors for treating several hereditary disorders, AIDS, and cancer. Theoretically, all viruses, because of their tropism for certain cells or their means of reproduc-

tion have the potential of becoming swords beaten into plowshares.—G.T.S.

See also ACQUIRED IMMUNE DEFICIENCY SYNDROME (AIDS); BACTERIA; GERM THEORY OF DISEASE; INFLUENZA; SMALLPOX ERADICATION

Vitalism

Vitalism is the belief that life depends on the action of a *special factor* that guides the physical and chemical processes occurring within the organism. In other words, the functioning of living things cannot be reduced to nothing more than these chemical and physical processes. The invocation of a special factor, however, does not explain the process; no vitalist has managed to provide a theory of the factor detailed enough to provide explanations or predictions. Nevertheless, vitalists are of the belief that it interacts causally with the material organism. Call this factor the *Life* (with a capital *L*) of the organism.

What, then, is this factor called Life? All vitalists agree that:

1. It is a substance capable of causal transaction with matter.
2. Its effects on matter are necessary for life.
3. It is nonmaterial; or if material, it is assigned properties and powers that do not appear outside living organisms.
4. It shows purposive action.

The belief that living things possess a Life (in our sense) rests on two main unsophisticated sources. The first is linguistic. There seems little difference between "It's alive," and "It has a life." And there is in common parlance a great number of metaphors, idioms, and fashions of speech that bewitch us into thinking that life is like a substance: "He squeezed her life out," "Get a life!," and so on. Naturally, these usages of *life* do not entail the existence of Life, but they pervasively suggest and reinforce it.

The second source is a rich background of folklore, myth, and religious imagery. The most compelling example is depicted on the ceiling of the Sistine Chapel: Jehovah putting Life into Adam through their fingertips. Similar in concept if not sublimity are Dracula sucking Life with blood, and Frankenstein's patchwork man drawing Life from an electric storm. So Life can be identified with a substance or postulated fluid (this latter was popular in the 18th century) such as blood, the breath, animal magnetism, electric fluid, the Galenic bodily humors, or the "vital spirits."

By the late 19th and early 20th centuries, the battle lines of an intellectual war had been drawn, with vitalism on one side and mechanism, materialism, and reductionism on the other. The leader on the vitalist side was the German embryologist and Kantian philosopher Hans Driesch (1867–1941).

One famous skirmish in this war was begun by Wilhelm Roux (1850–1924), an embryologist and avowed mechanist. In 1888, he published an experiment that he claimed showed that an organism is a machine. He took a fertilized sea urchin egg after its first division and killed one of the daughter cells with a hot needle. The other continued to develop into one-half an urchin larva. He had destroyed, Roux claimed, one-half of the urchin-building machine.

In 1891, Driesch countered with his own experiment. After the first division of an urchin egg, he simply separated the daughter cells. Both grew into complete larvae, proving, he thought, that whatever machinery was present in the egg, half of the machine was enough to build an organism. (These two intellectual efforts strike us today as whimsical. Roux's claim—that you can't get the whole from a part—is contradicted by every gardener who has grown a whole plant from a leaf or a root cutting. So Driesch did not need his experiment to defend his thesis, although the two experiments injected some drama into the debate.)

The Life, on Driesch's view, is a nonspatial, nonmaterial, purposive, and organizing entity. Its actions are never in conflict with the laws of nature, which put constraints on what can happen. For example, I can choose or not choose to raise my arm. Either way, all the principles of thermodynamics, chemical combination, etc., will be satisfied.

Meanwhile, the debate over vitalism had been raging on other fronts. As early as 1828, the German chemist Friedrich Wöhler (1800–1882) had synthesized urea; this was a significant discovery, for until then it had been assumed that organisms were necessary for the production of organic substances. In 1844, Hermann Kolbe (1818–

1884), another German chemist, synthesized acetic acid from nonorganic substances, and in the 1850s the French chemist Pierre Bertholet (1827–1907) synthesized a variety of other organic compounds: methyl and ethyl alcohol, benzene, acetylene, and methane.

On the other side, vitalism received a boost when the research of Louis Pasteur (1822–1895) disproved the theory of spontaneous generation; the prior existence of Life seemed to be necessary for the creation of new Life. Pasteur's research into fermentation also strengthened the vitalist position, but only temporarily. In 1897, the German chemist Eduard Buchner (1860–1917) demonstrated that living yeast cells were not necessary for fermentation. Rather, it was a nonliving substance, an enzyme, that was responsible for turning sugar into alcohol.

Vitalism has never regained the following it had in the 19th century. At the same time, however, the expectation of the materialists that all living processes can be reduced to known chemical and physical principles has not been fulfilled. The scientific community has almost totally abandoned this old war, and we now have better ways of conceptualizing, for example, the issues that separated Driesch and Roux. Today, the claim to be a vitalist or materialist is likely to be met with yawn.—M.B.

See also ENZYMES; FERMENTATION; SPONTANEOUS GENERATION; UREA

Further Reading: Ranier Schubert-Soldern, *Mechanism and Vitalism: Philosophical Aspects of Biology* (trans. C. E. Robin), 1962.

Vitamins

Vitamins are organic compounds essential to proper bodily functioning and good health. They play a key role in regulating the metabolic processes through which an organism converts food into energy and tissues. Most vitamins serve this purpose by acting as enzymes, chemical catalysts that produce or speed up chemical reactions while themselves remaining unchanged. Other vitamins perform different functions, such as protecting tissues by serving as antioxidants.

Long before vitamins were discovered, it was known that nutritional deficiencies were the cause of certain diseases. In 1747, the Scottish physician James Lind (1716–1794) showed that scurvy could be prevented and cured by giving citrus fruits to those afflicted with the disease. Christiaan Eijkman (1858–1930), a Dutch physician living in the East Indies, found in 1887 that beriberi, a disease that caused paralysis and sometimes death, could be traced to a diet based on polished rice (rice with the hulls milled off). Eijkman thought that the disease was caused by a toxin in the rice, and that the toxin was somehow neutralized by the rice hulls. In 1901, his colleague Gerrit Grijns (1865–1944) determined that rice hulls prevented beriberi, not because they contained an antitoxin but because they held an essential nutrient.

In 1906, the English biochemist Frederick Gowland Hopkins (1861–1947) conducted nutritional experiments with mice that convinced him that a diet of fats, carbohydrates, and proteins was not nutritionally adequate, and that what he called "accessory food factors" were necessary for health and growth. In 1911, Casimir Funk (1884–1967), a Polish biochemist, suggested that these accessory food factors be given the name *vitamines*, a contraction of "amine compounds vital for life." In 1920, it was determined that not all of these substances were amines, and the *e* was dropped to give the present-day term *vitamin.*

The years that followed saw the identification of a growing number of distinct vitamins. These were given alphanumeric designations that indicated the chronological order in which they were discovered. These designations became the source of considerable confusion, and the trend now is to use their chemical names instead. The first vitamin to be identified as such was, logically enough, given the name vitamin A (it is now also known as *retinol*). It was discovered in 1913 by Elmer McCollum (1879–1967) as a result of nutritional studies conducted with white rats. McCollum found that the rats failed to grow when lard or olive oil was their only source of fat, but they did well when butter or eggs were added to their diet. McCollum used the name "fat-soluble A" for the substance he discovered, and gave the name "water-soluble B" to the antiberiberi substance previously identified by Eijkman. As it turned out, B was not a single vitamin but a group of vitamins. The B vitamin group encompasses B_1 (thiamine), B_2 (riboflavin), B_3 (niacin or niacinimide), B_5 (pantothenic acid), B_6 (pyridoxine), and B_{12} (cyanocobalamin).

Vitamin C (ascorbic acid) does not act as an catalyst but serves to regulate the formation of collagen, a protein that makes up the connective tissue in the body. Vitamin C proved to be the critical ingredient in the citrus fruits that were first used to cure scurvy in the mid-18th century. It was extracted from cabbages, oranges, and paprikas by Albert Szent-György (1893–1986) in Hungary, and from oranges and lemons by Charles G. King (1896–1988) at the University of Pittsburgh. The two reported results of their research a few weeks apart in 1932. It was synthesized a year later through the efforts of Walter Haworth (1883–1950), who determined its chemical structure, and Tadeus Reichstein (1897–1996), who did the actual synthesis. Rather oddly, human beings and other primates are almost alone among animals in their inability to synthesize vitamin C within their bodies; consequently they must obtain it from the food they eat.

Vitamin D is essential for normal bone growth because it aids the absorption of calcium and phosphorous from the intestinal tract into the bloodstream. What is known as vitamin D is actually a group of vitamins that are similar to cholesterol in their chemical structures. The most important of these is vitamin D_3, which is produced on the skin's surface under the influence of ultraviolet light from the sun. Vitamin D deficiencies may lead to rickets, a bone disorder that is manifested in bowed legs and spinal curvature. In 1921, the English biochemist Edward Mellanby (1884– 1955) reported that rickets could be cured by adding certain fats to the diet that contained "a vitamin or accessory food factor." In 1922, he isolated this substance and gave it the name vitamin D.

Vitamin E (tocopherol) was discovered in 1922 when it was noticed that laboratory rats did not reproduce when lard was their only source of fat. A compound extracted from wheat germ and lettuce that was subsequently given the name vitamin E corrected the disorder, but its value for human beings was not evident for many years. Although its role in metabolism is unknown, it is thought that it helps maintain the structural integrity of muscle tissue and some elements of the nervous, vascular, and reproductive systems. It also is reasonably certain that it serves as an antioxidant that keeps tissues from breaking down as a result of the oxidation of unsaturated fatty acids.

The discovery of Vitamin K began with efforts by the Danish biochemist Carl Dam (1895–1976) to determine the cause of small hemorrhages in laboratory chickens. Dam correctly surmised that they were due to the absence of a substance that he called vitamin K (for "koagulation"). It was eventually isolated by an American group under the leadership of Edward Doisy (1893–1986). Unlike most other vitamins, vitamin K is synthesized by bacteria in the intestinal tract, so there usually is no need to take supplements unless one is on a long-term antibiotic treatment that destroys the bacteria. It is, however, often administered to newborn infants, who initially lack the bacteria in their intestinal tracts.

As was noted above, more recently discovered vitamins have not been given alphanumeric designations. Instead, they are known by their chemical names: folic acid, biotin, and choline. Two other compounds, inositol and para-aminopbenzoic acid (PABA), may also be significant, but their function in human beings has yet to be determined.

Some foods are particularly rich in particular vitamins. A diet that contains these foods will go a long way towards providing an adequate intake of vitamins.

VITAMIN	SOURCE
Vitamin A	liver, carrots, sweet potatoes, butter, margarine, leafy vegetables
Vitamin B1	pork, nuts
Vitamin B2	liver, dairy products
Vitamin B3	meat, fish, poultry, liver
Vitamin B5	legumes, meat, milk, whole-grain cereals, egg yolk
Vitamin B6	beans, beef, lover, bananas, whole-grain cereals
Vitamin B12	meat, fish, poultry, milk products
Vitamin C	citrus fruits, broccoli, strawberries
Vitamin D	fish liver oils, vitamin D–enriched milk
Vitamin E	vegetable oils, whole-grain cereal, butter, leafy green vegetables
Vitamin K	leafy green vegetables, yogurt
Folic acid	leafy green vegetables, whole-grain cereals, nuts, liver
Biotin	fresh vegetables, kidney, liver, eggs, milk
Choline	liver, yeast, eggs, grains

The diets of many people are often deficient in natural sources of at least some vitamins. This makes it advisable to take vitamin supplements on a regular basis. Suggested dosages are given as recommended daily allowances (RDA). These are measured in two ways: weights and units. Weights refer to the actual weight of the vitamin that should be taken daily, usually given in milligrams or micrograms. When the compound has not been isolated and hence cannot be weighed, animal studies are conducted to determine how much of a substance is necessary to produce the desired effect. These dosages are then converted into arbitrary measurements known as units. In some cases, most notably vitamin A, dosages are still expressed as units because the practice has been entrenched for many years.

How large a dosage is necessary is still a matter of controversy. Guidelines have been developed by government agencies and individual nutritionists, but these usually apply only to average adults, and very few people's needs are equivalent to a statistical average. A better approximation of one's vitamin requirements may be gained by consulting one or more of the many nutritional guides that have been produced for this purpose. There is generally no danger in taking more vitamin supplements than are required. Vitamin A and vitamin D can be toxic when taken at very high dosages for an extended period of time, but the effects disappear when the overdosing ceases. Excessive intake of vitamin C can result in diarrhea but produces no other known side effects. With the possible exception of vitamin E, vitamin supplements derived from natural sources seem to have no advantage over synthesized supplements.

Voice Mail

Voice mail is an electronic subsystem within a private branch exchange (PBX) switch, or in a private network served by a Centrex system in a telephone company's central office switch. The function of voice mail is to record, transfer, and store voice messages among subscribers within the private network. However, many systems also combine voice mail with other automated answering function to direct outside callers to the internal voice mail "mailboxes" of employees, where the caller can leave a recorded message.

Voice mail serves numerous telephone lines within organizations such as government agencies, hotels, hospitals, universities, telemarketers, and transportation systems. A voice mail system enables each authorized employee to create, send, and receive voice messages to or from coworkers within the network served by the PBX or Centrex system. To create and send, the employee accesses the system, speaks the message (to a recording medium), and enters the appropriate extension number or name. The same message can be sent to multiple extensions simultaneously, or broadcast to all the phones on the voice mail system. The privacy of each mailbox is protected by individual passwords assigned to the users. By dialing the voice mail code and entering the assigned password, the subscriber can listen to the messages, repeat them, or save or delete them.

Voice mail is especially valuable in enabling coworkers in different operating shifts (e.g., doctors and nurses in hospitals, detectives in police departments) to leave important or confidential messages for each other. In global communications systems, employees in different time zones can use voice mail to speed up information transfer without the inconvenience of a middle-of-the-night phone call.

Voice mail is one part of a larger PBX-based (or Centrex-based) system called *voice messaging*, which uses circuit packs and disk drives to provide up to 4,000 electronic mailboxes and storing up to 480 hours of digitally recorded voice messages. More than 30 calls can be handled simultaneously. The general public often uses the term *voice mail* to cover all the services provided by voice messaging. Besides voice mail, these include "call answering," "automated attendant," "Information service," and "announcements."

Call answering. This service automatically answers an incoming call to a specific telephone with a recorded greeting, then offers the caller the opportunity to leave a recorded voice message for that particular subscriber, if the person is away from his or her desk or busy on the phone with another caller. Call answering can be activated automatically, following a certain number of rings, or can be set for a specific time period.

Automated attendant. Incoming calls to a main PBX number are automatically greeted by a voice recording instead of a human attendant. The caller is then offered a "menu" from which to select how the call is to be directed, using the touchtone keys on the telephone; i.e., press 1 for sales, 2 for shipping, 3 for engineering, 4 for personnel. The call is then transferred to either a specific phone line or a new menu, with a narrower scope of choices. If nobody answers the phone selected from the menu, call answering may be activated so that the caller can leave a recorded message. Although the automated attendant is favored by PBX owners as a cost cutter that frees personnel for more productive activities, it can be a frustrating experience for people who refuse to "talk to machines." Even more frustrating for others is the seemingly endless series of menu after menu in some systems. Many PBX owners try to overcome that problem by providing an option to "press O for an operator."

Information service. A recorded greeting and menu provides callers with specific recorded information, such as location, hours of operation, product availability, prices, etc. The caller does not have the option to leave a message.

Announcements. Recorded announcements to be broadcast throughout the telephone system can be created by the system administrators or selected from among a package of prerecorded announcements in English or various other languages.

Voice messaging systems were introduced in the early 1980s. Prior to that, organizations with PBX systems relied on operators, receptionists, and secretaries to write down or memorize messages, which were later passed on to the recipients. However, this practice lacked privacy and was subject to mistakes, distortions, omissions, and even failure to deliver the message.—R.Q.H.

See also TELEPHONE ANSWERING MACHINE

Voice Synthesis and Recognition

Having a computer that can talk to you and, more importantly, recognize your speech as a source of commands, is one of the standard features of many science fiction movies and novels. It is, however, a practical goal for a large segment of the population—people with physical disabilities that prevent them from using a keyboard or seeing a display, as well as people from cultures where the ideographic writing style is not easily adapted to standard keyboards.

The output part of this problem (voice synthesis) is quite easily accomplished and has been a standard part of computer technology for many years. There is no problem in prerecording various words (such as the integers one, two, three, etc.), digitizing them, storing the individual words on a computer disk, and then playing them back when answers are required for simple questions. A typical example of such an application is when you put a call into a time service and are clearly told that it is now "two-thirty-nine." The individual words are simply extracted from the memory device and played back to you in the order that is required.

A much more difficult problem, and one that has not been completely solved, is to feed ordinary text into a computer and have it pronounce the words as would a human being reading the same material. It would be possible to simply record someone saying each word in the language and then put these together as the occasion demands, but there is much more to verbal communication than the basic words. For example, meaning is conveyed to the listener when a speaker provides some inflection in the words and stresses different parts of a sentence when asking a question, as opposed to giving a command. When this type of information is absent, it is difficult to understand the situation unless one has had a great deal of experience listening to this type of speech. Even when some processing of the text can be done to add inflection, the resulting speech synthesis systems are limited because not all variations in all possible words can be stored.

A possible solution to this problem would be to produce a "rule-based" system that examines the text and attempts to get the computer to pronounce the words according to the information stored in its database of rules. Although significant progress has been made with this approach, it is difficult to accommodate all possible variations of a language, particularly a language like English. Words that have similar spellings but different pronunciations (such as the pair *through* and *tough*) make such programs very difficult to produce. Some languages do not have this problem to the same degree as English, and much better text-to-speech programs have been created for them, but these languages have other difficulties that make synthesized speech very difficult to understand all the time.

The opposite side of this coin, getting a computer to understand someone's speech, is an enormously difficult problem. However, some very good programs have been produced in the last few years. The act of speaking creates very complex sound waves containing a huge amount of information. The fact that human beings can easily understand the same words when spoken by a big man with a bass voice and a small child with a high-pitched voice—to say nothing of the fact that we often understand words when spoken by someone with a speech defect—attests to the amazing properties of the human brain to filter out the extraneous information and to recognize the central core of information that carries meaning. If the human brain can perform such a task, then, at least in principle, a computer also can do it; the only problem is to figure out the best way of doing the job and to do it in "real time." There is very little point in having a speech recognition system that takes 20 minutes to analyze a 5-second utterance.

The majority of the speech recognition systems in use today rely on examining the digitized version of the sound waves produced and attempting to eliminate the unnecessary information at an early stage so that subsequent processing will have less data to wade through. The second processing stage is usually an attempt to determine the places at which words start and stop. This is not a simple problem. Even human beings, particularly in strained situations with unfamiliar vocabulary, often find it difficult to determine. If a computer system is limited to analyzing the situation in which a speaker will obviously pause between each word ("the cat ran up the hill"), then it can usually do a very good job of isolating each word, but even here it can sometimes be fooled by background noises such as air conditioners, noisy breathing, or others talking nearby. Once the end points of words have been identified, the signal is usually subjected to further processing to eliminate the differences in frequency of sound caused by variations in human anatomy between large and small people, men and women, etc.

After all this "signal processing" has been accomplished (the advent of faster computers is making this task a lot easier), the resulting data can be compared against a "dictionary" of word forms to see if the machine can match the spoken input against a pattern that will allow it to identify the word. This pattern is often stored in the system in a "training phase" of the program, during which time the user has to repeat specific words so that the system can learn to recognize their specific speaking characteristics. Often, several different patterns will match a spoken word closely enough to be potentially correct, and the computer will then attempt to use statistics to determine which of the potential words might be correct. These statistics can involve everything from a comparison of the speed at which the word was said to a recall of previously used words in an attempt to determine which word might be intended for this portion of the sentence. Of course, problems with homonyms (words such as *to*, *two*, and *too*) cause no end of difficulty with speech recognition systems, and most of the systems in use have some mechanism that will allow the user to choose from a list of alternatives when this situation arises.

Despite all the difficulties, there now exist computer systems that do a good job at recognizing speech, and they will undoubtedly get better. However, it must be understood that even a seemingly high level of accuracy may result in a large number of errors; a speech recognition program that is 95 percent accurate will average about 15 mistakes per page of typed text. Even so, effective voice recognition and synthesis technologies will meet a real need. While most people who are capable of using a keyboard or viewing a monitor may prefer these devices, voice recognition and synthesis systems do provide a viable, if slower, alternative to those who, for one reason or another, cannot use standard input and output devices.—M.R.W.

See also SIGNAL-TO-NOISE RATIO

Vostok and Voskhod

After surprising the world by orbiting a series of artificial satellites, the Soviet Union maintained their momentum by putting the first human beings into space. The vehicle that made this possible was the Vostok spacecraft. Before being used for human space travel, a Vostok prototype officially called Sputnik 5 housed two dogs during a flight that made 18 orbits around the Earth before safely returning to Earth. This flight cleared the way for the first human venture into space. On Apr. 12, 1961, Yuri Gagarin (1934–1968) made a 1-hour, 48-minute flight that included one orbit of the Earth at an altitude of 327 km (203 mi). On Aug. 6, 1961, Gagarin's epic flight was followed by Gherman Titov's (1935–) 17-orbit flight. On Aug. 11, 1962, the Soviet space program launched Vostok 3, followed by Vostok 4 a day later. The two missions remained in orbit for 4 and 5 days, respectively, before returning to Earth within 6 minutes and 190 miles of each other. Another double flight, Vostok 5 and Vostok 6 was noteworthy for putting the first woman, Valentina Tereshkova (1937–) into space on June 16, 1963, where she remained for 48 orbits.

Vostok spacecraft consisted of a 2.3-m (7.5-ft) spherical module that housed the cosmonaut and a cone-shaped equipment module that included the retro rockets, batteries, and instrumentation. This module had a diameter of 2.4 m (7.9 ft) and a length of 2.58 m (8.5 ft). Altogether, Vostok weighed 4,725 kg (10,417 lb). Unlike the Americans, the Soviets landed their spacecraft on solid ground rather than on water. Unsure of their ability to effect a soft landing, they separated the cosmonaut from the spacecraft during reentry. At 7,000 m (22,996 ft) the reentry capsule was separated from the equipment module. Two seconds after separation, the cosmonaut ejected from the capsule and was braked by a small parachute. At 4,000 m (13,123 ft), the seat was jettisoned, and another parachute was deployed, allowing a relatively slow return to Earth. Meanwhile, the capsule descended beneath its own parachute. The fact that Valentina Tereshkova had been a sport parachutist may have been a significant factor in her being chosen as the first woman cosmonaut.

Over a 26-month period, the Vostok program put six cosmonauts into space, one at a time. However, the long-term exploration of space would require multiple crews, docking with other spacecraft, and the ability of people to leave their spacecraft for sojourns in space. Moreover, Soviet Premier Nikita Krushchev was eager to exploit the propaganda value of the Soviet space program, and he demanded multiple-cosmonaut space flights that would upstage the flights of solitary Project Mercury astronauts. The result was the Voskhod ("sunrise") program. Voskhod required little development time, for in its basic layout it closely followed Vostok. By removing the ejection seat, and eliminating spacesuits and emergency gear, it was possible to fit three cosmonauts into the spacecraft. Unlike the procedures used for returning Vostok cosmonauts from space, Voskhod cosmonauts remained in the reentry capsule, which after separating from the equipment module parachuted back to Earth.

After two unmanned launches, Voskhod 1 was launched on Apr. 12, 1964, making 15 orbits in the course of a flight time of 24 hours, 17 minutes. The next (and last) mission, Vostok 2, contained only two cosmonauts; the third crewman's seat was replaced by a portable airlock. This was used by Alexei Leonov (1934–) for the world's first spacewalk on Mar. 18, 1965. Afterwards, the mission nearly went awry when the automatic reentry system failed. The cosmonauts were able to use manual controls to bring their ship back to Earth, but they landed far off course in the Ural Mountains region, where they spent a cold night before a rescue team arrived.

Vostok and Voskhod were key components of the Soviet space program, but subsequent efforts required larger and more sophisticated spacecraft. In 1967, the Soviets launched the first example of their next-generation spacecraft, Soyuz, which was to serve as the backbone of the Soviet manned space program for many years to come.

See also MERCURY PROJECT; SOYUZ AND SALYUT; SPUTNIK

Further Reading: Ray Spangenburg and Diane Moser, *Opening the Space Frontier*, 1989.

Vulcanized Rubber—see Rubber, Vulcanized

Washing Machines

Over the centuries, laundry has always been one of the most labor-intensive household tasks. Until the 1800s, only the rudimentary hand-washing techniques of pounding, rubbing, and rinsing in water were used to remove surface dirt from clothes and bedding. This laborious process, which was mostly done by women, was finally changed and refined with the development of simple machines during the first decades of the 19th century.

The earliest machines were wooden tubs containing a set of paddles manually driven by a flywheel to agitate the clothes. Some machines had a wringer attached to help squeeze out excess water, but this was not much of a labor-saving device since women had to hand feed the heavy, wet clothes through the rollers. Also, women still had to spend a lot of time and effort carrying, heating, and emptying all of the water required for washing and rinsing. To alleviate this burden, in the mid-1850s many middle-class American households began to adopt the European practice of sending the washing to local laundresses or commercial laundries. The lower classes, however, could not afford such services, and the upper classes continued to depend on inhouse servant labor to do the tedious and time-consuming work.

After the Civil War, inventions and innovations related to washing clothes rapidly increased. Women became particularly active in inventing washing machines and other household-related devices from the 1870s on. Margaret Plunkett Colvin (1828–1894) of Michigan, for example, patented her Triumph Rotary Washer in 1871 and presented it at the Philadelphia Centennial alongside other female inventors at the Women's Pavilion. Around 1884, Ella Goodwin of Illinois invented a washing machine that copied the washboard action of hand laundering; cranks moved the clothes and rubbed them between upper and lower boards.

The inventions that most dramatically influenced the development of the washing machine and household appliances in general were electricity and, consequently, the electric motor. In particular, Nikola Tesla's invention of a compact electric motor in 1899 allowed household appliances to become much more efficient and easy to use. One of the first companies to produce an electric washing machine was the Hurley Company, which introduced its Thor model in 1907. The same year, Alva Fisher patented a clothes washer that had a drum driven by an electric motor. Four years later, Maytag introduced its own electric washing machine, and during the next 16 years that company proceeded to manufacture 1 million machines.

During the 1920s, washing machines and other domestic appliances greatly benefited from a general consumer boom and the increasing manufacturing capacities of large companies such as General Electric. Also contributing to the expansion of the domestic-appliance industry was the need to make up for the loss of household labor as occupational opportunities for women expanded. Middle- and upper-class women who had depended on servants now became reliant on mechanical appliances to assist with housework. Lower-class women continued to do the same household tasks, but at least their work became a bit easier once appliances were available on the mass market at increasingly lower prices.

As competition among manufacturers increased in the 1930s, appliances finally began to be redesigned and were gradually streamlined to become extensions of each other in the kitchen. This streamline design was intended to convey the era's most important household values: modernity, cleanliness, and efficiency. Washing machines were among the last of the kitchen appliances to receive the streamlining treatment. Until the end of the decade, they remained much the same as they had always been: round tubs or drums on legs with a motor attached underneath. Maytag was the first company to square off the corners and design the washing machine more like its kitchen counterparts, with a cabinet enclosing the motor and eradicating the need for legs.

In terms of materials, washing machines were now made of steel rather than traditional wood, and were either galvanized or porcelain-enameled. Other important developments of the 1930s were the invention of automatic washing machines, with timing controls, variable cycles, and preset water levels, as well as of washer/dryer combinations. These major improvements were followed in the 1940s by the introduction of front-loading washers and the transition to the 'tumble action' method of washing clothes instead of the traditional "agitation" method, both changes first appearing in Bendix appliances.

Since World War II, innovations in household appliances, including washing machines, have continued at a slower pace. During the 1950s, developments tended to be more cosmetic than substantive as custom colors became fashionable, changing from year to year. Over the past several decades, the washing machine has become a common appliance in American homes, and issues of capacity and efficient use of energy and water have become the most important.—M.S.

See also MOTOR, ELECTRIC; STREAMLINING

Further Reading: Ruth Schwartz Cowan, *More Work for Mother: The Ironies of Household Technology from the Open Hearth to the Microwave*, 1983. Penny Sparke, *Electrical Appliances: Twentieth-Century Design*, 1987.

Waste Disposal

A universal characteristic of communities of human beings is the generation of waste. Settlements at all levels of size and technological sophistication have to get rid of their wastes, or at least learn how to cope with them. As populations have increased and economies have expanded, the problem of waste disposal has grown in magnitude. The biggest producer of waste is the United States, where farms, residences, industries, and businesses generate 5.75 billion metric tons (6.25 billion tons) of waste annually—70 percent of the world's waste. Global waste production is already prodigious, but if the rest of the world equaled the United States in this regard, the problems of waste disposal might become insurmountable.

Some forms of waste create special problems. About 4 percent of the waste in the United States is classified as hazardous because it is toxic, explosive, corrosive, or reactive. If not properly dealt with, these wastes can have serious consequences over a wide area and for a prolonged period of time. Nuclear wastes create unique problems of their own. Air pollution is the result of venting into the atmosphere the waste products of combustion. Although major strides have been taken in recent years, air pollution remains a major problem in many areas.

The collection and disposal of "ordinary" wastes are usually the responsibility of local governments, which also have the task of sewage collection and treatment. Collecting, transporting, and getting rid of garbage is a significant financial responsibility, usually costing local governments more than police and fire services. Like police work and firefighting, trash collection is a dangerous occupation; sanitation workers are four times more likely to be injured on the job than miners. The collection of rubbish is a noisy, unsightly process, but the real problem centers on what to do with the trash after it has been collected. In the past, a large amount of trash was incinerated, but burning trash has become less common due to escalating costs and tightening air pollution regulations. Unless they are strictly controlled, incinerators produce sulfur and nitrogen oxides, acid gases, furans and dioxins (highly toxic substances), and heavy metals like lead and cadmium. Even when efficient and clean-burning incinerators are installed (at considerable cost), the disposal problem remains, for incinerators only diminish the weight and volume of garbage, they do not eliminate it. Under normal conditions, incinerators reduce the weight of solid wastes by about 50 percent and their volume by about 60 percent; something still has to be done with the incinerated remains.

By far the most common way to dispose of wastes is to dump them somewhere. But simply depositing garbage in a pit is not permissible in most places, owing to the stench and vermin infestation typical of open dumps. Nearly three-quarters of the municipal refuse in the United States goes into landfills, open areas that are filled with garbage and covered over by earth-moving equipment. Landfills put wastes more or less out of sight, and in some cases landfills have been converted to public parks and other amenities. But landfills are not a perfect solution to the world's waste-disposal problems. Areas that can be filled with garbage are not in infinite supply, and existing ones are gradually being filled up. Also, covered-over garbage dumps may remove the visual evidence of garbage, but problems may lurk beneath the surface. Dangerous substances may leach into the soil, contaminating groundwater. This hazard can be overcome by using a lining that seals the pit from its surroundings, but this in turn creates a new problem because the waste materials residing in a sealed pit decompose very slowly, if at all. Even in an unlined pit, decomposition may take place at a very leisurely rate. Some objects are in a good state of preservation 40 years after they were dumped. A pit that is sealed off may serve as an excellent place for the permanent preservation of garbage.

Encouraging natural decomposition is one of the most effective ways of dealing with garbage. Many wastes break down under the influence of bacteria and other organisms, leaving behind a substance that can be used for fertilizer. This process can be used for more disposed items when they are consciously designed to be biodegradable, that is, capable of being naturally decomposed. For example, disposable diapers made from conventional plastics will not biodegrade, but diapers made from a cornstarch-based plastic will.

Many of the components of waste will resist deterioration under any circumstances. Plastics in particular have a long life expectancy. Fortunately, used plastic products have some value, for they can be recycled into new plastic products. Many other materials can be recycled, with subsequent savings in waste. Recycling, however, is not a perfect answer to the waste problem, for it entails significant expenditures of energy, labor, and capital. The only real way to arrest the growth in the volume of waste is to reduce the amount of waste being generated. This can be done through "source reduction," for example, the elimination of material that is destined to be thrown out immediately, such as elaborate packages for consumer goods. Wastes can also be lessened by repudiating the "throwaway society." This can be done by extending the "working lives" of the things we use through intelligent design, the use of durable materials, and an engineered-in capacity for future repair.

See also CARBON DIOXIDE; COMPOSTING; DIAPERS, DISPOSABLE; POLLUTION CHARGES AND CREDITS; RADIOACTIVITY AND RADIATION; RECYCLING; WATER SUPPLY AND TREATMENT

Water Supply and Treatment

During the first half of the 19th century, most major cities faced the daunting challenge of building technologically complex and capital-intensive waterworks. Until that time, local rivers and lakes supplied most of the water that cities used. These were supplemented by wells that provided access to subterranean water deposits and cisterns that captured rainwater. Nevertheless, water supplies remained limited, and the water that was available was generally of poor quality. Numerous private companies were chartered to provide water, but their efforts generally failed. High costs attended the construction of complex aqueduct systems, making it difficult or impossible to earn a profit from these ventures.

Despite these problems, the demand for greater supplies of water could not be ignored. Not only was water needed for personal consumption and a variety of industrial processes, it was also essential to physical safety. Fire posed a constant threat, as minor fires quickly became major conflagrations that burned large sections of cities. Without an ample water supply, people simply could not battle the blazes. Yellow fever, cholera, and typhus also regularly ravaged the population, and increased water supplies, while not promising to abate disease, would at least provide comfort to those who had fallen ill. In addition, many believed that an increased supply of water would encourage better personal hygiene and public sanitation, and this in turn would limit the impact of epidemics.

In most cities, the need for new sources of water was apparent, yet problems remained as to what sources should be tapped and how the water should be conveyed and distributed. In 1801, Philadelphia responded to this problem by building the first major municipal waterworks system in the United States. A tunnel conveyed water from the Schuylkill River to the city, about 1.6 km (1 mi) away. A network of pipes then distributed the water throughout the city, using steam engines to hoist and pump the water. Though these waterworks were expensive and prone to breakage, they remain noteworthy for the precedent that they set as the first public, capital-intensive waterworks system constructed in the United States. Unlike the self-contained wells and cisterns built earlier, these waterworks, and the ones subsequently built by other cities, used tunnels to convey water to reservoirs, from which networks of pipes distributed the water through the city. In the years that followed, New York City completed the Croton Aqueduct in 1842, and Boston's Cochituate Aqueduct brought water to that city in 1848.

Building municipal water systems demanded large financial outlays and the development of new funding methods. Most often, cities floated water loans that were to be redeemed after a fixed period of time. The presumption was that fees paid for water hookup and use would provide the funds to eventually retire the loans.

Developing new mechanisms of management and empowering them to build integrated water systems posed an even greater challenge than generating the requisite funds. First, cities and states had to sort out their respective roles. Although waterworks were constructed to serve a particular municipality, it was often the case that water sources and aqueduct routes extended far beyond city limits. As a result, the construction of water supply systems involved issues that went beyond a particular city or town. State legislatures usually had to settle these murky issues of management and policy. The Massachusetts State Legislature authorized Boston's government to appoint a water commission, while in New York State, the governor directly appointed the water commissioners. In the New York experience, significant hostility often surfaced between the state agency charged with constructing the facility and the city that the aqueduct was meant to serve.

Antagonism also grew and occasionally erupted between aqueduct engineers and city politicians, each seeking to fulfill their own particular agendas. For example, Croton's chief engineer, John B. Jervis (1795–

1885), faced off with New York City's leaders over the issue of who controlled construction of the distribution system once the aqueduct reached city limits. Jervis felt that his technical expertise and his cadre of engineers were the logical appointees. Of course, the city wanted to control construction within municipal boundaries, for the construction trade traditionally bore lucrative political fruit. New York City's politicians ultimately won this battle in the state legislature, and constructed the distribution network for water crossing city boundaries.

Despite the challenges of building and managing waterworks systems, everyone was delighted to see the various municipal water supplies improved. However, the demand for water was so great that it exceeded all estimates. As a more copious supply of water allowed more liberal use, the supply could not keep pace with the demand. By the end of the 19th century, most cities were forced to expand existing facilities and plan additional ones.

During the last decades of the 19th century, attention also turned to assessing and improving water quality. Though European scientists had addressed the issue of water purification since the 18th century, this had not yet attracted much attention in the United States. Until the 1870s, most aqueducts simply conveyed water to reservoirs, where the passive process of sedimentation allowed particulate matter to settle into chambers below, while the lighter and seemingly purer water remained above. Filtration could have complemented sedimentation and further clarified the water, but this technique remained largely undeveloped until the 1880s and 1890s. Filtration allowed water to pass through a medium consisting of various mixtures of gravel, sand, or charcoal that trapped suspended matter. In the United States, rapid filtration, where water is passed through the "sponge" under great pressure, came into widespread use at the turn of the century. Often, coagulation agents were introduced into the water supply to facilitate rapid filtration.

For the most part, the goal of late 19th-century research and experimentation in the United States was to improve the physical appearance of the water supply by clarifying it. In the 20th century, attention turned to purifying it. Acceptance of the germ theory of disease as well as contemporary trends in European bacteriological and chemical research influenced this shift in emphasis. Chlorine was first introduced to disinfect water supplies in 1908, and its use quickly became widespread. Since then, research into the effectiveness of other chemicals, as well as electricity, ozone, and ultraviolet rays, to improve water quality has continued.

In recent decades, the effort to improve water quality has extended beyond disinfection to include softening hard water, improving taste and color, and introducing medicinal features into drinking water. For example, by 1979, half of the water in the United States had been fluoridated to reduce tooth decay. As the quality and usefulness of water increases, we can only expect that the consumption of water will increase as well, further intensifying the search for pure water.—J.A.G.

See also EPIDEMICS IN HISTORY; FLUORIDATION; GERM THEORY; STEAM ENGINE; ULTRAVIOLET RADIATION

Further Reading: M. N. Baker, *The Quest for Pure Water*, 1981.

Waterwheel

The first reliable source of mechanical energy, the waterwheel, first appears in written records in the 1st-century B.C.E. in both the eastern Mediterranean basin and in China. Waterwheels were configured either horizontally or vertically. In China, the predominant form was the horizontal wheel, so called because its blades, or paddles, rotated in the horizontal plane. The horizontal wheel took several forms. In its most primitive configuration, a small stream of water was directed against either flat or curved paddles on one side of the wheel. The water's impact turned the paddles, which turned the wheel's axle, which turned a millstone or moved a power transmission apparatus. In other cases, the entire horizontal wheel was enclosed in a casing or even a circular tower to ensure that water acted effectively.

In Europe, the predominant form of waterwheel was the *vertical wheel*, which rotated in the vertical plane. This wheel, too, took several forms. The simplest was the *undershot wheel*, consisting of flat paddles mounted on the circumference of a wheel and rotated by the impact of a stream striking the blades. The *overshot wheel* was more complex. It had containers, or buckets, built into its circumference. A mill-

race directed water over the top of the wheel into the buckets, where the water's weight turned the wheel. Slower moving and more expensive than the undershot wheel, the overshot was, however, more efficient. An intermediate form of vertical wheel, called the *breast wheel*, received water at or near axle level and often had a casing to hold the water on the wheel.

The output of early waterwheels was small, prob-

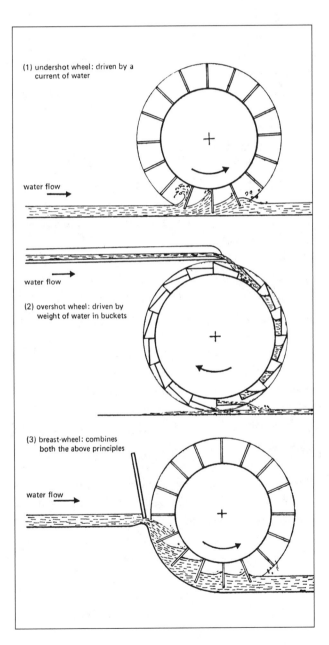

Three basic types of waterwheel (from A. Pacey, *The Maze of Ingenuity*, 1992; reprinted with permission).

ably no more than 3.7 kW (5 hp). In preindustrial civilizations, however, this was significant. A human being can average no more than .07 kW (0.1 hp) for an 8-hour day. Hence, a small vertical wheel could produce around 50 times more power—a huge increase in output. Moreover, a waterwheel could be worked 24 hours if necessary.

Little is known of the early history of the waterwheel. Archaeological evidence indicates that knowledge of waterwheels was widely spread throughout the Roman Empire. But before the 3d-century C.E. they seem to have been rarely used, perhaps because cheap labor made capital-intensive, labor-saving devices economically unattractive. The use of waterwheels accelerated, however, as Rome's expansion ceased and labor surpluses were replaced by labor shortages. By the 6th century, Rome itself had become totally dependent on watermills for grinding flour.

The fall of the Roman Empire only temporarily halted the expansion of waterwheels. The new civilization that crystallized in Western Europe was centered in a region where rainfall was heavier and steams flowed more regularly than in the heartlands of Mediterranean civilizations. Moreover, early Europe was chronically short of labor. These circumstances contributed to growing use of the waterwheel. Late 11th-century tax records compiled for William the Conqueror in England identified more than 6,000 watermills in his new realm.

One of the factors contributing to the wide diffusion of waterpower in medieval Europe was monasticism. Monasteries frequently used waterpower to secure self-sufficiency and promote piety. Often abbots could provide sufficient time for prayer and meditation only by using waterwheels for time-consuming tasks like grinding grain. Moreover, the missionary drive of Western monasteries spread knowledge of waterwheels all over Europe. For example, in the 12th century alone, the Cistercians built nearly 750 monasteries, most of which seem to have imitated the mother abbey at Clairvaux in making extensive use of waterpower.

In Europe, the Middle East, and China, waterpower was at first most heavily used for two tasks: grinding grain and raising water. But the advantages of mechanical energy were so evident that new applications of the wheel were found. Even before the beginning of the Christian era, Chinese millwrights had applied waterpower to blowing machines used to produce iron, and by the end of the Roman era, engineers had apparently adapted waterpower to sawing. But the Western medieval period saw the greatest diversification of waterpower. Applying first the cam (a small projection on the axis of a waterwheel) and then the crank to convert rotary motion to reciprocating motion, medieval millwrights adapted the waterwheel to a host of new tasks, including fulling cloth, preparing tanning bark, shaping metals, crushing ore, producing paper, sawing wood, and powering bellows.

Soon, location near falling water became important for many industries. When fulling was mechanized in England in the 13th century, the center of the English woolen industry shifted decisively from its traditional centers in southern and eastern England to northern and western England where steams were more widely available to power the trip hammers used in the fulling process. The introduction of the waterpowered hammer and water-powered bellows similarly pushed European and Chinese metallurgical operations to the banks of streams.

By 1500, the waterwheel had become the prime mover of choice in Western Europe. Although demographic, geographic, social, and political factors hindered the expansion of waterpower in other civilizations in the early modern period, European dependence continued to grow. By 1694, France alone had around 80,000 watermills. Moreover, by this period European engineers had begun constructing elaborate waterpower complexes. In the Harz mountains of Germany, for instance, between the years 1550 and 1800, mining engineers erected an elaborate system consisting of 60 water supply dams and reservoirs and 190 km (120 mi) of waterpower canals within a radius of 4 km (2.5 mi) of Clausthal. This system delivered water to 225 wheels to power mine pumps, hoisting engines, and stamp mills.

The growing importance of waterpower and the concurrent need to extract maximum power from existing sites as new ones became scarcer led to the first systematic, quantitative investigations of waterwheels beginning around 1700. These culminated in the extensive model experiments of John Smeaton (1724– 1792) in the 1750s that demonstrated conclusively that wheels utilizing the weight of water (overshot and breast wheels) were more efficient than those utilizing only its impact (undershot and many horizontal wheels). Smeaton also began to substitute iron for wood. By the early 1800s, thinner, stronger, more durable iron was rapidly replacing wood for the axles, spokes, and buckets of waterwheels in industrially advanced areas.

The period from around 1780 to 1860 saw the pinnacle of traditional waterwheel construction. In this period, millwrights and engineers erected metal and wood-metal hybrid wheels up to 24.4 m (80 ft) high and capable of producing nearly 225 kW (302 hp). Waterwheels provided the power for the first stages of Great Britain's Industrial Revolution. The growing use of the centralized factory for textile production created massive new demands for waterpower and led to the construction of even more elaborate waterpower systems. The most elaborate, that at Lowell, Mass., included a maze of locks and millraces that by 1850 supplied nearly 6,711 kW (9,000 hp) to 30 mills with more than 100 breast waterwheels and a few water turbines.

In the mid-19th century, the traditional waterwheel began to be supplanted by the water turbine. Even so, the waterwheel still held out in some areas: small direct-drive pumping plants, small factories, and some mines. For these applications, the waterwheel's higher efficiency at partial gate openings, ease of repair, reliability, and longevity remained central. But even this industrial toehold disappeared with the emergence and growing dependence on electric power between 1880 and 1910. In electric generators, high rotational velocity is crucial to cost, efficiency, and output. Turbines developed the high shaft speeds needed to power them much more easily than traditional waterwheels. By the early 20th century, the waterwheel was all but extinct as an industrial prime mover.—T.S.R.

See also INDUSTRIAL REVOLUTION; TURBINE, HYDRAULIC

Further Reading: Terry S. Reynolds, *Stronger Than a Hundred Men: A History of the Vertical Waterwheel*, 1983.

Wave Theory of Light

The nature of light has long been a source of speculation. In the 17th century, Christian Huygens (1629–1695) proposed that light was propagated as a longitudinal wave that undulated in the direction of its motion. According to this conception, light was similar to sound, which was known to travel in waves. But the analogy was not an exact one; sound waves were known to bend around obstacles, but light did not seem to exhibit this property. The rival theory of light, that it consisted of tiny particles or "corpuscles," presented no such difficulties, for it could be assumed that the particles traveled in straight lines.

In fact, the postulate that light always travels in straight lines had been called into question by a 17th-century experiment conducted by Francesco Grimaldi (1618–1663), an Italian physicist. In this experiment, a card pierced by a small aperture was placed behind another card with a similar aperture. When light passed through the two apertures and then fell on a flat surface, the resulting band of light was a bit larger than the band that had entered the first aperture. From this Grimaldi inferred that the beam had been bent slightly away from the edges of the aperture. He called this phenomenon *diffraction*.

Although Grimaldi's demonstration provided some support for the wave theory of light, the rival particle theory enjoyed a stronger basis of support. Isaac Newton's research had led him to the belief that light was a kind of particle, and during most of his lifetime and for decades thereafter, his immense prestige was sufficient to lend an unshakable legitimacy to the particle theory. A serious challenge did not come until the early 19th century, when the wave theory was revived by another English scientist, Thomas Young (1773–1829). A medical doctor with a wide range of interests and abilities, Young became interested in light as a result of his early research into the functioning of the eye. In 1803, Young performed an experiment that somewhat resembled Grimaldi's in that it used cards with small apertures. Young found that when light was passed through a very small aperture, distinct bands of light appeared where sharp shadows might have been expected. Young interpreted this phenomenon as the effect of light waves being diffracted around the corners formed by the edges of the apertures. An even more convincing demonstration involved a card with two pinholes. When a narrow beam of light passed through them and then fell upon a blank card, the result was a number of colored bands as well as some dark ones.

This was inexplicable if light were a particle, but it made good sense if light were a wave. As Young interpreted it, the image on the card was an interference pattern: The bright bands were the result of the peaks of some light waves coinciding; the dark bands were caused by the trough of one wave canceling out the peak of another wave. Young was even able to use his diffraction experiments to calculate the wavelength of light by determining what wavelength would allow the small amount of observed bending. Red waves, he found, had a wavelength of 0.8 millionths of a meter, while the wavelength of violet light was half as much.

The wave theory explained diffraction, but it could not account for the double refraction that occurs when an image is viewed through certain crystals like Iceland spar (crystal calcite). In 1808, the eminent French astronomer and mathematician Pierre Simon Laplace (1749–1827) was able to explain this phenomenon by arguing that if light was made up of particles, these particles could be split into two separate rays, each with a different velocity. Young naturally attacked this idea, but was unable to provide a better explanation based on the wave theory of light. In the hopes of resolving this scientific puzzle, the French Académie des Sciences (of which Young was a member) offered a prize to anyone able to provide a mathematical theory that might explain the phenomenon.

In late 1808, an adherent of the corpuscular theory of light, a French army engineer named Étienne-Lous Malus (1775–1812), won the prize by introducing the concept of polarized light to explain double diffraction. In his formulation, the rays of light refracted by the crystal were opposite poles of light, analogous to magnetic poles; the double refraction occurred because the rays were polarized perpendicularly to each other.

The wave theory seemed to be in retreat, and there was now the prospect of solidifying the position of the corpuscular theory if only it could be reconciled with what was known about diffraction. Accordingly, in 1817 the Académie offered another prize, this one for an adequate explanation of diffraction. As it turned out, the winner of the prize, Augustin Jean Fresnel (1788–1827), came up with an explanation that gave a strong boost to the wave theory. Young had been at a loss to explain double diffraction because he assumed that light waves oscillated back and forth in the direction

that they were traveling—that is, in the same manner that sound waves are propagated. If, however, light waves moved transversely to their direction, both diffraction and double refraction could be explained. By 1817, Young had embraced the idea that light waves were propagated transversely, and he communicated this belief in two letters he sent to Dominique Francois Arago (1786–1853), an influential French physicist. Arago was a friend of Fresnel, to whom he passed on Young's reconsideration of the nature of light waves. Building upon Young's reconceptualization, Fresnel was able to produce a well-developed wave-based theory of light. Not only was this an important contribution to science, it had practical consequences as well, for Fresnel was able to apply his theory to the design of highly efficient lenses for lighthouses.

This, however, was not the end of the matter. The propagation of waves seemed to require some sort of medium through which they could travel. Sound waves presented no problem, for it was known that they traveled through the air and other fluids. But what was the medium through which light traveled? To resolve this, 19th-century scientists had to postulate, as Huygens had, the existence of something they called the "luminiferous ether" to serve as this medium. But as efforts to detect it failed, the whole idea seemed to be more metaphysical than scientific. By the beginning of the 20th century, efforts to come to a more complete understanding of the nature of light played a major role in generating a revolution in physics. Ironically, this revolution reintroduced the idea that light was a particle—but a particle that exhibited wave-like qualities.

See also LIGHT, POLARIZED; LIGHT, SPEED OF; QUANTUM THEORY AND QUANTUM MECHANICS

Weather Forecasting

Weather forecasting is an estimation of the probable trend or future condition of atmospheric changes in a certain area. Forecasting involves measuring the present weather conditions over a large area, and extrapolating the information into the future, usually with the aid of high-powered computers and computer modeling techniques. A computer-based scientific model is an approximation of a real system, such as the atmosphere, which omits all but the most essential variables of the system. In atmospheric modeling, these essential variables consist of temperature, humidity, and wind speed and direction. Actual or theoretical information is input into the model to obtain present or probable trends in weather patterns. Key data are obtained by a network of over 10,000 land-based weather data stations, supplemented by aircraft and satellite observations that occur throughout the day. These data are sent to the National Meteorological Center (NMC) in Camp Springs, Md., where meteorological maps are published for international distribution. Forecast information is then obtained by extending the probable course of these variables into the future, taking account of their past performance.

In the United States, weather data are collected by a variety of agencies, depending on the type of weather experienced and its severity. Land-based day-to-day weather observations across the country on rain, temperature, winds, and cloud cover are made by automated stations located at airports, as well as by thousands of volunteer spotters. When weather information is received from these stations and ground observers, it is sent to 116 nationwide weather forecast offices (WFOs), branch offices of the National Weather Service. Buoy stations on lighthouses, offshore oil rigs, and other coastal locations provide information on severe storms such as hurricanes. Buoy stations send information to the National Hurricane Center (NHC) at Coral Gables, Fla., which tracks and forecasts severe storms. Data from these agencies, as well as information from worldwide, shipboard, buoy, and satellite locations are sent to the National Meteorological Center. This is also the location of the National Weather Service's Environmental Modeling Center, which uses a Cray C90 supercomputer to model atmospheric variables such as temperature, humidity, wind speed, and wind direction, and to project the outcome into the future. These forecasts are available nationally and internationally.

Weather is a global phenomenon that affects such diverse matters as airline flights, marine transport, agriculture, and sporting events. Weather is, therefore, the concern of the government agency concerned with commerce. The Department of Commerce is the administrative agency responsible for the activities of the National Atmospheric and Oceanic Administration (NOAA), the parent agency of the National Weather Service (NWS). NOAA conducts research and gathers data about the oceans, atmosphere, space, and sun through the use of orbiting satellites. The Geostationary Operational Environmental Satellite (GOES) satellite orbits the Earth at a fixed position 36,000 km (22,500 mi) above the Earth's equator, while a series of polar-orbiting satellites circle the Earth in a polar (meridional) direction. The GOES satellite is the cornerstone of the NWS modernization effort, which produces both images and atmospheric soundings of temperature, moisture, vertical moisture, and stratospheric ozone content. Satellite data are supplemented with data obtained by the NWS, which is responsible for weather data acquisition, data analysis, forecast dissemination, and storm watches and warnings. A storm watch is a notification by the NWS that conditions are favorable for the development of severe weather. A warning is issued when Weather Service Personnel use information from radar, forecast data, and other sources to inform people that an area is in imminent danger of severe thunderstorm or tornado activity. Severe storms are monitored by a system of 161 weather surveillance radar stations that use Doppler radar, the weather forecaster's most powerful new tool.

Radar is an acronym for *ra*dio *d*etection *a*nd *r*anging. It has been used since World War II to detect and track severe weather systems such as tornadoes, thunderstorms, and hurricanes. Since some of these weather systems are small, they have often eluded detection by the network of weather stations set up across the country. Doppler radar works by determining the velocity of falling precipitation either toward or away from the radar unit by using the principles of the Doppler shift and displaying these changes as a color-coded pattern on a screen. Two or more stations can give a three-dimensional picture of a storm. The Doppler shift or effect refers to the change in frequency, the number of waves to pass a fixed point in 1 second, that occurs when the emitter or the observer is moving toward or away from the other. It is as if you (the observer) were standing on a train platform and a train, several hundred meters away, starts its whistle (the emitter). The train continues past you with unchanging speed, pulling its whistle without stop as it passes you. The rising change in pitch as the train approaches is the increase in frequency, the increase in the number of sound waves as the train passes you (the waves compress as they approach you). The waves are at their greatest compression and highest pitch when the whistle passes you. As the train recedes, the pitch decreases as the frequency decreases and the waves expand. This principle can also be applied to microwaves, radio, and light waves. While ordinary Doppler radar uses microwaves, light waves are the basis of Doppler lidar (*l*ight *d*etection *a*nd *r*anging). Doppler lidar is an experimental system used to determine the change in frequency of falling precipitation, cloud particles, and dust that serve as indications of a tornado vortex signature, one of the most important uses for Doppler equipment. With this system, storms can be detected before they become severe, wind speeds can be charted at many levels, and rain can be measured as it falls into a stream basin. With the ability to measure rainfall over a basin area, warnings of flash floods are issued, which greatly aids in the prevention of loss of life and property. The National Weather Service is currently putting into effect 135 Doppler radar units called NEXRAD, an acronym for NEXt Generation Weather RADar. These units, called WSR-88D (Weather Service Radar, model 88D), will be tied to a set of computers that perform a variety of functions, including input of data, display of data on monitors, and running the data through algorithms, computer programs that simulate and predict severe weather.

The National Weather Service is in the process of modernizing its network of 116 weather forecast offices. The Advanced Weather Interactive Processing System (AWIPS) is the center of this modernization effort. It consists of new high-speed computer workstations and a communications system that will be capable of receiving, processing, and aiding forecasters in the analysis of vast amounts of weather data.

Weather monitoring and prediction took on international importance when world travel and trade became commonplace in the 19th century. In 1878, the International Meteorological Organization (IMO) was created for the purpose of expanding human knowledge about worldwide weather patterns. The name of this agency was changed in 1947 to the World Meteorological Organization (WMO); its headquarters is in Geneva, Switzerland. As an agency of the United Nations, the WMO coordinates weather data collection and analysis by its over 145 member nations and is responsible for the international exchange of weather data. Data are collected through the World Weather Watch (WWW), an international weather-monitoring network coordinated by the WMO. The WMO also maintains the integrity of weather data by certifying that observation procedures among member nations do not change. It also has standardized procedures for collecting information, as well as the symbols used by meteorologists. These data are sent by teletype, telephone, or satellite to the three WMO data collection centers (Moscow, Russia; Melbourne, Australia; and Washington, D.C.) From here, the information is transmitted to the National Meteorological Center (NMC), where the data are analyzed, weather charts are composed, and national and international forecasts are made.

Two of the most effective tools for disseminating weather data have been the use of new communications media such as the Weather Channel on cable television and the various Internet computer facilities. Weather information on the net is provided by several organizations: the National Weather Service, which is an interactive Weather Information Center giving all U.S. weather warnings (at http://www.nws.noaa.gov); the Weather Net of the University of Michigan, which links to nearly 300 other weather sites and 24 weathercams (at http://cirrus.sprl.umich.edu/wxnet); and the Daily Planet, the multimedia guide to meteorology of the University of Illinois (at http://www.atmos.uiuc.edu).—M.B.

See also ALGORITHM; DOPPLER EFFECT; INTERNET; MATHEMATICAL MODELING; MICROWAVE COMMUNICATION; OZONE LAYER; RADAR.

Further Reading: C. Donald Ahrens, *Meteorology Today: An Introduction to Weather, Climate, and the Environment*, 1994.

Weather Modification

Weather modification is defined as any change in atmospheric conditions that is induced by human activity, either intentionally or unintentionally. Cloud seeding to increase rainfall and fog dispersal are two of the major types of intentional weather modification, while atmospheric changes as a result of air pollution and deforestation are examples of unintentional weather modification. One example of a pollution-induced unintentional atmospheric change is global warming caused by increased levels of greenhouse gases such as carbon dioxide (CO_2), methane (CH_4), and nitrous oxide (N_2O) that result from the burning of fossil fuels and industrial activity. Another example of unintentional weather modification is an increase in rainfall resulting from acid rain caused by an increase in particulate matter. Deforestation, the massive worldwide loss of trees, is another unintentional cause of atmospheric change that produces a further increase of carbon dioxide into the atmosphere. Loss of trees has been aggravated by the highly acidic rain that falls on forested areas, burning leaves and increasing the acidity of soil and water.

Weather modification, especially relating to precipitation (rain, hail, and snow), is probably one of humankind's earliest attempts at influencing nature. In the absence of irrigation, crops depend on rain, and without crops a tribal group often perished. For this reason, the tribal rainmaker frequently held a position of prominence, especially among semiarid and desert-dwelling tribes. Later attempts to modify the weather occurred in grape-growing regions in 14th-century Europe, where church bells were rung in the hope of suppressing hailstorms. These efforts were intensified in the 19th century when M. Albert Stiger, a grape grower and city official of Windisch-Feistritz, Austria, designed and built a funnel-shaped cannon that he believed would suppress hail when it fired smoke particles into the air. He believed that the heated particles would melt the hail so that less damage would be done to the vineyards. Amazingly, when cannon were experimentally discharged in 1896–97, no hail was reported despite considerable damage in nearby areas. Many hail cannons were tried elsewhere in Europe in wine-manufacturing regions, but Stiger's successes were not duplicated, and by 1905 interest in his machine waned. After World War II, Stiger's ideas were tried again in Italy where, until the early 1970s when it was outlawed, farmers fired explosive rockets into thunderclouds in order to break up the formation of hailstones.

In the Soviet Union, silver iodide crystals were seeded into thunderclouds with the same objective. Because hail was responsible for approximately $700 million in annual losses, the United States was receptive to the Soviets' hail suppression techniques. In 1972, the federal government funded the National Hail Research Experiment to take place over eastern Colorado. After 3 years of cloud seeding with silver iodide, a lot of information was learned about hail storms, but there was little proven success in hail storm suppression. This research ended in 1979.

Rain has been another form of precipitation that has proved popular in weather modification efforts. Among technologically oriented societies, attempts at weather modification relating to rain (technically, droplets of water greater than 0.5 mm [.02 in.] in diameter surrounding a nucleus) occurred when cloud seeding was undertaken in the years since World War II. Cloud seeding is an attempt to stimulate natural precipitation processes by injecting nucleating agents, such as silver iodide or dry ice (solid carbon dioxide at a temperature of about –80°C [–112°F]), into cold clouds. Cold clouds are defined as those whose temperatures are below 0°C (32°F), and are composed of ice crystals or supercooled water droplets, or a mixture of both. Cloud seeding can be performed from an aircraft or from ground-based generators, though aircraft-based seeding increases the dispersal of nucleating material.

To test the effectiveness of cloud seeding, a rigorous set of experiments was conducted by the National Oceanic and Atmospheric Administration (NOAA) in the 1970s and early 1980s over south Florida using a blind test involving silver iodide (a chemically active nucleating agent) and sand grains (an inactive substance that doesn't attract water droplets). The Florida Area Cumulus Experiment (FACE) was conducted by seeding cumulus clouds (clouds resembling huge puffs of cotton formed by updrafts) with silver iodide crystals and with sand grains, and then measuring any differences in precipitation. In later stages of testing, results were determined to be inconclusive because they weren't replicated in successive tests; i.e., rain was increased by 25 percent in one experiment but not in others. Larger questions remain, however, such as how much rain would have fallen if the cloud hadn't been seeded. Also, if cloud seeding produces rain in one area of the country, does it deprive another area of rain?

Fog is a cloud with its base at the Earth's surface; its

dispersal is another area of intentional weather modification. Dense fog, which greatly lowers visibility, is responsible for many collisions by cars, airplanes, and ships at sea. Fog dispersal, the clearing of fog either by increasing the air temperature or by cloud seeding, occurs most often at airports, where fog-based delays are most prevalent and the cause of millions of dollars in losses. During World War II, the British used fuel burners set along runways to disperse fog by lowering the relative humidity, the amount of moisture the air is capable of absorbing at a particular temperature, causing the fog to dissipate. Their success triggered postwar efforts in using heat to disperse fog, but interest in this technique diminished because of the expense. Today, the only major airports to use heat as a fog-dispersal agent are at Orly and De Gaulle airports outside Paris. In addition to heat, cloud seeding has also been used to disperse fog.

In warm fogs, i.e., fogs above 0°C (32°F), hygroscopic substances (substances that accelerate water vapor condensation) have been used to reduce the relative humidity and cause the formation of raindrops that fall to the surface. In cold fogs, fogs whose temperature lies between 0°C (32°F) and –20°C (–4°F), pellets of dry ice are used. With this method, which is called the Bergeron process, the formation of precipitation in cold clouds causes ice crystals to grow at the expense of supercooled water droplets. Today, dry ice is most commonly used at airports for the dispersal of cold fogs. At present, however, fog dispersal using cloud seeding has had varying degrees of success, and many questions remain unanswered.

One of the most intriguing workers in the field of weather modification was the brilliant Serbo-Croatian inventor and mechanical engineer Nikola Tesla (1856–1943). Tesla, who was a contemporary and competitor of Thomas Edison, was not well known in the United States until recently, when his experiments and inventions have come to be better appreciated. The holder of over 140 U.S. patents (including some that are still thought to be classified by the U.S. government) and the man who was largely responsible for the early development of alternating current, Tesla formed a strong belief that weather could be controlled. To this end he built a device called a *magnifying transmitter*, which emitted powerful radio signals pulsed at very low frequencies. In his experiments, Tesla was able to use the magnifying

transmitter to create dense fogs in his laboratory. He felt that the device could be placed in arid regions to draw unlimited amounts of water from the ocean that could be transported inland for use in irrigation and power. Unfortunately, Tesla died before he could put his ideas to practical use.

One of the most complex issues facing humanity today is the unintentional effect that atmospheric heating is causing the global environment. Though carbon dioxide (CO_2) occurs naturally in the atmosphere, and, along with other greenhouse gases, traps heat, the amount present has steadily increased. Since the Industrial Revolution in the mid-19th century, the industrialized countries have greatly increased their use of fossil fuels, a byproduct of which is carbon dioxide. Carbon dioxide is also part of the biochemical reaction known as *photosynthesis*, whereby plants take in carbon dioxide and give off oxygen. Because of the enormous loss of trees due to logging operations, less CO_2 is being absorbed, leaving more in the atmosphere. Deforestation has, therefore, aggravated the global warming problem. The loss of trees in many parts of the world due to acid rain has made the problem even more severe.—M.B.

See also ACID RAIN; ALTERNATING CURRENT; CLIMATE CHANGE; DEFORESTATION; FUELS, FOSSIL; INDUSTRIAL REVOLUTION; MOTOR, ELECTRIC; PHOTOSYNTHESIS

Further Reading: J. M. Moran and M. D. Morgan, *Meteorology: The Atmosphere and the Science of Weather*, 4th ed., 1994.

Weaving

The first fabrics were made by matting fibers together to form a kind of felt. A superior means of making fabrics, weaving, was devised in the later Mesolithic era (c. 7000 B.C.E.), perhaps inspired by the making of baskets. In its most basic form, weaving begins with a number of parallel threads that are held in tension (the *warp*). These threads are then interlaced at right angles with a sequence of threads (the *weft* or *woof*). Weaving can be done with nothing more complicated than a frame to hold the warp threads and something to hold the weft thread as it is pushed through the warp.

Weaving can be done much more rapidly if some

kind of loom is used. The simplest of these is the warp-weighted loom. It can be little more than two posts joined at the top by a roller (the *cloth beam*) on which the cloth is wound as it is woven. The warp threads hang from the cloth beam and are kept vertical by weights attached to their ends. The weaver passes the weft thread from side to side, going over one warp thread, and then under the next one, repeating the sequence until the last thread is reached. After battening down the new weft thread, the weaver moves on to the next row, putting the thread under the warp that the previous thread had put over the warp, and so on for successive warp threads.

The process requires the weaver to raise or depress each succeeding warp thread with his or her fingers. Through the use of a more sophisticated loom, it is possible to mechanically raise a group of warp threads with a single movement. In a loom of this sort, alternating warp threads hang down in front of and in back of a rod that is fixed to the upright beam (the *shed rod*). A second rod, known as a *heddle*, is loosely attached to the warp threads situated behind the shed rod. This creates a space (known as a *shed*) through which the weft thread can be passed from one side to the other. The weaver then moves the heddle forward, pulling the rear warp threads forward and creating a new space through which the weft thread can be passed. By alternating the position of the heddle, the weft thread can be quickly woven through the weft. This produces a plain weave; increasing the number of heddles allows the weaving of more elaborate patterns.

Looms of this sort date back to antiquity. In Europe by the middle of 12th century, this vertical loom began to be supplanted by a horizontal loom. Possibly

Industrial loom used for the manufacture of rugs, c. 1950 (courtesy National Archives).

derived from the Chinese silk loom or from the looms used in the Islamic world, the horizontal loom had the advantage of allowing the weaver to remain seated as he or she worked. The major innovation of the horizontal loom was the use of foot-operated treadles to change the position of the warp threads. The weft thread was passed through the shed by a boat-shaped *shuttle* containing a bobbin (known as a *quill*) that had been wound with thread. The shuttle was weighted, which allowed it to be propelled with some force from one end to another.

A modern hand loom works on similar principles, except that instead of fastening the warp threads to the heddle, the ends of the individual threads pass through eyelets within a group of vertical wires, one eyelet per thread. The wires, in turn, are connected to the treadles. By depressing one or more of these treadles, the weaver causes a particular group of warp threads to rise, leaving a shed. A weft thread attached to a shuttle is then passed through the space, and then beaten down to lie tightly alongside the adjacent weft thread. By selecting a particular treadle or group of treadles, the weaver can control the movement of warp threads, thereby producing a particular pattern in the woven cloth.

By Renaissance times, looms had become quite elaborate, with thousands of moving parts involved in their operation. Along with the clock and the organ, these looms were the most elaborate mechanical devices built up to that time. They allowed the weaving of elaborate patterns, but weaving still required a great deal of human labor. Some improvements were wrought in the Renaissance era, such as the ribbon loom, invented in 1579, which could simultaneously weave several pieces of narrow fabric. By the early 17th century, Willem van Sonnevelt, a Dutch inventor, had put into operation a loom that could weave as many as 24 ribbons at once.

Stimulated in part by improvements in spinning, many 18th-century inventors worked on the mechanization of weaving. One of the most important advances was the *flying shuttle*, patented by John Kay (1704–1764) in 1733. In this invention, the shuttle was equipped with small wheels that ran in a groove. When the weaver pulled a cord, small hammers at either side of the loom struck the shuttle, forcibly propelling it from one end of the warp to the other. Weaving was still

a manual process until Edmund Cartwright (1743–1823) invented a power loom in 1785. He installed it in a factory in 1787, but 4 decades passed before all the mechanical difficulties were overcome. The first commercially successful power looms were run by waterpower, and a location near a stream or river was the first requirement of a textile mill during the late 18th and early l9th centuries. By the second half of the 19th century, this requirement was relaxed as steam engines provided the motive power for many power looms.

In the years that followed, much inventive effort went into making weaving a more automatic operation. Of particular interest was the use of punch cards to control the weaving of patterned fabrics; this technology went on to be of major significance for automatic control in general. The 1820s saw the introduction of looms with mechanisms that automatically stretched the fabric across the loom; it has been estimated that this mechanism saved 10 minutes of every hour that previously had been taken up by manual adjustment.

Automatically stretching the fabric was a substantial improvement, but power looms still had to be stopped when the shuttle's bobbin ran out of thread, a fairly frequent occurrence. This difficulty was overcome in 1894, when James H. Northrup, an English émigré working for a the Draper Loom Company in Massachusetts, put the first successful automatic loom into operation. Northrup's loom, which went under the commercial name of Model A Draper, used a circular hopper that held a supply of bobbins. If a feeler detected that a bobbin was about to run out of thread, the empty bobbin was knocked out and a new one was inserted. Another mechanism automatically stopped the loom when a warp or weft thread broke. With looms of this sort, labor productivity went up substantially; a single attendant could oversee the operation of as many as 24 looms at once. Improvements to the automatic loom continued to be made in the 20th century. Although a mature invention, the loom continued to benefit from many incremental improvements: a Model E Draper loom has a manufacturer's plate that lists no less than 134 patents granted from 1912 to 1927. One outstanding example of a perfected loom was made in Japan by the Toyoda loom works. Japan's rising technological prowess was evidenced by the fact that a well-established British manufacturer of textile machinery purchased the manufacturing rights to the Toyoda loom in 1929.

The technological advance of weaving has not been a completely unmixed blessing, however. From the 18th century onward, technical improvements in weaving brought diminished fortunes to some. Prior to its being mechanized, weaving had been a craft enterprise conducted by independent artisans. Many weavers earned a bare subsistence, but at least they had some control over their working lives. The development of mechanized looms and the aggregation of these looms in large factories threatened jobs and the traditional way of life. On occasion, hand-loom weavers fought bitterly to hold onto their livelihood, sometimes resorting to the smashing of mechanized looms. The effects of mechanization were not just confined to the industrialized world. The Indian weaving industry, which largely depended on hand labor, was decimated by the importation of cheap, machine-woven textiles. On the positive side, mechanized weaving improved productivity, lowered prices, and created a number of skilled occupations. But mechanized weaving remains a difficult working environment. A battery of powered looms produces immense amounts of noise, and the vibration can even cause significant structural damage.

See also FACTORY SYSTEM; LOOM, JACQUARD; LUDDISM; STEAM ENGINE; TURBINE, HYDRAULIC; WATERWHEEL

Weights and Measures, International System of

As early as 1747, the French astronomer Charles-Marie La Condamine (1701–1779) suggested the development of an international standard of linear measure based on a geodesically determined magnitude, the length of a pendulum with a 1 second period at the equator. Throughout the 18th century, French academicians argued in favor of establishing some uniform standards of weights and measures and fought about what those standards should be. Hopes for a universal system of measurements began to be realized on May 8, 1790, when the French Revolutionary Assembly voted to establish a set of standards. In September 1793, a *Commission des Poids et Mesures* was established to do the job, but wartime conditions dragged the process out, and it was not until 1799 that the National Assembly adopted the new metric standard.

Fundamental SI Units

Physical Quantity	Unit	Abbreviation
Length	meter	m
Mass	kilogram	kg
Time	second	s
Electric current	ampere	A
Temperature	kelvin	K
Luminous intensity	candela	cd
Amount of substance*	mole	mol

*Adopted since 1971.

SI Prefixes and Abbreviations

Prefix	Abbreviation	Factor
deka-	da	10^1
hecto-	h	10^2
kilo-	k	10^3
mega-	M	10^6
giga-	G	10^9
tera-	T	10^{12}
peta-	P	10^{15}
exa-	E	10^{18}
deci-	d	10^{-1}
centi-	c	10^{-2}
milli-	m	10^{-3}
micro-	μ	10^{-6}
nano-	n	10^{-9}
pico-	p	10^{-12}
femto-	f	10^{-15}
atto-	a	10^{-18}

The meter, the basic standard of length, now became 1/10,000,000 of a meridional quadrant of the Earth (i.e., 90 degrees of arc along a great circle through the two poles) at sea level. For practical purposes, the distance between two marks on a platinum (later platinum-iridium) bar at a specified pressure and temperature was defined as the international prototypal meter. In 1866, the U.S. Congress adopted the metric system as the only official system of measures that the country has ever had, but until recently most nonscientists (including most engineers) in the United States persisted in using customary British units of foot and pound measure.

Since 1875, an International Bureau of Weights and Measures, located near Paris, has served as a repository of prototype standards for both the primary measure of length and for secondary measures. In addition, it coordinates international measurement techniques.

In 1960, the meter was redefined as 1,650,763.73 times the wavelength in vacuum of the orange-red spectral line of light emitted by krypton-86. In 1971, the present International System of measures (SI), a highly refined version of the original metric system based on the six fundamental units and 16 basic prefixes (to indicate multiple and fractional powers of 10) was adopted by most nations. All other physical units are defined in terms of the fundamental ones. For example, force is defined in terms of mass times length divided by the square of the time.—R.O.

See also MEASURES OF ENERGY

Welding, Electric

Welding is the joining of two or more metal parts through the application of heat. It is possible to weld certain materials such as aluminum through the application of great pressure, but most welding operations require heat. When applied to a joint, heat causes the metal in that area to liquefy; when the heat is removed the cooled joints are permanently joined. For millennia, this was done by putting the pieces of metal to be joined in a forge and then hammering them together after they had reached the right temperature. It was a time-consuming process that called for a fair amount of skill and strength. It was not until the late 19th century that new ways of applying heat made welding a faster, less-laborious process.

For most applications, the heat required for welding is supplied by an electric current or by a direct flame. For specialized applications, electron beam welding and laser welding are being used, and the latter will likely take on greater importance in the years to come. But at the present time, the most common form of welding is arc welding, which essentially consists of putting the work to be welded in series with an

electric circuit that is completed when an electrode is brought close to the work. The resulting electrical arc produces heat that melts the portion of the metal to be welded. In most forms of arc welding, the electrode is made of a metal similar to the work to be welded. In the course of welding, it is deposited on the joint, forming a strengthening fillet. The electrode is often coated with materials that aid in the welding process and make for a stronger weld. Other kinds of arc welding use nonconsumable electrodes that are usually made from tungsten. When these are used, the area to be welded is shielded with an inert gas such as argon or helium; this allows the welding of metals that react readily with the atmosphere, such as titanium.

The other major form of electrical welding is resistance welding, a technology developed in the late 1880s by Elihu Thomson (1853–1937). This method is based on the fact that resistance to the passage of an electric current produces heat. When there is enough current, the heat is sufficient to melt metals and effect a weld. The pieces are clamped tightly together to ensure that a good electrical contact is maintained and that current is passed through the assembly. In another form of resistance welding known as *flash welding*, the pieces to be welded are held lightly together. The application of current causes rapid vaporization of metal at the contact points, which is then deposited on the surfaces to be welded; the pieces are then quickly clamped together to produce the weld. Resistance welding is used primarily for the welding of sheet metal, where the relative thinness of the material allows the rapid buildup of resistance. From its inception, this manner of welding has been of great value to industry, for it allowed the automation of a task that previously had required a high level of blacksmithing skill and effort.

Welding has to a significant degree replaced riveting as a way of making joints for large structures such as buildings, bridges, and ships, thereby saving time and labor costs. At first, the replacement of riveting with welding caused some problems, for riveted structures had the advantage of revealing flaws more easily, and the rivet holes served to stop developing cracks. In welded structures, problems tended to be hidden, and the structure could fail with no warning. Better materials, devices, and techniques, along with the design of structures to take full advantage of welding, have obviated these problems, and welding failures are rare today.

See also LASER; WELDING, GAS

Further Reading: Andrew D. Althouse et al., *Modern Welding*, 1980.

Welding, Gas

Welding is one of the most common means of joining metals. Although the majority of welding jobs are done through various kinds of electrical welding, gas welding is still widely employed for the fabrication and repair of metal assemblies. Although electrical welding equipment produces higher temperatures, gas welding requires a bit less skill than many forms of electrical welding, and it can be used where there is no access to electricity.

Most forms of gas welding require two cylinders of gas, one containing oxygen, the other a combustible substance, usually acetylene. The combustion of the two immediately outside the nozzle of a welding torch produces a flame with a temperature of 2,750 to 3,300°C (5,000–6,000°F), hot enough to easily melt any metal. For most purposes, the oxygen and acetylene are mixed such that 2.5 times more acetylene is consumed, but the mix may be varied for particular purposes. The operator of a welding torch can determine the proportion of the two gases by looking at the color and shape of the flame. Optimal combustion produces a flame with a blue inner cone, at the end of which is the hottest part of the flame. With an additional stream of oxygen, a torch can also be used to cut metal smoothly and rapidly.

The invention of the welding torch followed Karl von Linde's invention in 1895 of a commercially practical means of liquefying air and extracting oxygen from it. In 1903, Fouch and Picard created the first oxygen-acetylene torch. In the following years, gas welding supplanted traditional techniques of forge welding, greatly increasing the speed and ease of most welding operations.

See also ACETYLENE; WELDING, ELECTRICAL

Wheel

The wheel is often singled out as the most fundamental mechanical invention ever made. It is certainly true that

without the wheel a vast number of mechanisms would have been impossible. Life as it is lived today is inconceivable without the wheel. At the same time, however, some quite sophisticated societies have flourished without making any practical use of the wheel. As with all technologies, the significance of the wheel depends on the social context in which it is used.

The precise time and place of the wheel's invention is not known and probably never will be. The invention of the wheel required an imaginative leap of the first order, for circular motion of any sort is rarely found in the natural world. It is possible that the potter's wheel was the first application of rotary motion; a Mesopotamian clay tablet depicting such a wheel goes back to 3500 B.C.E. The oldest existing depiction of a wheeled cart, also from Mesopotamia, is dated around 3200 B.C.E., but the wheel may have been in use for many centuries before this date. There is no direct evidence that the potter's wheel gave rise to the idea of the cartwheel; the two may in fact have been independent inventions.

Whatever their ultimate source, wheels used for transportation diffused rapidly into Europe. Archeological sites have yielded wheels as much as a meter (3.28 ft) in diameter, often made from three or more planks held together with pegs or battens. It would seem that a wheel could be made most easily from a cross section of tree trunk, but a wheel of this sort would quickly split along the rays that run at right angles to the concentric rings. One of the reasons for humankind's rather recent invention of the wheel may be that its construction required fairly sophisticated carpentry tools. Such tools were certainly needed for the building of spoked wheels, which began to be used for war chariots around 2000 B.C.E. The rims of these wheels were made up of a number of C-shaped segments known as *fellies*, each of which had to be shaped with a fair degree of accuracy. Iron tires appeared around 700 B.C.E., followed by the Celtic invention about 3 centuries later of shrink fitting these tires.

Although wheeled vehicles facilitated land transportation, many of the great monuments of the ancient world were built without them. To take the most notable example, the constructors of the pyramids of Egypt made no use of the wheel. Rather, the huge blocks from which these structures were built were moved on large sledges. Friction was reduced by pouring oil or water on the pathway immediately ahead of the sledges' runners.

Early wheels either rolled freely around a fixed shaft (known as a *dead axle*) or were connected to a rotating shaft (known as a *live axle*). By the middle of the 2d-millennium B.C.E., tallow derived from animal fat and oil obtained from plants was used as a lubricant. Plain bearings made of metal also helped to reduce friction, and, in the late 18th century, ball bearings began to be used on vehicles. The construction of the wheels themselves underwent slow improvement. The most important of these was the *dished wheel* that first appeared in the 16th century. In wheels of this sort, the rim was not in the same plane as the hub, and the spokes were arranged like a shallow cone. This gave the wheels more lateral strength, and allowed a larger load to be carried between them. This was done by mounting the wheels so that their tops were tilted outwards, while the lower part of the wheel was situated directly below the hub.

Wheel technology stagnated in the years that followed, the next major improvement not coming until the late 19th century, when wire-spoke wheels began to be fitted to early bicycles. At first the spokes were arranged radially, but wheels using this arrangement were easily deformed. A better pattern was the tangentially spoked wheel, patented by James Starley (1801–1881) in 1874. The bicycle also created the market for the pneumatic tire, which did much to improve ride and handling. Like horse-drawn carts, many early automobiles had wheels with wooden spokes, although wire-spoke wheels were also common. In the 1930s, the pressed-steel wheel became increasingly popular.

As noted above, for all its seeming obviousness, the wheel was not used in all cultures. Among the societies that eschewed the wheel were the Precolumbian civilizations of Central and South America. Failure to use the wheel was not due to ignorance of the basic principle; wheeled figurines used as toys or ceremonial objects appeared in Mesoamerica beginning in the 4th-century C.E. Wheels, however, were not used for transport, a fact that has been explained by the lack of draft animals and the difficulty of building vehicular roads in rugged terrain. This explanation is not completely convincing, but nothing better has been offered.

Other cultures used the wheel but then largely abandoned it. In the Middle East, the invention of the camel saddle between 500 and 100 B.C.E. allowed camels to be used as pack animals. In many ways, a camel with a pack was superior to a cart pulled by another beast of burden, and the latter came to be supplanted by the former. But over the long run, unfamiliarity with the wheel may have contributed to subsequent technological retardation in the Middle East—ironically, the place where the wheel was perhaps first invented.

See also BEARINGS; BICYCLE; TIRE, PNEUMATIC

Wheelbarrow

A wheeled vehicle reduces effort by supporting the weight of a load and moving it with much less friction than would be produced by dragging it. The simplest of wheeled vehicles is the wheelbarrow. Small in size and endowed with but a single wheel, it might be thought that the wheelbarrow was one of the earliest wheeled vehicles; in fact, it was not widely used in Europe until the Middle Ages. There is some linguistic evidence that wheelbarrows were used around construction projects in ancient Greece, but they seem to have disappeared for many centuries thereafter. There is no pictorial or verbal indication that they were used in ancient Rome or in Europe in the centuries immediately following the collapse of the Roman Empire.

At this time, however, the wheelbarrow was widely employed in China, perhaps as early as the 1st-century B.C.E. and certainly no later than the 3d-century C.E.. Chinese wheelbarrows were (and sometimes still are) different from those familiar to Westerners. Rather than having a small wheel at the front, Chinese wheelbarrows had a substantially larger wheel located near the vehicle's midpoint. Goods or passengers were carried alongside the wheel, and if the load was not properly balanced the operator had to tilt the conveyance to the opposite side, often at a rather pronounced angle. In parts of north China where the roads were flat and the winds strong, wheelbarrows were sometimes fitted with sails for additional propulsion.

The wheelbarrow's reemergence in Western Europe during the second half of the 12th century may have been a case of "stimulus diffusion," whereby an artifact of another culture is not directly copied but provides the

The large wheel of the Chinese wheelbarrow allows it to carry much larger loads than the typical Western wheelbarrow (from R. Temple, *The Genius of China*, 1986).

inspiration for a rather different device serving the same purpose. Alternatively, the wheelbarrow may have been independently invented, perhaps through the simple expedient of adding a wheel to one end of a handbarrow (a handbarrow is a box or frame with shafts at opposite ends, and hence requires two people to carry it). Then too, the European wheelbarrow may have survived in the Byzantine Empire of Eastern Europe, eventually making its way back to the western part of the continent. It is also possible that its disappearance is more apparent than real, the result of inadequate documentation. In any event, by the 14th century it appears to be in common use, a reflection perhaps of the need to save labor in a continent depopulated by the Black Death.

In the following centuries, the wheelbarrow served as an indispensable conveyance for farmers, construction workers, miners, and anyone else who had to move fairly light loads. Yet as simple as its basic design was, there still remained room for improvement, such as the fitting of metal wheels and better bearings. Even the optimal dimensions of a wheelbarrow were a matter of some experimentation. When canals were being built in late 18th-century America, one construction supervisor had to ask an experienced English canal builder for the proper materials and dimensions so that he could have a wheelbarrow made to the most efficient design.

See also BEARINGS; EPIDEMICS IN HISTORY; WHEEL; WINDMILLS

Whistleblowing

Government and private organizations sometimes do things that are harmful. When they are misused, science and technology can multiply the damage these organizations do. For example, the production of a new material may result in the venting of toxic gases into the air, or a government agency may set up a secret surveillance system that intrudes on the privacy of law-abiding citizens. When things like these happen, the organization and its members may prefer that nobody knows about them. In order to draw attention to the problem, somebody from within the organization may have to take it upon himself or herself to reveal negligence, mistakes, or malfeasance. When they do so, they are said to have "blown the whistle."

A whistleblower may first go to his or her immediate superior in the hope of rectifying the problem, confident that the facts will speak for themselves. This sometimes works, but for various reasons the boss may prefer that the situation not be addressed. This leaves the whistleblower with the option of calling the issue to the attention of someone higher up in the organization's chain of command. However, attempting to circumvent the hierarchical order by going over the head of one's boss may not be successful. Higher-level officials usually expect employees to respect the authority structure by "going through channels." Just getting a hearing is likely to be difficult in these circumstances. Also, a higher-level manager will likely have doubts about an employee who has gone behind the back of his or her immediate superior, for loyalty is much prized in most organizations.

If the whistleblower fails to get a problem remedied or at least addressed within the organization, he or she has the option of calling the problem to the attention of the media or some other outside agency. This of course is disloyalty to the organization as a whole, and is likely to have serious, if not fatal, consequences for the whistleblower's career in that organization. One recent example of this is the experience of Roger Boisjoly, an engineer for Morton-Thiokol, the firm that built the solid booster rockets for the ill-fated *Challenger* space shuttle. His testimony before the special commission investigating the tragedy pinpointed the technical and managerial failures that caused the shuttle's fatal explosion. Isolated and ostracized at the firm he had served for many years, Boisjoly had little choice but to leave Morton-Thiokol, suffering considerable financial and psychological distress as a result.

Should an employee choose to take on the role of whistleblower, his or her struggle will most likely be a lonely one. The professional organizations of engineers and other science and technology workers usually draw many of their members from the ranks of management. Under these circumstances, professional societies are usually not inclined to lend their support to employees perceived as "disloyal." One highly publicized case of whistleblowing involved Ernest Fitzgerald, an engineer who was fired after making public vast cost overruns on a military aircraft built by his employer, Lockheed Aircraft. His professional association, the American Institute of Industrial Engineers, refused to come to his defense on the grounds that it was a "technical organization," not a "professional society." The influence wielded by defense

industry executives within the institute made this a predictable outcome.

In recent years, whistleblowers have received some help from the government. In the United States, the Civil Service Reform Act of 1978 includes a section that forbids reprisals against employees who disclose information about activities reasonably believed to pose "substantial and specific danger to the public health or safety." However, a law is only as good as its enforcement. The Department of Labor is supposed to provide hearings before outside examiners in these matters, but it has rarely displayed much enthusiasm in defending whistleblowers.

Whistleblowers have provided a public service in calling attention to reprehensible actions taken by both private and public organizations. At the same time, however, it cannot be assumed that every whistleblower acts out of a sense of public duty. Every organization has its disgruntled workers, and false accusations under the guise of whistleblowing can be a nuisance or worse. And sometimes well-intentioned whistleblowers may see a problem where none really exists. There is a fine line between an organizational and legal environment that encourages employees to "blow the whistle" when circumstances warrant it and an environment that leaves the way open for self-interested and irresponsible behavior.

See also *Challenger* DISASTER

Wind Shear

Wind shear, a technical term meaning a sharp variation in wind velocity at right angles to the prevalent wind direction, was first used in aeronautical terminology shortly before the middle of the 20th century. Vertical wind shear is more commonly called a *gust*, while the term *wind shear* itself is usually understood to refer to sharp horizontal changes in wind velocity. Both phenomena must be considered in the safe operation of airplanes.

Most people who have flown aboard passenger jet aircraft have experienced a sudden drop or, less commonly, a rise in the craft's altitude as it proceeds ahead at a steady horizontal velocity. This may cause some slight alarm and discomfort to the passengers but, more importantly, such gusts exert significant loads on the airplane's structure, and it is important that the craft be designed to resist such loads. In fact, vertical gusts have posed little danger to airplanes since the early days of flying when aeronautical engineers learned to design the machine's structure to successfully carry such loads. An exception to this is the danger from a strong downdraft close to the ground, since such an occurrence might dash an airplane into the ground with disastrous results. Fortunately such occurrences are exceedingly rare.

It is horizontal gusts, or wind shear, that is of greater concern to pilots, since this entails, especially if in the fore and aft direction, a change in the effective airspeed of an airplane. Wind shear from the craft's tail means the potential loss of lift, leading to a rapid and dangerous loss of altitude and control by the pilot, whereas wind shear from the nose of the airplane can lead to a rapid gain in altitude, with the pilot still having to alter control settings. The latter case is less likely to have dire consequences. Pilots are trained to make appropriate, rapid corrections in both situations. The greatest problem is the immediate detection of wind shear in order to avoid accidents.

Wind shear is most dangerous when an airplane is landing or taking off. In the vicinity of controlled airports, meteorologists monitor air conditions in order to have the air controllers warn pilots about the possibility of wind shear occurrences. Airplanes in the vicinity of uncontrolled airports receive no such assistance. Fortunately, such airports are used mostly by smaller airplanes that will respond more rapidly to corrective actions than will the larger jet-powered airplanes used by most passengers. Instruments to detect wind shear are now available and will warn pilots of dangerous wind shear conditions.—M.L.

Further Reading: David B. Thurston, *Design for Safety*, 1980.

Wind Tunnel

Wind tunnels are ductlike structures through which a controlled stream of air flows past an object, model or full scale, in order to ascertain the forces that the flow exerts on the object. Most commonly the test object is an aircraft or one of its components. Modern wind tunnels all descend from the primitive one designed by Britain's F. H. Wenham about 1870. With the aid of J. Browning, Wenham built the tunnel and conducted important aerodynamic research in 1871 with the support of the Aeronautical Society (later the Royal Aeronautical Society).

Before Wenham and Browning conducted their tests, the only way to obtain (much less accurate) aerodynamic data was by use of the whirling arm introduced by Benjamin Robins (1707–1751) and John Smeaton (1724–1792) in Great Britain during the middle of the 18th century. The wind tunnel superseded the whirling arm for aeronautical research in little more than a decade. Other important early wind tunnel work was done by H. F. Phillips in Great Britain during the early 1880s, the Wright brothers in the United States beginning in 1901, and Gustave Eiffel (1832–1923) in pre–World War I France. The tests that the Wright brothers performed on models with the wind tunnel they built in the autumn of 1901 were crucial to the success of their 1903 Wright Flyer.

Wind tunnels of the sort used through World War I were of the open-circuit type; i.e., air was drawn into the tunnel's fan from the atmosphere and then exhausted back into the atmosphere after passing through the test section where the model was located. After that war the closed-circuit wind tunnel was introduced. In a closed-circuit wind tunnel, the air was contained in a closed duct and recirculated after passing the test section, with the aid of vanes that helped to reduce air turbulence, the ever-present foe of wind tunnel designers and users. Among the advantages of the closed-circuit tunnel is the better control of the conditions at the test section that it allows. A more sophisticated version of the closed circuit wind tunnel, the variable-density tunnel, was first built and operated at the Langley Aeronautical Laboratory of the American National Advisory Committee for Aeronautics (NACA) in the early 1920s. The motivation for the NACA to build the variable-density tunnel (VDT) was that by using air compressed to as much as 20 atmospheres it was possible, by aerodynamic scaling, to conduct tests on small models that simulated tests on full-scale airplanes. The NACA's VDT, designed by the talented aerodynamicist Max M. Munk, a postwar immigrant from Germany, was very important for the NACA's early research on airfoil profiles (wing sections). By the mid-1920s, that organization had become the world leader in experimental aerodynamic research. Somewhat later, the NACA constructed a wind tunnel sufficiently large so that small airplanes could be tested at full scale; that tunnel became operational in 1931.

So long as airplane speeds remained below 175 mi/sec (400 mph), air could be considered to be an (approximately) incompressible fluid, and wind tunnel test data did not need to be corrected for compressibility effects. However, at subsonic speeds higher than that, such corrections have to be made when using conventional subsonic tunnels. For velocities higher than Mach 0.8 (0.8 times the velocity of sound), one must design a wind tunnel to match the desired velocity. Consequently, we now have transonic tunnels for Mach .8 to 1.2, supersonic tunnels for Mach 1 to 6, hypersonic tunnels for Mach 6 to 12, and hypervelocity tunnels for more than Mach 12. Each of these types of wind tunnels has its own design peculiarities and, in fact, requires modification with interchangeable components if it is to be used for tests at other than a specific Mach number. The power requirements to operate supersonic and faster wind tunnels is so great that the throat size of the test section must be kept rather small, a small fraction of a

square meter. The power required may be as high as several times 10^4 kW/m^2 (10^3 hp/ft^2). Consequently, it is impractical to run such tunnels continuously; they usually are operated in short bursts with a controlled release of highly compressed air lasting as little as a small fraction of a second. This is a far cry from the Wright brothers' continuous operation of their 1901 wind tunnel, which used a 1.5-kW (2-hp) gasoline engine to provide an airstream velocity of 12 m/sec (27 mph).

Special wind tunnels have been designed to test vertical takeoff and landing (VTOL) aircraft such as the V-22 Osprey, as well as automobiles, models of buildings, and even portions of cities. A famous early, nonaeronautical use of wind tunnels was the study of the 1940 failure of the Tacoma Narrows Bridge, which was conducted at the California Institute of Technology in 1942. That study, supervised by Theodore von Kármán (1881–1963), led to significant changes in the design of suspension bridges. Each of these types of subsonic wind tunnels poses challenging problems for the designer. Wind tunnels, usually thought of as a handmaiden to technology, constitute a fascinating set of technologies in their own right.—M.L.

See also AIRCRAFT, STOL, VTOL, AND VSTOL; AIRPLANE; BALLISTICS; BRIDGES, SUSPENSION; HYPER-

The wind tunnel used by the National Advisory Council on Aeronautics from the 1920s onwards (from R. E. Bilstein, *Flight in America*, 1984).

SONIC FLIGHT; NATIONAL AERONAUTICS AND SPACE ADMINISTRATION (NASA); SUPERSONIC FLIGHT

Further Reading: James R. Hansen, *Engineer in Charge: A History of the Langley Aeronautical Laboratory, 1917–1958*, 1987.

Windmills

The windmill and the waterwheel were the first devices to provide nonanimate sources of energy. In most places and at most times, the waterwheel has been the more widely used, for steady winds do not appear with the same regularity as running water. Windmills also tend to be more complex devices; consequently, their development and use occurred at a slower pace. At the same time, however, windmills have been—and still are—important sources of energy in some parts of the world.

The first documented evidence of windmills comes from 9th-century Persia and Afghanistan, arid regions lacking in running water. By the 12th century, windmills were in operation throughout the Islamic world. Windmills of this sort had a distinctive arrangement. Their blades were fixed to a vertical shaft, and the apparatus was enclosed in a walled structure open at both ends. Wind entered the structure through a wide opening that narrowed down toward the windmill's shaft, producing a funnel effect. Only one side of the blade-and-shaft assembly was open to the wind, which caused the shaft to rotate. Such an arrangement was obviously of use only when the wind blew in the proper direction, as was usually the case in this region.

It used to be thought that windmills were introduced to Europe after they came to the attention of the Crusaders. The windmills of the Middle East were probably a source of inspiration, but the windmills subsequently built in Europe were quite different in their basic design. The transference of windmill technology from the Middle East to Europe can be taken as an example of *stimulus diffusion*, a process through which exposure to a new idea or a device (in this case making use of the wind) motivates an innovation that is constructed in a different manner.

Whereas the windmills of the Islamic world used a vertical shaft, the windmills that began to appear for the first time in Europe during the 12th century had a horizontal shaft. This feature allowed them to be turned into the wind. The earliest European windmills were known as *post mills*, for the entire structure housing the shaft, blades, and other apparatus turned on a central post. The difficulty of turning the whole windmill limited their size. Larger windmills were made possible by the development of "tower" or "turret" mills, where only the top half of the structure was turned. The design for such a windmill can be found in Leonardo da Vinci's notebooks, but the bulk of the actual design and construction work was done by the Dutch in the late 16th century. Windmills made important contributions to the economic development of the Netherlands in the years that followed. Their most important use was to run the pumps that drained waterlogged areas and made them suitable for cultivation. Windmills were also used for grinding grain into flour, making paper pulp, and sawing lumber.

Windmills were improved by the installation of gearing and transmission systems derived from water mills. At the same time, windmills made their own contribution to mechanics through the development of self-regulating mechanisms. The first such device was the "flier fan," a small set of blades that actuated a set of gears that automatically turned the mill's blades into the wind. The speed at which the blades turned could also be regulated automatically. In the event of very high winds, the covering would be taken off the blades so that they did not rotate, but in less blustery conditions the blades' speed was sometimes regulated by a centrifugal governor that had been invented in the mid-18th century. This device was later adopted for use with steam engines, and it stands as one of the first cybernetic devices, that is, one that uses feedback for self-regulation.

In addition to being of considerable practical value, windmills stimulated a number of early scientific inquiries. In the mid-16th century, Jerome Cardan (1501–1576) analyzed the mechanics of the whirling blades and stressed the importance of setting the blades at an angle. In the 1780s, some experimenters endeavored to determine the optimal shape for the blades of windmills. As is sometimes the case, scientific inquiry served to affirm the validity of long-standing practice, for it was found that the optimal shape had been used by the Dutch for centuries.

A 22½-foot railroad pattern Standard windmill pumps water for livestock on the W. N. Fleck Ranch south of Alamogordo, N.M., c. 1923 (from T. Lindsay Baker, *Blades in Sky*, 1992).

Windmills of this sort provided an average of 10 horsepower, not a great deal by modern standards but a significant amount of power for a preindustrial economy. Even when other sources of power emerged, windmills continued to play an important role in many parts of the world. Some of their most notable contributions were made in the American West, where farming and ranching were dependent on water brought to the surface by pumps powered by windmills. Such devices are properly called *wind pumps*, but the old term continues in widespread usage.

Although a very old technology, windmills have taken on a new importance as producers of electricity. Thousands of wind turbines are already at work, most notably in California, where they supply 1.2 percent of that state's electrical energy. Windmills have great potential as a relatively nonpolluting source of energy. However, wind-generated electricity is more expensive than electricity generated by conventional means, and it has needed some government assistance to be competitive. Along with other renewable sources of energy, until 1987 wind-powered electrical generators benefited from a federal tax credit of 1.5 cents per kilowatt-hour for their first 10 years of operation.

It is likely that in the years to come, the cost of wind-generated electricity will go down as a result of advances in aerodynamics, electronics, and materials. At the same time, the cost of electricity from other sources will probably increase, making electricity produced by modern windmills an increasingly attractive alternative.

See also CYBERNETICS; TECHNOLOGICAL INNOVATIONS, DIFFUSION OF; UNIVERSAL JOINT; WATERWHEEL

Wood

Civilizations have relied on wood to produce an enormous range of objects from the intricate and beautiful to the rugged and utilitarian. Wood served as the primary material in houses, barns, fences, furniture, clocks, tools, barrels, pipes, wagons, ships, bridges, and even roads, while also providing a source of fuel and necessary products like tar, pitch, potash, turpentine, and charcoal for the production of iron. While sawmills date to the 15th century, most wooden objects were fashioned with hand tools and were built according to ancient methods. Woodworking required an intimate knowledge of various types of woods and of their strengths and physical properties. In craft production, a woodworker typically produced an entire object, whether it was a furniture-maker creating tables and chairs or a carpenter making the windows, doors, and trim for the houses he built.

During the 19th century, machines gradually supplanted hand labor, especially in the basic tasks of preparing materials for assembly and finishing. Samuel Bentham (1757–1831) has been called "the father of wood cutting machines." Bentham established a number of patents in Britain by 1793 for general-purpose planing, boring, molding, rebating, mortising, and sawing machines, as well as for special-purpose machines for producing window sash and carriage wheels. His designs included adjustable fences, sliding carriages, V tracks, and tilting tables to allow for bevel cuts. The use of machines led increasingly to the subdivision of labor that demanded greater precision from machines and less skill from operators. As inspector general for British naval works, Bentham introduced a series of machines designed by Marc Brunel (1769–1849) to manufacture ships' blocks (pulleys) at the Portsmouth Dockyard. The British Navy required about 100,000 blocks each year. Henry Maudslay (1771–1831) built 43 machines for the project. Widely celebrated, the manufacture of blocks at Portsmouth anticipated by 50 years the strategy of sequential operations for the manufacture of rifles by United States armories that became known as the "American system of manufactures." British arsenals did not adopt these methods. But Maudslay's "innate love of truth and accuracy" in machine design greatly influenced a generation of British tool builders.

Increasing demand for wood products spurred innovations in machine tools that in turn allowed for the growth of a woodworking-machine industry. By 1850, the United States assumed leadership in this industry. Since the colonial period, the great wealth of timber resources had provided cash for Americans in the Atlantic economy. Sawmills were always common in America, and for an extended period wood remained an abundant and inexpensive material. Indeed, the abundance of forests affected American attitudes toward what would later be a treasured natural resource. For American farmers, the forests were an obstacle to overcome. Trees were destroyed by simply burning them to the ground to allow for more efficient agriculture. In a society short on labor and capital, it made economic sense to exploit fully an abundant resource that was easily worked by simple machines.

American woodworking machines met very diverse needs. American machine builders provided individuals and small shops with simple, manually powered, hand-guided machines that resembled closely the hand tools they displaced. They also provided large manufacturers with expensive, steam-powered general- and special-purpose machinery employing rotary cutters at high speeds. In large plants, machine production raised new technical problems for woodworkers regarding power sources, shafting, bearings, machine placement, and work sequences.

To a great extent, the character of a particular woodworking industry depended on the type of product and the potential size of the market. The railroads used prodigious quantities of lumber, especially in the United States where wood was used for ties, bridges, and rolling stock. Because of the great demand, the railroads employed numerous woodworking machines, many of unique design for special purposes. By 1860, carriage and wagon manufacture had become a significant industry in the Midwest. However, even large manufacturers did not develop their own machinery since they could purchase standard and special-purpose machines from established machine builders. In contrast, the Singer Sewing Machine Company devoted considerable time and resources to the manufacture of wooden cabinets for sewing machines. By the 20th century, their plant in South Bend, Ind., employed 3,000 workers who made 2 million wooden cabinets a year. To meet this demand, they developed highly mechanized and capital-intensive methods of manufacture with machines designed inhouse.

For most of the 19th century, the typical woodworking shop was a relatively small enterprise that used only simple and inexpensive machinery. As late as 1870, despite the presence of numerous sash, door, and blind factories throughout the United States, most carpenters still made their own windows and doors in their workshops. Factory products increased rapidly as urban growth created tremendous demand for building materials. But small specialty shops remained a fixture of the market, because many jobs like commercial work in banks, hotels, and churches required special designs that could not be met by standardized factory products.

The furniture industry illustrates most clearly how the nature of the market set limits on the industrialization of woodworking. The industry adopted machinery at an early date, but machines were capable of only simple, basic tasks. Consequently, woodworking technology was widely diffused in an industry of small shops open to new competitors. Elaborate work continued to be a hand process and a basic component of an industry that promoted variety and extended product lines. The furniture industry remained more concerned with marketing and design than with simplifying the manufacturing process.

The introduction of ball bearings and electric motors led to the redesign of all woodworking machinery and plants by the 1920s. Markets opened for the manufacture of refrigerators and automobile bodies and for radio, record player, and, later, television cabinets. As the cost of timber products increased, manufacturers developed innovative uses for plywood, veneers, and pressed wood. But most markets disappeared or were severely reduced as plastics and metals replaced wood as the basic material for common objects. Wood no longer serves as the cheapest and most easily worked material. At the same time, in modern society conservation of natural resources has become a major economic and political issue affecting attitudes toward forest products.—J.C.B.

See also BEARINGS; MASS PRODUCTION; MORTISING AND TENONING, MACHINES FOR; MOTOR, ELECTRIC; PLYWOOD; RAILROAD; SEWING MACHINE; STEAM ENGINE

Wool

Wool is a textile fiber obtained from the coats of a number of domesticated animals. Although sheep provide the largest amount of wool used worldwide, wool is also obtained from goats (the source of cashmere and mohair), vicunas, llamas, alpacas, and camels. When examined microscopically, wool fibers exhibit a pattern of overlapping scales. The fibers also have a pronounced twist, known as *crimp*. These physical features give wool a number of attractive qualities. Woolen fabrics resist wrinkling yet hold a crease, and absorb moisture while at the same time resisting water. Wool fabrics stretch easily, making them resistant to tearing. At the same time, their elasticity allows them to return to their original shape after being stretched.

Depending on its texture, the coat of a single sheep will have as few as 10 million and as many as 100 million individual fibers. These fibers can be anywhere from 10 to 70 microns in diameter (a micron is one-millionth of a meter). Wool is classified according to the thickness of the fibers: coarse (31 microns and

above), medium (24–31 microns), fine (18–24 microns), and superfine (15–18 microns).

The use of wool goes back to Neolithic times and the domestication of sheep and goats. Over the centuries, selective breeding has produced animals that give far more wool than their ancestors. The first fabric made from wool was felt, which is made by applying heat, moisture, and pressure to wool fibers. This results in a thick fabric that has no grain and does not fray. However, garments made from felt do not hold their shape well.

Most woolen fibers are converted to cloth by either knitting or weaving. Before this occurs, the wool goes through a number of preparatory steps. After being sheared from the animal—done today with mechanical shearers that remove the wool coat as a single piece—the wool is cleaned, yielding lanolin as a useful by-product. The wool is then sorted. This is a necessary step, as the quality of the wool varies according to what part of the animal it comes from. The wool is then combed in order to separate long and short fibers. Alternatively, the wool may be carded, a process that puts the fibers parallel to one another. It is then spun into yarn that can be woven or knitted into fabrics. Often, the spinning process is set up to produce worsted, a compactly twisted yarn made from long fibers. If the yarn is woven, the resulting fabric is then fulled, a process that shrinks it and tightens the weave. Prior to the industrialization of wool processing, fulling was done by beating the fabric with mallets, or simply walking over it. Fulling often entailed the application of a fine powder known as *fuller's earth* (diatomaceous earth). The fabric is then washed, stretched, bleached, and dyed (sometimes this last step is done earlier in the production process). In the past, it was also necessary to crop the fabric, a process of raising the nap and cutting it level. In the late 18th and early 19th centuries, the introduction of mechanical croppers occasionally provoked outbreaks of machine-smashing by people who feared the loss of their jobs.

Wool has been the basis of many regional and national economies. In the late Middle Ages, much of the wealth of Venice and Florence was produced by the wool-based textile industries located in these cities. Their preeminent position eroded as a result of competition from the woolen industries of Flanders and England. So significant was the wool trade to England that by the end of the 17th century, nearly half of its exports consisted of wool and woolen cloth. Domestic demand was also high, encouraged by English laws such as a law that stipulated the burial of the dead in nothing but woolen fabrics.

The use of powered machinery for the production of textiles was a central component of the Industrial Revolution. Most of the advances, however, came in the cotton industry; wool production was much more slowly industrialized. Even in the middle of the 19th century, fewer than half of England's woolen textile workers labored in factories. The disparity in the tempo of industrialization in wool and cotton production was primarily due to the greater difficulty of adopting wool to mechanized processes, although tighter government regulations over the manufacture of woolens and the limited price elasticity of woolen goods also were significant.

Wool production remains an important industry in many countries today. The world's sheep population numbers over a billion, of which fewer than 9 million are in the United States. By contrast, Australia has 135 million sheep. New Zealand, long noted as a country with more sheep than people, is home to 60 million. Other nations with large sheep populations are Uruguay (25 million) and South Africa (35 million).

Wool is the primary material for many kinds of apparel and is used for a number of other products such as insulating materials and carpets. Although a large number of synthetic fabrics have been developed in recent decades, none combines all of the attractive qualities of wool. Wool is likely to remain an important textile material for many years to come.

See also ACRYLICS; ANIMAL BREEDING, SELECTIVE; COTTON; FACTORY SYSTEM; INDUSTRIAL REVOLUTION; LUDDISM; NYLON; RAYON

Word Processing

The invention of the typewriter brought fundamental changes to office work. Typewriters produced cleanly lettered documents at a rapid pace, but they had a number of shortcomings. The slightest typing mistake required erasure and retyping, which slowed things down and often left unsightly smudges. Any modification to a document required that it be retyped, even if only one word had to be changed. The typing of dupli-

cate documents entailed the use of carbon paper, a messy business that allowed the production of only three or four reasonably clear copies.

Efforts to overcome these difficulties date back to the early decades of the 20th century. During World War I, the Hooven Co. produced automatic typewriters that worked on the same principle as a player piano; the striking of the keys was governed by an embossed cylinder. In 1932, the American Automatic Typewriter Co. began to produce its Auto-Typist. This machine recorded up to 300 typed lines on a punched-paper memory tape that controlled the typewriter as it generated form letters at a rate of up to 150 words per minute. A punched paper tape also served as the basis of an automatic typewriter manufactured and marketed by Commercial Controls, Inc. The memory unit had been designed during World War II by International Business Machines (IBM), a firm that eventually would play a leading role in the development of automatic typewriters.

At about this time, the development of electronic computers was opening up the possibility of manipulating words as well as numbers. By the late 1950s, some academic institutions were using time-shared computers to experiment with what was then known as *text editing*. In the years that followed, many text-editing features were created in computer science departments, but the major source of development came through efforts to produce a better typewriter. In 1964, IBM began to market a memory typewriter, the magnetic-tape Selectric. Its memory resided in a 1.27-cm-wide (.5-in.) magnetic tape; 30.5 m (100 ft) of this tape could store up to 24,000 characters. The memory allowed multiple copies of the same document to be typed automatically. Equally important, it allowed text editing: If a word had to be changed, the operator did not have to start all over again. Corrections and changes could be made while the document was being composed and then stored in the memory, which in turn was used to produce one or more printed documents.

While pioneering the use of magnetic tape for typewriters, IBM coined the term *word processing*. At first, IBM assumed that the memory Selectric typewriter would be used primarily for the automatic writing of business letters. However, one of IBM's employees in Germany had a more expansive vision. As he saw it, automatic typewriters could serve as the basis for the centralized typing of dictated letters. In German, this was called *textverarbeitung*, which IBM translated into English as *word processing*, the verbal analog of data processing.

First-generation word processors were an improvement over the typewriters of the day, but they had significant limitations. For example, it was not possible to insert new lines of text that were longer than the deleted lines. The typist also had to enter editing codes along with the new text. The capabilities of early word processors were considerably expanded when they gained a second driver for magnetic cards or tape. This allowed the combination of data from two sources; for example, a list of mailing addresses could be used in conjunction with a form letter.

In early the 1970s, the second generation of word processors made its appearance. These machines displayed the text on a videoscreen, allowing the operator to see a portion of the text, along with any changes that had been made. Known at the time as *display text editors*, the machines incorporated a number of new capabilities, such as the ability to move lines and even whole pages (the "cut-and-paste" function), automatic margin justification, and searches for particular words. They also contained libraries of "boilerplate," stored verbiage that could be inserted into a new document as needed. These capabilities did not come cheap; a dedicated word processor could cost up to $20,000 in the mid- to late-1970s.

These machines enjoyed a brief vogue, but as personal computers became more powerful even as their prices were dropping, there was little reason to purchase a dedicated word processor. Beginning in the late 1970s, the rapid diffusion of the personal computer and the growing use of word processing occurred in a reciprocal fashion. The personal computer did much to advance the popularity of word processing, while at the same time, the availability of word-processing programs provided a strong impetus for the purchase of personal computers.

Personal computers doing word processing have become fixtures in many homes and most offices, relegating the conventional typewriter to the status of a museum piece. The ubiquity of word processing has altered routines in many offices. Executives and administrators who in the past relied on secretaries to do their typing now may do it themselves, although some still cling to the notion that typing is a demeaning task. On the secretarial side, word processing has accelerated the trend of centralizing the production of letters and other documents. For typists in this situation, word processing may mean less variation in the tasks they are required to do, making for a more monotonous work environment. Furthermore, typists may find themselves under closer supervision due to the use of word-processing programs that count the number of words typed in a given period of time.

See also COMPUTER, PERSONAL; TYPEWRITER

X Rays

In late 1895, the German physicist Wilhelm Konrad Roentgen (1845–1923) was performing a set of experiments with cathode rays. In particular, he was interested in a phenomenon reported by other experimenters, the fluorescing of certain materials when they were struck by the rays. In order to better observe faint fluorescence, Roentgen darkened the room and wrapped a sheet of thin, black cardboard around the cathode-ray tube. As the tube went into operation, he saw that a sheet of paper coated with a fluorescent substance (barium platinocyanide) was glowing, even though it was in another part of the room. Roentgen turned off the cathode-ray tube; the paper went dark. He turned it back on; the paper glowed. He took the paper into the next room and closed the door, and still the paper glowed.

Through further experimentation, Roentgen determined that the strange radiation passed through thin sheets of metal. He discovered that it ionized gasses, and that it was unaffected by magnetic or electrical fields. He also noted that the radiation moved in straight lines and that its absorption depended on the density of the material at which it was directed. Although he began to ascertain some of its properties and effects, Roentgen was mystified about the nature of the radiation. He therefore called it an X ray, since *X* is the mathematical symbol for an unknown.

In addition to publishing his findings in an academic journal, Roentgen gave a public demonstration of X rays on Jan. 23, 1896. At this demonstration, an X-ray photograph was taken of the hand of an 80-year-old volunteer, clearly revealing the bones that had been hidden underneath the skin. The medical value of X rays was grasped immediately. In the United States, X rays were used to locate a bullet lodged in a patient's leg—a mere 4 days after the arrival of news of Roentgen's discovery.

X rays allowed hidden objects like bones and bullets to be seen because the rays easily penetrate soft tissue but are largely stopped by hard substances like bones. Bones therefore block portions of an X-ray beam and cast a kind of shadow on a photographic plate. X rays have great penetrating power because they have short wavelengths and high frequencies. They are part of the electromagnetic spectrum that includes, among other things, radio waves and visible light. Whereas visible light encompasses wavelengths between 390 to 750 nanometers (a nanometer is 10^{-9}m), the wavelength of X rays extends from 0.001 to 10 nanometers.

The unique properties of X rays have led to their being used for much more than medical diagnosis. In 1909, Max von Laue (1879–1960) began to study X rays while at the University of Munich. At that time, the wavelengths of X rays was unknown, for the techniques that had been used to determine the wavelengths of visible light could not be applied. The wavelengths of visible light had been determined by passing the light through a diffraction grating with lines at known distances, but this was not feasible in the case of X rays. Von Laue's solution was to use the regular molecular structures of crystals to diffract the rays and hence allow the calculation of their wavelengths.

This technique was doubly valuable. On the one hand, it allowed the measurement of the wavelengths of X rays by passing them through a crystal with a known structure. On the other hand, through the use of X rays of known wavelength it became possible to study the hitherto unknown atomic structures of certain crystals. The technique could even be used to ascertain the atomic structure of noncrystalline substances like polymers that had internal regularities that were sufficient to diffract the X rays. X-ray diffraction played a key role in one of the most significant scientific events of the 20th century, the discovery of the molecular structure of DNA in 1953.

X rays are also widely used for many other scientific studies. X-ray spectrometry is used in a manner analogous to visible-light spectrometry for the quantitative analysis of a substance's elemental composition. X rays are used for the investigation of very small and very large objects. In the former case, X-ray spectroscopy is used to provide quantitative chemical information about samples as small as 10^{-14} g. In the latter case, X-ray astronomy is concerned with the X rays emanating from distant galaxies. Analyses of X rays that have originated in different parts of the universe are used to provide clues about energy production in the stars, the nature of pulsars and black holes, and the structure of the universe.

It should be evident that X rays have been of the greatest significance for the advancement of science. In an even larger sense, the discovery of X rays marked the initiation of what has been called "the second revolution in science" (the Galilean-Newtonian being the first). At the time Roentgen discovered X rays, physics seemed to have settled into a period of permanent maturity. All the big issues seemed to have been resolved, and all that remained to be done was the ironing out of some details. Suddenly, X rays came along to present a phenomenon that no one could explain. Moreover, less than a year later an experiment with X rays led directly to the discovery of radiation. This in turn initiated the process of analyzing matter into ever-smaller components, a process that continues to this day.

For most people, however, the primary relevance of X rays lies in their medical applications. Although technologies such as computerized axial tommography and nuclear magnetic resonance imaging have complemented and in some cases supplanted the use of X rays, the latter continue to be essential diagnostic aids. So prevalent has been the use of X rays that they have been identified as a safety hazard. In the not-too-distant past, many physicians, dentists, and hospitals made excessive use of X rays, and in so doing put pa-

tients at greater risk of contracting leukemia and other cancers. It is now recognized that a prudent course of action is to receive no more X-ray exposure than is truly necessary.

See also BLACK HOLES; CATHODE-RAY TUBE; COMPUTERIZED AXIAL TOMMOGRAPHY; DNA; MAGNETIC RESONANCE IMAGING; POLYMERS; RADIOACTIVITY AND RADIATION; SPECTROSCOPY; TELESCOPE, RADIO

Xerography

Xerography is a method of making copies of written materials through the use of light. It differs fundamentally from other copying technologies such as microfilming, which is derived from traditional photographic procedures. Unlike these, xerography is a dry process, which is reflected in its name—a derivation of the Greek words for *dry* and *writing*. The first commercially successful xerographic copier was sold under the trade name Xerox, and as sometimes happens with novel products, the name has become a generic term, like kleenex or jello.

As is often the case with inventions, the origins of xerography lay in the desire to save effort. Chester Carlson (1906–1968), the inventor of xerography, was employed to copy patent drawings and documents by P. R. Mallory & Company, a manufacturer of electrical components. Carlson had long wanted to be a professional inventor, and his tedious job motivated him to try to devise an alternative to hand copying. A graduate of the California Institute of Technology, Carlson was well versed in physics, and he knew that some materials changed their electrical charge when exposed to light. He also was aware of the discovery of Paul Selenyi (1884–1954), a Hungarian physicist, that charged particles could be attracted to an oppositely charged surface in a predetermined pattern. Carlson's plan was to use these phenomena for copying through a process he called *electrophotography*.

In 1937, Carlson obtained a patent on the basic principles involved, but electrophotography remained a paper invention. On Oct. 22, 1938, he and his assistant, Otto Kornei, used these principles to produce the first example of electrophotography in a workshop located in the Astoria section of Queens, New York. An electrostatic charge was created by rubbing a cotton handkerchief over a sulfur-coated zinc plate. A glass slide upon which was inscribed the date and the place of the experiment, "Astoria 10.-22.-38," was then laid on plate and exposed to a bright light. The slide was separated from the plate, which was then sprinkled with a vegetable-based powder. Finally, a piece of wax paper was pressed against the plate and the two were heated. When the paper was peeled off, it had on it a somewhat blurred image of the words originally written on the slide.

Carlson built a prototype copier and used it to demonstrate electrophotography to established firms such as IBM, Eastman Kodak, RCA, and General Electric, but none could see any value in a complicated apparatus that seemed to do no more than take the place of carbon paper. In 1944, he finally succeeded in getting a research organization, the Battelle Memorial Institute, to make an initial investment of $3,000 in return for receiving a portion of any royalties garnered through the sales of copiers (their share eventually came to about $350 million).

As with many radically new inventions, while the basic ideas behind electrophotography were correct, a great deal of development work was required in order to translate them into a commercially viable copying machine. Development, in turn, required a considerable infusion of cash, more than the Battelle Institute was able to put up. In 1947, Carlson and Battelle obtained financial support from the Haloid Company, a manufacturer of photographic papers and supplies. The original investment was $25,000, not a huge sum, but one that represented nearly a quarter of Haloid's income in 1946. In addition to putting up the money, the firm coined the term *xerography* for Carlson's process.

In 1949, Haloid produced its first copying machine, the XeroX Model A. Again, the response was subdued, and the Model A found no buyers. In retrospect, the lack of interest seems perverse, but there were reasons for the seeming obtuseness. The Model A was expensive, complicated, and difficult to use, costing around $400 and requiring 14 separate actions by its operator to produce copies. Although it failed as an office copier, the first Xerox machine eventually found a modest commercial niche for making masters used

for offset printing. By the mid-1950s, the company was also making some money with a machine that made prints from microfilm originals. Finally, in 1960 after an investment of $60 million, the firm marketed the first commercially successful copier. Known as the Model 914 because of the size of the paper it used—9 inches (22.9 cm) by 14 inches (35.6 cm)—it worked at a rate of seven copies per minute. Market surveys commissioned by Haloid prior to the introduction of the Model 914 had indicated that xerography was not likely to be a profit-making venture. Market surveys, however, are often inaccurate when a novel product is involved, a fact that was clearly demonstrated by the explosive growth of sales of copiers. By the end of the decade, Xerox machines and their competitors were churning out 10 billion copies per year in the United States alone.

Today's xerographic copiers work on the same basic principles first employed by Chester Carlson. In a typical xerography process, the image to be copied is illuminated, which causes reflected light from the white areas to strike a charged photoconductor belt or plate. The light neutralizes the negatively charged portions, leaving only the dark areas with a charge. Areas with this negative charge then attract particles of positively charged toner. After its charge is relaxed, the photoconductor belt or plate then comes in contact with a negatively charged sheet of paper that attracts the positively charged toner particles. The paper is then separated from the belt or plate, and carried to a place where the toner particles are softened by heat and fused onto the paper's surface, leaving an accurate copy of the original.

In the early 1990s, xerography machines in American offices were turning out an estimated 350 billion copies every year. Not all of these copies were directly relevant to the work being done in these offices; one study found that as many as 37 percent of the copies made were either unnecessary or served personal matters. Misuse of this sort costs American business as much as $2.6 billion annually. The overuse of xerography provides an excellent example of the tendency of some technologies to create a new set of needs. In an ironic reversal of the days when there seemed to be no demand for xerographic copying, there is now a strong tendency to engage in promiscuous copying. Where one additional copy for some other party might have been deemed sufficient in the past, the ability to make copies rapidly and inexpensively has created a situation where anyone involved, however remotely, is now furnished with copies of memos, letters, reports, and all other pieces of communication relating to the matter at hand.

Excessive copying occurs in cultures where information circulates with little hindrance. But in closed societies such as the Soviet Union, access to xerographic machines was strictly limited for fear that they would be used for the dissemination of material con-

sidered dangerous by the government. Consequently, the Soviet Union was the world's largest consumer of carbon paper and had some of the world's most inefficient bureaucracies. Finally, although xerographic copiers posed a serious threat to authoritarian governments, they can be a problem for democratic governments too, for they now produce such accurate copies that they can be used for counterfeiting and the forging of identity cards and other documents. Defeating misused xerography will require the development of new, copy-resistant technologies.

See also RESEARCH AND DEVELOPMENT

Z

Zipper

Zippers are used today on virtually every article of clothing: jackets, trousers, dresses, blouses, shirts, and skirts. Yet the first use of the principle embodied in the zipper was for footwear. In the late 19th century, men, women, and children were commonly shod with high-button shoes, highly practical for walking through mud and dust, but a pain to take on and off. In 1893, a young inventor named Whitcomb Judson (1846–1909) received

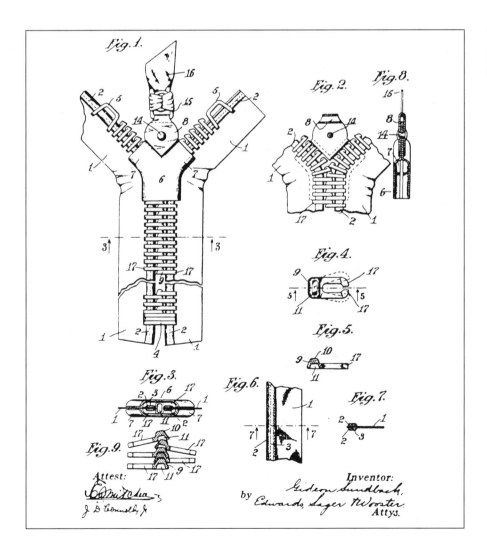

The patent application for Gideon Sundbeck's "Separable Fastener" (from Robert F. Friedel, *Zipper: An Exploration in Novelty,* 1994).

two patents for a "clasp locker or unlocker for shoes." In the first patent description, this device employed parallel rows of clasps that were engaged and disengaged by a slider; the second employed hooks and eyes. The slider was not permanently attached; it was removable. It was used from either end, depending on whether the fastener was to be opened or closed. Unlike many inventors, Judson was fortunate in having a financial backer, Lewis Walker (1855–1938), whose support was essential during the many years that the fastener failed to be either technically or financially successful.

In 1904, Judson modified his design so that the entire fastener was clamped to the edge of a cloth tape, obviating the need to tediously attach each clasp. At about this time the fastener was being marketed for use on women's shirts and dresses, but its unreliability prevented its widespread use.

The basic principle was finally employed for an effective fastener by a Swedish immigrant, Gideon Sundback (1880–1954). After first making some improvements to Judson's design, in 1913 Sundback came up with a completely redesigned fastener that employed interlocking cup-shaped members in the place of hooks and eyes. At least as important as the design was the machinery that produced it, for there was no way that the basic elements could be made by hand. For the fastener to function properly, each part had to be virtually identical to the others, as well as be exactly aligned. Yet this precision did not come at a high price; in 1914 Sundbeck's Hookless #2 was being sold at four for a dollar.

A tough marketing job remained. Clothing manufacturers were not convinced that customers would accept clothing with the new fastener, and balked at what they perceived as an excessively high cost. Instead of being used for clothing, the fastener caught on when employed for money belts sold to soldiers and sailors embarking for service in World War I. A rather surprising success was the Locktite tobacco pouch, which, as its makers proclaimed, used "no strings—no buttons." A major success came in 1923, when the B. F. Goodrich Company introduced a line of galoshes that used the fastener. These were called "zipper boots," named after the sound made when the boot was opened or closed. Soon the name *zipper* was being applied to the fastener itself.

By the 1930s, zippers were being used for a variety of articles of apparel, especially children's clothes. Toward the end of the decade, designers of high-fashion men's and women's clothing began to employ zippers, which fit perfectly with the sleek look in vogue at the time. Reinforced by intensive advertising by zipper manufacturers, these fashion trends culminated in the widespread use of zippers, 300 million of which were sold in 1939. By 1950, the figure had risen to more than a billion. The zipper had become commonplace, an item whose ubiquity masked an impressive level of precision in its design and manufacture.

Further Reading: Robert F. Friedel, *Zipper: An Exploration in Novelty*, 1994.

Bibliography

Adamson, A.W. *Textbook of Physical Chemistry*, New York: Academic Press, 1973.

Ahrens, C. Donald. *Meteorology Today: An Introduction to Weather, Climate, and the Environment.* St. Paul, MN: West, 1994.

Aitken, Hugh G.J. *The Continuous Wave: Technology and American Radio, 1900–1932.* Princeton, NJ: Princeton University Press, 1985.

———. *Syntony and Spark: The Origins of Radio.* Princeton, NJ: Princeton University Press, 1985.

Alic, John A., et al. *Beyond Spinoff: Military and Commercial Technologies in a Changing World.* Boston: Harvard Business School Press, 1992.

Althouse, Andrew D., et al. *Modern Welding.* South Holland, IL: Goodheart-Wilcox, 1980.

American Cancer Society. *Breast Cancer Facts and Figures, 1996.* ACS, Dec. 1995.

Anderson Jr., Oscar Edward. *Refrigeration in America: A History of a New Technology and Its Impact.* Princeton, NJ: Princeton University Press, 1953.

Anderson, Robert O. *Fundamentals of the Petroleum Industry.* Norman, OK: University of Oklahoma Press, 1984.

Andrews, Clinton J., ed. *Regulating Regional Power Systems.* Westport, CT: Quorum Books, 1995.

Apple, Rima. *Mothers and Medicine: A Social History of Infant Feeding 1890–1950.* Madison, WI: University of Wisconsin Press, 1987.

Arms Control Today. Start Supplement (Nov. 1991), *Start II Supplement* (Jan./Feb. 1993), and *Start II Resolution of Ratification* (Feb. 1996).

Armstrong, John. *The Railroad: What It Is, What It Does*, 3d ed. Omaha, NE: Simmons-Boardman, 1990.

Asimov, Isaac. *Eyes on the Universe: A History of the Telescope.* Boston: Houghton Mifflin, 1975.

Atkins, P.W. *The Second Law: Energy, Chaos, and Form.* San Francisco: Freeman, 1984.

Bachrach, Bernard S. "Charles Martel, Mounted Shock Combat, the Stirrup, and Feudal Origins." In *Studies in Medieval and Renaissance History*, vol. 7, 1970, pp. 47–76.

Baker, M.N. *The Quest for Pure Water*, 2 vols. New York: American Water Works Association, 1981.

Baker, T. Lindsay. *Blades in the Sky: Windmilling through the Eyes of B. H. "Tex" Burdick.* Lubbock, TX: Texas Tech University Press, 1992, p. 39.

Balcomb, J. Douglas, ed. *Passive Solar Buildings.* Cambridge, MA: MIT Press, 1992.

Barraclagh, K.C. *Steelmaking, 1850–1890.* London: Institute of Metals, 1990.

Bartky, Ian R. "The Adoption of Standard Time." *Technology and Culture*, vol. 30, no. 1 (Jan. 1989) pp. 25–56.

Basalla, George. *The Evolution of Technology.* Cambridge and New York: Cambridge University Press, 1988.

Belew, Leland F., ed. *Skylab, Our First Space Station.* Washington, DC: National Aeronautics and Space Administration, 1977.

Bell, Daniel. *The Coming of Post-Industrial Society: A Venture in Social Forecasting.* New York: Basic Books, 1973.

Belton, John. *Widescreen.* Cambridge, MA: Harvard University Press, 1992.

Benarde, Melvin A. *The Chemicals We Eat.* New York: McGraw-Hill, 1971.

Ben David, Joseph. *The Scientist's Role in Society.* Englewood Cliffs, NJ: Prentice Hall, 1971.

Bender, Barry, and Ralph M Stair, Jr. *Quantitative Analysis for Management.* Boston: Allyn & Bacon, 1991.

BerggrenT, W.A., and John A. Van Couvering. *Catastrophes and Earth History: The New Uniformitarianism.* Princeton, N.J: Princeton University Press, 1984.

Bijker, Wiebe E. "The Social Construction of Fluorescent Lighting: Or How an Artifact Was Invented in Its Diffusion Stage." *In* Weibe E. Bijker and John Law. *Shaping Technology/Building Society: Studies in Sociotechnical Change.* Cambridge, MA: MIT Press, 1992.

———. *Of Bicycles, Bakelite, and Bulbs: Toward a Theory of Sociotechnical Change.* Cambridge, MA: MIT Press, 1995.

Bilstein, Roger E. *Flight in America: From the Wrights to the Astronauts.* Baltimore: The Johns Hopkins University Press, 1984.

Blackmore, Howard L. *Guns and Rifles of the World.* New York: Viking, 1965.

Bolt, Bruce A. *Earthquakes: A Primer.* New York: Freeman, 1988, pp. 99–115.

Bott, Martin Harold Phillips. *The Interior of the Earth.* New York: St. Martin's Press, 1971.

Bowen, Robert. *Geothermal Resources,* 2d ed. New York: Elsevier Science, 1989.

Bowler, Peter J. *Evolution: The History of an Idea,* rev. ed. Berkeley, CA: University of California Press, 1989.

Bradbury, S. *The Evolution of the Microscope.* Oxford, UK: Pergamon Press, 1967.

Brandt, Allan M. "Racism and Research: The Case of the Tuskegee Syphilis Study." In *The Hastings Center Report,* vol. 54 (Dec. 1978): 21–29.

Braun, Ernest, and Stuart MacDonald. *Revolution in Miniature: The History and Impact of Semiconductor Electronics.* Cambridge and New York: Cambridge University Press, 1978.

Britten, Karen Gerhardt. *Bale o' Cotton: The Mechanical Art of Cotton Ginning.* College Station, TX: Texas A&M University Press, 1992.

Broad, William, and Nicholas Wade. *Betrayers of the Truth: Fraud and Deceit in the Halls of Science.* New York: Simon & Schuster, 1982.

Brock, William H. *The Norton History of Chemistry.* New York: Norton, 1993.

Brown, Michael E. *Flying Blind: The Politics of the U.S. Strategic Bomber Program.* Ithaca, NY: Cornell University Press, 1992.

Bruce, Alfred W. *The Steam Locomotive in America: Its Development in the Twentieth Century.* New York: Norton, 1952.

Burlingame, R. *March of the Iron Men.* New York: Grosset & Dunlop, 1938.

Burns, Russell, ed. *Radar Development to 1945.* London: Peter Peregrinus, 1988.

Bush, Donald J. *The Streamlined Decade.* New York: George Braziller, 1975.

Bushong, Stewart C. *Magnetic Resonance Imaging: Physical and Biological Principles.* St Louis: Mosby, 1988.

Butti, Ken, and John Perlin. *A Golden Thread: 2500 Years of Solar Architecture and Technology.* New York: Van Nostrand Reinhold, 1980.

Cambridge Encyclopedia of Space. Cambridge: Cambridge University Press, 1990, pp. 278–83.

Carson, Rachel. *Silent Spring.* Boston: Houghton Mifflin, 1987. Originally published in 1962.

Cassidy, David. *Einstein and Our World.* Atlantic Highlands, NJ: Humanities Press, 1995.

Cellular Telecommunications Industry Assoc., Washington, DC 20036.

Chaisson, Eric J. *The Hubble Wars.* New York: HarperCollins, 1994.

Chapman, S.D., and Chambers, J.D. *The Beginnings of Industrial Britain.* London: University Tutorial Press, 1970.

Charren, Peggy. "A Public Policy Perspective on Televised Violence and Youth." *Harvard Educational Review* (Summer 1995): 282-91.

Clancy, Tom. *Submarine: A Guided Tour Inside a Nuclear Warship.* New York: Berkeley Books, 1993.

Clark, Ezekail L. "Cogeneration—Efficient Energy Source." *Annual Review of Energy,* vol. II (Nov. 1986): 275–94.

Close Frank. *Too Hot to Handle: The Race for Cold Fusion.* Princeton, NJ: Princeton University Press, 1991.

———, Michael Marten, and Christine Sutton. *The Particle Explosion.* New York: Oxford University Press, 1987.

Coel, Margaret. "Keeping Time by Atom." *American Heritage of Invention and Technology,* vol. 3, no. 3 (Winter 1988).

Cohen, I. Bernard. *Revolution in Science.* Cambridge, MA: Harvard University Press, 1985.

Collins, Harry, and Trevor Pinch. *The Golem: What Everyone Should Know About Science.* Cambridge and New York: Cambridge University Press, 1993, pp. 79–90.

Commission on Life Sciences, National Research Council. *The Use of Laboratory Animals in Biomedical and Behavioral Research.* Washington, DC: National Academy Press, 1988.

Condon, W.U., and G.H. Shortley. *The Theory of Atomic Spectra.* Cambridge and New York: Cambridge University Press, 1957.

Constant, Edward. *Origins of the Turbojet Revolution.* Baltimore and London: Johns Hopkins University Press, 1980.

Coombs, Charles. *Soaring.* New York: Henry Holt, 1988.

Cooper, Bryan, and John Batchelor. *Fighter: A History of Fighter Aircraft.* New York: Ballantine, 1973.

Cooper, C.C. *Shaping Invention.* New York: Columbia University Press, 1991.

Cooper, Gail. *Air Conditioning in America: Engineers and The Controlled Environment 1900–1960.* Baltimore: Johns Hopkins University Press, 1998.

Cooper, Grace Rogers. *The Sewing Machine: Its Invention and Development.* Washington, DC: Smithsonian Institution Press, 1976.

Copp, Newton H., and Andrew W. Zanella. *Discovery, Innovation and Risk: Case Studies in Science and Technology.* Cambridge, MA: MIT Press, 1993, pp. 192–242.

Cortada, James W. *Before the Computer.* Princeton, NJ: Princeton University Press, 1993.

Cortwright, Edgar M., ed. *Apollo Expeditions to the Moon.* Washington, DC: National Aeronautics and Space Administration, 1975.

Cowan, Ruth Schwartz. *More Work for Mother: The Ironies of Household Technology from the Open Hearth to the Microwave.* New York: Basic Books, 1983.

Crispeels, Maarten J., and David Sadava. *Plants, Food, and People.* San Francisco: Freeman, 1977.

Crowe, Michael J. *The Extraterrestrial Life Debate 1750–1900: The Idea of a Plurality of Worlds from Kant to Lowell.* Cambridge, UK: Cambridge University Press, 1986.

Cummings, Michael R. *Human Heredity: Principles and Issues,* 3d ed. St. Paul, MN: West, 1994.

Curvin, Robert, and Bruce Porter. *Blackout Looting! New York City, July 13, 1977.* New York: Gardner Press, 1979.

Dalgleish, D. Douglas, and Larry Schweikart. *Trident.* Carbondale, IL: Southern Illinois University Press, 1984.

DeBoer, David J. *Piggyback and Containers: A History of Rail Intermodal on America's Steel Highway.* San Marino, CA: Golden West Books, 1992.

de Bono, Edward. *Eureka: An Illustrated History of Inventions.* London: Thames & Hudson, 1974.

Delany, Janice, Mary Jane Lupton, and Emily Toth. *The Curse: A Cultural History of Menstruation.* Urbana, IL: University of Illinois Press, 1988.

De Laquil, III, Pascal, et al. "Solar-Thermal Electric Technology." *In* Thomas B. Johansson et al., eds. *Renewable Energy: Sources for Fuels and Electricity.* Washington, DC: Island Press, 1993.

Denning, Peter J., ed. *Computers Under Attack: Intruders, Worms, and Viruses.* Reading, MA: Addison-Wesley, 1990.

DeVries, Kelly. *Medieval Military Technology.* Peterborough, Ontario: Broadview Press, 1992.

Dick, Steven J. *The Biological Universe: The Twentieth Century Extraterrestrial Life Debate and the Limits of Science.* Cambridge and New York: Cambridge University Press, 1996.

————. *Plurality of Worlds: The Origins of the Extraterrestrial Life Debate from Democritus to Kant.* Cambridge, MA: Cambridge University Press, 1982.

Dickerson, R.E. "The DNA Double Helix and How It Is Read." *Scientific American* (1983).

Donaldson, Barry, and Bernard Nagengast. *Heat and Cold: Mastering the Great Indoors: A Selective History of Heating, Ventilation, Refrigeration, and Air Conditioning.* Atlanta: American Society of Heating, Refrigeration, and Air-Conditioning Engineers, 1994.

Dowling, Harry F. *Fighting Infection: Conquests of the Twentieth Century.* Cambridge, MA: Harvard University Press, 1977, pp. 105–57.

Dunsheath, Percy. *A History of Electrical Power Engineering.* Cambridge, MA: MIT Press, 1962, pp. 89–122.

Dutton, Diana B. "DES and the Elusive Goal of Drug Safety." In *Worse Than the Disease: Pitfalls of Medical Progress.* Cambridge and New York: Cambridge University Press, 1988, pp. 31–90.

Eberly, Joseph H., and Peter W. Milonni. *Lasers.* New York: Wiley, 1988.

Edgerton, Harold E., and James R. Killian, Jr. *Moments of Vision: The Stroboscopic Revolution in Photography.* Cambridge, MA: MIT Press, 1979.

Einstein, Albert. *Relativity: The Special and General Theory.* London, Methuen, 1954.

Eisenstein, Elizabeth. *The Printing Revolution in Early Modern Europe.* Cambridge and New York: Cambridge University Press, 1984.

Elliott, Cecil D. *Technics and Architecture: The Development of Materials and Systems for Buildings.* Cambridge, MA: MIT Press, 1992.

Ellis, John. *The Social History of the Machine Gun.* New York: Pantheon, 1975.

Elzen, Boelie. "Two Ultracentrifuges: A Comparative Study of the Social Construction of Artifacts." *Social Studies of Science,* vol. 16 (1986): 621–62.

Enos, John Lawrence. *Petroleum Progress and Profits: A History of Process Innovation.* Cambridge, MA: MIT Press, 1962, pp. 1–130.

Evans, Hughes. "Losing Touch: The Controversy Over the Introduction of Blood Pressure Instruments into Medicine." *Technology and Culture,* vol. 34, no. 4 (Oct. 1993): 784–807.

Ezell, Edward C. *The Great Rifle Controversy.* Harrisburg, PA: Stackpole, 1984.

Fakhry, Ahmed. *The Pyramids.* Chicago: University of Chicago Press, 1969.

Farey, J.A. *A Treatise on the Steam Engine.* London, 1827.

Farrington, Benjamin. *Greek Science.* Baltimore: Penguin, 1953.

Fisher, David E., and Marshall, Jon Fisher. *Tube: The*

Invention of Television. San Diego, CA: Harcourt, Brace, 1997.

Fitzgerald, Deborah. *The Business of Breeding: Hybrid Corn in Illinois.* Ithaca, NY: Cornell University Press, 1990.

Flink, James. *The Automobile Age.* Cambridge, MA: MIT Press, 1988.

Foley, Vernard George Palmer, and Werner Soedel. "The Crossbow," *Scientific American*, vol. 252, no. 1 (Jan. 1985): 104–10.

Frank, Felix. *Polywater.* Cambridge, MA.: MIT Press, 1981.

Freight Car Design Manual. Milwaukee: William K. Walthers, 1946.

Friedel, Robert F. *Pioneer Plastic: The Making and Selling of Celluloid.* Madison, WI: University of Wisconsin Press, 1983.

———. *Zipper: An Exploration in Novelty.* New York: Norton, 1994.

Fritsche, P.A.. *A Nation of Fliers.* Cambridge and London: Harvard University Press, 1992.

Fuhrman, R.A. "The Fleet Ballistic Missile System: Polaris to Trident." *The Journal of Spacecraft*, vol. 15, no. 5. AIAA Paper 77–355, reprinted in Sep.–Oct. 1978, pp. 265–86.

Gamow, George. *Gravity.* Garden City, NY: Doubleday, 1962.

———. *One, Two, Three, Infinity.* New York: Viking, 1961.

Gardiner, Martin. *The Relativity Explosion.* New York: Vintage Books, 1976.

Gatland, Kenneth. *The Illustrated Encyclopedia of Space Technology: A Comprehensive History of Space Exploration.* New York: Harmony Books, 1981, pp. 182–89.

Gies, Francis, and Joseph Gies. *Cathedral, Forge and Waterwheel.* New York: HarperCollins, 1994.

Gilbertson, Roger G. *Muscle Wires Project Book.*, 3d ed. San Anselmo, CA: Mondo-tronics, 1994.

Gilder, George. *Life After Television.* Nashville, TN: Whittle Books, 1990.

Glaser, Anton. *History of Binary and other Nondecimal Numeration.* Los Angeles: Tomask, 1981.

Glischinski, Steve. *Santa Fe Railway.* Oscala, WI: Motorbooks International, 1997.

Gillessen, Klaus, and Werner Schairer. *Light Emitting Diodes: An Introduction.* Englewood Cliffs, NJ: Prentice-Hall, 1987.

Giordano, Frank R., and Maurice D. Weir. *A First Course on Mathematical Modeling.* Monterey, CA: Brooks-Cole, 1985.

Goldberg, Stanley. *Understanding Relativity: Original Impact of a Scientific Revolution.* Boston: Birkhaüser, 1984.

Goodrich, L.C., and Comeran, N. *The Face of China.* New York: Aperture, 1978

Gordon, John Steele. "The Chicken Story." *American Heritage*, vol. 47, no. 5 (Sep. 1996).

Gott, Philip G. *Changing Gears: The Development of the Automotive Transmission.* Warrendale, PA: Society of Automotive Engineers, 1991.

Graham, Margaret B. *RCA and the VideoDisc: The Business of Research.* Cambridge and New York: Cambridge University Press, 1986.

Grainger, A. *The Threatening Desert: Controlling Desertification.* London: Earthscan, 1990.

Gray, Leonard. *How to Navigate Today*, 6th ed. Cambridge, MD: Cornell Maritime Press, 1986.

Gray, Jeremy. *Ideas of Space: Euclidean, Non-Euclidean, and Relativistic.* Oxford, UK: Clarendon Press, 1989.

Greenleaf, William. *Monopoly on Wheels: Henry Ford and the Selden Automobile Patent.* Detroit, MI: Wayne State University Press, 1961.

Greenwood, Ted. *The Making of the MIRV: A Study of Defense Decision Making.* Lanham, MD: University Press of America, 1988.

Griffiths, Anthony, J.F., et al. *An Introduction to Genetic*

Analysis, 5th ed. New York: Freeman, 1993, p. 310.

Grumbine, R. Edward. *Ghost Bears: Exploring the Biodiversity Crisis.* Washington, DC: Island Press, 1992.

Gunston, Bill, and Mike Spick. *Modern Air Combat.* New York: Random House, 1988.

Gurcke, Karl. *Bricks and Brickmaking: A Handbook for Historical Archaeology.* Moscow ID: University of Idaho Press, 1987.

Guthke, Karl S. *The Last Frontier: Imagining Other Worlds, from the Copernican Revolution to Modern Science Fiction.* Ithaca, NY: Cornell University Press, 1990.

Hackmann, William. *Seek & Strike: Sonar, Anti-Submarine Warfare and the Royal Navy, 1914–1954.* London: Her Majesty's Stationery Office, 1984.

Hagar, Charles F. *Planetarium: Window to the Universe.* Oberkochen, W. Germany: Carl Zeiss, 1980.

Hagstrom, Warren. *The Scientific Community.* New York: Basic Books, 1965.

Haines, Charles. *The Industrialization of Wood: The Transformation of a Material.* Ph.D. dissertation, University of Delaware, 1990.

Haining, Peter. *The Dream Machines.* New York: World, 1973.

Hallion, Richard P. *The Hypersonic Revolution: Eight Case Studies in the History of Hypersonic Technology: vol. 1, From Max Valier to Project Prime.* Dayton, OH: USAF Special Staff Office, Aeronautical Systems Division, 1987, and *vol. 2: From Scramjet to the National Aero-Space Plane,* 1987.

Handel, S. *The Electronic Revolution.* Hammondsworth, UK: Penguin, 1967.

Hansen, James R. *Engineer in Charge: A History of the Langley Aeronautical Laboratory, 1917–1958.* Washington, DC: NASA, NASA History Series, 1987.

Hanson, David J. "New Agent Orange Study Links Herbicides to Diseases." *Chemical and Engineering News,* vol. 71 (Aug. 23, 1993).

Hard, M. *Machines are Frozen Spirit.* Boulder, CO: Westview, 1994.

Hardy, Robert. *Longbow: A Social and Military History,* 3d ed. London: Bois d'Arc Press, 1992.

Harris, J.R. *Industry and Technology in the Eighteenth Century, 1780–1850.* Hampshire, UK: Vaiorum, 1992.

Hawkes, Nigel, et al. *Chernobyl: The End of the Nuclear Dream.* New York: Vintage, 1986.

Headrick, Daniel R. *The Invisible Weapon: Telecommunications and International Politics, 1851–1945.* New York: Oxford University Press, 1991, pp. 11–115.

Heath, T.L. ed. *The Thirteen Books of Euclid's Elements.* Dover, NY: 1956.

Heppenheimer, T.A. "Flying Blind." *American Heritage of Invention and Technology,* vol. 10, no. 4 (Spring 1995): 54–63.

Herbert, Vernon, and Attilio Bisio. *Synthetic Rubber: A Project That Had to Succeed.* Westport, CT: Greenwood Press, 1985.

Herman, Robin. *Fusion: The Search for Endless Energy.* Cambridge and New York: Cambridge University Press, 1990.

Herring, Thomas A. "The Global Positioning System." *Scientific American,* vol. 274, no. 2 (Feb. 1996), 44–50.

Hickman, Ian. *Oscilloscopes: How to Use Them, How They Work,* 4th ed. Oxford, UK: Newnes, 1995.

Hilgartner, Stephen, Richard C. Gell, and Rory O'Connor. *Nukespeak: Nuclear Language, Vision, and Mindset.* San Francisco: Science Club Books, 1982.

Hill, Donald R. "Trebuchets." *Viator,* vol. 4 (1973): 99–114.

Hills, Richard L. *Power from Steam: A History of the Stationary Steam Engine.* Cambridge, UK: Cambridge University Press, 1989.

Hindle, Brooke, and Steven Lubar. *Engines of Change: The American Industrial Revolution 1790-1860.* Washington, DC: Smithsonian Institution Press, 1986.

Hirsh, Richard F. *Technology and Transformation in*

the Electric Utility Industry. Cambridge and New York: Cambridge University Press, 1989.

Hoffman, Banesh. *The Strange Story of the Quantum.* New York: Dover, 1959.

Hogan, William T., S.J. *An Economic History of the Iron and Steel Industry in the United States,* 5 vols. Lexington, MA: Lexington Books, 1971.

Hogg, Ian V. *Military Small Arms of the 20th Century,* 5th ed. Northfield, IL: DBI, 1985.

Holden, Alan, and Phylis Singer. *Crystals and Crystal Growing.* Garden City, NY: Doubleday, 1960.

Holton, Gerald. *Introduction to Concepts and Theories in Physical Science,* 2d ed., rev. and enlarged by Stephen Brush. Reading, MA: Addison-Wesley, 1973, ch. 25.

Hounshell, David. *From the American System to Mass Production, 1800–1932: The Development of Manufacturing Technology in the United States.* Baltimore: John Hopkins University Press, 1984.

———, and John K. Smith. *Science and Corporate Strategy: Du Pont R&D, 1902–1980.* Cambridge, UK: Cambridge University Press, 1989.

Hoy, Suellen. "The Garbage Disposer, the Public Health, and the Good Life." *Technology and Culture,* vol. 26, no 4 (Oct. 1985).

Hughes, Thomas P. *Elmer Sperry: Inventor and Engineer.* Baltimore: Johns Hopkins University Press, 1971.

Hutcheson, G. Dan, and Jerry D. Hutcheson. "Technology and Economics in the Semiconductor Industry." *Scientific American,* vol. 274, no. 1 (Jan. 1996): 54–62.

Jackson, Donald C. *Building the Ultimate Dam: John S. Eastwood and the Control of Water in the West.* Lawrence, KS: University Press of Kansas, 1995.

Jaffe, Bernard. *Michelson and the Speed of Light.* Garden City, NY: Doubleday, 1960.

Jansen, Niels. *Pictorial History of American Trucks.* Bideford, UK: Bag View Books, 1995.

Jenkins, Virginia Scott. *The Lawn: A History of an American Obsession.* Washington, DC: Smithsonian Institution Press, 1994.

Johansson, Thomas B., et al., ed. *Renewable Energy: Sources for Fuels and Electricity.* Washington, DC: Island Press, 1993, chs. 6–11.

Joravsky, David. *The Lysenko Affair.* Cambridge, MA: Harvard University Press, 1970.

Joseph, Lawrence E. *Gaia: The Growth of an Idea.* New York: St. Martin's Press, 1990.

Kandel, Eric R., and James H. Schwartz. *Principles of Neural Science.* New York: Elsevier Science Publishing, 1985.

Kaplan, Lawrence J., and Rosemarie Tong. *Controlling Our Reproductive Destiny.* Cambridge, MA: MIT Press, 1994.

Kearey, Philip, and Frederick J. Vine. *Global Tectonics.* Cambridge, MA: Blackwell Scientific Publications, 1990.

Kevles, Daniel J. *In the Name of Eugenics: Genetics and the Uses of Human Heredity.* New York: Knopf, 1985.

Kit Planes magazine and publications of the *Experimental Aircraft Association, Inc.* Information may be obtained from the latter at EAA, Wittman Airfield, Oshkosh, WI 54903–3086.

Klotz. Irving M. "The N-Ray Affair." *Scientific American,* vol. 242, no. 5 (May 5, 1980): 168–75.

Knights, David, and Hugh Willmott, eds. *New Technology and the Labour Process.* Basingstoke, UK: Macmillan, 1988.

Kotz, John C., and Keith F. Purcell. *Chemistry and Chemical Reactivity,* 2d ed. Philadelphia: Saunders College Publishing, 1991.

Kranakis, Eda Fowlkes. "The French Connection: Giffard's Injector and the Nature of Heat." *Technology and Culture,* vol. 23, no. 1 (Jan. 1982): 3–38.

Kranzberg, M., and C.W. Pursell, Jr. *Technolgy in Western Civilization,* vol. I. New York: Oxford University Press, 1967.

Krasnow, Erwin G., Lawrence D. Longley, and Herbert A. Terry. *The Politics of Broadcast Regulation*, 3d ed. New York: St. Martin's Press, 1982.

Kuhn, Thomas. "Energy Conservation as an Example of Simultaneous Discovery." *In* Marshall Clagett, ed. *Critical Problems in the History of Science.* Madison, WI: University of Wisconsin Press, 1969, pp. 321–56.

———. *The Structure of Scientific Revolutions*, 2d ed. Chicago: University of Chicago Press, 1970.

Kunkle, Gregory C. "Technology in the Seamless Web: Success and Failure in the History of the Electron Microscope." *Technology and Culture*, vol. 36, no. 1 (Jan. 1995).

Lambright, W. Henry. *Shooting Down the Nuclear Plane.* Indianapolis, IN: Bobbs-Merrill, 1967.

Landes, David. *Revolution in Time: Clocks and the Making of the Modern World.* Cambridge, MA: Harvard University Press, 1983.

Lardner, James. *Fast Forward: Hollywood, the Japanese, and the VCR Wars.* New York: Mentor, 1988.

Law, R.J. *James Watt and the Separate Condenser: An Account of the Invention.* London: Her Majesty's Stationery Office, 1969.

Lay, M.G. *Ways of the World: A History of the World's Roads and the Vehicles That Used Them.* New Brunswick, NJ: Rutgers University Press, 1992.

Leahey, Thomas H., and Richard J. Harris. *Learning and Cognition*, 4th ed. Upper Saddle River, NJ: Prentice-Hall, 1997.

Leiter, Robert D. *Featherbedding and Job Security.* New York: Twayne, 1964.

Leslie, Jacques. "Food Irradiation." *Atlantic Monthly* (Sep. 1990).

Lewin, R. *Human Evolution: An Illustrated Introduction.* New York: Freeman, 1984.

Lewis, Tom. *Empire of the Air: The Men Who Made Radio.* New York: HarperCollins, 1991.

Liebs, Chester A. *Main Street to Miracle Mile: American Roadside Architecture.* Baltimore: Johns Hopkins University Press, 1985.

Lifshey, Earl. *The Housewares Story: A History of the American Housewares Industry.* Chicago: National Housewares Manufacturers Association, 1973.

Lilienthal, David E. *TVA: Democracy on the March.* New York: Harper & Brothers, 1944.

Lubar, Steven. "Do Not Fold, Spindle, or Mutilate: The Cultural History of the Punch Card." *Journal of American Culture* (Winter 1992).

Macey, Samuel L. *Clocks and the Cosmos: Time in Western Life and Thought.* Hamden, CT: Archon Books, 1980.

McClellan, James E. III. *Science Reorganized: Scientific Societies in the Eighteenth Century.* New York: Columbia University Press, 1985.

McCurdy, Howard E. *Inside NASA: High Technology and Organizational Change in the U.S. Space Program.* Baltimore: Johns Hopkins University Press, 1993.

McDougall, Walter A. *The Heavens and the Earth: A Political History of the Space Age.* New York: Basic Books, 1985, pp. 141–56.

McGraw-Hill Encyclopedia of Science and Technology, 7th ed. New York: McGraw-Hill, 1992.

McKinney, Michael L., and Robert M. Schoch, *Environmental Science: Systems and Solutions.* Minneapolis/St. Paul, MN: West, 1996, ch. 19.

McNeill, William H. *Plagues and Peoples.* Garden City, NY: Anchor, 1976.

Mallove, Eugene. *Fire from Ice: Searching for the Truth Behind the Cold Fusion Furore.* New York: John Wiley, 1991.

Mann, Julia de L. "The Textile Industry: Machinery for Cotton, Flax and Wool, 1760 to 1860." *In* Charles Singer, ed. *A History of Technology*, vol. IV. Oxford, UK: Oxford University Press, 1958.

Manning, Peter. *Electronic and Computer Music*, 2d ed. Oxford, UK: Clarendon Press, 1993.

Mark, Robert. *Experiments in Gothic Structure*. Cambridge, MA: MIT Press, 1982.

Market, Linda Rae. *Contemporary Technology*. South Holland, Il: Goodheart–Willcox, 1989.

Marsden, E.W. *Greek and Roman Artillery: Historical Development*. Oxford, UK: Clarendon Press, 1969.

Marvin, Ursula B. *Continental Drift: The Evolution of a Concept*. Washington, DC: Smithsonian Institution Press, 1973.

Mayr, Otto. *The Origins of Feedback Control*. Cambridge, MA: MIT Press, 1970.

Meikle, Jeffrey I. *American Plastic: A Cultural History*. New Brunswick, NJ: Rutgers University Press, 1995.

Melosi, Martin V. *Coping with Abundance: Energy and Environment in Industrial America*. New York: Knopf, 1985.

Merrick, David. *Coal Combustion and Conversion Technology*. Amsterdam: Elsevier, 1984.

Micklos, David A., and Greg A. Freyer. *A First Course in Recombinant DNA Technology*. Cold Spring Harbor, NY: Cold Spring Harbor Laboratory Press, 1990.

———, and ———. *DNA Science*. Cold Spring Harbor, NY: Cold Spring Harbor Laboratory Press, 1990, p. 33.

Millard, Andre. *America on Record: A History of Recorded Sound*. Cambridge and New York: Cambridge University Press, 1995.

Miller, G. Tyler. *Energy and Environment*. Belmont, CA: Wadsworth, 1980.

Miller, Kenton, and Laura Tangley. *Trees of Life: Saving Tropical Forests and their Biological Wealth*. Boston: Beacon Press, 1991.

Minnich, Jerry, and Marjorie Hunt. *The Rodale Guide to Composting*. Emmaus, PA: Rodale Press, 1979.

Mintz, Sidney W. *Sweetness and Power: The Place of Sugar in Modern History*. New York: Viking Penguin, 1985.

Misa, Thomas J. *A Nation of Steel: The Making of Modern America, 1865–1925*. Baltimore: Johns Hopkins University Press, 1995.

Monda, R. "Shedding Light on Dark Matter." *Astronomy*, vol. 20, no. 2 (Feb. 1992).

Moore, David S. *The Basic Practice of Statistics*. New York: Freeman, 1995.

Moore, Patrick. *Mission to the Planets: An Illustrated History of Man's Exploration of the Solar System*. New York: Norton, 1990.

Moran, J.M., and M.D. Morgan. *Meteorology: The Atmosphere and the Science of Weather*, 4th ed. New York: Macmillan, 1994.

Morison, Elting W. *From Know-How to Nowhere*. New York: Basic Books, 1974.

Morris, Peter J.Y. *Polymer Pioneers*. Philadelphia: Beckman Center for the History of Chemistry, 1990.

Multhauf, Robert. *Neptune's Gift: A History of Common Salt*. Baltimore: Johns Hopkins University Press, 1978.

Munro, John. "Textile Technology" *In* Joseph Strayer, ed. *The Dictionary of the Middle Ages*, vol. XI. New York: Scribner, 1988.

NASA web page: http://station.nasa.gov/.

National Research Council. *Possible Health Effects of Exposure to Residential Electric and Magnetic Fields*. Washington, DC: National Academy Press, 1997.

Needham, Joseph. *Science and Civilization in China*. Cambridge and New York: Cambridge University Press, 1965.

Nelson, Daniel. *Frederick W. Taylor and the Rise of Scientific Management*. Madison, WI: University of Wisconsin Press, 1980.

Newhall, Beaumont. *The History of Photography:*.

From 1839 to the Present. New York: Museum of Modern Art, 1982, pp. 8–71, 268–79.

Nichols, Eve K. *Mobilizing Against AIDS.* Cambridge, MA: Harvard University Press, 1989.

Noble, David F. *Forces of Production: A Social History of Industrial Automation.* New York: Oxford University Press, 1986.

Noller, Carl N. *Chemistry of Organic Compunds*, 2d ed. Philadelphia: Saunders, 1957.

Norbye, Jan P. "Pulling to the Front: Front-Wheel Drive." *Automotive Quarterly*, vol. 35, no. 2 (May 1996): 24–39.

North, J.D. "The Astrolabe." *Scientific American*, vol. 230, no. 1 (Jan. 1974), 96–103.

Nye, David E. *Electrifying America: Social Meanings of a New Technology, 1880–1940.* Cambridge, MA: MIT Press, 1990.

Odum, Eugene P. *Ecology and Our Endangered Life-Support Systems.* Sunderland, MA: Sinauer Associates, 1993.

Olby, Robert C. *Origins of Mendelism.* New York: Schocken, 1966.

"On Shaky Ground: A History of Earthquake Resistant Building Design Codes and Safety Standards in the United States in the Twentieth Century." *Bulletin of Science, Technology, and Society*, vol. 16, nos. 5–6 (1996): 311–27.

Ord-Hume, Arthur W. *Perpetual Motion: The History of an Obsession.* New York: St. Martin's Press, 1980.

Overy, R.J. *The Air War, 1939–45.* New York: Stein & Day, 1981.

Pacey, Arnold. *Technology in World Civilization.* Cambridge, MA: MIT Press, 1990.

———. *The Maze of Ingenuity*, 2d ed. Cambridge, MA: MIT Press, 1992.

Parker, Geoffrey. *The Military Revolution.* Cambridge, UK: Cambridge University Press, 1988.

Paulos, John Allen. *Innumeracy: Mathematical Illiteracy and Its Consequences.* New York: Hill & Wang, 1988.

Peebles, Curtis. *Watch the Skies! Chronicle of the Flying Saucer Myth.* Washington. DC: Smithsonian Institution Press, 1994.

Penrose, James E. "Inventing Electrocution." *American Heritage of Invention and Technology*, vol. 9, no. 4 (Spring 1994): 34–44.

Perrow, Charles. *Normal Accidents: Living with High-Risk Technologies.* New York: Basic Books, 1984.

Perry, Tekla S., and John Voelker. "Of Mice and Menus: Designing the User-Friendly Interface." *IEEE Spectrum* (Sep. 1989): 46–51.

Petroski, Henry. *To Engineer Is Human: The Role of Failure in Successful Design.* New York: St. Martin's Press, 1985.

———. *The Origin of Useful Things.* New York: Knopf, 1992, pp.51–77.

Pierce, John R. *Electrons, Waves, and Messages.* Garden City, NY: Hanover House, 1956, pp. 211–18.

Pinkepad, Jerry A. *The Second Diesel Spotter's Guide.* Milwaukee, WI: Kalmbach Publishing Co., 1973.

Plummer, Charles C., and David McGeary. *Physical Geology*, 7th ed. Dubuque, IA: William C. Brown, 1996.

Pohlmann, Ken C. *The Compact Disc: A Handbook of Theory and Use.* Madison, WI: A-R Editions, 1989.

Porkert, Manfred, and Christian Ullmann. *Chinese Medicine.* New York: William Morrow, 1988.

Porter, Theodore M. *Trust in Numbers: The Pursuit of Objectivity in Science and Public Life.* Princeton: Princeton University Press, 1995.

Power to Produce. Washington, DC: U.S. Department of Agriculture, 1960, pp. 25–45.

President's Commission. *Accident at Three Mile Island.*

Washington, DC: U.S. Government Printing Office, 1979.

Pretty, Jules N. *Regenerating Agriculture: Politics and Practice for Sustainability and Self-Reliance.* Washington, DC: Joseph Henry Press, 1995.

Prokosch, Eric. *The Technology of Killing: A Military and Political History of Antipersonnel Weapons.* Atlantic Highlands, NJ: Zed Books, 1995.

"A Public Policy Perspective on Televised Violence and Youth." *Harvard Educational Review* (Summer 1995): 282–91.

Purcell, C.W., Jr., ed. *Technology in America.* Cambridge and London: MIT Press, 1982.

Reid, T.R. *The Chip: How Two Americans Invented the Microchip and Launched a Revolution.* New York: Simon & Schuster, 1985.

Reiser, Stanley Joel. "The Concept of Screening for Disease." *Millbank Memorial Fund Quarterly,* vol. 54, no. 6 (1978): 403–25.

———. *Medicine and the Reign of Technology.* Cambridge and New York: Cambridge University Press, 1978, pp. 23–44.

Reynolds, Clark. *The Fast Carriers: The Forging of an Air Navy.* New York: McGraw-Hill, 1968.

Reynolds, Terry S. "Defining Professional Boundaries: Chemical Engineering in the Early 20th Century." *Technology and Culture,* vol. 27, no. 4 (Oct. 1986): 694–716.

———. *Stronger Than a Hundred Men: A History of the Vertical Water Wheel.* Baltimore: Johns Hopkins University Press, 1983.

Rhodes, Richard. *Nuclear Renewal.* New York: Viking Penguin, 1993.

Rhodes, R.G., and B.E. Mulhall. *Magnetic Levitation for Rail Transport.* Oxford, UK: Clarenden Press, 1992.

Rindos, David. *The Origin of Agriculture: An Ecological Perspective.* Orlando, FL: Academic Press, 1984.

Ring, Malvin E. *An Illustrated History of Dentistry.* New York: Harry N. Abrams, 1985.

Ritzer, George. *Expressing America: A Critique of the Global Credit Card Society.* Thousand Oaks, CA: Pine Forge Press, 1995.

Roberts, Royston M. *Serendipity: Accidental Discoveries in Science.* New York: John Wiley, 1989, pp. 177–86.

Rogers, Everett. *The Diffusion of Innovations,* 3d ed. New York: Free Press, 1983.

Rohlf, James W. *Modern Physics from A to Z.* New York: John Wiley, 1994.

Roland, Alex. "Secrecy, Technology, and War: Greek Fire and the Defense of Byzantium, 678–1204." *Technology and Culture,* vol. 33, no. 4 (Oct. 1992): 655–79.

Ronan, Colin. *Science: Its History and Development Among the World's Cultures.* New York: Facts On File, 1982.

Rose, Mark H. *Cities of Light and Heat: Domesticating Gas and Electricity in Urban America.* University Park, PA: Pennsylvania State University Press, 1995.

———. *Interstate: Express Highway Politics, 1939–1989,* 2d ed. Knoxville, TN: University of Tennessee Press, 1990.

Royce, James E. *Alcohol Problems and Alcoholism: A Comprehensive Survey.* New York: Free Press, 1989.

Sabins, Floyd F. *Remote Sensing,* 3d ed. New York, Freeman, 1997.

Saltman, Paul, Joel Gurin, and Ira Mothner. *The University of California San Diego Nutrition Book.* Boston: Little Brook, 1993.

Sammet. Jean. *Programming Languages: History and Fundamentals.* Englewood Cliffs, NJ: Prentice-Hall, 1969.

Sawyer, W.W. *What Is Calculus About?* New York: Random House, 1993.

Saxby, Graham. *Practical Holography,* 2d ed. New York: Prentice-Hall, 1994.

Schiffer, Michael Brian. *Taking Charge: The Electric Automobile in America*. Washington, DC: Smithsonian Institution Press, 1994.

————. *The Portable Radio in American Life*. Tucson, AZ: University of Arizona Press, 1991.

Schnitter, Nicholas J. *A History of Dams: The Useful Pyramids*. Brookfield, VT: Balkema, 1994.

Schobert, Harold A. *Coal: The Energy Source of Past and Future*. Washington, DC: American Chemical Society, 1987.

Schreiber, G.R. *Concise History of Vending in the U.S.A.* Chicago: Vend, 1961.

Schroeder, Ralph. *Possible Worlds: The Social Dynamic of Virtual Reality Technology*. Boulder, CO: Westview, 1996.

Schroeer, Dietrich. *Science, Technology, and the Nuclear Arms Race*. New York: John Wiley, 1984.

Schubert-Soldern, Ranier. *Mechanism and Vitalism: Philosophical Aspects of Biology* (trans. C.E. Robin). Notre Dame, IN: Notre Dame University Press, 1962.

Schwartz, Frederic D. "The Epic of the TV Dinner." *American Heritage of Invention and Technology*, vol. 9, no. 4 (Spring 1994).

Seely, Bruce E. *Building the American Highway System: Engineers as Policy Makers*. Philadelphia: Temple University Press, 1987.

Seideman, Tony. "Bar Codes Sweep the World." *American Heritage of Invention and Technology*, vol. 8, no. 4 (Spring 1993).

Sendzimir, Vanda. "Black Box." *American Heritage of Invention and Technology*, vol. 12, no. 2 (Fall 1966): 26–36.

Serling, Robert J. *The Jet Age*. Alexandria, VA: Time-Life, 1982.

Seyfert, Carl K., and Leslie A. Sirkin. *Earth History and Plate Tectonics*. New York: Harper & Row 1973.

Shank, W.H. *Towpaths to Tugboats*. York, PA: American Canal & Transportation Center, 1982.

Sheff, David. *Game Over: How Nintendo Conquered the World*. New York: Vintage, 1994.

Shipman, Harry L. *Black Holes, Quasars, and the Universe*, 2d ed. Boston: Houghton Mifflin, 1980.

Shurkin, Joel. *Engines of the Mind: The Evolution of the Computer from Mainframes to Microprocessors*. New York: Norton, 1996.

Sicilia, David B. "A Most Invented Invention." *American Heritage of Invention and Technology*, vol. 6, no. 1 (Spring/Summer 1990).

Sinclair, Ian R. ed. *Audio & Hi-Fi Handbook*. Newton, MA: Butterworth-Heinemann, 1995.

Smallwood, James. "Automatic Teller Machines" In Larry Schweikart, ed. *Encyclopedia of American Business History and Biography: Banking and Finance from 1913 to 1989*. New York: Facts On File, 1990.

Smil, Vaclav. *Biomass Energies: Resources, Links, Constraints*. New York: Plenum Press, 1983.

————. *Energy in World History*. Boulder, CO: Westview Press, 1994, pp. 40–49.

Smith, Anthony, ed. *Television: An International History*. Oxford, UK; Oxford University Press, 1995.

Smith, Marvin. *Radio, TV, & Cable: A Telecommunications Approach*. New York: CBS College Publishing, 1985.

Smulyan, Susan. *Selling Radio: The Commercialization of American Broadcasting, 1920–1934*. Washington, DC: Smithsonian Institution Press, 1994.

Soaring Society of America, P.O. Box E, Hobbs, NM 88241.

Sobel, Dava. *Longitude*. New York: Walker, 1995.

Soedel, Werner, and Vernard Foley. "Ancient Catapults." *Scientific American*, vol. 240, no. 3 (Mar. 1979): 150–60.

Spangenburg, Ray, and Diane Moser. *Opening the Space Frontier*. New York: Facts On File, 1989, pp. 42–50, 52–55.

Sparke, Penny. *Electrical Appliances: Twentieth-Century Design*. New York: Dutton, 1987.

Spielberg, Nathan, and Bryon D. Anderson. *Seven Ideas that Shook the Universe*. New York: John Wiley, 1987, ch. 7.

Splinter, William C. "Center-Pivot Irrigation." *Scientific American* (June 1976): 234–36.

Stanley, Steven M. *The New Evolutionary Timetable: Fossils, Genes, and the Origin of Species*. New York: Basic Books, 1981.

Stearns, Peter N. *The Industrial Revolution in World History*. Boulder, CO: Westview Press, 1993.

Stern, Colin W., and Robert L. Carroll. *Paleontology: The Record of Life*. John Wiley, 1989.

Stern, Rudi. *Let There Be Neon*. New York: Harry N. Abrams, 1979.

Stewart, Frances. *Technology and Underdevelopment*. Boulder, CO: Westview, 1977.

Stockholm International Peace Research Institute. *Antipersonnel Weapons*. New York: Crane, Russak, 1978.

Strada, Gino. "The Horror of Land Mines." *Scientific American*, vol. 274, no. 5 (May 1996): 40–45.

Strasser, Susan. *Never Done: A History of American Housework*. New York: Pantheon Books, 1982.

Street, James H. *The New Revolution in the Cotton Economy*. Chapel Hill, NC: University of North Carolina Press, 1957.

Tarr, Joel A., James McCurley III, Francis C. McMichael, and Terry Yosie. "Water and Wastes: A Retrospective Assessment of Wastewater Technology in the United States, 1800–1932." *Technology and Culture*, vol. 25, no. 2 (Apr. 1984): 226–39.

Temple, R. *The Genius of China*. New York: Simon & Schuster, 1986.

Thomis, Malcolm I. *The Luddites: Machine-Breaking in Regency England*. New York: Schocken, 1972.

Thompson, Jim. *Machine Guns: A Pictorial, Tactical, and Practical History*. London: Greenhill, 1990.

Thorne, Kip S. *Black Holes and Time Warps: Einstein's Outrageous Legacy*. New York: Norton, 1994.

Thorpe, James. *The Gutenberg Bible*. San Marino, CA: Huntington Library, 1975.

Thurston, David B. *Design for Flying*. New York: McGraw-Hill, 1978.

———. *Design for Safety*. New York: McGraw-Hill, 1980, pp. 145–51.

Travis, Anthony S. *The Rainbow Makers: The Origins of the Synthetic Dyestuffs Industry in Western Europe*. Bethleham, PA: Lehigh University Press, 1993.

Traweek, Sharon. *Beamtimes and Lifetimes: The World of High Energy Physicists*. Cambridge, MA: Harvard University Press, 1988.

Turco, Richard P., et al. "Nuclear Winter: Global Consequences of Multiple Nuclear Explosions." *Science*, vol. 222 (1983): 1283–92.

Tyne, Gerald F.J. *Saga of the Vacuum Tube*. Indianapolis, IN: Howard W. Sams, 1977, pp. 40–51.

U.S. Department of Agriculture. *Home and Garden Bulletin*, no. 253–1 through 253–8. Washington, DC: Government Printing Office, 1993.

———. *Power to Produce*. Washington, DC: Government Printing Office, 1960, pp. 25–45.

Van Esterik, Penny. *Beyond the Breast-Bottle Controversy*. New Brunswick, NJ: Rutgers University Press, 1989.

van Rossum, Gerhard Dohrn. *History of the Hour: Clocks and Modern Temporal Orders*. Chicago: University of Chicago Press, 1996.

Van Vat, Dan. *Stealth at Sea: The History of the Submarine*. London: Werdenfeld & Nickolson, 1994.

Vaughn, Dione. *The Challenger Launch Decision: Riskiy Technology, Culture, and Deviance at NASA*.

Chicago: University of Chicago Press, 1996.

von Brandt, Andres. *Fish Catching Methods of the World*. London: Fishing News Books Ltd., 1964.

von Spronsen, J.W. Van. *The Periodic System of the Elements*. Amsterdam: Elsevier, 1969.

Walker, Graham. "The Stirling Engine." *Scientific American*, vol. 229, no. 2 (Aug. 1973).

Weinberg, Steven. *The First Three Minutes: A Modern View of the Origin of the Universe*, 2d ed. New York: Basic Books, 1993.

Wertime, Theodore A. *The Coming of the Age of Steel*. Leiden: Brill, 1961.

————, and James D. Muhly, eds. *The Coming of the Age of Iron*. New Haven: Yale University Press, 1980.

White, John A. *The Great Yellow Fleet*. San Marino, CA.: Golden West Books, 1986.

————. *The American Railroad Freight Car*. Baltimore: Johns Hopkins University Press, 1993, pp. 527–46.

White, J.H. *A History of the Phlogiston Theory*. London: E. Arnold, 1932.

White, Lynn, Jr. *Medieval Technology and Social Change*. New York: Oxford University Press, 1966.

————. "The Invention of the Parachute." *Technology and Culture*, vol. 9, no. 3 (July 1968): 462–67.

Wietzman, David. *Windmills, Bridges & Old Machines*. New York: Scribners, 1982.

Wilford, John Noble. *The Riddle of the Dinosaur*. New York: Random House, 1985.

Williams, J.E.D. *From Sails to Satellites: The Origin and Development of Navigational Science*. Oxford, UK: Oxford University Press, 1992, pp. 21–27, 128–44.

Williams, L. Pearce. *Michael Faraday: A Biography*. New York: Basic Books, 1965, ch. 4.

————, ed. *Relativity Theory: Its Origins and Impact on Modern Thought*. New York: John Wiley, 1968.

Williams, Michael R. *A History of Computing Technology*. Los Alamitos, CA: IEEE Computer Society Press, 1997.

Willloughby, Vic. *Back to Basics*. Newbury Park, CA: Haynes Publishing Group, 1981.

Wohleber, Curt. "How the Movies Learned to Talk." *American Heritage of Invention and Technology*, vol. 10, no. 3 (Winter 1995).

Womack, James, et al. *The Machine That Changed the World*. New York: Rawson Associates/Macmillan, 1990.

Wonnacott, Thomas H., and Ronald J. Wonnacott. *Introductory Statistics for Business and Economics*, 4th ed. New York: John Wiley, 1990.

Woodbury, Robert S. *History of the Gear-Cutting Machine: A Historical Study in Geometry and Machines*. Cambridge, MA: MIT Press, 1958.

————. *Studies in the History of Machine Tools*. Cambridge, MA: MIT Press, 1972.

Worthington, William Jr. "Early Risers." *American Heritage of Invention and Technology*, vol. 4, no. 3 (Winter 1989): 40–44.

Yannas, Simos. *Solar Energy and Housing Design*. London: Architectural Association, 1994.

Yenne, Bill. *100 Inventions*. San Francisco: Bluewood, 1993.

Young, Hugh D. *Fundamentals of Waves, Optics, and Modern Physics*, 2d ed. New York: McGraw-Hill, 1976.

Zuckerman, Harriet. *The Scientific Elite: Nobel Laureates in the United States*. New York: Free Press, 1977.

Index

A

Aachen Capitulary, 136–137

Abacus, 1–2

Abbe, Cleveland, 939

Abbe, Ernest, 631
  theory of, 631

ABC (American Broadcasting Company), 156

Abd Allah, 484

ABO groupings, 725

Academic American Encyclopedia, 182

Accademia dei Lincei, 884

Accademia del Cimento, 884

Accelerators, particle, 740–742

Accidents, normal, 698–699

Accretion disk, 817

Accumulator, 108

Acetylene, 2

Achromatic lens, 1021

Acid rain, 2–3, 441, 1121

Ackeret, Jacob, 1067

Ackerman, Thomas, 709

Acquired immunodeficiency syndrome (AIDS), 3–6, 256, 373, 1103

Acrylate rubber, 6

Acrylics, 6

Actinides, 748

Action for Children's Television, 1091

Action potential, 693

Active solar design, 913–916

Actualism, 180

Acupuncture, 6–8

Acute lymphoblastic leukemia, 200

Adams, Thomas, 1094

Adenosine deaminase (ADA), 456
  deficiency of, 459

Adenoviruses, 533

Adherends, 8

Adhesives, 8–9

Ad Hoc Committee on the Triple Revolution, 1007, 1008

Adhydroxydiamino-arsenobenzene hydrochloride, 100

Adobe, 144

Adolphus, Gustavus, 637

Adrenaline, 506–507, 957

Adrian, Edgar, 340

Ads for sanitary protection products, 873

Advanced Research Projects Agency (ARPA), 291

Advanced Research Projects Agency (ARPA) Network, 346

Advanced Technology Program (ATP), 683

Advanced Weather Interactive Processing System (AWIPS), 1120

Aeolipile (steam engine), 945, 1070–1071

Aerial Phenomena Research Organization (APRO), 1082

Aerojet General Corporation, 863

Aerosol sprays, 9–10

African sleeping disease, 199

Agent Orange, 10–11

Agent Orange Act (1991), 10

Aging, theories of, 580–581

Agricola (Georg Bauer), 556, 806

Agricultural Act (1949), 12

Agricultural Adjustment Acts (1933 and 1938), 12

Agricultural revolution, 11–13

Agriculture
  chemical fertilizers in, 405–406
  crop rotation in, 270–271

Green Revolution in, 479–481

irrigated, 13–15

Neolithic revolution in, 689–691

nitrogen cycle in, 696–697

and organic fertilizers, 406–407

slash-and-burn, 270

Soviet, 602

Agriculture, United States Department of (USDA), 290–291

AIDS (acquired immunodeficiency syndrome), 3–6, 256, 373, 1103

Aiken, Howard, 245

Airbag, 22–24

Airbrake, 24–25

Airbus Industrie, 417

Air conditioning, 15–18
  automotive, 18–19

Aircraft, 32–35
  and instrument flying, 540–541
  nuclear-powered, 28
  pressurized, 29
  short takeoff and landing (STOL), 30–31
  ultralight, 31–32
  vertical/short takeoff and landing (V/STOL), 30–31
  vertical takeoff and landing (VTOL), 30–31

Aircraft carrier, 25–28

Aircraft propeller, 797–798

Air cushion vehicle, 509

Air-launched cruise missiles (ALCMs), 350, 964

Airplane crashes, and flight data recorders, 412–413

Airplanes. *See* Aircraft

Air Force, *See* United States Air Force

Cage construction, 155

Cahill, Thaddeus, 675

Cai Lun, 734–736

Caisson, 158–159

Caisson disease, 159

Calcium carbonate (CaCO3), 582

Calcium cyclamate, 276

Calcium sulfide (CaSO4), 869

Calculating aids, abacus as, 1–2

Calculator, 159–160

Calculus

differential, 160–162

integral, 160–162

California Aqueduct, 67, 68

California Environmental Quality Act
(CEQA), 368

California Institute of Technology, 1132

Caligula, 112

Caloric, 162–163, 355, 493, 536, 878

Caloric fluid, 355, 536

Caloric theory, 536

Calories, 356–357

Calvin, Melvin, 760

Camera obscura, 657, 757

Cameron, George H., 260

Campbell, Angus, 264

Campbell-Swinton, Alan, 1023

Canadian Pacific Railroad, 828

Canal lock, 164–165

Canals, 165–167

Cancer, 167–169

bladder, 167

breast, 167, 506, 609–611

endometrial, 505–506

gene therapy for, 456

lung, 167

scrotal, 167

treatment for, 548

Cancer Control Supplement of the
National Health Interview Survey,
610

CANDU (Canadian Deuterium
Uranium), 705

Cann, Rebecca, 385

Canned food and beverages, 169–170

Cannizaro, Stanislao, 82, 84, 98, 197

Capacitor, 170

Čapek, Karel, 89

Capital punishment, 447–448

Capone, Al, 37

Capture effect, 434

Carbon-11, 1050

Carbon-14 dating, 826–827

Carbon dioxide, 170–171, 208

Carboniferous system, 463

Carbonization, 427

Carbon-pencil microphone, 628

Carburetor, 171–172

Carcinogens, 422

Cardan, Jerome, 1133

Cardano, Girolamo, 1088

Cardan shaft, 1088

Cardiac pacemaker, 172–173

Cardinal Polignac, 963

Carlile, Richard, 1094

Carlisle, Anthony, 341

Carlson, Chester, 1140

Carnegie, Andrew, 954

Carnegie Institution's Station for the
Experimental Study of Evolution,
458

Carnivores, 421

Carnot, Sadi, 163, 173, 355, 357, 366

Carnot cycle, 163, 164, 173–174, 357

Carolingian Empire, 136

Carothers, Wallace, 711, 778, 865

Carré, Ferdinand, 839

Carrel, Alexis, 725

Carrier, Willis H., 16

Carruthers, H. M., 914

Carson, Rachel, 284, 538

Carterfone Decision, 1014

Cartwright, Edmund, 1124

Casegrain, 1018

Caselli, Giovanni, 401

Cash, Richard, 724

Cash register, 175

Cast iron, 175–176, 554, 558

Cast-iron stoves, 962

Catalyst, 177

Catalytic converter, 176–177

Catalytic cracking, 177–178

Catalytic processes, 222

Catapult, torsion, 178–179

Catastrophism, 179–181

Caterpillar tractor, 400

Cathedral of Notre Dame, 417

Cathode-ray oscilloscope, 728

Cathode rays, 728

Cathode-ray tubes (CRTs), 181–182,
589, 668, 1099

CAT scans, 251–252, 1140

Cattell, James McKean, 544

Cauchy, 1050

Cause-and-effect, 881

Cavalieri, Bonaventura, 161

Cavendish, Henry, 104, 171, 478, 487

Caventou, Joseph Bienaimé, 760

Cavitation, 504, 798

Cavity magnetron, 729

Cayley, George, 967

C-band, 231

CBS (Columbia Broadcasting System),
156

CD-ROMs, 182–183, 912

Celestial globes, 766

Cell, 183–185

Cellini, Benvenuto, 112

Cellular telephone systems, 1015–1017

Celluloid, 101, 185–187, 778

Celluloid Manufacturing Company, 186

Cellulose acetate, 472

Celsius scale, 1039

Cement, 252

Center-pivot irrigation, 560

Centers for Disease Control and
Prevention (CDC), 804

Centigrade scale, 1039

Central heating, 187–189

Central limit theorem, 942

Central Pacific Railroad, 828

Centre Européen de Researche Nucléaire
(CERN), 814

Centrifugal supercharger, 981

Centrifuge, 189–190

Centrioles, 185

Ceramics engineering, 366

Cesare Borgia, 165

Cesium-137, 521

CFC-12, 19

Chadwick, James, 695, 800

Chaff, 944

Chagas' disease, 199

Chain, Ernst, 747

Chain drive, 190–191

Chain link fence, 191

Chain saw, 191–192

Eijkman, Christiaan, 1105

Eilmer, 472

Einstein, Albert, 78, 119, 466, 478, 494, 572, 621, 697, 703, 811–812, 844, 1054

Einthoven, Willem, 339, 340

Eisenhower, Dwight D.
   and doctrine of massive retaliation, 79
   and interstate highway system, 551
   and space exploration, 62, 681

Eisinga, Eise, 766

Elastomers, 8

Eldredge, Niles, 386

Electromagnet, 342

Electrical illumination, 525–527

Electrical-power systems, 336–337

Electric Boat Company, 974, 975

Electric chair, 332–333

Electric generator, 457–458

Electric guitar, 482–483

Electricity
   alternating current in, 44–45, 303
   direct current in, 303
   for heat, 188

Electric meter, 333

Electric-power blackouts, 333–335

Electric Power Research Institute, 761

Electric starter, 335–336

Electric Vehicle Company, 885

Electric welding, 1126–1127

Electrification, rural, 338–339

Electrocardiogram, 339–340

Electro-Dynamic Company, 975

Electroencephalogram (EEG), 340–341

Electrolysis, 341–342

Electromagnetic fields, 785–787

Electromagnetic induction, 45, 342–344, 619–620

Electromagnetic radiation, 813

Electromagnets, 1009

Electromechanical computers, 245

Electron, 344–345, 1050

Electron capture, 345

Electronic intelligence (ELINT) satellites, 875

Electronic mail (e-mail), 345–347, 549

Electronic music, 674–676

Electronics, 345

Electron microscope, 631–632

Electrons, 77, 1036

Electrophoresis, 998

Electrophotography, 1140

Electroscope, 260

Elements, periodic table of, 748–750

Elevated railways, 978

Elevator, 347–348

Elion, Gertrude, 200

Ellet, Charles, 148

Elliptical galaxies, 283

Ellsworth, Henry L., 11

Elwell-Parker Co., 425–426

Ely, Eugene, 25

e-mail. *See* Electronic mail

Embankment dams, 280

Emerson, James, 1070

Emissions controls, automotive, 348–350

Emissions-control technologies, 3

Empire State Building, 904, 905

Empirical adequacy, 1033

Empiricism, 350–351

Enamel, 351–352

Encke, W., 1067

Encryption, 1011

Endangered Species Act, 352–353, 519

Endeavor (space shuttle), 512, 925

Enders, John F., 1092

Endometrial cancer, 505–506

Endoplasmic reticulum (ER), 185

Endoscope, 409

Endothermy, evidence for, 300

End-stage renal disease, 566

Endurance limit, 625

Energy
   conservation of, 354–356
   fusion, 442–443
   geothermal, 467–468
   measures of, 356–357
   rest, 846

Energy efficiency, 353–354

Energy intensity, 354

Energy Policy Act (1992), 519

Engine
   diesel, 357–358
   four-stroke, 358–362
   steam, 945–948
   stirling, 362
   two-stroke, 362–364
   Wankel rotary, 365

Engineering
   ceramics, 366
   computer-aided, 250
   knowledge, 71
   reverse, 1002

Engineering bricks, 145

Engineering Research Associates (ERA), 246

Englebart, Douglas, 668

ENIAC (Electronic Numerical Integrator and Calculator), 246, 629

Enrichment, 561

Enterprise (space shuttle), 924

Entropy, 174, 366–367

Enumerative induction, 529

Environmental impact reports (EIRs), 367–369

Environmental Protection Agency (EPA), 291, 369–370, 537, 776, 909

Enzymes, 370–372

Epicenter, 326

Epicurus, 81, 392

Epicycloid, 451

Epidemics, 803
   in history, 372–374

Epinephrine, 506–507, 507

Epitrochoid, 365

Equants, 918

Equivalence, principle of, 843

Ercker, Lazarus, 557

Ergonomics and human factors, 374–375

Ergot fungus (*Claviceps purpurea*), 488

Ergot poisoning, 488

Erickson, C. J., 1014

Erickson, J., 1014

Ericsson, John, 913

Erie Canal, 166

Erosion, 375–376

Error sampling, 873

Erythroxylon coca, 223

Escalator, 377

Escape velocity, 389

Escherichia coli, 100, 216, 323, 990

Estrogen replacement therapy (ERT), 505

Estrogens, 507

Estrogen use, 505–506

Estrone, 507

Ethanol, 404, 449

Ethernet, 243

Fixed air, 171, 760

Fizeau, Armand, 583

Flamethrowers, 679

Flash welding, 1126

Flat-slab dams, 280. *See also* Dams

Flavr-Savr tomato, 1053

Fleischmann, Martin, 228

Fleming, Alexander, 199–200, 580, 746

Fleming, Ambrose, 1060

Fleming, John Ambrose, 49, 302, 842, 878, 1036

Fleming, Walther, 184

Flexible-production systems, 397

Flight data recorder (FDR), 412–413

Flint glass, 634

Floating caisson, 159

Floppy disk, 623–624

Florey, Howard, 580, 747

Florida Area Cumulus Experiment (FACE), 1121

Flourens, Pierre, 762

Flow, 366

Fluidized-bed combustion, 413–415

Fluorescent tubes, specialized, 1085

Fluoxetine (Prozac), 59

Flush, toilet, 1044–1045

Fluxion, 161

Flyballs, 476

Fly-by-wire technology, 416–417

Flying buttresses, 417–418

Flying saucer, 1081

Flying shuttle, 1124

Flywheel, 418–419

Fokker, Anthony, 606

Fokker scourge, 606

Fonzi, Giuseppangelo, 398

Food

  frozen, 436–437

  irradiated, 559–560

  spoilage of, 525

Food, Drug, and Cosmetic Act, 422, 559, 981

  Delany Clause Amendment of, 288

Food Additives Amendment (1958), 419

Food and Agriculture Organization (FAO), 538, 559, 804

  Code of Conduct, 539

Food and Drug Administration (FDA), 419–420, 422, 504, 559, 804

and diethylstilbestrol (DES), 293

Food chain, 331, 420–421

Food poisoning, *Escherichia coli* in, 100, 216, 323, 990

Food preservatives, 421–422

Football helmet, 422–423

Ford, Henry, 75, 92, 450, 744, 1000

Ford, Model-T, 614–615

Ford Motor Company, 400

Fore-and-aft sail, 896

Forest, Lee de, 49, 302, 660, 820, 842, 1036, 1060

Forest Products Laboratory (FPL), 775

Forests

  rain, 287

  temperate, 286

  tropical, 287

Forging, 423–425

Forklift and industrial trucks, 425–426

Forlanini, Enrico, 519

Formaldehyde (HCHO), 449

Forssmann, Werner, 53

Fortification, trace italienne style of, 1048

FORTRAN (FORmula TRANslation), 239–240, 250

Fossil fuels, 440–441

Fossils, 426–428

Foucault, Jean-Bernard Léon, 485, 540, 584, 1019

Fouch, 1127

Foundry technique, 963

Fourier, Joseph, 618–619

Fourneyron, Benôit, 1068

Four-stroke engine, 358–362

  See also Engine

Fowler, Lorenzo, 762

Fowler, Orson, 762

Foyle, Joseph W., 315

Fracastoro, Girolomo, 468

Fractal geometry, 195

Fractals, 428–430, 1048

Fractional distillation, 305

Fragmentation mines, 571

Framework Convention on Climate Change, 210

Francis, James B., 1069

Franco-Prussian War (1870), 142, 603

Franklin, Benjamin, 342, 395, 586, 963

  kite-flying experiment of, 170

Franklin, Rosalind, 306

Franklin stove, 963

Fraud in science, 430–432

Fraunhofer, Joseph von, 311, 929, 1021

Freedman, Wendy, 121

Free radicals, 825

Freeway, 432–433

Freeze drying, 433

French Academy of Sciences, 884

French Revolution, 482

Freon-12, 204

Frequency-division multiplexing (FDM), 673

Frequency modulation (FM), 433–435, 676, 821

Fresnel, Jean, 42

Fresnel, Augustin Jean, 582, 1118

Freud, Sigmund, 223, 715

Freyssinet, Eugene, 254

Friedmann, Alexander, 119

Frisch, Otto, 703

Fritts, Charles, 761

Froelich, John, 399

Fromont, Paul, 664

Front-disc brakes, 140

Front-wheel drive, 435–436

Frosch, Paul, 1102

Frozen food, 436–437

Frozen prepared meals, 437–438

Fructose, 979

Fry, William, 1083

Frye standard, 310

Fuel cell, 438–439

Fuel efficiency, 226

Fuel injection, 439–440

Fuel rail, 440

Fuels

  biomass, 125–127

  fossil, 440–441

Fuller, Buckminster, 311

Fuller, Calvin, 761

Fuller's earth, 1136

Fulton, Robert, 798, 949, 975

Functional Cargo Block, 925

Fundamental theorem of calculus, 160–161

Funk, Casimir, 1105

Furnace, reverberatory, 441–442

Furniture industry, 1135–1136

Halo effect, 289

Haloid Company, 1141

Halsted, William Stewart, 167, 990

Hamill brothers, 317

Hamilton, Alice, 580

Hamilton, W. D., 911

Hamiltonian kin selection, 911

Hammurabi's code, 532

Hancock, Thomas, 867

Handoff, 1015–1016

Handsaws, 876

Hang glider, 472

Hanmore, Hiram, 73

Hanna, G. C., 1046

Hansen, William, 790

Hapsburg-Valios Wars (1494–1559), 637, 1048

Harbors, 489–491

Hard disks, 304, 623

Hard driving, 954

Harden, Arthur, 371

Harder, Delmar, 88

Hardware, 912

Hardy, James, 726

Hargreaves, James, 931–932

Harington, John, 1045

Harriot, Thomas, 884

Harrison, John, 213, 596, 687

Harsanyi, John C., 617

Hartley, David, 622

Hartley, Harold, 834

Hartlib, Samuel, 882

Hartman, Louis, 1008

Hart-Parr Company, 399

Harvester, combine, 229–230

Harvey, E. Newton, 1083

Harvey, William, 54, 131–132

Hata, Sahachiro, 199

Hatch Act (1877), 290

Hauksbee, Francis, 457

Hauron, Louis Ducos de, 755

Hauy, René Just, 273

Hawking, Stephen, 128

Haworth, Walter, 1105

Hayashi, Chushiro, 119

Hayflick, Leonard, 581

Hayflick phenomenon, 581

al-Haytham, 465

Hazan, Al, 469

HDLs (high-density lipoproteins), 190, 205, 505

Head, Howard, 1031

Headers, 772

Health, 802

Hearing aid, 491–492

Hearst, William Randolph, 792

Heart, artificial, 492–493

Heart block, 172

Heart transplants, 492

Heat

central, 187–189

kinetic theory of, 493–494

latent, 573–574

stoves, 963

Heatley, Norman, 747

Heat pump, 366

Heberling, Emory, 210

Heddle, 1123

Hegel, G. W. F., 393

Heilig, Morton, 1100

Heilmeier, George, 589

Heilmeier, N. J., 589

Heisenberg, Werner, 494–495, 621, 813

Heisenberg uncertainty principle, 345, 494–495

HeLa cells, 168

Helfrich, Wolfgang, 589

Helical springs, 936

Helicopter, 495–497

in Korean War, 496

Heliocaminus, 916

Heliocaminus room, 916

Heliocentric, 919

Heliocentrism, 918

Helium, 929

Helmer, Olaf, 289

Helmet, football, 422–423

Helmholtz, Hermann von, 342, 356, 795–796

Helmont, Johannes Van Buren, Martin, 759

Helper T cells, 726

Helve hammer, 424

Heme, 898

Hemorrhagic fevers, 1103

Hemorrhoids, 1045

Henderson, Charles B., 864

Hendricks, Jimi, 483

HeNe laser, 572

Henle, Jakob, 468

Hennebique, François, 253, 254

Hennig, W., 998

Henry, Joseph, 342, 343, 344, 664, 1009

Henry repeating rifles, 142

Hepatitis B, vaccines for, 528

Herbicides, 497–498, 542–543

Herbivores, 420

Herd immunity, 528

Hering, Rudolf, 891

Heroin, 498–499

Heron (or Hero) of Alexandria, 85, 451, 945, 1032, 1071, 1094

Héroult, Paul, 46, 951

Herpes simplex virus reproduction, 200

Herrick, James B., 898

Herrick, Richard, 726

Herrick, Ronald, 726

Herschel, John, 757–758

Herschel, William, 535, 1018

Hertz, Heinrich, 620, 634, 811, 820

Hess, Harry, 771

Hess, Viktor Franz, 260

Hessen, Boris, 500

Heterozygote, 461

Hevesy, Gyorgy, 561

Hewitt, Peter Cooper, 836

Hickman, Henry, 51

Higgs Boson, 982

High-definition color television (HDTV), 1028–1029

High-density lipoproteins (HDLs), 190, 205, 505

High-fidelity recording, 834–835

High-level waste, 707

High Speed Ground Transportation Act (1965), 607

High-temperature glass, 470–471

Highway, 432

Highway Act (1956), 551

Highway Emergency Location Plan (HELP), 821

Higinbotham, Willy, 1095

Hildebrand brothers, 666

Hildebrandt, August, 51

Hill, Julian, 711

Hiller, Lejaren, 241

Hilton, W. F., 986

Mannheim, Amedee, 907

Manufacturing Extension Partnership, 683

Manville Covering Company, 73

Many-body problem, 611

Maple syrup urine disease, 459

Marat, Jean-Paul, 482

Maratha Empire of India, 636

Marburg fever, 1103

Marconi, Guglielmo, 818, 820
Wireless Telegraphy Company of, 878

Marco Polo bridge, 146

Margarine, 611–612

Margulis, Lynn, 444

Marine Corps, United States, and use of napalm, 679–680

Marine propeller, 798–799

Marine Spill Response Corporation, 721

Mark, Herman, 778, 779

Marriot, Fred, 96

Marsden, Ernest, 43

Marshall, George C. Space Flight Center, 681

Martel, Charles, 960

Martenot, Ondes, 675

Martian, 393

Martin, Archer John Porter, 206

Marvel, Carl, 778

Marx, Karl, 14, 296

Maser, 572

Maskelyne, Nevil, 596–597

Masonry, 612–613

Masonry gravity dams, 280

Mass, 621

Massachusetts Institute of Technology (MIT), 250

Massive compact halo objects, 283

Massive planets, 283

Massive retaliation, doctrine of, 79

Mass production, 396, 613–615. *See also* Assembly lines

Mass spectrograph, 561

Mass spectrometer, 71

Masterman, Margaret, 737

Materials, composite, 235

Mathaei, Heinrich, 455

Mathematical modeling, 615–617, 991

Mathematics, role in science, 617–619

Mathews, Max, 241, 242

Matrix mechanics, 813

Matthews, B. H. C., 340

Matthews, D. H., 771

Mauchly, John, 246

Maudslay, Harry, 574

Maudslay, Henry, 613, 1135

Mau-kuen Wu, 984

Maurice of Nassau, 637

Maxam, Alan, 455

Maxim, Hiram, 603

Maximum tolerated dose (MTD), 288

Max Planck Institut für Radioastronomie at Effelsberg, 1017

Maxson Food Systems, 438

Maxwell, James Clerk, 82, 446, 535, 611, 618, 619, 634, 755, 820, 944

Maxwell's equations, 619–620

Maybach, Wilhelm, 172, 360–361

Mayer, Julius Robert, 355–356, 795

Mayr, Ernest, 386

McAdam, John Louden, 860

McAuliffe, Christa, 194

McCarthy, John, 70

McCarthy, Joseph, 521

McCarty, M., 455

McClintock, Barbara, 455

McCollum, Elmer, 1105

McCormick, Cyrus H., 11, 832

McCormick, Katharine Dexter, 256

McCormick reaper, 833

McEvoy, C. A., 919

McFarlane, Robert, 964

McGaffey, Ives W., 1093

McKay, Frederick, 415

McKenzie, D. P., 771

McLean, Malcolm, 1057

McLouth Steel, 951

McMillan, Edwin, 277

McNair, Ronald, 192

Mean, 943

Measles, 1102

Measure of dispersion, 943

Mechanical clock, 451

Mechanics, Newtonian, 620–623

Medawar, Peter, 725

Median, 943

Medicine
alternative, 6–8
use of lasers in, 573

Megahallucinogen, 488

Megé-Mourièz, Hippolyte, 611

Meiosis, 207

Meissner, Walter, 983

Meister, Joseph, 528

Meitner, Lise, 703

Mekong Delta, 680

Melies, Geroges, 661

Mellanby, Edward, 1105

Memory
computer, 623–624
immunologic, 527

Memprobamate, 56

Menai Straits Bridge, 148

Mendel, Gregor, 207, 378, 454, 460–461, 769

Mendeleev, Dmitri Ivanovich, 80, 197, 748

Mendelian genetics, 460–461

Mendel's laws, 258

Menopause, 505

Mental age (MA), 545

Mental level, 544

Menten, M. L., 371

Menzel, Donald, 1081

Mercalli scale, 853

Mercury Project, 624

Mereng, Joseph von, 507

Merganthaler, Otto, 588

Mergesort, 40–41

Meridians, 7

Merry, Charles, 314

Merryman, Jerry, 160

Mersenne, Marin, 882

Mertz, Janet, 216

Mescaline, 488

Meson, 261

Mesophyllic bacteria, 122

Mesotron, 261

Metal fatigue, 624–625

Metallic-cartridge ammunition, 603

Metal Mike, 890

Metal stamping, 625–626

Methadone, 498, 626–627, 1001

Methane, 208, 222

Methanol, 38, 449

Methyl alcohol, 38

Methyl esters, 6

Methyl methacrylate, 6

Metric, 466, 843

Moving-target indication radars (MTIs), 819

Moxibustion, 7

MS-DOS, 248

Muddy Waters, 483

Mueller, Erwin, 453

Mueller, Karl, 984

Mughal Empire of India, 636

Muhammed ibn Musa Al-Khowarizmi, 39

Muir, John, 329, 519

Muldem, 673

Mulder, Gerardus, 697

Mule spinning, 669–670

Muller, Hermann, 454

Müller, Otto Freidrich, 99

Müller, Paul, 284, 538

MULTICS, 241

Multiphasic screening, 670–671

Multiple-arch dams, 280

Multiple independently targetable reentry vehicles (MIRVs), 671–672, 872

Multiplexing, 672–674

Mumford, Lewis, 556, 746

Mumps, vaccines for, 528

Munitions, precision-guided, 787–788

Munk, Max M., 1131

Munroe effect, 109–110

Munzer, Carl, 840

Muon neutrino, 815

Muons, 815

Murchison, Roderick, 463

Murdock, William, 585

Murray, Joseph E., 726

Murrell, K. F. H., 374

Muscarine, 488–489

Mushet, David, 954

Mushet, Robert, 952

Music
    computer, 241–243
    electronic, 674–676

Musical instrument digital interface (MIDI), 243

Musick, Mark, 537–538

Musket, 676–677

Musschenbroek, Pieter van, 170

Mustard gas, 448

Mutual inductance, 344

Mutual induction, 344

Mutually assured destruction (MAD), 79, 871, 965

Muybridge, Edweard, 658

Mycobacterium tuberculosis, 200

N

NAFTA, 290

Nagasaki, 79

Nägeli, Karl Wilhelm von, 183, 461, 778

Nails, finish, 678

Nalin, David, 724

Napalm, 678–680

Napier, John, 595, 906

Napier's bones, 906

Napoleon, 975

Napoleon III, 611

Nash, John F., 617

Nasir al-Dīn al-Tūsī, 464

Nasmyth, James, 424

Natalie O. Warren, 994

National Academy of Engineering (NAE), 680

National Academy of Sciences (NAS), 10–11, 680–681

National Advisory Committee for Aeronautics (NACA), 624, 681

National Advisory Council, 986

National Aeronautics and Space Administration (NASA), 194, 232, 393–394, 681–682, 761, 924, 1007. *See also* Space exploration
    Apollo Project, 61–63, 417, 541
    Challenger (space shuttle), 1130
        explosion of, 192–194, 511, 682, 925
    Goddard Space Flight Center, 511
    Langley Research Center, 522
    Mercury Project, 624

National Aerospace Plane, 983

National Air and Space Agency, 730

National Atmospheric and Oceanic Administration (NOAA), 1119

National Broadcasting Corporation (NBC), 152, 156, 1028

National Bureau of Standards (NBS), 210, 682, 941

National Cancer Institute Act (1937), 168

National Cancer Program, 168–169

National Cash Register Company (NCR), 175

National Center for Toxicological Research, 420

National Commission on Product Safety, 576

National Cooperative Research Act (NCRA), 1029

National Defense Research Committee (NDRC), 679

National Environmental Policy Act (NEPA), 368

National Football League (NFL), 423

National Hail Research Experiment, 1121

National Historic Landmarks, 158

National Hurricane Center (NHC), 1119

National Institute of Standards and Technology (NIST), 682–683

National Institutes of Health (NIH), 683–684, 804

National Investigations Committee on Aerial Phenomena (NICAP), 1082

National Library of Medicine, 684

National Maglev Institute, 607

National Marine Fisheries Service, 352

National Meteorological Center (NMC), 1119

National Oceanic and Atmospheric Administration (NOAA), 730, 731, 1121

National Organ Transplant Act (1984), 727

National Railroad Passenger Corporation (Amtrak), 829

National Research Council (NRC), 291, 680

National Research Defense Committee (NRDC), 291

National Science Foundation, 684–685

National Science Foundation Act, 684

National Security Act (1947), 291

National Sickle Cell Anemia Control Act (1972), 459

National System of Interstate and Defense Highways, 551

National Weather Service, 1119, 1120
    Environmental Modeling Center, 1119

Natta, Giulio, 778, 779

Office of Scientific Research and Development, 684

Office of Technology Assessment, 715–716

Offset, 590

Offset lithography, 591

Ogallala aquifer, 560

Ohain, Hans von, 1074

Ohio, 973–974, 977

Ohl, Russell, 761

Ohm, George Simon, 849

Ohm's law, 906

Oil exploration, 716–717

Oil pipelines, 717–718

  Alaskan, 718

Oil refining, 718–719

Oil shale, 720

Oil spills, 720–721

Oils, lubricating, 599–600

Oldenburg, Henry, 883

Oldendorf, William H., 251

Oldowan culture, 502

Olds, Ransom E., 92

Oleography, 590

Oliver, J., 771

Olmsted, Frederick Law, 432

Olson, Scott, 902

Olympus Mons, 922

Omnibus Budget Reconciliation Act (1981), 459

Onager, 179

123 Compound, 984

One-gene one-enzyme theory, 168

Onizuka, Ellison, 192

OnTyme, 346

Open caisson, 159

Open-cast mining, 640

Open-hearth process, 950

Open-loop system, 889

Open-pit mining, 220

Open universe, 120

Operating system programs, 912

Operation Ranch Hand, 10

Operations research, 721–723, 990–991

Opium, 723

Oppenheimer, J. Robert, 78, 521

Optical microscope, 632–634

Optical telescopes, 1017

Oral contraceptives, 256–257

Oral rehydration therapy, 724–725

Orbit, 345

Orbitals, 199

Orbiter missions, 922

Ordovician system, 463

Organ, pipe, 765–766

Organic fertilizer, 406–407

Organization of Petroleum Exporting Countries (OPEC), 259, 686

Organization of Standardization (ISO), 941

Organophosphates, 538–539

Organ transplantation, 725–727

Orientable surfaces, 1047

O-ring, 193–194

Ornithischia, 300

Orphan Drug Act (1983), 727

Orphan drugs, 727

Orthodontics, 727–728

Oscilloscope, 728–729

Ostromislensky, Ivan, 779

Otis, Elisha Graves, 347

Otis Elevator Company, 348, 377

Otophone, Marconi, 491

Otter boards, 692

Otto, Nikolaus, 92, 360, 361, 666

Ottoman Empire, 636

Oughtred, William, 906

Oven, microwave, 729–730

Overshot wheel, 1115

"Over the Horizon" (OTH) radar, 819

Owen, Richard, 299

Owens-Illinois Glass Company, 408

Oxidation, 162

Oxidation number, 198

Oxidative damage, 580

Oxygen-15, 1050

Oxygen-acetylene torch, 1127

Ozone, 909

Ozone depletion, 730

Ozone hole, 730–731

Ozone layer, 730, 1085

P

Pacemaker, cardiac, 172–173

Pacific Gas and Electric Company, 915

Pacinotti, Antonio, 458, 664

Packet switching, 243

Page, Charles Grafton, 664

Paine, Thomas, 392–393

Paint, 732–733

Pain transmission, gate control theory of, 8

Paleontology, 426

Paleotechnic era, 556

Palladio, Andrea, 150

Palmer, Timothy, 150

Palo Alto Research Center (PARC), 249, 668

Panama Canal, 166

Pan American World Airways, 562

Paneling, 895

Pangaea, 255

Panther Valley Television, 156

Panzerschreck (tank terror), 110

Papanicolaou, George N., 733

Paper, 734–736

Paper clip, 736

Papin, Denis, 789, 946, 948

Pap test, 733–734

Para-aminopbenzoic acid (PABA), 1106

Parachute, 736–737

Paradigm, 737–738

Parafoils, 737

Parallel postulate, 464

Parallel processing, 739–740

Parawings, 737

PARC, 668

Paris Academy of Sciences, 883, 884

Paris-Bordeaux-Paris automobile race, 1043

Paris Exposition, 691

Parker, R. L., 771

Parker, Zebulon, 1070

Parkes, Alexander, 186

Parkesine, 186

Parking meter, 740

Parkways, 432–433

Parsons, Charles A., 1066, 1071

Parsons, Ed, 156

Parsons, William, 1019

Particle accelerators, 740–742, 983

Partridge, Seth, 907

PASCAL, 241

Pascal, Blaise, 20, 107, 122, 159, 245, 445–446

Sedimenting centrifuges, 189

Seeberger, Charles, 377

Seguin, Marc, 148

Seikan tunnel, 1066

Seismic waves, 325, 466

Seismograms, 326

Seismological Society of America, 323

Seismometers, 326

Selden, George, 744, 885

Selden patent, 885–886

Selectavision, 1098–1099

Selenyi, Paul, 1140

Self-induction, 344

Selikoff, Irving, 73

Sellers, William, 940

Selten, Reinhard, 617

Semaphore system, 1008

Semiautomatic rifles, 887

Semiautomatic weapons, 886–888

Semiconductor, 888

Semi-elliptical, 936

Semmelweis, Ignaz, 988

Semon, Waldo L., 780

Senébier, Jean, 760

Seneca, 1065

Senefelder, Aloys, 590

Separate condenser, 888–889

Septicemic plague, 153

Serber, Robert, 814

Sericin, 900

Serotonergic hallucinogens, 488

Serpoller, Leon, 666

Sertraline (Zoloft), 59

Sertuerner, Friedrich, 655

Serum therapy, 528

Server, 243

Service industries, 1087

Servomechanism, 889–890

Severe combined immune deficiency, 456, 726, 1103

Sewers, 890–892

Sewing machine, 892–894

Sewing Machine Combination, 893

Shadoof, 14

Shaft mine, 643

Shallenberger, Oliver, 333

Shaped charges, 109–110

Shape memory alloys (SMAs), 894–895

Shape memory effect, 894

Shapin, Steven, 501

Shaping machines, 895–896

Shapley, Harlow, 388

Share, 773

Sharpey-Shaefer, Albert, 507

Sharps repeating rifles, 142

Shattering, 504

Shaw Communication Inc., 1091

Shed, 1123

Shed rod, 1123

Shells, 345

Sherrington, Charles, 694

Shiatsu, 7

Shibasaburo Kitasato, 528

Shields, 643

Shipman, Pat, 503

Ships, sailing, 896–898

Shockley, William, 1054

Sholes, Christopher Latham, 1078

Short bow, 137

Short takeoff and landing (STOL) aircraft, 30

Showtime, 156

Shrapnel, 60

Shreve, Henry M., 949

Shuey, Henry M., 863

Shull, George Harrison, 258, 769

Shultz, George, 964

Shuman, Frank, 913

Shuttle, 1124

Sick-building syndrome, 188

Sickle-cell anemia, 43, 124, 898–899
    and malaria, 898
    screening for, 459

Sickles, Frederick, 889

Sidereal period, 802

Siege warfare, 636

Siemens, William, 457, 950

Sigl, Georg, 590

Signal beam, 501

Signal-to-noise ratio, 899–900

Sign and Grudge, 1081

Sikorsky, Igor, 496

Silicon, 406, 888

Silicon-controlled rectifier (SCR), 836

Silicosis, 221

Silk, 900–901

Silkworms, raising, 901

Silliman, Benjamin, 1035

Silurian system, 463

Silver, Robert, 1089

Simon, Pierre (Marquis de Laplace), 127

Simon, Theodore, 544

Simple Mail Transfer Protocol, 346

Simpson, James Young, 51

Simpson, O. J., 310

Singer, Isaac Merritt, 892

Singer Sewing Machine Company, 1135

Singularity, 127

Sirius, 949

Skates, roller, 901–902

Skating, ice, 901

SKETCHPAD, 250, 668

Skinner, B. F., 114, 495

Skylab, 902

Skyscrapers, 155, 903–906

Slash-and-burn agriculture, 270

Slater, James, 190

Sleeping sickness, 100

Slide rule, 906–907

Sling psychrometer, 16–17

Slipher, Vesto Melvin, 389

Sloan, Howard, 917

Sludge disposal, 891

Small, James, 773

Smallpox, 527, 1102
    eradication of, 907–908
    vaccine for, 907–908

Smeaton, John, 253, 947, 1116, 1131

Smelling salts, 47

Smith, Cyril Stanley, 555

Smith, Francis, 1010

Smith, G. A., 661

Smith, Michael, 192

Smith, Robert Angus, 3

Smith, William, 462

Smog, photochemical, 908–909

Snail mail, 346

Snell, George Davis, 725

Snider-Pellergrine, Antonio, 771

Snow, George B., 398

Soap, 909–910

Social construction of technology, 1005–1006

Social ecology, 329–330

Social-psychological studies, 880

Society in Sunderland for Preventing

Thermals, 473

Thermodynamics, 1038

   First Law of, 163, 356, 750, 829

   Second Law of, 173, 260, 366, 367, 750, 829, 850

Thermoionic (vacuum) tube, 1036–1038

Thermometer, 1038–1039

Thermonuclear bombs, 520

Thermophyllic bacteria, 122

Thermoplastics, 8, 779

Thermosets, 8

Thermostat, 274, 1039–1040

Theropods, 300

Thessaloniki, siege of, 1058

Thiacril, 6

Thimonnier, Barthelemy, 892

Thiokol, 863

Thirty Years' War (1618-1648), 637, 1049

Thomas, Sidney Gilchrist, 954

Thomas W. Lawson (schooner), 993

Thompson, Benjamin (Count Rumford), 163, 355, 493

Thompson, Robert W., 1043

Thompson, William, 173

Thomson, C. J., 556

Thomson, Elihu, 1126

Thomson, George Paget, 812

Thomson, J. J., 43, 83, 116, 182, 344–345, 631, 799, 1036, 1060

Thomson, William (Lord Kelvin), 82, 163, 356, 381, 569, 826, 829, 971

Thoreau, Henry David, 329

Thorndike, Edward, 114

Thornycroft, John, 509

3-D motion picture, 657–658. *See also* Motion pictures

Three Gorges Dam, 282, 518–519

Three Mile Island, 704

   nuclear accident at, 1040–1041

Threshing, 229

Throwing, 901

Thurston, W., 1048

Thyristor, 553

Thyroid gland, 507, 1050

Tidal power, 1041

Tilden, William, 865

Tiles, 1041–1043

Tilt-up construction, 253

Time, standard, 939–940

Time dilation, 845

Time-Division Multiple Access (TDMA), 1016

Time-division multiplexing (TDM), 673

Time-rock units, 462

Time sharing, 346

Tire

   pneumatic, 1043–1044

   radial, 1044

Tissue, 183

Titan, 453

Titanic, 286

Titanic, sinking of, 818

Titov, Gherman, 624, 1109

T lymphocytes, 547

Tocopherol, 1105, 1106

Toilet, flush, 1044–1045

Tokamak Fusion Test Reactor, 442, 443

Tolerance, 450

Tolman, Edward C., 114

Tomato harvester, mechanical, 1046

Toon, Brian, 709

Topology, 1046–1048

Torque, 451

Torrey Canyon disaster, 993

Torricelli, Evangelista, 20, 106, 946

Torsion bar, 936

Touchtone dials, 1015

Touhy, Kevin, 255

Towed sonar arrays (TASS), 919

Townes, Charles, 572

Town lattice, 111

Toxic shock syndrome (TSS), 992

Toxic Substances Control Act (TSCA), 369

Toyo Kogyo, 365

Toyota Motor Company, 578

Trace fossils, 428

Trace italienne, 637, 1048–1050

Tracers, radioactive, 1050

Tracked vehicles, 1050–1051

Traction trebuchet, 1058. *See also* Trebuchet

Tractor, farm, 399–400

Traffic signals, 1051–1052

Trampoline effect, 1031

Tranquilizer, 56

Transducers, 45, 627

Transformation, 366

Transformer, 1052

Transform fault, 770

Transforming principle, 455

Transgenic organisms, 1053

Transistors, 160, 247, 630, 934, 1053–1056

   radio, 822–823

Transit, 1032

Transition metals, 748

Transition temperature, 983

Transit-mixed cement, 253

Translation, 308

Translocations, 207

Transplantation, organ, 725–727

Transportation, intermodal, 1056–1058

Trautwein, 675

Travers, Morris W., 691

Trawler, 692

Trebuchet, 1058–1059

Treponema pallidum, 199

Trésaguet, Pierre-Marie Jérôme, 859–860

Trevithick, Richard, 593

Triassic system, 463

Triangle, differential, 161–162

Triangulation, 1032

2,4,5-Trichlorophenoxyacetic acid (2,4,5-T), 497

Trident submarine, 973–975

Triode, 1059–1060

Trip hammers, 424

Triple junction, 770

Triplet code, 455

Trismegistos, Hermes, 35

Tritium, 442, 443

Trivers, R. L., 911

Trojan Horse, 245

Trolley. *See* Streetcar

Trolleybus, 1060–1061

Tropical forests, 287

Tropical rain forests, 287

Tropicana, 994

Tropsch, Hans, 223

Trucks, 1061–1063

   forklift and industrial, 425–426

Truman, Harry S, 521

   and interstate highway system, 550

   and labor relations, 951–952